Bending

Normal stress

$$\sigma = \frac{My}{I}$$

Unsymmetric bending

$$\sigma = -\frac{M_z y}{I_z} + \frac{M_y z}{I_y}, \qquad \tan \alpha = \frac{I_z}{I_y} \tan \theta$$

Shear

Average direct shear stress

$$\tau_{avg} = \frac{V}{A}$$

Transverse shear stress

$$\tau = \frac{VQ}{It}$$

Shear flow

$$q = \tau t = \frac{VQ}{I}$$

Stress in Thin-Walled Pressure Vessel

Cylinder $\qquad \sigma_1 = \dfrac{pr}{t} \qquad \sigma_2 = \dfrac{pr}{2t}$

Sphere $\qquad \sigma_1 = \sigma_2 = \dfrac{pr}{2t}$

Stress Transformation Equations

$$\sigma_{x'} = \frac{\sigma_x + \sigma_y}{2} + \frac{\sigma_x - \sigma_y}{2} \cos 2\theta + \tau_{xy} \sin 2\theta$$

$$\tau_{x'y'} = -\frac{\sigma_x - \sigma_y}{2} \sin 2\theta + \tau_{xy} \cos 2\theta$$

Principal Stress

$$\tan 2\theta_p = \frac{\tau_{xy}}{(\sigma_x - \sigma_y)/2}$$

$$\sigma_{1,2} = \frac{\sigma_x + \sigma_y}{2} \pm \sqrt{\left(\frac{\sigma_x - \sigma_y}{2}\right)^2 + \tau_{xy}^2}$$

Maximum in-plane shear stress

$$\tan 2\theta_s = -\frac{(\sigma_x - \sigma_y)/2}{\tau_{xy}}$$

$$\tau_{max} = \sqrt{\left(\frac{\sigma_x - \sigma_y}{2}\right)^2 + \tau_{xy}^2}, \qquad \sigma_{avg} = \frac{\sigma_x + \sigma_y}{2}$$

Absolute maximum shear stress

$$\tau_{\substack{abs \\ max}} = \frac{\sigma_{max} - \sigma_{min}}{2}, \qquad \sigma_{avg} = \frac{\sigma_{max} + \sigma_{min}}{2}$$

Material Property Relations

Poisson's ratio

$$\nu = -\frac{\epsilon_{lat}}{\epsilon_{long}}$$

Generalized Hooke's Law

$$\epsilon_x = \frac{1}{E}[\sigma_x - \nu(\sigma_y + \sigma_z)]$$

$$\epsilon_y = \frac{1}{E}[\sigma_y - \nu(\sigma_z + \sigma_x)]$$

$$\epsilon_z = \frac{1}{E}[\sigma_z - \nu(\sigma_x + \sigma_y)]$$

$$\gamma_{xy} = \frac{1}{G}\tau_{xy}, \quad \gamma_{yz} = \frac{1}{G}\tau_{yz}, \quad \gamma_{zx} = \frac{1}{G}\tau_{zx}$$

where

$$G = \frac{E}{2(1 + \nu)}$$

Relations Between w, V, M

$$\frac{dV}{dx} = -w(x), \qquad \frac{dM}{dx} = V$$

Elastic Curve

$$\frac{1}{\rho} = \frac{M}{EI}$$

$$EI\frac{d^4v}{dx^4} = -w(x)$$

$$EI\frac{d^3v}{dx^3} = V(x)$$

$$EI\frac{d^2v}{dx^2} = M(x)$$

Buckling

Critical axial load

$$P_{cr} = \frac{\pi^2 EI}{(KL)^2}$$

Critical stress

$$\sigma_{cr} = \frac{\pi^2 E}{(KL/r)^2}, \qquad r = \sqrt{I/A}$$

STATICS AND MECHANICS OF MATERIALS

STATICS

AND

MECHANICS OF MATERIALS

R. C. HIBBELER

PRENTICE HALL
Upper Saddle River, NJ 07458

LIBRARY OF CONGRESS CATALOGING IN PUBLICATION
Hibbeler, R.C.
    Statics and mechanics of materials / R. C. Hibbeler.
        p.   cm.
    "Portions of this book are reprinted from Mechanics of Materials by R. C. Hibbeler,
© 1991 by Macmillan Publishing Company, and Engineering Mechanics: Statics,
Sixth edition, by R. C. Hibbeler, copyright 1992 by Macmillan Publishing Company"
CIP galley.
    Includes index.
    ISBN 0-02-354091-5
    1. Strength of materials. 2. Statics. 3. Structural analysis (Engineering) I. Title.
TA405.h48   1993
620.l'l--dc20
                                                92-28506
                                                      CIP

Editors: David Johnstone, John Griffin
Production Supervisor: Margaret Comaskey
Production Manager: Sandra E. Moore

This book was set in Times Roman by York Graphic Services, Inc.
Portions of this book are reprinted from *Mechanics of Materials* by
R. C. Hibbeler, copyright © 1991 by Macmillan Publishing Company,
and *Engineering Mechanics: Statics*, Sixth Edition by R. C. Hibbeler,
copyright © 1992 by Macmillan Publishing Company.

 1993 by Prentice-Hall, Inc.
Upper Saddle River, NJ 07458

Printed in the United States of America

15 14 13 12 11

ISBN 0-02-354091-5

9 780023 540912

Prentice-Hall International (UK) Limited, *London*
Prentice-Hall of Australia Pty. Limited, *Sydney*
Prentice-Hall Canada, Inc., *Toronto*
Prentice-Hall Hispanoamericana, S.A., *Mexico*
Prentice-Hall of India Private Limited, *New Delhi*
Prentice-Hall of Japan, Inc., *Tokyo*
Simon & Schuster Asia Pte. Ltd., *Singapore*
Editora Prentice-Hall do Brasil, Ltda., *Rio de Janeiro*

Preface

This book represents a combined abridged version of two of the author's books, namely *Engineering Mechanics: Statics, 6th Edition* and *Mechanics of Materials*. It is intended for those students who do not need complete coverage of these subjects. Rather it provides a clear and thorough presentation of both the theory and application of the important fundamental topics of this material, which is often used in many engineering disciplines. Understanding is based on explaining the physical behavior of materials under load and then modeling this behavior to develop the theory. The development emphasizes the importance of satisfying equilibrium, compatibility of deformation, and material behavior requirements. The hallmark of the book, however, remains the same as the author's unabridged versions, and that is, strong emphasis is placed on drawing a free-body diagram, and the importance of selecting an appropriate coordinate system and an associated sign convention is stressed when the equations of mechanics are applied. Throughout the book, many analysis and design applications are presented which involve mechanical elements and structural members often encountered in engineering practice.

Organization and Approach

In order to aid both the instructor and the student, the contents of each chapter are organized into well-defined sections. Selected groups of sections contain an explanation of specific topics, followed by illustrative example problems and a set of homework problems. The topics within each section are often placed in subgroups denoted by boldface titles. The purpose of this is to present a structured method for introducing each new definition or concept and to make the book convenient for later reference and review.

As in the author's other textbooks, a "procedure for analysis" is used throughout the book, providing the student with a logical and orderly method to follow when applying the theory. The example problems are then solved using this outlined method in order to clarify its numerical application. It is to be understood, however, that once the relevant principles have been mastered and enough confidence and judgment have been acquired, the student can then develop his or her own procedures for solving problems. In most cases, it is felt that the first step in any procedure should be to draw a diagram. In doing so, the student forms the habit of tabulating the necessary data while focusing on the physical aspects of the problem and its associated geometry. If this step is correctly performed, applying the relevant equations becomes somewhat methodical, since the data can be taken directly from the diagram.

Contents

The book is divided into two parts, and the material is covered in the traditional manner.

Statics. The subject of statics is presented in 7 chapters. The text begins in Chapter 1 with an introduction to mechanics and a discussion of units. The notion of a vector and the properties of a concurrent force system are introduced in Chapter 2. Chapter 3 contains a general discussion of concentrated force systems and the methods used to simplify them. The principles of rigid-body equilibrium are developed in Chapter 4 and then applied to specific problems involving the equilibrium of trusses, frames, and machines in Chapter 5. Topics related to the center of gravity, centroid, and moment of inertia are treated in Chapter 6. Lastly, internal loadings in members is discussed in Chapter 7.

Mechanics of Materials. This portion of the text is covered in 10 chapters. Chapter 8 begins with a formal definition of both normal and shear stress, and a discussion of normal stress in axially loaded members and average shear stress caused by direct shear; finally, normal and shear strain are defined. In Chapter 9 a discussion of some of the important mechanical properties of materials is given. Separate treatments of axial load, torsion, bending, and transverse shear are presented in Chapters 10, 11, 12, and 13, respectively. Chapter 14 provides a partial review of the material covered in the previous chapters, in which the state of stress resulting from combined loadings is discussed. In Chapter 15 the concepts for transforming stress and strain are presented. Chapter 16 provides a means for a further summary and review of previous material by covering design applications of beams. Also, coverage is given for various methods for computing deflections of beams and finding the reactions on these members if they are statically indeterminate. Lastly, Chapter 17 provides a discussion of column buckling.

Sections of the book that contain more advanced material are indicated by a star (★). Time permitting, some of these topics may be included in the course. Furthermore, this material provides a suitable reference for basic principles when it is covered in other courses, and it can be used as a basis for assigning special projects.

Alternative Method for Coverage of Mechanics of Materials. Some instructors prefer to cover stress and strain transformations *first,* before discussing specific applications of axial load, torsion, bending, and shear. One possible method for doing this would be first to cover stress and strain and its transformations, Chapter 8 and Chapter 15. The discussion and example problems in Chapter 15 have been styled so that this is possible. Chapters 9 through 14 can then be covered with no loss in continuity.

Problems. Numerous problems in the book depict realistic situations encountered in engineering practice. It is hoped that this realism will both stimulate the student's interest in the subject and provide a means for developing the skill to reduce any such problem from its physical description to a model or symbolic representation to which the principles may be applied.

Throughout the text there is an approximate balance of problems using either SI or FPS units. Furthermore, in any set, an attempt has been made to arrange the problems in order of increasing difficulty. The answers to all but every fourth problem are listed in the back of the book. To alert the user to a problem without a reported answer, an asterisk (*) is placed before the problem number. Answers are reported to three significant figures, even though the data for material properties may be known with less accuracy. Although this might appear to be poor practice, it is done simply to be consistent and to allow the student a better chance to validate his or her solution. All the problems and their solutions have been independently checked for accuracy.

Computer Tutorial

Included with this book is a tutorial disk that contains several hundred review questions and problems on the major topics discussed in the book. It has been designed to run on an IBM or compatible computer having color graphics capability. The problems on this disk are very elementary and can be solved without the use of a pocket calculator. Many of my students have found it useful for self testing their understanding of the concepts and as a means for building their confidence prior to solving problems or taking examinations.

Acknowledgments

Preparation of the manuscript for this book has undergone several reviews and I owe the reviewers a personal debt of gratitude. Their encouragement and willingness to provide constructive criticism are very much appreciated; in particular, Professor William D. Webster, GMI Engineering Management Institute; Professor Hugh Currin, Oregon Institute of Technology; Professor George Staab, Ohio State University; Professor Alan Zhender, Cornell University; Professor Warren Campbell, University of Wisconsin—River Falls; Professor David McClellan, Mercer University; Professor Henry Petroski, Duke University; Professor Edward G. Lovell, University of Wisconsin—Madison; Professor Phillip Perdikaris, Case Western Reserve University; Professor Peter Wright, University of Toronto; Professor Stan Crawley, University of Utah; Professor Myron Snesrud, Winona State University; Lt. Col. Bob Pieri, U.S. Air Force Academy; and Professor John Bassani, University of Pennsylvania.

I would also like to thank all my students who have used the manuscript and the computer tutorial as well as their revisions, and made comments to improve their contents. A particular note of thanks goes to one of my former graduate students, Kai Beng Yap, who has been a great help to me in this regard. A special note of gratitude also goes to my editors and the staff at Macmillan, who have all been very supportive in allowing me to have a more creative and artistic license in the design and execution of this book. It has been a pleasure to work with all of them. Finally, appreciation goes to my wife, Conny, who has been a source of encouragement and has helped with the details of preparing the manuscript for publication.

Russell Charles Hibbeler

Contents

STATICS AND MECHANICS OF MATERIALS

STATICS

1 General Principles

This chapter provides an introduction to many of the fundamental concepts in mechanics. It includes a discussion of models or idealizations that are used to apply the theory, a statement of Newton's laws of motion, upon which this subject is based, and a general review of the principles for applying the SI system of units. Standard procedures for performing numerical calculations are then discussed. At the end of the chapter we will present a general guide that should be followed for solving problems.

1.1 Mechanics

Mechanics can be defined as that branch of the physical sciences concerned with the state of rest or motion of bodies that are subjected to the action of forces. In this book we will study two very important branches of mechanics, namely, statics and mechanics of materials. These subjects form a suitable basis for the design and analysis of many types of structural, mechanical, or electrical devices encountered in engineering.

Statics deals with the equilibrium of bodies, that is, it is used to determine the forces acting either external to the body or within it necessary to keep the body either at rest or moving with a constant velocity. *Mechanics of materials* studies the relationships between the external loads and the intensity of internal forces acting within the body. This subject is also concerned with computing the deformations of the body, and it provides a study of the body's stability when the body is subjected to external forces. In this book we will first study the principles of statics since for the design and analysis of any structural or mechanical element, it is *first* necessary to determine the forces acting both on and within its various members. Once these internal forces are determined, the size of the members, their deflection, and their stability can then be determined using the fundamentals of mechanics of materials, which will be covered later.

1.2 Fundamental Concepts

Before beginning our study of statics and mechanics of materials, it is important to understand the meaning of certain fundamental concepts and principles.

Basic Quantities.   The following four quantities are used throughout mechanics.

Length.   *Length* is needed to locate the position of a point in space and thereby describe the size of a physical system. Once a standard unit of length is defined, one can then quantitatively define distances and geometric properties of a body as multiples of the unit length.

Time.   *Time* is conceived as a succession of events. Although the principles of statics and mechanics of materials are time independent, this quantity does play an important role in the study of dynamics, which is a study of the accelerated motion of a body.

Mass.   *Mass* is a property of matter by which we can compare the action of one body with that of another. This property manifests itself as a gravitational attraction between two bodies and provides a quantitative measure of the resistance of matter to a change in velocity.

Force.   In general, *force* is considered as a "push" or "pull" exerted by one body on another. This interaction can occur when there is either direct contact between the bodies, such as a person pushing on a wall, or it can occur through a distance when the bodies are physically separated. Examples of the latter type include gravitational, electrical, and magnetic forces. In any case, a force is completely characterized by its magnitude, direction, and point of application.

Idealizations. Models or idealizations are used in mechanics in order to simplify application of the theory. A few of the more important idealizations will now be defined. Others that are noteworthy will be discussed at points where they are needed.

Particle. A *particle* has a mass but a size that can be neglected. When a body is idealized as a particle, the principles of mechanics reduce to a rather simplified form since the geometry of the body will not be involved in the analysis of the problem.

Rigid Body. A *rigid body* can be considered as a combination of a large number of particles in which all the particles remain at a fixed distance from one another both before and after applying a load. As a result, the material properties of any body that is assumed to be rigid will not have to be considered when analyzing the forces acting on the body. In most cases the actual deformations occurring in structures, machines, mechanisms, and the like are relatively small, and the rigid-body assumption is suitable for analysis.

Concentrated Force. A *concentrated force* represents the effect of a loading which is assumed to act at a point on a body. We can represent the effect of the loading by a concentrated force, provided the area over which the load is applied is *small* compared to the overall size of the body.

Newton's Three Laws of Motion.

The principles of mechanics are formulated, in part, on the basis of Newton's three laws of motion, the validity of which is based on experimental observation. These laws were originally stated for a particle, that is, a piece of matter having negligible size, which is moving in a nonaccelerating reference frame.

First Law. A particle originally at rest, or moving in a straight line with constant velocity, will remain in this state provided the particle is *not* subjected to an unbalanced force.

Second Law. A particle acted upon by an *unbalanced force* $\mathbf{F}$ experiences an acceleration $\mathbf{a}$ that has the same direction as the force and a magnitude that is directly proportional to the force.* If $\mathbf{F}$ is applied to a particle of mass m, this law may be expressed mathematically as

$$\mathbf{F} = m\mathbf{a} \qquad (1\text{–}1)$$

Third Law. The mutual forces of action and reaction between two particles are equal, opposite, and collinear.

---

*Stated another way, the unbalanced force acting on the particle is proportional to the time rate of change of the particle's linear momentum.

Newton's Law of Gravitational Attraction. Shortly after formulating his three laws of motion, Newton postulated a law governing the gravitational attraction between any two particles. Stated mathematically,

$$F = G\frac{m_1 m_2}{r^2} \tag{1-2}$$

where F = force of gravitation between the two particles
             G = universal constant of gravitation; according to experimental evidence, $G = 66.73(10^{-12})$ m³/(kg · s²)
      m_1, m_2 = mass of each of the two particles
             r = distance between the two particles

Weight. According to Eq. 1–2, any two particles or bodies have a mutual attractive (gravitational) force acting between them. In the case of a particle located at or near the surface of the earth, however, the only gravitational force having any sizable magnitude is that between the earth and the particle. Consequently, this force, termed the *weight,* will be the only gravitational force considered in our study of mechanics.

From Eq. 1–2, we can develop an approximate expression for finding the weight W of a particle having a mass $m_1 = m$. If we assume the earth to be a nonrotating sphere of constant density and having a mass m_2, then if r is the distance between the earth's center and the particle, we have

$$W = G\frac{mm_2}{r^2}$$

Letting $g = Gm_2/r^2$ yields

$$\boxed{W = mg} \tag{1-3}$$

By comparison with Eq. 1–1, we term g the acceleration due to gravity. Since it depends on r, it can be seen that the weight of a body is *not* an absolute quantity. Instead, its magnitude is determined from where the measurement was made. For most engineering calculations, however, g is determined at sea level and at a latitude of 45°, which is considered the "standard location."

1.3 Units of Measurement

The four basic quantities—length, time, mass, and force—are not all independent from one another; in fact, they are *related* by Newton's second law of motion, $\mathbf{F} = m\mathbf{a}$. Hence, the *units* used to define force, mass, length, and time cannot *all* be selected arbitrarily. The equality $\mathbf{F} = m\mathbf{a}$ is maintained only if three of the four units, called *base units,* are *arbitrarily defined* and the fourth unit is *derived* from the equation.

SI Units. The International System of units, abbreviated SI after the French "Système International d'Unités," is a modern version of the metric system which has received worldwide recognition. As shown in Table 1–1, the SI system specifies length in meters (m), time in seconds (s), and mass in kilograms (kg). The unit of force, called a newton (N), is *derived* from **F** = m**a.** Thus, 1 newton is equal to a force required to give 1 kilogram of mass an acceleration of 1 m/s² (N = kg · m/s²).

If the weight of a body located at the "standard location" is to be determined in newtons, then Eq. 1–3 must be applied. Here $g = 9.806\ 65$ m/s²; however, for calculations, the value $g = 9.81$ m/s² will be used. Thus,

$$W = mg \qquad (g = 9.81\ \text{m/s}^2) \tag{1–4}$$

Therefore, a body of mass 1 kg has a weight of 9.81 N, a 2-kg body weighs 19.62 N, and so on.

U.S. Customary. In the U.S. Customary system of units (FPS) length is measured in feet (ft), force in pounds (lb), and time in seconds (s), Table 1–1. The unit of mass, called a *slug,* is *derived* from **F** = m**a.** Hence, 1 slug is equal to the amount of matter accelerated at 1 ft/s² when acted upon by a force of 1 lb (slug = lb · s²/ft).

In order to determine the mass of a body having a weight measured in pounds, we must apply Eq. 1–3. If the measurements are made at the "standard location," then $g = 32.2$ ft/s² will be used for calculations. Therefore,

$$m = \frac{W}{g} \qquad (g = 32.2\ \text{ft/s}^2) \tag{1–5}$$

And so a body weighing 32.2 lb has a mass of 1 slug, a 64.4-lb body has a mass of 2 slugs, and so on.

Table 1–1 **Systems of Units**

Name	SI	FPS
Length	meter (m)	foot (ft)
Time	second (s)	second (s)
Mass	kilogram (kg)	slug* $\left(\dfrac{\text{lb} \cdot \text{s}^2}{\text{ft}}\right)$
Force	newton* (N) $\left(\dfrac{\text{kg} \cdot \text{m}}{\text{s}^2}\right)$	pound (lb)

*Derived unit.

Conversion of Units. In some cases it may be necessary to convert from one system of units to another. In this regard, Table 1–2 provides a set of direct conversion factors between FPS and SI units for the basic quantities. Also, in the FPS system, recall that 1 ft = 12 in. (inches), 5280 ft = 1 mi (mile), 1000 lb = 1 kip (kilo-pound), and 2000 lb = 1 ton.

Table 1–2 Conversion Factors

Quantity	Unit of Measurement (FPS)	Equals	Unit of Measurement (SI)
Force	lb		4.448 2 N
Mass	slug		14.593 8 kg
Length	ft		0.304 8 m

1.4 The International System of Units

The SI system of units is used extensively in this book since it is intended to become the worldwide standard for measurement. Consequently, the rules for its use and some of its terminology relevant to mechanics will now be presented.

Prefixes. When a numerical quantity is either very large or very small, the units used to define its size may be modified by using a prefix. Some of the prefixes used in the SI system are shown in Table 1–3. Each represents a multiple or submultiple of a unit which, if applied successively, moves the decimal point of a numerical quantity to every third place.* For example, 4 000 000 N = 4 000 kN (kilo-newton) = 4 MN (mega-newton), or 0.005 m = 5 mm (milli-meter). Notice that the SI system does not include the multiple deca (10) or the submultiple centi (0.01), which form part of the metric system. Except for some volume and area measurements, the use of these prefixes is to be avoided in science and engineering.

Table 1–3 Prefixes

	Exponential Form	Prefix	SI Symbol
Multiple			
1 000 000 000	10^9	giga	G
1 000 000	10^6	mega	M
1 000	10^3	kilo	k
Submultiple			
0.001	10^{-3}	milli	m
0.000 001	10^{-6}	micro	μ
0.000 000 001	10^{-9}	nano	n

*The kilogram is the only base unit that is defined with a prefix.

Rules for Use. The following rules are given for the proper use of the various SI symbols:

1. A symbol is *never* written with a plural "s," since it may be confused with the unit for second (s).

2. Symbols are always written in lowercase letters, with the following exceptions: symbols for the two largest prefixes shown in Table 1–3, giga and mega, are capitalized as G and M, respectively; and symbols named after an individual are also capitalized, e.g., N.

3. Quantities defined by several units which are multiples of one another are separated by a *dot* to avoid confusion with prefix notation, as indicated by $N = kg \cdot m/s^2 = kg \cdot m \cdot s^{-2}$. Also, m · s (meter-second), whereas ms (milli-second).

4. The exponential power represented for a unit having a prefix refers to both the unit *and* its prefix. For example, $\mu N^2 = (\mu N)^2 = \mu N \cdot \mu N$. Likewise, mm^2 represents $(mm)^2 = mm \cdot mm$.

5. Physical constants or numbers having several digits on either side of the decimal point should be reported with a *space* between every three digits rather than with a comma; e.g., 73 569.213 427. In the case of four digits on either side of the decimal, the spacing is optional; e.g., 8537 or 8 537. Furthermore, always try to use decimals and avoid fractions; that is, write 15.25, *not* $15\frac{1}{4}$.

6. When performing calculations, represent the numbers in terms of their *base or derived units* by converting all prefixes to powers of 10. The final result should then be expressed using a *single prefix*. Also, after calculation, it is best to keep numerical values between 0.1 and 1000; otherwise, a suitable prefix should be chosen. For example,

$$(50 \text{ kN})(60 \text{ nm}) = [50(10^3) \text{ N}][60(10^{-9}) \text{ m}]$$
$$= 3000(10^{-6}) \text{ N} \cdot \text{m} = 3(10^{-3}) \text{ N} \cdot \text{m} = 3 \text{ mN} \cdot \text{m}$$

7. Compound prefixes should not be used; e.g., $k\mu s$ (kilo-micro-second) should be expressed as ms (milli-second) since $1 \ k\mu s = 1(10^3)(10^{-6}) \ s = 1(10^{-3}) \ s = 1 \ ms$.

8. With the exception of the base unit the kilogram, in general avoid the use of a prefix in the denominator of composite units. For example, do not write N/mm, but rather kN/m; also, m/mg should be written as mm/kg.

9. Although not expressed in multiples of 10, the minute, hour, etc., are retained for practical purposes as multiples of the second. Furthermore, plane angular measurement is made using radians (rad). In this book, however, degrees will often be used, where $180° = \pi$ rad.

1.5 Numerical Calculations

Numerical work in engineering practice is most often performed by using hand-held calculators and computers. It is important, however, that the answers to any problem be reported with both justifiable accuracy and appropriate significant figures. In this section we will discuss these topics together with some other important aspects involved in all engineering calculations.

Dimensional Homogeneity. The terms of any equation used to describe a physical process must be *dimensionally homogeneous;* that is, each term must be expressed in the same units. Provided this is the case, all the terms of an equation can then be combined if numerical values are substituted for the variables. Consider, for example, the equation $s = vt + \frac{1}{2}at^2$, where, in SI units, s is the position in meters, m, t is time in seconds, s, v is velocity in m/s, and a is acceleration in m/s^2. Regardless of how this equation is evaluated, it maintains its dimensional homogeneity. In the form stated each of the three terms is expressed in meters [m, (m/s̸)s̸, (m/s̸^{2})s̸^{2}], or solving for a, $a = 2s/t^2 - 2v/t$, the terms are each expressed in units of m/s^2 [m/s^2, m/s^2, (m/s)/s].

Since problems in mechanics involve the solution of dimensionally homogeneous equations, the fact that all terms of an equation are represented by a consistent set of units can be used as a partial check for algebraic manipulations of an equation.

Significant Figures. The accuracy of a number is specified by the number of significant figures it contains. A *significant figure* is any digit, including a zero, provided it is not used to specify the location of the decimal point for the number. For example, the numbers 5604 and 34.52 each have four significant figures. When numbers begin or end with zeros, however, it is difficult to tell how many significant figures are in the number. Consider the number 40. Does it have one (4), or perhaps two (40) significant figures? In order to clarify this situation, the number should be reported using powers of 10. There are two ways of doing this. The format for *scientific notation* specifies one digit to the left of the decimal point, with the remaining digits to the right; for example, 40 expressed to one significant figure would be $4(10^1)$. Using *engineering notation,* which is preferred here, the exponent is displayed in multiples of three in order to facilitate conversion of SI units to those having an appropriate prefix. Thus, 40 expressed to one significant figure would be $0.04(10^3)$. Likewise, 2500 and 0.00546 expressed to three significant figures would be $2.50(10^3)$ and $5.46(10^{-3})$.

Rounding Off Numbers. For numerical calculations, the accuracy obtained from the solution of a problem generally can never be better than the accuracy of the problem data. This is what is to be expected, but often handheld calculators or computers involve more figures in the answer than the number of significant figures used for the data. For this reason, a calculated result should always be "rounded off" to an appropriate number of significant figures.

To ensure accuracy, the following rules for rounding off a number to n significant figures apply:

1. If the $n + 1$ digit is *less than 5,* the $n + 1$ digit and others following it are dropped. For example, 2.326 and 0.451 rounded off to $n = 2$ significant figures would be 2.3 and 0.45.

2. If the $n + 1$ digit is equal to 5 with zeros following it, then round off the nth digit to an *even number*. For example, 1245 and 0.8655 rounded off to $n = 3$ significant figures become 1240 and 0.866.

3. If the $n + 1$ digit is *greater than 5* or equal to 5 with any nonzero digits following it, then increase the nth digit by 1 and drop the $n + 1$ digit and others following it. For example, 0.723 87 and 565.500 3 rounded off to $n = 3$ significant figures become 0.724 and 566.

Calculations. As a general rule, to ensure accuracy of a final result when performing calculations with numbers of unequal accuracy, always retain one extra significant figure in the more accurate numbers than in the least accurate number *before* beginning the computations. Then round off the final result so that it has the same number of significant figures as the least accurate number. If possible, try to work out the computations so that numbers which are approximately equal are not subtracted, since accuracy is often lost from this calculation.

Most of the example problems in this book are solved with the assumption that any measured *data* are accurate to three significant figures.* Consequently, intermediate calculations are often worked out to four significant figures and the answers are generally reported to *three* significant figures.

The following examples illustrate application of the principles just discussed as related to the proper use and conversion of units.

*Of course, some numbers, such as π, e, or numbers used in derived formulas, are exact and are therefore accurate to an infinite number of significant figures.

■ Example 1–1

Convert 2 kN/m to lb/ft.

SOLUTION

Since 1 kN = 1000 N and from Table 1–2, 1 lb = 4.4482 N and 1 ft = 0.3048 m, the factors of conversion are arranged in the following order, so that a cancellation of the units can be applied:

$$2 \text{ kN/m} = \frac{2 \, \cancel{\text{kN}}}{\cancel{\text{m}}} \left(\frac{1000 \, \cancel{\text{N}}}{\cancel{\text{kN}}} \right) \left(\frac{\text{lb}}{4.4482 \, \cancel{\text{N}}} \right) \left(\frac{0.3048 \, \cancel{\text{m}}}{\text{ft}} \right)$$

$$= 137 \text{ lb/ft} \qquad\qquad Ans.$$

■ Example 1–2

Convert the quantity 300 lb · s to appropriate SI units.

SOLUTION

Using Table 1–2, 1 lb = 4.448 2 N.

$$300 \text{ lb} \cdot \text{s} = 300 \, \cancel{\text{lb}} \cdot \text{s} \left(\frac{4.448 \, 2 \text{ N}}{\cancel{\text{lb}}} \right)$$

$$= 1334.5 \text{ N} \cdot \text{s} = 1.33 \text{ kN} \cdot \text{s} \qquad\qquad Ans.$$

Example 1–3

Evaluate each of the following and express with SI units having an appropriate prefix: (a)   (50 mN)(6 GN),   (b)   (400 mm)(0.6 MN)^2, (c) 45 MN^3/900 Gg.

SOLUTION

First convert each number to base units, perform the indicated operations, then choose an appropriate prefix (see Rule 6 on p. 7).

Part (a)

$$(50 \text{ mN})(6 \text{ GN}) = [50(10^{-3}) \text{ N}][6(10^9) \text{ N}]$$

$$= 300(10^6) \text{ N}^2$$

$$= 300(10^6) \, \cancel{\text{N}}^2\left(\frac{1 \text{ kN}}{10^3 \, \cancel{\text{N}}}\right)\left(\frac{1 \text{ kN}}{10^3 \, \cancel{\text{N}}}\right)$$

$$= 300 \text{ kN}^2 \qquad\qquad \textit{Ans.}$$

Note carefully the convention kN^2 = (kN)^2 = 10^6 N^2 (Rule 4 on p. 7).

Part (b)

$$(400 \text{ mm})(0.6 \text{ MN})^2 = [400(10^{-3}) \text{ m}][0.6(10^6) \text{ N}]^2$$

$$= [400(10^{-3}) \text{ m}][0.36(10^{12}) \text{ N}^2]$$

$$= 144(10^9) \text{ m} \cdot \text{N}^2$$

$$= 144 \text{ Gm} \cdot \text{N}^2 \qquad\qquad \textit{Ans.}$$

We can also write

$$144(10^9) \text{ m} \cdot \text{N}^2 = 144(10^9) \text{ m} \cdot \cancel{\text{N}}^2\left(\frac{1 \text{ MN}}{10^6 \, \cancel{\text{N}}}\right)\left(\frac{1 \text{ MN}}{10^6 \, \cancel{\text{N}}}\right)$$

$$= 0.144 \text{ m} \cdot \text{MN}^2$$

Part (c)

$$45 \text{ MN}^3/900 \text{ Gg} = \frac{45(10^6 \text{ N})^3}{900(10^6) \text{ kg}}$$

$$= 0.05(10^{12}) \text{ N}^3/\text{kg}$$

$$= 0.05(10^{12}) \, \cancel{\text{N}}^3\left(\frac{1 \text{ kN}}{10^3 \, \cancel{\text{N}}}\right)^3 \frac{1}{\text{kg}}$$

$$= 0.05(10^3) \text{ kN}^3/\text{kg}$$

$$= 50 \text{ kN}^3/\text{kg} \qquad\qquad \textit{Ans.}$$

Here we have used rules 4 and 8 of p. 7.

1.6 General Procedure for Analysis

The most effective way of learning the principles of mechanics is to *solve problems*. To be successful at this, it is important always to present the work in a *logical* and *orderly manner,* as suggested by the following sequence of steps:

1. Read the problem carefully and try to correlate the actual physical situation with the theory studied.

2. Draw any necessary diagrams and tabulate the problem data.

3. Apply the relevant principles, generally in mathematical form.

4. Solve the necessary equations algebraically as far as practical, then, making sure they are dimensionally homogeneous, use a consistent set of units and complete the solution numerically. Report the answer with no more significant figures than the accuracy of the given data.

5. Study the answer with technical judgment and common sense to determine whether or not it seems reasonable.

6. Once the solution has been completed, review the problem. Try to think of other ways of obtaining the same solution.

In applying this general procedure, do the work as neatly as possible. Being neat generally stimulates clear and orderly thinking, and vice versa.

PROBLEMS

1–1. Round off the following numbers to three significant figures: (*a*) 3.45555 m, (*b*) 45.556 s, (*c*) 5555 N, (*d*) 4525 kg.

1–2. If a car weighs 3500 lb, determine its mass and express the result using SI units.

1–3. Represent each of the following combinations of units in the correct SI form using an appropriate prefix: (*a*) Mg/mm, (*b*) mN/μs, (*c*) μm · Mg.

*1–4. Convert 63 ft^2 · s to m^2 · s.

1–5. Represent each of the following quantities in the correct SI form using an appropriate prefix: (*a*) 0.000431 kg, (*b*) 35.3(10^3) N, (*c*) 0.00532 km.

1–6. Evaluate each of the following to three significant figures and express each answer in SI units using an appropriate prefix: (*a*) 0.631 Mm/(8.60 kg)^2, (*b*) (35 mm)^2(48 kg)^3.

1–7. Evaluate (204 mm)(0.004 57 kg)/(34.6 N) and express the answer in SI units using an appropriate prefix.

*1–8. Evaluate each of the following to three significant figures and express each answer in SI units using an appropriate prefix: (*a*) (212 mN)^2, (*b*) (52 800 ms)^2, (*c*) [548(10^6)]$^{1/2}$ ms.

1–9. Evaluate each of the following and express each answer in SI units using an appropriate prefix: (*a*) (684 μm)/43 ms, (*b*) (28 ms)(0.0458 Mm)/(348 mg), (*c*) (2.68 mm)(426 Mg).

1–10. Convert each of the following and express the answer using an appropriate prefix: (*a*) 175 lb/ft^3 to kN/m^3, (*b*) 6 ft/h to mm/s, (*c*) 835 lb · ft to kN · m.

1–11. The specific weight (wt./vol.) of brass is 520 lb/ft^3. Determine its density (mass/vol.) in SI units. Use an appropriate prefix.

*1–12. Determine the mass in kilograms of an object that has a weight of (*a*) 20 mN, (*b*) 150 kN, (*c*) 60 MN. Express each answer using an appropriate prefix.

2 Force Vectors

In this chapter we will introduce the concept of a concentrated force and give the procedures for adding forces, resolving them into components, and projecting them along an axis. Since force is a vector quantity, we must use the rules of vector algebra whenever forces are considered. We will begin our study by defining scalar and vector quantities and then develop some of the basic rules of vector algebra.

2.1 Scalars and Vectors

Most of the physical quantities in mechanics can be expressed mathematically by means of scalars and vectors.

Scalar. A quantity characterized by a positive or negative number is called a *scalar*. Mass, volume, and length are scalar quantities often used in statics. In this book, scalars are indicated by letters in italic type, such as the scalar A. The mathematical operations involving scalars follow the same rules as those of elementary algebra.

Vector. A *vector* is a quantity that has both a magnitude, direction, and sense, and obeys the parallelogram law of addition. This law, which will be described later, utilizes a form of construction that accounts for the combined magnitude and direction of the vector. Vector quantities commonly used in statics are position, force, and moment vectors.

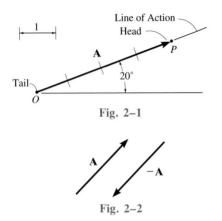

Fig. 2–1

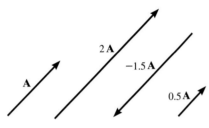

Fig. 2–2

A vector is represented graphically by an arrow, which is used to define its magnitude, direction, and sense. The *magnitude* of the vector is indicated by the length of the arrow, the *direction* is defined by the angle between a reference axis and the arrow's line of action, and the *sense* is indicated by the arrowhead. For example, the vector **A** shown in Fig. 2–1 has a magnitude of 4 units, a direction which is 20° measured clockwise from the horizontal axis, and a sense which is upward and to the right. The point *O* is called the *tail* of the vector, the point *P* is the *tip* or *head*.

For handwritten work, a vector is generally represented by a letter with an arrow written over it, such as $\vec{A}$. The magnitude is designated $|\vec{A}|$ or simply A. In this book vectors will be symbolized in boldface type; for example, **A** is used to designate the vector "A". Its magnitude, which is always a positive quantity, is symbolized in italic type, written as $|A|$, or simply A when it is understood that A is a positive scalar.

2.2 Basic Vector Operations

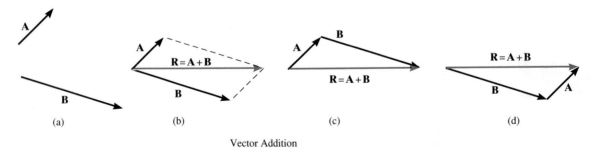

Scalar Multiplication and Division

Fig. 2–3

Multiplication and Division of a Vector by a Scalar. The product of vector **A** and scalar *a*, yielding *a***A,** is defined as a vector having a magnitude $|aA|$. The *sense* of *a***A** is the *same* as **A** provided *a* is *positive;* it is *opposite* to **A** if *a* is *negative*. Consequently, the negative of a vector is formed by multiplying the vector by the scalar (−1), Fig. 2–2. Division of a vector by a scalar can be defined using the laws of multiplication, since **A**/*a* = (1/*a*)**A,** *a* ≠ 0. Graphic examples of these operations are shown in Fig. 2–3.

Vector Addition. Two vectors **A** and **B** of the same type, Fig. 2–4*a*, may be added to form a "resultant" vector **R** = **A** + **B** by using the *parallelogram law*. To do this, **A** and **B** are joined at their tails, Fig. 2–4*b*. Parallel dashed lines drawn from the head of each vector intersect at a common point, thereby forming the adjacent sides of a parallelogram. As shown, the resultant **R** is the diagonal of the parallelogram, which extends from the tails of **A** and **B** to the intersection of the dashed lines.

(a)   (b)   (c)   (d)

Vector Addition

Fig. 2–4

We can also add **B** to **A** using a *triangle construction,* which is a special case of the parallelogram law, whereby vector **B** is added to vector **A** in a "head-to-tail" fashion, i.e., by connecting the head of **A** to the tail of **B,** Fig. 2–4c. The resultant **R** extends from the tail of **A** to the head of **B.** In a similar manner, **R** can also be obtained by adding **A** to **B,** Fig. 2–4d. By comparison, it is seen that vector addition is commutative; in other words, the vectors can be added in either order, i.e., **R = A + B = B + A.**

As a special case, if the two vectors **A** and **B** are *collinear,* i.e., both have the same line of action, the parallelogram law reduces to an *algebraic* or *scalar addition* $R = A + B$, as shown in Fig. 2–5.

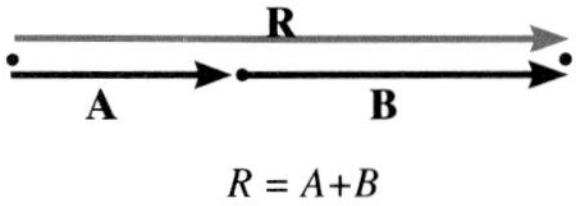

$R = A+B$
Fig. 2–5

Vector Subtraction.
The resultant *difference* between two vectors **A** and **B** of the same type may be expressed as

$$\mathbf{R'} = \mathbf{A} - \mathbf{B} = \mathbf{A} + (-\mathbf{B})$$

This vector sum is shown graphically in Fig. 2–6. Subtraction is therefore defined as a special case of addition, so the rules of vector addition also apply to vector subtraction.

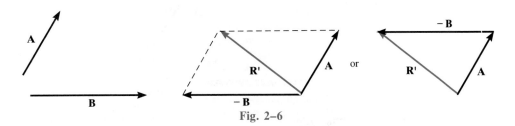

Fig. 2–6

Resolution of a Vector.
A vector may be resolved into two "components" having known lines of action by using the parallelogram law. For example, if **R** in Fig. 2–7a is to be resolved into components acting along the lines *a* and *b*, one starts at the *head* of **R** and extends a dashed line *parallel* to *a* until it intersects *b*. Likewise, a dashed line parallel to *b* is drawn from the *head* of **R** to the point of intersection with *a*, Fig. 2–7a. The two components **A** and **B** are then drawn such that they extend from the tail of **R** to the points of intersection, as shown in Fig. 2–7b.

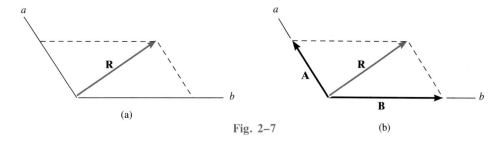

(a)

(b)

Fig. 2–7

2.3 Vector Addition of Forces

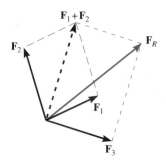

Fig. 2–8

Experimental evidence has shown that a force is a vector quantity since it has a specified magnitude, direction, and sense and it adds according to the parallelogram law. Two common problems in statics involve finding either the resultant force, knowing its components, or resolving a known force into two components. As described in Sec. 2–2, both of these problems require application of the parallelogram law.

If more than two forces are to be added, successive applications of the parallelogram law can be carried out in order to obtain the resultant force. For example, if three forces F_1, F_2, F_3 act at a point, Fig. 2–8, the resultant of any two of the forces is found—say, $F_1 + F_2$—and then this resultant is added to the third force, yielding the resultant of all three forces; i.e., $F_R = (F_1 + F_2) + F_3$. Using the parallelogram law to add more than two forces, as shown here, often requires extensive geometric and trigonometric calculation to determine the numerical values for the magnitude and direction of the resultant. Instead, problems of this type are easily solved by using the "rectangular-component method," which is explained in Sec. 2.4.

PROCEDURE FOR ANALYSIS

Problems that involve the addition of two forces and contain at most *two unknowns* can be solved by using the following procedure:

Parallelogram Law. Make a sketch showing the vector addition using the parallelogram law. If possible, determine the interior angles of the parallelogram from the geometry of the problem. Recall that the sum total of these angles is 360°. Unknown angles, along with known and unknown force magnitudes, should be clearly labeled on this sketch. Redraw a half portion of the constructed parallelogram to illustrate the triangular head-to-tail addition of the components.

Trigonometry. By using trigonometry, the two unknowns can be determined from the data listed on the triangle. If the triangle does *not* contain a 90° angle, the law of sines and/or the law of cosines may be used for the solution. These formulas are given in Fig. 2–9 for the triangle shown.

The following examples illustrate this method numerically.

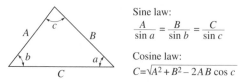

Sine law:
$$\frac{A}{\sin a} = \frac{B}{\sin b} = \frac{C}{\sin c}$$

Cosine law:
$$C = \sqrt{A^2 + B^2 - 2AB \cos c}$$

Fig. 2–9

Example 2–1

The screw eye in Fig. 2–10a is subjected to two forces, $\mathbf{F}_1$ and $\mathbf{F}_2$. Determine the magnitude and direction of the resultant force.

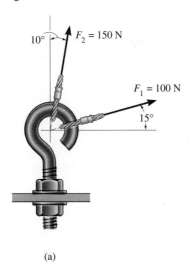

(a)

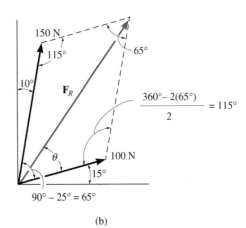

(b)

SOLUTION

Parallelogram Law. The parallelogram law of addition is shown in Fig. 2–10b. The two unknowns are the magnitude of $\mathbf{F}_R$ and the angle θ (theta). From Fig. 2–10b, the vector triangle, Fig. 2–10c, is constructed.

Trigonometry. F_R is determined by using the law of cosines:

$$F_R = \sqrt{(100)^2 + (150)^2 - 2(100)(150)\cos 115°}$$
$$= \sqrt{10\,000 + 22\,500 - 30\,000(-0.4226)} = 212.6 \text{ N}$$
$$= 213 \text{ N} \qquad\qquad\qquad\qquad Ans.$$

The angle θ is determined by applying the law of sines, using the computed value of F_R.

$$\frac{150}{\sin\theta} = \frac{212.6}{\sin 115°}$$

$$\sin\theta = \frac{150}{212.6}(0.9063)$$

$$\theta = 39.8°$$

Thus, the direction ϕ (phi) of $\mathbf{F}_R$, measured from the horizontal, is

$$\phi = 39.8° + 15.0° = 54.8° \quad \measuredangle^\phi \qquad\qquad Ans.$$

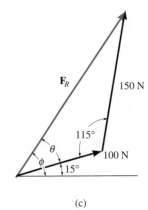

(c)

Fig. 2–10

Example 2-2

Resolve the 200-lb force shown acting on the pin, Fig. 2–11a, into components in the (a) x and y directions, and (b) x' and y directions.

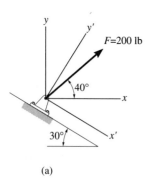

(a)

SOLUTION

In each case the parallelogram law is used to resolve **F** into its two components, and then the vector triangle is constructed to determine the numerical results by trigonometry.

Part (a). The vector addition $\mathbf{F} = \mathbf{F}_x + \mathbf{F}_y$ is shown in Fig. 2–11b. In particular, note that the length of the components is scaled along the x and y axes by first constructing dashed lines parallel to the axes in accordance with the parallelogram law. From the vector triangle, Fig. 2–11c,

$$F_x = 200 \cos 40° = 153 \text{ lb} \qquad Ans.$$
$$F_y = 200 \sin 40° = 129 \text{ lb} \qquad Ans.$$

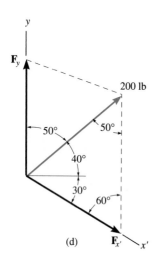

(d)

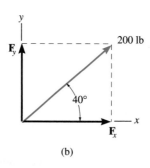

(b)

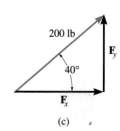

(c)

Fig. 2–11

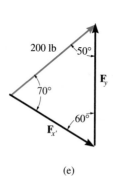

(e)

Part (b). The vector addition $\mathbf{F} = \mathbf{F}_{x'} + \mathbf{F}_y$ is shown in Fig. 2–11d. Note carefully how the parallelogram is constructed. Applying the law of sines and using the data listed on the vector triangle, Fig. 2–11e, yields

$$\frac{F_{x'}}{\sin 50°} = \frac{200}{\sin 60°}$$
$$F_{x'} = 200\left(\frac{\sin 50°}{\sin 60°}\right) = 177 \text{ lb} \qquad Ans.$$
$$\frac{F_y}{\sin 70°} = \frac{200}{\sin 60°}$$
$$F_y = 200\left(\frac{\sin 70°}{\sin 60°}\right) = 217 \text{ lb} \qquad Ans.$$

Example 2–3

The force **F** acting on the frame shown in Fig. 2–12a has a magnitude of 500 N and is to be resolved into two components acting along struts AB and AC. Determine the angle θ, measured *below* the horizontal, so that the component $\mathbf{F}_{AC}$ is directed from A toward C and has a magnitude of 400 N.

Fig. 2–12

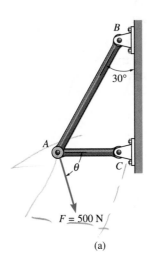

(a)

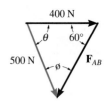

(b)

SOLUTION

By using the parallelogram law, the vector addition of the two components yielding the resultant is shown in Fig. 2–12b. Note carefully how the resultant force is resolved into the two components $\mathbf{F}_{AB}$ and $\mathbf{F}_{AC}$, which have specified lines of action. The corresponding vector triangle is shown in Fig. 2–12c. The angle ϕ can be determined by using the law of sines:

$$\frac{400}{\sin \phi} = \frac{500}{\sin 60^\circ}$$

$$\sin \phi = \left(\frac{400}{500}\right) \sin 60^\circ$$

$$\phi = 43.9^\circ$$

(c)

Hence,

$$\theta = 180^\circ - 60^\circ - 43.9^\circ = 76.1^\circ \quad \text{Ans.}$$

Using this value for θ, apply the law of cosines and show that $\mathbf{F}_{AB}$ has a magnitude of 560 N.

Notice that **F** can also be directed at an angle θ *above* the horizontal, as shown in Fig. 2–12d, and still produce the required component $\mathbf{F}_{AC}$. Show that in this case $\theta = 16.1^\circ$ and $F_{AB} = 161$ N.

(d)

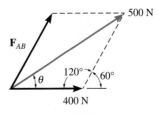

■ Example 2–4

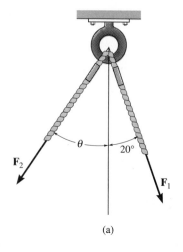

(a)

The ring shown in Fig. 2–13a is subjected to two forces, $\mathbf{F}_1$ and $\mathbf{F}_2$. If it is required that the resultant force have a magnitude of 1 kN and be directed vertically downward, determine (a) the magnitudes of $\mathbf{F}_1$ and $\mathbf{F}_2$ provided $\theta = 30°$, and (b) the magnitudes of $\mathbf{F}_1$ and $\mathbf{F}_2$ if F_2 is to be a minimum.

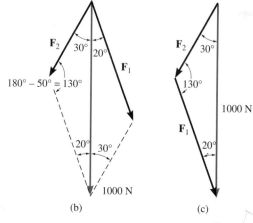

(b)   (c)

SOLUTION

Part (a).   A sketch of the vector addition according to the parallelogram law is shown in Fig. 2–13b. From the vector triangle constructed in Fig. 2–13c, the unknown magnitudes F_1 and F_2 can be determined by using the law of sines.

$$\frac{F_1}{\sin 30°} = \frac{1000}{\sin 130°}$$

$$F_1 = 653 \text{ N} \qquad \textit{Ans.}$$

$$\frac{F_2}{\sin 20°} = \frac{1000}{\sin 130°}$$

$$F_2 = 446 \text{ N} \qquad \textit{Ans.}$$

Part (b).   If θ is not specified, then by the vector triangle, Fig. 2–13d, $\mathbf{F}_2$ may be added to $\mathbf{F}_1$ in various ways to yield the resultant 1000-N force. In particular, the *minimum* length or magnitude of $\mathbf{F}_2$ will occur when its line of action is *perpendicular to* $\mathbf{F}_1$. Any other direction, such as OA or OB, yields a larger value for F_2. Hence, when $\theta = 90° - 20° = 70°$, F_2 is minimum. From the triangle shown in Fig. 2–13e, it is seen that

$$F_1 = 1000 \sin 70° = 940 \text{ N} \qquad \textit{Ans.}$$

$$F_2 = 1000 \sin 20° = 342 \text{ N} \qquad \textit{Ans.}$$

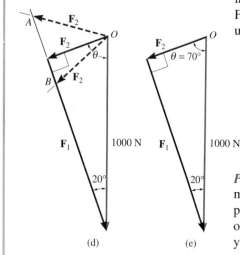

(d)   (e)

Fig. 2–13

PROBLEMS

2–1. Determine the magnitude of the resultant force $F_R = F_1 + F_2$ and its orientation θ, measured counterclockwise from the positive x axis.

2–2. Determine the magnitude of the resultant force $F_R = F_1 - F_2$ and its orientation θ, measured counterclockwise from the positive x axis.

2–5. Determine the magnitude of the resultant force $F_R = F_1 + F_2$ and its orientation θ, measured counterclockwise from the positive u axis.

2–6. Resolve the force F_1 into components acting along the u and v axes and determine the magnitudes of the components.

2–7. Resolve the force F_2 into components acting along the u and v axes and determine the magnitudes of the components.

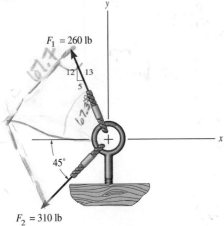

Probs. 2–1/2–2

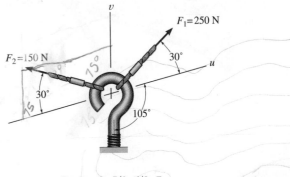

Probs. 2–5/2–6/2–7

2–3. Determine the magnitude of the resultant force $F_R = F_1 + F_2$ and its orientation θ, measured clockwise from the positive x axis.

*2–4. Determine the magnitude of the resultant force $F_R = F_1 + F_3$ and its orientation θ, measured counterclockwise from the positive x axis.

*2–8. The screw eye is subjected to the two forces shown. F is directed along the inclined plane and the 450-N force is horizontal. Determine the magnitude of F so that the resultant force F_R is directed normal to the plane, i.e., along the positive n axis. What is the magnitude of F_R?

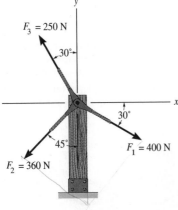

Probs. 2–3/2–4

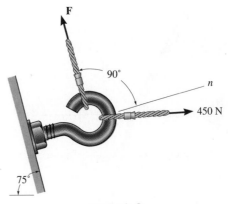

Prob. 2–8

2–9. Determine the design angle θ ($0° \leq \theta \leq 90°$) for member AB so that the 400-lb horizontal force has a component of 500 lb directed from A toward C. What is the component of force acting along member AB? Take $\phi = 40°$.

2–10. Determine the design angle ϕ ($0° \leq \phi \leq 90°$) between members AB and AC so that the 400-lb horizontal force has a component of 600 lb which acts up to the right, in the same direction as from B toward A. Also calculate the magnitude of the force component along AC. Take $\theta = 30°$.

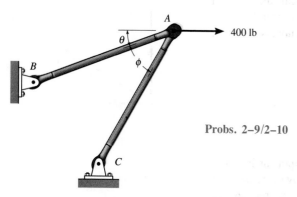

Probs. 2–9/2–10

2–11. The cable exerts a force of 600 N on the frame. Resolve this force into components acting *(a)* along the x and v axes and *(b)* along the y and u axes. What is the magnitude of each component?

*2–12. The cable exerts a force of 600 N on the frame. Resolve this force into components acting *(a)* along the x and y axes and *(b)* along the u and v axes. What is the magnitude of each component?

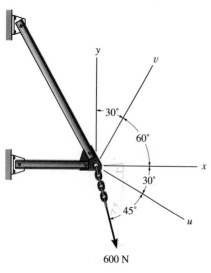

Probs. 2–11/2–12

2–13. Two forces, having a magnitude of 20 lb and 15 lb, act on the rod. If the resultant force has a magnitude of 30 lb, determine the angle θ between the forces.

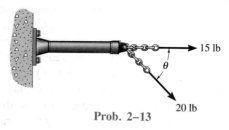

Prob. 2–13

2–14. The hipbone H is connected to the femur F at A using three different muscles, which exert the forces shown on the femur. Determine the resultant force on the femur and specify its orientation θ measured counterclockwise from the positive x axis.

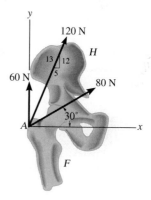

Prob. 2–14

2–15. The jet aircraft is being towed by two trucks B and C. Determine the magnitudes of the two towing forces $\mathbf{F}_B$ and $\mathbf{F}_C$ if the resultant force has a magnitude $F_R = 10$ kN and is directed along the positive x axis. Set $\theta = 15°$.

*2–16. If the resultant $\mathbf{F}_R$ of the two forces acting on the jet aircraft is to be directed along the positive x axis and have a magnitude of 10 kN, determine the angle θ of the cable attached to the truck at B such that the force $\mathbf{F}_B$ in this cable is a *minimum*. What is the magnitude of force in each cable when this occurs?

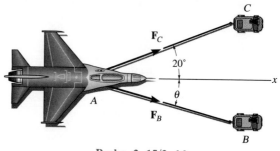

Probs. 2–15/2–16

2.4 Addition of a System of Coplanar Forces

When the resultant of more than two forces has to be obtained, it is easier to find the components of each force along specified axes, add these components algebraically, and then form the resultant, rather than form the resultant of the forces by successive application of the parallelogram law as discussed in Sec. 2.3. In this section we will resolve each force into its rectangular components $\mathbf{F}_x$ and $\mathbf{F}_y$, which lie along the x and y axes, respectively, Fig. 2–14a. Although the axes are shown here to be horizontal and vertical, they may in general be directed at any inclination, as long as they remain perpendicular to one another, Fig. 2–14b. In either case, by the parallelogram law, we require:

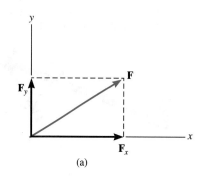

(a)

and

$$\mathbf{F} = \mathbf{F}_x + \mathbf{F}_y$$

$$\mathbf{F}' = \mathbf{F}'_x + \mathbf{F}'_y$$

As shown in Fig. 2–14, the sense of direction of each component is represented *graphically* by the *arrowhead*. For *analytical* work, however, we must establish a notation for representing the directional sense of the rectangular components for each coplanar vector. This can be done in one of two ways.

Scalar Notation. Since the x and y axes have designated positive and negative directions, the magnitude and directional sense of the rectangular components of a force can be expressed in terms of *algebraic scalars*. For example, the components of $\mathbf{F}$ in Fig. 2–14a can be represented by positive scalars F_x and F_y since their sense of direction is along the *positive* x and y axes, respectively. In a similar manner, the components of $\mathbf{F}'$ in Fig. 2–14b are F'_x and $-F'_y$. Here the y component is negative, since $\mathbf{F}'_y$ is directed along the negative y axis. It is important to keep in mind that this scalar notation is to be used only for computational purposes, not for graphical representations in figures. Throughout the text, the *head of a vector arrow* in any figure indicates the sense of the vector *graphically;* algebraic signs are not used for this purpose. Thus, the vectors in Figs. 2–14a and 2–14b are designated by using boldface (vector) notation.* Whenever italic symbols are written near vector arrows in figures, they indicate the *magnitude* of the vector, which is *always* a *positive* quantity.

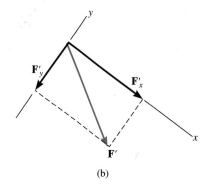

(b)

Fig. 2–14

---

*Negative signs are used only in figures with boldface notation when showing equal but opposite pairs of vectors as in Fig. 2–2.

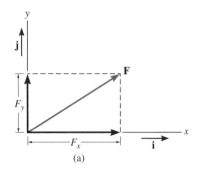

(a)

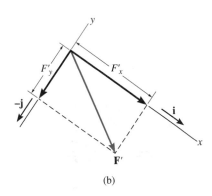

(b)

Fig. 2–15

Cartesian Vector Notation. It is also possible to represent the components of a force in terms of Cartesian unit vectors. By doing this the methods of vector algebra are easier to apply, and we will see that this becomes particularly advantageous for solving problems in three dimensions. In two dimensions the *Cartesian unit vectors* **i** and **j** are used to designate the *directions* of the x and y axes, respectively, Fig. 2–15a.* These vectors have a dimensionless magnitude of unity, and their sense (or arrowhead) will be described analytically by a plus or minus sign, depending on whether they are pointing along the positive or negative x or y axis.

As shown in Fig. 2–15a, the *magnitude* of each component of **F** is *always a positive quantity,* which is represented by the (positive) scalars F_x and F_y. Therefore, having established notation to represent the magnitude and the direction of each component, we can express **F** in Fig. 2–15a as a *Cartesian vector,* i.e.,

$$\mathbf{F} = F_x\mathbf{i} + F_y\mathbf{j}$$

And in the same way, **F′** in Fig. 2–15b can be expressed as

$$\mathbf{F}' = F_x'\mathbf{i} + F_y'(-\mathbf{j})$$

or simply

$$\mathbf{F}' = F_x'\mathbf{i} - F_y'\mathbf{j}$$

Coplanar Force Resultants. Either of the two methods just described for representing the rectangular components of a force can be used to determine the resultant of several *coplanar forces.* To do this, each force is first resolved into its x and y components and then the respective components are added using *scalar algebra* since they are collinear. The resultant force is then formed by adding the resultants of the x and y components using the parallelogram law. For example, consider the three forces in Fig. 2–16a, which have x and y components as shown in Fig. 2–16b. To solve this problem using *Cartesian vector notation,* each force is first represented as a Cartesian vector, i.e.,

$$\mathbf{F}_1 = F_{1x}\mathbf{i} + F_{1y}\mathbf{j}$$
$$\mathbf{F}_2 = -F_{2x}\mathbf{i} + F_{2y}\mathbf{j}$$
$$\mathbf{F}_3 = F_{3x}\mathbf{i} - F_{3y}\mathbf{j}$$

*For handwritten work, unit vectors are usually indicated using a circumflex, e.g., **î** and **ĵ**.

The vector resultant is therefore

$$\mathbf{F}_R = \mathbf{F}_1 + \mathbf{F}_2 + \mathbf{F}_3$$
$$= F_{1x}\mathbf{i} + F_{1y}\mathbf{j} - F_{2x}\mathbf{i} + F_{2y}\mathbf{j} + F_{3x}\mathbf{i} - F_{3y}\mathbf{j}$$
$$= (F_{1x} - F_{2x} + F_{3x})\mathbf{i} + (F_{1y} + F_{2y} - F_{3y})\mathbf{j}$$
$$= (F_{Rx})\mathbf{i} + (F_{Ry})\mathbf{j}$$

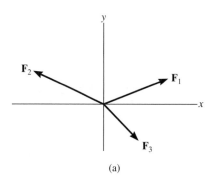

(a)

If *scalar notation* is used, then, from Fig. 2–16*b*, since x is positive to the right and y is positive upward, we have

$(\xrightarrow{+})$ $F_{Rx} = F_{1x} - F_{2x} + F_{3x}$

$(+\uparrow)$ $F_{Ry} = F_{1y} + F_{2y} - F_{3y}$

These results are the *same* as the $\mathbf{i}$ and $\mathbf{j}$ components of $\mathbf{F}_R$ determined above.

In the general case, the x and y components of the resultant of any number of coplanar forces can be represented symbolically by the algebraic sum of the x and y components of all the forces, i.e.,

$$\boxed{\begin{aligned} F_{Rx} &= \Sigma F_x \\ F_{Ry} &= \Sigma F_y \end{aligned}} \qquad (2\text{–}1)$$

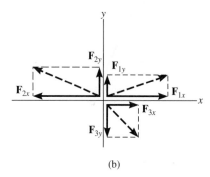

(b)

When applying these equations it is important to use the sign convention established for the components; and that is, components having a directional sense along the positive coordinate axes are considered positive scalars, whereas those having a directional sense along the negative coordinate axes are considered negative scalars. If this convention is followed, then the signs of the resultant components will specify the sense of these components. For example, a positive result indicates that the component has a directional sense which is in the positive coordinate direction.

Once the resultant components are determined, they may be sketched along the x and y axes in their proper directions, and the resultant force can be determined from vector addition, as shown in Fig. 2–16*c*. From this sketch, the magnitude of $\mathbf{F}_R$ is then found from the Pythagorean theorem; that is,

$$F_R = \sqrt{F_{Rx}^2 + F_{Ry}^2}$$

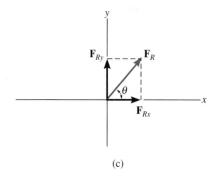

(c)

Also, the direction angle θ, which specifies the orientation of the force, is determined from trigonometry.

$$\theta = \tan^{-1}\left| \frac{F_{Ry}}{F_{Rx}} \right|$$

Fig. 2–16

The above concepts are illustrated numerically in the following examples.

Example 2–5

Determine the x and y components of $\mathbf{F}_1$ and $\mathbf{F}_2$ shown in Fig. 2–17a. Express each force as a Cartesian vector.

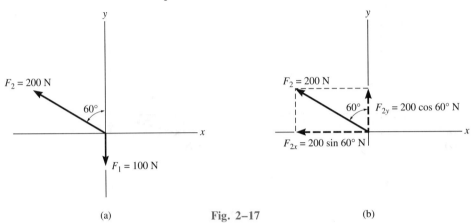

(a)   Fig. 2–17   (b)

SOLUTION

Scalar Notation. Since $\mathbf{F}_1$ acts along the negative y axis, and the magnitude of $\mathbf{F}_1$ is 100 N, the components written in scalar form are

$$F_{1x} = 0, \qquad F_{1y} = -100 \text{ N} \qquad \qquad Ans.$$

or, alternatively,

$$F_{1x} = 0, \qquad F_{1y} = 100 \text{ N} \downarrow \qquad \qquad Ans.$$

By the parallelogram law, $\mathbf{F}_2$ is resolved into x and y components, Fig. 2–17b. The magnitude of each component is determined by trigonometry. Since $\mathbf{F}_{2x}$ acts in the $-x$ direction, and $\mathbf{F}_{2y}$ acts in the $+y$ direction, we have

$$F_{2x} = -200 \sin 60° \text{ N} = -173 \text{ N} = 173 \text{ N} \leftarrow \qquad Ans.$$
$$F_{2y} = 200 \cos 60° \text{ N} = 100 \text{ N} = 100 \text{ N} \uparrow \qquad Ans.$$

Cartesian Vector Notation. Having computed the magnitudes of the components of $\mathbf{F}_2$, Fig. 2–17b, we can express each force as a Cartesian vector.

$$\mathbf{F}_1 = 0\mathbf{i} + 100 \text{ N}(-\mathbf{j})$$
$$= \{-100\mathbf{j}\} \text{ N} \qquad \qquad Ans.$$

and

$$\mathbf{F}_2 = 200 \sin 60° \text{ N}(-\mathbf{i}) + 200 \cos 60° \text{ N}(\mathbf{j})$$
$$= \{-173\mathbf{i} + 100\mathbf{j}\} \text{ N} \qquad \qquad Ans.$$

Example 2–6

Determine the x and y components of the force $\mathbf{F}$ shown in Fig. 2–18a.

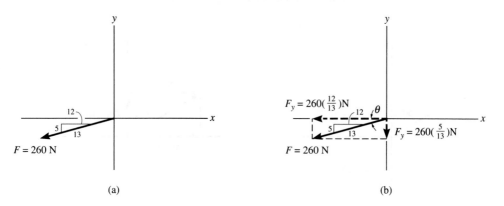

Fig. 2–18

SOLUTION

The force is resolved into its x and y components as shown in Fig. 2–18b. Here the *slope* of the line of action for the force is indicated. From this "slope triangle" we could obtain the direction angle θ, e.g., $\theta = \tan^{-1}\left(\frac{5}{12}\right)$, and then proceed to determine the magnitudes of the components in the same manner as for $\mathbf{F}_2$ in Example 2–5. An easier method, however, consists of using proportional parts of similar triangles, i.e.,

$$\frac{F_x}{260} = \frac{12}{13} \qquad F_x = 260\left(\frac{12}{13}\right) = 240 \text{ N}$$

Similarly,

$$F_y = 260\left(\frac{5}{13}\right) = 100 \text{ N}$$

Notice that the magnitude of the *horizontal component*, F_x, was obtained by multiplying the force magnitude by the ratio of the *horizontal leg* of the slope triangle divided by the hypotenuse; whereas the magnitude of the *vertical component*, F_y, was obtained by multiplying the force magnitude by the ratio of the *vertical leg* divided by the hypotenuse. Hence, using scalar notation,

$$F_x = -240 \text{ N} = 240 \text{ N} \leftarrow \qquad\qquad Ans.$$
$$F_y = -100 \text{ N} = 100 \text{ N} \downarrow \qquad\qquad Ans.$$

If $\mathbf{F}$ is expressed as a Cartesian vector, we have

$$\mathbf{F} = \{-240\mathbf{i} - 100\mathbf{j}\} \text{ N} \qquad\qquad Ans.$$

Example 2–7

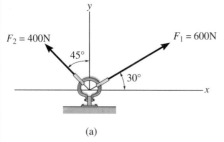

(a)

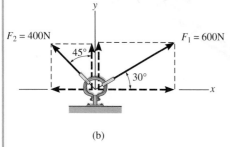

(b)

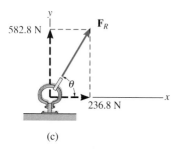

(c)

Fig. 2–19

The link in Fig. 2–19a is subjected to two forces F_1 and F_2. Determine the magnitude and orientation of the resultant force.

SOLUTION I

Scalar Notation.   This problem can be solved by using the parallelogram law; however, here we will resolve each force into its x and y components, Fig. 2–19b, and sum these components algebraically. Indicating the "positive" sense of the x and y force components alongside Eqs. 2–1, we have

$$\xrightarrow{+} F_{Rx} = \Sigma F_x; \qquad F_{Rx} = 600 \cos 30° - 400 \sin 45°$$
$$= 236.8 \text{ N} \rightarrow$$

$$+ \uparrow F_{Ry} = \Sigma F_y; \qquad F_{Ry} = 600 \sin 30° + 400 \cos 45°$$
$$= 582.8 \text{ N} \uparrow$$

The resultant force, shown in Fig. 2–19c, has a *magnitude* of

$$F_R = \sqrt{(236.8)^2 + (582.8)^2}$$
$$= 629 \text{ N} \qquad\qquad\qquad Ans.$$

From the vector addition, Fig. 2–19c, the direction angle θ is

$$\theta = \tan^{-1}\left(\frac{582.8}{236.8}\right) = 67.9° \qquad\qquad Ans.$$

SOLUTION II

Cartesian Vector Notation.   From Fig. 2–19b, each force expressed as a Cartesian vector is

$$\mathbf{F}_1 = 600 \cos 30°\mathbf{i} + 600 \sin 30°\mathbf{j}$$
$$\mathbf{F}_2 = -400 \sin 45°\mathbf{i} + 400 \cos 45°\mathbf{j}$$

Thus

$$\mathbf{F}_R = \mathbf{F}_1 + \mathbf{F}_2 = (600 \cos 30° - 400 \sin 45°)\mathbf{i}$$
$$+ (600 \sin 30° + 400 \cos 45°)\mathbf{j}$$
$$= \{236.8\mathbf{i} + 582.8\mathbf{j}\} \text{ N}$$

The magnitude and direction of $\mathbf{F}_R$ are determined in the same manner as shown above.

Comparing the two methods of solution, it is seen that use of scalar notation is more efficient, since the scalar components can be found *directly*, without first having to express each force as a Cartesian vector before adding the components. Cartesian vector analysis, however, will later be shown to be more advantageous for solving three-dimensional problems.

Example 2–8

The end of the boom O in Fig. 2–20a is subjected to three concurrent and coplanar forces. Determine the magnitude and orientation of the resultant force.

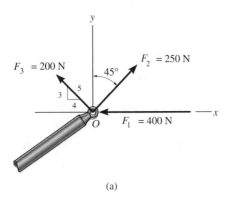

(a)

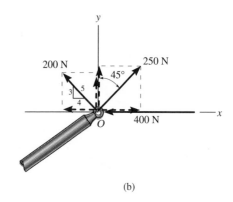

(b)

SOLUTION

Each force is resolved into its x and y components as shown in Fig. 2–20b. Summing the x components, we have

$$\xrightarrow{+} F_{Rx} = \Sigma F_x; \qquad F_{Rx} = -400 + 250 \sin 45° - 200(\tfrac{4}{5})$$
$$= -383.2 \text{ N} = 383.2 \text{ N} \leftarrow$$

The negative sign indicates that F_{Rx} acts to the left, i.e., in the negative x direction as noted by the small arrow. Summing the y components yields

$$+\uparrow F_{Ry} = \Sigma F_y; \qquad F_{Ry} = 250 \cos 45° + 200(\tfrac{3}{5})$$
$$= 296.8 \text{ N} \uparrow$$

The resultant force, shown in Fig. 2–20c, has a *magnitude* of

$$F_R = \sqrt{(-383.2)^2 + (296.8)^2}$$
$$= 485 \text{ N} \qquad\qquad Ans.$$

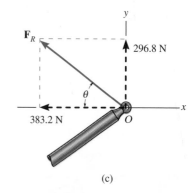

(c)

From the vector addition in Fig. 2–20c, the direction angle θ is

$$\theta = \tan^{-1}\left(\frac{296.8}{383.2}\right) = 37.8° \qquad Ans.$$

Realize that the single force $\mathbf{F}_R$ shown in Fig. 2–20c creates the *same effect* on the boom as the three forces in Fig. 2–20a.

Fig. 2–20

PROBLEMS

2–17. Express $\mathbf{F}_1$ and $\mathbf{F}_2$ as Cartesian vectors.

2–18. Determine the magnitude of the resultant force and its orientation measured clockwise from the positive x axis.

2–21. Express $\mathbf{F}_1$, $\mathbf{F}_2$, and $\mathbf{F}_3$ as Cartesian vectors.

2–22. Determine the magnitude of the resultant force and its orientation measured counterclockwise from the positive x axis.

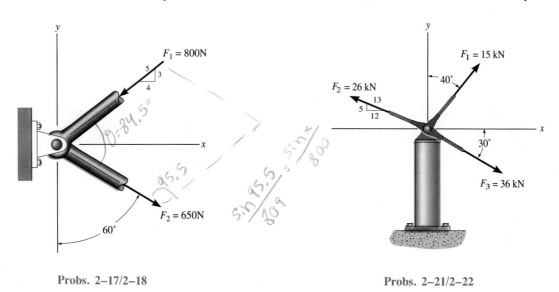

Probs. 2–17/2–18   Probs. 2–21/2–22

2–19. Express $\mathbf{F}_1$ and $\mathbf{F}_2$ as Cartesian vectors.

*2–20. Determine the magnitude of the resultant force and its orientation measured counterclockwise from the positive x axis.

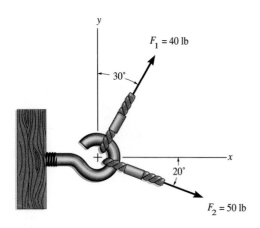

Probs. 2–19/2–20

2–23. Solve Prob. 2–3 by summing the rectangular or x, y components of the forces to obtain the resultant force.

*2–24. Solve Prob. 2–4 by summing the rectangular or x, y components of the forces to obtain the resultant force.

2–25. Solve Prob. 2–15 by summing the rectangular components of the forces to obtain the resultant force.

2-26. Express each force acting on the bracket in Cartesian vector form. What is the magnitude of the resultant force?

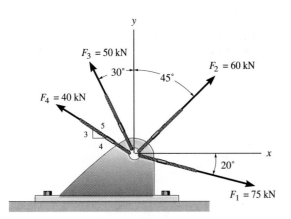

Prob. 2-26

2-27. Four concurrent forces act on the plate. Determine the magnitude of the resultant force and its orientation measured counterclockwise from the positive x axis.

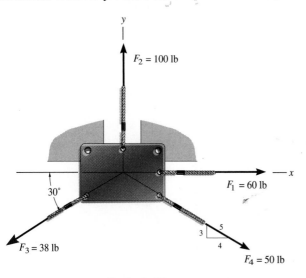

Prob. 2-27

*2-28. If $F_2 = 150$ lb and $\theta = 55°$, determine the magnitude and orientation, measured clockwise from the positive u axis, of the resultant force of the three forces acting on the bracket.

2-29. Three forces act on the bracket. Determine the magnitude and orientation θ of $\mathbf{F}_2$ so that the resultant force is directed along the positive u axis and has a magnitude of 50 lb.

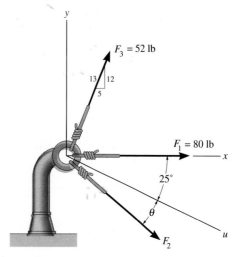

Probs. 2-28/2-29

2-30. Determine the magnitude and orientation θ of $\mathbf{F}_B$ so that the resultant force is directed along the positive y axis and has a magnitude of 1500 N.

2-31. Determine the magnitude and orientation, measured counterclockwise from the positive y axis, of the resultant force acting on the bracket, if $F_B = 600$ N and $\theta = 20°$.

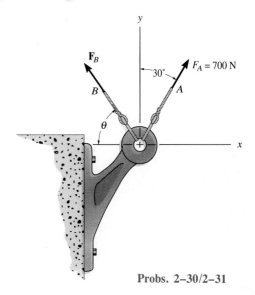

Probs. 2-30/2-31

2.5 Cartesian Vectors

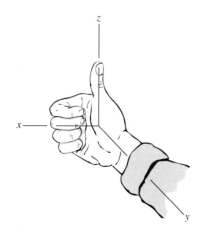

Fig. 2–21

The operations of vector algebra, when applied to solving problems in *three dimensions,* are greatly simplified if the vectors are first represented in Cartesian vector form. In this section we will present a general method for doing this, then in Sec. 2.6 we will apply this method to solving problems involving the addition of forces. Similar applications will be illustrated for the position and moment vectors given in later sections of the text.

Right-Handed Coordinate System. A right-handed coordinate system will be used for developing the theory of vector algebra that follows. A rectangular or Cartesian coordinate system is said to be *right-handed* provided the thumb of the right hand points in the direction of the positive z axis when the right-hand fingers are curled about this axis and directed from the positive x toward the positive y axis, Fig. 2–21. Furthermore, according to this rule, the z axis for a two-dimensional problem as in Fig. 2–20 would be directed outward, perpendicular to the page.

Rectangular Components of a Vector. A vector **A** may have one, two, or three rectangular components along the x, y, z coordinate axes, depending on how the vector is oriented relative to the axes. In general, though, when **A** is directed within an octant of the x, y, z frame, Fig. 2–22, then by two successive applications of the parallelogram law, we may resolve the vector into components as $\mathbf{A} = \mathbf{A}' + \mathbf{A}_z$ and then $\mathbf{A}' = \mathbf{A}_x + \mathbf{A}_y$. Combining these equations, **A** is represented by the vector sum of its *three* rectangular components,

$$\mathbf{A} = \mathbf{A}_x + \mathbf{A}_y + \mathbf{A}_z \tag{2–2}$$

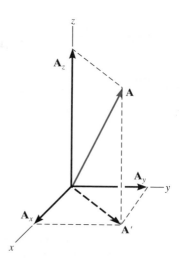

Fig. 2–22

Unit Vector. In general, a *unit vector* is a vector having a magnitude of 1. If **A** is a vector having a magnitude $A \neq 0$, then a unit vector having the *same direction* as **A** is represented by

$$\mathbf{u}_A = \frac{\mathbf{A}}{A} \qquad (2\text{--}3)$$

Rewriting this equation gives

$$\mathbf{A} = A\mathbf{u}_A \qquad (2\text{--}4)$$

Since vector **A** is of a certain type, e.g., a force vector, it is customary to use the proper set of units for its description. The magnitude A also has this same set of units; hence, from Eq. 2–3, the *unit vector will be dimensionless* since the units will cancel out. Equation 2–4 therefore indicates that vector **A** may be expressed in terms of both its magnitude and direction *separately;* i.e., A (a positive scalar) defines the *magnitude* of **A,** and $\mathbf{u}_A$ (a dimensionless vector) defines the *direction* and sense of **A,** Fig. 2–23.

Fig. 2–23

Cartesian Unit Vectors. In three dimensions, the set of Cartesian unit vectors, **i, j, k,** is used to designate the directions of the *x, y, z* axes respectively. As stated in Sec. 2–4, the *sense* (or arrowhead) of these vectors will be described analytically by a plus or minus sign, depending on whether they are pointing along the positive or negative *x, y,* or *z* axis. Thus the positive unit vectors are shown in Fig. 2–24.

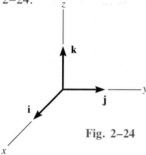

Fig. 2–24

Cartesian Vector Representation. Using Cartesian unit vectors, the three vector components of Eq. 2–2 may be written in "Cartesian vector form." Since the components act in the positive **i, j,** and **k** directions, Fig. 2–25, we have

$$\boxed{\mathbf{A} = A_x\mathbf{i} + A_y\mathbf{j} + A_z\mathbf{k}} \qquad (2\text{--}5)$$

There is a distinct advantage to writing vectors in terms of their Cartesian components. Since each of these components has the same form as Eq. 2–4, the *magnitude* and *direction* of each *component vector* are *separated,* and it will be shown that this will simplify the operations of vector algebra, particularly in three dimensions.

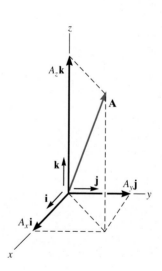

Fig. 2–25

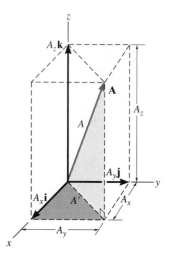

Fig. 2–26

Magnitude of a Cartesian Vector. It is always possible to obtain the magnitude of vector **A** provided the vector is expressed in Cartesian vector form. As shown in Fig. 2–26, from the colored right triangle, $A = \sqrt{A'^2 + A_z^2}$, and from the shaded right triangle, $A' = \sqrt{A_x^2 + A_y^2}$. Combining these equations yields

$$A = \sqrt{A_x^2 + A_y^2 + A_z^2} \qquad (2\text{–}6)$$

Hence, the magnitude of **A** *is equal to the positive square root of the sum of the squares of its components.*

Direction of a Cartesian Vector. The *orientation* of vector **A** is defined by the *coordinate direction angles* α (alpha), β (beta), and γ (gamma), measured between the *tail* of **A** and the *positive x, y, z axes* located at the tail of **A,** Fig. 2–27. Note that regardless of where **A** is directed, each of these angles will be between 0° and 180°. To determine α, β, and γ, consider the projection of **A** onto the *x, y, z* axes, Fig. 2–28. Referring to the colored right triangles shown in each figure, we have

$$\cos \alpha = \frac{A_x}{A} \qquad \cos \beta = \frac{A_y}{A} \qquad \cos \gamma = \frac{A_z}{A} \qquad (2\text{–}7)$$

These numbers are known as the *direction cosines* of **A.** Once they have been obtained, the coordinate direction angles α, β, γ can then be determined from the inverse cosines.

An easy way of obtaining the direction cosines of **A** is to form a unit vector in the direction of **A,** Eq. 2–3. Provided **A** is expressed in Cartesian

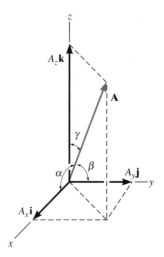

Fig. 2–27

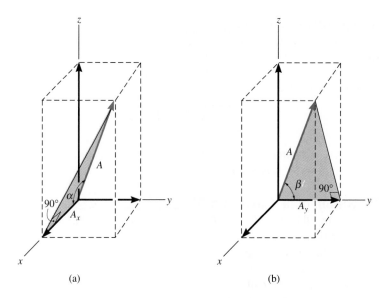

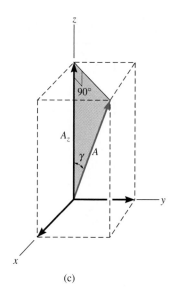

(a)   (b)   (c)

Fig. 2–28

vector form as $\mathbf{A} = A_x\mathbf{i} + A_y\mathbf{j} + A_z\mathbf{k}$ (Eq. 2–5), we have

$$\mathbf{u}_A = \frac{\mathbf{A}}{A} = \frac{A_x}{A}\mathbf{i} + \frac{A_y}{A}\mathbf{j} + \frac{A_z}{A}\mathbf{k} \qquad (2\text{–}8)$$

where $A = \sqrt{(A_x)^2 + (A_y)^2 + (A_z)^2}$ (Eq. 2–6). By comparison with Eqs. 2–7, it is seen that *the* $\mathbf{i}$, $\mathbf{j}$, *and* $\mathbf{k}$ *components of* $\mathbf{u}_A$ *represent the direction cosines of* $\mathbf{A}$, i.e.,

$$\mathbf{u}_A = \cos\alpha\mathbf{i} + \cos\beta\mathbf{j} + \cos\gamma\mathbf{k} \qquad (2\text{–}9)$$

Since the magnitude of a vector is equal to the positive square root of the sum of the squares of the magnitudes of its components, and $\mathbf{u}_A$ has a magnitude of 1, then from Eq. 2–9 an important relation between the direction cosines can be formulated as

$$\cos^2\alpha + \cos^2\beta + \cos^2\gamma = 1 \qquad (2\text{–}10)$$

Provided vector $\mathbf{A}$ lies in a known octant, this equation can be used to determine one of the coordinate direction angles if the other two are known. (See Example 2–10.)

Finally, if the magnitude and coordinate direction angles of $\mathbf{A}$ are given, $\mathbf{A}$ may be expressed in Cartesian vector form as

$$\begin{aligned}
\mathbf{A} &= A\mathbf{u}_A \\
&= A\cos\alpha\mathbf{i} + A\cos\beta\mathbf{j} + A\cos\gamma\mathbf{k} \qquad (2\text{–}11) \\
&= A_x\mathbf{i} + A_y\mathbf{j} + A_z\mathbf{k}
\end{aligned}$$

2.6 Addition and Subtraction of Cartesian Vectors

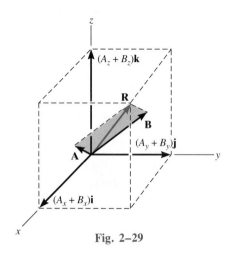

Fig. 2–29

The vector operations of addition and subtraction of two or more vectors are greatly simplified if the vectors are expressed in terms of their Cartesian components. For example, consider the two vectors **A** and **B,** both of which are directed within the positive octant of the x, y, z frame, Fig. 2–29. If $\mathbf{A} = A_x\mathbf{i} + A_y\mathbf{j} + A_z\mathbf{k}$ and $\mathbf{B} = B_x\mathbf{i} + B_y\mathbf{j} + B_z\mathbf{k}$, then the resultant vector, **R,** has components which represent the scalar sums of the **i, j,** and **k** components of **A** and **B,** i.e.,

$$\mathbf{R} = \mathbf{A} + \mathbf{B} = (A_x + B_x)\mathbf{i} + (A_y + B_y)\mathbf{j} + (A_z + B_z)\mathbf{k}$$

Vector subtraction, being a special case of vector addition, simply requires a scalar subtraction of the respective **i, j,** and **k** components of either **A** or **B.** For example,

$$\mathbf{R}' = \mathbf{A} - \mathbf{B} = (A_x - B_x)\mathbf{i} + (A_y - B_y)\mathbf{j} + (A_z - B_z)\mathbf{k}$$

Concurrent Force Systems. In particular, the above concept of vector addition may be generalized and applied to a system of several concurrent forces. In this case, the force resultant is the vector sum of all the forces in the system and can be written as

$$\boxed{\mathbf{F}_R = \Sigma\mathbf{F} = \Sigma F_x\mathbf{i} + \Sigma F_y\mathbf{j} + \Sigma F_z\mathbf{k}} \qquad (2\text{–}12)$$

Here ΣF_x, ΣF_y, and ΣF_z represent the algebraic sums of the respective x, y, z, or **i, j, k** components of each force in the system.

The following examples illustrate numerically the methods used to apply the above theory to the solution of problems involving force as a vector quantity.

Example 2–9

Determine the magnitude and the coordinate direction angles of the resultant force acting on the ring in Fig. 2–30a.

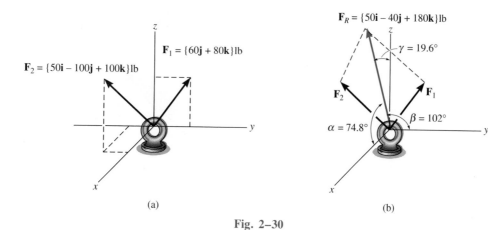

(a)

(b)

Fig. 2–30

SOLUTION

Since each force is represented in Cartesian vector form, the resultant force, shown in Fig. 2–30b, is

$$\mathbf{F}_R = \Sigma \mathbf{F} = \mathbf{F}_1 + \mathbf{F}_2 = (60\mathbf{j} + 80\mathbf{k}) + (50\mathbf{i} - 100\mathbf{j} + 100\mathbf{k})$$
$$= \{50\mathbf{i} - 40\mathbf{j} + 180\mathbf{k}\} \text{ lb}$$

The magnitude of $\mathbf{F}_R$ is found from Eq. 2–6, i.e.,

$$F_R = \sqrt{(50)^2 + (-40)^2 + (180)^2}$$
$$= 191.0 \text{ lb} \qquad \qquad \textit{Ans.}$$

The coordinate direction angles α, β, γ are determined from the components of the unit vector acting in the direction of $\mathbf{F}_R$.

$$\mathbf{u}_{F_R} = \frac{\mathbf{F}_R}{F_R} = \frac{50}{191.0}\mathbf{i} - \frac{40}{191.0}\mathbf{j} + \frac{180}{191.0}\mathbf{k}$$
$$= 0.2617\mathbf{i} - 0.2094\mathbf{j} + 0.9422\mathbf{k}$$

so that

$$\cos \alpha = 0.2617 \qquad \alpha = 74.8° \qquad \textit{Ans.}$$
$$\cos \beta = -0.2094 \qquad \beta = 102° \qquad \textit{Ans.}$$
$$\cos \gamma = 0.9422 \qquad \gamma = 19.6° \qquad \textit{Ans.}$$

These angles are shown in Fig. 2–30b. In particular, note that $\beta > 90°$ since the **j** component of $\mathbf{u}_{F_R}$ is negative.

Example 2–10

Express the force **F** shown in Fig. 2–31 as a Cartesian vector.

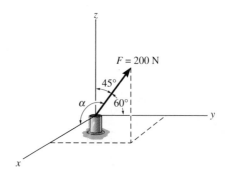

Fig. 2–31

SOLUTION

Since only two coordinate direction angles are specified, the third angle α is determined from Eq. 2–10; i.e.,

$$\cos^2 \alpha + \cos^2 \beta + \cos^2 \gamma = 1$$
$$\cos^2 \alpha + \cos^2 60° + \cos^2 45° = 1$$
$$\cos \alpha = \sqrt{1 - (0.707)^2 - (0.5)^2} = \pm 0.5$$

Hence,

$$\alpha = \cos^{-1}(0.5) = 60° \qquad \text{or} \qquad \alpha = \cos^{-1}(-0.5) = 120°$$

By inspection of Fig. 2–31, however, it is necessary that $\alpha = 60°$, since $\mathbf{F}_x$ is in the $+x$ direction.

Using Eq. 2–11, with $F = 200$ N, we have

$$\mathbf{F} = F \cos \alpha \mathbf{i} + F \cos \beta \mathbf{j} + F \cos \gamma \mathbf{k}$$
$$= 200 \cos 60° \mathbf{i} + 200 \cos 60° \mathbf{j} + 200 \cos 45° \mathbf{k}$$
$$= \{100.0\mathbf{i} + 100.0\mathbf{j} + 141.4\mathbf{k}\} \text{ N} \qquad \qquad \textit{Ans.}$$

By applying Eq. 2–6, note that indeed the magnitude of $F = 200$ N.

$$F = \sqrt{F_x^2 + F_y^2 + F_z^2}$$
$$= \sqrt{(100.0)^2 + (100.0)^2 + (141.4)^2} = 200 \text{ N}$$

Example 2–11

Express the force **F** shown acting on the hook in Fig. 2–32a as a Cartesian vector.

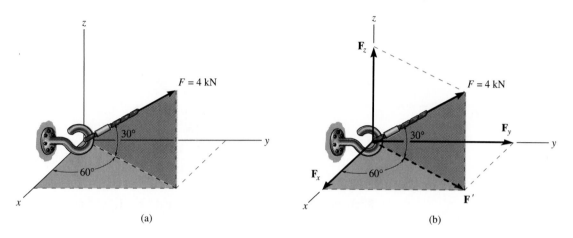

Fig. 2–32

SOLUTION

In this case the angles 60° and 30° defining the direction of **F** are *not* coordinate direction angles. Why? By two successive applications of the parallelogram law, however, **F** can be resolved into its x, y, z components as shown in Fig. 2–32b. First, from the colored triangle,

$$F' = 4 \cos 30° = 3.46 \text{ kN}$$
$$F_z = 4 \sin 30° = 2.00 \text{ kN}$$

Next, using **F**′ and the shaded triangle,

$$F_x = 3.46 \cos 60° = 1.73 \text{ kN}$$
$$F_y = 3.46 \sin 60° = 3.00 \text{ kN}$$

Thus,

$$\mathbf{F} = \{1.73\mathbf{i} + 3.00\mathbf{j} + 2.00\mathbf{k}\} \text{ kN} \qquad \textit{Ans.}$$

As an exercise, show that the magnitude of **F** is indeed 4 kN, and that the coordinate direction angle $\alpha = 64.3°$.

Example 2–12

Express force **F** shown in Fig. 2–33a as a Cartesian vector.

SOLUTION

As in Example 2–11, the angles of 60° and 45° defining the direction of **F** are *not* coordinate direction angles. The two successive applications of the parallelogram law needed to resolve **F** into its *x, y, z* components are shown in Fig. 2–33b. By trigonometry, the magnitudes of the components are

$$F_z = 100 \sin 60° = 86.6 \text{ lb}$$
$$F' = 100 \cos 60° = 50 \text{ lb}$$
$$F_x = 50 \cos 45° = 35.4 \text{ lb}$$
$$F_y = 50 \sin 45° = 35.4 \text{ lb}$$

Realizing that **F**$_y$ has a direction defined by $-\mathbf{j}$, we have

$$\mathbf{F} = \mathbf{F}_x + \mathbf{F}_y + \mathbf{F}_z$$
$$\mathbf{F} = \{35.4\mathbf{i} - 35.4\mathbf{j} + 86.6\mathbf{k}\} \text{ lb} \qquad \textit{Ans.}$$

To show that the magnitude of this vector is indeed 100 lb, apply Eq. 2–6,

$$F = \sqrt{F_x^2 + F_y^2 + F_z^2}$$
$$= \sqrt{(35.4)^2 + (-35.4)^2 + (86.6)^2} = 100 \text{ lb}$$

If needed, the coordinate direction angles of **F** can be determined from the components of the unit vector acting in the direction of **F**. Hence,

$$\mathbf{u} = \frac{\mathbf{F}}{F} = \frac{F_x}{F}\mathbf{i} + \frac{F_y}{F}\mathbf{j} + \frac{F_z}{F}\mathbf{k}$$

$$= \frac{35.4}{100}\mathbf{i} - \frac{35.4}{100}\mathbf{j} + \frac{86.6}{100}\mathbf{k}$$

$$= 0.354\mathbf{i} - 0.354\mathbf{j} + 0.866\mathbf{k}$$

so that

$$\alpha = \cos^{-1}(0.354) = 69.3°$$
$$\beta = \cos^{-1}(-0.354) = 111°$$
$$\gamma = \cos^{-1}(0.866) = 30.0°$$

These results are shown in Fig. 2–33c.

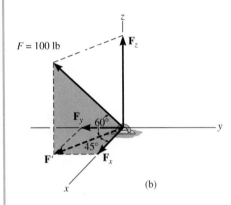

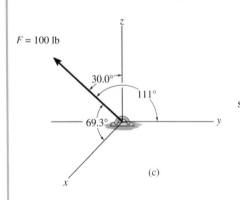

Fig. 2–33

Example 2–13

Two forces act on the hook shown in Fig. 2–34a. Specify the coordinate direction angles of $\mathbf{F}_2$ so that the resultant force $\mathbf{F}_R$ acts along the positive y axis and has a magnitude of 800 N.

SOLUTION

To solve this problem, the resultant force and its two components, $\mathbf{F}_1$ and $\mathbf{F}_2$, will each be expressed in Cartesian vector form. Then, as shown in Fig. 2–34b, it is necessary that $\mathbf{F}_R = \mathbf{F}_1 + \mathbf{F}_2$.

Applying Eq. 2–11,

$$\mathbf{F}_1 = F_1 \mathbf{u}_{F_1} = F_1 \cos \alpha_1 \mathbf{i} + F_1 \cos \beta_1 \mathbf{j} + F_1 \cos \gamma_1 \mathbf{k}$$
$$= 300 \cos 45°\mathbf{i} + 300 \cos 60°\mathbf{j} + 300 \cos 120°\mathbf{k}$$
$$= \{212.1\mathbf{i} + 150\mathbf{j} - 150\mathbf{k}\} \text{ N}$$
$$\mathbf{F}_2 = F_2 \mathbf{u}_{F_2} = F_{2x}\mathbf{i} + F_{2y}\mathbf{j} + F_{2z}\mathbf{k}$$

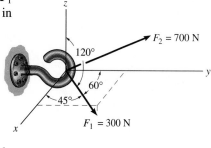

(a)

According to the problem statement, the resultant force $\mathbf{F}_R$ has a magnitude of 800 N and acts in the $+\mathbf{j}$ direction. Hence,

$$\mathbf{F}_R = (800 \text{ N})(+\mathbf{j}) = \{800\mathbf{j}\} \text{ N}$$

We require

$$\mathbf{F}_R = \mathbf{F}_1 + \mathbf{F}_2$$
$$800\mathbf{j} = 212.1\mathbf{i} + 150\mathbf{j} - 150\mathbf{k} + F_{2x}\mathbf{i} + F_{2y}\mathbf{j} + F_{2z}\mathbf{k}$$
$$800\mathbf{j} = (212.1 + F_{2x})\mathbf{i} + (150 + F_{2y})\mathbf{j} + (-150 + F_{2z})\mathbf{k}$$

To satisfy this equation, the corresponding $\mathbf{i}, \mathbf{j},$ and $\mathbf{k}$ components on the left and right sides must be equal. This is equivalent to stating that the x, y, z components of $\mathbf{F}_R$ be equal to the corresponding x, y, z components of $(\mathbf{F}_1 + \mathbf{F}_2)$. Hence,

$$0 = 212.1 + F_{2x} \qquad F_{2x} = -212.1 \text{ N}$$
$$800 = 150 + F_{2y} \qquad F_{2y} = 650 \text{ N}$$
$$0 = -150 + F_{2z} \qquad F_{2z} = 150 \text{ N}$$

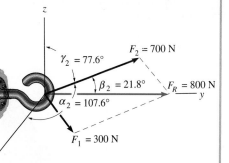

(b)

Since the magnitudes of $\mathbf{F}_2$ and its components are known, we can use Eq. 2–11 to determine α, β, γ.

$$-212.1 = 700 \cos \alpha_2; \qquad \alpha_2 = \cos^{-1}\left(\frac{-212.1}{700}\right) = 108° \quad \textit{Ans.}$$

$$650 = 700 \cos \beta_2; \qquad \beta_2 = \cos^{-1}\left(\frac{650}{700}\right) = 21.8° \quad \textit{Ans.}$$

$$150 = 700 \cos \gamma_2; \qquad \gamma_2 = \cos^{-1}\left(\frac{150}{700}\right) = 77.6° \quad \textit{Ans.}$$

Fig. 2–34

These results are shown in Fig. 2–34b.

PROBLEMS

*2–32. Express each force as a Cartesian vector and then determine the resultant force $\mathbf{F}_R$. Find the magnitude and coordinate direction angles of the resultant force and sketch this vector on the coordinate system.

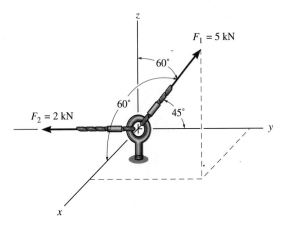

Prob. 2–32

2–33. Express each force as a Cartesian vector and then determine the resultant force $\mathbf{F}_R$. Find the magnitude and coordinate direction angles of the resultant force and sketch this vector on the coordinate system.

2–34. Express each force as a Cartesian vector and then determine the resultant force $\mathbf{F}_R$. Find the magnitude and coordinate direction angles of the resultant force and sketch this vector on the coordinate system.

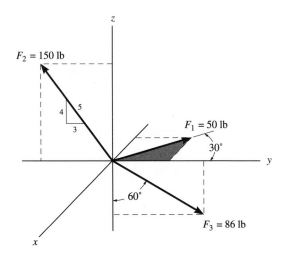

Prob. 2–34

2–35. The cable at the end of the beam exerts a force of 450 lb on the beam as shown. Express $\mathbf{F}$ as a Cartesian vector.

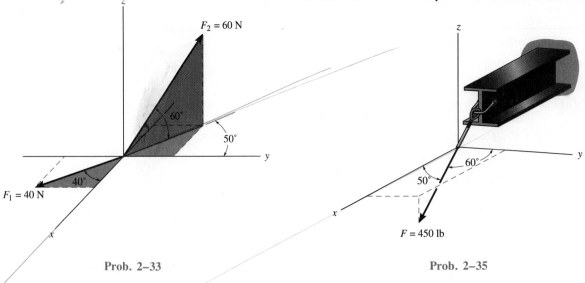

Prob. 2–33

Prob. 2–35

*2–36. The two forces $\mathbf{F}_1$ and $\mathbf{F}_2$ acting at the end of the pipe have a resultant force of $\mathbf{F}_R = \{120\mathbf{i}\}$ N. Determine the magnitude and coordinate direction angles of $\mathbf{F}_2$.

2–37. Determine the coordinate direction angles of the force $\mathbf{F}_1$ and indicate them on the figure.

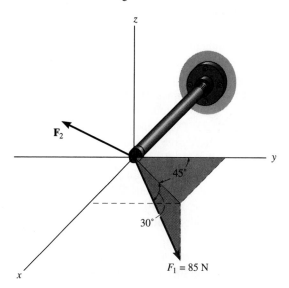

Probs. 2–36/2–37

2–38. Three forces act on the screw eye. If the resultant force $\mathbf{F}_R$ has a magnitude and orientation as shown, determine the magnitude and coordinate direction angles of force $\mathbf{F}_3$.

2–39. Determine the coordinate direction angles of $\mathbf{F}_1$ and $\mathbf{F}_R$.

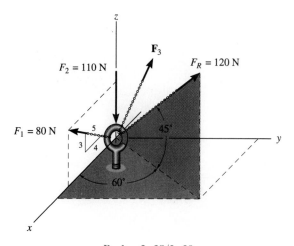

Probs. 2–38/2–39

*2–40. Determine the magnitude and coordinate direction angles of $\mathbf{F}_2$ so that the resultant of the two forces acts along the positive x axis and has a magnitude of 350 N.

2–41. Determine the magnitude and coordinate direction angles of $\mathbf{F}_2$ so that the resultant of the two forces is zero.

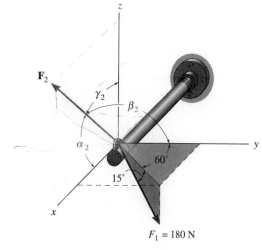

Probs. 2–40/2–41

2–42. The pipe is subjected to the force $\mathbf{F}$, which has components acting along the x, y, z axes as shown. If the magnitude of $\mathbf{F}$ is 12 kN, and $\alpha = 120°$ and $\gamma = 45°$, determine the magnitudes of its three components.

2–43. The pipe is subjected to the force $\mathbf{F}$ which has components $F_x = 1.5$ kN and $F_z = 1.25$ kN. If $\beta = 75°$, determine the magnitudes of $\mathbf{F}$ and $\mathbf{F}_y$.

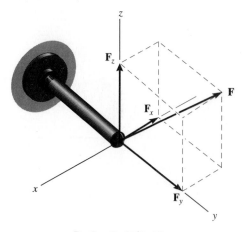

Probs. 2–42/2–43

2.7 Position Vectors

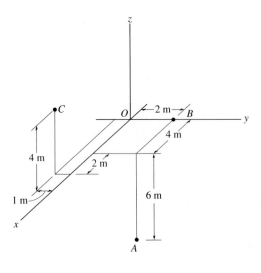

Fig. 2–35

In this section we will introduce the concept of a position vector. It will be shown in Sec. 2.8 that this vector is of importance in formulating a Cartesian force vector directed between any two points in space, and later, in Sec. 3–3, we will use it for finding the moment of a force.

***x, y, z* Coordinates.** Throughout this text we will use a *right-handed* coordinate system to reference the location of points in space. Furthermore, we will use the convention followed in many technical books, and that is to require the positive *z* axis to be directed *upward* (the zenith direction) so that it measures the height of an object or the altitude of a point. The *x, y* axes then lie in the horizontal plane, Fig. 2–35. Points in space are located relative to the origin of coordinates, *O*, by successive measurements along the *x, y, z* axes. For example, in Fig. 2–35 the coordinates of point *A* are obtained by starting at *O* and measuring $x_A = +4$ m along the *x* axis, $y_A = +2$ m along the *y* axis, and $z_A = -6$ m along the *z* axis. Thus, $A(4, 2, -6)$. In a similar manner, measurements along the *x, y, z* axes from *O* to *B* yield the coordinates of *B*, i.e., $B(0, 2, 0)$. Also notice that $C(6, -1, 4)$.

Position Vector. The *position vector* **r** is defined as a fixed vector which locates a point in space relative to another point. For example, if **r** extends from the origin of coordinates, *O*, to point $P(x, y, z)$, Fig. 2–36*a*, then **r** can be expressed in Cartesian vector form as

$$\mathbf{r} = x\mathbf{i} + y\mathbf{j} + z\mathbf{k}$$

In particular, note how the head-to-tail vector addition of the three components yields vector **r**, Fig. 2–36*b*. Starting at the origin *O*, one travels *x* in the $+\mathbf{i}$ direction, then *y* in the $+\mathbf{j}$ direction, and finally *z* in the $+\mathbf{k}$ direction to arrive at point $P(x, y, z)$.

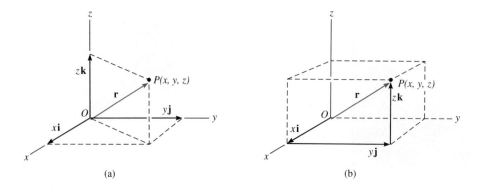

(a) (b)

Fig. 2–36

In the more general case, the position vector may be directed from point A to point B in space, Fig. 2–37a. As noted, this vector is also designated by the symbol **r.** As a matter of convention, however, we will *sometimes* refer to this vector with *two subscripts* to indicate from and to the point where it is directed, thus, **r** can also be designated as $\mathbf{r}_{AB}$. Also, note that $\mathbf{r}_A$ and $\mathbf{r}_B$ in Fig. 2–37a are referenced with only one subscript since they extend from the origin of coordinates.

From Fig. 2–37a, by the head-to-tail vector addition, we require

$$\mathbf{r}_A + \mathbf{r} = \mathbf{r}_B$$

Solving for **r** and expressing $\mathbf{r}_A$ and $\mathbf{r}_B$ in Cartesian vector form yields

$$\mathbf{r} = \mathbf{r}_B - \mathbf{r}_A = (x_B\mathbf{i} + y_B\mathbf{j} + z_B\mathbf{k}) - (x_A\mathbf{i} + y_A\mathbf{j} + z_A\mathbf{k})$$

or

$$\mathbf{r} = (x_B - x_A)\mathbf{i} + (y_B - y_A)\mathbf{j} + (z_B - z_A)\mathbf{k} \qquad (2\text{--}13)$$

Thus, the **i, j, k** *components of the position vector* **r** *may be formed by taking the coordinates of the tail of the vector, $A(x_A, y_A, z_A)$, and subtracting them from the corresponding coordinates of the head, $B(x_B, y_B, z_B)$.* Again note how the head-to-tail addition of these three components yields **r,** i.e., going from A to B, Fig. 2–37b, one first travels $(x_B - x_A)$ in the $+\mathbf{i}$ direction, then $(y_B - y_A)$ in the $+\mathbf{j}$ direction, and finally $(z_B - z_A)$ in the $+\mathbf{k}$ direction.

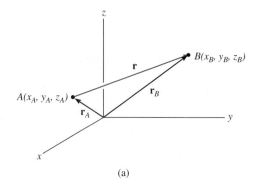

(a)

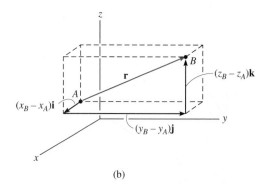

(b)

Fig. 2–37

Example 2–14

Determine the magnitude and direction of the position vector extending from A to B in Fig. 2–38a.

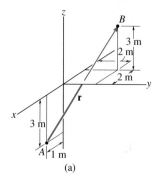

(a)

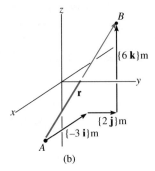

(b)

SOLUTION

In accordance with Eq. 2–13, the coordinates of the tail $A(1, 0, -3)$ are subtracted from the coordinates of the head $B(-2, 2, 3)$, which yields

$$\mathbf{r} = (-2 - 1)\mathbf{i} + (2 - 0)\mathbf{j} + [3 - (-3)]\mathbf{k}$$
$$= \{-3\mathbf{i} + 2\mathbf{j} + 6\mathbf{k}\} \text{ m}$$

As shown in Fig. 2–38b, the three components of $\mathbf{r}$ can also be obtained in a more *direct manner* by realizing that in going from A to B one must move along the x axis $\{-3\mathbf{i}\}$ m, along the y axis $\{2\mathbf{j}\}$ m, and finally along the z axis $\{6\mathbf{k}\}$ m.

The magnitude of $\mathbf{r}$ is thus

$$r = \sqrt{(-3)^2 + (2)^2 + (6)^2} = 7 \text{ m} \qquad \textit{Ans.}$$

Formulating a unit vector in the direction of $\mathbf{r}$, we have

$$\mathbf{u} = \frac{\mathbf{r}}{r} = \frac{-3}{7}\mathbf{i} + \frac{2}{7}\mathbf{j} + \frac{6}{7}\mathbf{k}$$

The components of this unit vector yield the coordinate direction angles

$$\alpha = \cos^{-1}\left(\frac{-3}{7}\right) = 115° \qquad \textit{Ans.}$$

$$\beta = \cos^{-1}\left(\frac{2}{7}\right) = 73.4° \qquad \textit{Ans.}$$

$$\gamma = \cos^{-1}\left(\frac{6}{7}\right) = 31.0° \qquad \textit{Ans.}$$

These angles are measured from the *positive axes* of a localized coordinate system placed at the tail of $\mathbf{r}$ as shown in Fig. 2–38c.

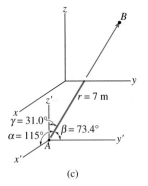

(c)

Fig. 2–38

2.8 Force Vector Directed Along a Line

Quite often in three-dimensional statics problems, the direction of a force is specified by two points through which its line of action passes. Such a situation is shown in Fig. 2–39, where the force **F** is directed along the cord *AB*. We can formulate **F** as a Cartesian vector by realizing that it has the *same direction* and *sense* as the position vector **r** directed from point *A* to point *B* on the cord. This common direction is specified by the *unit vector* **u** = **r**/r. Hence,

$$\mathbf{F} = F\mathbf{u} = F\left(\frac{\mathbf{r}}{r}\right)$$

Although we have represented **F** symbolically in Fig. 2–39, note that it has units of force, and unlike **r**, or coordinates *x*, *y*, *z*, which have units of length, **F** cannot be scaled along the coordinate axes.

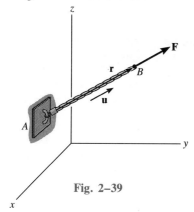

Fig. 2–39

PROCEDURE FOR ANALYSIS

When **F** is directed along a line which extends from point *A* to point *B*, then **F** can be expressed in Cartesian vector form as follows:

Position Vector. Determine the position vector **r** directed from *A* to *B*, and compute its magnitude *r*.

Unit Vector. Determine the unit vector **u** = **r**/r which defines the *direction* and *sense* of *both* **r** and **F**.

Force Vector. Determine **F** by combining its magnitude *F* and direction **u**, i.e., **F** = *F***u**.

This procedure is illustrated numerically in the following example problems.

Example 2–15

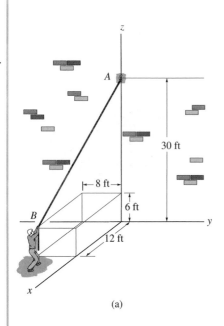

(a)

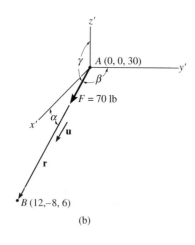

(b)

Fig. 2–40

The man shown in Fig. 2–40*a* pulls on the cord with a force of 70 lb. Represent this force, acting on the support *A*, as a Cartesian vector and determine its direction.

SOLUTION

Force **F** is shown in Fig. 2–40*b*. The *direction* of this vector, **u**, is determined from the position vector **r**, which extends from *A* to *B*, Fig. 2–40*b*. To formulate **F** as a Cartesian vector we use the following procedure.

Position Vector. The coordinates of the end points of the cord are $A(0, 0, 30)$ and $B(12, -8, 6)$. Forming the position vector by subtracting the corresponding x, y, and z coordinates of *A* from those of *B*, we have

$$\mathbf{r} = (12 - 0)\mathbf{i} + (-8 - 0)\mathbf{j} + (6 - 30)\mathbf{k}$$
$$= \{12\mathbf{i} - 8\mathbf{j} - 24\mathbf{k}\} \text{ ft}$$

Show on Fig. 2–40*a* how one can write **r** *directly* by going from *A* $\{12\mathbf{i}\}$ ft, then $\{-8\mathbf{j}\}$ ft, and finally $\{-24\mathbf{k}\}$ ft to get to *B*.

The magnitude of **r**, which represents the *length* of cord *AB*, is

$$r = \sqrt{(12)^2 + (-8)^2 + (-24)^2} = 28 \text{ ft}$$

Unit Vector. Forming the unit vector that defines the direction and sense of both **r** and **F** yields

$$\mathbf{u} = \frac{\mathbf{r}}{r} = \frac{12}{28}\mathbf{i} - \frac{8}{28}\mathbf{j} - \frac{24}{28}\mathbf{k}$$

Force Vector. Since **F** has a *magnitude* of 70 lb and a *direction* specified by **u,** then

$$\mathbf{F} = F\mathbf{u} = 70 \text{ lb} \left(\frac{12}{28}\mathbf{i} - \frac{8}{28}\mathbf{j} - \frac{24}{28}\mathbf{k} \right)$$
$$= \{30\mathbf{i} - 20\mathbf{j} - 60\mathbf{k}\} \text{ lb} \qquad \textit{Ans.}$$

As shown in Fig. 2–40*b*, the coordinate direction angles are measured between **r** (or **F**) and the *positive axes* of a localized coordinate system with origin placed at *A*. From the components of the unit vector:

$$\alpha = \cos^{-1}\left(\frac{12}{28}\right) = 64.6° \qquad \textit{Ans.}$$

$$\beta = \cos^{-1}\left(\frac{-8}{28}\right) = 107° \qquad \textit{Ans.}$$

$$\gamma = \cos^{-1}\left(\frac{-24}{28}\right) = 149° \qquad \textit{Ans.}$$

Example 2–16

The circular plate shown in Fig. 2–41a is partially supported by the cable AB. If the force of the cable on the hook at A is F = 500 N, express **F** as a Cartesian vector.

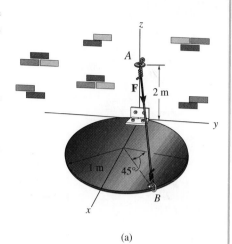

(a)

SOLUTION

As shown in Fig. 2–41b, **F** has the same direction and sense as the position vector **r**, which extends from A to B.

Position Vector.   The coordinates of the end points of the cable are A(0, 0, 2) and B(1.707, 0.707, 0), as indicated in the figure. Thus,

$$\mathbf{r} = (1.707 - 0)\mathbf{i} + (0.707 - 0)\mathbf{j} + (0 - 2)\mathbf{k}$$
$$= \{1.707\mathbf{i} + 0.707\mathbf{j} - 2\mathbf{k}\} \text{ m}$$

Note how one can calculate these components *directly* by going from A, $\{-2\mathbf{k}\}$ m along the z axis, then $\{1.707\mathbf{i}\}$ m along the x axis, and finally $\{0.707\mathbf{j}\}$ m along the y axis to get to B.

The magnitude of **r** is

$$r = \sqrt{(1.707)^2 + (0.707)^2 + (-2)^2} = 2.72 \text{ m}$$

Unit Vector.   Thus,

$$\mathbf{u} = \frac{\mathbf{r}}{r} = \frac{1.707}{2.72}\mathbf{i} + \frac{0.707}{2.72}\mathbf{j} - \frac{2}{2.72}\mathbf{k}$$
$$= 0.627\mathbf{i} + 0.260\mathbf{j} - 0.735\mathbf{k}$$

Force Vector.   Since F = 500 N and **F** has the direction **u**, we have

$$\mathbf{F} = F\mathbf{u} = 500 \text{ N}(0.627\mathbf{i} + 0.260\mathbf{j} - 0.735\mathbf{k})$$
$$= \{314\mathbf{i} + 130\mathbf{j} - 368\mathbf{k}\} \text{ N} \qquad\qquad Ans.$$

Using these components, notice that indeed the magnitude of **F** is 500 N; i.e.,

$$F = \sqrt{(314)^2 + (130)^2 + (-368)^2} = 500 \text{ N}$$

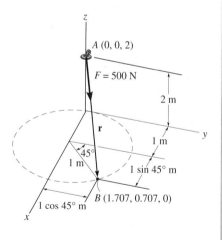

(b)

Fig. 2–41

Show that the coordinate direction angle $\gamma = 137°$, and indicate this angle on the figure.

Example 2–17

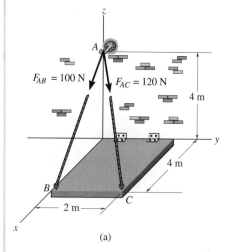

(a)

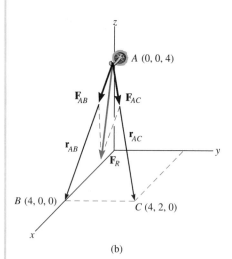

(b)

Fig. 2–42

The cables exert forces $F_{AB} = 100$ N and $F_{AC} = 120$ N on the ring at A as shown in Fig. 2–42a. Determine the magnitude of the resultant force acting at A.

SOLUTION

The resultant force $\mathbf{F}_R$ is shown graphically in Fig. 2–42b. We can express this force as a Cartesian vector by first formulating $\mathbf{F}_{AB}$ and $\mathbf{F}_{AC}$ as Cartesian vectors and then adding their components. The directions of $\mathbf{F}_{AB}$ and $\mathbf{F}_{AC}$ are specified by forming unit vectors $\mathbf{u}_{AB}$ and $\mathbf{u}_{AC}$ along the cables. These unit vectors are obtained from the associated position vectors $\mathbf{r}_{AB}$ and $\mathbf{r}_{AC}$. With reference to Fig. 2–42b for $\mathbf{F}_{AB}$ we have

$$\mathbf{r}_{AB} = (4 - 0)\mathbf{i} + (0 - 0)\mathbf{j} + (0 - 4)\mathbf{k}$$
$$= \{4\mathbf{i} - 4\mathbf{k}\} \text{ m}$$
$$r_{AB} = \sqrt{(4)^2 + (-4)^2} = 5.66 \text{ m}$$
$$\mathbf{F}_{AB} = 100 \text{ N}\left(\frac{\mathbf{r}_{AB}}{r_{AB}}\right) = 100 \text{ N}\left(\frac{4}{5.66}\mathbf{i} - \frac{4}{5.66}\mathbf{k}\right)$$
$$\mathbf{F}_{AB} = \{70.7\mathbf{i} - 70.7\mathbf{k}\} \text{ N}$$

For $\mathbf{F}_{AC}$ we have

$$\mathbf{r}_{AC} = (4 - 0)\mathbf{i} + (2 - 0)\mathbf{j} + (0 - 4)\mathbf{k}$$
$$= \{4\mathbf{i} + 2\mathbf{j} - 4\mathbf{k}\} \text{ m}$$
$$r_{AC} = \sqrt{(4)^2 + (2)^2 + (-4)^2} = 6 \text{ m}$$
$$\mathbf{F}_{AC} = 120 \text{ N}\left(\frac{\mathbf{r}_{AC}}{r_{AC}}\right) = 120 \text{ N}\left(\frac{4}{6}\mathbf{i} + \frac{2}{6}\mathbf{j} - \frac{4}{6}\mathbf{k}\right)$$
$$= \{80\mathbf{i} + 40\mathbf{j} - 80\mathbf{k}\} \text{ N}$$

The resultant force is therefore

$$\mathbf{F}_R = \mathbf{F}_{AB} + \mathbf{F}_{AC} = (70.7\mathbf{i} - 70.7\mathbf{k}) + (80\mathbf{i} + 40\mathbf{j} - 80\mathbf{k})$$
$$= \{150.7\mathbf{i} + 40\mathbf{j} - 150.7\mathbf{k}\} \text{ N}$$

The magnitude of $\mathbf{F}_R$ is thus

$$F_R = \sqrt{(150.7)^2 + (40)^2 + (-150.7)^2}$$
$$= 217 \text{ N} \qquad \qquad Ans.$$

PROBLEMS

*2–44. A position vector extends from the origin to the point (2 m, 3 m, 6 m). Determine the coordinate direction angles α, β, and γ which the tail of the vector makes with the x, y, and z axes, respectively.

2–45. Represent the position vector $\mathbf{r}$ acting from point $A(3$ m, 5 m, 6 m) to point $B(5$ m, -2 m, 1 m) in Cartesian vector form. Calculate its coordinate direction angles, and find the distance between points A and B.

2–46. Given the three position vectors

$$\mathbf{r}_1 = \{3\mathbf{i} - 4\mathbf{j} + 3\mathbf{k}\} \text{ m}$$
$$\mathbf{r}_2 = \{4\mathbf{i} - 5\mathbf{k}\} \text{ m}$$
$$\mathbf{r}_3 = \{3\mathbf{i} - 2\mathbf{j} + 5\mathbf{k}\} \text{ m}$$

determine the magnitude and coordinate direction angles of $\mathbf{r} = 2\mathbf{r}_1 - \mathbf{r}_2 + 3\mathbf{r}_3$.

2–47. Given the three position vectors

$$\mathbf{r}_1 = \{2\mathbf{i} + 5\mathbf{j} + 4\mathbf{k}\} \text{ m}$$
$$\mathbf{r}_2 = \{3\mathbf{i} + 2\mathbf{k}\} \text{ m}$$
$$\mathbf{r}_3 = \{-2\mathbf{i} + 4\mathbf{j}\} \text{ m}$$

determine the magnitude and coordinate direction angles of $\mathbf{r} = \mathbf{r}_1 - \mathbf{r}_2 + \frac{1}{2}\mathbf{r}_3$.

*2–48. Express the position vector $\mathbf{r}$ in Cartesian vector form; then determine its magnitude and coordinate direction angles.

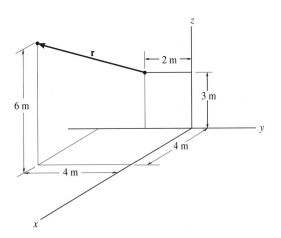

Prob. 2–48

2–49. Express the position vector $\mathbf{r}$ in Cartesian vector form; then determine its magnitude and coordinate direction angles.

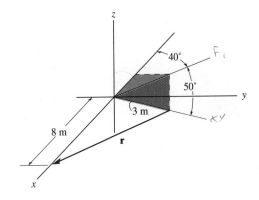

Prob. 2–49

2–50. Express the position vector $\mathbf{r}$ in Cartesian vector form; then determine its magnitude and coordinate direction angles.

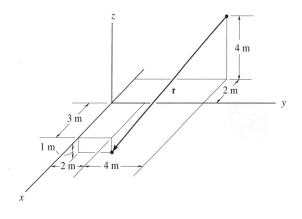

Prob. 2–50

2–51. Determine the length of the connecting rod AB by first formulating a Cartesian position vector from A to B and then determining its magnitude.

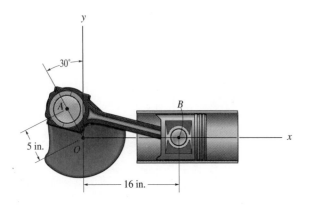

Prob. 2–51

*2–52.** The 80-ft-long cable is attached to the boom at A. If $x = 60$ ft, determine the coordinate location (y, z) to the end of the cable B. Take $z = 3y$.

2–53. The cable is attached to the boom at A and to point B, which has coordinates of $x = 80$ ft and $y = 40$ ft. If $z = 40$ ft, determine the length of the cable and the angle between the positive z axis and the cable.

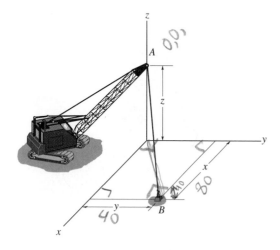

Probs. 2–52/2–53

2–54. Determine the lengths of wires AD, BD, and CD. The ring at D is midway between A and B.

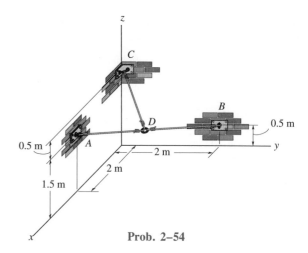

Prob. 2–54

2–55. Express the force as a Cartesian vector and then determine its coordinate direction angles.

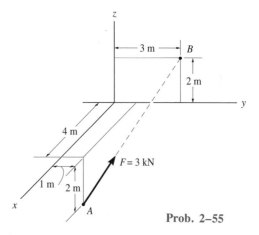

Prob. 2–55

*2–56.** Express the force as a Cartesian vector and then determine its coordinate direction angles.

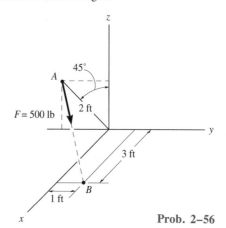

Prob. 2–56

2–57. Express the force as a Cartesian vector; then determine its coordinate direction angles.

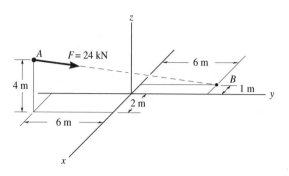

Prob. 2–57

2–59. Express each of the two forces in Cartesian vector form and then determine the magnitude and coordinate direction angles of the resultant force.

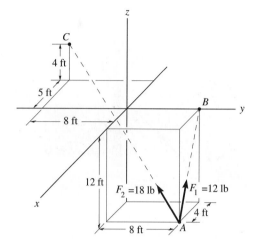

Prob. 2–59

2–58. Express each of the two forces in Cartesian vector form.

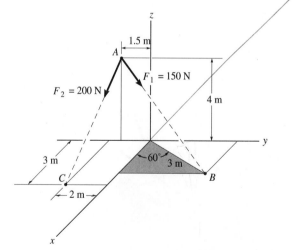

Prob. 2–58

*2–60. The cable, attached to the shear-leg derrick, exerts a force on the derrick of 350 lb. Express this force as a Cartesian vector.

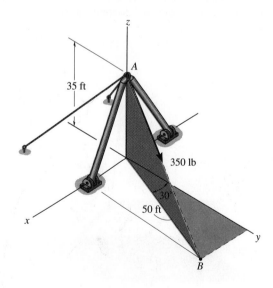

Prob. 2–60

2–61. Two tractors pull on the tree with the forces shown. Represent each force as a Cartesian vector and then determine the magnitude and coordinate direction angles of the resultant force.

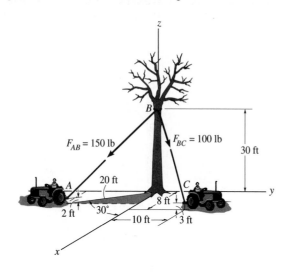

Prob. 2–61

2–62. The window is held open by chain AB. Determine the length of the chain, and express the 50-lb force acting at A along the chain as a Cartesian vector and determine its coordinate direction angles.

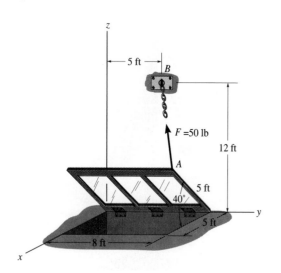

Prob. 2–62

2–63. The cable OA exerts a force on point O of $\mathbf{F} = \{-4\mathbf{i} + 3\mathbf{j} + 10\mathbf{k}\}$ lb, directed from O to A. If the length of the cable is 5 ft, what are the coordinates $(-x, y, z)$ of point A?

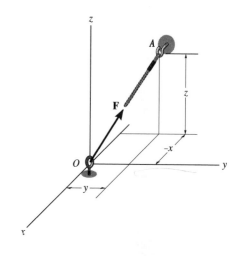

Prob. 2–63

***2–64.** The three supporting cables exert the forces shown on the sign. Represent each force as a Cartesian vector.

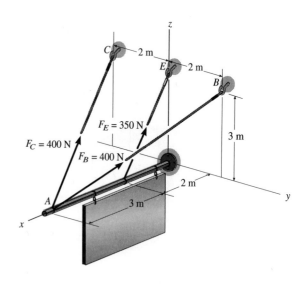

Prob. 2–64

2–65. The cylindrical vessel is supported by three cables which are concurrent at point D. Express each force which the cables exert on the vessel as a Cartesian vector, and determine the magnitude and coordinate direction angles of the resultant force.

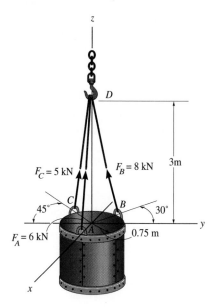

Prob. 2–65

2–66. The pole is held in place by three cables. If the force of each cable acting on the pole is shown, determine the magnitude and coordinate direction angles α, β, γ of the resultant force. Set $x = 4$ m, $y = 2$ m.

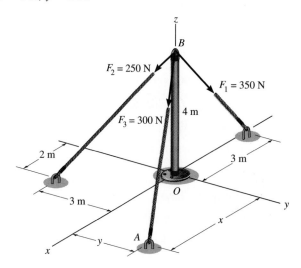

Prob. 2–66

2–67. The pole is held in place by three cables. If the force of each cable acting on the pole is shown, determine the position $(x, y, 0)$ for fixing cable BA so that the resultant force exerted on the pole is directed along its axis, from B toward O. Also, what is the magnitude of the resultant force?

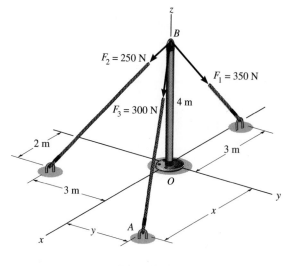

Prob. 2–67

*2–68. A force $\mathbf{F}$ is applied at the top of the tower at A. If it acts in the direction shown such that its vertical component has a magnitude of 80 lb, express $\mathbf{F}$ as a Cartesian vector and then determine its coordinate direction angles.

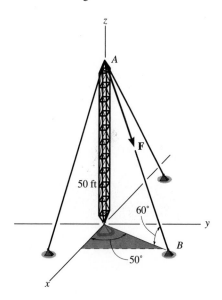

Prob. 2–68

2.9 Dot Product

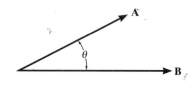

Fig. 2–43

Occasionally in statics one has to find the angle between two lines or the components of a force parallel and perpendicular to a line. In two dimensions, these problems can readily be solved by trigonometry since the geometry is easy to visualize. In three dimensions, however, this is often difficult, and consequently vector methods should be employed for the solution. The dot product defines a particular method for "multiplying" two vectors and is used to solve the above-mentioned problems.

The *dot product* of vectors **A** and **B**, written **A · B**, and read "A dot B," is defined as the product of the magnitudes of **A** and **B** and the cosine of the angle θ between their tails, Fig. 2–43. Expressed in equation form,

$$\boxed{\mathbf{A} \cdot \mathbf{B} = AB \cos \theta} \tag{2–14}$$

where $0° \leq \theta \leq 180°$. The dot product is often referred to as the *scalar product* of vectors, since the result is a *scalar* and not a vector.

Laws of Operation

1. Commutative law:

$$\mathbf{A} \cdot \mathbf{B} = \mathbf{B} \cdot \mathbf{A}$$

2. Multiplication by a scalar:

$$a(\mathbf{A} \cdot \mathbf{B}) = (a\mathbf{A}) \cdot \mathbf{B} = \mathbf{A} \cdot (a\mathbf{B}) = (\mathbf{A} \cdot \mathbf{B})a$$

3. Distributive law:

$$\mathbf{A} \cdot (\mathbf{B} + \mathbf{D}) = (\mathbf{A} \cdot \mathbf{B}) + (\mathbf{A} \cdot \mathbf{D})$$

It is easy to prove the first and second laws by using Eq. 2–14. The proof of the distributive law is left as an exercise (see Prob. 2–97).

Cartesian Vector Formulation. Equation 2–14 may be used to find the dot product for each of the Cartesian unit vectors. For example, $\mathbf{i} \cdot \mathbf{i} = (1)(1) \cos 0° = 1$ and $\mathbf{i} \cdot \mathbf{j} = (1)(1) \cos 90° = 0$. In a similar manner,

$$\mathbf{i} \cdot \mathbf{i} = 1 \qquad \mathbf{j} \cdot \mathbf{j} = 1 \qquad \mathbf{k} \cdot \mathbf{k} = 1$$
$$\mathbf{i} \cdot \mathbf{j} = 0 \qquad \mathbf{i} \cdot \mathbf{k} = 0 \qquad \mathbf{k} \cdot \mathbf{j} = 0$$

These results should not be memorized; rather, it should be clearly understood how each is obtained.

Consider now the dot product of two general vectors **A** and **B** which are expressed in Cartesian vector form. We have

$$\mathbf{A} \cdot \mathbf{B} = (A_x\mathbf{i} + A_y\mathbf{j} + A_z\mathbf{k}) \cdot (B_x\mathbf{i} + B_y\mathbf{j} + B_z\mathbf{k})$$
$$= A_xB_x(\mathbf{i} \cdot \mathbf{i}) + A_xB_y(\mathbf{i} \cdot \mathbf{j}) + A_xB_z(\mathbf{i} \cdot \mathbf{k})$$
$$+ A_yB_x(\mathbf{j} \cdot \mathbf{i}) + A_yB_y(\mathbf{j} \cdot \mathbf{j}) + A_yB_z(\mathbf{j} \cdot \mathbf{k})$$
$$+ A_zB_x(\mathbf{k} \cdot \mathbf{i}) + A_zB_y(\mathbf{k} \cdot \mathbf{j}) + A_zB_z(\mathbf{k} \cdot \mathbf{k})$$

Carrying out the dot-product operations, the final result becomes

$$\boxed{\mathbf{A} \cdot \mathbf{B} = A_x B_x + A_y B_y + A_z B_z} \qquad (2\text{–}15)$$

Thus, to determine the dot product of two Cartesian vectors, multiply their corresponding x, y, z components and sum their products algebraically. Since the result is a scalar, be careful *not* to include any unit vectors in the final result.

Applications. The dot product has two important applications in mechanics.

1. *The angle formed between two vectors or intersecting lines.* The angle θ between the tails of vectors **A** and **B** in Fig. 2–43 can be determined from Eq. 2–14 and written as

$$\theta = \cos^{-1}\left(\frac{\mathbf{A} \cdot \mathbf{B}}{AB}\right) \qquad 0° \leq \theta \leq 180°$$

Here $\mathbf{A} \cdot \mathbf{B}$ is computed from Eq. 2–15. In particular, notice that if $\mathbf{A} \cdot \mathbf{B} = 0$, $\theta = \cos^{-1} 0 = 90°$, so that **A** will be *perpendicular* to **B**.

2. *The components of a vector parallel and perpendicular to a line.* The component of vector **A** parallel to or collinear with the line aa' in Fig. 2–44 is defined by $\mathbf{A}_{\parallel}$, where $A_{\parallel} = A \cos \theta$. This component is sometimes referred to as the *projection* of **A** onto the line, since a right angle is formed in the construction. If the *direction* of the line is specified by the unit vector **u**, then, since $u = 1$, we can determine $A_{\parallel}$ directly from the dot product (Eq. 2–14); i.e.,

$$A_{\parallel} = A \cos \theta = \mathbf{A} \cdot \mathbf{u}$$

*Hence, the scalar projection of **A** along a line is determined from the dot product of **A** and the unit vector **u** which defines the direction of the line.* Notice that if this result is positive, then $\mathbf{A}_{\parallel}$ has a directional sense which is the same as **u**, whereas if $A_{\parallel}$ is a negative scalar, then $\mathbf{A}_{\parallel}$ has the opposite sense of direction to **u**. The component $\mathbf{A}_{\parallel}$ represented as a *vector* is therefore

$$\mathbf{A}_{\parallel} = A \cos \theta \,\mathbf{u} = (\mathbf{A} \cdot \mathbf{u})\mathbf{u}$$

Note that the component of **A** which is *perpendicular* to line aa' can also be obtained, Fig. 2–44. Since $\mathbf{A} = \mathbf{A}_{\parallel} + \mathbf{A}_{\perp}$ then $\mathbf{A}_{\perp} = \mathbf{A} - \mathbf{A}_{\parallel}$. There are two possible ways of obtaining $A_{\perp}$. The first would be to determine θ from the dot product, $\theta = \cos^{-1}(\mathbf{A} \cdot \mathbf{u}/A)$, then $A_{\perp} = A \sin \theta$. Alternatively, if $A_{\parallel}$ is known, then by the Pythagorean theorem we can also write $A_{\perp} = \sqrt{A^2 - A_{\parallel}^2}$.

The above two applications are illustrated numerically in the following example problems.

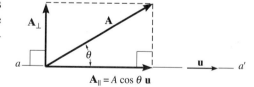

Fig. 2–44

Example 2–18

The frame shown in Fig. 2–45*a* is subjected to a horizontal force $\mathbf{F} = \{300\mathbf{j}\}$ N acting at its corner. Determine the magnitude of the components of this force parallel and perpendicular to member *AB*.

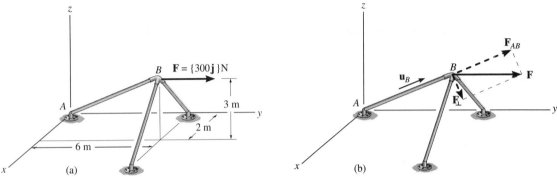

Fig. 2–45

SOLUTION

The magnitude of the component of $\mathbf{F}$ along *AB* is equal to the dot product of $\mathbf{F}$ and the unit vector $\mathbf{u}_B$ which defines the direction of *AB*, Fig. 2–45*b*. Since

$$\mathbf{u}_B = \frac{\mathbf{r}_B}{r_B} = \frac{2\mathbf{i} + 6\mathbf{j} + 3\mathbf{k}}{\sqrt{(2)^2 + (6)^2 + (3)^2}} = 0.286\mathbf{i} + 0.857\mathbf{j} + 0.429\mathbf{k}$$

Then

$$F_{AB} = F \cos\theta = \mathbf{F} \cdot \mathbf{u}_B = (300\mathbf{j}) \cdot (0.286\mathbf{i} + 0.857\mathbf{j} + 0.429\mathbf{k})$$
$$= (0)(0.286) + (300)(0.857) + (0)(0.429)$$
$$= 257.1 \text{ N}$$

Since the result is a positive scalar, $\mathbf{F}_{AB}$ has the same sense of direction as $\mathbf{u}_B$, Fig. 2–45*b*.

Expressing $\mathbf{F}_{AB}$ in Cartesian vector form, we have

$$\mathbf{F}_{AB} = F_{AB}\mathbf{u}_B = 257.1 \text{ N}(0.286\mathbf{i} + 0.857\mathbf{j} + 0.429\mathbf{k})$$
$$= \{73.5\mathbf{i} + 220\mathbf{j} + 110\mathbf{k}\} \text{ N} \qquad \textit{Ans.}$$

The perpendicular component, Fig. 2–45*b*, is therefore

$$\mathbf{F}_{\perp} = \mathbf{F} - \mathbf{F}_{AB} = 300\mathbf{j} - (73.5\mathbf{i} + 220\mathbf{j} + 110\mathbf{k})$$
$$= \{-73.5\mathbf{i} + 80\mathbf{j} - 110\mathbf{k}\} \text{ N}$$

Its magnitude can be determined either from this vector or from the Pythagorean theorem, Fig. 2–45*b*:

$$F_{\perp} = \sqrt{F^2 - F_{AB}^2}$$
$$= \sqrt{(300)^2 - (257.1)^2}$$
$$= 155 \text{ N} \qquad \textit{Ans.}$$

Example 2–19

The pipe in Fig. 2–46a is subjected to the force $F = 80$ lb at its end B. Determine the angle θ between $\mathbf{F}$ and the pipe segment BA, and the magnitudes of the components of $\mathbf{F}$, which are parallel and perpendicular to BA.

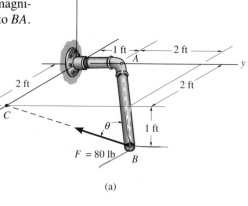

SOLUTION

Angle θ. Position vectors along BA and BC are first determined.

$$\mathbf{r}_{BA} = \{-2\mathbf{i} - 2\mathbf{j} + 1\mathbf{k}\} \text{ ft}$$
$$\mathbf{r}_{BC} = \{-3\mathbf{j} + 1\mathbf{k}\} \text{ ft}$$

Thus, the angle θ between their tails is

$$\cos\theta = \frac{\mathbf{r}_{BA} \cdot \mathbf{r}_{BC}}{r_{BA} r_{BC}} = \frac{(-2)(0) + (-2)(-3) + (1)(1)}{3\sqrt{10}}$$
$$= 0.7379$$
$$\theta = 42.5° \qquad\qquad\qquad Ans.$$

(a)

Components of F. We must first formulate the unit vector along BA and force $\mathbf{F}$ as Cartesian vectors.

$$\mathbf{u}_{BA} = \frac{\mathbf{r}_{BA}}{r_{BA}} = \frac{-2\mathbf{i} - 2\mathbf{j} + 1\mathbf{k}}{3} = -\frac{2}{3}\mathbf{i} - \frac{2}{3}\mathbf{j} + \frac{1}{3}\mathbf{k}$$

$$\mathbf{F} = 80 \text{ lb}\left(\frac{\mathbf{r}_{BC}}{r_{BC}}\right) = 80\left(\frac{-3\mathbf{j} + 1\mathbf{k}}{\sqrt{10}}\right) = -75.89\mathbf{j} + 25.30\mathbf{k}$$

Thus,

$$F_{BA} = \mathbf{F} \cdot \mathbf{u}_{BA} = (-75.89\mathbf{j} + 25.30\mathbf{k}) \cdot \left(-\frac{2}{3}\mathbf{i} - \frac{2}{3}\mathbf{j} + \frac{1}{3}\mathbf{k}\right)$$
$$= 0 + 50.60 + 8.43$$
$$= 59.0 \text{ lb} \qquad\qquad\qquad Ans.$$

Since θ was calculated in Fig. 2–46b, this same result can also be obtained directly from trigonometry.

$$F_{BA} = 80 \cos 42.5° = 59.0 \text{ lb} \qquad\qquad Ans.$$

Realize that both solutions are valid since, by the dot product,

$$F_{BA} = \mathbf{F} \cdot \mathbf{u}_{BA} = F(1) \cos\theta$$

Also,

$$F_{\perp} = F \sin\theta$$
$$= 80 \sin 42.5°$$
$$= 54.0 \text{ lb} \qquad\qquad\qquad Ans.$$

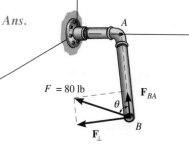

(b) Fig. 2–46

PROBLEMS

2–69. Determine the angle θ between the two position vectors.

2–70. Determine the projected component of vector $\mathbf{r}_1$ acting in the direction of $\mathbf{r}_2$. Express the result as a Cartesian vector.

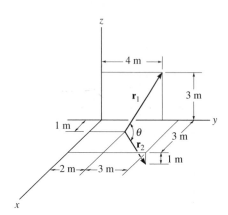

Probs. 2–69/2–70

2–71. Determine the angle θ between the two position vectors.

*2–72. Determine the projected component of vector $\mathbf{r}_1$ acting in the direction of $\mathbf{r}_2$. Express the result as a Cartesian vector.

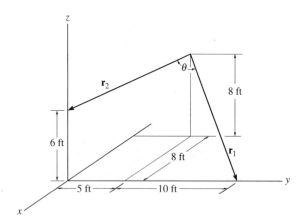

Probs. 2–71/2–72

2–73. Determine the magnitude of the projected component of the position vector $\mathbf{r}$ along the Oa axis.

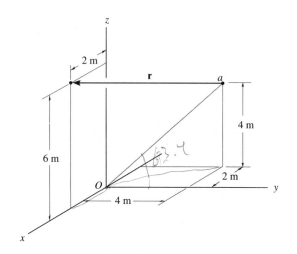

Prob. 2–73

2–74. Determine the magnitude of the projected component of the position vector $\mathbf{r}$ along the Oa axis.

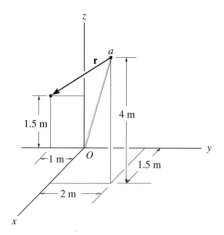

Prob. 2–74

2–75. A force of $F = 80$ N is applied to the handle of the wrench. Determine the angle θ between the tail of the force and the handle AB.

*2–76. A force of $F = 80$ N is applied to the handle of the wrench. Determine the magnitudes of the components of the force acting along the axis AB of the wrench handle and perpendicular to it.

2–78. Determine the magnitudes of the projected components of the force $\mathbf{F} = \{-80\mathbf{i} + 30\mathbf{j} + 20\mathbf{k}\}$ lb in the direction of the cables AB and AC.

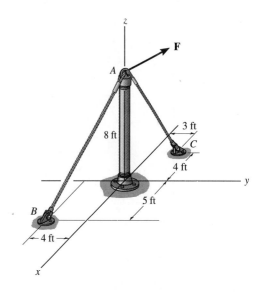

Prob. 2–78

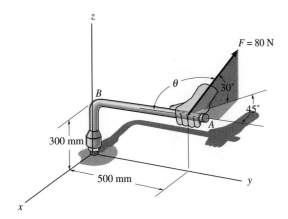

Probs. 2–75/2–76

2–77. A force of $\mathbf{F} = \{-40\mathbf{k}\}$ lb acts at the end of the pipe. Determine the magnitudes of the components $\mathbf{F}_1$ and $\mathbf{F}_2$ which act along the pipe's axis and perpendicular to it.

2–79. Determine the magnitude of the projected component of the 150-lb force acting along the axis BC of the pipe.

*2–80. Determine the angle θ between pipe segments BA and BC.

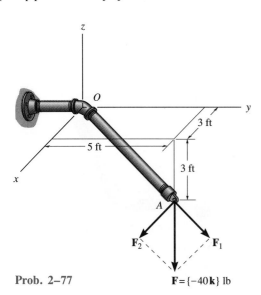

Prob. 2–77

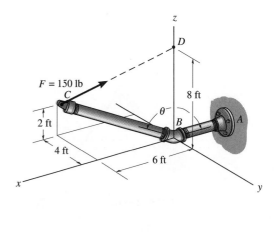

Probs. 2–79/2–80

2–81. Determine the angle θ between cables AB and AC.

2–82. If $\mathbf{F}$ has a magnitude of 55 lb, determine the magnitude of its projected component acting along the x axis and along cable AC.

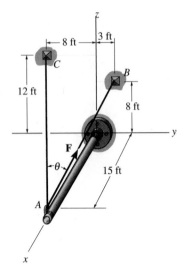

Probs. 2–81/2–82

2–83. Determine the angles θ and ϕ between the axis OA of the pole and each cable, AB and AC.

∗2–84. The two cables exert the forces shown on the pole. Determine the magnitude of the projected component of each force acting along the axis OA of the pole.

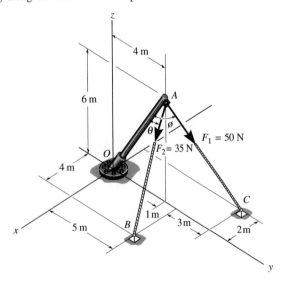

Probs. 2–83/2–84

2–85. Two cables exert forces on the pipe. Determine the magnitude of the projected component of $\mathbf{F}_1$ along the line of action of $\mathbf{F}_2$.

2–86. Determine the angle θ between the two cables attached to the pipe.

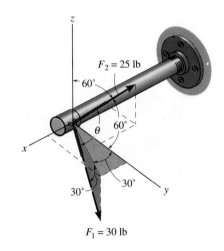

Probs. 2–85/2–86

2–87. The force $\mathbf{F} = \{25\mathbf{i} - 50\mathbf{j} + 10\mathbf{k}\}$ N acts at the end A of the pipe assembly. Determine the magnitudes of the components $\mathbf{F}_1$ and $\mathbf{F}_2$ which act along the axis of AB and perpendicular to it.

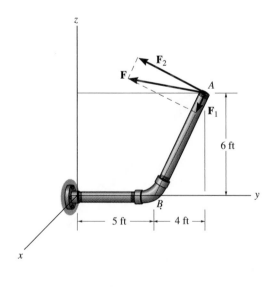

Prob. 2–87

REVIEW PROBLEMS

*2–88. The door is held open by means of two chains. If the tension in *AB* and *CD* is $F_A = 300$ N and $F_C = 250$ N, respectively, express each of these forces in Cartesian vector form.

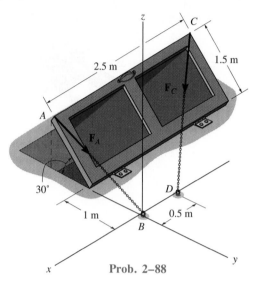

Prob. 2–88

2–89. Express each force acting on the screw eye as a Cartesian vector, and then determine the magnitude and orientation of the resultant force.

2–90. Use the parallelogram law to determine the magnitude of the resultant force acting on the screw eye. Determine its orientation, measured counterclockwise from the positive *x* axis.

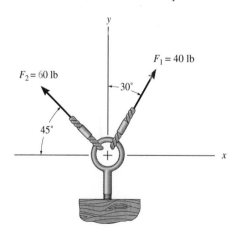

Probs. 2–89/2–90

2–91. The backbone exerts a vertical force of 110 lb on the lumbosacral joint of a person who is standing erect. Determine the magnitudes of the components of this force directed perpendicular to the surface of the sacrum, *v* axis, and parallel to it, *u* axis.

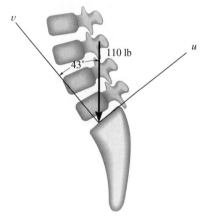

Prob. 2–91

*2–92. The cord *AB* exerts a force of $\mathbf{F} = \{12\mathbf{i} + 9\mathbf{j} - 8\mathbf{k}\}$ lb on the hook. If the cord is 8 ft long, determine the location *x*, *y* of the point of attachment *B*, and the height *z* of the hook.

2–93. The cord exerts a force of $F = 30$ lb on the hook. If the cord is 8 ft long, $z = 4$ ft, and the *x* component of the force is $F_x = 25$ lb, determine the location *x*, *y* of the point of attachment *B* of the cord to the ground.

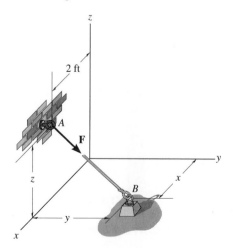

Probs. 2–92/2–93

2–94. Determine the magnitude of the projected component of force **F** along the *Oa* axis.

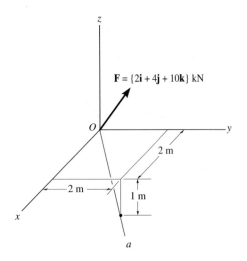

Prob. 2–94

2–95. Express each of the three forces acting on the column in Cartesian vector form and determine the magnitude of the resultant force.

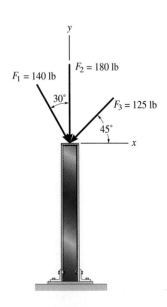

Prob. 2–95

*2–96. Determine the angle θ ($0° \le \theta \le 90°$) for cable *AB* so that the 500-N vertical force has a component of 200 N directed along cable *AB* from *A* toward *B*. What is the corresponding component of force acting along cable *CB*? Set $\phi = 60°$.

2–97. Determine the angle ϕ ($0° \le \phi \le 90°$) between cables *BA* and *BC* so that the 500-N vertical force has a component of 250 N which acts along cable *CB*, from *C* to *B*. Set $\theta = 60°$.

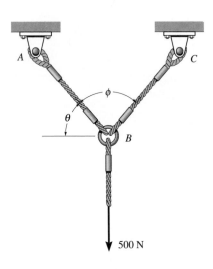

Probs. 2–96/2–97

3 Force System Resultants

In the previous chapter we developed the method for determining the resultant of a concurrent force system. Often the force system that acts on a body will not be concurrent, and, as a result, this will cause the body to have a tendency not only to translate, but also to rotate. To obtain the resultant of a nonconcurrent force system it is necessary to understand the concept of a *moment*. In this chapter a formal definition of a moment will be presented and ways of finding the moment of a force about a point or axis will be discussed. We will then use this concept to determine the resultant of a nonconcurrent force system. This development is important since application of the equations for force system simplification is similar to applying the equations of equilibrium for a rigid body, which will be discussed in the next chapter. Furthermore, the resultants of a force system will influence the state of equilibrium of a rigid body in the same way as the force system, and therefore we can study rigid-body behavior in a simpler manner by using the resultants.

3.1 Cross Product

The moment of a force will be formulated using Cartesian vectors in the next section. Before doing this, however, it is first necessary to expand our knowledge of vector algebra and introduce the cross-product method of vector multiplication.

The *cross product* of two vectors $\mathbf{A}$ and $\mathbf{B}$ yields the vector $\mathbf{C}$, which is written

$$\mathbf{C} = \mathbf{A} \times \mathbf{B}$$

and is read "$\mathbf{C}$ equals $\mathbf{A}$ cross $\mathbf{B}$."

Magnitude. The *magnitude* of **C** is defined as the product of the magnitudes of **A** and **B** and the sine of the angle θ between their tails ($0° \le \theta \le 180°$). Thus, $C = AB \sin \theta$.

Direction. Vector **C** has a *direction* that is perpendicular to the plane containing **A** and **B** such that the direction of **C** is specified by the right-hand rule; i.e., curling the fingers of the right hand from vector **A** (cross) to vector **B,** the thumb then points in the direction of **C,** as shown in Fig. 3–1.

Knowing both the magnitude and direction of **C,** we can write

$$\mathbf{C} = \mathbf{A} \times \mathbf{B} = (AB \sin \theta)\mathbf{u}_C \qquad (3\text{–}1)$$

where the scalar $AB \sin \theta$ defines the *magnitude* of **C** and the unit vector $\mathbf{u}_C$ defines the *direction* of **C.** The terms of Eq. 3–1 are illustrated graphically in Fig. 3–2.

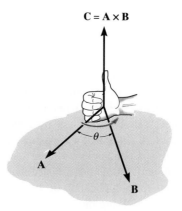

Fig. 3–1

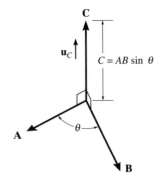

Fig. 3–2

Laws of Operation

1. The commutative law is *not* valid; i.e.,

$$\mathbf{A} \times \mathbf{B} \neq \mathbf{B} \times \mathbf{A}$$

Rather,

$$\mathbf{A} \times \mathbf{B} = -\mathbf{B} \times \mathbf{A}$$

This is shown in Fig. 3–3 by using the right-hand rule. The cross product $\mathbf{B} \times \mathbf{A}$ yields a vector that acts in the opposite direction to $\mathbf{C}$; i.e., $\mathbf{B} \times \mathbf{A} = -\mathbf{C}$.

2. Multiplication by a scalar:

$$a(\mathbf{A} \times \mathbf{B}) = (a\mathbf{A}) \times \mathbf{B} = \mathbf{A} \times (a\mathbf{B}) = (\mathbf{A} \times \mathbf{B})a$$

This property is easily shown, since the magnitude of the resultant vector ($|a|AB \sin \theta$) and its direction are the same in each case.

3. The distributive law:

$$\mathbf{A} \times (\mathbf{B} + \mathbf{D}) = (\mathbf{A} \times \mathbf{B}) + (\mathbf{A} \times \mathbf{D})$$

The proof of this identity is left as an exercise (see Prob. 3–1). It is important to note that *proper order* of the cross products must be maintained, since they are not commutative.

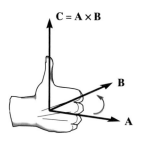

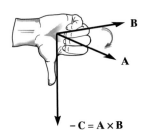

Fig. 3–3

Cartesian Vector Formulation. Equation 3–1 may be used to find the cross product of a pair of Cartesian unit vectors. For example, to find $\mathbf{i} \times \mathbf{j}$, the *magnitude* of the resultant vector is $(i)(j)(\sin 90°) = (1)(1)(1) = 1$, and its *direction* is determined using the right-hand rule. As shown in Fig. 3–4, the resultant vector points in the $+\mathbf{k}$ direction. Thus, $\mathbf{i} \times \mathbf{j} = (1)\mathbf{k}$. In a similar manner,

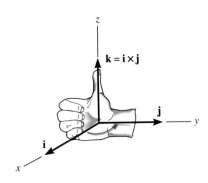

Fig. 3–4

$$\begin{array}{lll} \mathbf{i} \times \mathbf{j} = \mathbf{k} & \mathbf{i} \times \mathbf{k} = -\mathbf{j} & \mathbf{i} \times \mathbf{i} = 0 \\ \mathbf{j} \times \mathbf{k} = \mathbf{i} & \mathbf{j} \times \mathbf{i} = -\mathbf{k} & \mathbf{j} \times \mathbf{j} = 0 \\ \mathbf{k} \times \mathbf{i} = \mathbf{j} & \mathbf{k} \times \mathbf{j} = -\mathbf{i} & \mathbf{k} \times \mathbf{k} = 0 \end{array}$$

These results should *not* be memorized; rather, it should be clearly understood how each is obtained by using the right-hand rule and the definition of the cross product. A simple scheme shown in Fig. 3–5 is helpful for obtaining the same results when the need arises. If the circle is constructed as shown, then "crossing" two unit vectors in a *counterclockwise* fashion around the circle yields the *positive* third unit vector; e.g., $\mathbf{k} \times \mathbf{i} = \mathbf{j}$. Moving *clockwise*, a *negative* unit vector is obtained; e.g., $\mathbf{i} \times \mathbf{k} = -\mathbf{j}$.

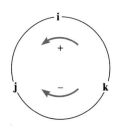

Fig. 3–5

Consider now the cross product of two general vectors **A** and **B** which are expressed in Cartesian vector form. We have

$$\mathbf{A} \times \mathbf{B} = (A_x\mathbf{i} + A_y\mathbf{j} + A_z\mathbf{k}) \times (B_x\mathbf{i} + B_y\mathbf{j} + B_z\mathbf{k})$$
$$= A_xB_x(\mathbf{i} \times \mathbf{i}) + A_xB_y(\mathbf{i} \times \mathbf{j}) + A_xB_z(\mathbf{i} \times \mathbf{k})$$
$$+ A_yB_x(\mathbf{j} \times \mathbf{i}) + A_yB_y(\mathbf{j} \times \mathbf{j}) + A_yB_z(\mathbf{j} \times \mathbf{k})$$
$$+ A_zB_x(\mathbf{k} \times \mathbf{i}) + A_zB_y(\mathbf{k} \times \mathbf{j}) + A_zB_z(\mathbf{k} \times \mathbf{k})$$

Carrying out the cross-product operations and combining terms yields

$$\mathbf{A} \times \mathbf{B} = (A_yB_z - A_zB_y)\mathbf{i} - (A_xB_z - A_zB_x)\mathbf{j} + (A_xB_y - A_yB_x)\mathbf{k} \qquad (3\text{–}2)$$

This equation may also be written in a more compact determinant form as

$$\mathbf{A} \times \mathbf{B} = \begin{vmatrix} \mathbf{i} & \mathbf{j} & \mathbf{k} \\ A_x & A_y & A_z \\ B_x & B_y & B_z \end{vmatrix} \qquad (3\text{–}3)$$

Thus, to find the cross product of any two Cartesian vectors **A** and **B,** it is necessary to expand a determinant whose first row of elements consists of the unit vectors **i, j,** and **k** and whose second and third rows represent the x, y, z components of the two vectors **A** and **B,** respectively.*

*A determinant having three rows and three columns can be expanded using three minors, each of which is multiplied by one of the three terms in the first row. There are four elements in each minor, e.g.,

$$\begin{vmatrix} A_{11} & A_{12} \\ A_{21} & A_{22} \end{vmatrix}$$

By *definition,* this notation represents the terms $(A_{11}A_{22} - A_{12}A_{21})$, which is simply the product of the two elements of the arrow slanting downward to the right $(A_{11}A_{22})$ *minus* the product of the two elements intersected by the arrow slanting downward to the left $(A_{12}A_{21})$. For a 3×3 determinant, such as Eq. 3–3, the three minors can be generated in accordance with the following scheme:

For element **i:**
$$\begin{vmatrix} \mathbf{i} & \mathbf{j} & \mathbf{k} \\ A_x & A_y & A_z \\ B_x & B_y & B_z \end{vmatrix} = \mathbf{i}(A_yB_z - A_zB_y)$$

For element **j:**
$$\begin{vmatrix} \mathbf{i} & \mathbf{j} & \mathbf{k} \\ A_x & A_y & A_z \\ B_x & B_y & B_z \end{vmatrix} = -\mathbf{j}(A_xB_z - A_zB_x)$$

For element **k:**
$$\begin{vmatrix} \mathbf{i} & \mathbf{j} & \mathbf{k} \\ A_x & A_y & A_z \\ B_x & B_y & B_z \end{vmatrix} = \mathbf{k}(A_xB_y - A_yB_x)$$

Adding the results and noting that the **j** element *must include the minus sign* yields the expanded form of **A** × **B** given by Eq. 3–2.

3.2 Moment of a Force—Scalar Formulation

The *moment* of a force about a point or axis provides a measure of the tendency of the force to cause a body to rotate about the point or axis. For example, consider the horizontal force F_x, which acts perpendicular to the handle of the wrench and is located a distance d_y from point O, Fig. 3–6a. It is seen that this force tends to cause the pipe to turn about the z axis and so the force creates a moment $(M_O)_z$ about the axis. The larger the force or the length d_y, the greater the effect. This tendency for rotation caused by F_x is sometimes called a *torque,* but most often it is called the *moment of a force* or simply the *moment* $(M_O)_z$. In particular, note that the moment axis (z) is perpendicular to the shaded plane (x–y) which contains both F_x and d_y and that this axis intersects the plane at point O. Now consider applying the force F_z to the wrench, Fig. 3–6b. This force will *not* rotate the pipe about the z axis. Instead, it tends to rotate it about the x axis. Keep in mind that although it may not be possible actually to "rotate" or turn the pipe in this manner, F_z still creates the *tendency* for rotation and so the moment $(M_O)_x$ is produced. As before, the force and distance d_y lie in the shaded plane (y–z) which is perpendicular to the moment axis (x). Lastly, if a force F_y is applied to the wrench, Fig. 3–6c, no moment is produced about point O. This lack of turning effect results, since the line of action of the force passes through O and therefore no tendency for rotation is possible.

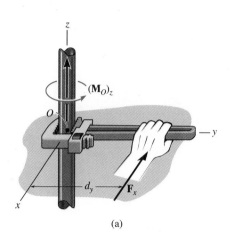

(a)

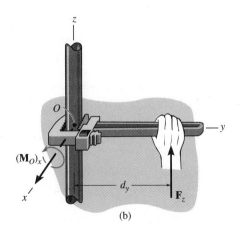

(b)

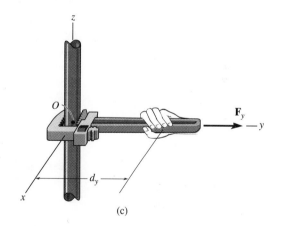

(c)

Fig. 3–6

For the general case, consider the force F and point O which lie in a shaded plane as shown in Fig. 3–7a. The moment M_O about point O, or about an axis passing through O and perpendicular to the plane, is a *vector quantity* since it has a specified magnitude and direction and adds according to the parallelogram law.

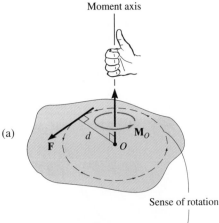

(a)

Moment axis

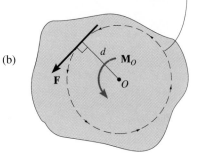

(b)

Sense of rotation

Fig. 3–7

Magnitude. The magnitude of $\mathbf{M}_O$ is

$$\boxed{M_O = Fd} \tag{3–4}$$

where d is referred to as the *moment arm* or perpendicular distance from the axis at point O to the line of action of the force. Units of moment magnitude consist of force times distance, e.g., N $\cdot$ m or lb $\cdot$ ft.

Direction. The direction of $\mathbf{M}_O$ will be specified by using the "right-hand rule." To do this, the fingers of the right hand are curled such that they follow the sense of rotation, which would occur if the force could rotate about point O, Fig. 3–7a. The *thumb* then *points* along the *moment axis* so that it gives the direction and sense of the moment vector, which is *upward* and *perpendicular* to the shaded plane containing $\mathbf{F}$ and d. By this definition, the moment $\mathbf{M}_O$ can be considered as a *sliding vector* and therefore acts at any point along the moment axis.

In three dimensions, $\mathbf{M}_O$ is illustrated by a vector arrow with a curl on it to *distinguish* it from a force vector, Fig. 3–7a. Many problems in mechanics, however, involve coplanar force systems that may be conveniently viewed in two dimensions. For example, a two-dimensional view of Fig. 3–7a is given in Fig. 3–7b. Here $\mathbf{M}_O$ is simply represented by the (counterclockwise) curl, which indicates the action of $\mathbf{F}$. The arrowhead on this curl is used to show the *sense of rotation* caused by $\mathbf{F}$. Using the right-hand rule, however, realize that the direction and sense of the moment vector in Fig. 3–7b are specified by the thumb, which points *out* of the page, since the fingers follow the curl. In particular, notice that *this curl or sense of rotation can always be determined by observing in which direction the force would "orbit" about point O* (counterclockwise in Fig. 3–7b). In two dimensions we will often refer to finding the moment of a force "about a point" (O). Keep in mind, however, that the moment *actually* acts about an axis which is perpendicular to the plane containing $\mathbf{F}$ and d, and this axis intersects the plane at the point (O), Fig. 3–7a.

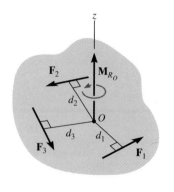

Fig. 3–8

Resultant Moment of a System of Coplanar Forces. If a system of forces all lie in the x–y plane, then the moment produced by each force about point O will be directed along the z axis, Fig. 3–8. Consequently, the resultant moment $\mathbf{M}_{R_O}$ of the system can be determined by simply adding the moments of each force *algebraically,* since all the moment vectors are collinear. We can write this vector sum symbolically as

$$\zeta + M_{R_O} = \Sigma Fd \tag{3–5}$$

Here the counterclockwise curl written alongside the equation indicates that by the scalar sign convention, the moment of any force will be positive if it is directed along the $+z$ axis, whereas a negative moment is directed along the $-z$ axis.

Example 3–1

Determine the moment of the 800-N force acting on the frame in Fig. 3–9 about points A, B, C, and D.

SOLUTION

In general, $M = Fd$, where d is the moment arm or *perpendicular distance* from the *point* on the moment axis to the *line of action* of the force. Hence,

$M_A = 800 \text{ N}(2.5 \text{ m}) = 2000 \text{ N} \cdot \text{m} \downarrow$ *Ans.*

$M_B = 800 \text{ N}(1.5 \text{ m}) = 1200 \text{ N} \cdot \text{m} \downarrow$ *Ans.*

$M_C = 800 \text{ N}(0) = 0$ (line of action of **F** passes through C)   *Ans.*

$M_D = 800 \text{ N}(0.5 \text{ m}) = 400 \text{ N} \cdot \text{m} \uparrow$ *Ans.*

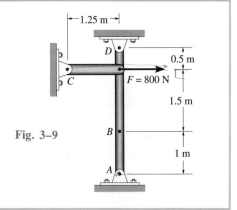

Fig. 3–9

Example 3–2

Determine the location of the point of application P and the direction of a 20-lb force that lies in the plane of the square plate shown in Fig. 3–10a, so that this force creates the greatest counterclockwise moment about point O. What is this moment?

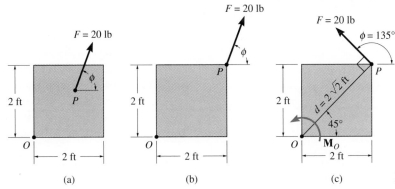

(a)   (b)   (c)   **Fig. 3–10**

SOLUTION

Since the maximum moment created by the force is required, the force must act on the plate at a distance *farthest* from point O. As shown in Fig. 3–10b, the point of application must therefore be at the diagonal corner. In order to produce *counterclockwise* rotation of the plate about O, **F** must act at an angle $45° < \phi < 225°$. The greatest moment is produced when the line of action of **F** is *perpendicular* to d, i.e., $\phi = 135°$, Fig. 3–10c. The maximum moment is therefore

$$M_O = Fd = (20 \text{ lb})(2\sqrt{2} \text{ ft}) = 56.6 \text{ lb} \cdot \text{ft} \uparrow \qquad \textit{Ans.}$$

By the right-hand rule, **M**$_O$ is directed out of the page.

Example 3–3

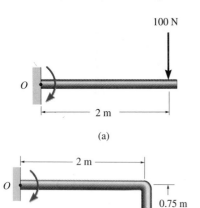

(a)

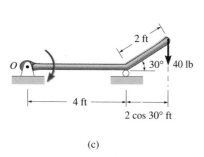

(b)

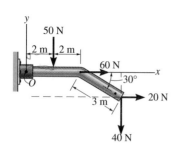

(c)

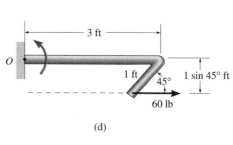

(d)

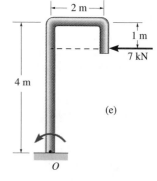

(e)

Fig. 3–11

For each case illustrated in Fig. 3–11, determine the moment of the force about point O.

SOLUTION

The line of action of each force is extended as a dashed line in order to establish the moment arm d. Also, the tendency of rotation caused by the force is shown as a colored curl. Thus,

Fig. 3–11a, $M_O = (100 \text{ N})(2 \text{ m}) = 200 \text{ N} \cdot \text{m}$ ↓   *Ans.*

Fig. 3–11b, $M_O = (50 \text{ N})(0.75 \text{ m}) = 37.5 \text{ N} \cdot \text{m}$ ↓   *Ans.*

Fig. 3–11c, $M_O = (40 \text{ lb})(4 \text{ ft} + 2 \cos 30° \text{ ft}) = 229 \text{ lb} \cdot \text{ft}$ ↓   *Ans.*

Fig. 3–11d, $M_O = (60 \text{ lb})(1 \sin 45° \text{ ft}) = 42.4 \text{ lb} \cdot \text{ft}$ ↑   *Ans.*

Fig. 3–11e, $M_O = (7 \text{ kN})(4 \text{ m} - 1 \text{ m}) = 21.0 \text{ kN} \cdot \text{m}$ ↑   *Ans.*

Example 3–4

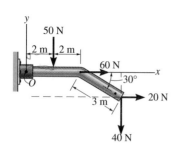

Fig. 3–12

Determine the resultant moment of the four forces acting on the member shown in Fig. 3–12 about point O.

SOLUTION

Here it is necessary to apply Eq. 3–5. Assuming that positive moments act in the $+\mathbf{k}$ direction, i.e., counterclockwise, we have

$$\downarrow + M_{R_O} = \Sigma Fd;$$

$$M_{R_O} = -50 \text{ N}(2 \text{ m}) + 60 \text{ N}(0) + 20 \text{ N}(3 \sin 30° \text{ m})$$
$$-40 \text{ N}(4 \text{ m} + 3 \cos 30° \text{ m})$$

$$M_{R_O} = -334 \text{ N} \cdot \text{m} = 334 \text{ N} \cdot \text{m}$$ ↓   *Ans.*

For this calculation, note how the moment-arm distances for the 20-N and 40-N forces are established from the extended (dashed) lines of action of each of these forces.

3.3 Moment of a Force—Vector Formulation

The moment of a force $\mathbf{F}$ about point O, or actually about an axis passing through O and perpendicular to the plane containing O and $\mathbf{F}$, Fig. 3–13a, can also be expressed using the vector cross product, namely,

$$\mathbf{M}_O = \mathbf{r} \times \mathbf{F} \qquad\qquad (3\text{–}6)$$

Here $\mathbf{r}$ represents a position vector drawn from O to *any point* lying on the line of action of $\mathbf{F}$. It will now be shown that indeed the moment $\mathbf{M}_O$, when determined by this cross product, has the proper magnitude and direction.

Magnitude. The magnitude of the above cross product is defined from Eq. 3–1 as $M_O = rF \sin\theta$. Here, the angle θ is measured between the *tails* of $\mathbf{r}$ and $\mathbf{F}$. Hence, $\mathbf{r}$ must be treated as a sliding vector so that θ can be constructed properly, Fig. 3–13b. As shown, the moment arm $d = r \sin\theta$, so that indeed

$$M_O = rF \sin\theta = F\,(r \sin\theta) = Fd$$

which agrees with Eq. 3–4.

Direction. The direction and sense of $\mathbf{M}_O$ in Eq. 3–6 are determined by the right-hand rule as it applies to the cross product. Thus, extending $\mathbf{r}$ to the dashed position and curling the right-hand fingers from $\mathbf{r}$ toward $\mathbf{F}$, "$\mathbf{r}$ cross $\mathbf{F}$," the thumb is directed upward or perpendicular to the plane containing $\mathbf{r}$ and $\mathbf{F}$ and this is in the *same direction* as $\mathbf{M}_O$, the moment of the force about point O, Fig. 3–13b. Note that the "curl" of the fingers, like the curl around the moment vector, indicates the sense of rotation caused by the force. Since the cross product is not commutative, it is important that the *proper order* of $\mathbf{r}$ and $\mathbf{F}$ be maintained in Eq. 3–6.

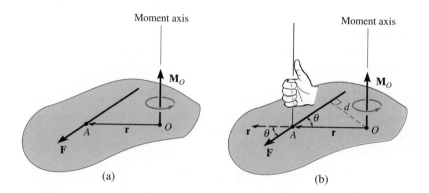

(a) (b)

Fig. 3–13

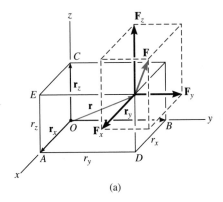

(a)

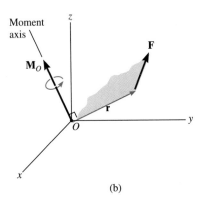

(b)

Fig. 3–14

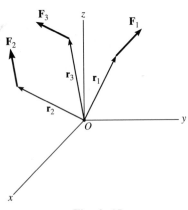

Fig. 3–15

Cartesian Vector Formulation. If the position vector **r** and force **F** are expressed in Cartesian vector form, Fig. 3–14, then from Eq. 3–6 we have

$$\mathbf{M}_O = \mathbf{r} \times \mathbf{F} = \begin{vmatrix} \mathbf{i} & \mathbf{j} & \mathbf{k} \\ r_x & r_y & r_z \\ F_x & F_y & F_z \end{vmatrix} \qquad (3\text{–}7)$$

where r_x, r_y, r_z represent the x, y, z components of the position vector drawn from point O to *any point* on the line of action of the force

F_x, F_y, F_z represent the x, y, z components of the force vector

If the determinant is expanded, then like Eq. 3–2 we have

$$\mathbf{M}_O = (r_y F_z - r_z F_y)\mathbf{i} - (r_x F_z - r_z F_x)\mathbf{j} + (r_x F_y - r_y F_x)\mathbf{k} \qquad (3\text{–}8)$$

The physical meaning of these three moment components becomes evident by studying Fig. 3–14a. For example, the **i** component of $\mathbf{M}_O$ is determined from the moments of $\mathbf{F}_x$, $\mathbf{F}_y$, and $\mathbf{F}_z$ about the x axis. In particular, note that $\mathbf{F}_x$ does *not* create a moment or tendency to cause turning about the x axis, since this force is *parallel* to the x axis. The line of action of $\mathbf{F}_y$ passes through point E, and so the magnitude of the moment of $\mathbf{F}_y$ about point A on the x axis is $r_z F_y$. By the right-hand rule this component acts in the negative **i** direction. Likewise, $\mathbf{F}_z$ contributes a moment component of $r_y F_z \mathbf{i}$. Thus, $(M_O)_x = (r_y F_z - r_z F_y)$ as shown in Eq. 3–7. As an exercise, establish the **j** and **k** components of $\mathbf{M}_O$ in this manner, and show that indeed the expanded form of the determinant, Eq. 3–8, represents the moment of $\mathbf{F}$ about point O. Once determined, realize that $\mathbf{M}_O$ will be *perpendicular* to the shaded plane containing vectors **r** and **F,** Fig. 3–14b.

It will be shown in Example 3–5 that the computation of the moment using the cross product has a distinct advantage over the scalar formulation when solving problems in three dimensions. This is because it is generally easier to establish the position vector **r** to the force, rather than determining the moment-arm distance d that must be directed *perpendicular* to the line of action of the force.

Resultant Moment of a System of Forces. The resultant moment of a system of forces about point O can be determined by vector addition resulting from successive applications of Eq. 3–7. This resultant can be written symbolically as

$$\mathbf{M}_{R_O} = \Sigma \mathbf{r} \times \mathbf{F} \qquad (3\text{–}9)$$

and is shown in Fig. 3–15.

Example 3–6 illustrates numerical application of Eqs. 3–7 and 3–9.

Example 3–5

The pole in Fig. 3–16a is subjected to a 60-N force that is directed from C to B. Determine the magnitude of the moment created by this force about point A.

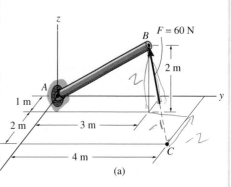

SOLUTION

As shown in Fig. 3–16b, either one of two position vectors can be used for the solution, since $\mathbf{M}_A = \mathbf{r}_B \times \mathbf{F}$ or $\mathbf{M}_A = \mathbf{r}_C \times \mathbf{F}$. The position vectors are represented as

$$\mathbf{r}_B = \{1\mathbf{i} + 3\mathbf{j} + 2\mathbf{k}\} \text{ m} \quad \text{and} \quad \mathbf{r}_C = \{3\mathbf{i} + 4\mathbf{j}\} \text{ m}$$

The force has a magnitude of 60 N and a direction specified by the unit vector $\mathbf{u}_F$, directed from C to B. Thus,

$$\mathbf{F} = (60 \text{ N})\mathbf{u}_F = (60) \left[\frac{(1 - 3)\mathbf{i} + (3 - 4)\mathbf{j} + (2 - 0)\mathbf{k}}{\sqrt{(-2)^2 + (-1)^2 + (2)^2}} \right]$$

$$= \{-40\mathbf{i} - 20\mathbf{j} + 40\mathbf{k}\} \text{ N}$$

Substituting into the determinant formulation, Eq. 3–7, and following the scheme for determinant expansion as stated in the footnote on page 72, we have

$$\mathbf{M}_A = \mathbf{r}_B \times \mathbf{F} = \begin{vmatrix} \mathbf{i} & \mathbf{j} & \mathbf{k} \\ 1 & 3 & 2 \\ -40 & -20 & 40 \end{vmatrix}$$

$$= [3(40) - 2(-20)]\mathbf{i} - [1(40) - 2(-40)]\mathbf{j} + [1(-20) - 3(-40)]\mathbf{k}$$

or

$$\mathbf{M}_A = \mathbf{r}_C \times \mathbf{F} = \begin{vmatrix} \mathbf{i} & \mathbf{j} & \mathbf{k} \\ 3 & 4 & 0 \\ -40 & -20 & 40 \end{vmatrix}$$

$$= [4(40) - 0(-20)]\mathbf{i} - [3(40) - 0(-40)]\mathbf{j} + [3(-20) - 4(-40)]\mathbf{k}$$

In both cases,

$$\mathbf{M}_A = \{160\mathbf{i} - 120\mathbf{j} + 100\mathbf{k}\} \text{ N} \cdot \text{m}$$

The *magnitude* of $\mathbf{M}_A$ is therefore

$$M_A = \sqrt{(160)^2 + (-120)^2 + (100)^2} = 224 \text{ N} \cdot \text{m} \qquad Ans.$$

As expected, $\mathbf{M}_A$ acts perpendicular to the shaded plane containing vectors $\mathbf{F}$, $\mathbf{r}_B$, and $\mathbf{r}_C$, Fig. 3–16c. (How would you find its coordinate direction angles $\alpha = 44.3°$, $\beta = 122°$, $\gamma = 63.4°$?) Had this problem been worked using a scalar approach, where $M_A = Fd$, notice the difficulty that might arise in obtaining the moment arm d.

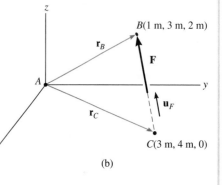

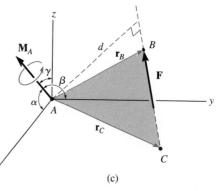

Fig. 3–16

Example 3–6

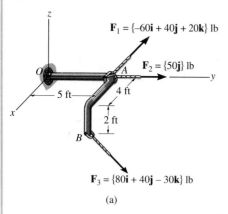

$\mathbf{F}_1 = \{-60\mathbf{i} + 40\mathbf{j} + 20\mathbf{k}\}$ lb

$\mathbf{F}_2 = \{50\mathbf{j}\}$ lb

5 ft

4 ft

2 ft

$\mathbf{F}_3 = \{80\mathbf{i} + 40\mathbf{j} - 30\mathbf{k}\}$ lb

(a)

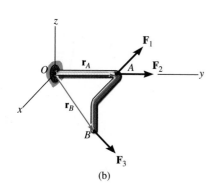

(b)

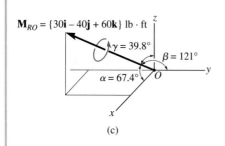

$\mathbf{M}_{RO} = \{30\mathbf{i} - 40\mathbf{j} + 60\mathbf{k}\}$ lb · ft

$\gamma = 39.8°$

$\beta = 121°$

$\alpha = 67.4°$

(c)

Fig. 3–17

Three forces act on the rod shown in Fig. 3–17a. Determine the resultant moment they create about the flange at O, and determine the direction of this moment axis.

SOLUTION

Here we must apply Eq. 3–9. Position vectors are directed from point O to each force as shown in Fig. 3–17b. These vectors are

$$\mathbf{r}_A = \{5\mathbf{j}\}\ \text{ft}$$
$$\mathbf{r}_B = \{4\mathbf{i} + 5\mathbf{j} - 2\mathbf{k}\}\ \text{ft}$$

Since $\mathbf{F}_2 = \{50\mathbf{i}\}$ lb, and the Cartesian components of the other forces are given, the resultant moment about O is therefore

$$\mathbf{M}_{R_O} = \Sigma \mathbf{r} \times \mathbf{F}$$
$$= \mathbf{r}_A \times \mathbf{F}_1 + \mathbf{r}_A \times \mathbf{F}_2 + \mathbf{r}_B \times \mathbf{F}_3$$

$$= \begin{vmatrix} \mathbf{i} & \mathbf{j} & \mathbf{k} \\ 0 & 5 & 0 \\ -60 & 40 & 20 \end{vmatrix} + \begin{vmatrix} \mathbf{i} & \mathbf{j} & \mathbf{k} \\ 0 & 5 & 0 \\ 0 & 50 & 0 \end{vmatrix} + \begin{vmatrix} \mathbf{i} & \mathbf{j} & \mathbf{k} \\ 4 & 5 & -2 \\ 80 & 40 & -30 \end{vmatrix}$$

$$= [5(20) - 40(0)]\mathbf{i} - [0\mathbf{j}] + [0(40) - (-60)(5)]\mathbf{k} + [0\mathbf{i} + 0\mathbf{j} + 0\mathbf{k}]$$
$$+ [5(-30) - (40)(-2)]\mathbf{i} - [4(-30) - 80(-2)]\mathbf{j} + [4(40) - 80(5)]\mathbf{k}$$
$$= \{30\mathbf{i} - 40\mathbf{j} + 60\mathbf{k}\}\ \text{lb} \cdot \text{ft} \qquad\qquad Ans.$$

The moment axis is directed along the line of action of $\mathbf{M}_{R_O}$. Since the magnitude of this moment is

$$M_{R_O} = \sqrt{(30)^2 + (-40)^2 + (60)^2} = 78.10\ \text{lb} \cdot \text{ft}$$

the unit vector which defines the direction of the moment axis is

$$\mathbf{u} = \frac{\mathbf{M}_{R_O}}{M_{R_O}} = \frac{30\mathbf{i} - 40\mathbf{j} + 60\mathbf{k}}{78.10} = 0.3841\mathbf{i} - 0.5121\mathbf{j} + 0.7682\mathbf{k}$$

Therefore, the coordinate direction angles of the axis are

$$\cos \alpha = 0.3841; \qquad \alpha = 67.4° \qquad Ans.$$
$$\cos \beta = -0.5121; \qquad \beta = 121° \qquad Ans.$$
$$\cos \gamma = 0.7682; \qquad \gamma = 39.8° \qquad Ans.$$

These results are shown in Fig. 3–17c. Realize that the three forces tend to cause the pipe to rotate about this axis in the manner shown by the curl indicated on the moment vector.

3.4 Transmissibility of a Force and the Principle of Moments

Using the definition of a moment formalized by the cross product, we will now present two important concepts often used throughout our study of mechanics.

Transmissibility of a Force. Consider the force $\mathbf{F}$ applied at point A in Fig. 3–18. The moment created by $\mathbf{F}$ about point O is $\mathbf{M}_O = \mathbf{r}_A \times \mathbf{F}$; however, it was shown that the position vector "$\mathbf{r}$" can extend from O to *any point* on the line of action of $\mathbf{F}$. Consequently, $\mathbf{F}$ may be applied at point B and the same moment $\mathbf{M}_O = \mathbf{r}_B \times \mathbf{F}$ will be computed. As a result, $\mathbf{F}$ has the properties of a *sliding vector* and can therefore act at *any point along its line of action and still create the same moment about point O.* We refer to $\mathbf{F}$ in this regard as being "transmissible," and we will discuss this property further in Sec. 3.7.

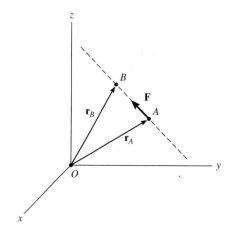

Fig. 3–18

Principle of Moments. A concept often used in mechanics is the *principle of moments,* which is sometimes referred to as *Varignon's theorem* since it was originally developed by the French mathematician Varignon (1654–1722). It states that *the moment of a force about a point is equal to the sum of the moments of the force's components about the point.* The proof follows directly from the distributive law of the vector cross product. To show this, consider the force $\mathbf{F}$ and two of its components, where $\mathbf{F} = \mathbf{F}_1 + \mathbf{F}_2$, Fig. 3–19. We have

$$\mathbf{M}_O = \mathbf{r} \times \mathbf{F}_1 + \mathbf{r} \times \mathbf{F}_2 = \mathbf{r} \times (\mathbf{F}_1 + \mathbf{F}_2) = \mathbf{r} \times \mathbf{F}$$

This concept has important applications to the solution of problems and proofs of theorems that follow, since it is often easier to determine the moments of a force's components rather than the moment of the force itself.

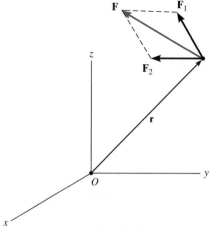

Fig. 3–19

Example 3–7

A 200-N force acts on the bracket shown in Fig. 3–20a. Determine the moment of the force about point A.

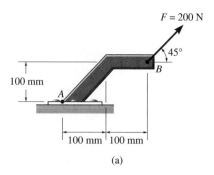

(a)

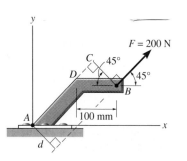

(b)

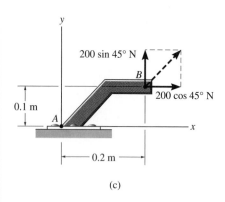

(c)

Fig. 3–20

SOLUTION I

The moment arm d can be found by trigonometry, using the construction shown in Fig. 3–16b. From triangle BCD,

$$CB = d = 100 \cos 45° = 70.71 \text{ mm} = 0.070\ 71 \text{ m}$$

Thus,

$$M_A = Fd = 200 \text{ N}(0.070\ 71 \text{ m}) = 14.1 \text{ N} \cdot \text{m} \curvearrowleft$$

According to the right-hand rule, $\mathbf{M}_A$ is directed in the $+\mathbf{k}$ direction since the force tends to rotate *counterclockwise* about point A. Hence, reporting the moment as a Cartesian vector, we have

$$\mathbf{M}_A = \{14.1\mathbf{k}\} \text{ N} \cdot \text{m} \qquad\qquad Ans.$$

SOLUTION II

The 200-N force may be resolved into x and y components, as shown in Fig. 3–20c. In accordance with the principle of moments, the moment of $\mathbf{F}$ computed about point A is equivalent to the sum of the moments produced by the two force components. Assuming counterclockwise rotation as positive, i.e., in the $+\mathbf{k}$ direction, we can apply Eq. 3–5, in which case

$$\curvearrowleft +M_A = (200 \sin 45°)(0.20) - (200 \cos 45°)(0.10)$$
$$= 14.1 \text{ N} \cdot \text{m} \curvearrowleft$$

Thus

$$\mathbf{M}_A = \{14.1\mathbf{k}\} \text{ N} \cdot \text{m} \qquad\qquad Ans.$$

By comparison, it is seen that Solution II provides a more *convenient method* for analysis than Solution I, since the moment arm for each component force is easier to establish.

Example 3–8

The force $\mathbf{F}$ acts at the end of the beam shown in Fig. 3–21a. Determine the moment of the force about point O.

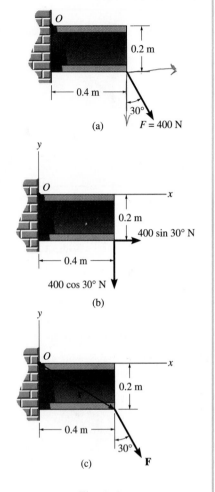

Fig. 3–21

SOLUTION I (SCALAR ANALYSIS)

The force is resolved into its x and y components as shown in Fig. 3–21b and the moments of the components are computed about point O. Taking positive moments as counterclockwise, i.e., in the $+\mathbf{k}$ direction, we have

$$\zeta + M_O = 400 \sin 30°(0.2) - 400 \cos 30°(0.4)$$
$$= -98.6 \text{ N} \cdot \text{m} = 98.6 \text{ N} \cdot \text{m} \, \downarrow$$

or

$$\mathbf{M}_O = \{-98.6\mathbf{k}\} \text{ N} \cdot \text{m} \qquad Ans.$$

SOLUTION II (VECTOR ANALYSIS)

Using a Cartesian vector approach, the force and position vectors shown in Fig. 4–21c can be represented as

$$\mathbf{r} = \{0.4\mathbf{i} - 0.2\mathbf{j}\} \text{ m}$$
$$\mathbf{F} = \{400 \sin 30°\mathbf{i} - 400 \cos 30°\mathbf{j}\} \text{ N}$$
$$= \{200.0\mathbf{i} - 346.4\mathbf{j}\} \text{ N}$$

The moment is therefore

$$\mathbf{M}_O = \mathbf{r} \times \mathbf{F} = \begin{vmatrix} \mathbf{i} & \mathbf{j} & \mathbf{k} \\ 0.4 & -0.2 & 0 \\ 200.0 & -346.4 & 0 \end{vmatrix}$$
$$= 0\mathbf{i} - 0\mathbf{j} + [0.4(-346.4) - (-0.2)(200.0)]\mathbf{k}$$
$$= \{-98.6\mathbf{k}\} \text{ N} \cdot \text{m} \qquad Ans.$$

By comparison, it is seen that the scalar analysis (Solution I) provides a more *convenient method* for analysis than Solution II, since the direction of the moment and the moment arm for each component force are easy to establish. Hence, this method is generally recommended for solving problems displayed in two dimensions. On the other hand, Cartesian vector analysis is generally recommended only for solving three-dimensional problems, where the moment arms and force components are often more difficult to determine.

PROBLEMS

3.1   Determine the magnitude and directional sense of the moment of the force at *A* about point *P*.

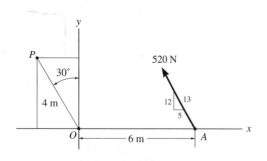

Prob. 3–1

3–2.   Determine the magnitude and directional sense of the resultant moment of the forces at *A* and *B* about point *O*.

3–3.   Determine the magnitude and directional sense of the resultant moment of the forces at *A* and *B* about point *P*.

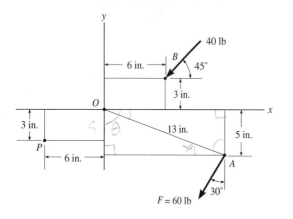

Probs. 3–2/3–3

*3–4.   Determine the magnitude and directional sense of the resultant moment of the forces at *A* and *B* about point *O*.

3–5.   Determine the magnitude and directional sense of the resultant moment of the forces at *A* and *B* about point *P*.

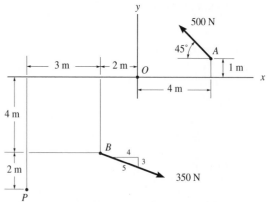

Probs. 3–4/3–5

3–6.   A force of 30 lb is applied to the handle of the wrench. Determine the moment of this force about point *O*.

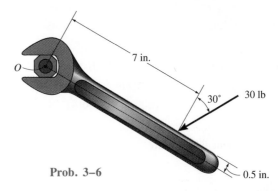

Prob. 3–6

3–7.   Using one hand, the man attempts to lift the sledge hammer, which has a head having a weight of 10 lb. Determine the greatest angle θ to which he can hold it if the maximum moment he can develop at his wrist is 18 lb · ft. Neglect the weight of the handle.

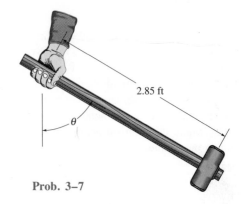

Prob. 3–7

*3–8. The crane can be adjusted for any angle $0° \leq \theta \leq 90°$ and any extension $0 \leq x \leq 5$ m. For a suspended mass of 150 kg, determine the moment developed at A as a function of x and θ. What values of both x and θ develop the maximum possible moment at A? Compute this moment. Neglect the size of the pulley at B.

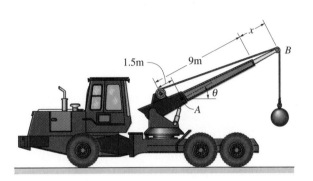

Prob. 3–8

3–9. Determine the resultant moment about point A of the three forces acting on the beam.

3–10. Determine the resultant moment about point B of the three forces acting on the beam.

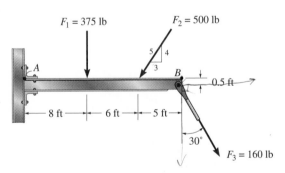

Probs. 3–9/3–10

3–11. Determine the magnitude of force $\mathbf{F}_2$ that must be applied perpendicular to the handle so that the resultant moment of both $\mathbf{F}_1$ and $\mathbf{F}_2$ at O is zero.

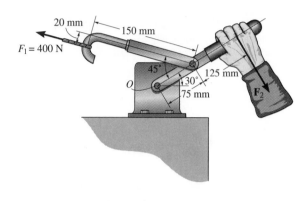

Prob. 3–11

*3–12. Determine the orientation $\theta (0° \leq \theta \leq 180°)$ of the 40-lb force $\mathbf{F}$ so that $\mathbf{F}$ produces (a) the maximum moment about point A and (b) no moment about point A. Compute the moment in each case.

3–13. Determine the moment of the force $\mathbf{F}$ about point A as a function of θ. Plot the results of M (ordinate) versus θ (abscissa) for $0° \leq \theta \leq 180°$.

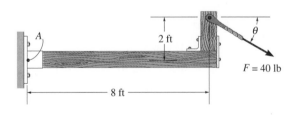

Probs. 3–12/3–13

3–14. The tongs are used to grip the ends of the drilling pipe P. Determine the torque (moment) M_P that the applied force $F = 150$ lb exerts on the pipe as a function of θ. Plot this moment M_P versus θ for $0 \le \theta \le 90°$.

3–15. The tongs are used to grip the ends of the drilling pipe P. If a torque (moment) of $M_P = 800$ lb · ft is needed at P to turn the pipe, determine the cable force F that must be applied to the tongs. Set $\theta = 30°$.

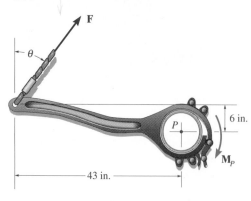

Probs. 3–14/3–15

3–18. Determine the resultant moment of the forces about point O. Express the result as a Cartesian vector.

3–19. Determine the resultant moment of the forces about point P. Express the result as a Cartesian vector.

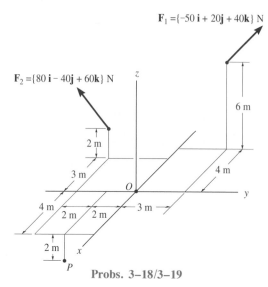

$F_1 = \{-50\,\mathbf{i} + 20\mathbf{j} + 40\mathbf{k}\}$ N

$F_2 = \{80\,\mathbf{i} - 40\mathbf{j} + 60\mathbf{k}\}$ N

Probs. 3–18/3–19

*3–16. Determine the moment of the force $\mathbf{F}$ at A about point O. Express the result as a Cartesian vector.

3–17. Determine the moment of the force $\mathbf{F}$ at A about point P. Express the result as a Cartesian vector.

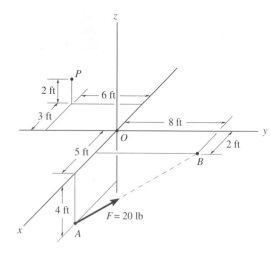

Probs. 3–16/3–17

*3–20. The cable exerts a 140-N force on the telephone pole as shown. Determine the moment of this force at the base A of the pole. Solve the problem two ways, i.e., by using a position vector from A to C, then A to B.

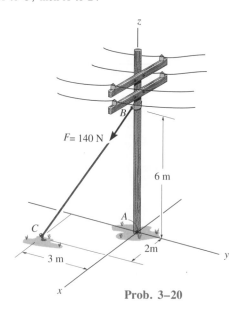

$F = 140$ N

Prob. 3–20

3–21. Using Cartesian vectors, compute the moment of each of the two forces acting on the pipe assembly about point O. Add these moments and calculate the magnitude and coordinate direction angles of the resultant moment.

3–22. Using Cartesian vectors, determine the moment of each of the two forces acting on the pipe assembly about point A. Add these moments and calculate the magnitude and coordinate direction angles of the resultant moment.

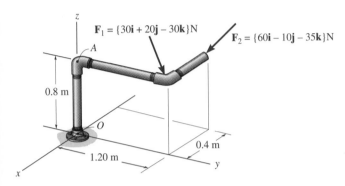

Probs. 3–21/3–22

3–23. The curved pipe has a radius of 5 ft. If a force of 80 lb acts at its end as shown, determine the moment of this force about point C. Solve the problem by using two different position vectors.

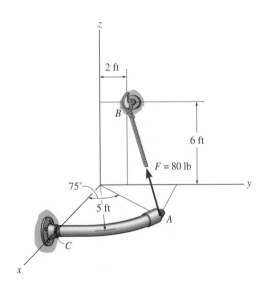

Prob. 3–23

*3–24. A force $\mathbf{F}$ having a magnitude of $F = 100$ N acts along the diagonal of the parallelepiped. Using Cartesian vectors, determine the moment of $\mathbf{F}$ about point A where $\mathbf{M}_A = \mathbf{r}_B \times \mathbf{F}$ and $\mathbf{M}_A = \mathbf{r}_C \times \mathbf{F}$.

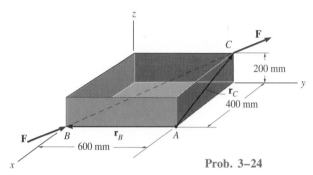

Prob. 3–24

3–25. Find the resultant moment of the two forces acting at the end of the pipe assembly about each of the joints at A, B, and C.

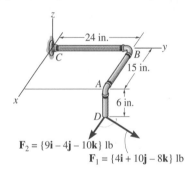

Prob. 3–25

3–26. If $\mathbf{F} = \{50\mathbf{i} + 60\mathbf{j} + 30\mathbf{k}\}$ lb, determine the magnitude and coordinate direction angles of the moment of $\mathbf{F}$ about point A.

3–27. Determine the coordinate direction angles of the force $\mathbf{F}$ applied at the end of the pipe such that the moment created by $\mathbf{F}$ about point A is zero.

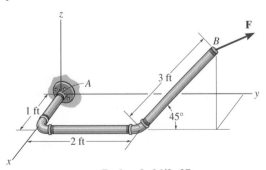

Probs. 3–26/3–27

3.5 Moment of a Force About a Specified Axis

Recall that when the moment of a force is computed about a point, the moment and its axis are *always* perpendicular to the plane containing the force and the moment arm. In some problems it is important to find the *component* of this moment along a specified axis that passes through the point. To solve this problem either a scalar or vector analysis can be used.

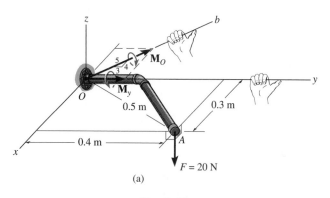

(a)

Fig. 3–22

Scalar Analysis. As a numerical example of this problem, consider the pipe assembly shown in Fig. 3–22a, which lies in the horizontal plane and is subjected to the vertical force of $F = 20$ N applied at point A. The moment of this force about point O has a *magnitude* of $M_O = (20 \text{ N})(0.5 \text{ m}) = 10 \text{ N} \cdot \text{m}$ and a *direction* defined by the right-hand rule, as shown in Fig. 3–22a. This moment tends to turn the pipe about the Ob axis. For practical reasons, however, it may be necessary to determine the *component* of M_O about the y axis, M_y, since this component tends to unscrew the pipe from the flange at O. From Fig. 3–22a, M_y has a magnitude of $M_y = \frac{3}{5}(10 \text{ N} \cdot \text{m}) = 6 \text{ N} \cdot \text{m}$ and a sense of direction shown by the vector resolution. Rather than performing this *two-step* process of first finding the moment of the force about point O and then resolving the moment along the y axis, it is also possible to solve this problem *directly*. To do so, it is necessary to determine the perpendicular or moment-arm distance from the line of action of $\mathbf{F}$ to the y axis. From Fig. 3–22a this distance is 0.3 m. Thus the *magnitude* of the moment of the force about the y axis is again $M_y = 0.3(20 \text{ N}) = 6 \text{ N} \cdot \text{m}$, and the *direction* is determined by the right-hand rule as shown.

In general, then, *if the line of action of a force* **F** *is perpendicular to an axis aa*, the magnitude of the moment of **F** about the axis can be determined from the equation

$$M_a = Fd_a$$ (3–10)

Here d_a is the *perpendicular or shortest distance* from the force line of action to the axis. The direction is determined from the thumb of the right hand when the fingers are curled in accordance with the direction of rotation as produced by the force. In particular, realize that a *force will not contribute a moment about an axis if its line of action is parallel to the axis or its line of action passes through the axis.*

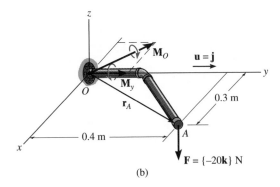

(b)

Vector Analysis.   The previous two-step solution of first finding the moment of the force about a point on the axis and then finding the projected component of the moment about the axis can also be performed using a vector analysis, Fig. 3–22b. Here the moment about point O is first determined from $\mathbf{M}_O = \mathbf{r}_A \times \mathbf{F} = (0.3\mathbf{i} + 0.4\mathbf{j}) \times (-20\mathbf{k}) = \{-8\mathbf{i} + 6\mathbf{j}\}$ N · m. The component or projection of this moment along the y axis is then determined from the dot product (Sec. 2.9). Since the unit vector for the axis (or line) is $\mathbf{u} = \mathbf{j}$, then $M_y = \mathbf{M}_O \cdot \mathbf{u} = (-8\mathbf{i} + 6\mathbf{j}) \cdot \mathbf{j} = 6$ N · m. This result, of course, is to be expected, since it represents the $\mathbf{j}$ component of $\mathbf{M}_O$.

A vector analysis such as this is particularly advantageous for finding the moment of a force about an axis when the force components or the appropriate moment arms are difficult to determine. For this reason, the above two-step process will now be generalized and applied to a body of arbitrary shape. To do so, consider the body in Fig. 3–23, which is subjected to the force **F** acting at point A. Here we wish to determine the effect of **F** in tending to rotate the body about the aa' axis. This tendency for rotation is measured by the moment component $\mathbf{M}_a$. To determine $\mathbf{M}_a$ we first compute the moment of **F**

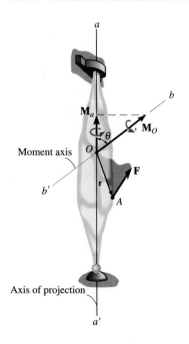

Moment axis

Axis of projection

Fig. 3–23

about an *arbitrary point O* that lies on the axis. In this case, $\mathbf{M}_O$ is expressed by the cross product $\mathbf{M}_O = \mathbf{r} \times \mathbf{F}$, where $\mathbf{r}$ is directed from O to A. Since $\mathbf{M}_O$ acts along the moment axis bb', which is perpendicular to the plane containing $\mathbf{r}$ and $\mathbf{F}$, the component or projection of $\mathbf{M}_O$ onto the aa' axis is then represented by $\mathbf{M}_a$. The *magnitude* of $\mathbf{M}_a$ is determined by the dot product, $M_a = M_O \cos\theta = \mathbf{M}_O \cdot \mathbf{u}_a$, where $\mathbf{u}_a$ is a unit vector used to define the direction of the aa' axis. Combining these two steps as a general expression, we have $M_a = (\mathbf{r} \times \mathbf{F}) \cdot \mathbf{u}_a$. Since the dot product is commutative, we can also write

$$M_a = \mathbf{u}_a \cdot (\mathbf{r} \times \mathbf{F})$$

In vector algebra, this combination of dot and cross product yielding the scalar M_a is termed the *triple scalar product*. Provided the Cartesian components of each of the vectors are known, the mixed triple product may be written in determinant form as

$$M_a = (u_{a_x}\mathbf{i} + u_{a_y}\mathbf{j} + u_{a_z}\mathbf{k}) \cdot \begin{vmatrix} \mathbf{i} & \mathbf{j} & \mathbf{k} \\ r_x & r_y & r_z \\ F_x & F_y & F_z \end{vmatrix}$$

or simply

$$M_a = \mathbf{u}_a \cdot (\mathbf{r} \times \mathbf{F}) = \begin{vmatrix} u_{a_x} & u_{a_y} & u_{a_z} \\ r_x & r_y & r_z \\ F_x & F_y & F_z \end{vmatrix} \tag{3-11}$$

where $u_{a_x}, u_{a_y}, u_{a_z}$ represent the x, y, z components of the unit vector defining the direction of the aa' axis

r_x, r_y, r_z represent the x, y, z components of the position vector drawn from any point O on the aa' axis to any point A on the line of action of the force

F_x, F_y, F_z represent the x, y, z components of the force vector.

When M_a is evaluated from Eq. 3–11, it will yield a positive or negative scalar. The sign of this scalar indicates the sense of direction of $\mathbf{M}_a$ along the aa' axis. If it is positive, then $\mathbf{M}_a$ will have the same sense as $\mathbf{u}_a$, whereas if it is negative, then $\mathbf{M}_a$ will act opposite to $\mathbf{u}_a$.

Once M_a is determined, we can then express $\mathbf{M}_a$ as a Cartesian vector, namely,

$$\mathbf{M}_a = M_a\mathbf{u}_a = [\mathbf{u}_a \cdot (\mathbf{r} \times \mathbf{F})]\mathbf{u}_a \tag{3-12}$$

Finally, if the resultant moment of a series of forces is to be computed about the axis, then the moment components of each force are added together *algebraically,* since each component lies along the same axis.

The following examples illustrate a numerical application of the above concepts.

Example 3–9

The force $\mathbf{F} = \{-40\mathbf{i} + 20\mathbf{j} + 10\mathbf{k}\}$ N acts at point A shown in Fig. 3–24a. Determine the moments of this force about the x and Oa axes.

SOLUTION I (*VECTOR ANALYSIS*)

We can solve this problem by using the position vector $\mathbf{r}_A$. Why? Since $\mathbf{r}_A = \{-3\mathbf{i} + 4\mathbf{j} + 6\mathbf{k}\}$ m, and $\mathbf{u}_x = \mathbf{i}$, then applying Eq. 3–11,

$$M_x = \mathbf{i} \cdot (\mathbf{r}_A \times \mathbf{F}) = \begin{vmatrix} 1 & 0 & 0 \\ -3 & 4 & 6 \\ -40 & 20 & 10 \end{vmatrix}$$

$$= 1[4(10) - 6(20)] - 0[(-3)(10) - 6(-40)] + 0[(-3)(20) - 4(-40)]$$

$$= -80 \text{ N} \cdot \text{m} \qquad\qquad Ans.$$

The negative sign indicates that the sense of $\mathbf{M}_x$ is opposite to $\mathbf{i}$.

Realize that this result also represents the $\mathbf{i}$ component of the moment of $\mathbf{F}$ about point O, i.e., $\mathbf{M}_O = \mathbf{r}_A \times \mathbf{F}$.

We can compute M_{Oa} using $\mathbf{r}_A$ and $\mathbf{u}_{Oa} = -\frac{3}{5}(\mathbf{i}) + \frac{4}{5}(\mathbf{j})$. Thus,

$$M_{Oa} = \mathbf{u}_{Oa} \cdot (\mathbf{r}_A \times \mathbf{F}) = \begin{vmatrix} -\frac{3}{5} & \frac{4}{5} & 0 \\ -3 & 4 & 6 \\ -40 & 20 & 10 \end{vmatrix}$$

$$= -\tfrac{3}{5}[4(10) - 6(20)] - \tfrac{4}{5}[(-3)(10) - 6(-40)] + 0[(-3)(20) - 4(-40)]$$

$$= -120 \text{ N} \cdot \text{m} \qquad\qquad Ans.$$

What does the negative sign indicate?

The moment components are shown in Fig. 3–24b.

SOLUTION II (*SCALAR ANALYSIS*)

Since the force components and moment arms are easy to determine for computing M_x a scalar analysis can be used to solve this problem. Referring to Fig. 3–24c, only the 10-N and 20-N forces contribute moments about the x axis. (The line of action of the 40-N force is *parallel* to this axis and hence its moment about the x axis is zero.) Using the right-hand rule, the algebraic sum of the moment components about the x axis is therefore

$$M_x = (10 \text{ N})(4 \text{ m}) - (20 \text{ N})(6 \text{ m}) = -80 \text{ N} \cdot \text{m} \qquad Ans.$$

Although not required here, note also that

$$M_y = (10 \text{ N})(3 \text{ m}) - (40 \text{ N})(6 \text{ m}) = -210 \text{ N} \cdot \text{m}$$

$$M_z = (40 \text{ N})(4 \text{ m}) - (20 \text{ N})(3 \text{ m}) = 100 \text{ N} \cdot \text{m}$$

If we were to determine M_{Oa} by this scalar method it would require much more effort, since the force components of 40 N and 20 N are *not* perpendicular to the direction of Oa. The vector analysis yields a more direct solution.

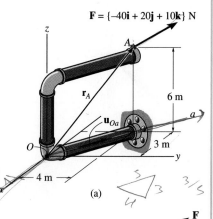

(a)

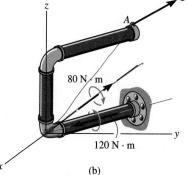

(b)

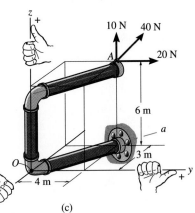

(c)

Fig. 3–24

Example 3–10

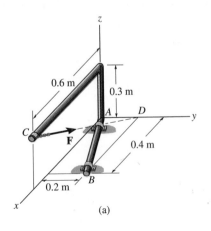

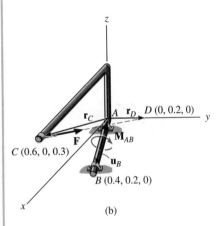

Fig. 3–25

The rod shown in Fig. 3–25a is supported by two brackets at A and B. Determine the moment $\mathbf{M}_{AB}$ produced by $\mathbf{F} = \{-600\mathbf{i} + 200\mathbf{j} - 300\mathbf{k}\}$ N, which tends to rotate the rod about the AB axis.

SOLUTION

A vector analysis using Eq. 3–11 will be considered for the solution since the moment arm is difficult to determine. In general, $M_{AB} = \mathbf{u}_B \cdot (\mathbf{r} \times \mathbf{F})$. Each of the terms in this equation will now be identified.

Unit vector $\mathbf{u}_B$ defines the direction of the AB axis of the rod, Fig. 3–25b, where

$$\mathbf{u}_B = \frac{\mathbf{r}_B}{r_B} = \frac{0.4\mathbf{i} + 0.2\mathbf{j}}{\sqrt{(0.4)^2 + (0.2)^2}} = 0.894\mathbf{i} + 0.447\mathbf{j}$$

Vector $\mathbf{r}$ is directed from *any point* on the AB axis to *any point* on the line of action of the force. For example, position vectors $\mathbf{r}_C$ and $\mathbf{r}_D$ are suitable, Fig. 3–25b. (Although not shown, $\mathbf{r}_{BC}$ or $\mathbf{r}_{BD}$ can also be used.) For simplicity, we choose $\mathbf{r}_D$, where

$$\mathbf{r}_D = \{0.2\mathbf{j}\} \text{ m}$$

The force is

$$\mathbf{F} = \{-600\mathbf{i} + 200\mathbf{j} - 300\mathbf{k}\} \text{ N}$$

Substituting these vectors into the determinant form and expanding, we have

$$M_{AB} = \mathbf{u}_B \cdot (\mathbf{r}_D \times \mathbf{F}) = \begin{vmatrix} 0.894 & 0.447 & 0 \\ 0 & 0.2 & 0 \\ -600 & 200 & -300 \end{vmatrix}$$

$$= 0.894[0.2(-300) - 0(200)] - 0.447[0(-300) - 0(-600)]$$
$$+ 0[0(200) - 0.2(-600)]$$

$$= -53.64 \text{ N} \cdot \text{m}$$

The negative sign indicates that the sense of $\mathbf{M}_{AB}$ is opposite to that of $\mathbf{u}_B$. Expressing $\mathbf{M}_{AB}$ as a Cartesian vector yields

$$\mathbf{M}_{AB} = M_{AB}\mathbf{u}_B = (-53.64 \text{ N} \cdot \text{m})(0.894\mathbf{i} + 0.447\mathbf{j})$$
$$= \{-48.0\mathbf{i} - 24.0\mathbf{j}\} \text{ N} \cdot \text{m} \qquad\qquad Ans.$$

The result is shown in Fig. 3–25b.

It should be mentioned that if axis AB is defined using a unit vector directed from B toward A, then in the above formulation $-\mathbf{u}_B$ would have to be used. This would lead to $M_{AB} = +53.64$ N · m. Consequently, $\mathbf{M}_{AB} = M_{AB}(-\mathbf{u}_B)$, and the above result would again be determined.

PROBLEMS

*3–28.   Determine the moment of the force **F** about the *Oa* axis. Express the result as a Cartesian vector.

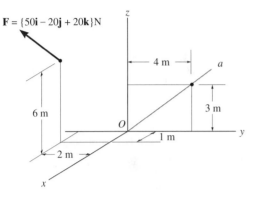

F = {50**i** – 20**j** + 20**k**}N

4 m

a

3 m

6 m

O

1 m

y

2 m

x

Prob. 3–28

3–30.   Determine the moment of the force **F** about the *aa* axis. Express the result as a Cartesian vector.

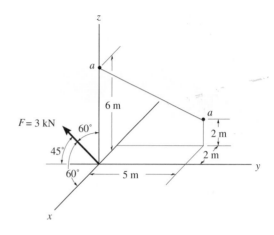

a

6 m

F = 3 kN   60°

a

2 m

45°

2 m

60°   5 m

y

x

Prob. 3–30

3–31.   Determine the moments of the force **F** about the *x*, *y*, and *z* axes. Solve a problem (*a*) using a Cartesian vector approach and (*b*) using a scalar approach.

*3–32.   Determine the moment of the force **F** about an axis extending between *O* and *A*. Express the result as a Cartesian vector.

3–29.   Determine the resultant moment of two forces about the *aa* axis. Express the result as a Cartesian vector.

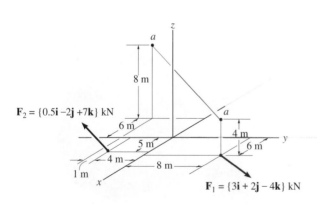

a

8 m

$F_2 = \{0.5\mathbf{i} - 2\mathbf{j} + 7\mathbf{k}\}$ kN

6 m

a

5 m

4 m

4 m

6 m

y

4 m

8 m

1 m   *x*

$F_1 = \{3\mathbf{i} + 2\mathbf{j} - 4\mathbf{k}\}$ kN

Prob. 3–29

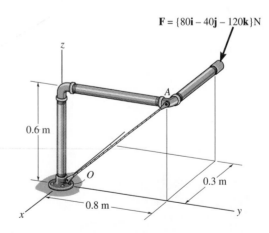

F = {80**i** – 40**j** – 120**k**}N

A

0.6 m

O

0.3 m

x

0.8 m

y

Probs. 3–31/3–32

3–33. Determine the magnitude of the resultant moment of the three forces about the axis AB. Solve the problem (a) using a Cartesian vector approach and (b) using a scalar approach.

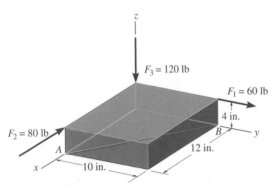

Prob. 3–33

3–34. A force of $\mathbf{F} = \{8\mathbf{i} - 1\mathbf{j} + 1\mathbf{k}\}$ lb is applied to the handle of the box wrench. Determine the component of the moment of this force about the z axis which is effective in tightening the bolt.

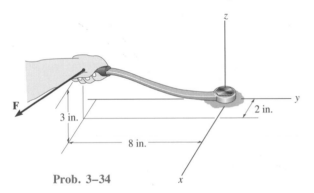

Prob. 3–34

3–35. The 50-lb force acts on the gear in the direction shown. Determine the moment of this force about the y axis.

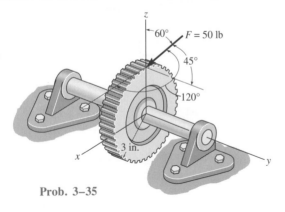

Prob. 3–35

*3–36. The chain AB exerts a force of 20 lb on the door at B. Determine the magnitude of the moment of this force along the hinged axis x of the door.

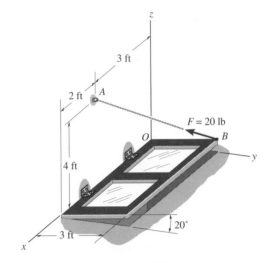

Prob. 3–36

3–37. The lug and box wrenches are used in combination to remove the lug nut from the wheel hub. If the applied force on the end of the box wrench is $\mathbf{F} = \{4\mathbf{i} - 12\mathbf{j} + 2\mathbf{k}\}$ N, determine the magnitude of the moment of this force about the x axis which is effective in unscrewing the lug nut.

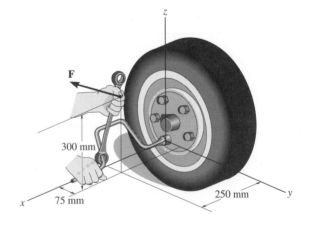

Prob. 3–37

3–38. Determine the moment that the force **F** exerts about the *y* axis of the shaft. Solve the problem using a Cartesian vector approach and using a scalar approach. Express the result as a Cartesian vector.

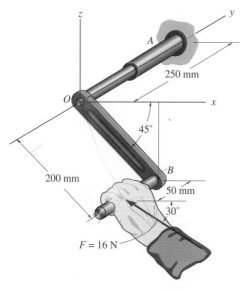

Prob. 3–38

3–39. The bracket is acted upon by a 600-N force at *A*. Determine the moment of this force about the *y* axis.

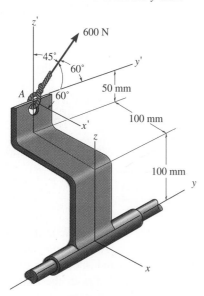

Prob. 3–39

*3–40. The force of *F* = 80 lb acts along the edge *DB* of the tetrahedron. Determine the magnitude of the moment of this force about the edge *AC*.

3–41. If the moment of the force **F** about the edge *AC* of the tetrahedron has a magnitude of *M* = 200 lb · ft and is directed from *C* toward *A*, determine the magnitude of **F**.

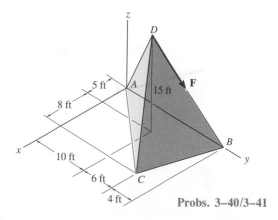

Probs. 3–40/3–41

3–42. A vertical force of *F* = 60 N is applied to the handle of the pipe wrench. Determine the moment that this force exerts along the axis *AB* (*x* axis) of the pipe assembly. Both the wrench and pipe assembly *ABC* lie in the *x–y* plane. *Suggestion:* Use a scalar analysis.

3–43. Determine the magnitude of the vertical force **F** acting on the handle of the wrench so that this force produces a component of moment along the *AB* axis (*x* axis) of the pipe assembly of $(\mathbf{M}_A)_x = \{-5\mathbf{i}\}$ N · m. Both the pipe assembly *ABC* and the wrench lie in the *x–y* plane. *Suggestion:* Use a scalar analysis.

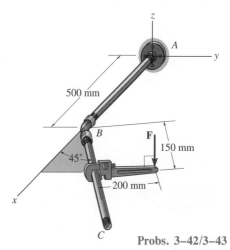

Probs. 3–42/3–43

3.6 Moment of a Couple

A *couple* is defined as two parallel forces that have the same magnitude, opposite directions, and are separated by a perpendicular distance d, Fig. 3–26. Since the resultant force of the two forces composing the couple is zero, the only effect of a couple is to produce a rotation or tendency of rotation in a specified direction.

Fig. 3–26

The moment produced by a couple, called a *couple moment,* is equivalent to the sum of the moments of both couple forces, determined about *any* arbitrary point O in space. To show this, consider position vectors r_A and r_B, directed from O to points A and B lying on the line of action of $-F$ and F, Fig. 3–27. The couple moment computed about O is therefore

$$M = r_A \times (-F) + r_B \times (F)$$
$$= (r_B - r_A) \times F$$

By the triangle law of vector addition, $r_A + r = r_B$ or $r = r_B - r_A$, so that

$$M = r \times F \qquad (3–13)$$

This result indicates that a couple moment is a *free vector,* i.e., it can act at *any point,* since M depends *only* upon the position vector directed *between* the forces and *not* the position vectors r_A and r_B, directed from point O to the forces. This concept is therefore unlike the moment of a force, which requires a definite point (or axis) about which moments are determined.

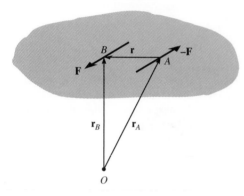

Fig. 3–27

Scalar Formulation.   The moment of a couple, M, Fig. 3–28, is defined as having a *magnitude* of

$$\boxed{M = Fd} \qquad (3–14)$$

where F is the magnitude of one of the forces and d is the perpendicular distance or moment arm between the forces. The *direction* and sense of the couple moment are determined by the right-hand rule, where the thumb indicates the direction when the fingers are curled with the sense of rotation caused by the two forces. In all cases, M acts perpendicular to the plane containing these forces.

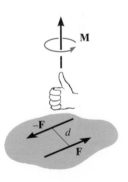

Fig. 3–28

Vector Formulation. The moment of a couple can also be expressed by the vector cross product using Eq. 3–13,

$$M = r \times F \qquad (3\text{--}15)$$

Application of this equation is easily remembered if one thinks of taking the moments of both forces about a point lying on the line of action of one of the forces. For example, if moments are taken about point A in Fig. 3–27 the moment of $-F$ is zero about this point and the moment of F is defined from Eq. 3–15. Therefore, in the formulation r is crossed with the force F to which it is directed.

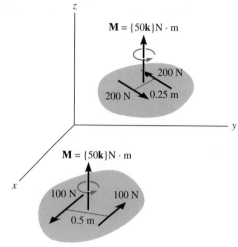

Fig. 3–29

Equivalent Couples. Two couples are said to be equivalent if they produce the same moment. Since the moment produced by a couple is always perpendicular to the plane containing the couple forces, it is therefore necessary that the forces of equal couples lie either in the same plane or in planes that are *parallel* to one another. In this way, the direction of each couple moment will be the same, that is, perpendicular to the parallel planes. For example, the two couples shown in Fig. 3–29 are equivalent. One couple is produced by a pair of 100-N forces separated by a distance of $d = 0.5$ m, and the other is produced by a pair of 200-N forces separated by a distance of 0.25 m. Since the planes in which the forces act are parallel to the x–y plane, the moment produced by each of the couples may be expressed as $M = \{50k\}$ N · m.

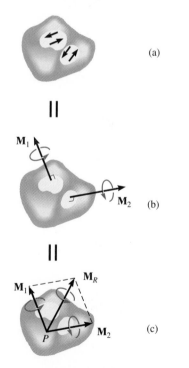

Fig. 3–30

Resultant Couple Moment Since couple moments are free vectors, they may be applied at any point P on a body and added vectorially. For example, the two couples acting on different planes of the rigid body in Fig. 3–30a may be replaced by their corresponding couple moments M_1 and M_2, Fig. 3–30b, and then these free vectors may be moved to the *arbitrary point P* and added to obtain the resultant couple moment $M_R = M_1 + M_2$, shown in Fig. 3–30c.

If more than two couple moments act on the body, we may generalize this concept and write the vector resultant as

$$M_R = \Sigma r \times F \qquad (3\text{--}16)$$

where each couple moment is computed in accordance with Eq. 3–15.

The following examples illustrate these concepts numerically. In general, problems projected in two dimensions should be solved using a scalar analysis, since the moment arms and force components are easy to compute.

Example 3–11

A couple acts at the end of the beam shown in Fig. 3–31a. Replace it by an equivalent couple having a pair of forces that act through (a) points A and B, and (b) points D and E.

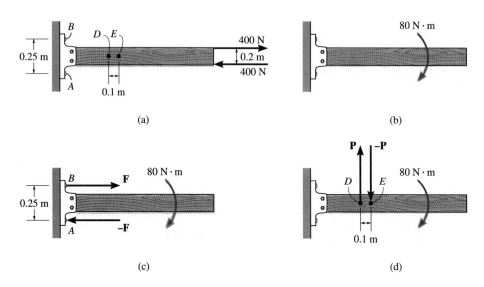

Fig. 3–31

SOLUTION (SCALAR ANALYSIS)

The couple has a magnitude of $M = Fd = 400(0.2) = 80$ N · m and a direction that is into the page since the forces tend to rotate clockwise. **M** is a free vector so that it can be placed at any point on the beam, Fig. 3–31b.

Part (a).   To preserve the direction of **M,** *horizontal* forces acting through points A and B must be directed as shown in Fig. 3–31c. The magnitude of each force is

$$M = Fd$$
$$80 = F(0.25)$$
$$F = 320 \text{ N} \qquad\qquad Ans.$$

Part (b).   To generate the required clockwise rotation, forces acting through points D and E must be *vertical* and directed as shown in Fig. 3–31d. The magnitude of each force is

$$M = Pd$$
$$80 = P(0.1)$$
$$P = 800 \text{ N} \qquad\qquad Ans.$$

Example 3–12

Determine the moment of the couple acting on the beam shown in Fig. 3–32*a*.

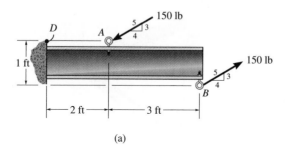

(a)

SOLUTION (*SCALAR ANALYSIS*)

Here it is somewhat difficult to determine the perpendicular distance between the forces and compute the couple moment as $M = Fd$. Instead, we can resolve each force into its horizontal and vertical components, $F_x = \frac{4}{5}(150) = 120$ lb and $F_y = \frac{3}{5}(150) = 90$ lb, Fig. 3–32*b*, and then use the principle of moments. The couple moment can be determined about *any point*. For example, if point *D* is chosen, we have for all four forces,

$$\zeta+M = 120 \text{ lb}(0 \text{ ft}) - 90 \text{ lb}(2 \text{ ft}) + 90 \text{ lb}(5 \text{ ft}) + 120 \text{ lb}(1 \text{ ft})$$
$$= 390 \text{ lb} \cdot \text{ft} \uparrow \qquad\qquad\qquad Ans.$$

It is easier, however, to determine the moments about point *A* or *B* in order to *eliminate* the moment of the forces acting at the moment point. For point *A*, Fig. 3–28*b*, we have

$$\zeta+M = 90 \text{ lb}(3 \text{ ft}) + 120 \text{ lb}(1 \text{ ft})$$
$$= 390 \text{ lb} \cdot \text{ft} \uparrow \qquad\qquad\qquad Ans.$$

Show that one obtains this same result if moments are summed about point *B*. Notice also that the couple in Fig. 3–32*a* has been replaced by *two* couples in Fig. 3–32*b*. Using $M = Fd$, one couple has a moment of $M_1 = 90 \text{ lb}(3 \text{ ft}) = 270 \text{ lb} \cdot \text{ft}$ and the other has a moment of $M_2 = 120 \text{ lb}(1 \text{ ft}) = 120 \text{ lb} \cdot \text{ft}$. By the right-hand rule, both couple moments are counterclockwise and are therefore directed out of the page. Since these couples are free vectors, they can be moved to any point and added, which yields $M = 270 \text{ lb} \cdot \text{ft} + 120 \text{ lb} \cdot \text{ft} = 390 \text{ lb} \cdot \text{ft} \uparrow$, the same result determined above. **M** is a free vector and can therefore act at any point on the beam, Fig. 3–32*c*.

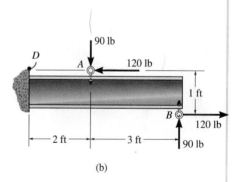

(b)

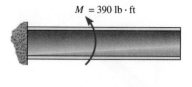

(c)

Fig. 3–32

Example 3–13

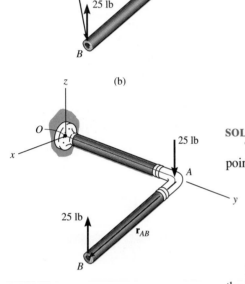

(b)

Determine the couple moment acting on the pipe shown in Fig. 3–33a. Segment AB is directed $30°$ below the x–y plane.

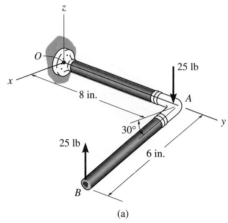

(a)

SOLUTION I (VECTOR ANALYSIS)

The moment of the two couple forces can be found about *any point*. If point O is considered, Fig. 3–33b, we have

$$\mathbf{M} = \mathbf{r}_A \times (-25\mathbf{k}) + \mathbf{r}_B \times (25\mathbf{k})$$
$$= (8\mathbf{j}) \times (-25\mathbf{k}) + (6 \cos 30°\mathbf{i} + 8\mathbf{j} - 6 \sin 30°\mathbf{k}) \times (25\mathbf{k})$$
$$= -200\mathbf{i} - 129.9\mathbf{j} + 200\mathbf{i}$$
$$= \{-130\mathbf{j}\} \text{ lb} \cdot \text{in.} \qquad \qquad \text{Ans.}$$

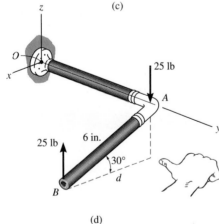

(c)

It is *easier* to take moments of the couple forces about a point lying on the line of action of one of the forces, e.g., point A, Fig. 3–33c. In this case the moment of the force at A is zero, so that

$$\mathbf{M} = \mathbf{r}_{AB} \times (25\mathbf{k})$$
$$= (6 \cos 30°\mathbf{i} - 6 \sin 30°\mathbf{k}) \times (25\mathbf{k})$$
$$= \{-130\mathbf{j}\} \text{ lb} \cdot \text{in.} \qquad \qquad \text{Ans.}$$

SOLUTION II (SCALAR ANALYSIS)

Although this problem is shown in three dimensions, the geometry is simple enough to use the scalar equation $M = Fd$. The perpendicular distance between the lines of action of the forces is $d = 6 \cos 30° = 5.20$ in., Fig. 3–33d. Hence, taking moments of the forces about either point A or B yields

$$M = Fd = 25(5.20) = 129.9 \text{ lb} \cdot \text{in.}$$

Applying the right-hand rule, $\mathbf{M}$ acts in the $-\mathbf{j}$ direction. Thus,

$$\mathbf{M} = \{-130\mathbf{j}\} \text{ lb} \cdot \text{in.} \qquad \qquad \text{Ans.}$$

(d)

Fig. 3–33

Example 3–14

Replace the two couples acting on the triangular block in Fig. 3–34a by a resultant couple moment.

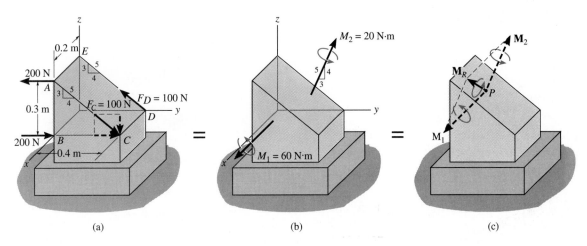

(a)   (b)   (c)

Fig. 3–34

SOLUTION (*VECTOR ANALYSIS*)

The couple moment M_1, developed by the forces at A and B, can easily be determined from a scalar formulation.

$$M_1 = Fd = 200(0.3) = 60 \text{ N} \cdot \text{m}$$

By the right-hand rule, M_1 acts in the $+i$ direction, Fig. 3–34b. Hence,

$$M_1 = \{60i\} \text{ N} \cdot \text{m}$$

Vector analysis will be used to determine M_2, caused by forces at C and D. If moments are computed about point D, Fig. 3–34a, $M_2 = r_{DC} \times F_C$, then

$$M_2 = r_{DC} \times F_C = (0.2i) \times [100(\tfrac{4}{5})j - 100(\tfrac{3}{5})k]$$
$$= (0.2i) \times [80j - 60k] = 16(i \times j) - 12(i \times k)$$
$$= \{12j + 16k\} \text{ N} \cdot \text{m}$$

Try to establish M_2 by using a scalar formulation, Fig. 3–34b.

Since M_1 and M_2 are free vectors, they may be moved to some arbitrary point P on the block and added vectorially, Fig. 3–34c. The resultant couple moment becomes

$$M_R = M_1 + M_2 = \{60i + 12j + 16k\} \text{ N} \cdot \text{m} \qquad Ans.$$

PROBLEMS

*3–44. Determine the magnitude and direction of the couple moment.

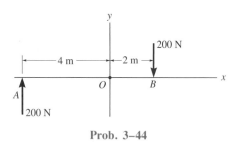

Prob. 3–44

3–45. Determine the magnitude and direction of the couple moment.

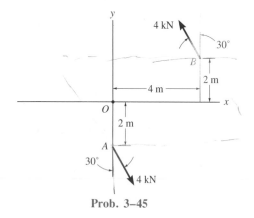

Prob. 3–45

3–46. Determine the magnitude and direction of the couple moment.

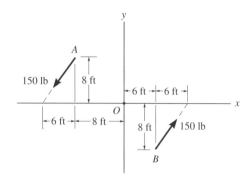

Prob. 3–46

3–47. The road reacts with a torque of $M_A = 400$ N · m and $M_B = 200$ N · m on the brushes of the road sweeper. Determine the magnitude of the couple forces that are developed by the road on the rear wheels of the sweeper, so that the resultant couple moment on the sweeper is zero. What is the magnitude of these forces if the brush at B is turned off?

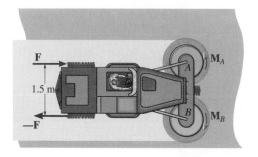

Prob. 3–47

*3–48. A clockwise couple moment having a magnitude of $M = 40$ lb · ft is resisted by the shaft of an electric motor. Determine the magnitudes of the reactive forces **R** and $-$**R**, which act at the supports A and B, so that the resultant of the two couples is zero.

3–49. Determine the magnitude of the couple moment **M** acting on the electric motor if the reactive forces **R** and $-$**R** each have a magnitude of 20 lb and the resultant of the two couples is zero.

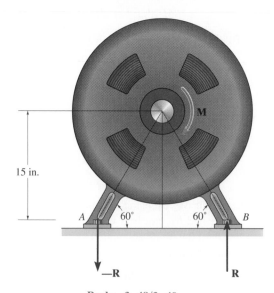

Probs. 3–48/3–49

3–50. The cord passing over the two small pegs A and B of the board is subjected to a tension of 10 lb. Determine the *minimum* tension P and the orientation θ of the cord passing over the pegs C and D, so that the resultant couple moment produced by the two cords is 20 lb · in., clockwise.

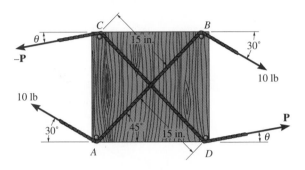

Prob. 3–50

3–51. Two couples act on the beam. Determine the magnitude of F so that the resultant couple moment is 450 lb · ft, counterclockwise. Where on the beam does the resultant couple moment act?

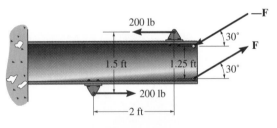

Prob. 3–51

*3–52. Two couples act on the frame. If the resultant couple moment is to be zero, determine the distance d between the 100-lb couple forces.

3–53. Two couples act on the frame. If $d = 4$ ft, determine the resultant couple moment. Compute the result by resolving each force into x and y components and (*a*) finding the moment of each couple (Eq. 3–14) and (*b*) summing the moments of all the force components about point A.

3–54. Two couples act on the frame. If $d = 6$ ft, determine the resultant couple moment. Compute the result by resolving each force into x and y components and (*a*) finding the moment of each couple (Eq. 3–14) and (*b*) summing the moments of all the force components about point B.

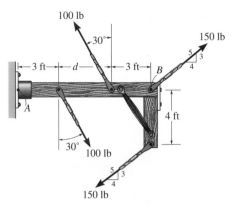

Probs. 3–52/3–53/3–54

3–55. Determine the resultant couple moment acting on the beam. Solve the problem two ways: (*a*) sum moments about point O; and (*b*) sum moments about point A.

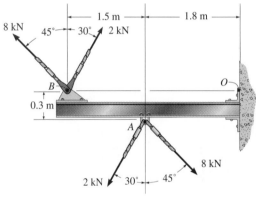

Prob. 3–55

*3–56. Determine the couple moment. Express the result as a Cartesian vector.

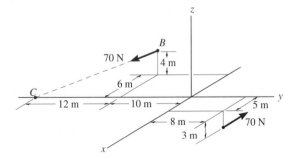

Prob. 3–56

3–57. The gear reducer is subjected to the couple moments shown. Determine the resultant couple moment and specify its magnitude and coordinate direction angles.

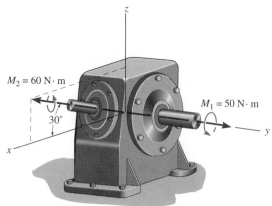

Prob. 3–57

3–58. Express the moment of the couple acting on the pipe assembly in Cartesian vector form. What is the magnitude of the couple moment?

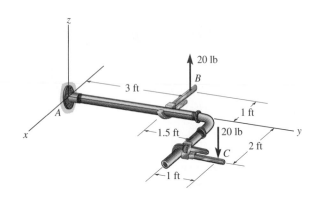

Prob. 3–58

3–59. Determine the resultant couple moment of the two couples that act on the pipe assembly. The distance from A to B is $d = 400$ mm. Express the result as a Cartesian vector.

*3–60. Determine the distance d between A and B so that the resultant couple moment has a magnitude of $M_R = 20$ N · m.

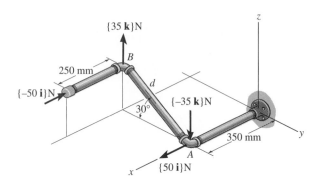

Probs. 3–59/3–60

3–61. If the resultant couple moment of the three couples acting on the triangular block is to be zero, determine the magnitude of forces $\mathbf{F}$ and $\mathbf{P}$.

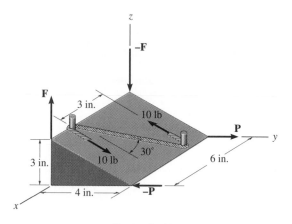

Prob. 3–61

3.7 Movement of a Force on a Rigid Body

If a force acts on a body, it generally has the effect of causing the body to both translate and rotate. The supports for the body, however, are intended to stop this motion from occurring. For example, consider the effect of holding the end of a stick of negligible weight, when a force $\mathbf{F}$ is applied at its other end,

Fig. 3–35a. Experience tells us that **F** produces both a downward force at the grip and a clockwise twist. This force and moment can be determined by first applying equal but opposite forces **F** and **–F** at the grip, Fig. 3–35b. This step in no way alters the effect at the grip, however, the two forces indicated by the slash across them form a couple which is clockwise and has a magnitude of $M = Fd$. This couple moment is a *free vector* and can actually act *at any point* on the stick. In addition to this, the force **F** acts at the grip, Fig. 3–35c. In other words, the force **F** and couple moment **M** acting as shown in Fig. 3–35c, create the same effect on the hand as the force **F** acting at the end of the stick in Fig. 3–35a. By using this construction procedure, an *equivalent system* has been maintained between each of the diagrams, as indicated by the equal sign.

If the stick is held vertically, Fig. 3–35d, we can repeat the above procedure, by first applying equal but opposite forces at the grip, Fig. 3–35e, then noting that the forces with the slash through them can be canceled, leaving the force **F** at the grip, Fig. 3–35f. Here the couple moment is zero, and the force has simply been "transmitted" along its line of action, from one end of the stick to the other. In other words, the force can be considered as a *sliding vector* since it can be applied at any point along its line of action and produce the same effect at the grip. This concept is referred to as the *principle of transmissibility,* and it was initially discussed in Sec. 3.4. It can be formally stated as follows: *The external effects on a body remain unchanged when a force, acting at a given point on the body, is applied to another point lying on the line of action of the force.* It is important to realize that only the external effects caused by the force remain unchanged after moving it. Certainly the internal effects depend on where **F** is located. For example, when **F** acts at the end of the stick, Fig. 3–35a, the internal forces within the stick at its end have a high intensity around its point of application, whereas movement of **F** away from this point will cause these internal forces to decrease.

The above two examples indicate how a force **F** may be moved from point A to another point O on a body and yet produce an equivalent *external* effect, Fig. 3–36a. This simply requires moving the force to point O and adding a couple moment to the body, determined by calculating the moment of the force about point O, in which case **M** = **r** × **F**, Fig. 3–36b. Here the position vector **r** extends from O to A, Fig. 3–36a. Remember that **M** is a free vector and can act at any point P on the body.

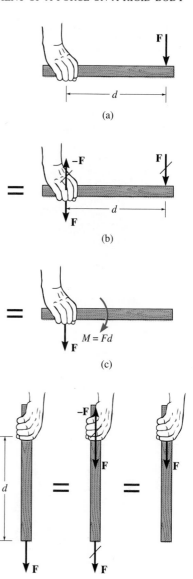

Fig. 3–35

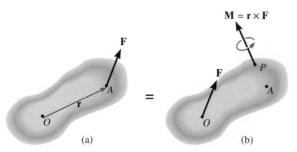

Fig. 3–36

3.8 Resultants of a Force and Couple System

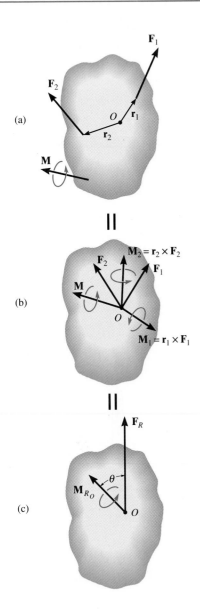

(a)

(b)

(c)

Fig. 3–37

When a body is subjected to a system of forces and couple moments, it is often simpler to study the external effects on the body by using the force and couple moment resultants, rather than the force and couple moment system. To show how to simplify a system of forces and couple moments to their resultants, consider the rigid body in Fig. 3–37a. The force and couple moment system acting on it will be simplified by moving the forces and couple moments to the arbitrary point O. In this regard, the couple moment $\mathbf{M}$ is simply moved to O, since it is a free vector. Forces $\mathbf{F}_1$ and $\mathbf{F}_2$ are sliding vectors, and since O does not lie on the line of action of these forces, each must be moved to O in accordance with the procedure stated in Sec. 3.7. For example, when $\mathbf{F}_1$ is applied at O, a corresponding couple moment $\mathbf{M}_1 = \mathbf{r}_1 \times \mathbf{F}_1$ must also be applied to the body, Fig. 3–37b. By vector addition, the force and couple moment system shown in Fig. 3–37b can now be reduced to an *equivalent* resultant force $\mathbf{F}_R = \mathbf{F}_1 + \mathbf{F}_2$ and resultant couple moment $\mathbf{M}_{R_O} = \mathbf{M} + \mathbf{M}_1 + \mathbf{M}_2$ as shown in Fig. 3–37c. Note that both the magnitude and direction of $\mathbf{F}_R$ are *independent* of the location of point O. On the other hand, $\mathbf{M}_{R_O}$ depends upon this location, since the moments $\mathbf{M}_1$ and $\mathbf{M}_2$ are computed using the position vectors $\mathbf{r}_1$ and $\mathbf{r}_2$. Realize also that $\mathbf{M}_{R_O}$ is a free vector and can act at *any point* on the body, although point O is generally chosen as its point of application.

PROCEDURE FOR ANALYSIS

The above method for simplifying any force and couple moment system to a resultant force acting at point O and a resultant couple moment will now be stated in general terms. To apply this method it is first necessary to establish an x, y, z coordinate system.

Three-Dimensional Systems. A Cartesian vector analysis is generally used to solve problems involving three-dimensional force and couple systems for which the force components and moment arms are difficult to determine.

Force Summation. The resultant force is equivalent to the vector sum of all the forces in the system; i.e.,

$$\boxed{\mathbf{F}_R = \Sigma \mathbf{F}}$$

$$(3\text{–}17)$$

Moment Summation. The resultant couple moment is equivalent to the vector sum of all the couple moments in the system plus the moments about point O of all the forces in the system; i.e.,

$$\boxed{\mathbf{M}_{R_O} = \Sigma \mathbf{M}_O}$$

$$(3\text{–}18)$$

Coplanar Force Systems. Since force components and moment arms are easy to determine in two dimensions, a scalar analysis provides the most convenient solution to problems involving coplanar force systems. Assuming that the forces lie in the x–y plane and any couple moments are perpendicular to this plane (along the z axis), then the resultants are determined as follows:

Force Summation. The *resultant force* $\mathbf{F}_R$ is equivalent to the vector sum of its two components $\mathbf{F}_{R_x}$ and $\mathbf{F}_{R_y}$. Each component is found from the scalar (algebraic) sum of the components of all the forces in the system that act in the same direction; i.e.,

$$\boxed{\begin{aligned} F_{R_x} &= \Sigma F_x \\ F_{R_y} &= \Sigma F_y \end{aligned}} \qquad\qquad (3\text{–}19)$$

Moment Summation. The *resultant couple moment* $\mathbf{M}_{R_O}$ is perpendicular to the plane containing the forces and is equivalent to the scalar (algebraic) sum of all the couple moments in the system *plus* the moments about point O of all the forces in the system; i.e.,

$$\boxed{\mathbf{M}_{R_O} = \Sigma M_O} \qquad\qquad (3\text{–}20)$$

When determining the moments of the forces about O, it is generally advantageous to use the *principle of moments;* i.e., determine the moments of the *components* of each force rather than the moment of the force itself.

It is important to remember that, when applying any of these equations, attention should be paid to the sense of direction of the force components and the moments of the forces. If they are along the positive coordinate axes, they represent positive scalars; whereas if these components have a directional sense along the negative coordinate axes, they are negative scalars. By following this convention, a positive result, for example, indicates that the resultant vector has a sense of direction along the positive coordinate axis.

The following two examples illustrate these procedures numerically.

Example 3–15

Replace the forces acting on the pipe shown in Fig. 3–38a by an equivalent resultant force and couple moment acting at point A.

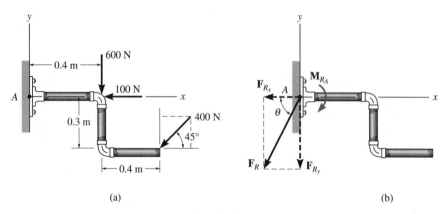

(a)   (b)

Fig. 3–38

SOLUTION (*SCALAR ANALYSIS*)

The principle of moments will be applied to the 400-N force, whereby the moments of its two rectangular components will be considered.

Force Summation.   The resultant force has x and y components of

$$\xrightarrow{+}F_{R_x} = \Sigma F_x; \quad F_{R_x} = -100 - 400 \cos 45° = -382.8 \text{ N} = 382.8 \text{ N} \leftarrow$$
$$+\uparrow F_{R_y} = \Sigma F_y; \quad F_{R_y} = -600 - 400 \sin 45° = -882.8 \text{ N} = 882.8 \text{ N} \downarrow$$

As shown in Fig. 3–38b, $\mathbf{F}_R$ has a magnitude of

$$F_R = \sqrt{(F_{R_x})^2 + (F_{R_y})^2} = \sqrt{(382.8)^2 + (882.8)^2} = 962 \text{ N} \quad \textit{Ans.}$$

and a direction defined from the vector sketch of

$$\theta = \tan^{-1}\left(\frac{F_{R_y}}{F_{R_x}}\right) = \tan^{-1}\left(\frac{882.8}{382.8}\right) = 66.6° \quad \theta \nearrow \qquad \textit{Ans.}$$

Moment Summation.   The resultant couple moment $\mathbf{M}_{R_A}$ is determined by summing the moments of the forces about point A. Assuming that positive moments act counterclockwise, i.e., in the $+\mathbf{k}$ direction, we have

$$\downarrow+M_{R_A} = \Sigma M_A;$$
$$M_{R_A} = 100(0) - 600(0.4) - (400 \sin 45°)(0.8) - (400 \cos 45°)(0.3)$$
$$= -551 \text{ N} \cdot \text{m} = 551 \text{ N} \cdot \text{m} \downarrow \qquad\qquad\qquad \textit{Ans.}$$

In conclusion, when $\mathbf{M}_{R_A}$ and $\mathbf{F}_R$ act on the pipe at point A, Fig. 3–38b, they will produce the *same* external effect or reaction at the support A as that produced by the force system in Fig. 3–38a.

Example 3–16

A beam is subjected to a couple moment $\mathbf{M}$ and forces $\mathbf{F}_1$ and $\mathbf{F}_2$ as shown in Fig. 3–39a. Replace this system by an equivalent resultant force and couple moment acting at point O.

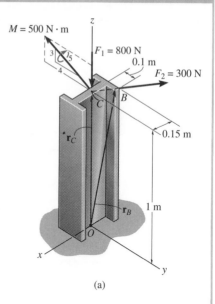

(a)

SOLUTION (*VECTOR ANALYSIS*)

Expressing the forces and couple moment as Cartesian vectors, we have

$$\mathbf{F}_1 = \{-800\mathbf{k}\} \text{ N}$$

$$\mathbf{F}_2 = (300 \text{ N})\mathbf{u}_{CB} = (300 \text{ N}) \left(\frac{\mathbf{r}_{CB}}{r_{CB}} \right)$$

$$= 300\left[\frac{-0.15\mathbf{i} + 0.1\mathbf{j}}{\sqrt{(-0.15)^2 + (0.1)^2}} \right] = \{-249.6\mathbf{i} + 166.4\mathbf{j}\} \text{ N}$$

$$\mathbf{M} = -500(\tfrac{4}{5})\mathbf{j} + 500(\tfrac{3}{5})\mathbf{k} = \{-400\mathbf{j} + 300\mathbf{k}\} \text{ N} \cdot \text{m}$$

Force Summation

$$\mathbf{F}_R = \Sigma\mathbf{F}; \qquad \mathbf{F}_R = \mathbf{F}_1 + \mathbf{F}_2 = -800\mathbf{k} - 249.6\mathbf{i} + 166.4\mathbf{j}$$

$$= \{-249.6\mathbf{i} + 166.4\mathbf{j} - 800\mathbf{k}\} \text{ N} \qquad\qquad \textbf{\textit{Ans.}}$$

Moment Summation

$$\mathbf{M}_{R_O} = \Sigma\mathbf{M}_O;$$

$$\mathbf{M}_{R_O} = \mathbf{M} + \mathbf{r}_C \times \mathbf{F}_1 + \mathbf{r}_B \times \mathbf{F}_2$$

$$= (-400\mathbf{j} + 300\mathbf{k}) + (1\mathbf{k}) \times (-800\mathbf{k}) + \begin{vmatrix} \mathbf{i} & \mathbf{j} & \mathbf{k} \\ -0.15 & 0.1 & 1 \\ -249.6 & 166.4 & 0 \end{vmatrix}$$

$$= (-400\mathbf{j} + 300\mathbf{k}) + (\mathbf{0}) + (-166.4\mathbf{i} - 249.6\mathbf{j})$$

$$= \{-166.4\mathbf{i} - 649.6\mathbf{j} + 300\mathbf{k}\} \text{ N} \cdot \text{m} \qquad\qquad \textbf{\textit{Ans.}}$$

The results are shown in Fig. 3–39b.

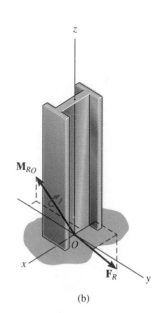

(b)

Fig. 3–39

3.9 Further Reduction of a Force and Couple System

Simplification to a Single Resultant Force. Consider now a special case for which the system of forces and couple moments acting on a body, Fig. 3–40*a*, reduces at point O to a resultant force $\mathbf{F}_R = \Sigma\mathbf{F}$ and resultant couple moment $\mathbf{M}_{R_O} = \Sigma\mathbf{M}_O$, which are *perpendicular* to one another, Fig. 3–40*b*. Whenever this occurs, we can further simplify the force and couple moment system by moving $\mathbf{F}_R$ to another point P, located either on or off the body so that no resultant couple moment has to be applied to the body, Fig. 3–40*c*. In other words, if the force and couple moment system in Fig. 3–40*a* is reduced to a resultant system at point P, only the force resultant will have to be applied to the body, Fig. 3–40*c*.

The location of point P, measured from point O, can always be determined provided $\mathbf{F}_R$ and $\mathbf{M}_{R_O}$ are known, Fig. 3–40*b*. As shown in Fig. 3–40*c*, P must lie on the *bb* axis, which is perpendicular to the line of action of $\mathbf{F}_R$ and the *aa* axis. This point is chosen such that the distance d satisfies the scalar equation $M_{R_O} = F_R d$ or $d = M_{R_O}/F_R$. With $\mathbf{F}_R$ so located, it will produce the same external effects on the body as the force and couple moment system in Fig. 3–40*a*, or the force and couple moment resultants in Fig. 3–40*b*. We refer to the force and couple moment system in Fig. 3–40*a* as being *equivalent or equipollent* to the single force "system" in Fig. 3–40*c*, because each system produces the *same* resultant force and resultant moment when replaced at point O.

If a system of forces is either concurrent, coplanar, or parallel, it can always be reduced, as in the above case, to a single resultant force $\mathbf{F}_R$ acting through a unique point P. This is because in each of these cases $\mathbf{F}_R$ and $\mathbf{M}_{R_O}$ will always be perpendicular to each other when the force system is simplified at *any* point O.

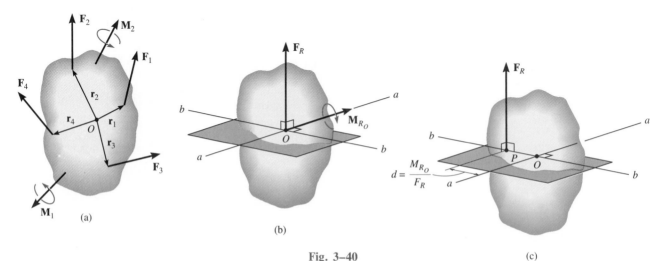

(a)

(b)

(c)

Fig. 3–40

Concurrent force systems have been treated in detail in Chapter 2. Obviously, all the forces act at a point for which there is no resultant couple moment, so the point P is automatically specified, Fig. 3–41.

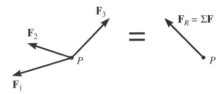

Fig. 3–41

Coplanar force systems, which may include couple moments directed perpendicular to the plane of the forces, as shown in Fig. 3–42a, can be reduced to a single resultant force, because when each force in the system is moved to any point O in the x–y plane, it produces a couple moment that is *perpendicular* to the plane, i.e., in the $\pm\mathbf{k}$ direction. The resultant moment $\mathbf{M}_{R_O} = \Sigma\mathbf{M} + \Sigma\mathbf{r} \times \mathbf{F}$ is thus perpendicular to the resultant force $\mathbf{F}_R$, Fig. 3–42b.

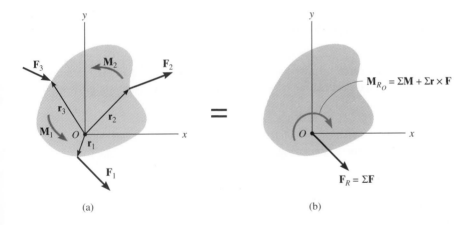

(a)   (b)

Fig. 3–42

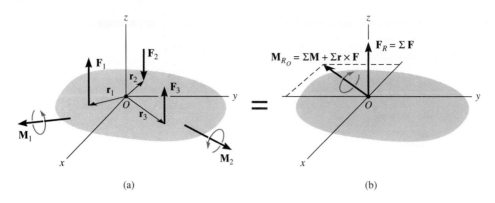

Fig. 3–43

Parallel force systems, which can include couple moments which are perpendicular to the forces, as shown in Fig. 3–43*a*, can be reduced to a single resultant force, because when each force is moved to any point *O* in the *x–y* plane, it produces a couple moment that has components only about the *x* and *y* axes. The resultant moment $\mathbf{M}_{R_O} = \Sigma \mathbf{M} + \Sigma \mathbf{r} \times \mathbf{F}$ is thus perpendicular to the resultant force $\mathbf{F}_R$, Fig. 3–43*b*.

PROCEDURE FOR ANALYSIS

The technique used to reduce a coplanar or parallel force system to a single resultant force follows the general procedure outlined in the previous section. First establish an *x*, *y*, *z* coordinate system. Then simplification requires the following two steps:

Force Summation. The resultant force $\mathbf{F}_R$ equals the sum of all the forces of the system, Fig. 3–40*a* and 3–40*c*; i.e.,

$$\mathbf{F}_R = \Sigma \mathbf{F}$$

Moment Summation. The distance *d* from the arbitrary point *O* to the line of action of $\mathbf{F}_R$ is determined by equating the moment of $\mathbf{F}_R$ about *O*, $\mathbf{M}_{R_O}$, Fig. 3–40*c*, to the sum of the moments about point *O* of all the couple moments and forces in the system, $\Sigma \mathbf{M}_O$, Fig. 3–40*a*; i.e.,

$$\mathbf{M}_{R_O} = \Sigma \mathbf{M}_O$$

Most often a scalar analysis can be used to apply these equations, since the force components and the moment arms are easily determined for either coplanar or parallel force systems.

Example 3–17

Replace the system of forces acting on the beam shown in Fig. 3–44a by an equivalent resultant force. Specify the distance the force acts from point A.

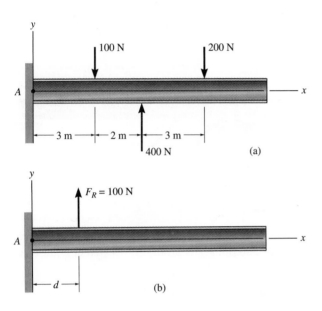

(a)

(b)

Fig. 3–44

SOLUTION

Here the force system is parallel and coplanar. We will use the established x, y, z axes to solve this problem.

Force Summation.   From Fig. 3–44a the force resultant F_R is

$$+\uparrow F_R = \Sigma F; \quad F_R = -100 \text{ N} + 400 \text{ N} - 200 \text{ N} = 100 \text{ N} \uparrow \qquad \textit{Ans.}$$

Moment Summation.   Moments will be summed about point A. Considering counterclockwise rotations as positive, i.e., positive moment vectors act in the $+\mathbf{k}$ direction, then from Figs. 3–44a and 3–44b we require the moment of $\mathbf{F}_R$ about A to equal the moments of the force system about A; i.e.,

$$\zeta + M_{R_A} = \Sigma M_A;$$
$$100 \text{ N}(d) = -(100 \text{ N})(3 \text{ m}) + (400 \text{ N})(5 \text{ m}) - (200 \text{ N})(8 \text{ m})$$
$$(100)(d) = 100$$
$$d = 1 \text{ m} \qquad \textit{Ans.}$$

Note that using a clockwise sign convention would yield the same result. Since d is *positive*, $\mathbf{F}_R$ acts to the right of A as shown in Fig. 3–44b.

Example 3-18

The beam AE in Fig. 3-45a is subjected to a system of coplanar forces. Determine the magnitude, direction, and location on the beam of a resultant force which is equivalent to the given system of forces.

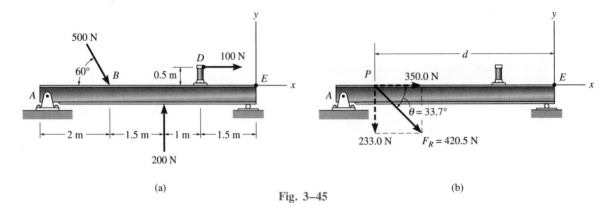

(a)   (b)

Fig. 3-45

SOLUTION

The origin of coordinates is arbitrarily located at point E as shown in Fig. 3-45a.

Force Summation. Resolving the 500-N force into x and y components, and summing the force components, yields

$$\xrightarrow{+} F_{R_x} = \Sigma F_x; \qquad F_{R_x} = 500 \cos 60° + 100 = 350.0 \text{ N} \rightarrow$$
$$+ \uparrow F_{R_y} = \Sigma F_y; \qquad F_{R_y} = -500 \sin 60° + 200 = -233.0 \text{ N}$$
$$= 233.0 \text{ N} \downarrow$$

The magnitude and direction of the resultant force are established from the vector addition shown in Fig. 3-45b. We have

$$F_R = \sqrt{(350.0)^2 + (233.0)^2} = 420.5 \text{ N} \qquad \qquad Ans.$$

$$\theta = \tan^{-1}\left(\frac{233.0}{350.0}\right) = 33.7° \ \ ^{\searrow}_{\theta} \qquad \qquad Ans.$$

Moment Summation. Moments will be summed about point E. Hence, from Figs. 3-45a and 3-45b, we require the moments of the components of $\mathbf{F}_R$ (or the moment of $\mathbf{F}_R$) about point E to equal the moments of the force system about E; i.e.,

$$\zeta + M_{R_E} = \Sigma M_E; \ 233.0(d) + 350.0(0)$$
$$= (500 \sin 60°)(4) + (500 \cos 60°)(0) - (100)(0.5) - (200)(2.5)$$

$$d = \frac{1182.1}{233.0} = 5.07 \text{ m} \qquad \qquad Ans.$$

■ Example 3–19

The frame shown in Fig. 3–46a is subjected to three coplanar forces. Replace this loading by an equivalent resultant force and specify where the resultant's line of action intersects members AB and BC.

SOLUTION

Force Summation. Resolving the 75-lb force into x and y components and summing the force components yields

$$\xrightarrow{+} F_{R_x} = \Sigma F_x; \quad F_{R_x} = -75(\tfrac{3}{5}) - 15 = -60 \text{ lb} = 60 \text{ lb} \leftarrow$$

$$+ \uparrow F_{R_y} = \Sigma F_y; \quad F_{R_y} = -75(\tfrac{4}{5}) - 90 = -150 \text{ lb} = 150 \text{ lb} \downarrow$$

As shown by the vector addition in Fig. 3–46b,

$$F_R = \sqrt{(60)^2 + (150)^2} = 161.6 \text{ lb} \qquad\qquad \textit{Ans.}$$

$$\theta = \tan^{-1}\left(\frac{150}{60}\right) = 68.2° \quad \theta\nearrow \qquad\qquad \textit{Ans.}$$

Moment Summation. Moments will be summed about point A. If the line of action of $\mathbf{F}_R$ intersects AB, Fig. 3–46b, we require the moment of the components of $\mathbf{F}_R$ in Fig. 3–46b about A to equal the moments of the force system in Fig. 3–46a about A; i.e.,

$$\zeta + M_{R_A} = \Sigma M_A; \quad 0(150) + y(60) = 4(15) + 0(90) + 7(75)(\tfrac{3}{5}) - 2(75)(\tfrac{4}{5})$$

$$y = 4.25 \text{ ft} \qquad\qquad \textit{Ans.}$$

By the principle of transmissibility, $\mathbf{F}_R$ can also intersect BC, Fig. 3–46b, in which case we have

$$\zeta + M_{R_A} = \Sigma M_A; \quad 7(60) - x(150) = 4(15) + 0(90) + 7(75)(\tfrac{3}{5}) - 2(75)(\tfrac{4}{5})$$

$$x = 1.10 \text{ ft} \qquad\qquad \textit{Ans.}$$

We can also solve for these positions by assuming $\mathbf{F}_R$ acts at the arbitrary point (x, y) on its line of action, Fig. 3–46b. Summing moments about point A yields

$$\zeta + M_{R_A} = \Sigma M_A; \quad y(60) - x(150) = 4(15) + 0(90) + 7(75)(\tfrac{3}{5}) - 2(75)(\tfrac{4}{5})$$

$$60y - 150x = 255$$

which is the equation of the colored dashed line in Fig. 3–46b. To find the points of intersection with the frame, set x = 0, then y = 4.25 ft, and set y = 7 ft, then x = 1.10 ft.

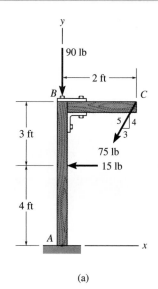

(a)

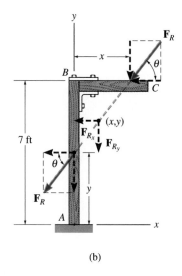

(b)

Fig. 3–46

Example 3–20

The slab in Fig. 3–47a is subjected to four parallel forces. Determine the magnitude and direction of a resultant force equivalent to the given force system, and locate its point of application on the slab.

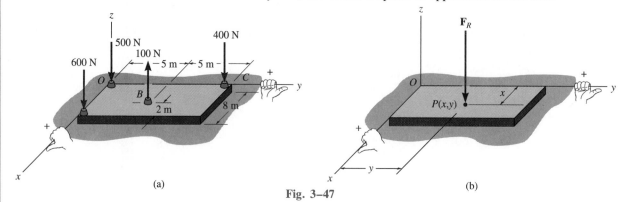

(a)

(b)

Fig. 3–47

SOLUTION (*SCALAR ANALYSIS*)

Force Summation. From Fig. 3–47a, the resultant force is

$$+\uparrow F_R = \Sigma F; \qquad F_R = -600 + 100 - 400 - 500$$
$$= -1400 \text{ N} = 1400 \text{ N} \downarrow \qquad \qquad \textit{Ans.}$$

Moment Summation. We require the moment about the x axis of the resultant force, Fig. 3–47b, to be equal to the sum of the moments about the x axis of all the forces in the system, Fig. 3–47a. The moment arms are determined from the y coordinates since these coordinates represent the *perpendicular distances* from the x axis to the lines of action of the forces. Using the right-hand rule, where positive moments act in the $+\mathbf{i}$ direction, we have

$$M_{R_x} = \Sigma M_x;$$
$$-1400y = 600(0) + 100(5) - 400(10) + 500(0)$$
$$-1400y = -3500 \qquad y = 2.50 \text{ m} \qquad \qquad \textit{Ans.}$$

In a similar manner, assuming that positive moments act in the $+\mathbf{j}$ direction, a moment equation can be written about the y axis using moment arms defined by the x coordinates of each force.

$$M_{R_y} = \Sigma M_y;$$
$$1400x = 600(8) - 100(6) + 400(0) + 500(0)$$
$$1400x = 4200 \qquad x = 3.00 \text{ m} \qquad \qquad \textit{Ans.}$$

Hence, a force of $F_R = 1400$ N placed at point $P(3.00$ m, 2.50 m) on the slab, Fig. 3–47b, is equivalent to the parallel force system acting on the slab in Fig. 3–47a.

■ Example 3–21 ▬▬▬▬▬▬

Three parallel forces act on the rim of the circular plate in Fig. 3–48a. Determine the magnitude and direction of a resultant force equivalent to the given force system and locate its point of application, P, on the plate.

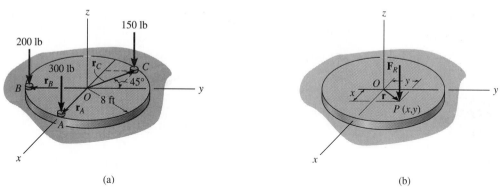

(a)   (b)

Fig. 3–48

SOLUTION (*VECTOR ANALYSIS*)

Force Summation.   From Fig. 3–48a, the force resultant $\mathbf{F}_R$ is

$$\mathbf{F}_R = \Sigma \mathbf{F}; \qquad \mathbf{F}_R = -300\mathbf{k} - 200\mathbf{k} - 150\mathbf{k}$$
$$= \{-650\mathbf{k}\} \text{ lb} \qquad\qquad Ans.$$

Moment Summation.   Choosing point O as a reference for computing moments and assuming that $\mathbf{F}_R$ acts at a point $P(x, y)$, Fig. 3–48b, we require

$$\mathbf{M}_{R_O} = \Sigma \mathbf{M}_O; \quad \mathbf{r} \times \mathbf{F}_R = \mathbf{r}_A \times (-300\mathbf{k}) + \mathbf{r}_B \times (-200\mathbf{k}) + \mathbf{r}_C \times (-150\mathbf{k})$$
$$(x\mathbf{i} + y\mathbf{j}) \times (-650\mathbf{k}) = (8\mathbf{i}) \times (-300\mathbf{k}) + (-8\mathbf{j}) \times (-200\mathbf{k})$$
$$+ (-8 \sin 45°\mathbf{i} + 8 \cos 45°\mathbf{j}) \times (-150\mathbf{k})$$
$$650x\mathbf{j} - 650y\mathbf{i} = 2400\mathbf{j} + 1600\mathbf{i} - 848.5\mathbf{j} - 848.5\mathbf{i}$$

Equating the corresponding $\mathbf{j}$ and $\mathbf{i}$ components yields

$$650x = 2400 - 848.5 \qquad\qquad (1)$$
$$-650y = 1600 - 848.5 \qquad\qquad (2)$$

Solving these equations, we obtain the coordinates of point P,

$$x = 2.39 \text{ ft} \qquad y = -1.16 \text{ ft} \qquad\qquad Ans.$$

The negative sign indicates that it was wrong to have assumed a $+y$ position for $\mathbf{F}_R$ as shown in Fig. 3–48b.

As a review, try to establish Eqs. 1 and 2 by using a scalar analysis; i.e., apply the sum of moments about the x and y axes, respectively.

PROBLEMS

3–62. Replace the force at *A* by an equivalent force and couple moment at *P*.

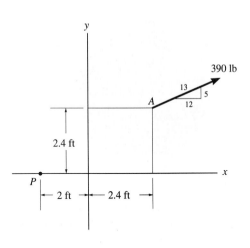

Prob. 3–62

*3–64. Replace the force and couple moment system by an equivalent force and couple moment acting at point *P*.

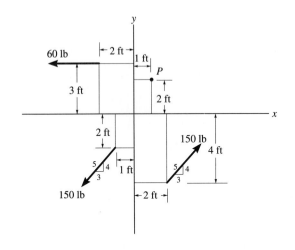

Prob. 3–64

3–65. Replace the force and couple moment system by an equivalent force and couple moment acting at point *P*.

3–63. Replace the force at *A* by an equivalent force and couple moment at *P*.

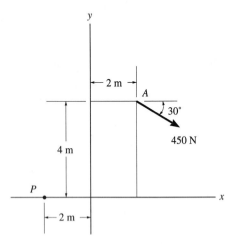

Prob. 3–63

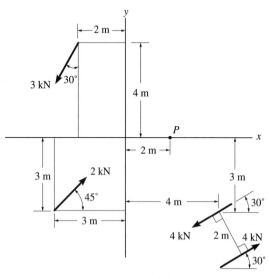

Prob. 3–65

3–66. Replace the force system by an equivalent resultant force and specify its point of application measured along the *x* axis from point *O*.

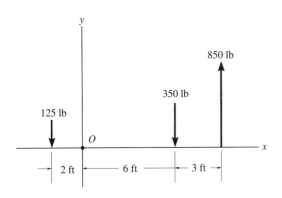

Prob. 3–66

3–67. Replace the coplanar force system by an equivalent resultant force and specify where its line of action intersects the *x* axis.

*3–68. Replace the coplanar force system by an equivalent resultant force and specify where its line of action intersects the *y* axis.

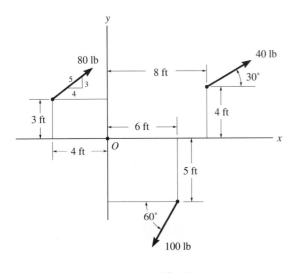

Probs. 3–67/3–68

3–69. The gear is subjected to the two forces shown. Replace these forces by an equivalent resultant force and couple moment acting at point *O*.

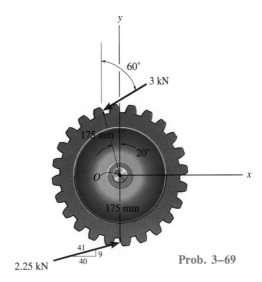

Prob. 3–69

3–70. Determine the magnitude of the vertical force **F** acting tangent to the gear so that **F** and **M** can be replaced by an equivalent resultant force at point *A*.

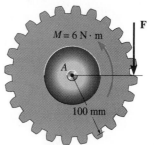

Prob. 3–70

3–71. Replace the two forces by an equivalent resultant force and couple moment at point *O*. Set $F = 20$ lb.

*3–72. Replace the two forces by an equivalent resultant force and couple moment at point *O*. Set $F = 15$ lb.

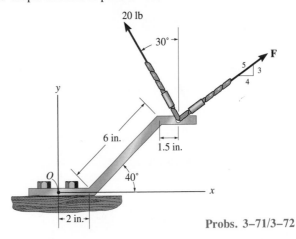

Probs. 3–71/3–72

3–73.   Replace the force system acting on the beam by an equivalent resultant force and couple moment at point A.

3–74.   Replace the force system acting on the beam by an equivalent resultant force and couple moment at point O.

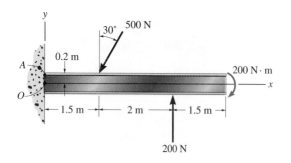

Probs. 3–73/3–74

3–75.   Replace the force system acting on the beam by an equivalent resultant force and couple moment at point B.

*3–76.   Replace the force system acting on the beam by an equivalent resultant force and couple moment at point A.

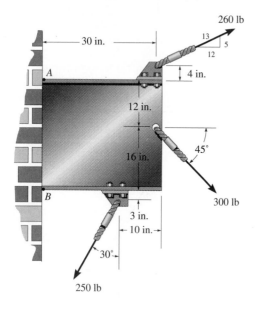

Probs. 3–75/3–76

3–77.   Replace the force system acting on the frame by an equivalent resultant force, and specify where the resultant's line of action intersects member AB measured from point A. Neglect the thickness of the members.

3–78.   Replace the force system acting on the frame by an equivalent resultant force and couple moment acting at point C. Neglect the thickness of the members.

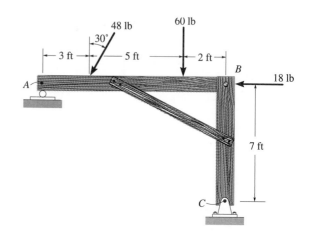

Probs. 3–77/3–78

3–79.   Determine the magnitude and orientation θ of force $\mathbf{F}$ and its placement d on the beam so that the loading system is equivalent to a resultant force of 15 kN acting vertically downward at point O and a clockwise couple moment of 60 kN · m.

*3–80.   Determine the magnitude and orientation θ of force $\mathbf{F}$ and its placement d on the beam so that the loading system is equivalent to a resultant force of 20 kN acting vertically downward at point O, and a clockwise couple moment of 80 kN · m.

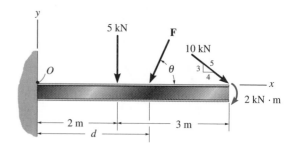

Probs. 3–79/3–80

3–81. The frame is subjected to the coplanar system of loads. Replace this system by an equivalent resultant force and couple moment acting at point *A*.

3–82. The frame is subjected to the coplanar system of loads. Replace this system by an equivalent resultant force and couple moment acting at point *B*.

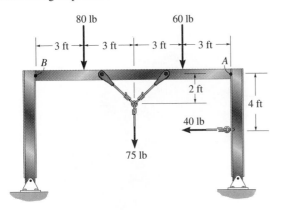

Probs. 3–81/3–82

3–83. Replace the force system acting on the frame by an equivalent resultant force and couple moment acting at point *A*.

*3–84. Replace the force system acting on the frame by an equivalent resultant force and specify where the resultant's line of action intersects member *AB* measured from point *A*.

3–85. Replace the force system acting on the frame by an equivalent resultant force and specify where the resultant's line of action intersects member *BC* measured from point *B*.

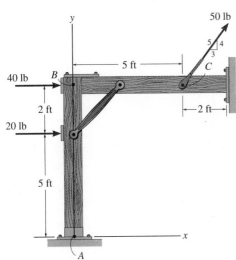

Probs. 3–83/3–84/3–85

3–86. The system of parallel forces acts on the top cord of the truss. Determine the equivalent resultant force of the system and specify its location measured from point *A*.

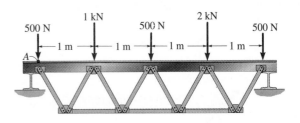

Prob. 3–86

3–87. Replace the force at *A* by an equivalent resultant force and couple moment at point *O*. Express the results in Cartesian vector form.

*3–88. Replace the force at *A* by an equivalent resultant force and couple moment at point *P*. Express the results in Cartesian vector form.

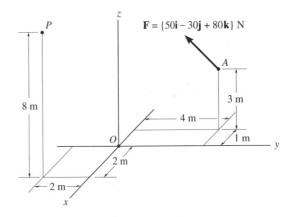

Probs. 3–87/3–88

3–89.   Replace the force and couple moment system by an equivalent resultant force and couple moment at point O. Express the results in Cartesian vector form.

3–90.   Replace the force and couple moment system by an equivalent resultant force and couple moment at point P. Express the results in Cartesian vector form.

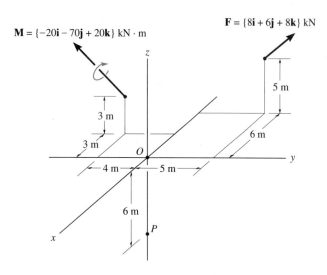

Probs. 3–89/3–90

3–91.   Replace the force $\mathbf{F}$ having a magnitude of $F = 70$ lb and acting at point A by an equivalent resultant force and couple moment at point O.

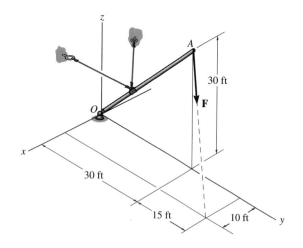

Prob. 3–91

*3–92.   The tube supports the four parallel forces. Determine the magnitudes of forces $\mathbf{F}_C$ and $\mathbf{F}_D$ acting at C and D so that the equivalent resultant force of the system acts through the midpoint O of the tube.

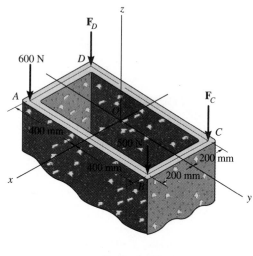

Prob. 3–92

3–93.   The building slab is subjected to four parallel column loadings. Determine the equivalent resultant force and specify its location (x, y) on the slab.

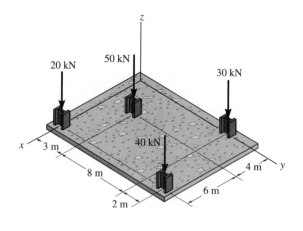

Prob. 3–93

3–94.   A force and couple act on the pipe assembly. Replace this system by an equivalent resultant force. Specify the location of the resultant force along the x axis, measured from A. The pipe lies in the x–y plane and the three forces are vertical.

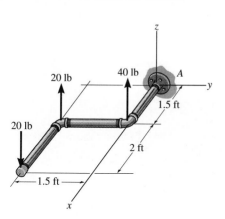

Prob. 3–94

3–95.   The crate is to be hoisted using the three slings shown. Replace the system of forces acting on the slings by an equivalent resultant force and couple moment at point O. The force F_1 is vertical.

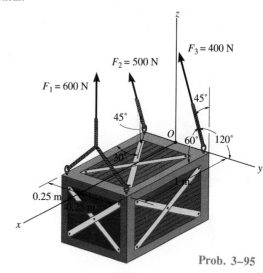

Prob. 3–95

REVIEW PROBLEMS

3–96.   Determine the moment about point A of each of the three forces acting on the beam. What is the resultant moment of all the forces about point A?

3–97.   Determine the moment about point B of each of the three forces acting on the beam. What is the resultant moment of all the forces about point B?

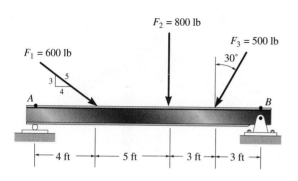

Probs. 3–96/3–97

3–98.   Determine the moment of the force **F** which is directed along ED of the block about the axis BA. Solve the problem by using two different position vectors **r**. Express the result as a Cartesian vector.

3–99.   Determine the moment of the force **F** which is directed along ED of the block about the axis OC. Solve the problem by using two different position vectors **r**. Express the result as a Cartesian vector.

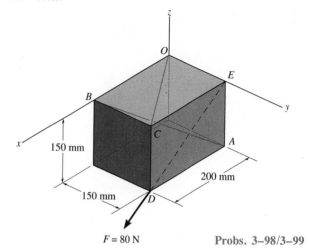

Probs. 3–98/3–99

*3–100. The main beam along the wing of an airplane is swept back at an angle of 25°. From load calculations it is determined that the beam is subjected to couple moments $M_x = 17$ kip · ft and $M_y = 25$ kip · ft. Determine the resultant couple moments created about the x' and y' axes. The axes all lie in the same horizontal plane.

3–103. The weights of the various components of the truck are shown. Replace this system of forces by an equivalent resultant force and couple moment acting at point A.

*3–104. The weights of the various components of the truck are shown. Replace this system of forces by an equivalent resultant force and specify its location measured from point A.

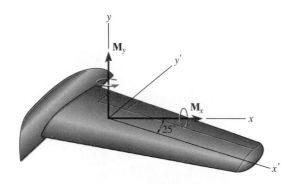

Prob. 3–100

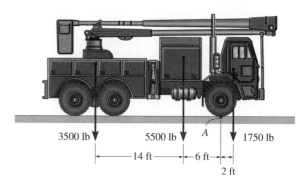

Probs. 3–103/3–104

3–101. The curved rod lies in the x–y plane and has a radius of 3 m. If a force of $F = 80$ N acts at its end as shown, determine the moment of this force about point O.

3–102. The curved rod lies in the x–y plane and has a radius of 3 m. If a force of $F = 80$ N acts at its end as shown, determine the moment of this force about point B.

3–105. The belt passing over the pulley is subjected to forces $\mathbf{F}_1$ and $\mathbf{F}_2$, each having a magnitude of 40 N. $\mathbf{F}_1$ acts in the $-\mathbf{k}$ direction. Replace these forces by an equivalent resultant force and couple moment at point A. Express the results in Cartesian vector form. Set $\theta = 0°$ so that $\mathbf{F}_2$ acts in the $-\mathbf{j}$ direction.

3–106. The belt passing over the pulley is subjected to two forces $\mathbf{F}_1$ and $\mathbf{F}_2$, each having a magnitude of 40 N. $\mathbf{F}_1$ acts in the $-\mathbf{k}$ direction. Replace these forces by an equivalent force and couple moment at point A. Express the results in Cartesian vector form. Take $\theta = 45°$.

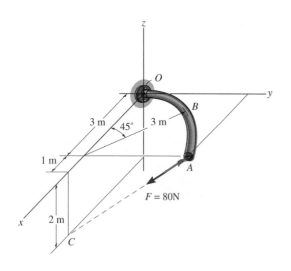

Probs. 3–101/3–102

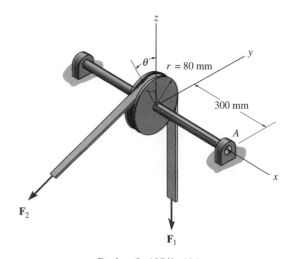

Probs. 3–105/3–106

4 Equilibrium

In this chapter the fundamental concepts of rigid-body equilibrium will be discussed. It will be shown that equilibrium requires both a *balance of forces,* to prevent the body from translating with accelerated motion, and a *balance of moments,* to prevent the body from rotating.

Many types of engineering problems involve symmetric loadings and can be solved by projecting all the forces acting on a body onto a single plane. Hence, in the first part of this chapter, the equilibrium of a body subjected to a *coplanar* or *two-dimensional force system* will be considered. Ordinarily the geometry of such problems is not very complex, so a scalar solution is suitable for analysis. The more general discussion of rigid bodies subjected to *three-dimensional force systems* is given in the next part of this chapter. It will be seen that many of these types of problems can best be solved by using vector analysis. Lastly, we will consider a special case of problems that involve frictional forces.

4.1 Conditions for Equilibrium

Equilibrium refers to a condition in which an object is at rest if originally at rest, or has a constant velocity if originally in motion. Most often, however, the term ''equilibrium'' or, more specifically, ''static equilibrium'' is used to describe an object at rest. To maintain a state of equilibrium, it is *necessary* to satisfy Newton's first law of motion, which states that if the *resultant force* acting on a particle is *zero,* then the particle is in equilibrium. This condition may be stated mathematically as

$$\Sigma \mathbf{F} = \mathbf{0}$$

where $\Sigma \mathbf{F}$ is the vector *sum of all the forces* acting on the particle.

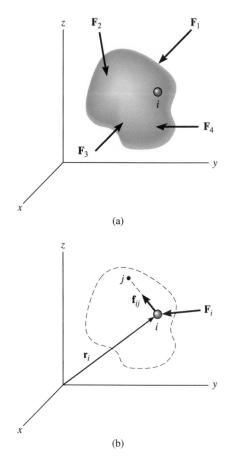

(a)

(b)

Fig. 4–1

Not only is this equation a necessary condition for equilibrium, it is also a *sufficient* condition. This follows from Newton's second law of motion, which can be written as $\Sigma \mathbf{F} = m\mathbf{a}$. Since the force system satisfies the above equation, then $m\mathbf{a} = \mathbf{0}$, and therefore the particle's acceleration $\mathbf{a} = \mathbf{0}$, and consequently the particle indeed moves with constant velocity or remains at rest.

Using the fact that $\Sigma \mathbf{F} = \mathbf{0}$ for a particle, we will now develop the conditions required to maintain equilibrium for a body. To do this, consider the body in Fig. 4–1*a*, which is fixed in the x, y, z reference and is either at rest or moves with the reference at constant velocity. The arbitrary ith particle of the body is shown in Fig. 4–1*b*. There are two types of forces which act on it. The *internal forces,* represented symbolically as

$$\sum_{\substack{j=1 \\ (j \neq 1)}}^{n} \mathbf{f}_{ij} = \mathbf{f}_i$$

are forces which all the other particles exert on the ith particle and produce the resultant $\mathbf{f}_i$. Although only one of these particles is shown in Fig. 4–1*b*, the summation extends over all n particles composing the body. For this summation note that it is meaningless for $i = j$ since the ith particle cannot exert a force on itself. The resultant *external force* $\mathbf{F}_i$ represents, for example, the effects of gravitational, electrical, magnetic, or contact forces between the ith particle and adjacent bodies or particles *not* included within the body. If the particle is in equilibrium, then applying Newton's first law we have

$$\mathbf{F}_i + \mathbf{f}_i = \mathbf{0}$$

When the equation of equilibrium is applied to each of the other particles of the body, similar equations will result. If all these equations are added together *vectorially,* we obtain

$$\Sigma \mathbf{F}_i + \Sigma \mathbf{f}_i = \mathbf{0}$$

The summation of the internal forces if carried out will equal zero since the internal forces between particles within the body will occur in equal but opposite collinear pairs, Newton's third law. Consequently, only the sum of the *external forces* will remain; and therefore, letting $\Sigma \mathbf{F}_i = \Sigma \mathbf{F},$ the above equation can be written as

$$\Sigma \mathbf{F} = \mathbf{0}$$

Let us now consider the moments of the forces acting on the ith particle about the arbitrary point O, Fig. 4–1*b*. Using the particle equilibrium equation and the distributive law of the vector cross product yields

$$\mathbf{r}_i \times (\mathbf{F}_i + \mathbf{f}_i) = \mathbf{r}_i \times \mathbf{F}_i + \mathbf{r}_i \times \mathbf{f}_i = \mathbf{0}$$

Similar equations can be written for the other particles of the body, and adding them together vectorially, we obtain

$$\Sigma \mathbf{r}_i \times \mathbf{F}_i + \Sigma \mathbf{r}_i \times \mathbf{f}_i = \mathbf{0}$$

The second term is zero since, as stated above, the internal forces occur in equal but opposite collinear pairs, and by the transmissibility of a force, as discussed in Sec. 3.4, the moment of each pair of forces about point O is therefore zero. Hence, using the notation $\Sigma \mathbf{M}_O = \Sigma \mathbf{r}_i \times \mathbf{F}_i$, we can write the previous equation as

$$\Sigma \mathbf{M}_O = \mathbf{0}$$

Hence the *equations of equilibrium* for a body can be summarized as follows:

$$\boxed{\begin{aligned} \Sigma \mathbf{F} &= \mathbf{0} \\ \Sigma \mathbf{M}_O &= \mathbf{0} \end{aligned}} \tag{4-1}$$

These equations require that a body will be in equilibrium provided the sum of all the *external forces* acting on the body is equal to zero and the sum of the moments of the external forces about a point is equal to zero. The fact that these conditions are *necessary* for equilibrium has now been proven. They are also *sufficient* conditions. To show this, let us assume that the body is *not* in equilibrium, and yet the force system acting on it satisfies Eqs. 4–1. Suppose that an *additional force* $\mathbf{F}'$ is required to hold the body in equilibrium. As a result, the equilibrium equations become

$$\Sigma \mathbf{F} + \mathbf{F}' = \mathbf{0}$$

$$\Sigma \mathbf{M}_O + \mathbf{M}'_O = \mathbf{0}$$

where $\mathbf{M}'_O$ is the moment of $\mathbf{F}'$ about O. Since $\Sigma \mathbf{F} = \mathbf{0}$ and $\Sigma \mathbf{M}_O = \mathbf{0}$, then we require $\mathbf{F}' = \mathbf{0}$ (also $\mathbf{M}'_O = \mathbf{0}$). Consequently, the additional force $\mathbf{F}'$ is not required for holding the body, and indeed Eqs. 4–1 are also sufficient conditions for equilibrium.

It is important to realize that Eqs. 4–1 also apply to any type of *deformable body*. However, application will require knowing the *deformed shape* of the body once it has been loaded since the orientation and location of each force must be specified in order to properly use the equations. To obtain the body's deformed shape is usually a difficult task. Fortunately, most engineering materials, such as steel and concrete, can be approximated as rigid bodies, and so the *original* dimensions and location of each force are thereby maintained, and these measurements can be used when applying Eqs. 4–1 without introducing significant errors. For this reason, unless otherwise stated, we will always assume the body is rigid when applying the equations of equilibrium throughout this text.

Equilibrium in Two Dimensions

4.2 Free-Body Diagrams

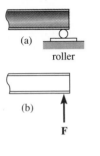

(a)

roller

(b)

F

Fig. 4–2

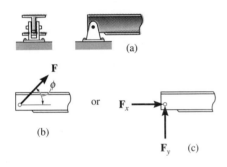

(a)

F

ϕ

or **F**$_x$

(b)

F$_y$ (c)

Fig. 4–3

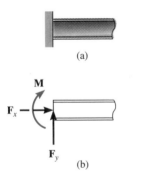

(a)

M

F$_x$

F$_y$ (b)

Fig. 4–4

Successful application of the equations of equilibrium requires a complete specification of *all* the known and unknown external forces that act *on* the body. The best way to account for these forces is to draw the body's *free-body diagram*. This diagram is a sketch of the outlined shape of the body, which represents it as being *isolated* or "free" from its surroundings. On this sketch it is necessary to show *all* the forces and couple moments that the surroundings exert *on the body*. By using this diagram the effects of all the applied forces and couple moments acting on the body can be accounted for when the equations of equilibrium are applied. For this reason, *a thorough understanding of how to draw a free-body diagram is of primary importance for solving problems in mechanics*.

Support Reactions. Before presenting a formal procedure as to how to draw a free-body diagram, we will first consider the various types of reactions that occur at supports and points of support between bodies subjected to coplanar force systems.

The principles involved for determining these reactions can be illustrated by considering three ways in which a horizontal member, such as a beam, is commonly supported at its end. The first method of support consists of a *roller* or cylinder, Fig. 4–2*a*. Since this type of support only prevents the beam from translating in the vertical direction, it is necessary that the roller exerts a force on the beam in this direction, Fig. 4–2*b*.

The beam can be supported in a more restrictive manner by using a *pin* as shown in Fig. 4–3*a*. The pin passes through holes in the beam and two leaves which are fixed to the ground. Here the pin will prevent translation of the beam in *any direction* ϕ, Fig. 4–3*b*, and so it must exert a force **F** on the beam in this direction. For purposes of analysis, it is generally easier to represent this effect by its two components **F**$_x$ and **F**$_y$, Fig. 4–3*c*.

The most restrictive way to support the beam would be to use a *fixed support* as shown in Fig. 4–4*a*. This support will prevent both translation and rotation of the beam, and so to do this a force and couple moment must be developed on the beam at its point of connection, Fig. 4–4*b*. As in the case of the pin, the force is usually represented by its components **F**$_x$ and **F**$_y$.

From this discussion, we can now make a general statement as to how to determine the types of reactions that are developed on a contacting member by *any kind* of support. In this regard, *if a support prevents translation in a given direction, then a force is developed on the contacting member in that direction. Likewise, if rotation is prevented, a couple moment is exerted on the member*. Table 4–1 lists some common types of supports used to support bodies subjected to coplanar force systems. (In all cases the angle θ is assumed to be known.) Carefully study each of the symbols used to represent these supports and the types of reactions they exert on their contacting members. Although concentrated forces and couple moments are shown in this table, they actually represent the *resultants* of *distributed surface loads* that exist between each support and its contacting member. It is these *resultants* which are determined in practice, and it is generally not important to determine the actual distribution of the load, since the surface area over which it acts is considerably *smaller* than the *total surface area* of the connected member.

Springs.
If a *linear elastic spring* is used to support a body, the length of the spring will change in direct proportion to the force acting on it. A characteristic that defines the "elasticity" of a spring is the *spring constant* or *stiffness k*. Specifically, the magnitude of force developed by a linear elastic spring which has a stiffness k, and is deformed (elongated or compressed) a distance s measured from its unloaded position, is

$$\boxed{F = ks} \tag{4–2}$$

Note that s is determined from the difference in the spring's deformed length l and its undeformed length l_o, i.e., $s = l - l_o$. Thus, if s is positive, **F** "pulls" on the spring; whereas if s is negative, **F** must "push" on it. For example, the spring shown in Fig. 4–5 has an undeformed length $l_o = 0.4$ m and stiffness $k = 500$ N/m. To stretch it so that $l = 0.6$ m, a force $F = ks = (500 \text{ N/m})(0.6 \text{ m} - 0.4 \text{ m}) = 100$ N is needed. Likewise, to compress it to a length $l = 0.2$ m, a force $F = ks = (500 \text{ N/m})(0.2 \text{ m} - 0.4 \text{ m}) = -100$ N is required, Fig. 4–5.

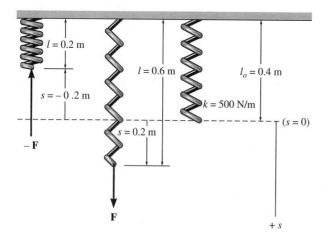

Fig. 4–5

Table 4–1 Supports for Bodies Subjected to Two-Dimensional Force Systems

Types of Connection	Reaction	Number of Unknowns
(1) cable		One unknown. The reaction is a tension force which acts away from the member in the direction of the cable.
(2) weightless link	or	One unknown. The reaction is a force which acts along the axis of the link.
(3) roller		One unknown. The reaction is a force which acts perpendicular to the surface at the point of contact.
(4) roller or pin in confined smooth slot	or	One unknown. The reaction is a force which acts perpendicular to the slot.
(5) rocker		One unknown. The reaction is a force which acts perpendicular to the surface at the point of contact.
(6) smooth contacting surface		One unknown. The reaction is a force which acts perpendicular to the surface at the point of contact.
(7) member pin connected to collar on smooth rod	or	One unknown. The reaction is a force which acts perpendicular to the rod.

Table 4–1 (Contd.)

Types of Connection	Reaction	Number of Unknowns
(8) smooth pin or hinge		Two unknowns. The reactions are two components of force, or the magnitude and direction ϕ of the resultant force. Note that ϕ and θ are not necessarily equal [usually not, unless the rod shown is a link as in (2)].
(9) member fixed connected to collar on smooth rod		Two unknowns. The reactions are the couple moment and the force which acts perpendicular to the rod.
(10) fixed support		Three unknowns. The reactions are the couple moment and the two force components, or the couple moment and the magnitude and direction ϕ of the resultant force.

External and Internal Forces. Since a body is a composition of particles, both *external* and *internal* loadings may act on it. It is important to realize, however, that if the free-body diagram for the body is drawn, the forces that are *internal* to the body are *not represented* on the free-body diagram. As discussed in Sec. 4.1, these forces always occur in equal but opposite collinear pairs, and therefore their *net effect* on the body is zero.

In some problems, a free-body diagram for a "system" of connected bodies may be used for an analysis. An example would be the free-body diagram of an entire automobile (system) composed of its many parts. Obviously, the connecting forces between its parts would represent *internal forces* which would *not* be included on the free-body diagram of the automobile. To summarize, then, internal forces act between particles which are located *within* a specified system which is contained within the boundary of the free-body diagram. Particles or bodies outside this boundary exert external forces on the system, and these alone must be shown on the free-body diagram.

Weight and the Center of Gravity. When a body is subjected to a gravitational field, each of its particles has a specified weight as defined by Newton's law of gravitation, $F = Gm_1m_2/r^2$, Eq. 1–2. If we assume the size of the body to be "small" in relation to the size of the earth, then it is appropriate to consider these gravitational forces to be represented as a *system* *of parallel forces* acting on the particles contained within the boundary of the body. It was shown in Sec. 3.9 that such a system can be reduced to a single resultant force acting through a specified point. We refer to this force resultant as the *weight* **W** of the body, and to the location of its point of application as the *center of gravity G*. The methods used for its calculation will be developed in Chapter 6.

In the examples and problems that follow, if the weight of the body is important for the analysis, this force will then be reported in the problem statement. Also, when the body is *uniform* or made of homogeneous material, the center of gravity will be located at the body's *geometric center* or *centroid;* however, if the body is nonhomogeneous or has an unusual shape, then its center of gravity will be given.

PROCEDURE FOR DRAWING A FREE-BODY DIAGRAM

To construct a free-body diagram for a body or group of bodies considered as a single system, the following steps should be performed:

Step 1. Imagine the body to be *isolated* or cut "free" from its constraints and connections, and draw (sketch) its outlined shape.

Step 2. Identify all the external forces and couple moments that act on the body. Those generally encountered are due to (1) applied loadings, (2) reactions occurring at the supports or at points of contact with other bodies (see Table 4–1), and (3) the weight of the body. To account for all these effects, it may help to trace over the boundary, carefully noting each force or couple moment acting on it.

Step 3. Indicate the dimensions of the body necessary for computing the moments of forces. The forces and couple moments that are known should be labeled with their proper magnitudes and directions. Letters are used to represent the magnitudes and direction angles of forces and couple moments that are *unknown*. In particular, if a force or couple moment has a known line of action but unknown magnitude, the arrowhead which defines the sense of the vector can be assumed. The correctness of the assumed sense will become apparent after solving the equilibrium equations for the unknown magnitude. By definition, the *magnitude* of a vector is *always positive,* so that if the solution yields a "negative" scalar, the *minus sign* indicates that the vector's sense is *opposite* to that which was originally assumed.

Before proceeding, review this section; then carefully study the following examples. Afterward, attempt to draw the free-body diagrams for the objects in Fig. 4–6 through 4–10 without "looking" at the solutions. Further practice in drawing free-body diagrams should be gained by solving *all* the problems given at the end of this section.

Example 4–1

The crate in Fig. 4–6a has a weight of 20 lb. Draw a free-body diagram of the crate, the cord *BD*, and the ring at *B*.

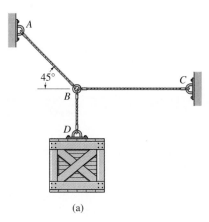

(a)

SOLUTION

If we imagine the crate to be isolated from its surroundings, then by inspection there are only two forces acting on it, namely, the gravitational force or weight of 20 lb, and the force of the cord *BD*. Thus the free-body diagram is shown in Fig. 4–6b.

If the cord *BD* is isolated, then there are only two forces acting on it, Fig. 4–6c, namely, the force of the connection on the top of the crate, $\mathbf{F}_D$, and the force $\mathbf{F}_B$ at *B* which is caused by the ring. Since these forces tend to *stretch* the cord, we can state that the cord is in *tension*. (This must be the case, since compressive forces would cause the cord to collapse.)

When the ring at *B* is isolated from its surroundings, it is realized that three forces act on it. All these forces are caused by the attached cords, Fig. 4–6d. Notice that $\mathbf{F}_B$ shown here is equal but opposite to that shown in Fig. 4–6c, a consequence of Newton's third law.

$\mathbf{F}_D$ (Force of cord on crate)

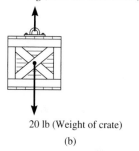

20 lb (Weight of crate)

(b)

$\mathbf{F}_B$ (Force of ring on cord)

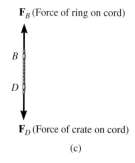

$\mathbf{F}_D$ (Force of crate on cord)

(c)

$\mathbf{F}_A$ (Force of cord *BA* on ring)

45°

$\mathbf{F}_C$ (Force of cord *BC* on ring)

$\mathbf{F}_B$ (Force of cord *BD* on ring)

(d)

Fig. 4–6

Example 4–2

Draw the free-body diagram of the uniform beam shown in Fig. 4–7a. The beam has a mass of 100 kg.

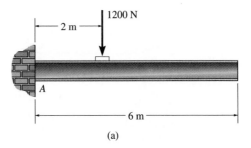

(a)

SOLUTION

The free-body diagram of the beam is shown in Fig. 4–7b. Since the support at A is a fixed wall, there are three reactions acting *on the beam* at A, denoted as $\mathbf{A}_x$, $\mathbf{A}_y$, and $\mathbf{M}_A$. The magnitudes of these vectors are *unknown,* and their sense has been *assumed.* (How does one obtain the *correct* sense of these vectors?) The weight of the beam, $W = 100(9.81) = 981$ N, acts through the beam's center of gravity G, 3 m from A since the beam is uniform.

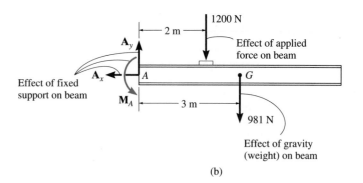

(b)

Fig. 4–7

Example 4–3

Draw the free-body diagram for the bell crank *ABC* shown in Fig. 4–8*a*.

SOLUTION

The free-body diagram is shown in Fig. 4–8*b*. The pin support at *B* exerts force components **B**$_x$ and **B**$_y$ *on the crank,* each having a known line of action but unknown magnitude. The link at *C* exerts a force **F**$_C$ acting in the direction of the link and having an unknown magnitude. The dimensions of the crank are also labeled on the free-body diagram, since this information will be useful in computing the moments of the forces. As usual, the sense of the three unknown forces has been assumed. The correct sense will become apparent after solving the equilibrium equations.

Although not part of this problem, three-dimensional views of the free-body diagrams of the pin and two pin leaves at *B* are shown in Fig. 4–8*c*. Since the leaves are *fixed-connected* to the wall, there are three unknowns that the wall exerts on each leaf; namely, **B**$_x''$, **B**$_y''$, **M**$_B''$. These reactions are shown to be equal in magnitude and direction on each leaf due to the symmetry of the loading and geometry. Note carefully how the principle of action—equal but opposite collinear reaction is used when applying the forces **B**$_x'$ and **B**$_y'$ to each leaf and the pin. All of these unknowns can be obtained from the equations of equilibrium once **B**$_x$ and **B**$_y$ are obtained.

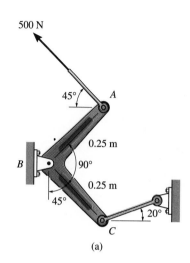

(a)

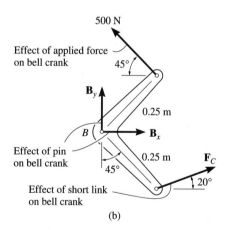

(b)

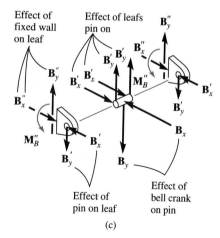

(c)

Fig. 4–8

Example 4–4

Two smooth tubes A and B, each having a mass of 2 kg, rest between the inclined planes shown in Fig. 4–9a. Draw the free-body diagrams for tube A, tube B, and tubes A and B together.

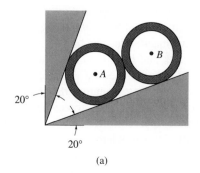

(a)

SOLUTION

The free-body diagram for tube A is shown in Fig. 4–9b. Its weight is $W = 2(9.81) = 19.62$ N. Since all contacting surfaces are *smooth,* the reactive forces $\mathbf{T}, \mathbf{F}, \mathbf{R}$ act in a direction *normal* to the tangent at their surfaces of contact.

The free-body diagram of tube B is shown in Fig. 4–9c. Can you identify each of the three forces acting *on the tube?* In particular, note that $\mathbf{R}$, representing the force of tube A on tube B, Fig. 4–9c, is equal and opposite to $\mathbf{R}$ representing the force of tube B on tube A, Fig. 4–9b. This is a consequence of Newton's third law of motion.

The free-body diagram of both tubes combined (''system'') is shown in Fig. 4–9d. Here the contact force $\mathbf{R}$, which acts between A and B, is considered as an *internal* force and hence is not shown on the free-body diagram. That is, it represents a pair of equal but opposite collinear forces which cancel each other.

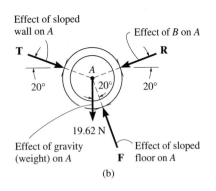

(b)

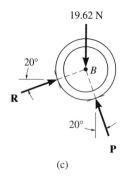

(c)

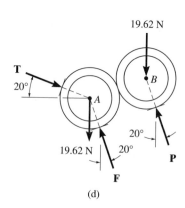

(d)

Fig. 4–9

Example 4–5

The free-body diagram of each object in Fig. 4–10 is drawn and the forces acting on the object are identified. The weight of the objects is neglected except where indicated.

SOLUTION

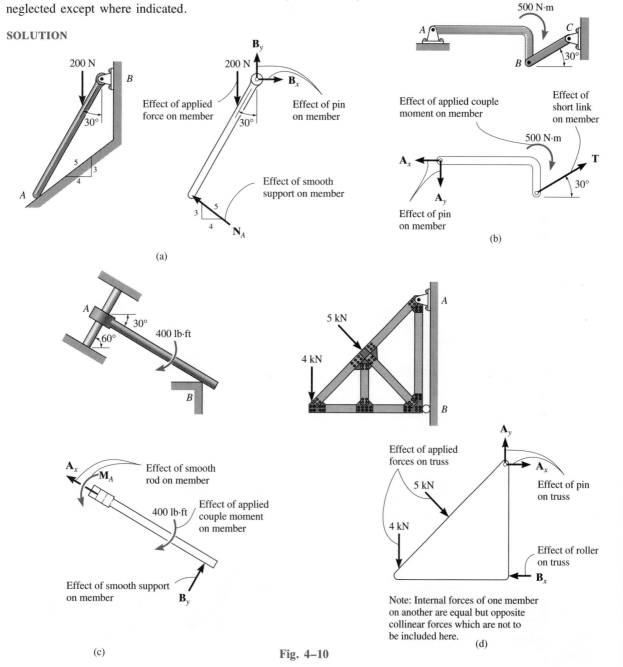

Fig. 4–10

PROBLEMS

4–1. The sphere has a mass of 6 kg and is supported as shown. Draw a free-body diagram of the sphere and the knot at C. Explain the significance of each force on the diagram. (See Fig. 4–6d.)

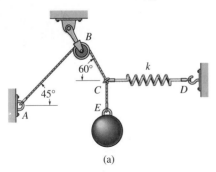

(a)

Prob. 4–1

4–2. Draw the free-body diagram of the beam. The support at B is smooth. Explain the significance of each force on the diagram.

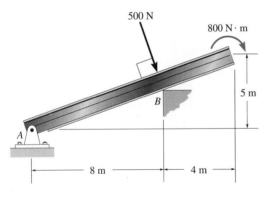

Prob. 4–2

4–3. Draw the free-body diagram of the automobile, which has a mass of 5 Mg and center of mass at G. The tires are free to roll, so rolling resistance can be neglected. Explain the significance of each force on the diagram.

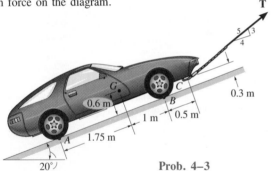

Prob. 4–3

*4–4. Draw the free-body diagram of the beam which supports the 80-kg load and is supported by the pin at A and a cable which wraps around the pulley at D. Explain the significance of each force on the diagram.

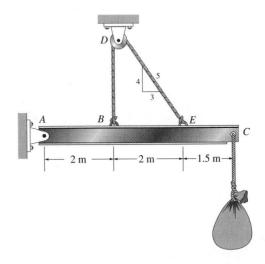

Prob. 4–4

4–5. Draw the free-body diagram of the spanner wrench which is subjected to a force of 20 lb at the grip. The wrench is pinned at A, and the surface of contact at B is smooth. Explain the significance of each force on the diagram.

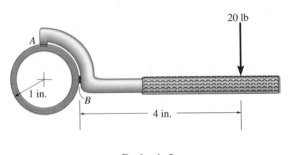

Prob. 4–5

4–6. Draw a free-body diagram of the crane boom *ABC*, which has a mass of 45 kg, center of gravity at *G*, and supports a load of 30 kg. The boom is pin-connected to the frame at *B* and to a vertical chain *CD*. The chain supporting the load is attached to the boom at *A*.

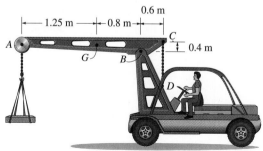

Prob. 4–6

4–7. Draw the free-body diagram of the beam. The incline at *B* is smooth.

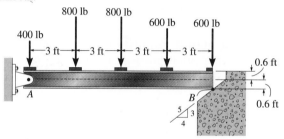

Prob. 4–7

*4–8. Draw the free-body diagram of the winch, which consists of a drum of radius 4 in. It is pin-connected at its center *C*, and at its outer rim is a ratchet gear having a mean radius of 6 in. The pawl *AB* serves as a two-force member (short link) and holds the drum from rotating.

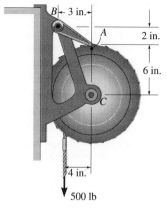

Prob. 4–8

4–9. Draw the free-body diagram of the smooth bar, which has points of contact at *A*, *B*, and *C*.

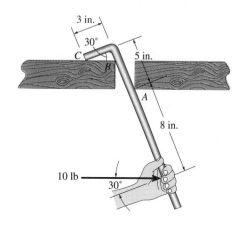

Prob. 4–9

4–10. Draw the free-body diagram of the uniform beam, which has a mass of 100 kg and is supported by making contact with the smooth surfaces at *A*, *B*, and *C*.

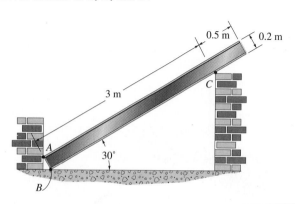

Prob. 4–10

4.3 Equations of Equilibrium

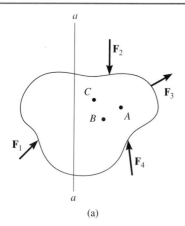

(a)

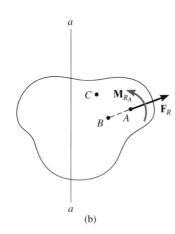

(b)

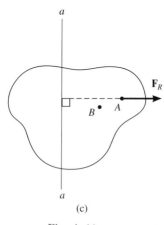

(c)

Fig. 4–11

In Sec. 4.1 we developed the two equations which are both necessary and sufficient for the equilibrium of a body, namely, $\Sigma \mathbf{F} = \mathbf{0}$ and $\Sigma \mathbf{M}_O = \mathbf{0}$. When the body is subjected to a system of forces, which all lie in the x–y plane, then the forces can be resolved into their x and y components. Consequently, the conditions for equilibrium in two dimensions are

$$\Sigma F_x = 0$$
$$\Sigma F_y = 0 \qquad (4\text{–}3)$$
$$\Sigma M_O = 0$$

Here ΣF_x and ΣF_y represent, respectively, the algebraic sums of the x and y components of all the forces acting on the body, and ΣM_O represents the algebraic sum of the couple moments and the moments of all the force components about an axis perpendicular to the x–y plane and passing through the arbitrary point O, which may lie either on or off the body.

Alternative Sets of Equilibrium Equations. Although Eqs. 4–3 are *most often* used for solving equilibrium problems involving coplanar force systems, two *alternative* sets of three independent equilibrium equations may also be used. One such set is

$$\Sigma F_a = 0$$
$$\Sigma M_A = 0 \qquad (4\text{–}4)$$
$$\Sigma M_B = 0$$

When using these equations it is required that the moment points A and B do *not* lie on a line that is *perpendicular* to the a axis. To prove that Eqs. 4–4 provide the *conditions* for equilibrium, consider the free-body diagram of an arbitrarily shaped body shown in Fig. 4–11a. Using the methods of Sec. 3.8, the loading acting on the body may be replaced by a single resultant force $\mathbf{F}_R = \Sigma \mathbf{F}$, acting at point A, and a resultant couple moment $\mathbf{M}_{R_A} = \Sigma \mathbf{M}_A$, Fig. 4–11$b$. For equilibrium it is required that $\mathbf{F}_R = \mathbf{0}$ and $\mathbf{M}_{R_A} = \mathbf{0}$. Hence, if $\Sigma M_A = 0$ is satisfied, it is necessary that $\mathbf{M}_{R_A} = \mathbf{0}$. Furthermore, in order that $\mathbf{F}_R$ satisfy $\Sigma F_a = 0$, it must have no component along the a axis, and therefore its line of action must be perpendicular to the a axis, Fig. 4–11c. Finally, if it is required that $\Sigma M_B = 0$, where B does not lie on the line of action of $\mathbf{F}_R$, then $\mathbf{F}_R = \mathbf{0}$, and indeed the body shown in Fig. 4–11a must be in equilibrium.

A second alternative set of equilibrium equations is

$$\Sigma M_A = 0$$
$$\Sigma M_B = 0 \qquad (4\text{–}5)$$
$$\Sigma M_C = 0$$

Here it is necessary that points A, B, and C do not lie on the same line. To prove that these equations when satisfied ensure equilibrium, consider again the free-body diagram in Fig. 4–11b. If $\Sigma M_A = 0$ is to be satisfied, then the resultant couple moment $\mathbf{M}_{R_A} = \mathbf{0}$. $\Sigma M_B = 0$ is satisfied if the line of action of $\mathbf{F}_R$ passes through point B as shown, and finally, if we require $\Sigma M_C = 0$, where C does not lie on line AB, it is necessary that $\mathbf{F}_R = \mathbf{0}$, and the body in Fig. 4–11a must then be in equilibrium.

PROCEDURE FOR ANALYSIS

The following procedure provides a method for solving coplanar force equilibrium problems:

Free-Body Diagram.   Draw a free-body diagram of the body as discussed in Sec. 4.2. Briefly, this requires showing all the external forces and couple moments acting *on the body*. The magnitudes of these vectors must be labeled and their directions specified. Dimensions of the body, necessary for computing the moments of forces, are also included on the free-body diagram. Identify the unknowns. The sense of a force or couple moment having an *unknown* magnitude but known line of action can be *assumed*.

Equations of Equilibrium.   Establish the x, y, z axes and apply the equations of equilibrium: $\Sigma F_x = 0$, $\Sigma F_y = 0$, $\Sigma M_O = 0$ (or the alternative sets of Eqs. 4–4 or 4–5). To *avoid* having to solve simultaneous equations, apply the moment equation $\Sigma M_O = 0$ about a point (O) *that lies at the intersection of the lines of action of two unknown forces*. In this way, the moments of these unknowns are *zero* about O, and one can obtain a *direct solution* for the third unknown. When applying the force equations $\Sigma F_x = 0$ and $\Sigma F_y = 0$, orient the x and y axes along lines that will provide the simplest resolution of the forces into their x and y components. If the solution of the equilibrium equations yields a *negative* scalar for an unknown force or couple moment, it indicates that the sense is *opposite* to that which was assumed on the free-body diagram.

The following example problems illustrate this procedure numerically.

Example 4–6

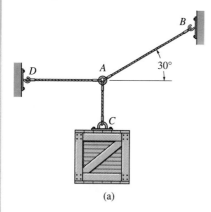

(a)

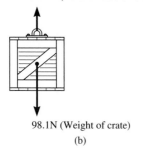

98.1N (Force of cord on crate)

98.1N (Weight of crate)

(b)

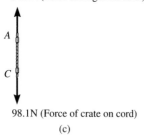

98.1N (Force of ring A on cord)

98.1N (Force of crate on cord)

(c)

Fig. 4–12

Determine the tension in cords AB and AD for equilibrium of the 10-kg crate shown in Fig. 4–12a.

SOLUTION

The free-body diagram of the crate is shown in Fig. 4–12b. Here the weight of the crate is $(10 \text{ kg})(9.81 \text{ m/s}^2) = 98.1 \text{ N}$. As a result, the force of cord CA on the crate must also be 98.1 N in order to hold the crate in equilibrium. By Newton's third law, this force acts in an equal but opposite manner on the cord CA, Fig. 4–12c, and so the cord is held in equilibrium by the 98.1-N force of the ring. The force in cords AB and AD can now be obtained by investigating the equilibrium of the ring at A, because this "particle" is subjected to the force of both of these cords.

Free-Body Diagram.   As shown in Fig. 4–12d, there are three concurrent forces *acting on the ring*. The cord tension forces $\mathbf{T}_B$ and $\mathbf{T}_D$ have unknown magnitudes but known directions. Cord AC exerts a downward force on A equal to 98.1 N. Why?

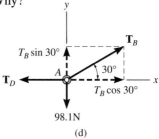

(d)

Equations of Equilibrium.   Since the equilibrium equations require a summation of the x and y components of each force, $\mathbf{T}_B$ must be resolved into x and y components. These components, shown dashed on the free-body diagram, have magnitudes of $T_B \cos 30°$ and $T_B \sin 30°$, respectively. Applying the equations of equilibrium, we have

$$\xrightarrow{+} \Sigma F_x = 0; \qquad\qquad T_B \cos 30° - T_D = 0 \qquad\qquad (1)$$

$$+\uparrow \Sigma F_y = 0; \qquad\qquad T_B \sin 30° - 98.1 = 0 \qquad\qquad (2)$$

Solving Eq. 2 for T_B and substituting into Eq. 1 to obtain T_D yields

$$T_B = 196 \text{ N} \qquad\qquad\qquad Ans.$$
$$T_D = 170 \text{ N} \qquad\qquad\qquad Ans.$$

The accuracy of these results, of course, depends on the accuracy of the data, i.e., measurements of geometry and loads. For most engineering work involving a problem such as this, the data as measured to three significant figures would be sufficient. Also, note that here we have neglected the weights of the cords, a reasonable assumption since they would be small in comparison with the weight of the crate.

Example 4–7

Determine the required length of cord AC in Fig. 4–13a so that the 8-kg lamp is suspended in the position shown. The *undeformed* length of the spring AB is $l'_{AB} = 0.4$ m, and the spring has a stiffness of $k_{AB} = 300$ N/m.

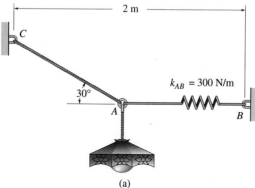

(a)

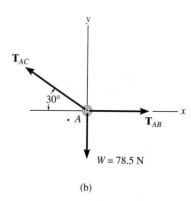

(b)

Fig. 4–13

SOLUTION

If the force in spring AB is known, the stretch of the spring can be found $(F = ks)$. Using the problem geometry, it is then possible to calculate the required length of AC.

Free-Body Diagram.   The lamp has a weight $W = 8(9.81) = 78.5$ N. The free-body diagram of the ring at A is shown in Fig. 4–13b.

Equations of Equilibrium

$$\xrightarrow{+}\Sigma F_x = 0; \qquad T_{AB} - T_{AC} \cos 30° = 0$$
$$+\uparrow \Sigma F_y = 0; \qquad T_{AC} \sin 30° - 78.5 = 0$$

Solving, we obtain

$$T_{AC} = 157.0 \text{ N}$$
$$T_{AB} = 136.0 \text{ N}$$

The stretch of spring AB is therefore

$$T_{AB} = k_{AB} s_{AB}; \qquad 136.0 \text{ N} = 300 \text{ N/m}(s_{AB})$$
$$s_{AB} = 0.453 \text{ m}$$

so the stretched length is

$$l_{AB} = l'_{AB} + s_{AB}$$
$$l_{AB} = 0.4 \text{ m} + 0.453 \text{ m} = 0.853 \text{ m}$$

The horizontal distance from C to B, Fig. 4–13a, requires

$$2 \text{ m} = l_{AC} \cos 30° + 0.853 \text{ m}$$
$$l_{AC} = 1.32 \text{ m} \qquad\qquad\qquad Ans.$$

■ **Example 4–8**

Determine the horizontal and vertical components of reaction for the beam loaded as shown in Fig. 4–14a. Neglect the weight and thickness of the beam in the calculations.

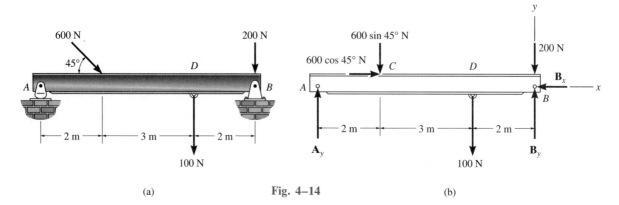

(a)   **Fig. 4–14**   (b)

SOLUTION

Free-Body Diagram. Can you identify each of the forces shown on the · free-body diagram of the beam, Fig. 4–14b? For simplicity in applying the equilibrium equations, the 600-N force is represented by its x and y components as shown. Also, note that a 200-N force acts on the beam at B, and is independent of the force components $\mathbf{B}_x$ and $\mathbf{B}_y$ which represent the effect of the pin on the beam.

Equations of Equilibrium. Summing forces in the x direction yields

$$\xrightarrow{+}\Sigma F_x = 0; \qquad\qquad 600 \cos 45° \text{ N} - B_x = 0$$

$$B_x = 424 \text{ N} \qquad\qquad\qquad Ans.$$

A direct solution for $\mathbf{A}_y$ can be obtained by applying the moment equation $\Sigma M_B = 0$ about point B. For the calculation, it should be apparent that forces 200 N, 600 cos 45° N, $\mathbf{B}_x$, and $\mathbf{B}_y$ all create zero moment about B. Assuming counterclockwise rotation about B to be positive (in the $+\mathbf{k}$ direction), Fig. 4–14b, we have

$$\zeta+\Sigma M_B = 0; \quad 100 \text{ N}(2 \text{ m}) + (600 \sin 45° \text{ N})(5 \text{ m}) - A_y(7 \text{ m}) = 0$$

$$A_y = 332 \text{ N} \qquad\qquad\qquad Ans.$$

Summing forces in the y direction, using the result A_y = 332 N, gives

$$+\uparrow \Sigma F_y = 0; \quad 332 \text{ N} - 600 \sin 45° \text{ N} - 100 \text{ N} - 200 \text{ N} + B_y = 0$$

$$B_y = 393 \text{ N} \qquad\qquad\qquad Ans.$$

Example 4–9

The cord shown in Fig. 4–15a supports a force of 100 lb and wraps over the frictionless pulley. Determine the tension in the cord at C and the horizontal and vertical components of reaction at pin A.

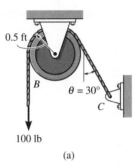

(a)

SOLUTION

Free-Body Diagrams. The free-body diagrams of the cord and pulley are shown in Fig. 4–15b. Note that the principle of action, equal but opposite reaction, must be carefully observed when drawing each of these diagrams: the cord exerts an unknown load distribution p along part of the pulley's surface, whereas the pulley exerts an equal but opposite effect on the cord. For the solution, however, it is simpler to *combine* the free-body diagrams of the pulley and the contacting portion of the cord, so that the distributed load becomes *internal* to the system and is therefore eliminated from the analysis, Fig. 4–15c.

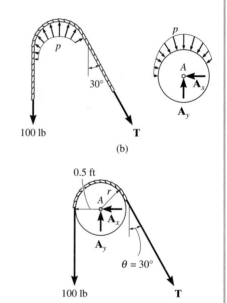

(b)

(c)

Fig. 4–15

Equations of Equilibrium. Summing moments about point A to eliminate $\mathbf{A}_x$ and $\mathbf{A}_y$, Fig. 4–15c, we have

$$\zeta + \Sigma M_A = 0; \qquad 100 \text{ lb}(0.5 \text{ ft}) - T(0.5 \text{ ft}) = 0$$

$$T = 100 \text{ lb} \qquad \textit{Ans.}$$

It is seen that the tension remains *constant* as the cord passes over the pulley. (This of course is true for *any angle* θ at which the cord is directed and for *any radius* r of the pulley.) Using the result for T, a force summation is applied to determine the components of reaction at pin A.

$$\xrightarrow{+} \Sigma F_x = 0; \qquad -A_x + 100 \sin 30° \text{ lb} = 0$$

$$A_x = 50.0 \text{ lb} \qquad \textit{Ans.}$$

$$+ \uparrow \Sigma F_y = 0; \qquad A_y - 100 \text{ lb} - 100 \cos 30° \text{ lb} = 0$$

$$A_y = 187 \text{ lb} \qquad \textit{Ans.}$$

Example 4–10

Determine the reactions at the fixed support A for the loaded frame shown in Fig. 4–16a.

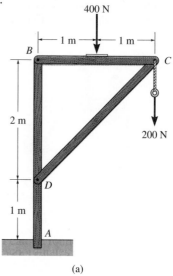

(a)

SOLUTION

Free-Body Diagram. The free-body diagram for the *entire frame* is shown in Fig. 4–16b. There are three unknowns at the fixed support, represented by the magnitudes of A_x, A_y, and M_A.

Equations of Equilibrium

$$\xrightarrow{+}\Sigma F_x = 0; \qquad\qquad A_x = 0 \qquad\qquad\qquad Ans.$$
$$+\uparrow \Sigma F_y = 0; \qquad A_y - 400\text{ N} - 200\text{ N} = 0$$
$$A_y = 600\text{ N} \qquad\qquad Ans.$$
$$\zeta+\Sigma M_A = 0; \qquad M_A - 400\text{ N(1 m)} - 200\text{ N(2 m)} = 0$$
$$M_A = 800\text{ N} \cdot \text{m} \qquad\qquad Ans.$$

Point A was chosen for summing moments since the lines of action of the *unknown* forces A_x and A_y pass through this point, and therefore these forces were not included in the moment summation. Realize, however, that M_A must be *included* in the moment summation. This couple moment is a free vector and represents the effect of the fixed support on the frame.

Although only *three* independent equilibrium equations can be written for a body, it is a good practice to *check* the calculations using a fourth equilibrium equation. For example, the above computations may be verified by summing moments about point C:

$$\zeta+\Sigma M_C = 0; \quad 400\text{ N(1 m)} - 600\text{ N(2 m)} + 800\text{ N} \cdot \text{m} \equiv 0$$
$$400\text{ N} \cdot \text{m} - 1200\text{ N} \cdot \text{m} + 800\text{ N} \cdot \text{m} \equiv 0$$

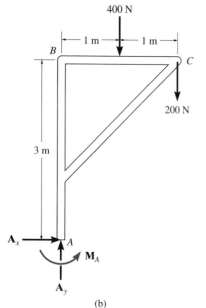

(b)

Fig. 4–16

Example 4–11

The uniform rod shown in Fig. 4–17*a* is subjected to a force and couple moment. If the rod is supported at *A* by a smooth wall and at *B* and *C* either at the top or bottom by smooth contacts, determine the reactions at these supports. Neglect the weight of the rod.

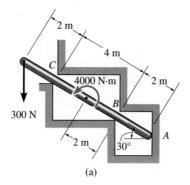

(a)

SOLUTION

Free-Body Diagram. As shown in Fig. 4–17*b*, all the support reactions act normal to the surface of contact since the contacting surfaces are smooth. The reactions at *B* and *C* are shown acting in the positive *y'* direction. This assumes that only the contacts located on the bottom of the rod are used for support.

Equations of Equilibrium. Using the *x*, *y* coordinate system in Fig. 4–17*b*, we have

$$\xrightarrow{+}\Sigma F_x = 0; \qquad C_{y'} \sin 30° + B_{y'} \sin 30° - A_x = 0 \qquad (1)$$

$$+\uparrow \Sigma F_y = 0; \quad -300 \text{ N} + C_{y'} \cos 30° + B_{y'} \cos 30° = 0 \qquad (2)$$

$$\underset{+}{\curvearrowright}\Sigma M_A = 0; \qquad -B_{y'}(2 \text{ m}) + 4000 \text{ N} \cdot \text{m} - C_{y'}(6 \text{ m})$$
$$+ (300 \cos 30° \text{ N})(8 \text{ m}) = 0 \qquad (3)$$

When writing the moment equation, it should be noticed that the line of action of the force component 300 sin 30° N passes through point *A*, and therefore this force is not included in the moment equation.

Solving Eqs. 2 and 3 simultaneously, we obtain

$$B_{y'} = -1000.0 \text{ N} \qquad \qquad Ans.$$
$$C_{y'} = 1346.4 \text{ N} \qquad \qquad Ans.$$

Since $B_{y'}$ is a negative scalar, the sense of $\mathbf{B}_{y'}$ is opposite to that shown on the free-body diagram in Fig. 4–17*b*. Therefore, the top contact at *B* serves as the support rather than the bottom one. Retaining the negative sign for $B_{y'}$ (Why?) and substituting the results into Eq. 1, we obtain

$$1346.4 \sin 30° \text{ N} - 1000.0 \sin 30° \text{ N} - A_x = 0$$
$$A_x = 173.2 \text{ N} \qquad \qquad Ans.$$

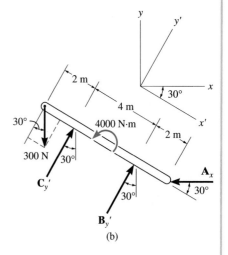

(b)

Fig. 4–17

Example 4–12

The beam shown in Fig. 4–18a is pin-connected at A and rests against a smooth support at B. Compute the horizontal and vertical components of reaction at the pin A.

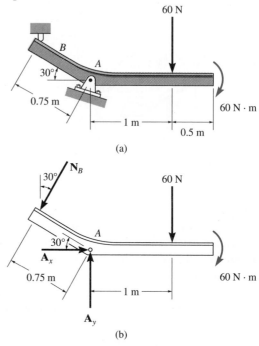

(a)

(b)

Fig. 4–18

SOLUTION

Free-Body Diagram. As shown in Fig. 4–18b, the reaction $\mathbf{N}_B$ is perpendicular to the beam at B. Also, horizontal and vertical components of reaction are represented at A, even though the base of the pin support is tilted.

Equations of Equilibrium. Summing moments about A, we obtain a direct solution for N_B,

$$\zeta + \Sigma M_A = 0; \quad -60 \text{ N} \cdot \text{m} - 60 \text{ N}(1 \text{ m}) + N_B(0.75 \text{ m}) = 0$$
$$N_B = 160 \text{ N}$$

Using this result,

$$\xrightarrow{+} \Sigma F_x = 0; \qquad A_x - 160 \sin 30° \text{ N} = 0$$
$$A_x = 80.0 \text{ N} \qquad \qquad \textit{Ans.}$$

$$+ \uparrow \Sigma F_y = 0; \qquad A_y - 160 \cos 30° \text{ N} - 60 \text{ N} = 0$$
$$A_y = 199 \text{ N} \qquad \qquad \textit{Ans.}$$

Example 4–13

A force of 150 lb acts on the end of the beam shown in Fig. 4–19a. Determine the magnitude and direction of the reaction at the pin A and the tension in the cable.

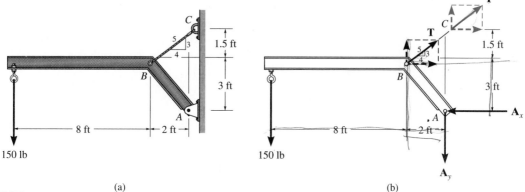

(a)   (b)

Fig. 4–19

SOLUTION

Free-Body Diagram.   The forces acting on the beam are shown in Fig. 4–19b.

Equations of Equilibrium.   Summing moments about point A to obtain a direct solution for the cable tension yields

$$\zeta+\Sigma M_A = 0; \quad -(\tfrac{3}{5}T)(2 \text{ ft}) - (\tfrac{4}{5}T)(3 \text{ ft}) + 150 \text{ lb}(10 \text{ ft}) = 0$$

$$-3.6T + 150 \text{ lb}(10 \text{ ft}) = 0 \qquad\qquad (1)$$

$$T = 416.7 \text{ lb} \qquad\qquad\qquad Ans.$$

Using the principle of transmissibility it is also possible to locate $\mathbf{T}$ at C, even though this point is not on the beam, Fig. 4–19b. In this case, the vertical component of $\mathbf{T}$ creates *zero moment* about A and the moment arm of the horizontal component ($\tfrac{4}{5}T$) becomes 4.5 ft. Hence, $\Sigma M_A = 0$ yields Eq. 1 directly since $(\tfrac{4}{5}T)(4.5) = 3.6T$.

Summing forces to obtain A_x and A_y, using the result for T, we have

$$\xrightarrow{+}\Sigma F_x = 0; \qquad\qquad -A_x + (\tfrac{4}{5})(416.7 \text{ lb}) = 0$$

$$A_x = 333.3 \text{ lb} \rightarrow$$

$$+\uparrow \Sigma F_y = 0; \qquad (\tfrac{3}{5})416.7 \text{ lb} - 150 \text{ lb} - A_y = 0$$

$$A_y = 100 \text{ lb} \downarrow$$

Thus,

$$F_A = \sqrt{(333.3 \text{ lb})^2 + (100 \text{ lb})^2}$$

$$= 348.0 \text{ lb} \qquad\qquad\qquad Ans.$$

$$\theta = \tan^{-1}\frac{100 \text{ lb}}{333.3 \text{ lb}} = 16.7° \qquad\qquad Ans.$$

■ Example 4–14

The 100-kg uniform beam AB shown in Fig. 4–20a is supported at A by a pin and at B and C by a continuous cable which wraps around a frictionless pulley located at D. If a maximum tension force of 800 N can be developed in the cable before it breaks, determine the greatest distance d at which the 6-kN force can be placed on the beam. What are the horizontal and vertical components of reaction at A just before the cable breaks? Neglect the thickness of the beam in the calculation.

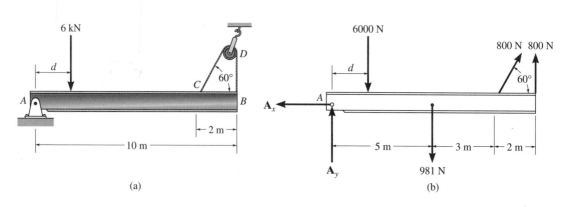

Fig. 4–20

(a)

(b)

SOLUTION

Free-Body Diagram. Since the cable is continuous and passes over a frictionless pulley, the entire cable is subjected to its maximum tension of 800 N when the 6-kN force acts at d. Hence, the cable exerts an 800-N force at points C and B *on the beam* in the direction of the cable, Fig. 4–20b.

Equations of Equilibrium. By summing moments of the force system about point A it is possible to obtain a direct solution for the dimension d. Why?

$$\zeta+\Sigma M_A = 0;$$
$$-(6000 \text{ N})(d) - 981 \text{ N}(5 \text{ m}) + (800 \sin 60° \text{ N})(8 \text{ m}) + 800 \text{ N}(10 \text{ m}) = 0$$
$$d = 1.44 \text{ m} \qquad\qquad Ans.$$

Summing forces in the x and y directions, we have

$$\xrightarrow{+}\Sigma F_x = 0; \qquad -A_x + 800 \cos 60° \text{ N} = 0$$
$$A_x = 400 \text{ N} \qquad\qquad Ans.$$
$$+\uparrow \Sigma F_y = 0; \quad A_y - 6000 \text{ N} - 981 \text{ N} + 800 \sin 60° \text{ N} + 800 \text{ N} = 0$$
$$A_y = 5.49 \text{ kN} \qquad\qquad Ans.$$

4.4 Two- and Three-Force Members

The solution to some equilibrium problems can be simplified if one is able to recognize members that are subjected to only two or three forces.

Two-Force Members. When a member is subject to *no couple moments* and forces are applied at only two points on a member, the member is called a *two-force member*. An example of this situation is shown in Fig. 4–21a. The forces at A and B are first summed to obtain their respective *resultants* $\mathbf{F}_A$ and $\mathbf{F}_B$, Fig. 4–21b. These two forces will maintain *translational or force equilibrium* ($\Sigma \mathbf{F} = \mathbf{0}$) provided $\mathbf{F}_A$ is of equal magnitude and opposite direction to $\mathbf{F}_B$. Furthermore, *rotational or moment equilibrium* ($\Sigma \mathbf{M}_O = \mathbf{0}$) is satisfied if $\mathbf{F}_A$ is *collinear* with $\mathbf{F}_B$. As a result, the line of action of both forces is known, since it always passes through A and B. Hence, only the force magnitude must be determined or stated. Other examples of two-force members held in equilibrium are shown in Fig. 4–22.

Three-Force Members. If a member is subjected to three coplanar forces, then it is necessary that the forces be either *concurrent* or *parallel* if the member is to be in equilibrium. To show this, consider the body in Fig. 4–23 and suppose that any two of the three forces acting on the body have lines of action that intersect at point O. To satisfy moment equilibrium about O, i.e., $\Sigma M_O = 0$, the third force must also pass through O, which then makes the force system *concurrent*. If two of the three forces are parallel, the point of concurrency, O, is considered to be at "infinity" and the third force must be parallel to the other two forces to intersect at this "point." Provided the three forces are concurrent at a point, only force equilibrium ($\Sigma \mathbf{F} = \mathbf{0}$) must be satisfied.

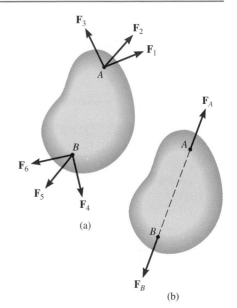

(a)

(b)

Two-force member

Fig. 4–21

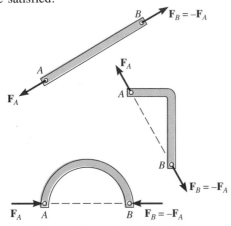

Two-force members

Fig. 4–22

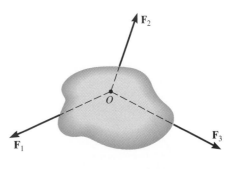

Three-force member

Fig. 4–23

■ Example 4–15

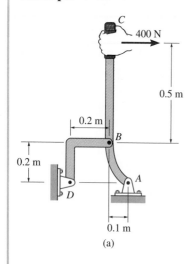

0.5 m

0.2 m

0.2 m

0.1 m

(a)

The lever *ABC* is pin-supported at *A* and connected to a short link *BD* as shown in Fig. 4–24a. If the weight of the members is negligible, determine the force of the pin on the lever at *A*.

SOLUTION

Free-Body Diagrams. As shown by the free-body diagram, Fig. 4–24b, the short link *BD* is a *two-force member*, so the *resultant forces* at pins *D* and *B* must be equal, opposite, and collinear. Although the magnitude of the force is unknown, the line of action is known, since it passes through *B* and *D*.

Lever *ABC* is a *three-force member,* and therefore, in order to satisfy moment equilibrium, the three nonparallel forces acting on it must be concurrent at *O*, Fig. 4–24c. In particular, note that the force **F** on the lever is equal but opposite to **F** acting at *B* on the link. Why? The distance *CO* must be 0.5 m, since the lines of action of **F** and the 400-N force are known.

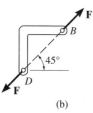

(b)

Equations of Equilibrium. By requiring the force system to be concurrent at *O*, so that $\Sigma M_O = 0$, the angle θ which defines the line of action of $\mathbf{F}_A$ can be determined from trigonometry,

$$\theta = \tan^{-1}\left(\frac{0.7}{0.4}\right) = 60.3° \quad \measuredangle\theta \qquad \textit{Ans.}$$

Applying the force equilibrium equations, we can obtain F and F_A.

$\xrightarrow{+} \Sigma F_x = 0; \qquad F_A \cos 60.3° - F \cos 45° + 400 \text{ N} = 0$

$+ \uparrow \Sigma F_y = 0; \qquad F_A \sin 60.3° - F \sin 45° = 0$

Solving, we get

$$F_A = 1075 \text{ N} \qquad \textit{Ans.}$$
$$F = 1320 \text{ N}$$

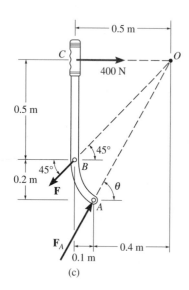

(c)

Fig. 4–24

Note: We can also solve this problem by representing the force at *A* by its two components $\mathbf{A}_x$ and $\mathbf{A}_y$ and applying $\Sigma M_A = 0$, $\Sigma F_x = 0$, $\Sigma F_y = 0$. Once A_x and A_y are determined, how would you find F_A and θ?

PROBLEMS

4–11. Determine the magnitudes of $\mathbf{F}_1$ and $\mathbf{F}_2$ so that the particle is in equilibrium.

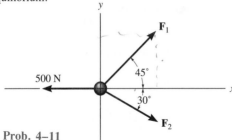

Prob. 4–11

*4–12. Determine the magnitude and orientation θ of $\mathbf{F}$ so that the particle is in equilibrium.

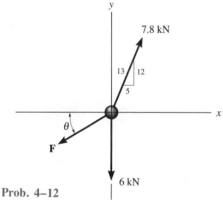

Prob. 4–12

4–13. The members of a truss are pin-connected at joint O as shown. Determine the magnitudes of $\mathbf{F}_1$ and $\mathbf{F}_2$ for equilibrium.

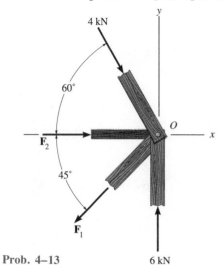

Prob. 4–13

4–14. Determine the forces in cables AC and AB needed to hold the 20-kg cylinder D in equilibrium. Set $F = 300$ N and $d = 1$ m.

4–15. The cylinder D has a mass of 20 kg. If a force of $F = 100$ N is applied horizontally to the ring at A, determine the largest dimension d so that the force in cable AC is zero.

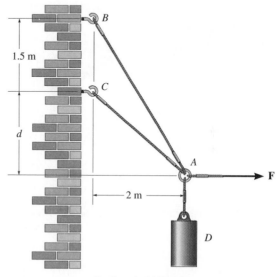

Probs. 4–14/4–15

*4–16. The pan and contents on the scale have a weight of 10 lb. If each spring has an unstretched length of 4 ft and a stiffness of $k = 5$ lb/ft, determine the angle θ for equilibrium.

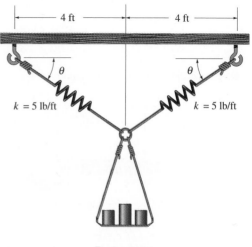

Prob. 4–16

4–17. Determine the stretch of each spring for equilibrium of the 20-kg block. The springs are shown in their equilibrium position.

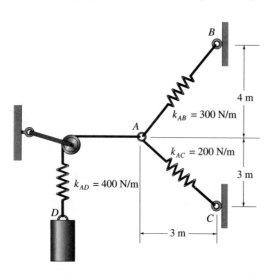

Prob. 4–17

4–18. Determine the reactions at the supports.

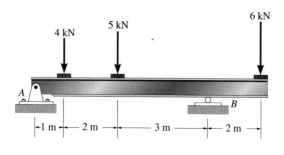

Prob. 4–18

4–19. Determine the reactions at the supports.

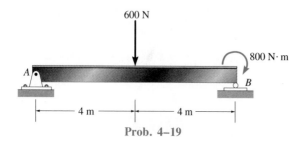

Prob. 4–19

*4–20. Determine the reactions at the pins A and B. The spring has an unstretched length of 80 mm.

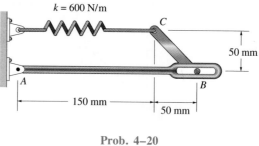

Prob. 4–20

4–21. Determine the reactions at the fixed support A.

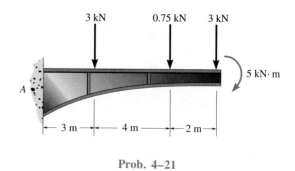

Prob. 4–21

4–22. Determine the reactions at the roller A and the pin at B.

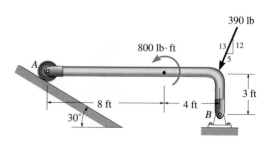

Prob. 4–22

4–23. The uniform rod *AB* has a weight of 15 lb. Determine the force in the cable when the rod is in the position shown.

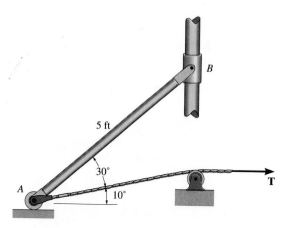

Prob. 4–23

*4–24. The wheel is subjected to a torque of 500 N · m. Determine the force **F** that acts along the axis *AB* of the toggle needed to prevent motion. Also, determine the magnitude of the resultant force that acts at the pinned axis *C*.

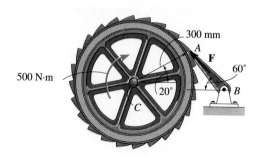

Prob. 4–24

4–25. Determine the horizontal and vertical components of reaction at the pin *A* and the force in the cable *BC*.

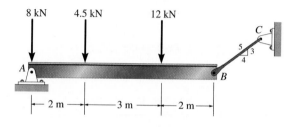

Prob. 4–25

4–26. Determine the horizontal and vertical components of reaction at the pin *A* and the force in the short link *BD*.

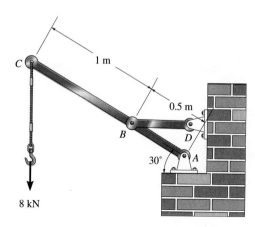

Prob. 4–26

4–27. Determine the reactions at *A* and *B* acting on the 50-kg roll of paper, which has a center of mass at *G* and rests on the smooth blade of the paper hauler.

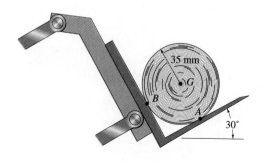

Prob. 4–27

*4–28. Determine the support reactions on the beam.

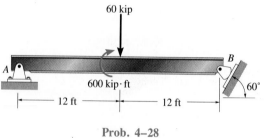

Prob. 4–28

4–29. Determine the support reactions on the beam.

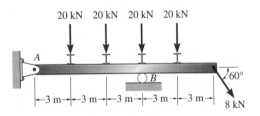

Prob. 4–29

4–30. The cutter is subjected to a horizontal force of 580 lb and a normal force of 350 lb. Determine the horizontal and vertical components of force acting on the pin A and the force along the hydraulic cylinder BC (a two-force member).

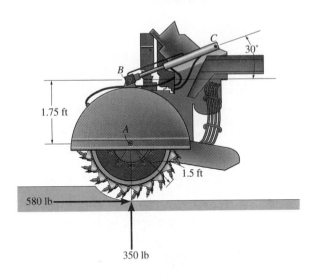

Prob. 4–30

4–31. The crane consists of three parts, which have weights of $W_A = 3500$ lb, $W_B = 900$ lb, $W_C = 1500$ lb and centers of gravity at A, B, and C, respectively. Neglecting the weight of the boom, determine (*a*) the reactions on each of the four tires if the load is hoisted at constant velocity and has a weight of 800 lb, and (*b*), with the boom held in the position shown, the maximum load the crane can lift without tipping over.

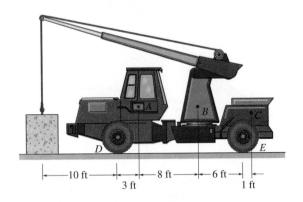

Prob. 4–31

***4–32.** Determine the resultant normal force acting on *each set* of wheels of the airplane. There is a set of wheels in the front, A, and a set of wheels under each wing, B. Both wings have a total weight of 50 kip and center of gravity at G_w, the fuselage has a weight of 180 kip and center of gravity at G_f, and both engines (one on each side) have a weight of 22 kip and center of gravity at G_e.

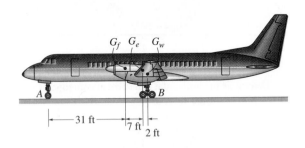

Prob. 4–32

4–33. The crane lifts a 400-kg load L. If the primary boom AB has a mass of 1.20 Mg and a center of mass at G_1, whereas the secondary boom BC has a mass of 0.6 Mg and a center of mass at G_2, determine the tension in the cable BD and the horizontal and vertical components of reaction at the pin A.

4–35. The uniform door has a weight of 100 lb and a center of gravity at G. Determine the reactions at the hinges if the hinge at A supports only a horizontal reaction on the door; whereas hinge at B exerts both horizontal and vertical reactions.

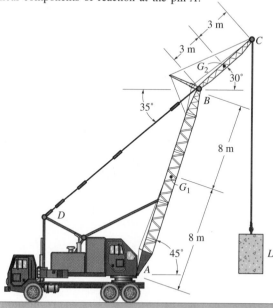

Prob. 4–33

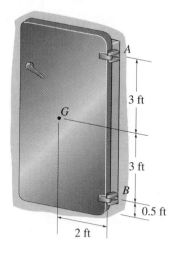

Prob. 4–35

4–34. The picnic table has a weight of 50 lb and a center of gravity at G. If a man weighing 225 lb has a center of gravity at G_M and sits down in the centered position shown, determine the vertical reaction at each of the two legs at B. Neglect the thickness of the legs. What can you conclude from the results?

*4–36. The hold-down clamp exerts a compressive force of 400 N on the wood block at C. If the clamp is loosely bolted to the bench using two symmetrically placed bolts B (one of which is shown), determine the force along the axis of each bolt and the vertical reaction of the bench on the clamp at A.

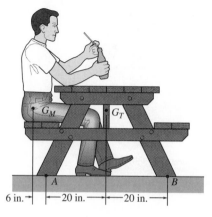

Prob. 4–34

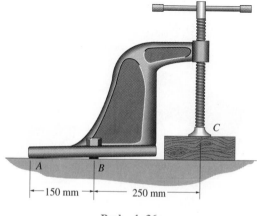

Prob. 4–36

4–37.   The 25-kg lawn roller is to be lifted over the 100-mm-high step. Compare the magnitudes of force **P** required to (a) push it and (b) pull it over the step if in each case the force is directed at θ = 30° along member AB as shown.

4–38.   Determine the minimum magnitude of force **P** and its associated angle θ required to pull the 25-kg roller over the 100-mm step.

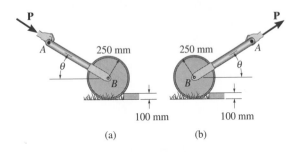

(a)   (b)

Probs. 4–37/4–38

4–39   The dumpster D of the truck has a weight of 5000 lb and a center of gravity at G. Determine the horizontal and vertical components of reaction at the pin A and along the hydraulic cylinder BC (two-force member).

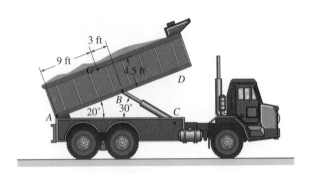

Prob. 4–39

*4–40.   The boom supports the two vertical loads. Neglect the size of the collars at D and B and the thickness of the boom, and determine the horizontal and vertical components of force at the pin A and the force in cable CB.

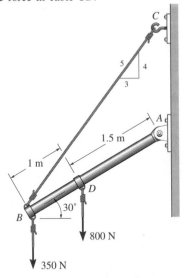

Prob. 4–40

4–41.   The davit is used to suspend the lifeboat over the side of a ship. If the boat exerts a force of 800 lb on the cable at D, determine the force acting along the hydraulic cylinder BC (two-force member) and the horizontal and vertical components of reaction at the pin A.

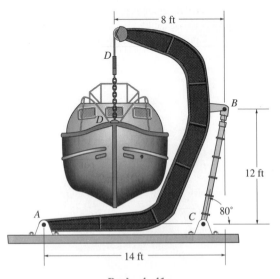

Prob. 4–41

4–42. Determine the horizontal and vertical components of reaction at the pin A and the reaction at the roller B needed to support the semicircular arch.

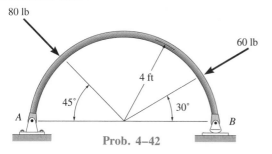

Prob. 4–42

4–43. The motor has a weight of 850 lb. Determine the force that each of the chains exerts on the supporting hooks at A, B, and C. Neglect the size of the hooks and the thickness of the beam.

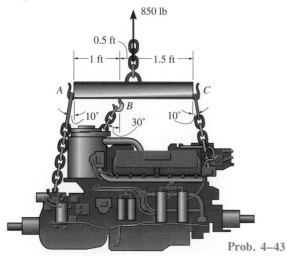

Prob. 4–43

*4–44. The man has a weight W and stands at the center of the plank. If the planes at A and B are smooth, determine the tension in the cord in terms of W and θ.

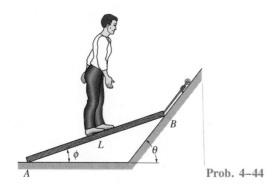

Prob. 4–44

4–45. Determine the magnitude of $\mathbf{F}_1$ and $\mathbf{F}_2$ if the supports at A and B exert forces of 4 kN and 4.5 kN, respectively, on the beam.

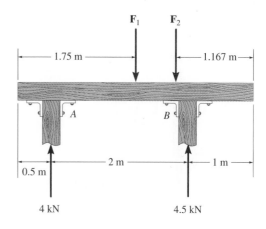

Prob. 4–45

4–46. The rigid metal strip of negligible weight is used as part of an electromagnetic switch. If the stiffness of the springs at A and B is $k = 5$ N/m, and the strip is originally horizontal when the springs are unstretched, determine the smallest vertical force needed to close the contact gap at C.

4–47. The rigid metal strip of negligible weight is used as part of an electromagnetic switch. Determine the maximum stiffness k of the springs at A and B so that the contact at C closes when the vertical force developed there is 0.5 N. Originally the strip is horizontal as shown.

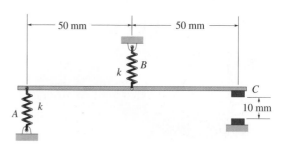

Probs. 4–46/4–47

*4–48.   Determine the magnitude of force at the pin *A* and in the cable *BC* needed to support the 500-lb load. Neglect the weight of the boom *AB*.

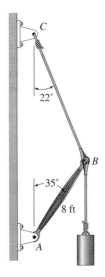

Prob. 4–48

4–49.   Determine the angle θ at which the link *ABC* is held in equilibrium if member *BD* moves 2 in. to the right. The springs are originally unstretched when $\theta = 0°$. Each spring has the stiffness shown. The springs remain horizontal since they are attached to roller guides.

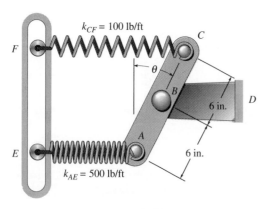

Prob. 4–49

4–50.   A horizontal force **P** is applied to the end of the uniform bar of weight *W*. When the bar is in equilibrium, the force is displaced a distance *d* as shown. Determine the magnitude of the reaction at the pin *A* in terms of *W*, *d*, and *l*.

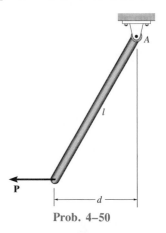

Prob. 4–50

4–51.   The rigid beam of negligible weight is supported horizontally by two springs and a pin. If the springs are unstretched, when the force **F** is removed, determine the force in each spring when **F** is applied. Also, compute the vertical deflection of end *C*. Assume that the spring stiffness *k* is large enough so that only small deflections occur.

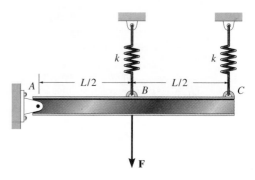

Prob. 4–51

*4–52.   The uniform rod *AB* has a weight of 15 lb and the spring is unstretched when $\theta = 0°$. If $\theta = 30°$, determine the stiffness *k* of the spring.

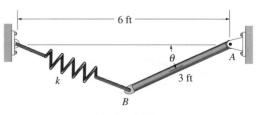

Prob. 4–52

Equilibrium in Three Dimensions

4.5 Free-Body Diagrams

The first step in solving three-dimensional equilibrium problems, as in the case of two dimensions, is to draw a free-body diagram of the body (or group of bodies considered as a system). Before we show this, however, it is necessary to discuss the types of reactions that can occur at the supports.

Support Reactions. The reactive forces and couple moments acting at various types of supports and connections, when the members are viewed in three dimensions, are listed in Table 4–2. It is important to recognize the symbols used to represent each of these supports and to understand clearly how the forces and couple moments are developed by each support. As in the two-dimensional case, *a force is developed by a support that restricts the translation of the attached member, whereas a couple moment is developed when rotation of the attached member is prevented.* For example, in Table 4–2, the ball-and-socket joint (4) prevents any translation of the connecting member; therefore, a force must act on the member at the point of connection. This force has three components having unknown magnitudes, F_x, F_y, F_z. Provided these components are known, one can obtain the magnitude of force, $F = \sqrt{F_x^2 + F_y^2 + F_z^2}$, and the force's orientation defined by the coordinate direction angles α, β, γ, Eqs. 2–7.* Since the connecting member is allowed to rotate freely about *any* axis, no couple moment is resisted by a ball-and-socket joint.

It should be noted that the *single* bearing supports (5) and (7), the *single* pin (8), and the *single* hinge (9) are shown to support both force and couple-moment components. If, however, these supports are used in conjunction with *other* bearings, pins, or hinges to hold the body in equilibrium, and provided the physical body maintains its *rigidity* when loaded and the supports are *properly aligned* when connected to the body, then the *force reactions* at these supports may *alone* be adequate for supporting the body. In other words, the couple moments become redundant and may be neglected on the free-body diagram. The reason for this will be clear after studying the examples which follow, but essentially the couple moments will not be developed at these supports since the rotation of the body is prevented by the reactions developed at the other supports and not by the supporting couple moments.

---

*The three unknowns may also be represented as an unknown force magnitude F and two unknown coordinate direction angles. The third direction angle is obtained using the identity $\cos^2 \alpha + \cos^2 \beta + \cos^2 \gamma = 1$, Eq. 2–10.

Table 4–2 Supports for Bodies Subjected to Three-Dimensional Force Systems

Types of Connection	Reaction	Number of Unknowns
(1) cable	**F**	One unknown. The reaction is a force which acts away from the member in the direction of the cable.
(2) smooth surface support	**F**	One unknown. The reaction is a force which acts perpendicular to the surface at the point of contact.
(3) roller	**F**	One unknown. The reaction is a force which acts perpendicular to the surface at the point of contact.
(4) ball and socket	$\mathbf{F}_z$ $\mathbf{F}_y$ $\mathbf{F}_x$	Three unknowns. The reactions are three rectangular force components.
(5) single journal bearing	$\mathbf{M}_z$ $\mathbf{F}_z$ $\mathbf{M}_x$ $\mathbf{F}_x$	Four unknowns. The reactions are two force and two couple-moment components which act perpendicular to the shaft.
(6) single journal bearing with square shaft	$\mathbf{M}_z$ $\mathbf{F}_z$ $\mathbf{M}_y$ $\mathbf{M}_x$ $\mathbf{F}_x$	Five unknowns. The reactions are two force and three couple-moment components.

Table 4–2 (Contd.)

Types of Connection	Reaction	Number of Unknowns
(7) single thrust bearing		Five unknowns. The reactions are three force and two couple-moment components.
(8) single smooth pin		Five unknowns. The reactions are three force and two couple-moment components.
(9) single hinge		Five unknowns. The reactions are three force and two couple-moment components.
(10) fixed support		Six unknowns. The reactions are three force and three couple-moment components.

Free-Body Diagrams. The general procedure for establishing the free-body diagram of a body has been outlined in Sec. 4.2. Essentially it requires first "isolating" the body by drawing its outlined shape. This is followed by a careful *labeling* of *all* the forces and couple moments in reference to an established x, y, z coordinate system. As a general rule, *components of reaction* having an *unknown magnitude* are shown acting on the free-body diagram in the *positive sense*. In this way, if any negative values are obtained, they will indicate that the components act in the negative coordinate directions.

Example 4–16

Three examples of objects along with their associated free-body diagrams are shown in Fig. 4–25. In all cases, the x, y, z axes are established and the unknown reaction components are indicated in the positive sense. The weight of the objects is neglected.

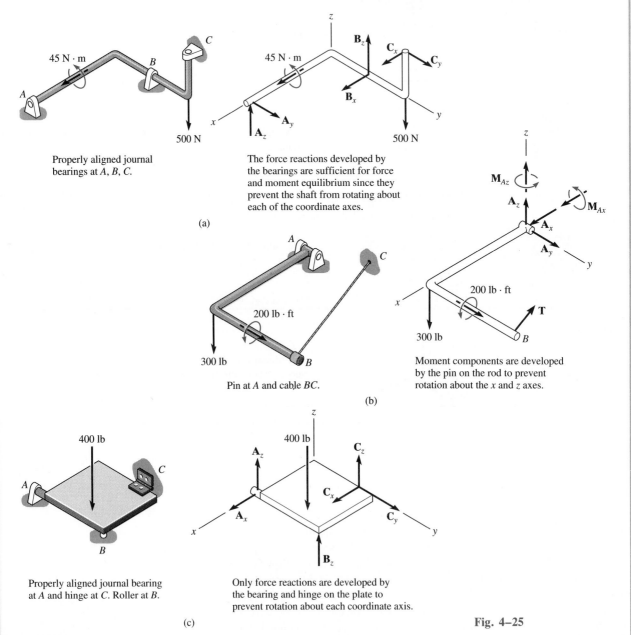

Properly aligned journal bearings at A, B, C.

The force reactions developed by the bearings are sufficient for force and moment equilibrium since they prevent the shaft from rotating about each of the coordinate axes.

(a)

Pin at A and cable BC.

Moment components are developed by the pin on the rod to prevent rotation about the x and z axes.

(b)

Properly aligned journal bearing at A and hinge at C. Roller at B.

Only force reactions are developed by the bearing and hinge on the plate to prevent rotation about each coordinate axis.

(c)

Fig. 4–25

4.6 Equations of Equilibrium

As stated in Sec. 4.1, the conditions for equilibrium of a body subjected to a three-dimensional force system require that both the *resultant* force and *resultant* couple moment acting on the body be equal to *zero*.

Vector Equations of Equilibrium. The two conditions for equilibrium of a body may be expressed mathematically in vector form as

$$\Sigma \mathbf{F} = \mathbf{0}$$
$$\Sigma \mathbf{M}_O = \mathbf{0}$$

(4–6)

where $\Sigma \mathbf{F}$ is the vector sum of all the external forces acting on the body and $\Sigma \mathbf{M}_O$ is the sum of the couple moments and the moments of all the forces about any point O located either on or off the body.

Scalar Equations of Equilibrium. If all the applied external forces and couple moments are expressed in Cartesian vector form and substituted into Eqs. 4–6, we have

$$\Sigma \mathbf{F} = \Sigma F_x \mathbf{i} + \Sigma F_y \mathbf{j} + \Sigma F_z \mathbf{k} = \mathbf{0}$$
$$\Sigma \mathbf{M}_O = \Sigma M_x \mathbf{i} + \Sigma M_y \mathbf{j} + \Sigma M_z \mathbf{k} = \mathbf{0}$$

Since the $\mathbf{i}, \mathbf{j},$ and $\mathbf{k}$ components are independent from one another, the above equations are satisfied provided

$$\Sigma F_x = 0$$
$$\Sigma F_y = 0$$
$$\Sigma F_z = 0$$

(4–7a)

and

$$\Sigma M_x = 0$$
$$\Sigma M_y = 0$$
$$\Sigma M_z = 0$$

(4–7b)

These *six scalar equilibrium equations* may be used to solve for at most six unknowns shown on the free-body diagram. Equations 4–7a express the fact that the sum of the external force components acting in the x, y, and z directions must be zero, and Eqs. 4–7b require the sum of the moment components about the x, y, and z axes to be zero.

PROCEDURE FOR ANALYSIS

The following procedure provides a method for solving three-dimensional equilibrium problems.

Free-Body Diagram. Construct the free-body diagram for the body. Be sure to include *all* the forces and couple moments that act *on* the body. These interactions are commonly caused by the externally applied loadings, contact forces exerted by adjacent bodies, support reactions, and the weight of the body if it is significant compared to the magnitudes of the other applied forces. Establish the origin of the x, y, z axes at a convenient point, and orient the axes so that they are parallel to as many of the external forces and moments as possible. Identify the unknowns, and in general show all the unknown components having a positive sense if the sense cannot be determined. Dimensions of the body, necessary for calculating the moments of forces, are also included on the free-body diagram.

Equations of Equilibrium. Apply the equations of equilibrium. In many cases, problems can be solved by *direct application* of the six scalar equations $\Sigma F_x = 0$, $\Sigma F_y = 0$, $\Sigma F_z = 0$, $\Sigma M_x = 0$, $\Sigma M_y = 0$, $\Sigma M_z = 0$, Eqs. 4–7; however, if the force components or moment arms seem difficult to determine, it is recommended that the solution be obtained by using vector equations: $\Sigma \mathbf{F} = \mathbf{0}$, $\Sigma \mathbf{M}_O = \mathbf{0}$, Eqs. 4–6. In any case, it is *not necessary* that the set of axes chosen for force summation *coincide* with the set of axes chosen for moment summation. Instead, it is recommended that one *choose the direction of an axis for moment summation such that it intersects the lines of action of as many unknown forces as possible*. The moments of forces passing through points on this axis or forces which are parallel to the axis will then be zero. Furthermore, *any set of three nonorthogonal axes* may be chosen for either the force or moment summations. By the proper choice of axes, it may be possible to solve directly for an unknown quantity, or at least reduce the need for solving a large number of simultaneous equations for the unknowns.

The following example problems illustrate this procedure numerically.

Example 4–17

The homogeneous plate shown in Fig. 4–26a has a mass of 100 kg and is subjected to a force and couple moment along its edges. If it is supported in the horizontal plane by means of a roller at A, a ball-and-socket joint at B, and a cord at C, determine the components of reaction at the supports.

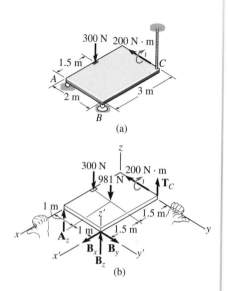

(a)

SOLUTION (*SCALAR ANALYSIS*)

Free-Body Diagram. There are five unknown reactions acting on the plate, as shown in Fig. 4–26b. Each of these reactions is assumed to act in a positive coordinate direction.

Equations of Equilibrium. Since the three-dimensional geometry is rather simple, a *scalar analysis* provides a *direct solution* to this problem. A force summation along each axis yields

$\Sigma F_x = 0;$ $B_x = 0$ *Ans.*

$\Sigma F_y = 0;$ $B_y = 0$ *Ans.*

$\Sigma F_z = 0;$ $A_z + B_z + T_C - 300\ \text{N} - 981\ \text{N} = 0$ (1)

Recall that the moment of a force about an axis is equal to the product of the force magnitude and the perpendicular distance (moment arm) from the line of action of the force to the axis. The sense of the moment is determined by the right-hand rule. Hence, summing moments of the forces on the free-body diagram, with positive moments acting along the positive x or y axis, we have

$\Sigma M_x = 0;$ $T_C(2\ \text{m}) - 981\ \text{N}(1\ \text{m}) + B_z(2\ \text{m}) = 0$ (2)

$\Sigma M_y = 0;$

$300\ \text{N}(1.5\ \text{m}) + 981\ \text{N}(1.5\ \text{m}) - B_z(3\ \text{m}) - A_z(3\ \text{m}) - 200\ \text{N} \cdot \text{m} = 0$

(3)

The components of force at B can be eliminated if the x', y', z' axes are used. We obtain

$\Sigma M_{x'} = 0;$ $981\ \text{N}(1\ \text{m}) + 300\ \text{N}(2\ \text{m}) - A_z(2\ \text{m}) = 0$ (4)

$\Sigma M_{y'} = 0;$

$-300\ \text{N}(1.5\ \text{m}) - 981\ \text{N}(1.5\ \text{m}) - 200\ \text{N} \cdot \text{m} + T_C(3\ \text{m}) = 0$ (5)

Solving Eqs. 1 through 3 or the more convenient Eqs. 1, 4, and 5 yields

$A_z = 790\ \text{N}$ $B_z = -217\ \text{N}$ $T_C = 707\ \text{N}$ *Ans.*

The negative sign indicates that $\mathbf{B}_z$ acts downward.

Note that the solution of this problem does not require the use of a summation of moments about the z axis. The plate is partially constrained since the supports will not prevent it from turning about the z axis if a force is applied to it in the x–y plane.

Fig. 4–26

Example 4–18

The windlass shown in Fig. 4–27a is supported by a thrust bearing at A and a smooth journal bearing at B, which are properly aligned on the shaft. Determine the magnitude of the vertical force **P** that must be applied to the handle to maintain equilibrium of the 100-kg crate. Also calculate the reactions at the bearings.

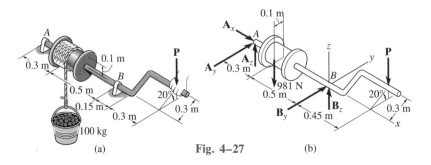

Fig. 4–27

(a)   (b)

SOLUTION (*SCALAR ANALYSIS*)
Free-Body Diagram.   Since the bearings at A and B are aligned correctly, *only* force reactions occur at these supports, Fig. 4–27b. Why are there no moment reactions?

Equations of Equilibrium.   Summing moments about the x axis yields a direct solution for **P.** Why? For a scalar moment summation, it is necessary to determine the moment of each force as the product of the force magnitude and the *perpendicular distance* from the x axis to the line of action of the force. Using the right-hand rule and assuming positive moments act in the +**i** direction, we have

$\Sigma M_x = 0$;   981 N(0.1 m) − P(0.3 cos 20° m) = 0

$$P = 348.0 \text{ N} \qquad\qquad Ans.$$

Using this result and summing moments about the y and z axes yields

$\Sigma M_y = 0$;

−981 N(0.5 m) + A_z(0.8 m) + (348.0 N)(0.45 m) = 0

$$A_z = 417.4 \text{ N} \qquad\qquad Ans.$$

$\Sigma M_z = 0$;   −A_y(0.8 m) = 0 $A_y = 0$ *Ans.*

The reactions at B are determined by a force summation, using the results obtained above.

$\Sigma F_x = 0$; $A_x = 0$ *Ans.*

$\Sigma F_y = 0$;   0 + B_y = 0 $B_y = 0$ *Ans.*

$\Sigma F_z = 0$;   417.4 − 981 + B_z − 348.0 = 0 $B_z = 911.6$ N   *Ans.*

Example 4–19

Determine the tension in cables BC and BD and the reactions at the ball-and-socket joint A for the mast shown in Fig. 4–28a.

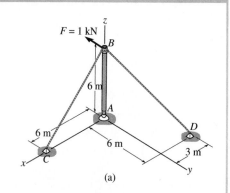

(a)

SOLUTION (*VECTOR ANALYSIS*)
Free-Body Diagram. There are five unknown force magnitudes shown on the free-body diagram, Fig. 4–28b.

Equations of Equilibrium. Expressing each force in Cartesian vector form, we have

$$\mathbf{F} = \{-1000\mathbf{j}\} \text{ N}$$
$$\mathbf{F}_A = A_x\mathbf{i} + A_y\mathbf{j} + A_z\mathbf{k}$$
$$\mathbf{T}_C = 0.707T_C\mathbf{i} - 0.707T_C\mathbf{k}$$
$$\mathbf{T}_D = T_D\left(\frac{\mathbf{r}_{BD}}{r_{BD}}\right) = -0.333T_D\mathbf{i} + 0.667T_D\mathbf{j} - 0.667T_D\mathbf{k}$$

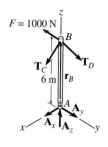

(b)

Applying the force equation of equilibrium gives

$$\Sigma\mathbf{F} = \mathbf{0}; \qquad \mathbf{F} + \mathbf{F}_A + \mathbf{T}_C + \mathbf{T}_D = \mathbf{0}$$
$$(A_x + 0.707T_C - 0.333T_D)\mathbf{i} + (-1000 + A_y + 0.667T_D)\mathbf{j}$$
$$+ (A_z - 0.707T_C - 0.667T_D)\mathbf{k} = \mathbf{0}$$
$$\Sigma F_x = 0; \qquad A_x + 0.707T_C - 0.333T_D = 0 \qquad (1)$$
$$\Sigma F_y = 0; \qquad A_y + 0.667T_D - 1000 = 0 \qquad (2)$$
$$\Sigma F_z = 0; \qquad A_z - 0.707T_C - 0.667T_D = 0 \qquad (3)$$

Fig. 4–28

Summing moments about point A, we have

$$\Sigma\mathbf{M}_A = \mathbf{0}; \qquad \mathbf{r}_B \times (\mathbf{F} + \mathbf{T}_C + \mathbf{T}_D) = \mathbf{0}$$
$$6\mathbf{k} \times (-1000\mathbf{j} + 0.707T_C\mathbf{i} - 0.707T_C\mathbf{k}$$
$$-0.333T_D\mathbf{i} + 0.667T_D\mathbf{j} - 0.667T_D\mathbf{k}) = \mathbf{0}$$

Evaluating the cross product and combining terms yields

$$(-4T_D + 6000)\mathbf{i} + (4.24T_C - 2T_D)\mathbf{j} = \mathbf{0}$$
$$\Sigma M_x = 0; \qquad -4T_D + 6000 = 0 \qquad (4)$$
$$\Sigma M_y = 0; \qquad 4.24T_C - 2T_D = 0 \qquad (5)$$

The moment equation about the z axis, $\Sigma M_z = 0$, is automatically satisfied. Why? Solving Eqs. 1 through 5 we have

$$T_C = 707 \text{ N} \qquad T_D = 1500 \text{ N} \qquad\qquad Ans.$$
$$A_x = 0 \text{ N} \qquad A_y = 0 \text{ N} \qquad A_z = 1500 \text{ N} \qquad Ans.$$

Since the mast is a two-force member, note that the value $A_x = A_y = 0$ could have been determined *by inspection*.

Example 4-20

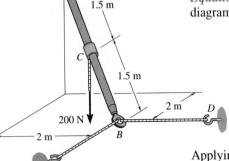

200 N

2 m

1.5 m

1.5 m

2 m

(a)

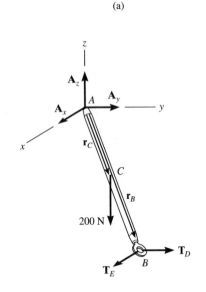

(b)

Fig. 4-29

Rod AB shown in Fig. 4–29a is subjected to the 200-N force. Determine the reactions at the ball-and-socket joint A and the tension in cables BD and BE.

SOLUTION (*VECTOR ANALYSIS*)
Free-Body Diagram. Fig. 4–29b.

Equations of Equilibrium. Representing each force on the free-body diagram in Cartesian vector form, we have

$$\mathbf{F}_A = A_x\mathbf{i} + A_y\mathbf{j} + A_z\mathbf{k}$$
$$\mathbf{T}_E = T_E\mathbf{i}$$
$$\mathbf{T}_D = T_D\mathbf{j}$$
$$\mathbf{F} = \{-200\mathbf{k}\}\ \text{N}$$

Applying the force equation of equilibrium,

$$\Sigma\mathbf{F} = \mathbf{0}; \qquad \mathbf{F}_A + \mathbf{T}_E + \mathbf{T}_D + \mathbf{F} = \mathbf{0}$$
$$(A_x + T_E)\mathbf{i} + (A_y + T_D)\mathbf{j} + (A_z - 200)\mathbf{k} = \mathbf{0}$$

$$\Sigma F_x = 0; \qquad\qquad A_x + T_E = 0 \qquad\qquad (1)$$
$$\Sigma F_y = 0; \qquad\qquad A_y + T_D = 0 \qquad\qquad (2)$$
$$\Sigma F_z = 0; \qquad\qquad A_z - 200 = 0 \qquad\qquad (3)$$

Summing moments about point A yields

$$\Sigma\mathbf{M}_A = \mathbf{0}; \qquad \mathbf{r}_C \times \mathbf{F} + \mathbf{r}_B \times (\mathbf{T}_E + \mathbf{T}_D) = \mathbf{0}$$

Since $\mathbf{r}_C = \frac{1}{2}\mathbf{r}_B$, then

$$(1\mathbf{i} + 1\mathbf{j} - 0.5\mathbf{k}) \times (-200\mathbf{k}) + (2\mathbf{i} + 2\mathbf{j} - 1\mathbf{k}) \times (T_E\mathbf{i} + T_D\mathbf{j}) = \mathbf{0}$$

Expanding and rearranging terms gives

$$(T_D - 200)\mathbf{i} + (-T_E + 200)\mathbf{j} + (2T_D - 2T_E)\mathbf{k} = \mathbf{0}$$

$$\Sigma M_x = 0; \qquad\qquad T_D - 200 = 0 \qquad\qquad (4)$$
$$\Sigma M_y = 0; \qquad\qquad -T_E + 200 = 0 \qquad\qquad (5)$$
$$\Sigma M_z = 0; \qquad\qquad 2T_D - 2T_E = 0 \qquad\qquad (6)$$

Solving Eqs. 1 through 6, we get

$$A_x = A_y = -200\ \text{N} \qquad\qquad\qquad \textit{Ans.}$$
$$A_z = T_E = T_D = 200\ \text{N} \qquad\qquad \textit{Ans.}$$

The negative sign indicates that $\mathbf{A}_x$ and $\mathbf{A}_y$ have a sense which is opposite to that shown on the free-body diagram, Fig. 4–29b.

PROBLEMS

4–53. Determine the x, y, z components of reaction at the fixed wall A.

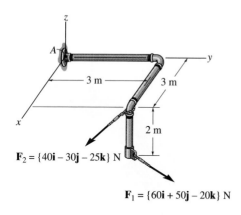

$F_2 = \{40i - 30j - 25k\}$ N

$F_1 = \{60i + 50j - 20k\}$ N

Prob. 4–53

4–54. The nonhomogeneous door of a large pressure vessel has a weight of 125 lb and a center of gravity at G. Determine the magnitudes of the resultant force and resultant couple moment developed at the hinge A, needed to support the door in any open position.

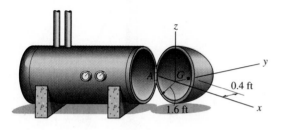

Prob. 4–54

4–55. The power drill is subjected to the forces shown acting on the grips. Determine the x, y, z components of force and the y and z components of moment reaction acting on the drill bit at A.

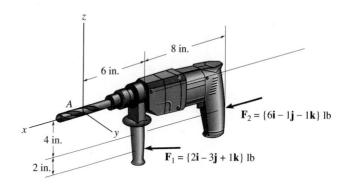

$F_2 = \{6i - 1j - 1k\}$ lb

$F_1 = \{2i - 3j + 1k\}$ lb

Prob. 4–55

*****4–56.** Determine the floor reaction on each wheel of the engine stand. The engine weighs 750 lb and has a center of gravity at G.

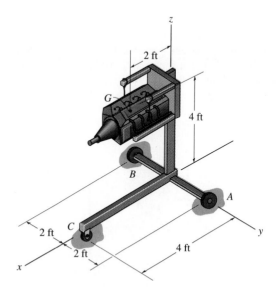

Prob. 4–56

4–57. The triangular plate is supported by a ball-and-socket joint at B and rollers at A and C. Determine the x, y, z components of reaction at these supports due to the loading shown.

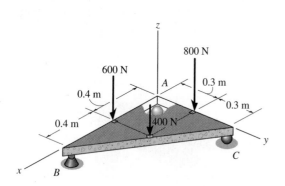

Prob. 4–57

4–58. The concrete slab is being hoisted with constant velocity using the cable arrangement shown. If the slab has a weight of 32,000 lb and a center of gravity at G, determine the tension in the supporting cables AC, BC, and DE.

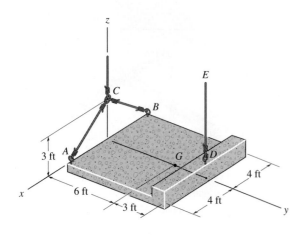

Prob. 4–58

4–59. The plate has a mass of 50 kg/m². Determine the force in each cable if it is suspended in the horizontal plane.

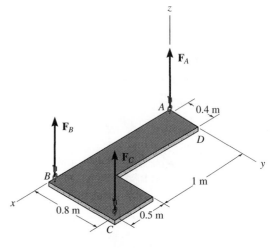

Prob. 4–59

*4–60. The boom is supported by a ball-and-socket joint at A and a guy wire at B. If the loads in the cables are each 5 kN and they lie in a plane which is parallel to the x–y plane, determine the x, y, z components of reaction at A and the tension in the cable at B.

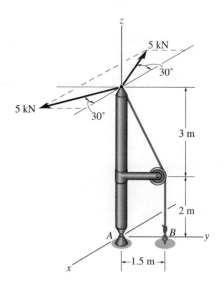

Prob. 4–60

4–61. The shaft is supported by a thrust bearing at A and a journal bearing at B. Determine the x, y, z components of reaction at these supports and the magnitude of force acting on the gear at C necessary to hold the shaft in equilibrium. The bearings are in proper alignment and exert only force reactions on the shaft.

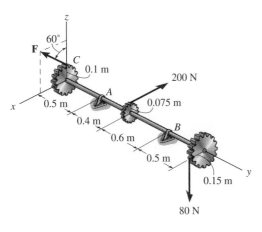

Prob. 4–61

4–62. The windlass is subjected to a load of 150 lb. Determine the horizontal force **P** needed to hold the handle in the position shown, and the x, y, z components of reaction at the ball-and-socket A and the smooth journal bearing B. The bearing is in proper alignment and exerts only force reactions on the shaft.

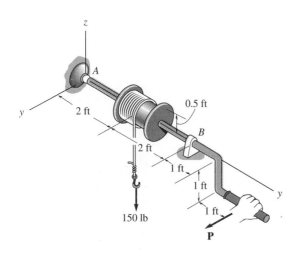

Prob. 4–62

4–63. The shaft is supported by journal bearings at A and B. A key is inserted into the bearing at B in order to prevent the shaft from rotating about and translating along its axis. Determine the x, y, z components of reaction at the bearings when the 600-N force is applied to the arm. The bearings are in proper alignment and exert only force reactions on the shaft.

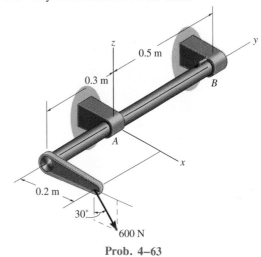

Prob. 4–63

*4–64. The bent rod is supported at A, B, and C by journal bearings. Determine the x, y, z reaction components at the bearings if the rod is subjected to a 200-lb vertical force and a 30-lb·ft couple moment as shown. The bearings are in proper alignment and exert only force reactions on the rod.

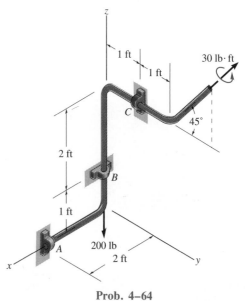

Prob. 4–64

4–65. The rod is supported by journal bearings at A, B, and C. Determine the x, y, z components of reaction at these supports due to the loading shown. The bearings are in proper alignment and exert only force reactions on the rod.

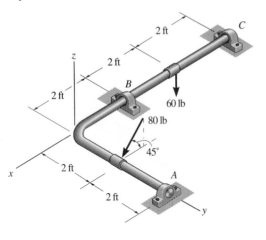

Prob. 4–65

4–66. The uniform 20-kg lid on a chest is propped open by the light rod CD. If the hinge at A prevents sliding of the lid along the x axis, whereas the hinge at B does not offer resistance in this direction, calculate the compressive force in CD and the x, y, z components of reaction at the hinges A and B. The hinges are in proper alignment and exert only force reactions on the lid.

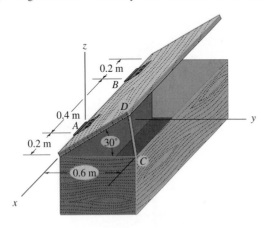

Prob. 4–66

4–67. Member AB is supported by a cable BC and at A by a smooth fixed *square* rod which fits loosely through the square hole of the collar. If $\mathbf{F} = \{20\mathbf{i} - 40\mathbf{j} - 75\mathbf{k}\}$ lb, determine the x, y, z components of reaction at A and the tension in the cable.

*4–68. Member AB is supported by a cable BC and at A by a smooth fixed *square* rod which fits loosely through the square hole of the collar. If the force $\mathbf{F} = -\{45\mathbf{k}\}$ lb, determine the tension in cable BC and the x, y, z components of reaction at A.

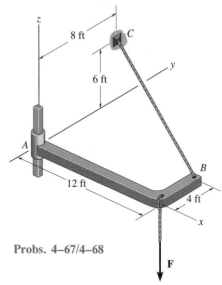

Probs. 4–67/4–68

4–69. The boom supports a load having a weight of $W = 850$ lb. Determine the x, y, z components of reaction at the ball-and-socket joint A and the tension in cables BC and DE.

4–70. Cable BC or DE can support a maximum tension of 700 lb before it breaks. Determine the greatest weight W that can be suspended from the end of the boom. Also, determine the x, y, z components of reaction at the ball-and-socket joint A.

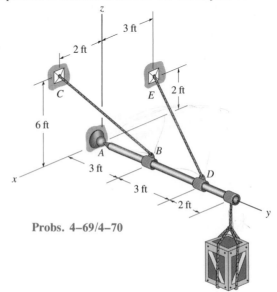

Probs. 4–69/4–70

4.7 Friction

In the previous sections the surfaces of contact between two bodies were considered to be perfectly *smooth*. Because of this, the force of interaction between the bodies always acts *normal* to the surface at points of contact. In reality, however, all surfaces are *rough,* and depending on the nature of the problem, the ability of a body to support a *tangential* as well as a *normal* force at its contacting surface must be considered. The tangential force is caused by friction, and in this section we will show how to analyze problems involving frictional forces.

Friction may be defined as a force of resistance acting on a body which prevents or retards slipping of the body relative to a second body or surface with which it is in contact. This force always acts *tangent* to the surface at points of contact with other bodies and is directed so as to oppose the possible or existing motion of the body relative to these points.

Theory of Dry Friction.
The theory of friction can best be explained by considering what effects are caused by pulling horizontally on a block of uniform weight **W** which is resting on a rough horizontal surface, Fig. 4–30a. To properly develop a full understanding of the nature of friction, it is necessary to consider the surfaces of contact to be *nonrigid or deformable*. The other portion of the block, however, will be considered rigid. As shown on the free-body diagram of the block, Fig. 4–30b, the floor exerts a *distribution* of both *normal force* $\Delta \mathbf{N}_n$ and *frictional force* $\Delta \mathbf{F}_n$ along the contacting surface. For equilibrium, the normal forces must act *upward* to balance the block's weight **W,** and the frictional forces act to the left to prevent the applied force **P** from moving the block to the right. Close examination of the contacting surfaces between the floor and block reveals how these frictional and normal forces develop, Fig. 4–30c. It can be seen that many microscopic irregularities exist between the two surfaces and, as a result, reactive forces $\Delta \mathbf{R}_n$ are developed at each of the protuberances.* These forces act at all points of contact and, as shown, each reactive force contributes both a frictional component $\Delta \mathbf{F}_n$ and a normal component $\Delta \mathbf{N}_n$.

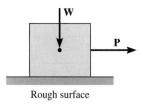

Rough surface

(a)

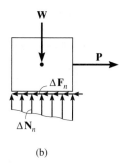

(b)

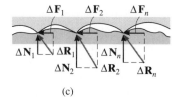

(c)

Fig. 4–30

---

*Besides mechanical interactions as explained here, a detailed treatment of the nature of frictional forces must also include the effects of temperature, density, cleanliness, and atomic or molecular attraction between the contacting surfaces. See D. Tabor, *Journal of Lubrication Technology,* 103, 169, 1981.

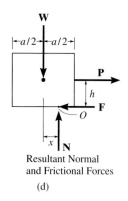

Resultant Normal
and Frictional Forces

(d)

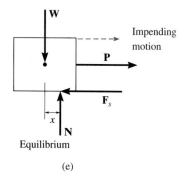

Equilibrium

(e)

Fig. 4–30 (Cont'd)

Equilibrium. For simplicity in the following analysis, the effect of the distributed normal and frictional loadings will be indicated by their *resultants* **N** and **F**, which are represented on the free-body diagram of the block as shown in Fig. 4–30d. Clearly, the distribution of ΔF_n in Fig. 4–30b indicates that **F** always acts *tangent to the contacting surface, opposite* to the direction of **P.** The normal force **N** is determined from the distribution of ΔN_n in Fig. 4–30b and is directed upward to balance the block's weight **W.** Notice that it acts a distance x to the right of the line of action of **W,** Fig. 4–30d. This location is necessary in order to balance the "tipping effect" caused by **P.** For example, if **P** is applied at a height h from the surface, Fig. 4–30d, then moment equilibrium about point O is satisfied if $Wx = Ph$ or $x = Ph/W$. In particular, note that the block will be on the verge of *tipping* if $x = a/2$.

Impending Motion. In cases where h is small or the surfaces of contact are rather "slippery," the frictional force **F** may *not* be great enough to balance the magnitude of **P,** and consequently the block will tend to slip *before* it can tip. In other words, as the magnitude of **P** is slowly increased, the magnitude of **F** correspondingly increases until it attains a certain *maximum value F_s,* called the *limiting static frictional force,* Fig. 4–30e. When this value is reached, the block is in *unstable equilibrium,* since any further increase in P will cause deformations and fractures at the points of surface contact and consequently the block will begin to move. Experimentally, it has been determined that the magnitude of the limiting static frictional force F_s is *directly proportional* to the magnitude of the resultant normal force **N.** This may be expressed mathematically as

$$F_s = \mu_s N \qquad (4\text{–}8)$$

where the constant of proportionality, μ_s (mu "sub" s), is called the *coefficient of static friction.*

Typical values for μ_s, found in many engineering handbooks, are given in Table 4–3. Although this coefficient is generally less than 1, be aware that in some cases it is possible, as in the case of aluminum on aluminum, for μ_s to be greater than 1. Physically this means, of course, that in this case the frictional force is greater than the corresponding normal force. Furthermore, it should be noted that μ_s is dimensionless and depends only on the characteristics of the two surfaces in contact. A wide range of values is given for each value of μ_s, since experimental testing was done under variable conditions of roughness and cleanliness of the contacting surfaces. For applications, therefore, it is important that both caution and judgment be exercised when selecting a coefficient of friction for a given set of conditions. When an exact calculation of F_s is required, the coefficient of friction should be determined directly by an experiment that involves the two materials to be used.

Table 4–3 **Typical Values for μ_s**

Contact Materials	Coefficient of Static Friction (μ_s)
Metal on ice	0.03–0.05
Wood on wood	0.30–0.70
Leather on wood	0.20–0.50
Leather on metal	0.30–0.60
Aluminum on aluminum	1.10–1.70

Motion. If the magnitude of **P** acting on the block is increased so that it becomes greater than F_s, the frictional force at the contacting surfaces drops slightly to a smaller value F_k, called the *kinetic frictional force*. The block will *not* be held in equilibrium ($P > F_k$); instead, it will begin to slide with increasing speed, Fig. 4–30*f*. The drop made in the frictional force magnitude, from F_s (static) to F_k (kinetic), can be explained by again examining the surfaces of contact, Fig. 4–30*g*. Here it is seen that when $P > F_s$, then P has the capacity to shear off the peaks at the contact surfaces and cause the block to "lift" somewhat out of its settled position and "ride" on top of the peaks. Once the block begins to slide, high local temperatures at the points of contact cause momentary adhesion (welding) of these points. The continued shearing of these welds is the dominant mechanism creating friction. Since the resultant contact forces $\Delta \mathbf{R}_n$ are aligned slightly more in the vertical direction than before, Fig. 4–30*c*, they thereby contribute *smaller* frictional components, $\Delta \mathbf{F}_n$, as when the irregularities are meshed.

Experiments with sliding blocks indicate that the magnitude of the resultant frictional force $\mathbf{F}_k$ is directly proportional to the magnitude of the resultant normal force $\mathbf{N}$. This may be expressed mathematically as

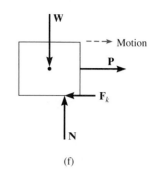

(f)

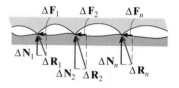

(g)

Fig. 4–30 (Cont'd)

$$F_k = \mu_k N \qquad (4\text{–}9)$$

where the constant of proportionality, μ_k, is called the *coefficient of kinetic friction*. Typical values for μ_k are approximately 25 percent *smaller* than those listed in Table 4–3 for μ_s.

The above effects regarding friction can be summarized by reference to the graph in Fig. 4–31, which shows the variation of the frictional force **F** versus the applied load **P.** Here the frictional force is categorized in three different ways: namely, **F** is a *static-frictional force* if equilibrium is maintained; **F** is a *limiting static-frictional force* $\mathbf{F}_s$ when its magnitude reaches a maximum value needed to maintain equilibrium; and finally, **F** is termed a *kinetic-frictional force* $\mathbf{F}_k$ when sliding occurs at the contacting surface. Notice also from the graph that for very large values of **P** or for high speeds, because of aerodynamic effects, $\mathbf{F}_k$ and likewise μ_k begin to decrease.

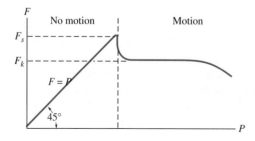

Fig. 4–31

Characteristics of Dry Friction. As a result of *experiments* that pertain to the foregoing discussion, the following rules which apply to bodies subjected to dry friction may be stated.

1. The frictional force acts *tangent* to the contacting surfaces in a direction *opposed* to the *relative motion* or tendency for motion of one surface against another.
2. The magnitude of the maximum static frictional force $\mathbf{F}_s$ that can be developed is independent of the area of contact, provided the normal pressure is not very low nor great enough to severely deform or crush the contacting surfaces of the bodies.
3. The magnitude of the maximum static frictional force is generally greater than the magnitude of the kinetic frictional force for any two surfaces of contact. However, if one of the bodies is moving with a *very low velocity* over the surface of another, F_k becomes approximately equal to F_s, i.e., $\mu_s \approx \mu_k$.
4. When *slipping* at the surface of contact is *about to occur,* the magnitude of the maximum static frictional force is proportional to the magnitude of the normal force, such that $F_s = \mu_s N$, Eq. 4–8.
5. When *slipping* at the surface of contact is *occurring,* the magnitude of the kinetic frictional force is proportional to the magnitude of the normal force, such that $F_k = \mu_k N$, Eq. 4–9.

Angle of Friction. It should be observed that Eqs. 4–8 and 4–9 have a specific yet *limited* use in the solution of friction problems. In particular, the frictional force acting at a contacting surface is determined from $F_k = \mu_k N$ *only* if *relative motion* is occurring between the two surfaces. Furthermore, if two bodies are *stationary,* the magnitude of the frictional force, F, *does not necessarily* equal $\mu_s N$; instead, F must satisfy the inequality $F \leq \mu_s N$. Only when *impending motion* occurs does F reach its upper limit, $F = F_s = \mu_s N$. This situation may be better understood by considering the block shown in Fig. 4–32a, which is acted upon by a force $\mathbf{P}$. In this case consider $P = F_s$, so that the block is on the *verge of sliding.* For equilibrium, the normal force $\mathbf{N}$ and frictional force $\mathbf{F}_s$ combine to create a resultant $\mathbf{R}_s$. The angle ϕ_s that $\mathbf{R}_s$ makes with $\mathbf{N}$ is called the *angle of static friction.* From the figure,

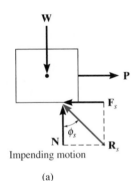

Impending motion

(a)

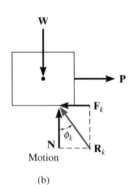

Motion

(b)

Fig. 4–32

$$\phi_s = \tan^{-1}\left(\frac{F_s}{N}\right) = \tan^{-1}\left(\frac{\mu_s N}{N}\right) = \tan^{-1}\mu_s$$

Provided the block is *not in motion,* any horizontal force $P < F_s$ causes a resultant $\mathbf{R}$ which has a line of action directed at an angle ϕ from the vertical such that $\phi \leq \phi_s$. If $\mathbf{P}$ creates uniform *motion* of the block, then $P = F_k$. In this case, the resultant $\mathbf{R}_k$ has a line of action defined by ϕ_k, Fig. 4–32b. This angle is referred to as the *angle of kinetic friction,* where

$$\phi_k = \tan^{-1}\left(\frac{F_k}{N}\right) = \tan^{-1}\left(\frac{\mu_k N}{N}\right) = \tan^{-1}\mu_k$$

By comparison, $\phi_s \geq \phi_k$.

Types of Friction Problems.
If a body is in equilibrium when it is subjected to a system of forces that includes the effect of friction, the force system must satisfy not only the equations of equilibrium but *also* the laws that govern the frictional forces. In general, there are three types of mechanics problems involving dry friction. They can easily be classified once the free-body diagrams are drawn and the total number of unknowns are identified and compared with the total number of available equilibrium equations. Each type of problem will now be explained and illustrated graphically by examples. In all these cases the geometry and dimensions for the problem are assumed to be known.

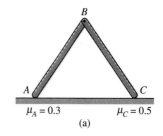

(a)

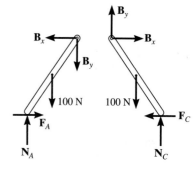

(b)

Fig. 4–33

Equilibrium.
Problems in this category are strictly equilibrium problems which require *the total number of unknowns to be equal to the total number of available equilibrium equations*. Once the frictional forces are determined from the solution, however, their numerical values must be checked to be sure they satisfy the inequality $F \leq \mu_s N$; otherwise, slipping will occur and the body will not remain in equilibrium. A problem of this type is shown in Fig. 4–33a. Here we must determine the frictional forces at A and C to check if the equilibrium position of the bars can be maintained. If the bars are uniform and have known weights of 100 N each, then the free-body diagrams are as shown in Fig. 4–33b. There are six unknown force components which can be determined *strictly* from the six equilibrium equations (three for each member). Once $\mathbf{F}_A$, $\mathbf{N}_A$, $\mathbf{F}_C$, and $\mathbf{N}_C$ are determined, then the bars will remain in equilibrium provided $F_A \leq 0.3N_A$ and $F_C \leq 0.5N_C$ are satisfied.

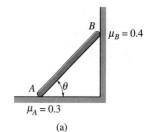

(a)

Impending Motion at All Points.
In this case *the total number of unknowns will equal the total number of available equilibrium equations plus the total number of available frictional equations, $F = \mu N$*. In particular, if *motion is impending* at the points of contact, then $F_s = \mu_s N$; whereas if the body is *slipping*, then $F_k = \mu_k N$. For example, consider the problem of finding the smallest angle θ at which the 100-N bar in Fig. 4–34a can be placed against the wall without slipping. The free-body diagram is shown in Fig. 4–34b. Here there are *five* unknowns: F_A, N_A, F_B, N_B, θ. For the solution there are *three* equilibrium equations and *two* static frictional equations which apply at *both* points of contact, so that $F_A = 0.3N_A$ and $F_B = 0.4N_B$. (It should also be noted that the bar will not be in a state where motion impends *unless* the bar slips at *both* points A and B simultaneously.)

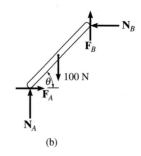

(b)

Fig. 4–34

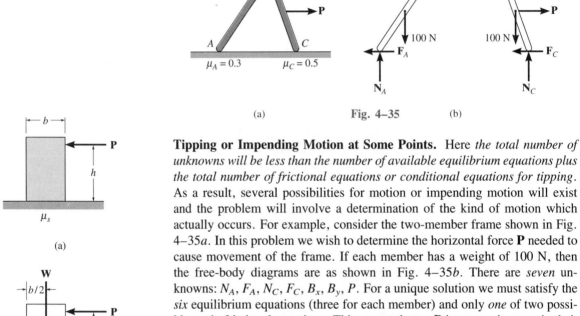

(a)   **Fig. 4–35**   (b)

Fig. 4–36

Tipping or Impending Motion at Some Points. Here *the total number of unknowns will be less than the number of available equilibrium equations plus the total number of frictional equations or conditional equations for tipping.* As a result, several possibilities for motion or impending motion will exist and the problem will involve a determination of the kind of motion which actually occurs. For example, consider the two-member frame shown in Fig. 4–35a. In this problem we wish to determine the horizontal force **P** needed to cause movement of the frame. If each member has a weight of 100 N, then the free-body diagrams are as shown in Fig. 4–35b. There are *seven* unknowns: N_A, F_A, N_C, F_C, B_x, B_y, P. For a unique solution we must satisfy the *six* equilibrium equations (three for each member) and only *one* of two possible static frictional equations. This means that as **P** increases its magnitude it will either cause slipping at A and no slipping at C, so that $F_A = 0.3N_A$ and $F_C \leq 0.5N_C$; or slipping occurs at C and no slipping at A, in which case $F_C = 0.5N_C$ and $F_A \leq 0.3N_A$. The actual situation can be determined by calculating P for each case and then choosing the case for which P is *smallest*. If in both cases the *same value* for P is calculated, which in practice would be highly improbable, then slipping at both points occurs simultaneously; i.e., the *seven unknowns* will satisfy *eight equations*. As a second example, consider a block having a width b, height h, and weight W which is resting on a rough surface, Fig. 4–36a. The force **P** needed to cause motion is to be determined. Inspection of the free-body diagram, Fig. 4–36b, indicates that there are *four unknowns*, namely, P, F, N, and x. For a unique solution, however, we must satisfy the *three* equilibrium equations and either *one* static friction equation or *one* conditional equation which requires the block not to tip. Hence two possibilities of motion exist. Either the block will *slip*, Fig. 4–36b, in which case $F = \mu_s N$ and the value obtained for x must satisfy $0 \leq x \leq b/2$; or the block will *tip*, Fig. 4–36c, in which case $x = b/2$ and the frictional force will satisfy the inequality $F \leq \mu_s N$. The solution yielding the *smallest* value of P will define the type of motion the block undergoes. If it happens that the same value of P is calculated for both cases, although this would be very improbable, then slipping and tipping will occur simultaneously; i.e., the *four unknowns* will satisfy *five equations*.

Equilibrium Versus Frictional Equations. It was stated earlier that the frictional force *always* acts so as to either oppose the relative motion or impede the motion of a body over its contacting surface. Realize, however, that we can *assume* the sense of the frictional force in problems which require F to be an "equilibrium force" and satisfy the inequality $F < \mu_s N$. The correct sense is made known *after* solving the equations of equilibrium for F. For example, if F is a negative scalar, the sense of $\mathbf{F}$ is the reverse of that which was assumed. This convenience of *assuming* the sense of $\mathbf{F}$ is possible because the equilibrium equations equate to zero the *components of vectors* acting in the *same direction*. In cases where the frictional equation $F = \mu N$ is used in the solution of a problem, however, the convenience of *assuming* the sense of $\mathbf{F}$ is *lost*, since the frictional equation relates only the *magnitudes* of two perpendicular vectors. Consequently, $\mathbf{F}$ *must always* be shown acting with its *correct sense* on the free-body diagram whenever the frictional equation is used for the solution of a problem.

PROCEDURE FOR ANALYSIS

The following procedure provides a method for solving equilibrium problems involving dry friction.

Free-Body Diagrams.   Draw the necessary free-body diagrams and determine the number of unknowns or equations required for a complete solution. Unless stated in the problem, *always* show the frictional forces as *unknowns;* i.e., *do not assume that* $F = \mu N$. Recall that only three equations of coplanar equilibrium can be written for each body. Consequently, if there are more unknowns than equations of equilibrium, it will be necessary to apply the frictional equation at some, if not all, points of contact to obtain the extra equations needed for a complete solution.

Equations of Friction and Equilibrium.   Apply the equations of equilibrium and the necessary frictional equations (or conditional equations if tipping is involved) and solve for the unknowns. If the problem involves a three-dimensional force system such that it becomes difficult to obtain the force components or the necessary moment arms, apply the equations of equilibrium using Cartesian vectors.

The following example problems illustrate this procedure numerically.

Example 4–21

The uniform crate shown in Fig. 4–37a has a mass of 20 kg. If a force $P = 80$ N is applied to the crate, determine if it remains in equilibrium. The coefficient of static friction is $\mu_s = 0.3$.

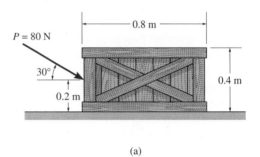

(a)

Fig. 4–37

SOLUTION

Free-Body Diagram.   As shown in Fig. 4–37b, the *resultant* normal force N_C must act a distance x from the crate's center line in order to counteract the tipping effect caused by **P**. There are *three unknowns: F, N_C, and x,* which can be determined strictly from the *three* equations of equilibrium.

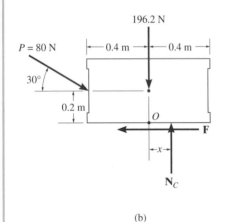

(b)

Equations of Equilibrium

$$\xrightarrow{+}\Sigma F_x = 0; \qquad 80 \cos 30° - F = 0$$
$$+\uparrow\Sigma F_y = 0; \qquad -80 \sin 30° + N_C - 196.2 = 0$$
$$\zeta+\Sigma M_O = 0; \qquad 80 \sin 30°(0.4) - 80 \cos 30°(0.2) + N_C(x) = 0$$

Solving,

$$F = 69.3 \text{ N}$$
$$N_C = 236 \text{ N}$$
$$x = -0.00908 \text{ m} = -9.08 \text{ mm}$$

Since x is negative it indicates the *resultant* normal force acts (slightly) to the *left* of the crate's center line. No tipping will occur since $x \leq 0.4$ m. Also, the *maximum* frictional force which can be developed at the surface of contact is $F_{max} = \mu_s N_C = 0.3(236 \text{ N}) = 70.8$ N. Since $F = 69.3$ N < 70.8 N, the crate will *not slip,* although it is very close to doing so.

Example 4–22

The pipe shown in Fig. 4–38a is gripped between two levers that are pinned together at C. If the coefficient of static friction between the levers and the pipe is $\mu = 0.3$, determine the maximum angle θ at which the pipe can be gripped without slipping. Neglect the weight of the pipe.

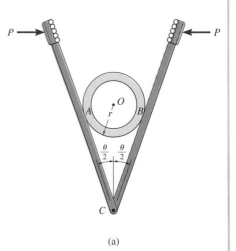

(a)

SOLUTION

Free-Body Diagram. As shown in Fig. 4–38b, there are five unknowns: N_A, F_A, N_B, F_B, and θ. The *three* equations of equilibrium and *two* frictional equations at A and B apply. The frictional forces act toward C to prevent upward motion of the pipe.

Equations of Friction and Equilibrium. The frictional equations are

$$F_s = \mu_s N; \qquad\qquad F_A = \mu N_A$$
$$F_B = \mu N_B$$

Using these results, and applying the equations of equilibrium, yields

$\xrightarrow{+} \Sigma F_x = 0;$

$$N_A \cos\left(\frac{\theta}{2}\right) + \mu N_A \sin\left(\frac{\theta}{2}\right) - N_B \cos\left(\frac{\theta}{2}\right) - \mu N_B \sin\left(\frac{\theta}{2}\right) = 0 \quad (1)$$

$\zeta + \Sigma M_O = 0;$

$$-\mu N_B(r) + \mu N_A(r) = 0 \qquad\qquad (2)$$

$+\uparrow \Sigma F_y = 0;$

$$N_A \sin\left(\frac{\theta}{2}\right) - \mu N_A \cos\left(\frac{\theta}{2}\right) + N_B \sin\left(\frac{\theta}{2}\right) - \mu N_B \cos\left(\frac{\theta}{2}\right) = 0 \quad (3)$$

From either Eq. 1 or 2 it is seen that $N_A = N_B$. This could also have been determined directly from the symmetry of *both* geometry and loading. Substituting the result into Eq. 3, we obtain

$$\sin\left(\frac{\theta}{2}\right) - \mu \cos\left(\frac{\theta}{2}\right) = 0$$

so that

$$\tan\left(\frac{\theta}{2}\right) = \frac{\sin(\theta/2)}{\cos(\theta/2)} = \mu = 0.3$$

$$\theta = 2 \tan^{-1} 0.3 = 33.4° \qquad\qquad Ans.$$

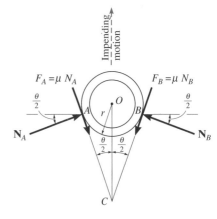

(b)

Fig. 4–38

■ **Example 4–23**

The uniform stone has a mass of 500 kg and is held in the horizontal position using a wedge at B as shown in Fig. 4–39a. If the coefficient of static friction is $\mu_s = 0.3$ at the surface in contact with the wedge, determine the force **P** needed to remove the wedge. Assume that the stone does not slip at A.

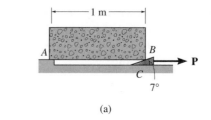

(a)

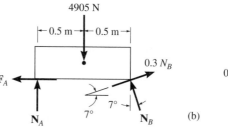

(b)

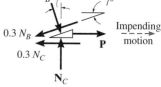

Fig. 4–39

SOLUTION

Since the wedge is to be removed, slipping is about to occur at the surfaces of contact. Thus, $F = \mu_s N$, and the free-body diagrams are shown in Fig. 4–39b. Note that on the stone at A, $F_A \leq \mu_s N_A$, since slipping does not occur there. From the free-body diagram of the stone,

$$\zeta + \Sigma M_A = 0; \quad -4905(0.5) + (N_B \cos 7°)(1) + (0.3 N_B \sin 7°)(1) = 0$$
$$N_B = 2383.1 \text{ N}$$

Using this result for the wedge, we have

$$\xrightarrow{+} \Sigma F_x = 0; \quad 2383.1 \sin 7° - 0.3(2383.1) \cos 7° + P - 0.3 N_C = 0$$
$$+\uparrow \Sigma F_y = 0; \quad N_C - 2383.1 \cos 7° - 0.3(2383.1) \sin 7° = 0$$
$$N_C = 2452.5 \text{ N}$$
$$P = 1154.9 \text{ N} = 1.15 \text{ kN} \qquad\qquad Ans.$$

Since P is positive, indeed the wedge must be pulled out.

Example 4–24

The homogeneous block shown in Fig. 4–40*a* has a weight of 20 lb and rests on the incline for which $\mu_s = 0.55$. Determine the largest angle of tilt, θ, of the plane before the block moves.

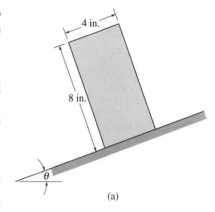

(a)

SOLUTION

Free-Body Diagram. As shown in Fig. 4–40*b*, the dimension *x* is used to locate the position of the resultant normal force $\mathbf{N}_B$ under the block. There are *four* unknowns, θ, N_B, F_B, and *x*. *Three* equations of equilibrium are available. The *fourth* equation is obtained by investigating the conditions for tipping or sliding of the block.

Equations of Equilibrium. Applying the equations of equilibrium yields

$$+\swarrow \Sigma F_x = 0; \qquad 20 \sin\theta - F_B = 0 \qquad (1)$$
$$+\nwarrow \Sigma F_y = 0; \qquad N_B - 20 \cos\theta = 0 \qquad (2)$$
$$\zeta + \Sigma M_O = 0; \qquad 20 \sin\theta(4) - 20 \cos\theta(x) = 0 \qquad (3)$$

(Impending Motion of Block.) This requires use of the frictional equation

$$F_s = \mu_s N; \qquad F_B = 0.55 N_B \qquad (4)$$

Solving Eqs. 1 through 4 yields

$$N_B = 17.5 \text{ lb} \qquad F_B = 9.64 \text{ lb} \qquad \theta = 28.8° \qquad x = 2.2 \text{ in.}$$

Since $x = 2.2$ in. > 2 in., the block will tip *before* sliding.

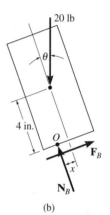

(b)

Fig. 4–40

(Tipping of Block.) This requires

$$x = 2 \text{ in.} \qquad (5)$$

Solving Eqs. 1 through 3 using Eq. 5 yields

$$N_B = 17.9 \text{ lb} \qquad F_B = 8.94 \text{ lb}$$
$$\theta = 26.6° \qquad \qquad \textit{Ans.}$$

Note: If we *first* assumed that the block tips, then the results for F_B would have to be checked with the maximum *possible* static frictional force; i.e.,

$$F_B = 8.94 \overset{?}{<} (0.55)(17.9 \text{ lb}) = 9.84 \text{ lb}$$

Since the inequality holds, indeed the block will tip before it slips.

Example 4–25

Beam AB is subjected to a concentrated force of 800 N and is supported at B by a post BC, Fig. 4–41a. If the coefficients of static friction at B and C are $\mu_B = 0.2$ and $\mu_C = 0.5$, determine the force $\mathbf{P}$ needed to pull the post out from under the beam. Neglect the weight of the members and the thickness of the post.

SOLUTION

Free-Body Diagrams. The free-body diagram of beam AB is shown in Fig. 4–41b. Applying $\Sigma M_A = 0$, we obtain $N_B = 400$ N. This result is shown on the free-body diagram of the post, Fig. 4–41c. Referring to this member, the *four* unknowns F_B, P, F_C, and N_C are determined from the *three* equations of equilibrium and *one* frictional equation applied either at B or C.

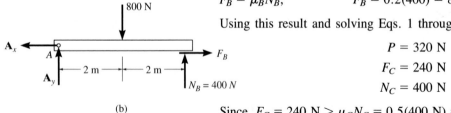

(a)

Equations of Equilibrium and Friction

$$\xrightarrow{+}\Sigma F_x = 0; \qquad\qquad P - F_B - F_C = 0 \qquad\qquad (1)$$

$$+\uparrow\Sigma F_y = 0; \qquad\qquad N_C - 400 = 0 \qquad\qquad (2)$$

$$\zeta+\Sigma M_C = 0; \qquad\qquad -P(0.25) + F_B(1) = 0 \qquad\qquad (3)$$

(Post Slips Only at B.) This requires $F_C \leq \mu N_C$ and

$$F_B = \mu_B N_B; \qquad\qquad F_B = 0.2(400) = 80 \text{ N}$$

Using this result and solving Eqs. 1 through 3, we obtain

$$P = 320 \text{ N}$$
$$F_C = 240 \text{ N}$$
$$N_C = 400 \text{ N}$$

Since $F_C = 240$ N $> \mu_C N_C = 0.5(400$ N$) = 200$ N, the other case of movement must be investigated.

(Post Slips Only at C.) Here $F_B \leq \mu_B N_B$ and

$$F_C = \mu_C N_C; \qquad\qquad F_C = 0.5 N_C \qquad\qquad (4)$$

Solving Eqs. 1 through 4 yields

$$P = 267 \text{ N} \qquad\qquad\qquad Ans.$$

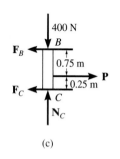

(c)

Fig. 4–41

Obviously, this case occurs first, since it requires a *smaller* value for P.

Example 4–26

Determine the normal force that must be exerted on the 100-kg spool shown in Fig. 4–42a to push it up the 20° incline at constant velocity. The coefficients of static and kinetic friction at the points of contact are $(\mu_s)_A = 0.18$, $(\mu_k)_A = 0.15$ and $(\mu_s)_B = 0.45$, $(\mu_k)_B = 0.4$.

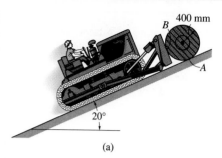

(a)

SOLUTION

Free-Body Diagram. As shown in Fig. 4–42b, there are four unknowns N_A, F_A, N_B, and F_B acting on the spool. These can be determined from the *three* equations of equilibrium and *one* frictional equation, which applies either at A or B. If slipping only occurs at B, the spool *rolls* up the incline; whereas if slipping only occurs at A, the spool will *slide* up the incline. Here we must calculate N_B.

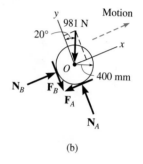

(b)

Fig. 4–42

Equations of Equilibrium and Friction

$$+\nearrow\Sigma F_x = 0; \qquad -F_A + N_B - 981 \sin 20° = 0 \qquad (1)$$
$$+\nwarrow\Sigma F_y = 0; \qquad N_A - F_B - 981 \cos 20° = 0 \qquad (2)$$
$$\zeta+\Sigma M_O = 0; \qquad F_B(400 \text{ mm}) - F_A(400 \text{ mm}) = 0 \qquad (3)$$

(Spool Rolls up Incline.) In this case $F_A \leq 0.18 N_A$ and

$$(F_k)_B = (\mu_k)_B N_B; \qquad F_B = 0.40 N_B \qquad (4)$$

The direction of the frictional force at B must be specified correctly. Why? Since the spool is being forced up the plane, $\mathbf{F}_B$ acts downward to prevent the clockwise rolling motion of the spool, Fig. 4–42b. Solving Eqs. 1 through 4, we have

$$N_A = 1146 \text{ N} \qquad F_A = 224 \text{ N} \qquad N_B = 559 \text{ N} \qquad F_B = 224 \text{ N}$$

The assumption regarding no slipping at A should be checked.

$$F_A \leq (\mu_s)_A N_A; \qquad 224 \text{ N} \overset{?}{\leq} 0.18(1146) = 206 \text{ N}$$

The inequality does *not apply,* and therefore slipping occurs at A and not at B. Hence, the other case of motion must be investigated.

(Spool Slides up Incline.) In this case, $F_B \leq 0.45 N_B$ and

$$(F_k)_A = (\mu_k)_A N_A; \qquad F_A = 0.15 N_A \qquad (5)$$

Solving Eqs. 1 through 3 and 5 yields

$$N_A = 1084 \text{ N} \qquad F_A = 163 \text{ N} \qquad N_B = 498 \text{ N} \qquad F_B = 163 \text{ N}$$

The validity of the solution ($N_B = 498$ N) can be checked by testing the assumption that indeed no slipping occurs at B.

$$F_B \leq (\mu_s)_B N_B; \qquad 163 \text{ N} < 0.45(498 \text{ N}) = 224 \text{ N} \qquad \text{(check)}$$

PROBLEMS

4–71. The footing of a ladder rests on a concrete surface for which $\mu_s = 0.8$ for the two materials. If the magnitude of a force directed along the axis of the ladder is 600 lb, determine the frictional force acting at the bottom of the ladder. What is the minimum coefficient of static friction that will prevent the ladder from slipping?

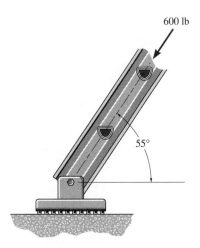

600 lb

55°

Prob. 4–71

*4–72. Determine the horizontal force P needed to just start moving the 300-lb crate up the plane. Take $\mu_s = 0.3$.

4–73. Determine the range of values for which the horizontal force P will prevent the crate from slipping down or up the inclined plane. Take $\mu_s = 0.1$.

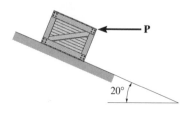

P

20°

Probs. 4–72/4–73

4–74. The refrigerator has a weight of 200 lb and a center of gravity at G. Determine the force **P** required to move it. Will the refrigerator tip or slip? Take $\mu_s = 0.4$.

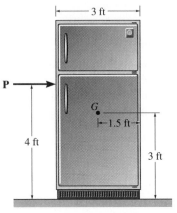

3 ft

P

G

1.5 ft

4 ft

3 ft

Prob. 4–74

4–75. The spool has a mass of 20 kg. If a cord is wrapped around its inner core and is attached to the wall, and the coefficient of static friction at A is $\mu_s = 0.15$, determine if the spool remains in equilibrium when it is released.

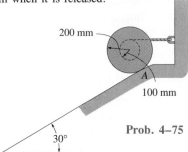

200 mm

A

100 mm

30°

Prob. 4–75

*4–76. The 100-lb cylinder rests between the two inclined planes. When $P = 15$ lb, the cylinder is on the verge of impending motion. Determine the coefficient of static friction μ_s between the surfaces of contact and the cylinder.

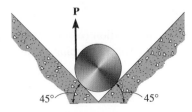

P

45°   45°

Prob. 4–76

4–77. The bracket can move freely along the pole until a load P is placed on it. When this occurs it contacts the pole at A and B. If the coefficient of static friction is $\mu_s = 0.6$, determine the smallest distance d at which the load can be applied so that the bracket does not slip. Neglect the weight of the bracket.

4–79. The square block has a weight W and rests on the board. If the board is tipped at an angle θ, the block will tip over. Determine θ and the smallest coefficient of static friction that will allow this to happen.

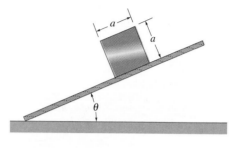

Prob. 4–79

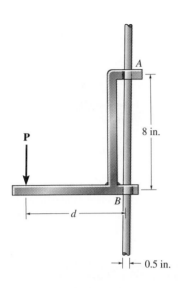

Prob. 4–77

4–78. The car has a weight of 4000 lb and a center of gravity at G. If it pulls off the side of a road, determine the greatest angle of tilt, θ, it can have without slipping or tipping over. The coefficient of static friction between its wheels and the ground is $\mu_s = 0.4$.

***4–80.** The uniform ladder has a weight W and rests against the wall at B and the floor at A. If the coefficient of static friction is μ, determine the smallest angle θ at which the ladder can be placed against the wall and not slip.

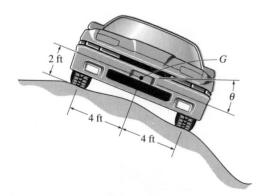

Prob. 4–78

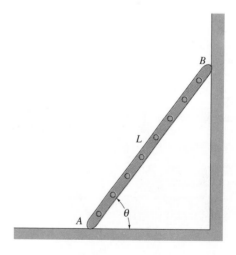

Prob. 4–80

4–81. If the coefficient of static friction between the incline and the 200-lb crate is $\mu_s = 0.3$, determine the force **P** that must be applied to the rope to begin moving the crate up the incline.

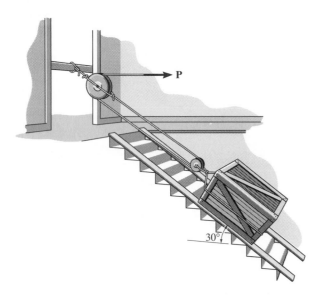

Prob. 4–81

4–82. The machine has a weight of 500 lb. Determine the force **P** that must be applied to the handle of the crowbar to lift it. The coefficient of static friction at A and B is $\mu_s = 0.3$.

4–83. The machine has a weight of 500 lb. Determine the force **P** that must be applied to the handle of the crowbar to lift it. The coefficient of static friction at A is $\mu_A = 0.3$ and at B, $\mu_B = 0.2$.

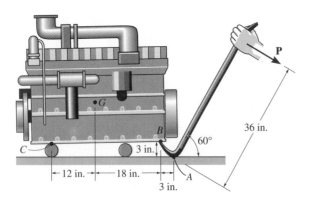

Probs. 4–82/4–83

4–84. The uniform roll of paper has a mass of 120 kg and rests on a 30° wedge of negligible mass. If the coefficient of static friction at all contacting surfaces is $\mu_s = 0.3$, determine the horizontal force **P** that must be applied to the wedge in order to life the paper. Neglect the weight of the wedge.

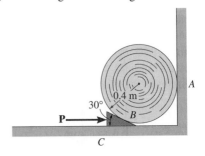

Prob. 4–84

4–85. The two stone blocks have weights of $W_A = 600$ lb and $W_B = 500$ lb. Determine the smallest horizontal force **P** that must be applied to block A in order to move it. The coefficient of static friction between the blocks is $\mu_s = 0.3$ and between the floor and each block $\mu_s' = 0.5$.

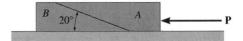

Prob. 4–85

4–86. Determine the smallest force P needed to lift the 3000-lb load. The coefficient of static friction between A and C and B and D is $\mu_s = 0.3$, and between A and B, $\mu_s' = 0.4$. Neglect the weight of each wedge.

4–87. Determine the reversed horizontal force $-\mathbf{P}$ needed to pull out wedge A. The coefficient of static friction between A and C and B and D is $\mu_s = 0.2$, and between A and B, $\mu_s' = 0.1$. Neglect the weight of each wedge.

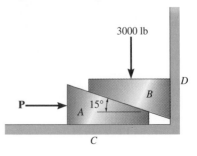

Probs. 4–86/4–87

*4–88.   Column D is subjected to a vertical load of 8000 lb. It is supported on two identical wedges A and B, for which the coefficient of static friction at the contacting surfaces AB and BC is $\mu_s = 0.4$. Determine the force **P** needed to move wedge B to the right and the smallest equilibrium force **P′** needed to hold wedge A stationary. The contacting surfaces AD are smooth. Neglect the weight of the wedges.

4–89.   Column D is subjected to a vertical load of 8000 lb. It is supported on two identical wedges A and B, for which the coefficient of static friction at the contacting surfaces AB and BC is $\mu_s = 0.4$. If the forces **P** and **P′** are removed, are the wedges self-locking? The contacting surfaces AD are smooth. Neglect the weight of the wedges.

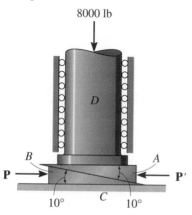

Probs. 4–88/4–89

4–90.   The truck has a mass of 1.25 Mg and a center of mass at G. Determine the greatest load it can pull if (*a*) the truck has rear-wheel drive while the front wheels are free to roll, and (*b*) the truck has four-wheel drive. The coefficient of static friction between the wheels and the ground is $\mu_s = 0.5$, and between the crate and the ground, it is $\mu_s' = 0.4$.

4–91.   Solve Prob. 4–90 if the truck and crate are traveling up a 10° incline.

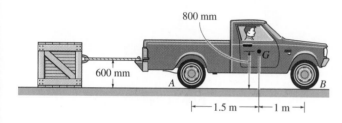

Probs. 4–90/4–91

*4–92.   The uniform dresser has a weight of 80 lb and rests on a tile floor for which $\mu_s = 0.25$. If the man pushes on it in the direction shown, determine the smallest magnitude of force **F** needed to move the dresser. Also, if the man has a weight of 150 lb, determine the smallest coefficient of static friction between his shoes and the floor so he doesn't slip.

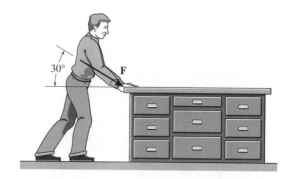

Prob. 4–92

4–93.   The tractor pulls on the fixed tree stump. Determine the torque that must be applied by the engine to the rear wheels to cause them to slip. The front wheels are free to roll. The tractor weighs 3500 lb and has a center of gravity at G. The coefficient of static friction between the rear wheels and the ground is $\mu_s = 0.5$.

4–94.   The tractor pulls on the fixed tree stump. If the coefficient of static friction between the rear wheels and the ground is $\mu_s = 0.6$, determine if the rear wheels slip or the front wheels lift off the ground as the engine provides torque to the rear wheels. What is the torque needed to cause the motion? The front wheels are free to roll. The tractor weighs 2500 lb and has a center of gravity at G.

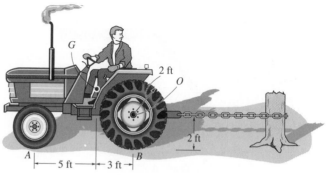

Probs. 4–93/4–94

4–95. The cabinet has a weight of 300 lb and due to the items inside the drawers, a center of gravity at G. If the coefficient of static friction at A and B is $\mu_s = 0.2$, determine if the 200-lb man can push the cabinet to the left. The coefficient of static friction between his shoes and the floor is $\mu'_s = 0.4$. Assume the man exerts only a horizontal force on the cabinet.

***4–96.** The cabinet has a weight of 300 lb and due to the items inside the drawers, a center of gravity at G. If the coefficient of static friction at A and B is $\mu_s = 0.2$, determine if the 200-lb man can pull the cabinet to the right. The coefficient of static friction between his shoes and the floor is $\mu'_s = 0.35$. Assume the man exerts only a horizontal force on the cabinet.

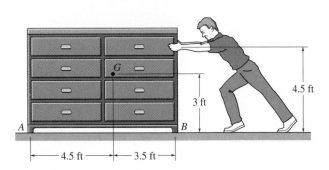

Probs. 4–95/4–96

4–97. The 2-m-long uniform rod has a weight W and is placed along the ledge. If the smallest angle at which it can lean without slipping is $\theta = 60°$, determine the coefficient of static friction μ between the contact surfaces A and B and the rod.

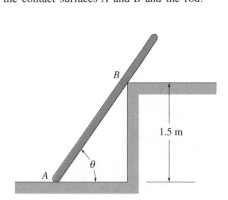

Prob. 4–97

4–98. Determine the greatest load W the boy can pull on the sled S, if the sled has a weight of 8 lb and $(\mu_S)_s = 0.1$ between the runners and the ice. The boy weighs 70 lb and $(\mu_B)_s = 0.4$ between his boots and the ice. Set $\theta = 30°$.

4–99. Determine the angle θ at which the cord must be directed so that the boy can pull the sled S, which supports a load of 400 lb. The sled weighs 8 lb and $(\mu_S)_s = 0.1$ between the runners and the ice. The boy weighs 70 lb, and $(\mu_B)_s = 0.4$ between his boots and the ice.

Probs. 4–98/4–99

***4–100.** Two boys, each weighing 60 lb, sit at the ends of a uniform board, which has a weight of 30 lb. If the board rests at its center on a post having a coefficient of static friction of $\mu_s = 0.6$ with the board, determine the greatest angle of tilt θ before slipping occurs. Neglect the size of the post and the thickness of the board in the calculations.

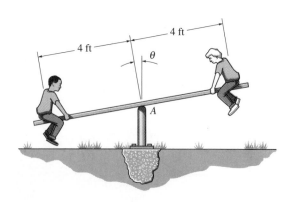

Prob. 4–100

4–101. Determine the minimum force **F** needed to push the tube *E* up the incline. The tube has a mass of 75 kg and the roller *D* has a mass of 100 kg. The force acts parallel to the plane and the coefficients of static friction at the contacting surfaces are $\mu_A = 0.3$, $\mu_B = 0.25$, and $\mu_C = 0.4$. Each cylinder has a radius of 150 mm.

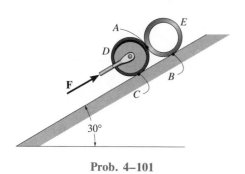

Prob. 4–101

4–103. The device is used to hold brooms and mops. It consists of two small rollers having a negligible weight that can slide freely along the smooth inclines. If the coefficient of static friction between the broomstick and each roller is $\mu_s = 0.5$, determine the angle of inclination θ at which the inclines should be designed so that any weight of broom can be supported.

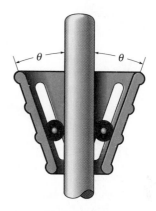

Prob. 4–103

4–102. The man has a mass of 60 kg and the crate has a mass of 100 kg. If the coefficient of static friction between his shoes and the ground is $\mu_S = 0.4$ and between the crate and the ground $\mu_C = 0.3$, determine if the man is able to move the crate using the rope-and-pulley system shown.

***4–104.** Determine the maximum weight *W* the man can lift using the pulley system, without and then with the "leading block" or pulley at *A*. The man has a weight of 200 lb and the coefficient of static friction between his feet and the ground is $\mu_s = 0.6$.

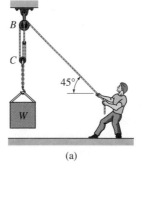

(a)

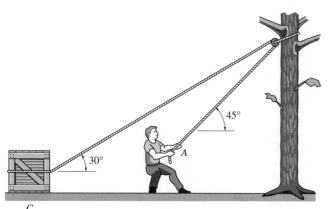

Prob. 4–102

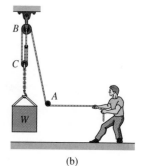

(b)

Prob. 4–104

4–105. The refrigerator has a mass of 75 kg and a center of mass at G. Determine the force **P** required to move it. Does it tip or slide? The coefficient of static friction between the refrigerator and the incline is $\mu_s = 0.3$.

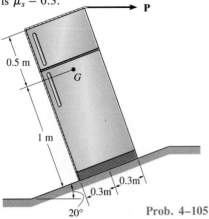

Prob. 4–105

4–106. Two blocks A and B, each having a mass of 6 kg, are connected by the linkage shown. If the coefficients of static friction at the contacting surfaces are $\mu_A = 0.2$ and $\mu_B = 0.3$, determine the largest vertical force **P** that may be applied to pin C without causing the blocks to slip. Neglect the weight of the links.

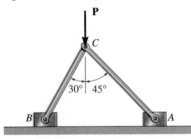

Prob. 4–106

4–107. The uniform rod has a weight W and rests on a smooth peg at A and against a wall at B for which $\mu_s = 0.25$. Determine the greatest angle θ for placement of the rod so that it does not slip.

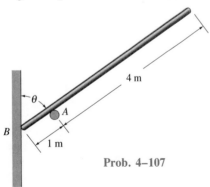

Prob. 4–107

*4–108. The beam has a mass of 20 kg and is subjected to a force of 250 N. It is supported at one end by a pin and at the other end by a spool having a mass of 35 kg. Determine the minimum cable force **P** needed to move the spool from under the beam. The coefficients of static friction at B and D are $\mu_B = 0.4$ and $\mu_D = 0.2$, respectively.

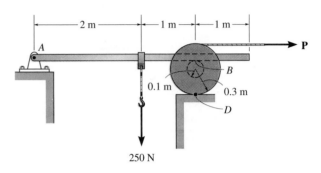

Prob. 4–108

4–109. The log of weight W is to be pulled up the inclined plane of slope α using a force **P**. If **P** acts at the angle ϕ as shown, show that for slipping to occur $P = W \sin (\alpha + \theta)/\cos (\phi - \theta)$, where θ is the angle of static friction; $\theta = \tan^{-1} \mu_s$.

4–110. Determine the angle ϕ at which the applied force **P** should act on the log so that the magnitude of **P** is as small as possible for pulling the log up the incline. What is the corresponding value of P? The log weighs W and the slope α is known. Express the answer in terms of the angle of kinetic friction, $\theta = \tan^{-1} \mu$.

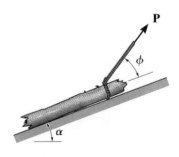

Probs. 4–109/4–110

REVIEW PROBLEMS

4–111. The horizontal beam is supported by springs at its ends. Each spring has a stiffness of $k = 5$ kN/m and is originally unstretched so that the beam is in the horizontal position. Determine the angle of tilt of the beam if a load of 800 N is applied at point C as shown.

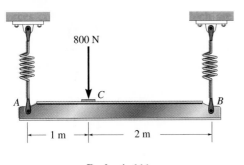

Prob. 4–111

4–113. A vertical force of 80 lb acts on the crankshaft. Determine the horizontal equilibrium force **P** that must be applied to the handle and the x, y, z components of force at the smooth journal bearing A and the thrust bearing B. The bearings are properly aligned and exert only force reactions on the shaft.

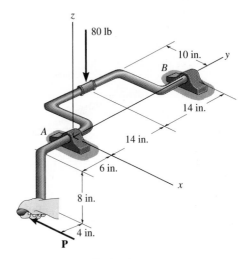

Prob. 4–113

4–114. The spring has a stiffness of $k = 80$ N/m and an unstretched length of 2 m. Determine the force in cables BC and BD when the spring is held in the position shown.

*4–112. The horizontal beam is supported by springs at its ends. If the stiffness of the spring at A is $k_A = 5$ kN/m, determine the required stiffness of the spring at B so that if the beam is loaded with the 800-N force it remains in the horizontal position both before and after loading.

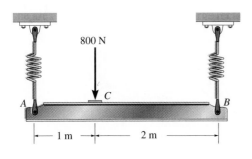

Prob. 4–112

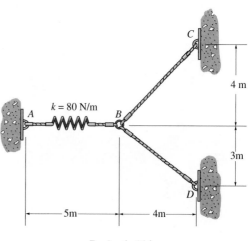

Prob. 4–114

4–115. A skeletal diagram of the lower leg is shown in the lower figure. Here it can be noted that this portion of the leg is lifted by the quadriceps muscle attached to the hip at A and to the patella bone at B. This bone slides freely over cartilage at the knee joint. The quadriceps is further extended and attached to the tibia at C. Using the mechanical system shown in the upper figure to model the lower leg, determine the tension $\mathbf{T}$ in the quadriceps and the magnitude of the resultant force at the femur (pin), D, in order to hold the lower leg in the position shown. The lower leg has a mass of 3.2 kg and a mass center at G_1, the foot has a mass of 1.6 kg and a mass center at G_2.

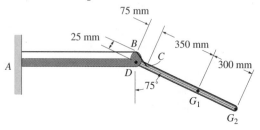

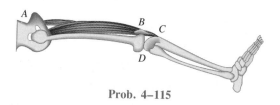

Prob. 4–115

***4–116.** The man has a mass of 40 kg. He plans to scale the vertical crevice using the method shown. If the coefficient of static friction between his shoes and the rock is $\mu_s = 0.4$ and between his backside and the rock, $\mu_s' = 0.3$, determine the smallest horizontal force his body must exert on the rock in order to do this.

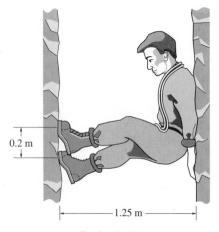

Prob. 4–116

4–117. Determine the reactions at the supports A and B of the frame.

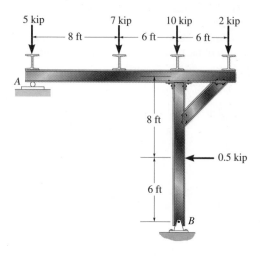

Prob. 4–117

4–118. The stiff-leg derrick used on ships is supported by a ball-and-socket joint at D and two cables BA and BC. The cables are attached to a smooth collar ring at B, which allows rotation of the derrick about the z axis. If the derrick supports a crate having a mass of 100 kg, determine the tension in the supporting cables and the x, y, z components of reaction at D.

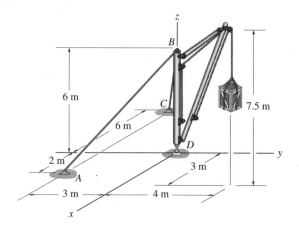

Prob. 4–118

5 Structural Analysis

In this chapter we will use the equations of equilibrium to analyze structures composed of pin-connected members. The analysis is based on the principle that if a structure is in equilibrium, then each of its members is also in equilibrium. By applying the equations of equilibrium to the various parts of a simple truss, frame, or machine, we will be able to determine all the forces acting at the connections.

The topics in this chapter are very important since they provide practice in drawing free-body diagrams, using the principle of action, equal but opposite collinear force reaction, and applying the equations of equilibrium.

5.1 Simple Trusses

A *truss* is a structure composed of slender members joined together at their end points. The members commonly used in construction consist of wooden struts or metal bars. The joint connections are usually formed by bolting or welding the ends of the members to a common plate, called a *gusset plate,* as shown in Fig. 5–1a, or by simply passing a large bolt or pin through each of the members, Fig. 5–1b.

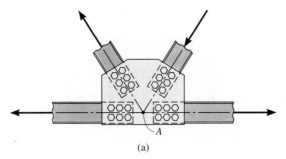

(a)

Fig. 5–1

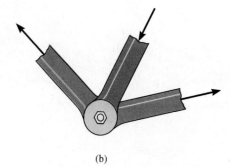

(b)

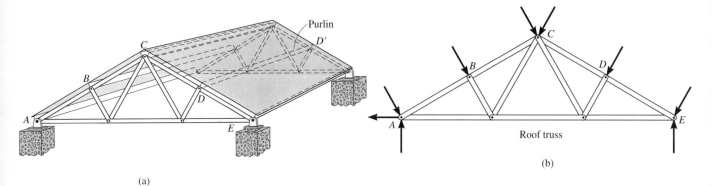

(a)

(b)

Roof truss

Fig. 5–2

Planar Trusses. *Planar* trusses lie in a single plane and are often used to support roofs and bridges. The truss *ABCDE*, shown in Fig. 5–2a, is an example of a typical roof-supporting truss. In this figure, the roof load is transmitted to the truss *at the joints* by means of a series of *purlins,* such as *DD′*. Since the imposed loading acts in the same plane as the truss, Fig. 5–2b, the analysis of the forces developed in the truss members is two-dimensional.

In the case of a bridge, such as shown in Fig. 5–3a, the load on the deck is first transmitted to *stringers,* then to *floor beams,* and finally to the *joints B, C,* and *D* of the two supporting side trusses. Like the roof truss, the bridge truss loading is also coplanar, Fig. 5–3b.

When bridge or roof trusses extend over large distances, a rocker or roller is commonly used for supporting one end, joint *E* in Figs. 5–2a and 5–3a. This type of support allows freedom for expansion or contraction of the members due to temperature or application of loads.

Fig. 5–3

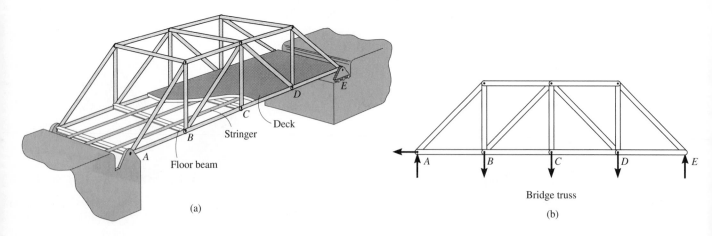

(a)

Bridge truss

(b)

Assumptions for Design. To design both the members and the connections of a truss, it is first necessary to determine the *force* developed in each member when the truss is subjected to a given loading. In this regard, two important assumptions will be made:

1. *All loadings are applied at the joints.* In most situations, such as for bridge and roof trusses, this assumption is true. Frequently in the force analysis the weights of the members are neglected, since the forces supported by the members are usually large in comparison with their weights. If the member's weight is to be included in the analysis, it is generally satisfactory to apply it as a vertical force, half of its magnitude applied at each end of the member.
2. *The members are joined together by smooth pins.* In cases where bolted or welded joint connections are used, this assumption is satisfactory provided the center lines of the joining members are *concurrent,* as in the case of point A in Fig. 5–1a.

Because of these two assumptions, *each truss member acts as a two-force member,* and therefore the forces at the ends of the member must be directed along the axis of the member. If the force tends to *elongate* the member, it is a *tensile force* (**T**), Fig. 5–4a; whereas if it tends to *shorten* the member, it is a *compressive force* (**C**), Fig. 5–4b. In the actual design of a truss it is important to state whether the nature of the force is tensile or compressive. Most often, compression members must be made *thicker* than tension members, because of the buckling or column effect that occurs when a member is in compression.

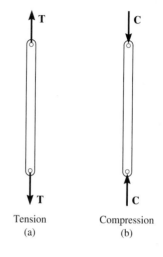

Tension
(a)

Compression
(b)

Fig. 5–4

Simple Truss. To prevent collapse, the framework of a truss must be rigid. Obviously, the four-bar frame *ABCD* in Fig. 5–5 will collapse unless a diagonal, such as *AC*, is added for support. The simplest framework which is rigid or stable is a *triangle.* Consequently, a *simple truss* is constructed by *starting* with a basic triangular element, such as *ABC* in Fig. 5–6, and connecting two members (*AD* and *BD*) to form an additional element. Thus it is seen that as each additional element of two members is placed on the truss, the number of joints for a simple truss is increased by one.

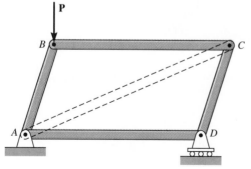

Fig. 5–5

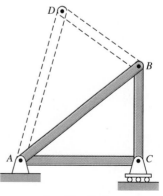

Fig. 5–6

5.2 The Method of Joints

If a truss is in equilibrium, then each of its joints must also be in equilibrium. The method of joints is based on this fact, since it consists of satisfying the equilibrium conditions for the forces exerted *on the pin* at each joint of the truss. Because the truss members are all straight two-force members lying in the same plane, the force system acting at each pin is *coplanar and concurrent*. Consequently, rotational or moment equilibrium is automatically satisfied at the joint (or pin), and it is only necessary to satisfy $\Sigma F_x = 0$ and $\Sigma F_y = 0$ to ensure translational or force equilibrium.

When using the method of joints, it is *first* necessary to draw the joint's free-body diagram before applying the equilibrium equations. To do this, recall that the *line of action* of each member force acting on the joint is *specified* from the geometry of the truss, since the force in a member passes along the axis of the member. As an example, consider the pin at joint B of the truss in Fig. 5–7a. Three forces act on the pin, namely, the 500-N force and the forces exerted by members BA and BC. The free-body diagram is shown in Fig. 5–7b. As shown, $\mathbf{F}_{BA}$ is "pulling" on the pin, which means that member BA is in *tension;* whereas $\mathbf{F}_{BC}$ is "pushing" on the pin, and consequently member BC is in *compression*.

In all cases, the analysis should start at a joint having at least one known force and at most two unknown forces, as in Fig. 5–7b. In this way, application of $\Sigma F_x = 0$ and $\Sigma F_y = 0$ yields two algebraic equations which can be

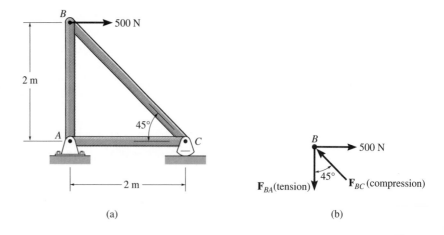

(a)

(b)

Fig. 5–7

solved for the two unknowns. When applying these equations, the correct sense of an unknown member force can be determined using one of two possible methods.

1. *Always assume* the *unknown member forces* acting on the joint's free-body diagram to be in *tension,* i.e., "pulling" on the pin. If this is done, then numerical solution of the equilibrium equations will yield *positive scalars for members in tension and negative scalars for members in compression.* Once an unknown member force is found, use its *correct* magnitude and sense (T or C) on subsequent joint free-body diagrams.

2. The *correct* sense of direction of an unknown member force can, in many cases, be determined "by inspection." For example, $\mathbf{F}_{BC}$ in Fig. 5–7b must push on the pin (compression) since its horizontal component, $F_{BC} \sin 45°$, must balance the 500-N force ($\Sigma F_x = 0$). Likewise, $\mathbf{F}_{BA}$ is a tensile force since it balances the vertical component, $F_{BC} \cos 45°$ ($\Sigma F_y = 0$). In more complicated cases, the sense of an unknown member force can be *assumed;* then, after applying the equilibrium equations, the assumed sense can be verified from the numerical results. A *positive* answer indicates that the sense is *correct,* whereas a *negative* answer indicates that the sense shown on the free-body diagram must be *reversed.* This is the method we will use in the example problems which follow.

PROCEDURE FOR ANALYSIS

The following procedure provides a typical means for analyzing a truss using the method of joints.

  Draw the free-body diagram of a joint having at least one known force and at most two unknown forces. (If this joint is at one of the supports, it generally will be necessary to know the external reactions at the truss support.) Use one of the two methods described above for establishing the sense of an unknown force. Orient the x and y axes such that the forces on the free-body diagram can be easily resolved into their x and y components and then apply the two force equilibrium equations $\Sigma F_x = 0$ and $\Sigma F_y = 0$. Solve for the two unknown member forces and verify their correct sense.

  Continue to analyze each of the other joints, where again it is necessary to choose a joint having at most two unknowns and at least one known force. Realize that once the force in a member is found from the analysis of a joint at one of its ends, the result can be used to analyze the forces acting on the joint at its other end. Strict adherence to the principle of action, equal but opposite reaction must, of course, be observed. Remember, a member in *compression* "pushes" on the joint and a member in *tension* "pulls" on the joint.

  Once the force analysis of the truss has been completed, the size of the members and their connections can be determined using the theory of mechanics of materials along with information given in engineering design codes.

Example 5–1

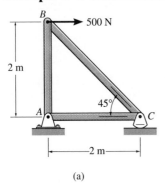

(a)

Determine the force in each member of the truss shown in Fig. 5–8a and indicate whether the members are in tension or compression.

SOLUTION

By inspection of Fig. 5–8a, there are two unknown member forces at joint B, two unknown member forces and an unknown reaction force at joint C, and two unknown member forces and two unknown reaction forces at joint A. Since we must have no more than two unknowns at the joint and at least one known force acting there, we must begin the analysis at joint B.

(b)

Joint B. The free-body diagram of the pin at B is shown in Fig. 5–8b. Three forces act on the pin: the external force of 500 N and the *two* unknown forces developed by members BA and BC. Applying the equations of joint equilibrium, we have

$$\xrightarrow{+} \Sigma F_x = 0; \qquad 500 - F_{BC} \sin 45° = 0 \qquad F_{BC} = 707.1 \text{ N} \quad (C) \quad Ans.$$

$$+ \uparrow \Sigma F_y = 0; \quad F_{BC} \cos 45° - F_{BA} = 0 \qquad F_{BA} = 500 \text{ N} \quad (T) \qquad Ans.$$

Since the force in member BC has been calculated, we can proceed to analyze joint C in order to determine the force in member AC and the support reaction at the rocker.

(c)

Joint C. From the free-body diagram of joint C, Fig. 5–8c, we have

$$\xrightarrow{+} \Sigma F_x = 0; \qquad -F_{AC} + 707.1 \cos 45° = 0 \qquad F_{AC} = 500 \text{ N} \quad (T) \quad Ans.$$

$$+ \uparrow \Sigma F_y = 0; \qquad C_y - 707.1 \sin 45° = 0 \qquad C_y = 500 \text{ N} \qquad Ans.$$

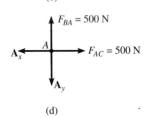

(d)

Joint A. Although not necessary, we can determine the support reactions at joint A using the results of $F_{AC} = 500$ N and $F_{AB} = 500$ N. From the free-body diagram, Fig. 5–8d, we have

$$\xrightarrow{+} \Sigma F_x = 0; \qquad 500 - A_x = 0 \qquad A_x = 500 \text{ N}$$

$$+ \uparrow \Sigma F_y = 0; \qquad 500 - A_y = 0 \qquad A_y = 500 \text{ N}$$

The results of the analysis are summarized in Fig. 5–8e. Note that the free-body diagram of each pin shows the effects of all the connected members and external forces applied to the pin, whereas the free-body diagram of each member shows only the effects of the end pins on the member.

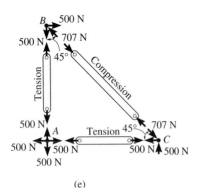

(e)

Fig. 5–8

Example 5–2

Determine the forces acting in all the members of the truss shown in Fig. 5–9a.

SOLUTION

By inspection, there are more than two unknowns at each joint. Consequently, the support reactions on the truss must first be determined. Show that they have been correctly calculated on the free-body diagram in Fig. 5–9b. We can now begin the analysis at joint C. Why?

Joint C. From the free-body diagram, Fig. 5–9c,

$$\xrightarrow{+}\Sigma F_x = 0; \qquad -F_{CD}\cos 30° + F_{CB}\sin 45° = 0$$
$$+\uparrow\Sigma F_y = 0; \qquad 1.5 + F_{CD}\sin 30° - F_{CB}\cos 45° = 0$$

These two equations must be solved *simultaneously* for each of the two unknowns. Note, however, that a *direct solution* for one of the unknown forces may be obtained by applying a force summation along an axis that is *perpendicular* to the direction of the other unknown force. For example, summing forces along the y' axis, which is perpendicular to the direction of F_{CD}, Fig. 5–9d, yields a direct solution for F_{CB}.

$$+\nearrow\Sigma F_{y'} = 0;$$
$$1.5\cos 30° - F_{CB}\sin 15° = 0 \qquad F_{CB} = 5.02 \text{ kN} \quad \text{(C)} \qquad Ans.$$

In a similar fashion, summing forces along the y'' axis, Fig. 5–9e, yields a direct solution for F_{CD}.

$$+\nearrow\Sigma F_{y''} = 0;$$
$$1.5\cos 45° - F_{CD}\sin 15° = 0 \qquad F_{CD} = 4.10 \text{ kN} \quad \text{(T)} \qquad Ans.$$

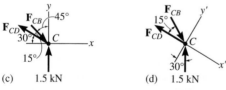

(c)   1.5 kN   (d)   1.5 kN

Joint D. We can now proceed to analyze joint D. The free-body diagram is shown in Fig. 5–9f.

$$\xrightarrow{+}\Sigma F_x = 0; \qquad -F_{DA}\cos 30° + 4.10\cos 30° = 0$$
$$F_{DA} = 4.10 \text{ kN} \quad \text{(T)} \qquad\qquad Ans.$$
$$+\uparrow\Sigma F_y = 0; \qquad F_{DB} - 2(4.10\sin 30°) = 0$$
$$F_{DB} = 4.10 \text{ kN} \text{ (T)} \qquad\qquad Ans.$$

The force in the last member, BA, can be obtained from joint B or joint A. As an exercise, draw the free-body diagram of joint B, sum the forces in the horizontal direction, and show that $F_{BA} = 0.777$ kN (C).

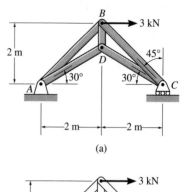

(a)

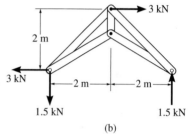

(b)

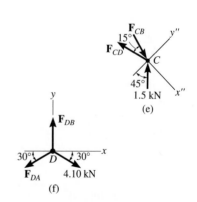

(e)

(f)

Fig. 5–9

Example 5–3

Determine the force in each member of the truss shown in Fig. 5–10a. Indicate whether the members are in tension or compression.

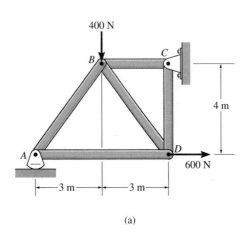

(a)

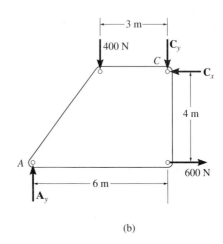

(b)

SOLUTION

Support Reactions. No joint can be analyzed until the support reactions are determined. Why? A free-body diagram of the entire truss is given in Fig. 5–10b. Applying the equations of equilibrium, we have

$$\xrightarrow{+}\Sigma F_x = 0; \qquad\qquad 600 - C_x = 0 \qquad C_x = 600 \text{ N}$$
$$\zeta+\Sigma M_C = 0; \quad -A_y(6) + 400(3) + 600(4) = 0 \qquad A_y = 600 \text{ N}$$
$$+\uparrow \Sigma F_y = 0; \qquad\qquad 600 - 400 - C_y = 0 \qquad C_y = 200 \text{ N}$$

The analysis can now start at either joint A or C. The choice is arbitrary, since there are one known and two unknown member forces acting on the pin at each of these joints.

Joint A (Fig. 5–10c). As shown on the free-body diagram, there are three forces that act on the pin at joint A. The inclination of $\mathbf{F}_{AB}$ is determined from the geometry of the truss. By inspection, can you see why this force is assumed to be compressive and $\mathbf{F}_{AD}$ tensile? Applying the equations of equilibrium, we have

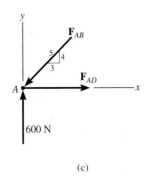

(c)

Fig. 5–10(a–c)

$$+\uparrow \Sigma F_y = 0; \quad 600 - \tfrac{4}{5}F_{AB} = 0 \qquad F_{AB} = 750 \text{ N} \quad (C) \qquad\qquad Ans.$$
$$\xrightarrow{+}\Sigma F_x = 0; \quad F_{AD} - \tfrac{3}{5}(750) = 0 \qquad F_{AD} = 450 \text{ N} \quad (T) \qquad\qquad Ans.$$

Joint D (Fig. 5–10d). The pin at this joint is chosen next since, by inspection of Fig. 5–10a, the force in *AD* is known and the unknown forces in *DB* and *DC* can be determined. Summing forces in the horizontal direction, Fig. 5–10d, we have

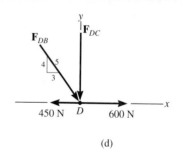

(d)

$$\xrightarrow{+}\Sigma F_x = 0; \quad -450 + \tfrac{3}{5}F_{DB} + 600 = 0 \qquad F_{DB} = -250 \text{ N}$$

The negative sign indicates that $\mathbf{F}_{DB}$ acts in the *opposite sense* to that shown in Fig. 5–10d.* Hence,

$$F_{DB} = 250 \text{ N} \quad \text{(T)} \qquad\qquad\qquad Ans.$$

To determine $\mathbf{F}_{DC}$, we can either correct the sense of $\mathbf{F}_{DB}$ and then apply $\Sigma F_y = 0$, or apply this equation and retain the negative sign for F_{DB}, i.e.,

$$+\uparrow \Sigma F_y = 0; \quad -F_{DC} - \tfrac{4}{5}(-250) = 0 \qquad F_{DC} = 200 \text{ N} \quad \text{(C)} \qquad Ans.$$

Joint C (Fig. 5–10e)

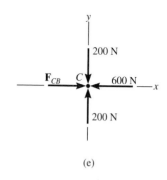

(e)

$$\xrightarrow{+}\Sigma F_x = 0; \qquad F_{CB} - 600 = 0 \qquad F_{CB} = 600 \text{ N} \quad \text{(C)} \qquad Ans.$$
$$+\uparrow \Sigma F_y = 0; \qquad\qquad 200 - 200 \equiv 0 \quad \text{(check)}$$

The analysis is summarized in Fig. 5–10f, which shows the correct free-body diagram for each pin and member.

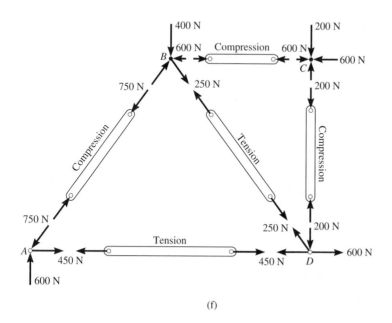

(f)

Fig. 5–10(d–f)

*The proper sense could have been determined by inspection, prior to applying $\Sigma F_x = 0$.

5.3 Zero-Force Members

Truss analysis using the method of joints is greatly simplified if one is able to first determine those members which support *no loading*. These *zero-force members* are used to increase the stability of the truss during construction and to provide support if the applied loading is changed.

The zero-force members of a truss can generally be determined *by inspection* of each of its joints. For example, consider the truss shown in Fig. 5–11*a*. If a free-body diagram of the pin at joint *A* is drawn, Fig. 5–11*b*, it is seen that members *AB* and *AF* are zero-force members. On the other hand, notice that we could not have come to this conclusion if we had considered the free-body diagrams of joints *F* or *B*, simply because there are five unknowns at each of these joints. In a similar manner, consider the free-body diagram of joint *D*, Fig. 5–11*c*. Here again it is seen that *DC* and *DE* are zero-force members. As a general rule, then, *if only two members form a truss joint and no external load or support reaction is applied to the joint, the members must be zero-force members*. The load on the truss in Fig. 5–11*a* is therefore supported by only five members as shown in Fig. 5–11*d*.

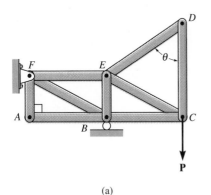

(a)

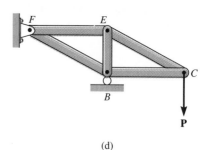

(d)

$$\xrightarrow{+} \Sigma F_x = 0; \; F_{AB} = 0$$
$$+\uparrow \; \Sigma F_y = 0; \; F_{AF} = 0$$

(b)

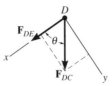

$$+ \searrow \Sigma F_y = 0; \; F_{DC} \sin\theta = 0; \; F_{DC} = 0 \text{ since } \sin\theta \neq 0$$
$$+ \swarrow \Sigma F_x = 0; \; F_{DE} + 0 = 0; \; F_{DE} = 0$$

(c)

Fig. 5–11

Now consider the truss shown in Fig. 5–12a. The free-body diagram of the pin at joint D is shown in Fig. 5–12b. By orienting the y axis along members DC and DE and the x axis along member DA, it is seen that DA is a zero-force member. Note that this is also the case for member CA, Fig. 5–12c. In general, then, *if three members form a truss joint for which two of the members are collinear, the third member is a zero-force member provided no external force or support reaction is applied to the joint*. The truss shown in Fig. 5–12d is therefore suitable for supporting the load **P**.

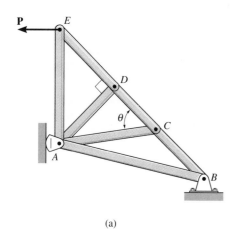

(a)

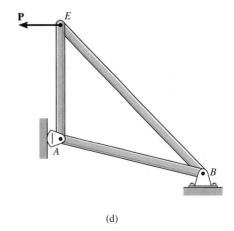

(d)

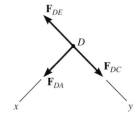

$+\swarrow \Sigma F_x = 0; \quad F_{DA} = 0$

$+\searrow \Sigma F_y = 0; \quad F_{DC} = F_{DE}$

(b)

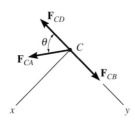

$+\swarrow \Sigma F_x = 0; \quad F_{CA} \sin \theta = 0; \quad F_{CA} = 0 \text{ since } \sin \theta \neq 0;$

$+\searrow \Sigma F_y = 0; \quad F_{CB} = F_{CD}$

(c)

Fig. 5–12

Example 5–4

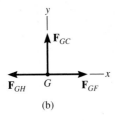

(b)

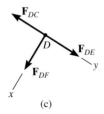

(c)

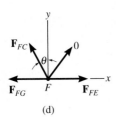

(d)

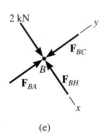

(e)

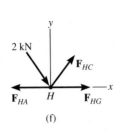

(f)

Using the method of joints, determine all the zero-force members of the *Fink truss* shown in Fig. 5–13a.

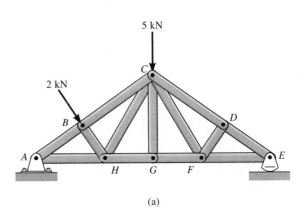

(a)

Fig. 5–13

SOLUTION

Looking for joint geometries that are similar to those outlined in Figs. 5–11 and 5–12, we have

Joint G (Fig. 5–13b)

$$+\uparrow \Sigma F_y = 0; \qquad\qquad F_{GC} = 0 \qquad\qquad\qquad Ans.$$

Realize that we could not conclude that GC is a zero-force member by considering joint C, where there are five unknowns. The fact that GC is a zero-force member means that the 5-kN load at C must be supported by members CB, CH, CF, and CD.

Joint D (Fig. 5–13c)

$$+\swarrow \Sigma F_x = 0; \qquad\qquad F_{DF} = 0 \qquad\qquad\qquad Ans.$$

Joint F (Fig. 5–13d)

$$+\uparrow \Sigma F_y = 0; \quad F_{FC} \cos \theta = 0, \text{ since } \theta \neq 90°; \quad F_{FC} = 0 \qquad Ans.$$

Note that if joint B is analyzed, Fig. 5–13e,

$$+\searrow \Sigma F_x = 0; \qquad 2 - F_{BH} = 0 \qquad F_{BH} = 2 \text{ kN} \quad (C)$$

Consequently, the numerical value of F_{HC} must satisfy $\Sigma F_y = 0$, Fig. 5–13f, and therefore HC is *not* a zero-force member.

PROBLEMS

5–1. Determine the force in each member of the truss and indicate whether the members are in tension or compression.

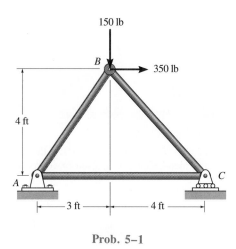

Prob. 5–1

5–2. Determine the force in each member of the truss and indicate whether the members are in tension or compression.

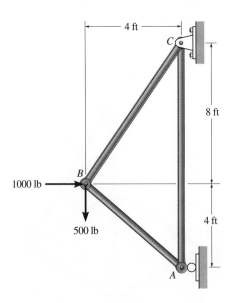

Prob. 5–2

5–3. Determine the force in each member of the truss and indicate whether the members are in tension or compression. Assume that all the members are pin-connected.

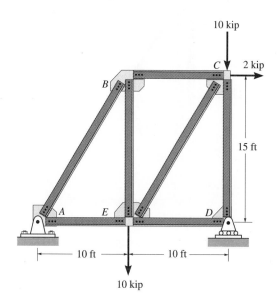

Prob. 5–3

*5–4. Determine the force in each member of the *A*-frame truss which is used to support the bridge-deck loading shown, and indicate whether the members are in tension or compression.

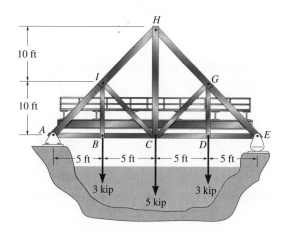

Prob. 5–4

5–5. The sign is partially supported using the side truss shown. The wind load on the face of the sign transmits the loading of 5 kN to joints D and E of the truss. Determine the force in each member and indicate whether the members are in tension or compression.

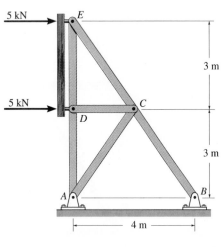

Prob. 5–5

5–6. Determine the force in each member of the truss and indicate whether the members are in tension or compression. Assume that all members are pin-connected.

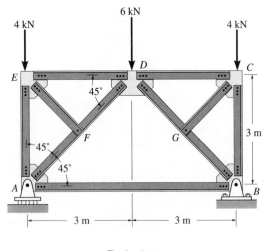

Prob. 5–6

5–7. Determine the force in each member of the truss and indicate whether the members are in tension or compression.

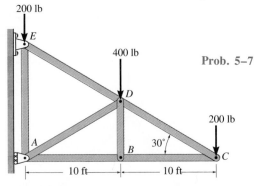

Prob. 5–7

*5–8. Determine the force in each member of the truss and indicate whether the members are in tension or compression.

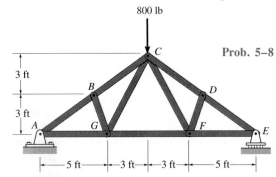

Prob. 5–8

5–9. Determine the force in each member of the truss in terms of the load P, and indicate whether the members are in tension or compression.

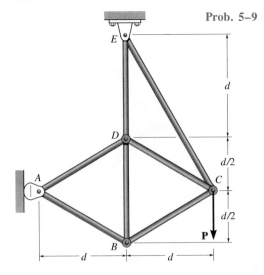

Prob. 5–9

5.4 The Method of Sections

The *method of sections* is used to determine the loadings acting within a body. It is based on the principle that if a body is in equilibrium, then any part of the body is also in equilibrium. To apply this method, one passes an *imaginary section* through the body, thus cutting it into two parts. When a free-body diagram of one of the parts is drawn, the loads acting at the section must be *included* on the free-body diagram. One then applies the equations of equilibrium to the part in order to determine the loading at the section. For example, consider the two truss members shown colored in Fig. 5–14. The internal loads at the section indicated by the dashed line can be obtained using one of the free-body diagrams shown on the right. Clearly, it can be seen that equilibrium requires that the member in tension be subjected to a "pull" **T** at the section, whereas the member in compression is subjected to a "push" **C**.

The method of sections can also be used to "cut" or section several members of an entire truss. If either of the two parts of the truss is isolated as a free-body diagram, we can then apply the equations of equilibrium to that part to determine the member forces at the "cut section." Since only *three* independent equilibrium equations ($\Sigma F_x = 0$, $\Sigma F_y = 0$, $\Sigma M_O = 0$) can be applied to the isolated part of the truss, one should try to select a section that, in general, passes through not more than *three* members in which the forces are unknown. For example, consider the truss in Fig. 5–15a. If the force in member *GC* is to be determined, section *aa* would be appropriate. The free-body diagrams of the two parts are shown in Figs. 5–15b and 5–15c. In particular, note that the line of action of each cut member force is specified from the *geometry* of the truss, since the force in a member passes along its axis. Also, the member forces acting on one part of the truss are equal but opposite to those acting on the other part—Newton's third law. As noted above, members assumed to be in *tension* (*BC* and *GC*) are subjected to a "pull," whereas the member in *compression* (*GF*) is subjected to a "push."

Fig. 5–14

Fig. 5–15

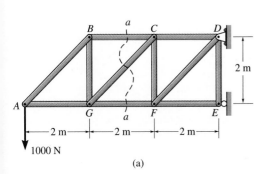

(a)

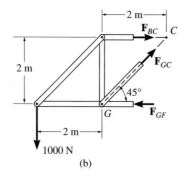

(b)

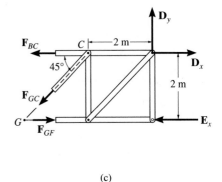

(c)

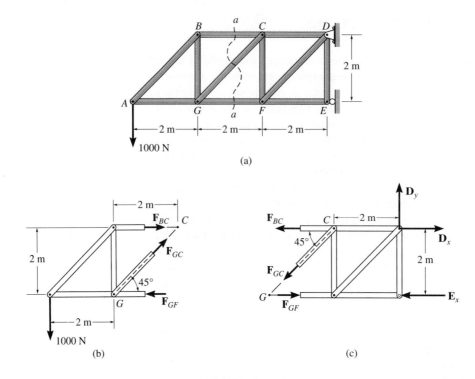

(a)

(b)

(c)

Fig. 5–15 *(Repeated)*

The three unknown member forces $\mathbf{F}_{BC}$, $\mathbf{F}_{GC}$, and $\mathbf{F}_{GF}$ can be obtained by applying the three equilibrium equations to the free-body diagram in Fig. 5–15b. If, however, the free-body diagram in Fig. 5–15c is considered, the three support reactions $\mathbf{D}_x$, $\mathbf{D}_y$, and $\mathbf{E}_x$ will have to be determined *first*. Why? (This, of course, is done in the usual manner by considering a free-body diagram of the *entire truss*.) When applying the equilibrium equations, one should consider ways of writing the equations so as to yield a *direct solution* for each of the unknowns, rather than having to solve simultaneous equations. For example, summing moments about C in Fig. 5–15b would yield a direct solution for $\mathbf{F}_{GF}$ since $\mathbf{F}_{BC}$ and $\mathbf{F}_{GC}$ create zero moment about C. Likewise, $\mathbf{F}_{BC}$ can be directly obtained by summing moments about G. Finally, $\mathbf{F}_{GC}$ can be found directly from a force summation in the vertical direction since $\mathbf{F}_{GF}$ and $\mathbf{F}_{BC}$ have no vertical components. This ability to *determine directly* the force in a particular truss member is one of the main advantages of using the method of sections.*

---

*By comparison, if the method of joints were used to determine, say, the force in member GC, it would be necessary to analyze joints A, B, and G in sequence.

As in the method of joints, there are two ways in which one can determine the correct sense of an unknown member force.

1. *Always assume* that the unknown member forces at the cut section are in *tension,* i.e., "pulling" on the member. By doing this, the numerical solution of the equilibrium equations will yield *positive scalars for members in tension and negative scalars for members in compression.*

2. The correct sense of an unknown member force can in many cases be determined "by inspection." For example, $\mathbf{F}_{BC}$ is a tensile force as represented in Fig. 5–15b, since moment equilibrium about G requires that $\mathbf{F}_{BC}$ create a moment opposite to that of the 1000-N force. Also, $\mathbf{F}_{GC}$ is tensile since its vertical component must balance the 1000-N force acting downward. In more complicated cases, the sense of an unknown member force may be *assumed.* If the solution yields a *negative* scalar, it indicates that the force's sense is *opposite* to that shown on the free-body diagram. This is the method we will use in the example problems which follow.

PROCEDURE FOR ANALYSIS

The following procedure provides a means for applying the method of sections to determine the forces in the members of a truss.

Free-Body Diagram. Make a decision as to how to "cut" or section the truss through the members where forces are to be determined. Before isolating the appropriate section, it may first be necessary to determine the truss's *external* reactions, so that the three equilibrium equations are used *only* to solve for member forces at the cut section. Draw the free-body diagram of that part of the sectioned truss which has the least number of forces acting on it. Use one of the two methods described above for establishing the sense of an unknown member force.

Equations of Equilibrium. Try to apply the three equations of equilibrium such that simultaneous solution of equations is avoided. In this regard, moments should be summed about a point that lies at the intersection of the lines of action of two unknown forces, so that the third unknown force is determined directly from the moment equation. If two of the unknown forces are *parallel,* forces may be summed *perpendicular* to the direction of these unknowns to determine *directly* the third unknown force.

The following examples illustrate these concepts numerically.

Example 5–5

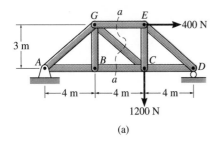

(a)

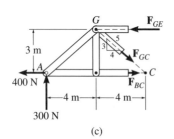

(b)

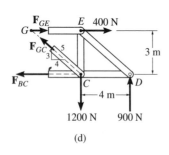

(c)

(d)

Fig. 5–16

Determine the force in members GE, GC, and BC of the truss shown in Fig. 5–16a. Indicate whether the members are in tension or compression.

SOLUTION

Section aa in Fig. 5–16a has been chosen since it cuts through the *three* members whose forces are to be determined. In order to use the method of sections, however, it is *first* necessary to determine the external reactions at A or D. Why? A free-body diagram of the entire truss is shown in Fig. 5–16b. Applying the equations of equilibrium, we have

$$\xrightarrow{+} \Sigma F_x = 0; \qquad 400 \text{ N} - A_x = 0 \qquad A_x = 400 \text{ N}$$
$$\curvearrowright + \Sigma M_A = 0; \quad -1200 \text{ N}(8 \text{ m}) - 400 \text{ N}(3 \text{ m}) + D_y(12 \text{ m}) = 0$$
$$D_y = 900 \text{ N}$$
$$+\uparrow \Sigma F_y = 0; \qquad A_y - 1200 \text{ N} + 900 \text{ N} = 0 \qquad A_y = 300 \text{ N}$$

Free-Body Diagrams. The free-body diagrams of the sectioned truss are shown in Figs. 5–16c and 5–16d. For the analysis the free-body diagram in Fig. 5–16c will be used since it involves the least number of forces.

Equations of Equilibrium. Summing moments about point G eliminates $\mathbf{F}_{GE}$ and $\mathbf{F}_{GC}$ and yields a direct solution for F_{BC}.

$$\curvearrowright + \Sigma M_G = 0; \quad -300 \text{ N}(4 \text{ m}) - 400 \text{ N}(3 \text{ m}) + F_{BC}(3 \text{ m}) = 0$$
$$F_{BC} = 800 \text{ N} \quad \text{(T)} \qquad\qquad Ans.$$

In the same manner, by summing moments about point C we obtain a direct solution for F_{GE}.

$$\curvearrowright + \Sigma M_C = 0; \qquad -300 \text{ N}(8 \text{ m}) + F_{GE}(3 \text{ m}) = 0$$
$$F_{GE} = 800 \text{ N} \quad \text{(C)} \qquad\qquad Ans.$$

Since $\mathbf{F}_{BC}$ and $\mathbf{F}_{GE}$ have no vertical components, summing forces in the y direction directly yields F_{GC}, i.e.,

$$+\uparrow \Sigma F_y = 0; \qquad\qquad 300 \text{ N} - \tfrac{3}{5} F_{GC} = 0$$
$$F_{GC} = 500 \text{ N} \quad \text{(T)} \qquad\qquad Ans.$$

Obtain these results by applying the equations of equilibrium to the free-body diagram shown in Fig. 5–16d.

Example 5–6

Determine the force in member *CF* of the truss shown in Fig. 5–17*a*. Indicate whether the member is in tension or compression.

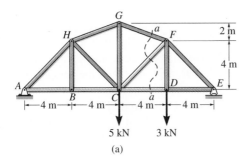

(a)

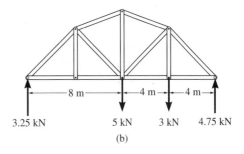

(b)

SOLUTION

Free-Body Diagram.   Section *aa* in Fig. 5–17*a* will be used since this section will "expose" the internal force in member *CF* as "external" on the free-body diagram of either the right or left portion of the truss. It is first necessary, however, to determine the external reactions on either the left or right side of the truss. Verify the results shown on the free-body diagram in Fig. 5–17*b*.

The free-body diagram of the right portion of the truss, which is the easiest to analyze, is shown in Fig. 5–17*c*. There are three unknowns, F_{FG}, F_{CF}, and F_{CD}.

Equations of Equilibrium.   The most direct method for solving this problem requires application of the moment equation about a point that eliminates two of the unknown forces. Hence, to obtain F_{CF}, we will eliminate F_{FG} and F_{CD} by summing moments about point *O*, Fig. 5–17*c*. Note that the location of point *O* measured from *E* is determined from proportional triangles, i.e., $4/(4 + x) = 6/(8 + x)$, $x = 4$ m. Or, stated in another manner, the slope of member *GF* has a drop of 2 m to a horizontal distance of $CD = 4$ m. Since *FD* is 4 m, Fig. 5–17*c*, then from *D* to *O* the distance must be 8 m.

An easy way to determine the moment of F_{CF} about point *O* is to resolve F_{CF} into its two rectangular components and then use the principle of transmissibility to move F_{CF} to point *C*. We have

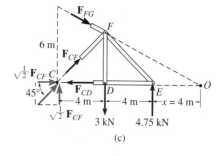

(c)

Fig. 5–17

$$\zeta + \Sigma M_O = 0; \quad -\frac{1}{\sqrt{2}} F_{CF}(12 \text{ m}) + (3 \text{ kN})(8 \text{ m}) - (4.75 \text{ kN})(4 \text{ m}) = 0$$

$$F_{CF} = 0.589 \text{ kN} \quad (C) \qquad\qquad Ans.$$

Example 5–7

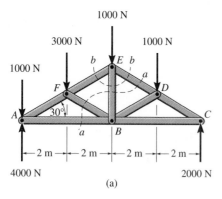

1000 N

3000 N　　　1000 N

1000 N

b　　*E b*

　　　　a

F　　　　　*D*

A　30°

　　a　　*B*　　　　　*C*

├─2 m─┼─2 m─┼─2 m─┼─2 m─┤

4000 N　　　　　　　　2000 N

(a)

Determine the force in member EB of the truss shown in Fig. 5–18a. Indicate whether the member is in tension or compression.

SOLUTION

Free-Body Diagrams. By the method of sections, any imaginary vertical section that cuts through EB, Fig. 5–18a, will also have to cut through three other members for which the forces are unknown. For example, section aa cuts through ED, EB, FB, and AB. If the components of reaction at A are calculated first ($A_x = 0$, $A_y = 4000$ N) and a free-body diagram of the left side of this section is considered, Fig. 5–18b, it is possible to obtain $\mathbf{F}_{ED}$ by summing moments about B to eliminate the other three unknowns; however, $\mathbf{F}_{EB}$ cannot be determined from the remaining two equilibrium equations. One possible way of obtaining $\mathbf{F}_{EB}$ is first to determine $\mathbf{F}_{ED}$ from section aa, then use this result on section bb, Fig. 5–18a, which is shown in Fig. 5–18c. Here the force system is concurrent and our sectioned free-body diagram is the same as the free-body diagram for the pin at E (method of joints).

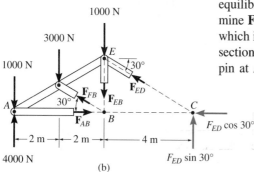

1000 N

3000 N

　　　E

1000 N　　　　　30°

　　　$\mathbf{F}_{FB}$　$\mathbf{F}_{ED}$

A　30°　$\mathbf{F}_{EB}$

　　　$\mathbf{F}_{AB}$　*B*　　　*C*

　　　　　　　　　$F_{ED} \cos 30°$

├─2 m─┼─2 m─┼──4 m──┤

4000 N　　　$F_{ED} \sin 30°$

(b)

Equations of Equilibrium. In order to determine the moment of $\mathbf{F}_{ED}$ about point B, Fig. 5–18b, we will resolve the force into its rectangular components and, by the principle of transmissibility, extend it to point C as shown. The moments of 1000 N, F_{AB}, F_{FB}, F_{EB}, and $F_{ED} \cos 30°$ are all zero about B. Therefore,

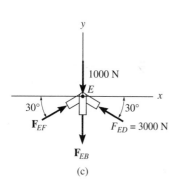

　　　y

1000 N

　　　E

30°　　　　30°　　　*x*

$\mathbf{F}_{EF}$　　$F_{ED} = 3000$ N

　　$\mathbf{F}_{EB}$

(c)

Fig. 5–18

$\zeta + \Sigma M_B = 0;\quad 1000(4) + 3000(2) - 4000(4) + F_{ED} \sin 30°(4) = 0$

$$F_{ED} = 3000 \text{ N}\quad (C)$$

Considering now the free-body diagram of section bb, Fig. 5–18c, we have

$\xrightarrow{+}\Sigma F_x = 0;\qquad F_{EF} \cos 30° - 3000 \cos 30° = 0$

$$F_{EF} = 3000 \text{ N}\quad (C)$$

$+\uparrow \Sigma F_y = 0;\qquad 2(3000 \sin 30°) - 1000 - F_{EB} = 0$

$$F_{EB} = 2000 \text{ N}\quad (T)\qquad\qquad Ans.$$

PROBLEMS

5–10. Determine the force in members GF, FC, and CD of the bridge truss and indicate whether the members are in tension or compression.

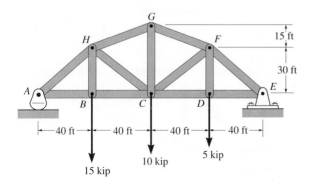

Prob. 5–10

5–11. Determine the force in members HG, GC, and HC of the truss and indicate whether the members are in tension or compression.

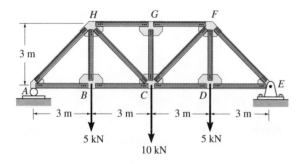

Prob. 5–11

*5–12. Determine the force in members CF and GC of the truss and indicate whether the members are in tension or compression.

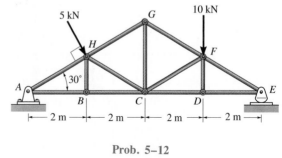

Prob. 5–12

5–13. Determine the force in members BE and EF of the truss and indicate whether the members are in tension or compression. After sectioning the truss, solve for each force *directly*, using a single equation of equilibrium to determine each force.

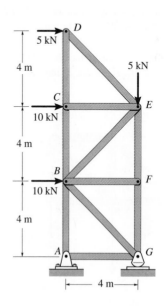

Prob. 5–13

5–14. The *Warren* truss is used to support a staircase. Determine the force in members *CE*, *ED*, and *DF*, and indicate whether the members are in tension or compression. Assume that all joints are pinned.

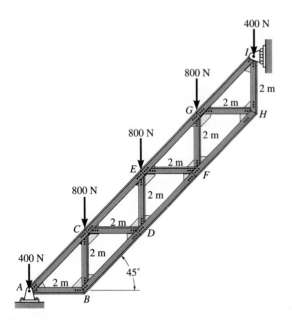

Prob. 5–14

5–15. Determine the force in members *CE*, *FE*, and *CD*, and indicate whether the members are in tension or compression.

*5–16. Determine the force in members *BC*, *EB*, and *AF*, and indicate whether the members are in tension or compression.

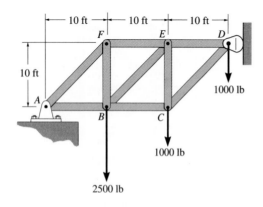

Probs. 5–15/5–16

5–17. Determine the force in members *FE*, *FB*, and *BC* of the truss and indicate whether the members are in tension or compression. After sectioning the truss, solve for each force *directly* using a single equilibrium equation to obtain each force. Assume that all joints are pin-connected.

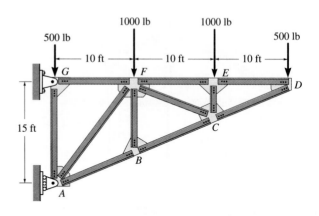

Prob. 5–17

5–18. For the given loading, determine the force in members *CD*, *CJ*, and *KJ* of the *Howe roof truss*. Indicate whether the members are in tension or compression.

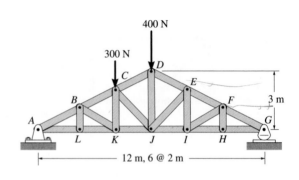

Prob. 5–18

5–19. Indicate all zero-force members, then determine the force in members *IC* and *CG* of the truss and indicate whether the members are in tension or compression.

*5–20. Indicate all zero-force members, then determine the force in members *JE* and *GF* of the truss and indicate whether the members are in tension or compression.

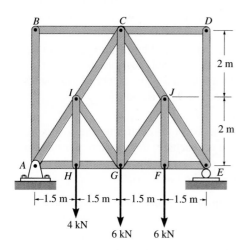

Probs. 5–19/5–20

5–21. Determine the force in members *BC*, *HC*, and *HG*, and indicate whether the members are in tension or compression. After the truss is sectioned, use a single equation of equilibrium to obtain each force.

5–22. Determine the force in members *CD*, *CF*, and *CG*, and indicate whether the members are in tension or compression.

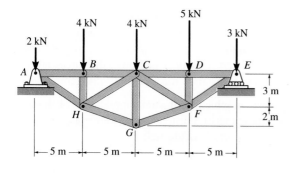

Probs. 5–21/5–22

5–23. Determine the force in members *KJ*, *JN*, and *CD*, and indicate whether the members are in tension or compression.

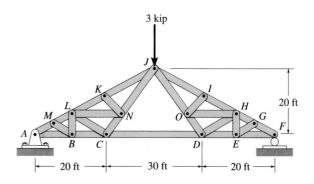

Prob. 5–23

*5–24. Determine the force in members *DE*, *JI*, and *DO* of the *K* truss, and indicate whether the members are in tension or compression. *Hint:* Use sections *aa* and *bb*.

5–25. Determine the force in members *CD* and *KJ* of the *K* truss, and indicate whether the members are in tension or compression. *Hint:* Note section *aa* can be used to find the force in members *DE* and *JI*.

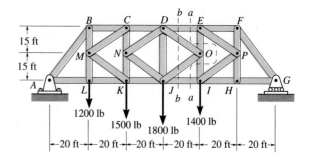

Probs. 5–24/5–25

5.5 Frames and Machines

Frames and machines are two common types of structures which are often composed of pin-connected *multiforce members,* i.e., members that are subjected to more than two forces. *Frames* are generally stationary and are used to support loads, whereas *machines* contain moving parts and are designed to transmit and alter the effect of forces. Provided a frame or machine is properly constrained and contains no more supports or members than are necessary to prevent collapse, the forces acting at the joints and supports can be determined by applying the equations of equilibrium to each member. Once the forces at the joints are obtained, it is then possible to *design* the size of the members, connections, and supports using the theory of mechanics of materials and an appropriate engineering design code.

Free-Body Diagrams. In order to determine the forces acting at the joints and supports of a frame or machine, the structure must be disassembled and the free-body diagrams of its parts must be drawn. In this regard, the following important points *must* be observed:

1. Isolate each part by drawing its *outlined shape*. Then show all the forces and/or couple moments that act on the part. Make sure to *label* or *identify* each known and unknown force and couple moment and indicate any dimensions used for taking moments. Most often the equations of equilibrium are easier to apply if the forces are represented by their rectangular components. As usual, the sense of an unknown force or couple moment can be assumed.
2. Identify all the two-force members in the structure, and represent their free-body diagrams as having two equal but opposite forces acting at their points of application. The line of action of the forces is defined by the line joining the two points where the forces act (see Sec. 4.4). By recognizing the two-force members, we can avoid solving an unnecessary number of equilibrium equations. (See Example 5–13.)
3. Forces common to any two *contacting* members act with equal magnitudes but opposite sense on the respective members. If the two members are treated as a ''*system*'' *of connected members,* then these forces are ''*internal*'' and are *not shown* on the *free-body diagram of the system;* however, if the free-body diagram of *each member* is drawn, the forces are ''*external*'' and *must* be shown on each of the free-body diagrams.

The following examples graphically illustrate application of these points in drawing the free-body diagrams of a dismembered frame or machine. In all cases, the weight of the members is neglected.

Example 5–8

For the frame shown in Fig. 5–19a, draw the free-body diagram of (a) each member, (b) the pin at B, and (c) the two members connected together.

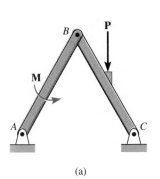

(a)

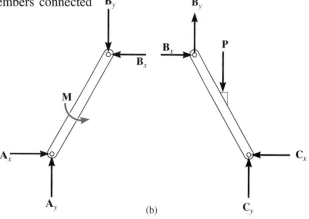

(b)

SOLUTION

Part (a).   By inspection, the members are *not* two-force members. Instead, as shown on the free-body diagrams, Fig. 5–19b, BC is subjected to *not* five but *three forces,* namely, the *resultants* of the two components of reaction at pins B and C and the external force **P.** Likewise, AB is subjected to the *resultant* pin-reactive forces at A and B and the external couple moment **M.**

Part (b).   It can be seen in Fig. 5–19a that the pin at B is subjected to only *two forces,* ii.e., the force of member BC on the pin and the force of member AB on the pin. For *equilibrium* these forces and therefore their respective components must be equal but opposite, Fig. 5–19c. Notice carefully how Newton's third law is applied between the pin and its contacting members, i.e., the effect of the pin on the two members, Fig. 5–19b, and the equal but opposite effect of the two members on the pin, Fig. 5–19c. Also note that B_x and B_y shown equal but opposite in Fig. 5–19b on members AB and BC is *not* the effect of Newton's third law; instead, this results from the *equilibrium* analysis of the pin, Fig. 5–19c.

Part (c).   The free-body diagram of both members connected together, yet removed from the supporting pins at A and C, is shown in Fig. 5–19d. The force components B_x and B_y are *not shown* on this diagram since they form equal but opposite collinear pairs of *internal* forces (Fig. 5–19b) and therefore cancel out.* Also, to be consistent when later applying the equilibrium equations, the unknown force components at A and C must act in the *same sense* as those shown in Fig. 5–19b.

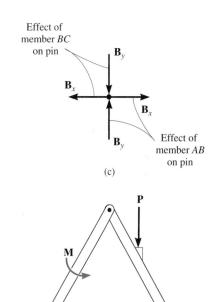

Effect of member BC on pin

Effect of member AB on pin

(c)

(d)

*This is similar to not including internal forces exerted between adjacent particles of a rigid body when drawing the free-body diagram of the entire rigid body.

Fig. 5–19

Example 5–9

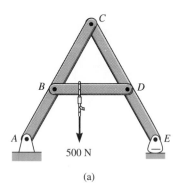

(a)

For the frame shown in Fig. 5–20*a*, draw the free-body diagrams of (a) each of the three members, and (b) members *ABC* and *BD* together.

SOLUTION

Part (a). By inspection, none of the three members of the frame are two-force members. Instead, each is subjected to *three* forces. The components of these forces are shown on the free-body diagrams in Fig. 5–20*b*. Notice that equal but opposite force reactions occur at *B*, *C*, and *D*. Draw a free-body diagram of one of the pins at *B*, *C*, or *D* and show why this is so.

Part (b). The free-body diagram of *ABC* and *BD* together is shown in Fig. 5–20*c*. Since the entire frame is in equilibrium, the force system on these two members also satisfies the equilibrium equations. Why not show the force components B_x and B_y on this diagram?

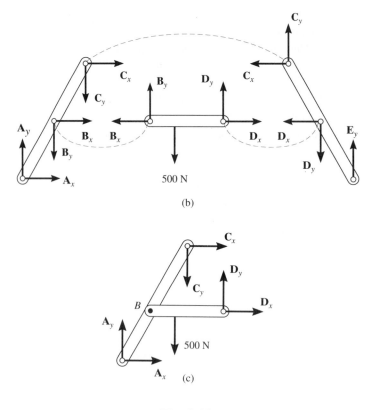

Fig. 5–20

Example 5–10

Draw the free-body diagram of each part of the smooth piston and link mechanism shown in Fig. 5–21a.

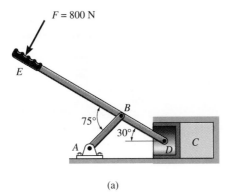

(a)

SOLUTION

By inspection, member *AB* is a two-force member. The free-body diagrams of the parts are shown in Fig. 5–21b. Since the pins at *B* and *D* *connect only two parts together,* the forces there are shown as equal but opposite on the separate free-body diagrams of their connected members. In particular, four components of force act on the piston: $\mathbf{D}_x$ and $\mathbf{D}_y$ represent the effect of the pin (or lever *EBD*), $\mathbf{N}_w$ is the *resultant force* of the cylinder's wall, and $\mathbf{P}$ is the resultant compressive force caused by the material within the cylinder *C*.

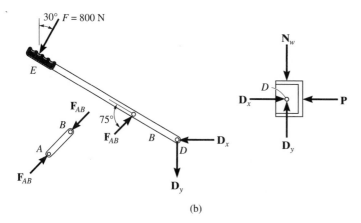

(b)

Fig. 5–21

Example 5–11

For the frame shown in Fig. 5–22*a*, draw the free-body diagrams of (a) the entire frame including the pulleys and cords, (b) the frame without the pulleys and cords, and (c) each of the pulleys.

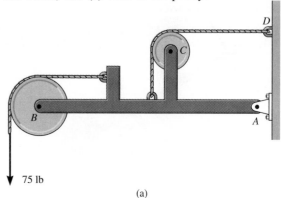

75 lb

(a)

SOLUTION

Part (a). When the entire frame including the pulleys and cords is considered, the interactions at the points where the pulleys and cords are connected to the frame become pairs of *internal forces* which cancel each other and therefore are not shown on the free-body diagram, Fig. 5–22*b*.

Part (b). When the cords and pulleys are removed, their effect *on the frame* must be shown, Fig. 5–22*c*.

Part (c). The force components B_x, B_y, C_x, C_y of the pins on the pulleys, Fig. 5–22*d*, are equal but opposite to the force components exerted by the pins on the frame, Fig. 5–22*c*. Why?

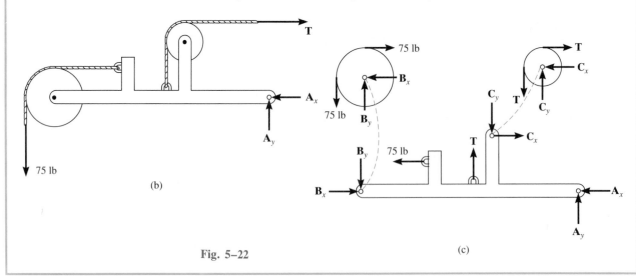

Fig. 5–22

Example 5–12

The hydraulic truck-mounted crane shown in Fig. 5–23a is used to lift a beam that has a mass of 1 Mg. Draw the free-body diagrams of each of its parts, including the pins at A and C.

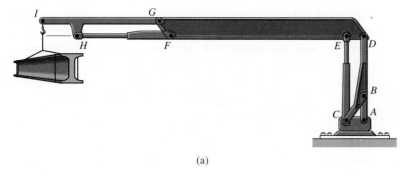

(a)

SOLUTION

By inspection, HF, EC, and AB are all two-force members. The free-body diagrams are shown in Fig. 5–23b. The pin at A is subjected to only *two* forces, namely, the force of the link AB and the force of the support. For equilibrium, these forces must be equal in magnitude but opposite in direction. The pin at C, however, is subjected to *three* forces. The force $\mathbf{F}_{EC}$ is caused by the hydraulic cylinder, the force components $\mathbf{C}_x$ and $\mathbf{C}_y$ are caused by member CBD, and finally, $\mathbf{C}'_x$ and $\mathbf{C}'_y$ are caused by the support. These components can be related by the equations of force equilibrium applied to the pin.

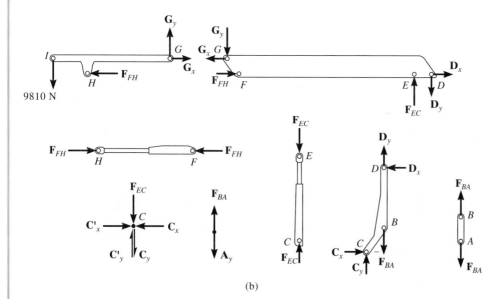

(b)

Fig. 5–23

Before proceeding, it is recommended to cover the solutions to the previous examples and attempt to draw the requested free-body diagrams. When doing so, make sure the work is neat and that all the forces and couple moments are properly labeled.

Equations of Equilibrium. Provided the structure (frame or machine) is properly supported and contains no more supports or members than are necessary to prevent its collapse, then the unknown forces at the supports and connections can be determined from the equations of equilibrium. If the structure lies in the $x-y$ plane, then for *each* free-body diagram drawn the loading must satisfy $\Sigma F_x = 0$, $\Sigma F_y = 0$, and $\Sigma M_O = 0$. The selection of the free-body diagrams used for the analysis is *completely arbitrary*. They may represent each of the members of the structure, a portion of the structure, or its entirety. For example, consider finding the six components of the pin reactions at A, B, and C for the frame shown in Fig. 5–24a. If the frame is dismembered, Fig. 5–24b, these unknowns can be determined by applying the three equations of equilibrium to each of the two members (total of six equations). The free-body diagram of the *entire frame* can also be used for part of the analysis, Fig. 5–24c. Hence, if so desired, all six unknowns can be determined by applying the three equilibrium equations to the entire frame, Fig. 5–24c, and also to either one of its members. Furthermore, the answers can be checked in part by applying the three equations of equilibrium to the remaining "second" member. In general, then, this problem can be solved by writing *at most* six equilibrium equations using free-body diagrams of the members and/or the combination of connected members. Any more than six equations written would *not* be unique from the original six and would serve only to check the results.

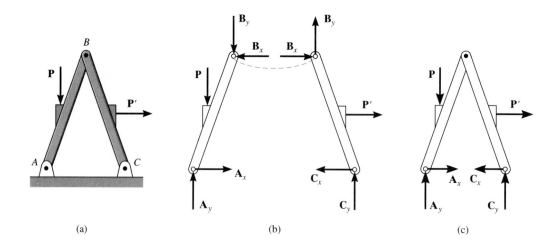

(a)   (b)   (c)

Fig. 5–24

PROCEDURE FOR ANALYSIS

The following procedure provides a method for determining the *joint reactions* of frames and machines (structures) composed of multiforce members.

Free-Body Diagrams.   Draw the free-body diagram of the entire structure, a portion of the structure, or each of its members. The choice should be made so that it leads to the most direct solution to the problem.

Forces common to two members which are in contact act with equal magnitude but opposite sense on the respective free-body diagrams of the members. Recall that all *two-force members,* regardless of their shape, have equal but opposite collinear forces acting at the ends of the member. The unknown forces acting at the joints of multiforce members should be represented by their rectangular components. In many cases it is possible to tell by inspection the proper sense of the unknown forces; however, if this seems difficult, the sense can be assumed.

Equations of Equilibrium.   Count the total number of unknowns to make sure that an equivalent number of equilibrium equations can be written for solution. Recall that in general three equilibrium equations can be written for each rigid body represented in two dimensions. Many times, the solution for the unknowns will be straightforward if moments are summed about a point that lies at the intersection of the lines of action of as many unknown forces as possible. If after obtaining the solution an unknown force magnitude is found to be negative, it means the sense of the force is the reverse of that shown on the free-body diagrams.

The examples that follow illustrate this procedure. All these examples should be *thoroughly understood* before proceeding to solve the problems.

Example 5–13

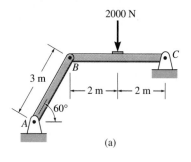

2000 N

(a)

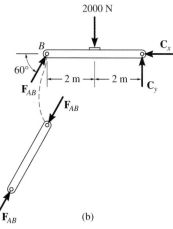

2000 N

(b)

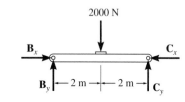

2000 N

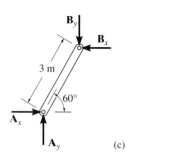

(c)

Fig. 5–25

Determine the horizontal and vertical components of force which the pin at C exerts on member CB of the frame in Fig. 5–25a.

SOLUTION I
Free-Body Diagrams. By inspection it can be seen that AB is a two-force member. The free-body diagrams are shown in Fig. 5–25b.

Equations of Equilibrium. The *three unknowns*, C_x, C_y, and F_{AB}, can be determined by applying the three equations of equilibrium to member CB.

$\zeta+\Sigma M_C = 0;$ $2000(2) - (F_{AB} \sin 60°)(4) = 0$ $F_{AB} = 1154.7$ N

$\xrightarrow{+}\Sigma F_x = 0;$ $1154.7 \cos 60° - C_x = 0$ $C_x = 577$ N   *Ans.*

$+\uparrow \Sigma F_y = 0;$ $1154.7 \sin 60° - 2000 + C_y = 0$ $C_y = 1000$ N   *Ans.*

SOLUTION II
Free-Body Diagrams. If one does not recognize that AB is a two-force member, then more work is involved in solving this problem. The free-body diagrams are shown in Fig. 5–25c.

Equations of Equilibrium. The *six unknowns*, A_x, A_y, B_x, B_y, C_x, C_y, are determined by applying the three equations of equilibrium to each member.
     Member AB

$\zeta+\Sigma M_A = 0;$ $B_x(3 \sin 60°) - B_y(3 \cos 60°) = 0$ (1)

$\xrightarrow{+}\Sigma F_x = 0;$ $A_x - B_x = 0$ (2)

$+\uparrow \Sigma F_y = 0;$ $A_y - B_y = 0$ (3)

     Member BC

$\zeta+\Sigma M_C = 0;$ $2000(2) - B_y(4) = 0$ (4)

$\xrightarrow{+}\Sigma F_x = 0;$ $B_x - C_x = 0$ (5)

$+\uparrow \Sigma F_y = 0;$ $B_y - 2000 + C_y = 0$ (6)

The results for C_x and C_y can be determined by solving these equations in the following sequence: 4, 1, 5, then 6. The results are

$$B_y = 1000 \text{ N}$$
$$B_x = 577 \text{ N}$$
$$C_x = 577 \text{ N} \qquad Ans.$$
$$C_y = 1000 \text{ N} \qquad Ans.$$

By comparison, Solution I is simpler since the requirement that $\mathbf{F}_{AB}$ in Fig. 5–25b be equal, opposite, and collinear at the ends of member AB automatically satisfies Eqs. 1, 2, and 3 above and therefore eliminates the need to write these equations. *As a result, always identify the two-force members before starting the analysis!*

Example 5–14

The beam shown in Fig. 5–26a is pin-connected at B. Determine the reactions at its supports. Neglect its weight and thickness.

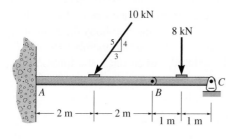

(a)

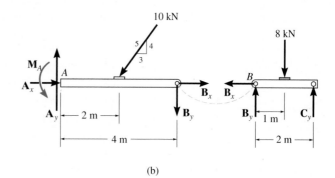

(b)

Fig. 5–26

SOLUTION

Free-Body Diagrams. By inspection, if we consider a free-body diagram of the entire beam ABC, there will be three unknown reactions at A and one at C. These four unknowns cannot all be obtained from the three equations of equilibrium, and so it will become necessary to dismember the beam into its two segments as shown in Fig. 5–26b.

Equations of Equilibrium. The six unknowns are determined as follows:

Segment BC

$$\xrightarrow{+}\Sigma F_x = 0; \qquad\qquad B_x = 0$$

$$\zeta+\Sigma M_B = 0; \qquad -8 \text{ kN}(1 \text{ m}) + C_y(2 \text{ m}) = 0$$

$$+\uparrow \Sigma F_y = 0; \qquad\qquad B_y - 8 \text{ kN} + C_y = 0$$

Segment AB

$$\xrightarrow{+}\Sigma F_x = 0; \qquad A_x - 10 \text{ kN}(\tfrac{3}{5}) + B_x = 0$$

$$\zeta+\Sigma M_A = 0; \quad M_A - 10 \text{ kN}(\tfrac{4}{5})(2 \text{ m}) - B_y(4 \text{ m}) = 0$$

$$+\uparrow \Sigma F_y = 0; \qquad A_y - 10 \text{ kN}(\tfrac{4}{5}) - B_y = 0$$

Solving each of these equations successively, using previously calculated results, we obtain

$$A_x = 6 \text{ kN} \qquad A_y = 12 \text{ kN} \qquad M_A = 32 \text{ kN} \cdot \text{m} \qquad Ans.$$

$$B_x = 0 \qquad\qquad B_y = 4 \text{ kN}$$

$$C_y = 4 \text{ kN} \qquad\qquad\qquad\qquad\qquad Ans.$$

Example 5–15

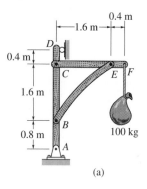

(a)

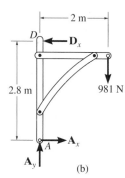

(b)

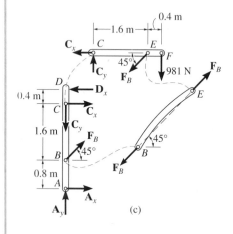

(c)

Fig. 5–27

Determine the horizontal and vertical components of force which the pin at C exerts on member $ABCD$ of the frame shown in Fig. 5–27a.

SOLUTION

Free-Body Diagrams. By inspection, the three components of reaction that the supports exert on $ABCD$ can be determined from a free-body diagram of the entire frame, Fig. 5–27b. Also, the free-body diagram of each frame member is shown in Fig. 5–27c. Notice that member BE is a two-force member. As shown by the colored dashed lines, the forces at B, C, and E have equal magnitudes but opposite directions on the separate free-body diagrams.

Equations of Equilibrium. The six unknowns A_x, A_y, F_B, C_x, C_y, and D_x will be determined from the equations of equilibrium applied to the entire frame and then to member CEF. We have

Entire Frame

$$\curvearrowright + \Sigma M_A = 0; \quad -981 \text{ N}(2 \text{ m}) + D_x(2.8 \text{ m}) = 0 \quad D_x = 700.7 \text{ N}$$

$$\xrightarrow{+} \Sigma F_x = 0; \qquad A_x - 700.7 \text{ N} = 0 \qquad A_x = 700.7 \text{ N}$$

$$+ \uparrow \Sigma F_y = 0; \qquad A_y - 981 \text{ N} = 0 \qquad A_y = 981 \text{ N}$$

Member CEF

$$\curvearrowright + \Sigma M_C = 0; \quad -981 \text{ N}(2 \text{ m}) - (F_B \sin 45°)(1.6 \text{ m}) = 0$$

$$F_B = -1734.2 \text{ N}$$

$$\xrightarrow{+} \Sigma F_x = 0; \qquad -C_x - (-1734.2 \cos 45° \text{ N}) = 0$$

$$C_x = 1230 \text{ N} \qquad \qquad Ans.$$

$$+ \uparrow \Sigma F_y = 0; \quad C_y - (-1734.2 \sin 45° \text{ N}) - 981 \text{ N} = 0$$

$$C_y = -245 \text{ N} \qquad \qquad Ans.$$

Since the magnitudes of forces $\mathbf{F}_B$ and $\mathbf{C}_y$ were calculated as negative quantities, they were assumed to be acting in the wrong sense on the free-body diagrams, Fig. 5–27c. The correct sense of these forces might have been determined "by inspection" *before* applying the equations of equilibrium to member CEF. As shown in Fig. 5–27c, moment equilibrium about point E on member CEF indicates that $\mathbf{C}_y$ must actually act *downward* to counteract the moment created by the 981-N force about point E. Similarly, summing moments about point C, it is seen that the vertical component of force $\mathbf{F}_B$ must actually act *upward*, and so $\mathbf{F}_B$ must act upward to the right.

The above calculations can be checked by applying the three equilibrium equations to member $ABCD$, Fig. 5–27c.

Example 5–16

The smooth disk shown in Fig. 5–28a is pinned at D and has a weight of 20 lb. Neglecting the weights of the other members, determine the horizontal and vertical components of reaction at pins B and D.

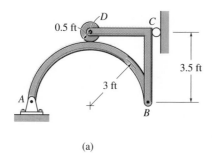

(a)

SOLUTION

Free-Body Diagrams.   By inspection, the three components of reaction at the supports can be determined from a free-body diagram of the entire frame, Fig. 5–28b. Also, free-body diagrams of the members are shown in Fig. 5–28c.

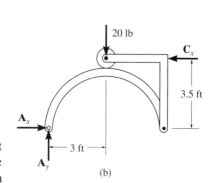

(b)

Equations of Equilibrium.   The eight unknowns can of course be obtained by applying the eight equilibrium equations to each member—three to member AB, three to member BCD, and two to the disk. (Moment equilibrium is automatically satisfied for the disk.) If this is done, however, all the results can be obtained only from a simultaneous solution of some of the equations. (Try it and find out.) To avoid this situation, it is best to first determine the three support reactions on the *entire* frame; then, using these results, the remaining five equilibrium equations can be applied to two other parts in order to solve successively for the other unknowns.

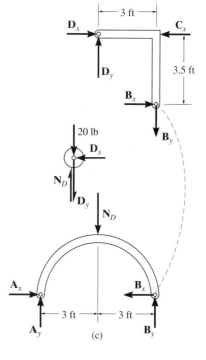

Entire Frame

$\zeta+\Sigma M_A = 0;$ $-20 \text{ lb}(3 \text{ ft}) + C_x(3.5 \text{ ft}) = 0$ $C_x = 17.1 \text{ lb}$

$\xrightarrow{+}\Sigma F_x = 0;$ $A_x - 17.1 \text{ lb} = 0$ $A_x = 17.1 \text{ lb}$

$+\uparrow\Sigma F_y = 0;$ $A_y - 20 \text{ lb} = 0$ $A_y = 20 \text{ lb}$

Member AB

$\xrightarrow{+}\Sigma F_x = 0;$ $17.1 \text{ lb} - B_x = 0$ $B_x = 17.1 \text{ lb}$ *Ans.*

$\zeta+\Sigma M_B = 0;$ $-20 \text{ lb}(6 \text{ ft}) + N_D(3 \text{ ft}) = 0$ $N_D = 40 \text{ lb}$

$+\uparrow\Sigma F_y = 0;$ $20 \text{ lb} - 40 \text{ lb} + B_y = 0$ $B_y = 20 \text{ lb}$ *Ans.*

Disk

$\xrightarrow{+}\Sigma F_x = 0;$ $D_x = 0$ *Ans.*

$+\uparrow\Sigma F_y = 0;$ $40 \text{ lb} - 20 \text{ lb} - D_y = 0$ $D_y = 20 \text{ lb}$ *Ans.*

Fig. 5–28

Example 5–17

Determine the tension in the cables and also the force **P** required to support the 600-N force using the frictionless pulley system shown in Fig. 5–29a.

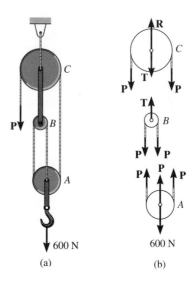

Fig. 5–29

SOLUTION

Free-Body Diagrams.   A free-body diagram of each pulley *including* its pin and a portion of the contacting cable is shown in Fig. 5–29b. Since the cable is *continuous* and the pulleys are frictionless, the cable has a *constant tension P* acting throughout its length (see Example 4–9). The link connection between pulleys *B* and *C* is a two-force member and therefore it has an unknown tension *T* acting on it. Notice that the *principle of action, equal but opposite reaction* must be carefully observed for forces **P** and **T** when the *separate* free-body diagrams are drawn.

Equations of Equilibrium.   The three unknowns are obtained as follows:

Pulley A

$+\uparrow \Sigma F_y = 0;$ $3P - 600 \text{ N} = 0$ $P = 200 \text{ N}$ *Ans.*

Pulley B

$+\uparrow \Sigma F_y = 0;$ $T - 2P = 0$ $T = 400 \text{ N}$ *Ans.*

Pulley C

$+\uparrow \Sigma F_y = 0;$ $R - 2P - T = 0$ $R = 800 \text{ N}$ *Ans.*

Example 5–18

The hand exerts a force of 8 lb on the grip of the spring compressor shown in Fig. 5–30a. Determine the force in the spring needed to maintain equilibrium of the mechanism in the position shown.

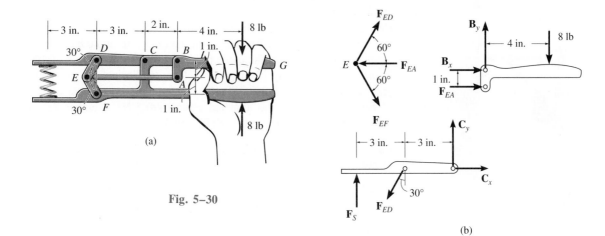

Fig. 5–30

SOLUTION

Free-Body Diagrams. By inspection, members EA, ED, and EF are all two-force members. The free-body diagrams for parts DC and ABC are shown in Fig. 5–30b. Also, the pin at E has been included here since *three* force interactions occur on this pin. They represent the effects of members ED, EA, and EF. Note carefully how equal and opposite force reactions occur between each of the parts.

Equations of Equilibrium. By studying the free-body diagrams, the most direct way to obtain the spring force is to apply the equations of equilibrium in the following sequence:

Lever ABG

$$\zeta + \Sigma M_B = 0; \qquad F_{EA}(1) - 8 \text{ lb}(4 \text{ in.}) = 0 \qquad F_{EA} = 32 \text{ lb}$$

Pin E

$$+ \uparrow \Sigma F_y = 0; \quad F_{ED} \sin 60° - F_{EF} \sin 60° = 0 \quad F_{ED} = F_{EF} = F$$
$$\stackrel{+}{\rightarrow} \Sigma F_x = 0; \qquad 2F \cos 60° - 32 \text{ lb} = 0 \qquad F = 32 \text{ lb}$$

Arm EC

$$\zeta + \Sigma M_C = 0; \qquad -F_s(6 \text{ in.}) + 32 \text{ lb} \cos 30°(3 \text{ in.}) = 0$$
$$F_s = 13.9 \text{ lb} \qquad\qquad\qquad \textit{Ans.}$$

Example 5–19

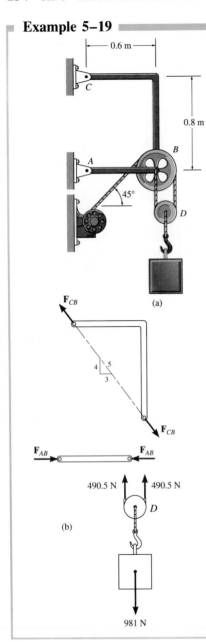

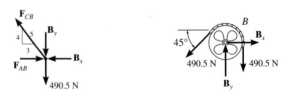

Fig. 5–31

The 100-kg block is held in equilibrium by means of the pulley and continuous cable system shown in Fig. 5–31a. If the cable is attached to the pin at B, compute the forces which this pin exerts on each of its connecting members.

SOLUTION

Free-Body Diagrams.   A free-body diagram of each member of the frame is shown in Fig. 5–31b. By inspection, members AB and CB are two-force members. Furthermore, the cable must be subjected to a force of 490.5 N in order to hold pulley D and the block in equilibrium. A free-body diagram of the pin at B is needed, since *four interactions* occur at this pin. These are caused by the attached cable (490.5 N), member AB ($\mathbf{F}_{AB}$), member CB ($\mathbf{F}_{CB}$), and pulley $B(\mathbf{B}_x$ and $\mathbf{B}_y)$.

Equations of Equilibrium.   Applying the equations of force equilibrium to pulley B, we have

$$\xrightarrow{+}\Sigma F_x = 0; \qquad B_x - 490.5 \cos 45° \text{ N} = 0 \qquad B_x = 346.8 \text{ N} \qquad Ans.$$
$$+\uparrow \Sigma F_y = 0; \qquad B_y - 490.5 \sin 45° \text{ N} - 490.5 \text{ N} = 0$$
$$B_y = 837.3 \text{ N} \qquad Ans.$$

Using these results, equilibrium of the pin requires that

$$+\uparrow \Sigma F_y = 0; \quad \tfrac{4}{5}F_{CB} - 837.3 \text{ N} - 490.5 \text{ N} = 0 \qquad F_{CB} = 1660 \text{ N} \quad Ans.$$
$$\xrightarrow{+}\Sigma F_x = 0; \quad F_{AB} - \tfrac{3}{5}(1660 \text{ N}) - 346.8 \text{ N} = 0 \qquad F_{AB} = 1343 \text{ N} \quad Ans.$$

It may be noted that the two-force member CB is subjected to bending as caused by the force $\mathbf{F}_{CB}$. From the standpoint of design, it would be better to make this member *straight* (from C to B) so that the force $\mathbf{F}_{CB}$ would only create tension in the member.

Before solving the following problems, it is suggested that a brief review be made of all the previous examples. This may be done by covering each solution and trying to locate the two-force members, drawing the free-body diagrams, and conceiving ways of applying the equations of equilibrium to obtain the solution.

PROBLEMS

5–26. Determine the reactive force at pins A and C of the two-member frame.

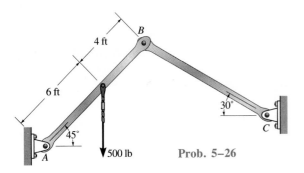

Prob. 5–26

5–27. Determine the horizontal and vertical components of force at pins A and C of the two-member frame.

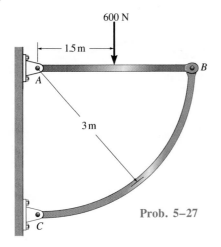

Prob. 5–27

*5–28. Determine the horizontal and vertical components of force acting at A, B, and C of the frame.

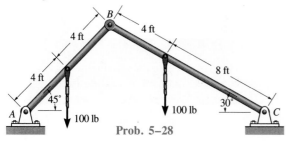

Prob. 5–28

5–29. Determine the force in members AB and AC of the overhang jack which is used to support the forms needed to cast the concrete slab. The forms exert a uniform load of 400 N/m on the top member of the jack assembly. There is a roller support at A and pins at A, B, C, and D.

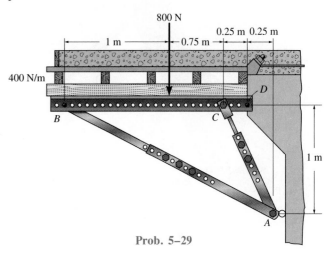

Prob. 5–29

5–30. The two insulators AB and BC are used to support the powerline P. If the weight of the line is 450 N, determine the force in each insulator if they are pin-connected at their ends. Also, what are the components of reaction at the fixed support D?

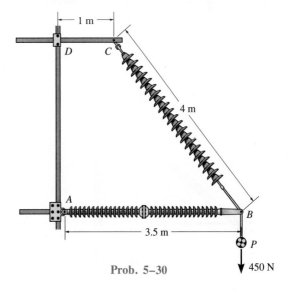

Prob. 5–30

5–31. Determine the force that must be developed in the hydraulic cylinder AB of the metal shear in order to develop a normal force of 4 MN in the grip. The jaws are pin-connected at C. Also, determine the magnitude of the force developed in the pin at C.

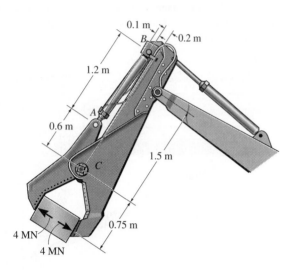

Prob. 5–31

5–34. Determine the clamping force exerted on the smooth pipe at B if a force of 100 N is applied to the handles of the pliers. The pliers are pinned together at A.

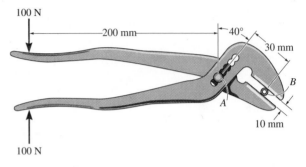

Prob. 5–34

5–35. Determine the compressive force exerted on the specimen by a vertical load of 50 N applied to the toggle press.

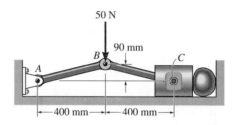

Prob. 5–35

*5–32. Determine the horizontal and vertical components of force which the pins at C and D exert on member ECD of the frame.

5–33. Determine the horizontal and vertical components of force which the pins at B and D exert on member ABD of the frame.

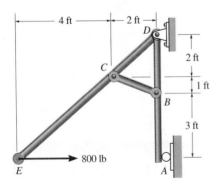

Probs. 5–32/5–33

*5–36. If block A weighs 50 lb and the platform B weighs 30 lb, determine the tension in the cables and the normal force which the block exerts on the platform.

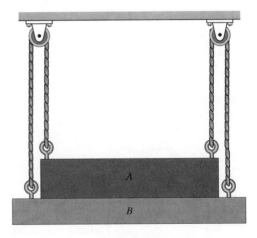

Prob. 5–36

5–37. Determine the tension in member BC of the frame and the horizontal and vertical components of force acting on the pin at D.

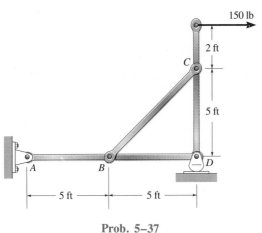

Prob. 5–37

5–38. Determine the reactions at the supports A and B of the compound beam. There is a pin at C.

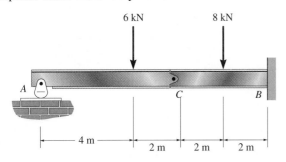

Prob. 5–38

5–39. The bridge frame consists of three segments which can be considered pinned at A, D, and E, rocker-supported at C and F, and roller-supported at B. Determine the horizontal and vertical components of reaction at all these supports due to the loading shown.

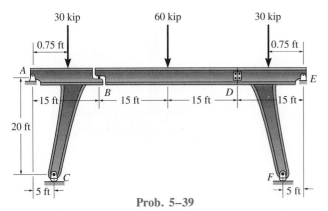

Prob. 5–39

*5–40. The compound beam is fixed at E and supported by a roller at A and B. Determine the reactions at these supports. There are hinges (pins) at C and D.

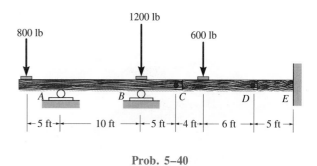

Prob. 5–40

5–41. Each of the clamshell buckets has a weight of 350 lb and a center of gravity at G. Determine the force in link AB and the reaction at the pin C and smooth contact D. Neglect the weights of the other members.

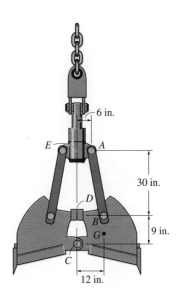

Prob. 5–41

5–42. A 5-lb force is applied to the handles of the vice grip. Determine the compressive force developed on the smooth bolt shank *A* at the jaws.

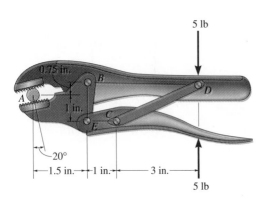

5 lb

0.75 in.

B

D

A

1 in.

C

E

20°

1.5 in. | 1 in. | 3 in.

5 lb

Prob. 5–42

5–43. If a force of 10 lb is applied to the grip of the clamp, determine the compressive force *F* that the wood block exerts on the clamp.

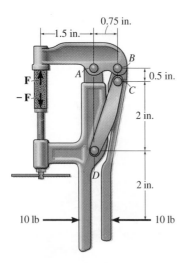

0.75 in.

1.5 in.

B

A

0.5 in.

C

F

2 in.

– F

D

2 in.

10 lb

10 lb

Prob. 5–43

*5–44. A force of 4 lb is applied to the handle of the toggle clamp. Determine the normal force *F* that the bolt exerts on the clamp.

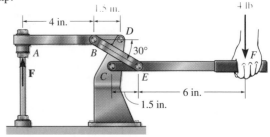

1.5 in.

4 in.

D

4 lb

A

B

30°

F

C

E

F

6 in.

1.5 in.

Prob. 5–44

5–45. Determine the horizontal and vertical components of force acting on the pins *A* and *C* of the frame. The pulley at *D* is frictionless.

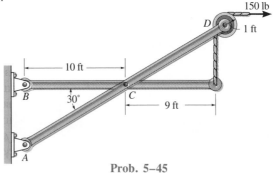

150 lb

D

1 ft

10 ft

B

30°

C

9 ft

A

Prob. 5–45

5–46. Determine the horizontal and vertical components of force which the pins exert on member *ABD*.

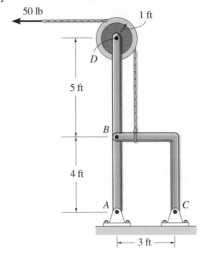

50 lb

1 ft

D

5 ft

B

4 ft

A

C

3 ft

Prob. 5–46

5–47. The shovel of the tractor supports the 400-kg load which has a center of mass at *G*. Determine the horizontal and vertical components of reaction at pin *A* and the force in the hydraulic cylinder *EF*. All labeled points are pin connections. The mechanism is the same on both sides and supports the load equally on both sides.

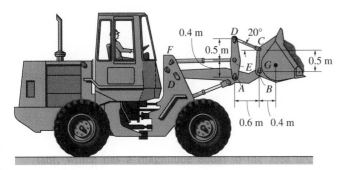

Prob. 5–47

5–49. Determine the horizontal and vertical components of force which the connecting pins at *B*, *E*, and *D* exert on member *BED*.

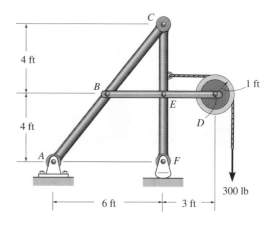

Prob. 5–49

*5–48. Determine the horizontal and vertical components of force at *C* which member *ABC* exerts on member *CEF*.

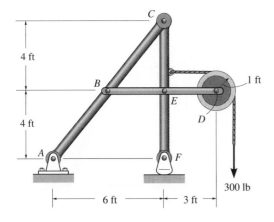

Prob. 5–48

5–50. The clamp has a rated load capacity of 1500 lb. Determine the compressive force this creates in the portion *AB* of the screw and the magnitude of force exerted at the pin *C*. The screw is pin-connected at its end *B* and passes through the pin-connected (swivel) block at *A*.

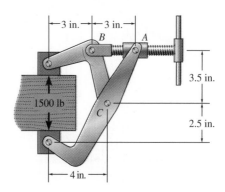

Prob. 5–50

5–51.   The crane is pin-connected at A and fits through the smooth collar at B. Determine the horizontal and vertical components of reaction at these supports and the resultant force that the pin at C exerts on the crane when it supports the crate, which has a mass of 175 kg.

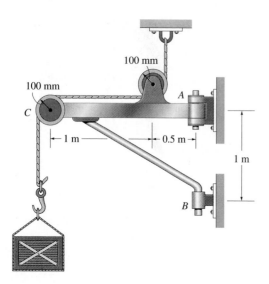

Prob. 5–51

*5–52.   Determine the reactions at pins A and B of the frame needed to support the 200-lb load. The large pulley is pinned at C and has a radius of 0.5 ft. The small pulley at D has a radius of 0.25 ft.

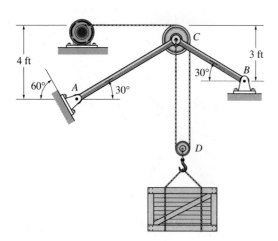

Prob. 5–52

5–53.   Determine the horizontal and vertical components of force which the pins exert on member $ABCD$. Note that the pin at C is attached to member $ABCD$ and passes through the smooth slot in member ECF.

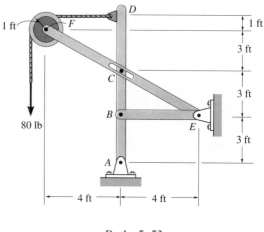

Prob. 5–53

5–54.   Determine the vertical reactions at the smooth ground and the horizontal and vertical components of reaction at the pin B when the man, who weighs 170 lb, stands upright on the ladder in the position shown. The two segments of the ladder are connected together by the cable DE.

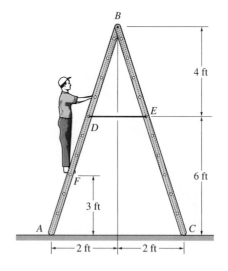

Prob. 5–54

5–55. Determine the tension T in the cord and the angle θ that the pulley-supporting link AB makes with the vertical. Neglect the mass of the pulleys and the link. The block has a mass of 30 kg, and the cord is attached to the pin at B.

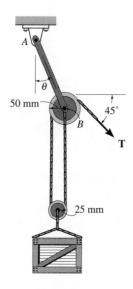

Prob. 5–55

5–57. The piston C moves vertically between the two smooth walls. If the spring has a stiffness of $k = 15$ lb/in. and is unstretched when $\theta = 0°$, determine the couple moment $\mathbf{M}$ that must be applied to AB to hold the mechanism in equilibrium when $\theta = 30°$.

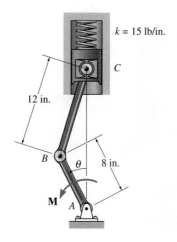

Prob. 5–57 ·

*5–56. If each of the three links of the mechanism has a weight of 25 lb, determine the angle θ for equilibrium. The spring, which always remains vertical, is unstretched when $\theta = 0°$.

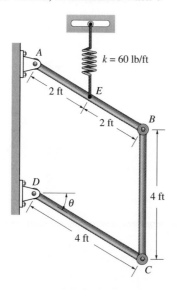

Prob. 5–56

5–58. The weight W is placed on the table such that its center of gravity G is directly over the pin at F. Determine the compressive force in the hydraulic cylinder DE in terms of W and the geometry of the platform. Each member has a length $2L$ and is pinned at its center.

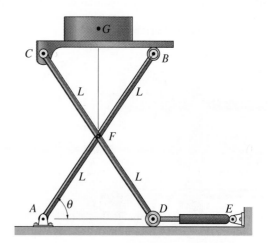

Prob. 5–58

5–59. The lever mechanism for a machine press serves as a toggle which develops a large force at E when a small force is applied at the handle H. To show that this is the case, determine the force at E if someone applies a vertical force of 80 N at H. The smooth head at D is able to slide freely downward. All members are pin-connected.

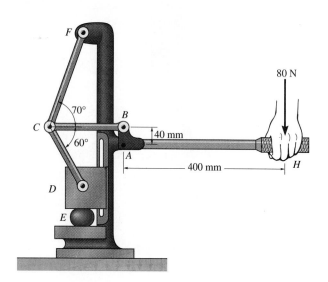

Prob. 5–59

*5–60. The spring mechanism is used as a shock absorber for a load applied to the drawbar AB. Determine the equilibrium length of each spring when the 50-N force is applied. Each spring has an unloaded length of 200 mm, and the drawbar slides along the smooth guide posts CG and EF. The ends of all springs are attached to their respective members.

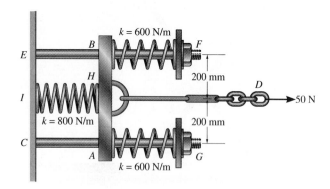

Prob. 5–60

5–61. The tongs consist of two jaws pinned to links at A, B, C, and D. Determine the horizontal and vertical components of force exerted on the 500-lb stone at F and G in order to lift it.

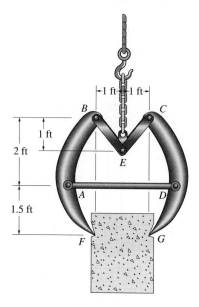

Prob. 5–61

5–62. The man weighs 150 lb and is standing on the uniform plank, which weighs 40 lb. Determine the force he exerts on the plank if he pulls with just enough force to lift the plank off the support at B. The plank is pin-connected at A.

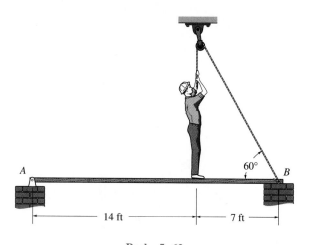

Prob. 5–62

REVIEW PROBLEMS

5–63. Determine the reactions at the supports of the compound beam. There is a short vertical link at C.

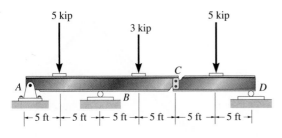

Prob. 5–63

5–65. Determine the force in each member of the truss and indicate whether the members are in tension or compression.

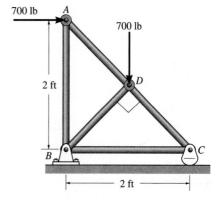

Prob. 5–65

***5–64.** The two-bar mechanism consists of a lever arm AB and smooth link CD, which has a fixed collar at its end C and a roller at the other end D. Determine the force **P** needed to hold the lever in the position θ. The spring has a stiffness k and unstretched length $2L$. The roller contacts either the top or bottom portion of the horizontal guide.

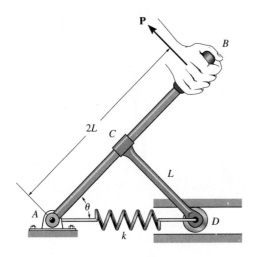

Prob. 5–64

5–66. The *Howe bridge truss* is subjected to the loading shown. Determine the force in members HD, CD, and GD, and indicate whether the members are in tension or compression.

5–67. The *Howe bridge truss* is subjected to the loading shown. Determine the force in members HI, HB, and BC, and indicate whether the members are in tension or compression.

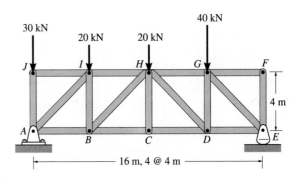

Probs. 5–66/5–67

*5–68. Determine the horizontal and vertical components of force at pins A, B, and C of the two-member frame.

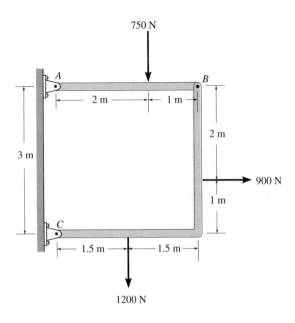

Prob. 5–68

5–69. The compound beam is supported by a rocker at B and fixed to the wall at A. If it is hinged (pinned) together at C, determine the reactions at the supports.

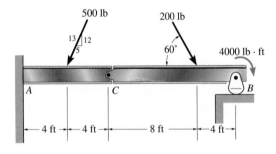

Prob. 5–69

5–70. Determine the horizontal and vertical components of reaction at A and B. The pin at C is fixed to member AE and fits through a smooth slot in member BD.

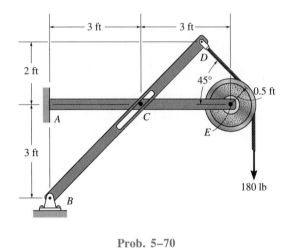

Prob. 5–70

5–71. In each case, determine the force P required to maintain equilibrium.

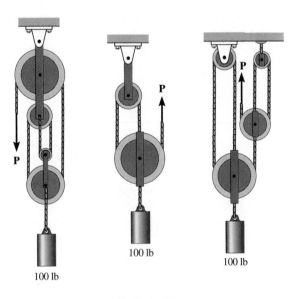

Prob. 5–71

6 Geometric Properties and Distributed Loadings

In this chapter we will discuss the method used to determine the location of the center of gravity of a system of particles, and then we will expand its application to include a body of arbitrary shape. The same method of analysis will also be used to determine the geometric center, or centroid, of areas and volumes. Once the centroid has been located, we will then show how to obtain the resultants of various types of distributed loadings. In the last part of the chapter we will discuss a method for determining the moment of inertia for an area. An important property used in mechanics of materials to analyze or design a structural member or mechanical element.

6.1 Center of Gravity and Center of Mass for a System of Particles

Center of Gravity. Consider the system of n particles fixed within a region of space as shown in Fig. 6–1a. The weights of the particles comprise a system of parallel forces* which can be replaced by a single (equivalent) resultant weight and a defined point of application. This point is called the *center of gravity G*. To find its $\bar{x}$, $\bar{y}$, $\bar{z}$ coordinates, we must use the principles outlined in Sec. 3.8. This requires that the resultant weight be equal to the

---

*This is not true in the exact sense, since the weights are not parallel to each other; rather they are all *concurrent* at the earth's center. Furthermore, the acceleration of gravity g is actually different for each particle, since it depends on the distance from the earth's center to the particle. For all practical purposes, however, both of these effects can be neglected.

total weight of all n particles; that is,

$$W_R = \Sigma W$$

The sum of the moments of the weights of all the particles about the x, y, and z axes is then equal to the moment of the resultant weight about these axes. Thus, to determine the $\bar{x}$ coordinate of G, we can sum moments about the y axis. This yields

$$\bar{x} W_R = \widetilde{x}_1 W_1 + \widetilde{x}_2 W_2 + \cdots + \widetilde{x}_n W_n$$

Likewise, summing moments about the x axis, we can obtain the $\bar{y}$ coordinate; i.e.,

$$\bar{y} W_R = \widetilde{y}_1 W_1 + \widetilde{y}_2 W_2 + \cdots + \widetilde{y}_n W_n$$

Although the weights do not produce a moment about the z axis, we can obtain the z coordinate of G by imagining the coordinate system, with the particles fixed in it, as being rotated 90° about either the x (or y) axis, Fig. 6–1b. Summing moments about the x axis, we have

$$\bar{z} W_R = \widetilde{z}_1 W_1 + \widetilde{z}_2 W_2 + \cdots + \widetilde{z}_n W_n$$

We can generalize these formulas, and write them symbolically in the form

$$\bar{x} = \frac{\Sigma \widetilde{x} W}{\Sigma W} \qquad \bar{y} = \frac{\Sigma \widetilde{y} W}{\Sigma W} \qquad \bar{z} = \frac{\Sigma \widetilde{z} W}{\Sigma W} \qquad (6\text{–}1)$$

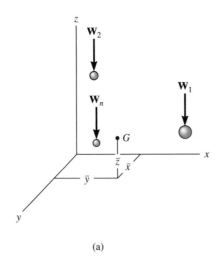

(a)

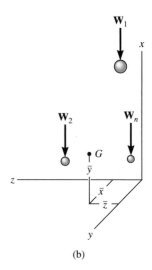

(b)

Fig. 6–1

Here

$\bar{x}, \bar{y}, \bar{z}$ represent the coordinates of the center of gravity G of the system of particles, regardless of the orientation of the x, y, z axes.

$\tilde{x}, \tilde{y}, \tilde{z}$ represent the coordinates of each particle in the system.

$W_R = \Sigma W$ is the total sum of the weights of all the particles in the system.

Formulas having this same form will be presented throughout this chapter to represent other "quantities" for a system. In all cases, however, keep in mind that they simply represent a balance between the sum of the moments of the "quantity" for *each part* of the system and the moment of the *resultant* "quantity" for the system.

6.2 Center of Gravity and Centroid for a Body

Center of Gravity. When the principles used to determine Eqs. 6–1 are applied to a system of particles composing a body, one obtains the same form as these equations except that each particle located at $(\tilde{x}, \tilde{y}, \tilde{z})$ is thought to have a *differential weight dW*. As a result, *integration* is required rather than a discrete summation of the terms. The resulting equations are

$$\bar{x} = \frac{\int \tilde{x}\, dW}{\int dW} \qquad \bar{y} = \frac{\int \tilde{y}\, dW}{\int dW} \qquad \bar{z} = \frac{\int \tilde{z}\, dW}{\int dW} \qquad (6\text{–}2)$$

In order to use these equations properly, the differential weight dW must be expressed in terms of its associated volume dV. If γ represents the *specific weight* of the body, measured as a weight per unit volume, then $dW = \gamma\, dV$ and therefore

$$\bar{x} = \frac{\int_V \tilde{x}\gamma\, dV}{\int_V \gamma\, dV} \qquad \bar{y} = \frac{\int_V \tilde{y}\gamma\, dV}{\int_V \gamma\, dV} \qquad \bar{z} = \frac{\int_V \tilde{z}\gamma\, dV}{\int_V \gamma\, dV} \qquad (6\text{–}3)$$

Here integration must be performed throughout the entire volume of the body.

Centroid. The *centroid* is a point which defines the *geometric center* of an object. Its location can be determined from formulas similar to those used to determine the body's center of gravity or center of mass. In particular, if the material composing a body is uniform or *homogeneous*, the *specific weight* will be *constant* throughout the body, and therefore this term will factor out of the integrals and *cancel* from both the numerators and denominators of Eqs. 6–3. The resulting formulas define the centroid of the body since they are independent of the body's weight and instead depend only on the body's geometry. Three specific cases will be considered.

Volume. If an object is subdivided into volume elements dV, Fig. 6–2, the location of the centroid $C(\bar{x}, \bar{y}, \bar{z})$ for the volume of the object can be determined by computing the "moments" of the elements about the coordinate axes. The resulting formulas are

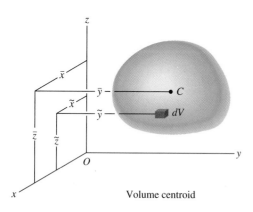

Volume centroid

Fig. 6–2

$$\bar{x} = \frac{\displaystyle\int_V \tilde{x}\, dV}{\displaystyle\int_V dV} \qquad \bar{y} = \frac{\displaystyle\int_V \tilde{y}\, dV}{\displaystyle\int_V dV} \qquad \bar{z} = \frac{\displaystyle\int_V \tilde{z}\, dV}{\displaystyle\int_V dV} \tag{6-4}$$

Area. In a similar manner, the centroid for the surface area of an object, such as a plate or shell, Fig. 6–3, can be found by subdividing the area into differential elements dA and computing the "moments" of these area elements about the coordinate axes, namely,

$$\bar{x} = \frac{\displaystyle\int_A \tilde{x}\, dA}{\displaystyle\int_A dA} \qquad \bar{y} = \frac{\displaystyle\int_A \tilde{y}\, dA}{\displaystyle\int_A dA} \qquad \bar{z} = \frac{\displaystyle\int_A \tilde{z}\, dA}{\displaystyle\int_A dA} \tag{6-5}$$

Notice that in the above cases the location of C does not necessarily have to be within the object; rather, it can be located off the object in space. Also, the *centroids* of some shapes may be partially or completely specified by

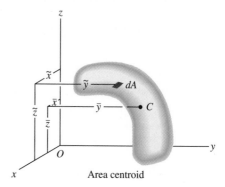

Fig. 6–3

Area centroid

using conditions of *symmetry*. In cases where the shape has an axis of symmetry, the centroid of the shape will lie along that axis. For example, the centroid C for the area shown in Fig. 6–4 must lie along the y axis, since for every elemental area dA at a distance $+\widetilde{x}$ to the right of the y axis, there is an identical element at a distance $-\widetilde{x}$ to the left. The total moment for all the elements about the axis of symmetry will therefore cancel; i.e., $\int \widetilde{x}\, dA = 0$ (Eq. 6–5), so that $\bar{x} = 0$. In cases where a shape has two or three axes of symmetry, it follows that the centroid lies at the intersection of these axes, Fig. 6–5.

It is important to remember that the terms $\widetilde{x}, \widetilde{y}, \widetilde{z}$ in Eqs. 6–2 through 6–5 refer to the "moment arms" or coordinates of the *center of gravity or centroid for the differential element* used in the equations. If possible, this differential element should be chosen such that it has a differential size or thickness in only *one direction*. When this is done, only a single integration is required to cover the entire region.

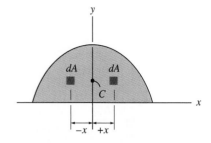

Fig. 6–4

PROCEDURE FOR ANALYSIS

The following procedure provides a method for determining the center of gravity or centroid of an object or shape using a single integration.

Differential Element. Specify the coordinate axes and choose an appropriate differential element for integration. For areas the element dA is generally a rectangle having a finite length and differential width; and for volumes the element dV is either a circular disk having a finite radius and differential thickness, or a shell having a finite length and radius and differential thickness. Locate the element so that it intersects the *boundary* of the shape at an *arbitrary point* (x, y, z).

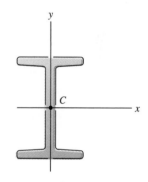

Fig. 6–5

Size and Moment Arms. Express the area dA, or volume dV of the element in terms of the coordinates used to define the boundary of the shape. Determine the coordinates or moment arms $\widetilde{x}, \widetilde{y}, \widetilde{z}$ for the centroid or center of gravity of the element.

Integrations. Substitute the data computed above into the appropriate equations (Eqs. 6–2 through 6–5) and perform the integrations.* Note that integration can be accomplished only when the function in the integrand is expressed in terms of the *same variable as the differential thickness of the element*. The limits of the integral are then defined from the two extreme locations of the element's differential thickness, so that when the elements are "summed" or the integration performed, the entire region is covered.

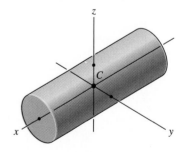

Fig. 6–6

The following examples illustrate this procedure numerically.

*Formulas for integration are given in Appendix A.

Example 6–1

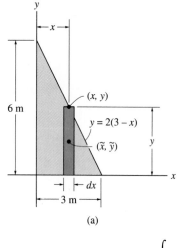

(a)

Locate the centroid of the triangular area shown in Fig. 6–7a.

SOLUTION I

Differential Element. A rectangular differential element of thickness dx is shown in Fig. 6–7a. Since the element intersects the curve at the *arbitrary point* (x, y), it has a height y.

Area and Moment Arms. The area of the element is $dA = y\,dx$, and its centroid is located at $\tilde{x} = x$, $\tilde{y} = \tfrac{1}{2}y$.

Integrations. Applying Eqs. 6–5 and integrating with respect to x yields

$$\bar{x} = \frac{\displaystyle\int_A \tilde{x}\,dA}{\displaystyle\int_A dA} = \frac{\displaystyle\int_0^3 xy\,dx}{\displaystyle\int_0^3 y\,dx} = \frac{\displaystyle\int_0^3 x\,2(3-x)\,dx}{\displaystyle\int_0^3 2(3-x)\,dx} = \frac{\left. \tfrac{6}{2}x^2 - \tfrac{2}{3}x^3 \right|_0^3}{\left. 6x - \tfrac{2}{2}x^2 \right|_0^3} = 1 \text{ m} \quad Ans.$$

$$\bar{y} = \frac{\displaystyle\int_A \tilde{y}\,dA}{\displaystyle\int_A dA} = \frac{\displaystyle\int_0^3 (\tfrac{1}{2}y)y\,dx}{\displaystyle\int_0^3 y\,dx} = \frac{\displaystyle\int_0^3 \tfrac{4}{2}(3-x)^2\,dx}{\displaystyle\int_0^3 2(3-x)\,dx} = \frac{\left. 2(9x - \tfrac{6}{2}x^2 + \tfrac{1}{3}x^3) \right|_0^3}{\left. 6x - \tfrac{2}{2}x^2 \right|_0^3} = 2 \text{ m} \quad Ans.$$

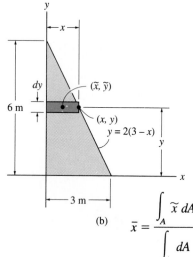

(b)

Fig. 6–7

SOLUTION II

Differential Element. A rectangular differential element of thickness dy is shown in Fig. 6–7b. The element intersects the curve at the *arbitrary point* (x, y), and so it has a length x.

Area and Moment Arms. The area of the element is $dA = x\,dy$, and its centroid is located at $\tilde{x} = \tfrac{1}{2}x$, $\tilde{y} = y$.

Integrations. Applying Eqs. 6–5 and integrating with respect to y yields

$$\bar{x} = \frac{\displaystyle\int_A \tilde{x}\,dA}{\displaystyle\int_A dA} = \frac{\displaystyle\int_0^6 (\tfrac{1}{2}x)x\,dy}{\displaystyle\int_0^6 x\,dy} = \frac{\displaystyle\int_0^6 \tfrac{1}{2}(3-\tfrac{1}{2}y)^2\,dy}{\displaystyle\int_0^6 (3-\tfrac{1}{2}y)\,dy} = \frac{\left. \tfrac{1}{2}(9y - \tfrac{3}{2}y^2 + \tfrac{1}{12}y^3) \right|_0^6}{\left. 3y - \tfrac{1}{4}y^2 \right|_0^6} = 1 \text{ m} \quad Ans.$$

$$\bar{y} = \frac{\displaystyle\int_A \tilde{y}\,dA}{\displaystyle\int_A dA} = \frac{\displaystyle\int_0^6 yx\,dy}{\displaystyle\int_0^6 x\,dy} = \frac{\displaystyle\int_0^6 y(3-\tfrac{1}{2}y)\,dy}{\displaystyle\int_0^6 (3-\tfrac{1}{2}y)\,dy} = \frac{\left. \tfrac{3}{2}y^2 - \tfrac{1}{6}y^3 \right|_0^6}{\left. 3y - \tfrac{1}{4}y^2 \right|_0^6} = 2 \text{ m} \quad Ans.$$

Example 6–2

Locate the centroid of the area shown in Fig. 6–8a.

SOLUTION I

Differential Element.   A differential element of thickness dx is shown in Fig. 6–8a. The element intersects the curve at the *arbitrary point* (x, y), and so it has a height y.

Area and Moment Arms.   The area of the element is $dA = y\,dx$, and its centroid is located at $\tilde{x} = x$, $\tilde{y} = y/2$.

Integrations.   Applying Eqs. 6–5 and integrating with respect to x yields

$$\bar{x} = \frac{\displaystyle\int_A \tilde{x}\,dA}{\displaystyle\int_A dA} = \frac{\displaystyle\int_0^1 xy\,dx}{\displaystyle\int_0^1 y\,dx} = \frac{\displaystyle\int_0^1 x^3\,dx}{\displaystyle\int_0^1 x^2\,dx} = \frac{0.250}{0.333} = 0.75 \text{ m} \qquad Ans.$$

$$\bar{y} = \frac{\displaystyle\int_A \tilde{y}\,dA}{\displaystyle\int_A dA} = \frac{\displaystyle\int_0^1 (y/2)y\,dx}{\displaystyle\int_0^1 y\,dx} = \frac{\displaystyle\int_0^1 (x^2/2)x^2\,dx}{\displaystyle\int_0^1 x^2\,dx} = \frac{0.100}{0.333} = 0.3 \text{ m} \quad Ans.$$

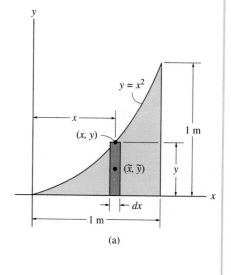

(a)

SOLUTION II

Differential Element.   The differential element of thickness dy is shown in Fig. 6–8b. The element intersects the curve at the *arbitrary point* (x, y), and so it has a length $(1 - x)$.

Area and Moment Arms.   The area of the element is $dA = (1 - x)\,dy$, and its centroid is located at

$$\tilde{x} = x + \left(\frac{1 - x}{2}\right) = \frac{1 + x}{2}, \quad \tilde{y} = y$$

Integrations.   Applying Eqs. 6–5 and integrating with respect to y, we obtain

$$\bar{x} = \frac{\displaystyle\int_A \tilde{x}\,dA}{\displaystyle\int_A dA} = \frac{\displaystyle\int_0^1 [(1 + x)/2](1 - x)\,dy}{\displaystyle\int_0^1 (1 - x)\,dy} = \frac{\dfrac{1}{2}\displaystyle\int_0^1 (1 - y)\,dy}{\displaystyle\int_0^1 (1 - \sqrt{y})\,dy} = \frac{0.250}{0.333} = 0.75 \text{ m} \quad Ans.$$

$$\bar{y} = \frac{\displaystyle\int_A \tilde{y}\,dA}{\displaystyle\int_A dA} = \frac{\displaystyle\int_0^1 y(1 - x)\,dy}{\displaystyle\int_0^1 (1 - x)\,dy} = \frac{\displaystyle\int_0^1 (y - y^{3/2})\,dy}{\displaystyle\int_0^1 (1 - \sqrt{y})\,dy} = \frac{0.100}{0.333} = 0.3 \text{ m} \quad Ans.$$

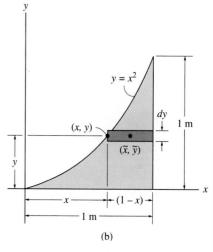

(b)

Fig. 6–8

Example 6–3

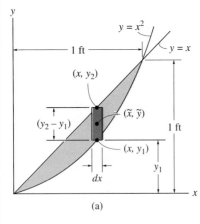

(a)

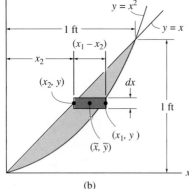

(b)

Fig. 6–9

Locate the $\bar{x}$ centroid of the shaded area bounded by the two curves $y = x$ and $y = x^2$, Fig. 6–9.

SOLUTION I

Differential Element. A differential element of thickness dx is shown in Fig. 6–9a. The element intersects the curves at *arbitrary points* (x, y_1) and (x, y_2), and so it has a height $(y_2 - y_1)$.

Area and Moment Arm. The area of the element is $dA = (y_2 - y_1)\, dx$, and its centroid is located at $\tilde{x} = x$.

Integrations. Applying Eq. 6–5, we have

$$\bar{x} = \frac{\displaystyle\int_A \tilde{x}\, dA}{\displaystyle\int_A dA} = \frac{\displaystyle\int_0^1 x(y_2 - y_1)\, dx}{\displaystyle\int_0^1 (y_2 - y_1)\, dx} = \frac{\displaystyle\int_0^1 x(x - x^2)\, dx}{\displaystyle\int_0^1 (x - x^2)\, dx} = \frac{\frac{1}{12}}{\frac{1}{6}} = 0.5 \text{ ft} \quad \textit{Ans.}$$

SOLUTION II

Differential Element. A differential element having a thickness dy is shown in Fig. 6–9b. The element intersects the curves at the *arbitrary points* (x_2, y) and (x_1, y), and so it has a length $(x_1 - x_2)$.

Area and Moment Arm. The area of the element is $dA = (x_1 - x_2)\, dy$, and its centroid is located at

$$\tilde{x} = x_2 + \frac{x_1 - x_2}{2} = \frac{x_1 + x_2}{2}$$

Integrations. Applying Eq. 6–5, we have

$$\bar{x} = \frac{\displaystyle\int_A \tilde{x}\, dA}{\displaystyle\int_A dA} = \frac{\displaystyle\int_0^1 [(x_1 + x_2)/2](x_1 - x_2)\, dy}{\displaystyle\int_0^1 (x_1 - x_2)\, dy} = \frac{\displaystyle\int_0^1 [(\sqrt{y} + y)/2](\sqrt{y} - y)\, dy}{\displaystyle\int_0^1 (\sqrt{y} - y)\, dy}$$

$$= \frac{\dfrac{1}{2}\displaystyle\int_0^1 (y - y^2)\, dy}{\displaystyle\int_0^1 (\sqrt{y} - y)\, dy} = \frac{\dfrac{1}{12}}{\dfrac{1}{6}} = 0.5 \text{ ft} \qquad\qquad \textit{Ans.}$$

Example 6–4

Locate the $\bar{y}$ centroid for the paraboloid of revolution, which is generated by revolving the shaded area shown in Fig. 6–10a about the y axis.

SOLUTION I

Differential Element. An element having the shape of a *thin disk* is chosen, Fig. 6–10a. This element has a thickness dy. In this "disk" method of analysis, the element of planar area, dA, is always taken *perpendicular* to the axis of revolution. Here the element intersects the generating curve at the *arbitrary point* $(0, y, z)$, and so its radius is $r = z$.

Volume and Moment Arm. The volume of the element is $dV = (\pi z^2) \, dy$, and its centroid is located at $\widetilde{y} = y$.

Integrations. Applying the second of Eqs. 6–4 and integrating with respect to y yields

$$\bar{y} = \frac{\displaystyle\int_V \widetilde{y} \, dV}{\displaystyle\int_V dV} = \frac{\displaystyle\int_0^{100} y(\pi z^2) \, dy}{\displaystyle\int_0^{100} (\pi z^2) \, dy} = \frac{100\pi \displaystyle\int_0^{100} y^2 \, dy}{100\pi \displaystyle\int_0^{100} y \, dy} = 66.7 \text{ mm} \quad \textit{Ans.}$$

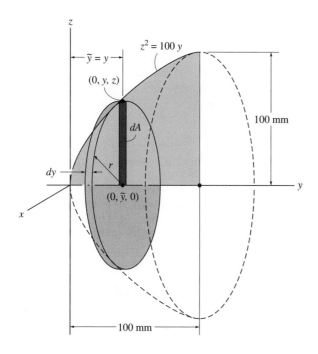

Fig. 6–10(a)

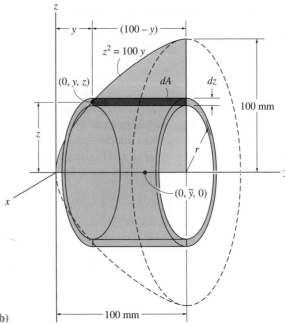

Fig. 6–10(b)

SOLUTION II

Differential Element. As shown in Fig. 6–10*b*, the volume element can be chosen in the form of a *thin cylindrical shell,* where the shell's thickness is *dz*. In this "shell" method of analysis, the element of planar area, *dA*, is always taken *parallel* to the axis of revolution. Here the element intersects the generating curve at point $(0, y, z)$, and so the radius of the shell is $r = z$.

Volume and Moment Arm. The volume of the element is $dV = 2\pi r\, dA = 2\pi z(100 - y)\, dz$, and its centroid is located at $\widetilde{y} = y + (100 - y)/2 = (100 + y)/2$.

Integrations. Applying the second of Eqs. 6–4 and integrating with respect to *z* yields

$$\bar{y} = \frac{\displaystyle\int_V \widetilde{y}\, dV}{\displaystyle\int_V dV} = \frac{\displaystyle\int_0^{100} [(100 + y)/2]2\pi z(100 - y)\, dz}{\displaystyle\int_0^{100} 2\pi z(100 - y)\, dz}$$

$$= \frac{\pi \displaystyle\int_0^{100} z(10^4 - 10^{-4}z^4)\, dz}{2\pi \displaystyle\int_0^{100} z(100 - 10^{-2}z^2)\, dz} = 66.7 \text{ mm} \qquad \textit{Ans.}$$

Example 6–5

Determine the location of the center of gravity of the cylinder shown in Fig. 6–11a if its specific gravity varies directly with its distance from the base, such that $\gamma = 200z$ lb/ft^3.

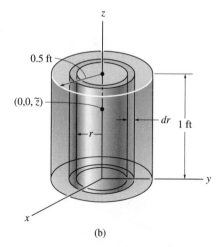

(a)

SOLUTION

For reasons of material symmetry,

$$\bar{x} = \bar{y} = 0 \qquad \textit{Ans.}$$

Differential Element. A disk element of radius 0.5 ft and thickness dz is chosen for integration, Fig. 6–11a, since γ *for the entire element is constant* for a given value of z. The element is located along the z axis at the *arbitrary point* $(0, 0, z)$.

Volume and Moment Arm. The volume of the element is $dV = \pi(0.5)^2\, dz$, and its centroid is located at $\tilde{z} = z$.

Integrations. Using the third of Eqs. 6–3 and integrating with respect to z, noting that $\gamma = 200z$, we have

$$\bar{z} = \frac{\displaystyle\int_V \tilde{z}\gamma\, dV}{\displaystyle\int_V \gamma\, dV} = \frac{\displaystyle\int_0^1 z(200z)\pi(0.5)^2\, dz}{\displaystyle\int_0^1 (200z)\pi(0.5)^2\, dz}$$

$$= \frac{\displaystyle\int_0^1 z^2\, dz}{\displaystyle\int_0^1 z\, dz} = 0.667 \text{ m} \qquad \textit{Ans.}$$

(b)

Fig. 6–11

Note: It is not possible to use a shell element for integration such as shown in Fig. 6–11b, since the specific weight of the material composing the shell would *vary* along the shell's height, and hence the location of $\tilde{z}$ for the element cannot be specified.

PROBLEMS

6–1. Determine the location $(\bar{x}, \bar{y})$ of the centroid of the triangular area.

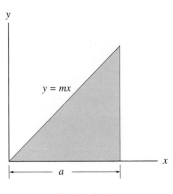

Prob. 6–1

6–2. Determine the location $(\bar{x}, \bar{y})$ of the centroid of the shaded area.

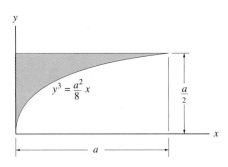

Prob. 6–2

6–3. Determine the location $(\bar{x}, \bar{y})$ of the centroid of the shaded area.

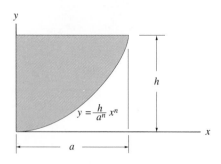

Prob. 6–3

*6–4. Determine the location $(\bar{x}, \bar{y})$ of the centroid of the shaded area.

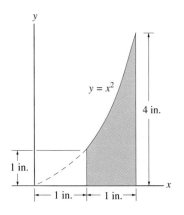

Prob. 6–4

6–5. Determine the location $(\bar{x}, \bar{y})$ of the centroid of the shaded area.

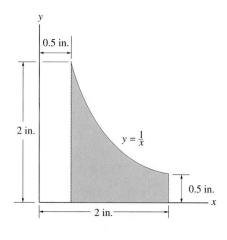

Prob. 6–5

6–6. Determine the location $(\bar{x}, \bar{y})$ of the centroid of the shaded area.

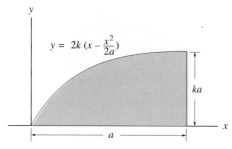

Prob. 6–6

6–7. Determine the location $(\bar{x}, \bar{y})$ of the centroid of the quarter-circular plate. Also compute the force in each of the supporting wires. The plate has a weight of 5 lb/ft^2.

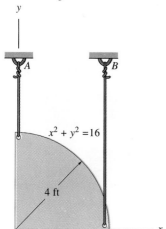

Prob. 6–7

*6–8. Determine the location $(\bar{x}, \bar{y})$ of the centroid of the quarter-elliptical plate.

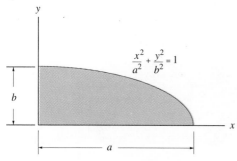

Prob. 6–8

6–9. Determine the location $(\bar{x}, \bar{y})$ of the centroid of the shaded area.

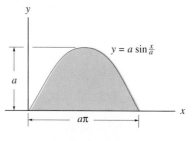

Prob. 6–9

6–10. Determine the location $(\bar{x}, \bar{y})$ of the centroid of the shaded area.

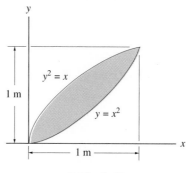

Prob. 6–10

6–11. Determine the location $(\bar{x}, \bar{y})$ of the centroid of the shaded area.

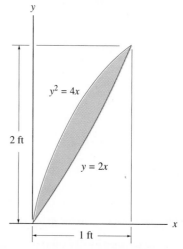

Prob. 6–11

*6–12. Determine the location $(\bar{x}, \bar{y})$ of the centroid of the shaded area. *Hint:* Choose elements of thickness dy and length $[(2 - y) - y^2]$.

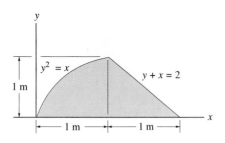

Prob. 6–12

6–13. Determine the location $(\bar{x}, \bar{y})$ of the centroid of the shaded area.

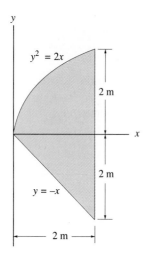

Prob. 6–13

6–14. Determine the distance $\bar{y}$ to the centroid of the cone.

6–15. Determine the distance $\bar{y}$ to the center of mass of the cone. The density of the material varies linearly from zero at the origin O to ρ_0 at $x = h$.

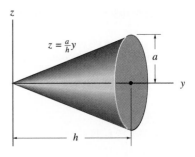

Probs. 6–14/6–15

*6–16. Determine the distance $\bar{y}$ to the centroid of the "bell-shaped" volume.

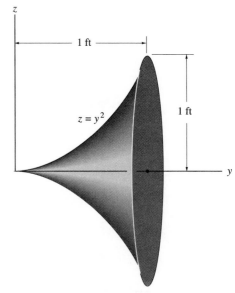

Prob. 6–16

6–17. Determine the distance $\bar{z}$ to the centroid of the volume.

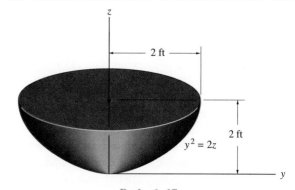

Prob. 6–17

6–18. Determine the distance $\bar{y}$ to the centroid of the paraboloid.

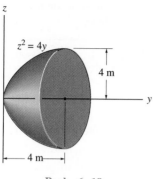

Prob. 6–18

6–19. Determine the distance $\bar{y}$ to the center of gravity of the volume. The material is homogeneous.

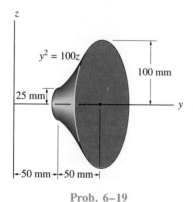

Prob. 6–19

*6–20. Determine the distance $\bar{z}$ to the centroid of the hemisphere.

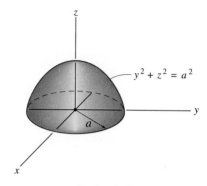

Prob. 6–20

6–21. Determine the distance $\bar{y}$ to the centroid of the semi-ellipsoid.

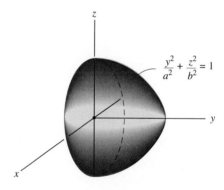

Prob. 6–21

6.3 Composite Bodies

A *composite body* consists of a series of connected "simpler" shaped bodies, which may be rectangular, triangular, semicircular, etc. Such a body can often be sectioned or divided into its composite parts, and provided the *weight* and location of the center of gravity of each of these parts are known, we can eliminate the need for integration to determine the center of gravity for the entire body. The method for doing this requires treating each composite part as a particle and following the procedure outlined in Sec. 6.1. Formulas analogous to Eqs. 6–1 result, since we must account for a finite number of weights. Rewriting these formulas, we have

$$\bar{x} = \frac{\Sigma \tilde{x} W}{\Sigma W} \qquad \bar{y} = \frac{\Sigma \tilde{y} W}{\Sigma W} \qquad \bar{z} = \frac{\Sigma \tilde{z} W}{\Sigma W} \qquad (6\text{--}6)$$

Here

$\bar{x}, \bar{y}, \bar{z}$ represent the coordinates of the center of gravity G of the composite body.

$\tilde{x}, \tilde{y}, \tilde{z}$ represent the coordinates of the center of gravity of each composite part of the body.

ΣW is the sum of the weights of all the composite parts of the body, or simply the total weight of the body.

When the body has a *constant density or specific weight,* the center of gravity *coincides* with the centroid of the body. The centroid for composite areas, and volumes can be found using relations analogous to Eqs. 6–6; however, the W's are replaced by A's, and V's, respectively. Centroids for common shapes of areas, and volumes are given in Appendix B.

PROCEDURE FOR ANALYSIS

The following procedure provides a method for determining the center of gravity of a body or the centroid of a composite geometrical object represented by an area or volume.

Composite Parts. Using a sketch, divide the body or object into a finite number of composite parts. If a composite part has a *hole,* or geometric region having no material, then consider the composite part without the hole, and the hole as an *additional* composite part having *negative* weight or size.

Moment Arms. Establish the coordinate axes on the sketch and determine the coordinates $\tilde{x}, \tilde{y}, \tilde{z}$ of the center of gravity or centroid of each part.

Summations. Determine $\bar{x}, \bar{y}, \bar{z}$ by applying the center of gravity equations, Eqs. 6–6, or the analogous centroid equations. If an object is *symmetrical* about an axis, recall that the centroid of the object lies on this axis.

If desired, the calculations can be arranged in tabular form, as indicated in the following three examples.

Example 6–6

Locate the centroid C of the cross-sectional area for the T-beam shown in Fig. 6–12a.

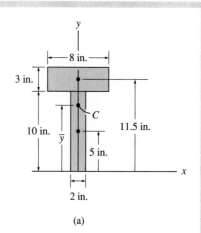

(a)

SOLUTION I

The y axis is placed along the axis of symmetry so that $\bar{x} = 0$, Fig. 6–12a. To obtain $\bar{y}$ we will establish the x axis (reference axis) through the base of the area. The area is segmented into two rectangles as shown, and the centroidal location $\bar{y}$ for each is established. Applying Eq. 6–6, we have

$$\bar{y} = \frac{\Sigma \tilde{y} A}{\Sigma A} = \frac{[5 \text{ in.}](10 \text{ in.})(2 \text{ in.}) + [11.5 \text{ in.}](3 \text{ in.})(8 \text{ in.})}{(10 \text{ in.})(2 \text{ in.}) + (3 \text{ in.})(8 \text{ in.})}$$

$$= 8.55 \text{ in.} \qquad\qquad\qquad Ans.$$

SOLUTION II

Using the same two segments, the x axis can be located at the top of the area as shown in Fig. 6–12b. Here

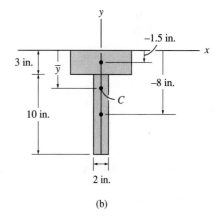

(b)

$$\bar{y} = \frac{\Sigma \tilde{y} A}{\Sigma A} = \frac{[-1.5 \text{ in.}](3 \text{ in.})(8 \text{ in.}) + [-8 \text{ in.}](10 \text{ in.})(2 \text{ in.})}{(3 \text{ in.})(8 \text{ in.}) + (10 \text{ in.})(2 \text{ in.})}$$

$$= -4.45 \text{ in.} \qquad\qquad\qquad Ans.$$

The negative sign indicates that C is located *below* the origin, which is to be expected. Also note that from the two answers 8.55 in. + 4.45 in. = 13.0 in., which is the depth of the beam as expected.

SOLUTION III

It is also possible to consider the cross-sectional area to be one large rectangle *less* two small rectangles, Fig. 6–12c. Hence we have

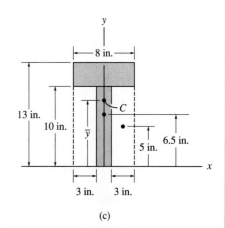

(c)

$$\bar{y} = \frac{\Sigma \tilde{y} A}{\Sigma A} = \frac{[6.5 \text{ in.}](13 \text{ in.})(8 \text{ in.}) - 2[5 \text{ in.}](10 \text{ in.})(3 \text{ in.})}{(13 \text{ in.})(8 \text{ in.}) - 2(10 \text{ in.})(3 \text{ in.})}$$

$$= 8.55 \text{ in.} \qquad\qquad\qquad Ans.$$

Fig. 6–12

Example 6–7

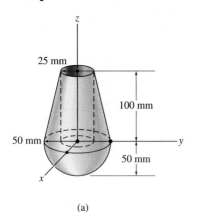

25 mm

100 mm

50 mm

50 mm

(a)

Fig. 6–13

Locate the center of mass of the composite assembly shown in Fig. 6–13a. The conical frustum has a density of $\rho_c = 8$ Mg/m^3, and the hemisphere has a density of $\rho_h = 4$ Mg/m^3.

SOLUTION

Composite Parts. The assembly can be thought of as consisting of four segments as shown in Fig. 6–13b. For the calculations, ③ and ④ must be considered as ''negative'' volumes in order that the four segments, when added together, yield the total composite shape shown in Fig. 6–13a.

Moment Arm. Using Appendix B, the computations for the centroid $\tilde{z}$ of each piece are shown in the figure.

Summations. Because of *symmetry*, note that

$$\bar{x} = \bar{y} = 0 \qquad \textit{Ans.}$$

Since $W = mg$ and g is constant, the third of Eqs. 6–6 becomes $\bar{z} = \Sigma \tilde{z} m / \Sigma m$. The mass of each piece can be computed from $m = \rho V$ and used for the calculations. Also, 1 Mg/m^3 = 10^{-6} kg/mm^3, so that

Segment	m (kg)	$\tilde{z}$ (mm)	$\tilde{z}m$ (kg · mm)
1	$8(10^{-6})(\frac{1}{3})\pi(50)^2(200) = 4.189$	50	209.440
2	$4(10^{-6})(\frac{2}{3})\pi(50)^3 = 1.047$	-18.75	-19.635
3	$-8(10^{-6})(\frac{1}{3})\pi(25)^2(100) = -0.524$	$100 + 25 = 125$	-65.450
4	$-8(10^{-6})\pi(25)^2(100) = -1.571$	50	-78.540
	$\Sigma m = 3.141$		$\Sigma \tilde{z}m = 45.815$

Thus,

$$\bar{z} = \frac{\Sigma \tilde{z}m}{\Sigma m} = \frac{45.815}{3.141} = 14.6 \text{ mm} \qquad \textit{Ans.}$$

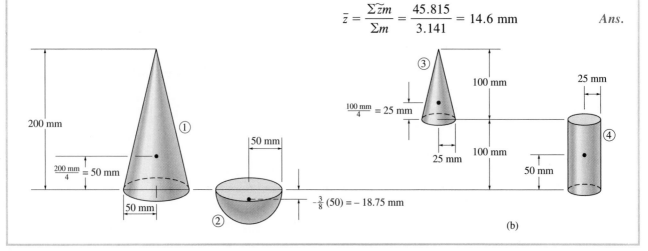

(b)

PROBLEMS

6–22. Determine the distance $\bar{y}$ to the centroid of the area.

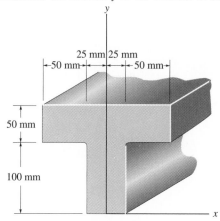

Prob. 6–22

6–23. Determine the distance $\bar{y}$ to the centroid of the beam's cross-sectional area.

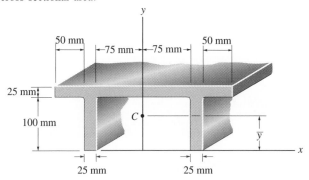

Prob. 6–23

*6–24. Determine the distance $\bar{y}$ to the centroid of the beam's cross-sectional area. Neglect the size of the corner welds for the calculation.

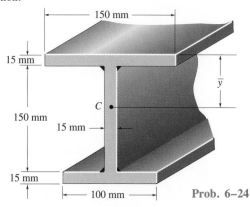

Prob. 6–24

6–25. Determine the location $(\bar{x}, \bar{y})$ of the centroid of the area.

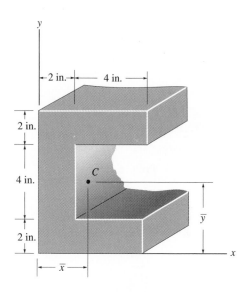

Prob. 6–25

6–26. Determine the location $(\bar{x}, \bar{y})$ of the centroid of the area.

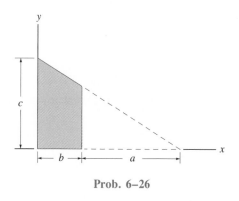

Prob. 6–26

6–27. Determine the location $(\bar{x}, \bar{y})$ of the centroid of the cross-sectional area.

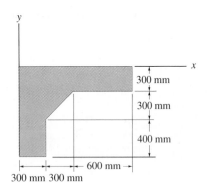

Prob. 6–27

*6–28. Determine the distance $\bar{y}$ to the centroid of the trapezoidal area in terms of the dimensions shown.

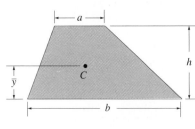

Prob. 6–28

6–29. The gravity wall is made of concrete. Determine the location $(\bar{x}, \bar{y})$ of the center of gravity G for the wall.

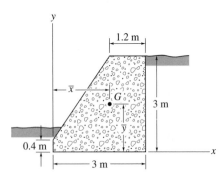

Prob. 6–29

6–30. Determine the distance $\bar{y}$ to the centroid of the shaded area.

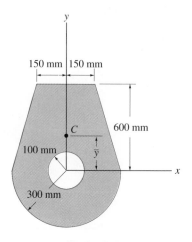

Prob. 6–30

6–31. Determine the location $(\bar{x}, \bar{y})$ of the centroid of the shaded area.

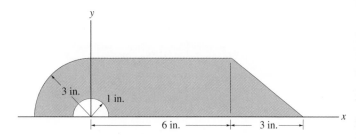

Prob. 6–31

*6–32. Determine the distance $\bar{y}$ to the centroid C for the circular sector having a radius $r = 0.25$ m.

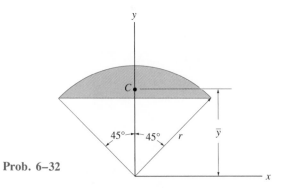

Prob. 6–32

6–33. The girder is to be lifted by a single cable attached to a point directly over the girder's center of gravity G. To locate this point, the girder is supported at end A and end B using two scales. The scale at A reads 5685 lb, and the one at B reads 4864 lb. Determine the distance $\bar{x}$ to G, measured from end A.

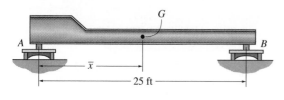

Prob. 6–33

6–34. Determine the distance $\bar{x}$ to the center of gravity of the gas dehydrator assembly. The weight and the center of gravity of each of the various components are indicated below. What are the vertical reactions at A and B needed to support the assembly?

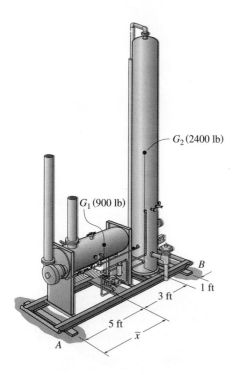

Prob. 6–34

6–35. Determine the distance $\bar{x}$ to the center of gravity of the generator assembly. The weight and the center of gravity of each of the various components are indicated below. What are the vertical reactions at blocks A and B needed to support the assembly?

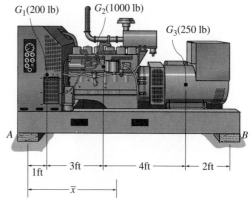

Prob. 6–35

*6–36. The wooden table is made from a square board having a weight of 15 lb. Each of the legs weighs 2 lb and is 3 ft long. Determine how high its center of gravity is from the floor. Also, what is the greatest angle, measured from the horizontal, through which its top surface can be tilted on two of its legs before it begins to overturn? Neglect the thickness of each leg.

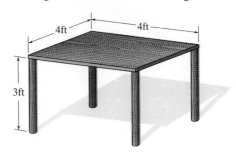

Prob. 6–36

6–37. Determine the distance $\bar{x}$ to the centroid of the volume.

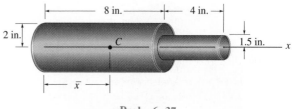

Prob. 6–37

6–38. Determine the distance $\bar{z}$ to the center of the casting that is formed from a hollow cylinder having a density of 8 Mg/m^3 and a hemisphere having a density of 3 Mg/m^3.

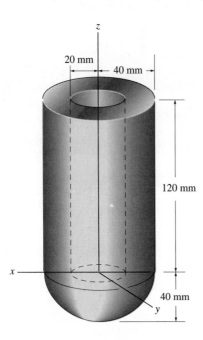

Prob. 6–38

6–39. Determine the dimension h of the block so that the centroid of the assembly lies at the base of the cylinder as shown.

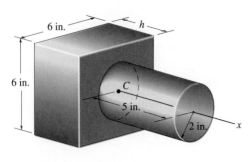

Prob. 6–39

*6–40. Determine the distance $\bar{y}$ to the center of mass of the assembly, which has a hole bored through the center. The material has a density of $\rho = 3 \text{ Mg/m}^3$.

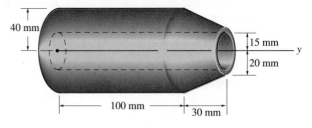

Prob. 6–40

6–41. The solid is formed by boring a conical hole into the cylinder. Determine the distance $\bar{z}$ to the center of gravity.

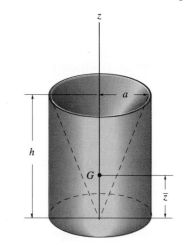

Prob. 6–41

6–42. The solid is formed by boring a conical hole into the hemisphere. Determine the distance $\bar{z}$ to the center of gravity.

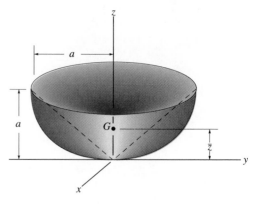

Prob. 6–42

6–43. Determine the distance h to which a 100-mm-diameter hole must be bored into the cone so that the center of mass G of the resulting shape is located at $\bar{z} = 115$ mm. The material has a density of 8 Mg/m³.

*6–44. A hole having a radius r is to be drilled in the center of the homogeneous block. Determine the depth h of the hole so that the center of gravity G is as low as possible.

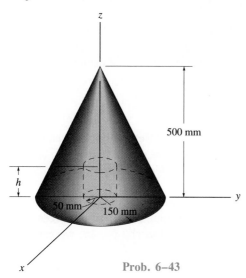

Prob. 6–43

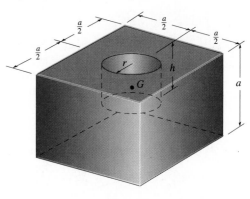

Prob. 6–44

6.4 Resultant of a Distributed Force System

In Chapter 4 we considered ways of simplifying a system of concentrated forces which act on a body. In many practical situations, however, the body may be subjected to loadings distributed over its surface. We have already encountered this situation in Sec. 4.7, while studying frictional and normal forces acting on the bottom of a block resting on a flat surface. Other examples of distributed loadings result from wind and hydrostatic pressure. The effects of these loadings can be studied in a simple manner if we replace them by their resultants. Here we will use the methods of Sec. 3–8 and show how to compute the resultant force of a distributed loading and specify its line of action. Two special cases will be considered.

Pressure Distribution over a Flat Surface. When a body supports a loading over its surface, its distribution is described by a *loading function* which defines the intensity of the load measured as a *force per unit area* or *pressure* acting on the body's surface. For example, the loading distribution or pressure acting on the plate shown in Fig. 6–14a is described by the loading function $p = p(x, y)$ Pa.* Knowing this function, we can determine the *magnitude* of the infinitesimal force $d\mathbf{F}$ acting on the differential area dA m² of the plate, located at the arbitrary point (x, y). This force magnitude is simply $dF = [p(x, y) \ \text{N/m}^2](dA \ \text{m}^2) = [p(x, y) \ dA]$ N. The entire loading

*A pascal, Pa = 1 N/m².

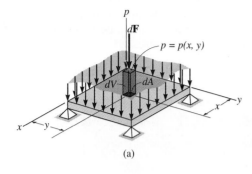

(a)

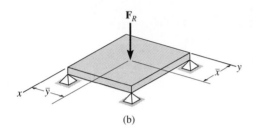

(b)

Fig. 6–14

on the plate is therefore represented as a system of *parallel forces* infinite in number and each acting on separate differential areas dA. This system of parallel forces will now be simplified to a single resultant force $\mathbf{F}_R$ acting through a unique point $(\bar{x}, \bar{y})$ on the plate, Fig. 6–14b.

Magnitude of Resultant Force. To determine the *magnitude* of $\mathbf{F}_R$, it is necessary to sum each of the differential forces $d\mathbf{F}$ acting over the plate's *entire surface area A*. This sum may be expressed mathematically as an integral.

$$F_R = \Sigma F; \qquad\qquad F_R = \int_A p(x, y)\ dA \qquad\qquad (6\text{–}7)$$

Note that $p(x, y)\ dA = dV$, the colored differential *volume element* shown in Fig. 6–14a. Therefore, the result indicates that the *magnitude of the resultant force is equal to the total volume under the distributed-loading diagram.*

Location of Resultant Force. The location $(\bar{x}, \bar{y})$ of $\mathbf{F}_R$ is determined by setting the moments of $\mathbf{F}_R$ equal to the moments of all the forces $d\mathbf{F}$ about the respective y and x axes. From Fig. 6–14a and 6–14b, we get

$$\bar{x} = \frac{\displaystyle\int_A xp(x, y)\ dA}{\displaystyle\int_A p(x, y)\ dA} \qquad \bar{y} = \frac{\displaystyle\int_A yp(x, y)\ dA}{\displaystyle\int_A p(x, y)\ dA} \qquad (6\text{–}8)$$

Setting $dV = p(x, y)\ dA$, we can also write

$$\bar{x} = \frac{\displaystyle\int_V x\ dV}{\displaystyle\int_V dV} \qquad \bar{y} = \frac{\displaystyle\int_V y\ dV}{\displaystyle\int_V dV} \qquad (6\text{–}9)$$

Hence, it can be seen that the *line of action of the resultant force passes through the geometric center or centroid of the volume under the distributed loading diagram.*

Linear Distribution of Load Along a Straight Line. In many cases a pressure distribution acting on a flat surface will be symmetric about an axis of the surface upon which it acts. An example of such a loading is shown in Fig. 6–15a. Here the loading function $p = p(x)$ Pa is only a function of x since the pressure is uniform along the y axis. If we multiply $p = p(x)$ by the width of a m of the plate, we obtain $w = [p(x)\ \text{N/m}^2](a\ \text{m}) = w(x)$ N/m. This loading function, shown in Fig. 6–15b, is a measure of load distribution along the line $y = 0$, which is in the plane of symmetry of the loading, Fig. 6–15c. As noted, it is measured as a force per unit length, rather than a force per unit area. Consequently, the load intensity diagram for $w = w(x)$ represents a system of *coplanar* parallel forces, shown in two dimensions in Fig. 6–15b. Each infinitesimal force $d\mathbf{F}$ acts over an element of length dx such that at any point x, $dF = (w(x)\ \text{N/m})(dx\ \text{m}) = w(x)\ dx$ N.

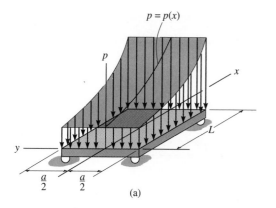

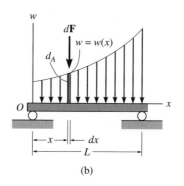

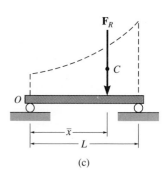

Fig. 6–15

Using the same method as above, we will now simplify the loading to a single resultant force and specify its location $\bar{x}$.

Magnitude of Resultant Force. The magnitude of $\mathbf{F}_R$ is determined from the sum

$$F_R = \int_L w(x)\, dx \qquad (6\text{–}10)$$

Note that $w(x)\, dx = dA$, where dA is the colored differential area shown in Fig. 6–15b. Thus we can state *the magnitude of the resultant force is equal to the total area under the distributed-loading diagram.*

Location of the Resultant Force. The location $\bar{x}$ of the line of action of $\mathbf{F}_R$ is determined by equating the moments of the force resultant and the force distribution about point O, Fig. 6–15b. This yields

$$\bar{x} = \frac{\displaystyle\int_L xw(x)\, dx}{\displaystyle\int_L w(x)\, dx} \qquad (6\text{–}11)$$

Setting $dA = w(x)\, dx$, we can also write

$$\bar{x} = \frac{\displaystyle\int_A x\, dA}{\displaystyle\int_A dA} \qquad (6\text{–}12)$$

This equation locates the $\bar{x}$ coordinate for the geometric center or centroid for the area. Hence, *the line of action of the resultant force passes through the geometric center or centroid of the area under the distributed-loading diagram.* Once $\bar{x}$ is determined, by symmetry, $\mathbf{F}_R$ passes through point $(\bar{x}, a/2)$ on the surface of the plate, Fig. 6–15c.

Example 6–8

In each case, determine the magnitude and location of the resultant of the distributed load acting on the beams in Fig. 6–16.

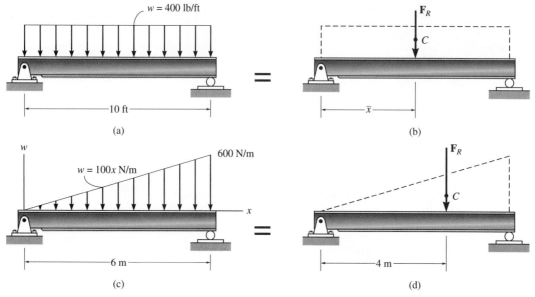

Fig. 6–16

SOLUTION

Uniform Loading.   As indicated $w = 400$ lb/ft, which is constant over the entire beam, Fig. 6–16a. This loading forms a rectangle, the area of which is equal to the resultant force, Fig. 6–16b; i.e.,

$$F_R = (400 \text{ lb/ft})(10 \text{ ft}) = 4000 \text{ lb} \qquad \qquad \textit{Ans.}$$

The location of $\mathbf{F}_R$ passes through the geometric center or centroid C of this rectangular area, so

$$\bar{x} = 5 \text{ ft} \qquad \qquad \textit{Ans.}$$

Triangular Loading.   Here the loading varies uniformly in intensity from 0 to 600 N/m, Fig. 6–16c. These values can be verified by substitution of $x = 0$ and $x = 6$ m into the loading function $w = 100x$ N/m. The area of this triangular loading is equal to $\mathbf{F}_R$, Fig. 6–16d. From Appendix B, $A = \frac{1}{2}bh$, so that

$$F_R = \frac{1}{2}(6 \text{ m})(600 \text{ N/m}) = 1800 \text{ N} \qquad \qquad \textit{Ans.}$$

The line of action of $\mathbf{F}_R$ passes through the centroid C of the triangle. Using Appendix B, this point lies at a distance of one third the length of the beam, measured from the right side. Hence,

$$\bar{x} = 6 \text{ m} - \frac{1}{3}(6 \text{ m}) = 4 \text{ m} \qquad \qquad \textit{Ans.}$$

Example 6–9

Determine the magnitude and location of the resultant of the distributed load acting on the beam shown in Fig. 6–17a.

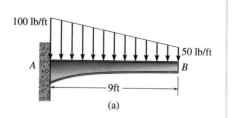

(a)

SOLUTION

The area of the loading diagram is a *trapezoid*, and therefore the solution can be obtained directly from the area and centroid formulas for a trapezoid listed in Appendix B. Since these formulas are not easily remembered, instead we will solve this problem by using "composite" areas. In this regard, we can divide the trapezoidal loading into a rectangular and triangular loading as shown in Fig. 6–17b. The magnitude of the force represented by each of these loadings is equal to its associated *area*,

$$F_1 = \tfrac{1}{2}(9 \text{ ft})(50 \text{ lb/ft}) = 225 \text{ lb}$$
$$F_2 = (9 \text{ ft})(50 \text{ lb/ft}) = 450 \text{ lb}$$

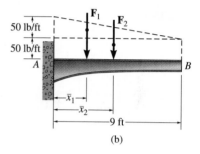

(b)

The lines of action of these parallel forces act through the *centroid* of their associated areas and therefore intersect the beam at

$$\bar{x}_1 = \tfrac{1}{3}(9 \text{ ft}) = 3 \text{ ft}$$
$$\bar{x}_2 = \tfrac{1}{2}(9 \text{ ft}) = 4.5 \text{ ft}$$

The two parallel forces F_1 and F_2 can be reduced to a single resultant F_R. The magnitude of F_R is

$$+\downarrow F_R = \Sigma F; \qquad F_R = 225 + 450 = 675 \text{ lb} \qquad Ans.$$

With reference to point A, Fig. 6–17b and c, we can define the location of F_R. We require

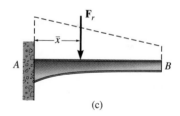

(c)

$$\uparrow + M_{R_A} = \Sigma M_A; \qquad \bar{x}(675) = 3(225) + 4.5(450)$$
$$\bar{x} = 4 \text{ ft} \qquad Ans.$$

Note: The trapezoidal area in Fig. 6–17a can also be divided into two triangular areas as shown in Fig. 6–17d. In this case

$$F_1 = \tfrac{1}{2}(9 \text{ ft})(100 \text{ lb/ft}) = 450 \text{ lb}$$
$$F_2 = \tfrac{1}{2}(9 \text{ ft})(50 \text{ lb/ft}) = 225 \text{ lb}$$

and

$$\bar{x}_1 = \tfrac{1}{3}(9 \text{ ft}) = 3 \text{ ft}$$
$$\bar{x}_2 = \tfrac{2}{3}(9 \text{ ft}) = 3 \text{ ft}$$

Using these results, show that again $F_R = 675 \text{ lb}$ and $\bar{x} = 4 \text{ ft}$.

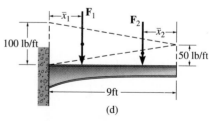

(d)

Fig. 6–17

Example 6–10

Determine the magnitude and location of the resultant force acting on the beam in Fig. 6–18a.

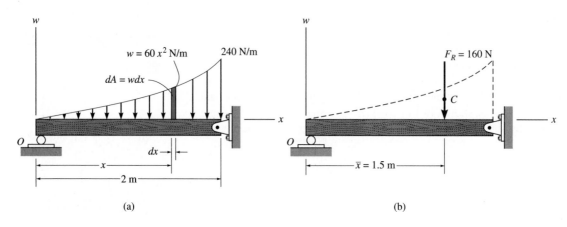

(a)   (b)

Fig. 6–18

SOLUTION

Since $w = w(x)$ is given, this problem will be solved by integration. The colored differential area element $dA = w\,dx = 60x^2\,dx$. Applying Eq. 6–10, by summing these elements from $x = 0$ to $x = 2$ m, we obtain the resultant force $\mathbf{F}_R$,

$$F_R = \Sigma F;$$

$$F_R = \int_L w(x)\,dx = \int_0^2 60x^2\,dx = 60\left[\frac{x^3}{3}\right]_0^2 = 60\left[\frac{2^3}{3} - \frac{0^3}{3}\right]$$

$$= 160\text{ N} \hspace{3cm} Ans.$$

Since the element of area dA is located an arbitrary distance x from O, the location $\bar{x}$ of $\mathbf{F}_R$ *measured from O,* Fig. 6–18b, is determined from Eq. 6–11.

$$\bar{x} = \frac{\int_L x\,w(x)\,dx}{\int_L w(x)\,dx} = \frac{\int_0^2 x(60x^2)\,dx}{160} = \frac{60\left[\frac{x^4}{4}\right]_0^2}{160} = \frac{60\left[\frac{2^4}{4} - \frac{0^4}{4}\right]}{160}$$

$$= 1.5\text{ m} \hspace{3cm} Ans.$$

These results may be checked by using Appendix B, where it is shown that for an exparabolic area of length a, height b, and shape shown in Fig. 6–18a,

$$A = \frac{ab}{3} = \frac{2(240)}{3} = 160\text{ N} \quad\text{and}\quad \bar{x} = \frac{3}{4}a = \frac{3}{4}(2) = 1.5\text{ m}$$

Example 6–11

A distributed loading of $p = 800x$ Pa acts over the top surface of the beam shown in Fig. 6–19a. Determine the magnitude and location of the resultant force.

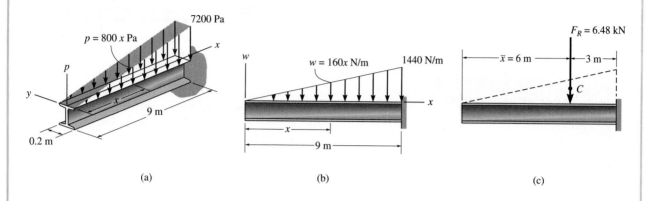

Fig. 6–19

SOLUTION

The loading function $p = 800x$ Pa indicates that the load intensity varies uniformly from $p = 0$ at $x = 0$ to $p = 7200$ Pa at $x = 9$ m. Since the intensity is uniform along the width of the beam (the y axis), the loading may be viewed in two dimensions as shown in Fig. 6–19b. Here

$$w = (800x \text{ N/m}^2)(0.2 \text{ m})$$
$$= (160x) \text{ N/m}$$

At $x = 9$ m, note that $w = 1440$ N/m. Although we may again apply Eqs. 6–10 and 6–11 as in Example 6–10, it is simpler to use Appendix B.

The magnitude of the resultant force is

$$F_R = \tfrac{1}{2}(9 \text{ m})(1440 \text{ N/m}) = 6480 \text{ N} = 6.48 \text{ kN} \qquad Ans.$$

The line of action of $\mathbf{F}_R$ passes through the *centroid C* of the triangle. Hence,

$$\bar{x} = 9 \text{ m} - \tfrac{1}{3}(9 \text{ m}) = 6 \text{ m} \qquad Ans.$$

The results are shown in Fig. 6–19c.

We may also view the resultant $\mathbf{F}_R$ as *acting* through the *centroid* of the *volume* of the loading diagram $p = p(x)$ in Fig. 6–19a. Hence $\mathbf{F}_R$ intersects the x–y plane at the point (6 m, 0). Furthermore, the *magnitude* of $\mathbf{F}_R$ is equal to the *volume* under the loading diagram; i.e.,

$$F_R = V = \tfrac{1}{2}(7200 \text{ N/m}^2)(9 \text{ m})(0.2 \text{ m}) = 6.48 \text{ kN} \qquad Ans.$$

PROBLEMS

6–45. Determine the resultant moment of both the 100-lb force and the triangular distributed load about point O.

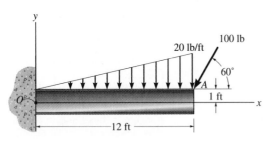

Prob. 6–45

6–46. Replace the loading by an equivalent resultant force and specify the location of the force on the beam, measured from point B.

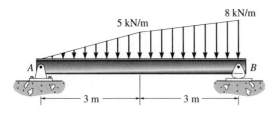

Prob. 6–46

6–47. Replace the distributed loading by an equivalent resultant force, and specify its location on the beam measured from the pin at C.

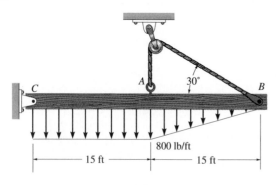

Prob. 6–47

*6–48. Determine the magnitude of the equivalent resultant force of the distributed loading and specify its location on the beam, measured from point A.

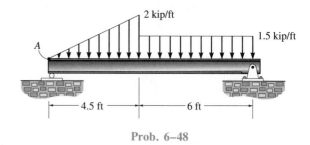

Prob. 6–48

6–49. The column is used to support the floor which exerts a force of 3000 lb on the top of the column. The effect of soil pressure along the side of the column is distributed as shown. Replace this loading by an equivalent resultant force and specify where it acts along the column measured from its base A.

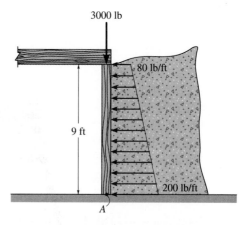

Prob. 6–49

6–50. The beam is subjected to the distributed loading. Determine the length b of the uniform load and its position a on the beam such that the resultant force and couple moment acting on the beam are zero.

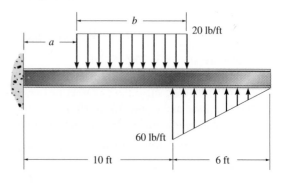

Prob. 6–50

6–51. Wet concrete exerts a pressure distribution along the wall of the form. Determine the resultant force of this distribution and specify the height h where the bracing strut should be placed so that it lies through the line of action of the resultant force. The wall has a width of 5 m.

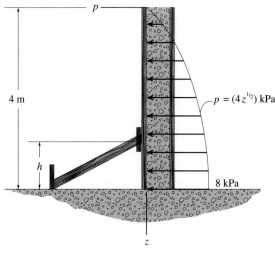

$p = (4z^{1/2})$ kPa

Prob. 6–51

*6–52. Determine the magnitude of the resultant force of the loading acting on the beam and specify where it acts from point O.

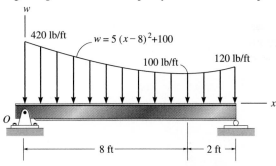

Prob. 6–52

6–53. The load over the plate varies linearly along the sides of the plate such that $p = \frac{2}{3}[x(4-y)]$ kPa. Determine the resultant force and its position $(\bar{x}, \bar{y})$ on the plate.

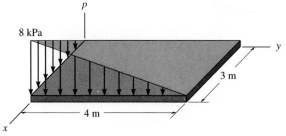

Prob. 6–53

6–54. Determine the magnitude and location of the resultant force of the parabolic loading acting on the plate.

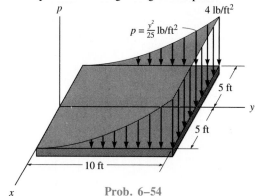

Prob. 6–54

6–55. The loading acting on a square plate is represented by a parabolic pressure distribution. Determine the magnitude and location of the resultant force. Also, what are the reactions at the rollers B and C and the ball-and-socket joint A? Neglect the weight of the plate.

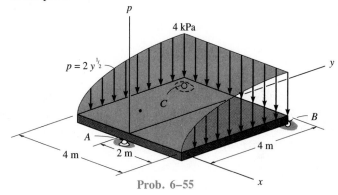

Prob. 6–55

*6–56. The pressure loading on the plate is described by the function $p = [-240/(x + 1) + 340]$ Pa. Determine the magnitude of the resultant force and coordinates of the point where the line of action of the force intersects the plate.

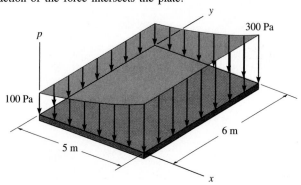

Prob. 6–56

6.5 Fluid Pressure

According to Pascal's law, a fluid at rest creates a pressure p at a point that is the *same* in *all* directions. The magnitude of p, measured as a force per unit area, depends on the specific weight γ or mass density ρ of the fluid and the depth z of the point from the fluid surface.* The relationship can be expressed mathematically as

$$p = \gamma z = \rho g z \qquad (6\text{--}13)$$

where g is the acceleration of gravity. Equation 6–13 is valid only for fluids that are assumed *incompressible,* as in the case of most liquids. Gases are compressible fluids, and since their density changes significantly with both pressure and temperature, Eq. 6–13 cannot be used.

To illustrate how Eq. 6–13 is applied, consider the submerged plate shown in Fig. 6–20. Three points on the plate have been specified. Since points A and B are both at depth z_2 from the liquid surface, the *pressure* at these points has a magnitude $p_2 = \gamma z_2$. Likewise, point C is at depth z_1; hence, $p_1 = \gamma z_1$. In all cases, the pressure acts *normal* to the surface area dA located at the specified point, Fig. 6–20. Using Eq. 6–13 and the results of Sec. 6.4, it is possible to determine the resultant force of a liquid pressure distribution, and specify its location on the surface of a submerged plate. Three different shapes of plates will now be considered.

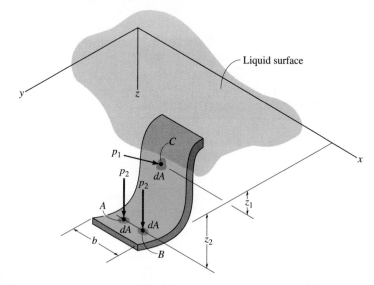

Fig. 6–20

*In particular, for water $\gamma = 62.4$ lb/ft^3, or $\gamma = 9810$ N/m^3, since $\rho = 1000$ kg/m^3 and $g = 9.81$ m/s^2.

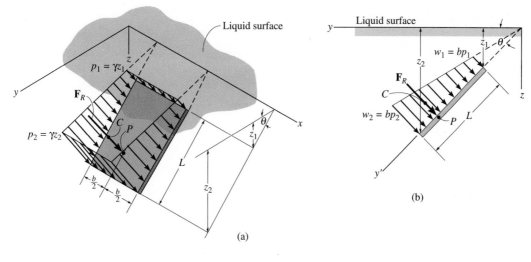

Fig. 6–21

Flat Plate of Constant Width. A flat rectangular plate of constant width, which is submerged in a liquid having a specific weight γ, is shown in Fig. 6–21a. The plane of the plate makes an angle θ with the horizontal, such that its top edge is located at a depth z_1 from the liquid surface and its bottom edge is located at a depth z_2. Since pressure varies linearly with depth, Eq. 6–13, the distribution of pressure over the plate's surface is represented by a trapezoidal volume having an intensity of $p_1 = \gamma z_1$ at depth z_1 and $p_2 = \gamma z_2$ at depth z_2. As noted in Sec. 6.4, the magnitude of the *resultant force* $\mathbf{F}_R$ is equal to the *volume* of this loading diagram and $\mathbf{F}_R$ has a *line of action* that passes through the volume's centroid C. Hence $\mathbf{F}_R$ does *not* act at the centroid of the plate; rather, it acts at point P, called the *center of pressure*.

Since the plate has a *constant width,* the loading distribution may also be viewed in two dimensions, Fig. 6–21b. Here the loading intensity is measured as force/length and varies linearly from $w_1 = bp_1 = b\gamma z_1$ to $w_2 = bp_2 = b\gamma z_2$. The magnitude of $\mathbf{F}_R$ in this case equals the trapezoidal *area,* and $\mathbf{F}_R$ has a *line of action* that passes through the area's *centroid C.* For numerical applications, the area and location of the centroid for a trapezoid are tabulated in Appendix B.

Curved Plate of Constant Width. When the submerged plate is curved, the pressure acting normal to the plate continually changes direction, and therefore calculation of the magnitude of $\mathbf{F}_R$ and its location P is more difficult than for a flat plate. Three- and two-dimensional views of the loading distribution are shown in Figs. 6–22a and 6–22b, respectively. Here integration can be used to determine both F_R and the location of the centroid C or center of pressure P.

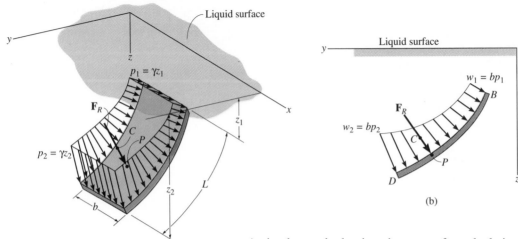

(a)

Fig. 6-22

A simpler method exists, however, for calculating the magnitude of $\mathbf{F}_R$ and its location along a curved (or flat) plate having a *constant width*. This method requires separate calculations for the horizontal and vertical *components* of $\mathbf{F}_R$. For example, the distributed loading acting on the curved plate *DB* in Fig. 6–22b can be represented by the *equivalent loading* shown in Fig. 6–23. Here the plate supports the weight of liquid W_f contained within the block *BDA*. This force has a magnitude $W_f = (\gamma b)(\text{area}_{BDA})$ and acts through the centroid of *BDA*. In addition, there are the pressure distributions caused by the liquid acting along the vertical and horizontal sides of the block. Along the vertical side *AD*, the force $\mathbf{F}_{AD}$ has a magnitude that equals the area under the trapezoid and acts through the centroid C_{AD} of this area. The distributed loading along the horizontal side *AB* is constant, since all points lying in this plane are at the same depth from the surface of the liquid. The magnitude of $\mathbf{F}_{AB}$ is simply the area of the rectangle. This force acts through the area's centroid C_{AB} or the midpoint of *AB*. Summing the three forces in Fig. 6–23 yields $\mathbf{F}_R = \Sigma\mathbf{F} = \mathbf{F}_{AD} + \mathbf{F}_{AB} + \mathbf{W}_f$, which is shown in Fig. 6–22. Finally, the location of the center of pressure *P* on the plate is determined by applying the equation $M_{R_O} = \Sigma M_O$, which states that the moment of the resultant force about a convenient reference point, Fig. 6–22, is equal to the sum of the moments of the three forces in Fig. 6–23 about the same point.

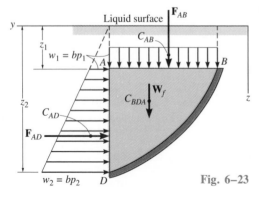

Fig. 6-23

Example 6–12

Determine the magnitude and location of the resultant hydrostatic force acting on the submerged rectangular plate AB shown in Fig. 6–24a. The plate has a width of 1.5 m; $\rho_w = 1000$ kg/m³.

SOLUTION

Since the plate has a constant width, the distributed loading can be viewed in two dimensions as shown in Fig. 6–24b. The intensity of the load at A and B is

$$w_A = b\rho_w g z_A = (1.5 \text{ m})(1000 \text{ kg/m}^3)(9.81 \text{ m/s}^2)(2 \text{ m})$$
$$= 29.4 \text{ kN/m}$$
$$w_B = b\rho_w g z_B = (1.5 \text{ m})(1000 \text{ kg/m}^3)(9.81 \text{ m/s}^2)(5 \text{ m})$$
$$= 73.6 \text{ kN/m}$$

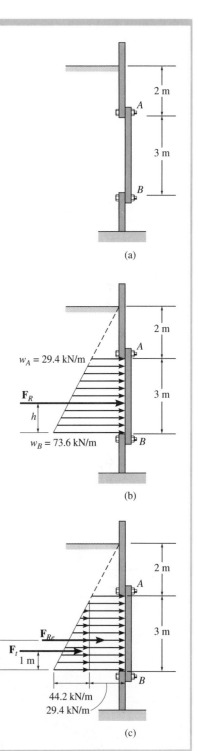

(a)

From Appendix B, the magnitude of the resultant force $\mathbf{F}_R$ created by the distributed load is

$$F_R = \text{area of trapezoid}$$
$$= \tfrac{1}{2}(3)(29.4 + 73.6) = 154.5 \text{ kN} \qquad Ans.$$

This force acts through the centroid of the area,

$$h = \frac{1}{3}\left(\frac{2(29.4) + 73.6}{29.4 + 73.6}\right)(3) = 1.29 \text{ m} \qquad Ans.$$

measured upward from B, Fig. 6–24b.

The same results can be obtained by considering two components of $\mathbf{F}_R$ defined by the triangle and rectangle shown in Fig. 6–24c. Each force acts through its associated centroid and has a magnitude of

$$F_{Re} = (29.4 \text{ kN/m})(3 \text{ m}) = 88.2 \text{ kN}$$
$$F_t = \tfrac{1}{2}(44.2 \text{ kN/m})(3 \text{ m}) = 66.3 \text{ kN}$$

Hence,

$$F_R = F_{Re} + F_t = 88.2 + 66.3 = 154.5 \text{ kN} \qquad Ans.$$

The location of $\mathbf{F}_R$ is determined by summing moments about B, Fig. 6–24b and c, i.e.,

$$\uparrow + (M_R)_B = \Sigma M_B; \quad (154.5)h = 88.2(1.5) + 66.3(1)$$
$$h = 1.29 \text{ m} \qquad Ans.$$

(b)

(c)

Fig. 6–24

Example 6–13

Determine the magnitude of the resultant hydrostatic force acting on the surface of a seawall shaped in the form of a parabola as shown in Fig. 6–25a. The wall is 5 m long; $\rho_w = 1020$ kg/m^3.

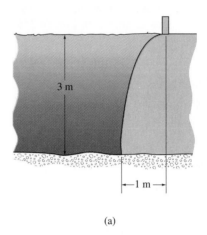

(a)

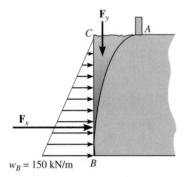

(b)

Fig. 6–25

SOLUTION

The horizontal and vertical components of the resultant force will be calculated, Fig. 6–25b. Since

$$w_B = b\rho_w g z_B = 5 \text{ m}(1020 \text{ kg/m}^3)(9.81 \text{ m/s}^2)(3 \text{ m}) = 150.0 \text{ kN/m}$$

then

$$F_x = \tfrac{1}{2}(3 \text{ m})(150.0 \text{ kN/m}) = 225.0 \text{ kN}$$

The area of the parabolic sector ABC can be determined using Appendix B. Hence, the weight of water within this region is

$$F_y = (\rho_w g b)(\text{area}_{ABC})$$
$$= (1020 \text{ kg/m}^3)(9.81 \text{ m/s}^2)(5 \text{ m})[\tfrac{1}{3}(1 \text{ m})(3 \text{ m})] = 50.0 \text{ kN}$$

The resultant force is therefore

$$F_R = \sqrt{F_x^2 + F_y^2} = \sqrt{(225.0)^2 + (50.0)^2}$$
$$= 230 \text{ kN} \qquad\qquad\qquad Ans.$$

PROBLEMS

6–57. Determine the magnitude of the resultant hydrostatic force acting on the dam and its location measured from the top surface of the water. The width of the dam is 8 m; $\rho_w = 1.0$ Mg/m³.

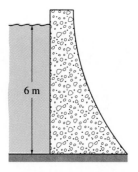

Prob. 6–57

6–59. The symmetric concrete "gravity" dam is held in place by its own weight. If the density of concrete is $\rho_c = 2.5$ Mg/m³, and water has a density of $\rho_w = 1.0$ Mg/m³, determine the smallest width d at its base that will prevent the dam from overturning about its end A. The dam has a width of 8 m.

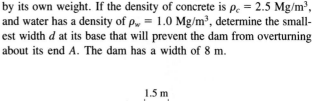
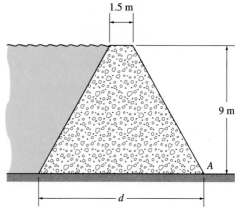

Prob. 6–59

6–58. The tank is filled with water to a depth of $d = 4$ m. Determine the resultant force the water exerts on side A and side B of the tank. If oil instead of water is placed in the tank, to what depth d should it reach so that it creates the same resultant forces? $\rho_o = 900$ kg/m³ and $\rho_w = 1000$ kg/m³.

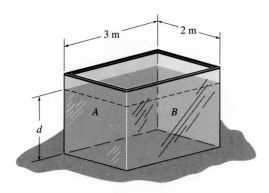

Prob. 6–58

*6–60. The gate AB is 8 m wide. Determine the horizontal and vertical components of force acting on the pin at B and the vertical reaction at the smooth support A. $\rho_w = 1.0$ Mg/m³.

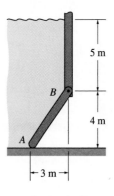

Prob. 6–60

6–61. Determine the resultant horizontal and vertical force components that the water exerts on the side of the dam. The dam is 25 ft long and $\gamma_w = 62.4$ lb/ft^3.

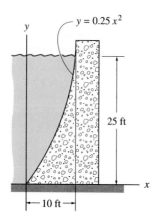

$y = 0.25\,x^2$

25 ft

10 ft

Prob. 6–61

6–62. Determine the resultant hydrostatic force acting per foot of length on the sea wall; $\gamma_w = 62.4$ lb/ft^3.

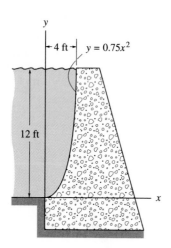

4 ft $y = 0.75x^2$

12 ft

Prob. 6–62

6–63. The tank is used to store a liquid having a specific weight of $\gamma = 80$ lb/ft^3. If it is filled to the top, determine the magnitude of force the liquid exerts on side $ABCD$.

*__6–64.__ The tank is used to store a liquid having a specific weight of $\gamma = 80$ lb/ft^3. If it is filled to the top, determine the magnitude of force the liquid exerts on side $ABEF$.

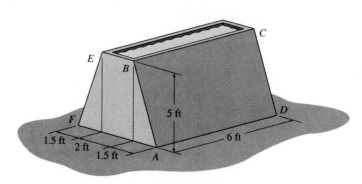

E
B
C
5 ft
F
D
1.5 ft
2 ft
1.5 ft
A
6 ft

Probs. 6–63/6–64

6–65. The tank is used to store a liquid having a density of 80 lb/ft^3. If it is filled to the top, determine the magnitude of force the liquid exerts on each of its two sides $ABDC$ and $BDFE$.

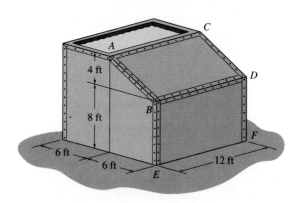

C
A
4 ft
D
B
8 ft
6 ft   6 ft
E
F
12 ft

Prob. 6–65

6.6 Moments of Inertia for Areas

In Sec. 6-2 we determined the centroid for an area by considering the first moment of the area about an axis; that is, for the computation we had to evaluate an integral of the form $\int x\,dA$. Integrals of the second moment of an area, such as $\int x^2\,dA$, are referred to as the *moment of inertia* for the area. The terminology "moment of inertia" as used here is actually a misnomer; however, it has been adopted because of the similarity with integrals of the same form related to mass. Since integrals of this form often arise in formulas used in mechanics of materials, structural mechanics, and machine design, the engineer should become familiar with the methods used for their computation.

Moment of Inertia. Consider the area A, shown in Fig. 6–26, which lies in the x–y plane. By definition, the moments of inertia of the differential planar area dA about the x and y axes are $dI_x = y^2\,dA$ and $dI_y = x^2\,dA$, respectively. For the entire area the *moments of inertia* are determined by integration; i.e.,

$$
\begin{aligned}
I_x &= \int_A y^2\,dA \\
I_y &= \int_A x^2\,dA
\end{aligned}
\tag{6–14}
$$

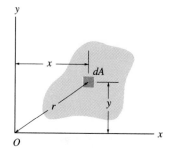

Fig. 6–26

We can also formulate the second moment of the differential area dA about the pole O or z axis, Fig. 6–26. This is referred to as the polar moment of inertia, $dJ_O = r^2\,dA$. Here r is the perpendicular distance from the pole (z axis) to the element dA. For the entire area the *polar moment of inertia* is

$$
J_O = \int_A r^2\,dA = I_x + I_y
\tag{6–15}
$$

The relationship between J_O and I_x, I_y is possible since $r^2 = x^2 + y^2$, Fig. 6–26.

From the above formulations it is seen that I_x, I_y, and J_O will *always* be *positive*, since they involve the product of distance squared and area. Furthermore, the units for moment of inertia involve length raised to the fourth power, e.g., m^4, mm^4, or ft^4, in^4.

6.7 Parallel-Axis Theorem

If the moment of inertia for an area is known about an axis passing through its centroid, it is convenient to determine the moment of inertia of the area about a corresponding parallel axis using the *parallel-axis theorem*. To derive this

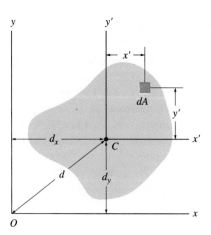

Fig. 6–27

theorem, consider finding the moment of inertia of the shaded area shown in Fig. 6–27 about the x axis. In this case, a differential element dA is located at an arbitrary distance y' from the *centroidal* x' axis, whereas the *fixed distance* between the parallel x and x' axes is defined as d_y. Since the moment of inertia of dA about the x axis is $dI_x = (y' + d_y)^2 \, dA$, then for the entire area,

$$I_x = \int_A (y' + d_y)^2 \, dA$$

$$= \int_A y'^2 \, dA + 2d_y \int_A y' \, dA + d_y^2 \int_A dA$$

The first integral represents the moment of inertia of the area about the centroidal axis, $\bar{I}_{x'}$. The second integral is zero since the x' axis passes through the area's centroid C; i.e., $\int y' \, dA = \bar{y} \int dA = 0$ since $\bar{y} = 0$. Realizing that the third integral represents the total area A, the final result is therefore

$$\boxed{I_x = \bar{I}_{x'} + Ad_y^2} \tag{6–16}$$

A similar expression can be written for I_y; i.e.,

$$\boxed{I_y = \bar{I}_{y'} + Ad_x^2} \tag{6–17}$$

And finally, for the polar moment of inertia about an axis perpendicular to the x–y plane and passing through the pole O (z axis), Fig. 6–27, we have

$$\boxed{J_O = \bar{J}_C + Ad^2} \tag{6–18}$$

The form of each of these equations states that *the moment of inertia of an area about an axis is equal to the moment of inertia of the area about a parallel axis passing through the area's centroid plus the product of the area and the square of the perpendicular distance between the axes.*

6.8 Moments of Inertia for an Area by Integration

When the boundaries for a planar area are expressed by mathematical functions, Eqs. 6–14 may be integrated to determine the moments of inertia for the area. If the element of area chosen for integration has a differential size in two directions as shown in Fig. 6–26, a double integration must be performed to evaluate the moment of inertia. Most often, however, it is easier to perform only a single integration by choosing an element having a differential size or thickness in only one direction.

PROCEDURE FOR ANALYSIS

If a single integration is performed to determine the moment of inertia of an area about an axis, it will first be necessary to specify the differential element dA. Most often this element will be rectangular, such that it will have a finite length and differential width. The element should be located so that it intersects the boundary of the area at the *arbitrary point (x, y)*. There are two possible ways to orient the element with respect to the axis about which the moment of inertia is to be determined.

Case 1.   The *length* of the element can be oriented *parallel* to the axis. This situation occurs when the rectangular element shown in Fig. 6–28 is used to determine I_y for the area. Direct application of Eq. 6–14, i.e., $I_y = \int x^2 \, dA$, can be made in this case, since the element has an infinitesimal thickness dx and therefore *all parts* of the element lie at the *same* moment-arm distance x from the y axis.*

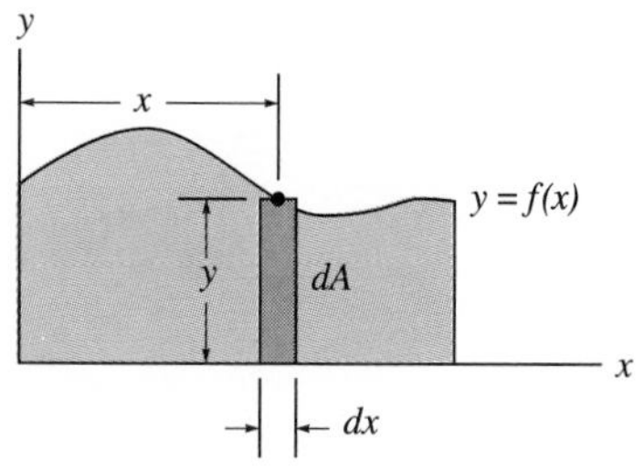

Case 2.   The *length* of the element can be oriented *perpendicular* to the axis. Here Eq. 6–14 *does not apply,* since all parts of the element will *not* lie at the same moment-arm distance from the axis. For example, if the rectangular element in Fig. 6–28 is used for determining I_x for the area, it will first be necessary to calculate the moment of inertia of the *element* about a horizontal axis passing through the element's centroid and then determine the moment of inertia of the *element* about the x axis by using the parallel-axis theorem. Integration of this result will yield I_x.

Fig. 6–28

Application of these cases is illustrated in the following examples.

---

*In the case of the element $dA = dx \, dy$, Fig. 6–26, the moment arms y and x are appropriate for the formulation of I_x and I_y (Eq. 6–14) since the *entire* element, because of its infinitesimal size, lies at the specified y and x perpendicular distances from the x and y axes.

Example 6–14

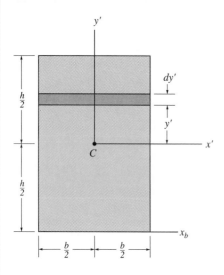

Fig. 6–29

Determine the moment of inertia for the rectangular area shown in Fig. 6–29 with respect to (a) the centroidal x' axis, (b) the axis x_b passing through the base of the rectangle, and (c) the pole or z' axis perpendicular to the $x'-y'$ plane and passing through the centroid C.

SOLUTION (CASE 1)

Part (a). The differential element shown in Fig. 6–29 is chosen for integration. Because of its location and orientation, the *entire element* is at a distance y' from the x' axis. Here it is necessary to integrate from $y' = -h/2$ to $y' = h/2$. Since $dA = b\, dy'$, then

$$\bar{I}_{x'} = \int_A y'^2\, dA = \int_{-h/2}^{h/2} y'^2 (b\, dy') = b \int_{-h/2}^{h/2} y'^2\, dy'$$

$$= \frac{1}{12} bh^3 \qquad\qquad Ans.$$

Part (b). The moment of inertia about an axis passing through the base of the rectangle can be obtained by using the result of part (a) and applying the parallel-axis theorem, Eq. 6–16.

$$I_{x_b} = \bar{I}_{x'} + Ad_y^2$$

$$= \frac{1}{12} bh^3 + bh\left(\frac{h}{2}\right)^2 = \frac{1}{3} bh^3 \qquad\qquad Ans.$$

Part (c). To obtain the polar moment of inertia about point C, we must first obtain $\bar{I}_{y'}$, which may be found by interchanging the dimensions b and h in the result of part (a), i.e.,

$$\bar{I}_{y'} = \frac{1}{12} hb^3$$

Using Eq. 6–15, the polar moment of inertia about C is therefore

$$\bar{J}_C = \bar{I}_{x'} + \bar{I}_{y'} = \frac{1}{12} bh(h^2 + b^2) \qquad\qquad Ans.$$

Example 6–15

Determine the moment of inertia of the shaded area shown in Fig. 6–30a about the x axis.

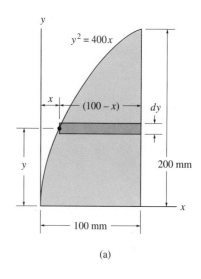

(a)

SOLUTION I (CASE 1)

A differential element of area that is *parallel* to the x axis, as shown in Fig. 6–30a, is chosen for integration. Since the element has a thickness dy and intersects the curve at the *arbitrary point* (x, y), the area is $dA = (100 - x)\, dy$. Furthermore, all parts of the element lie at the same distance y from the x axis. Hence, integrating with respect to y, from $y = 0$ to $y = 200$ mm, yields

$$I_x = \int_A y^2\, dA = \int_A y^2(100 - x)\, dy$$

$$= \int_0^{200} y^2\left(100 - \frac{y^2}{400}\right) dy = 100 \int_0^{200} y^2\, dy - \frac{1}{400} \int_0^{200} y^4\, dy$$

$$= 107(10^6)\ \text{mm}^4$$

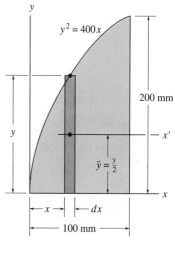

(b)

Fig. 6–30

SOLUTION II (CASE 2)

A differential element *parallel* to the y axis, as shown in Fig. 6–30b, is chosen for integration. It intersects the curve at the *arbitrary point* (x, y). In this case, all parts of the element do *not* lie at the same distance from the x axis, and therefore the parallel-axis theorem must be used to determine the *moment of inertia of the element* with respect to this axis. For a rectangle having a base b and height h, the moment of inertia about its centroidal axis has been determined in part (a) of Example 6–14. There it was found that $\bar{I}_{x'} = \frac{1}{12}bh^3$. For the differential element shown in Fig. 6–30b, $b = dx$ and $h = y$, and thus $d\bar{I}_{x'} = \frac{1}{12}\, dx\, y^3$. Since the centroid of the element is at $\tilde{y} = y/2$ from the x axis, the moment of inertia of the element about this axis is

$$dI_x = d\bar{I}_{x'} + dA\, \tilde{y}^2 = \frac{1}{12}\, dx\, y^3 + y\, dx\left(\frac{y}{2}\right)^2 = \frac{1}{3}\, y^3\, dx$$

[This result can also be concluded from part (b) of Example 6–14].
Integrating with respect to x, from $x = 0$ to $x = 100$ mm, yields

$$I_x = \int dI_x = \int_A \frac{1}{3}y^3\, dx = \int_0^{100} \frac{1}{3}(400x)^{3/2}\, dx$$

$$= 107(10^6)\ \text{mm}^4 \qquad\qquad Ans.$$

Example 6–16

Determine the moment of inertia with respect to the x axis of the circular area shown in Fig. 6–31a.

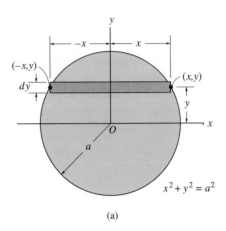

(a)

SOLUTION I (*CASE 1*)

Using the differential element shown in Fig. 6–31a, since $dA = 2x\,dy$, we have

$$I_x = \int_A y^2\,dA = \int_A y^2(2x)\,dy$$

$$= \int_{-a}^{a} y^2(2\sqrt{a^2 - y^2})\,dy = \frac{\pi a^4}{4} \qquad Ans.$$

SOLUTION II (*CASE 2*)

When the differential element is chosen as shown in Fig. 6–31b, the centroid for the element happens to lie on the x axis, and so, applying Eq. 6–16, noting that $d_y = 0$, we have

$$dI_x = \frac{1}{12}\,dx\,(2y)^3$$

$$= \frac{2}{3}y^3\,dx$$

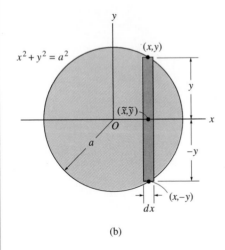

(b)

Fig. 6–31

Integrating with respect to x yields

$$I_x = \int_{-a}^{a} \frac{2}{3}(a^2 - x^2)^{3/2}\,dx = \frac{\pi a^4}{4} \qquad Ans.$$

PROBLEMS

6–66. The irregular area has a moment of inertia about the A–A axis of 500 in^4. If the total area is 12 in^2, determine the moment of inertia of the area about the B–B axis. The C–C axis passes through the centroid of the area.

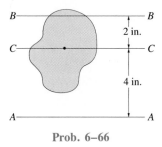

Prob. 6–66

6–67. Determine the moment of inertia of the triangular area about the x axis.

*6–68. Determine the moment of inertia of the triangular area about the y axis.

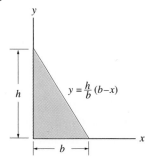

$$y = \frac{h}{b}(b-x)$$

Probs. 6–67/6–68

6–69. Determine the moments of inertia I_x and I_y of the shaded area.

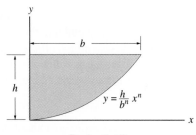

$$y = \frac{h}{b^n}x^n$$

Prob. 6–69

6–70. Determine the moment of inertia about the x axis of the area which is bounded by the lines $x = 0$, $y = 1$ in., and the parabola $y = 3x^2$.

6–71. Determine the moments of inertia about the x and y axes of the area bounded by the curve $y^2 = 2x$ and the lines $x = 2$ in. and $y = 0$.

*6–72. Determine the polar moments of inertia J_O for the cross-sectional area of a solid shaft having a radius of 20 mm and a tube having an outer radius of 20 mm and inner radius of 15 mm. What percentage of J_O is contributed by the tube to that of the solid shaft?

6–73. Determine the moment of inertia of the area about the x axis.

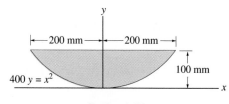

$$400\,y = x^2$$

Prob. 6–73

6–74. Determine the moment of inertia of the shaded area about the x axis.

6–75. Determine the moment of inertia of the shaded area about the y axis.

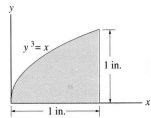

$$y^3 = x$$

Probs. 6–74/6–75

*6–76. Determine the moment of inertia of the shaded area about the x axis.

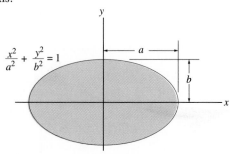

$$\frac{x^2}{a^2} + \frac{y^2}{b^2} = 1$$

Prob. 6–76

6–77. Determine the moment of inertia of the shaded area about the y axis.

6–78. Determine the moment of inertia of the shaded area about the x axis.

6–79. Determine the moment of inertia of the shaded area about the x axis.

*6–80. Determine the moment of inertia of the shaded area about the y axis.

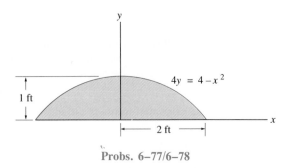

$4y = 4 - x^2$

1 ft

2 ft

Probs. 6–77/6–78

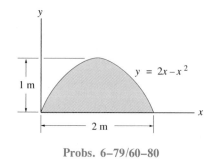

$y = 2x - x^2$

1 m

2 m

Probs. 6–79/60–80

6.9 Moments of Inertia for Composite Areas

A composite area consists of a series of connected "simpler" parts or shapes, such as semicircles, rectangles, and triangles. Provided the moment of inertia of each of these parts is known or can be determined about a common axis, then the moment of inertia of the composite area equals the *algebraic sum* of the moments of inertia of all its parts.

PROCEDURE FOR ANALYSIS

The following procedure provides a method for determining the moment of inertia of a composite area about a reference axis.

Composite Parts. Using a sketch, divide the area into its composite parts and indicate the perpendicular distance from the *centroid* of each part to the reference axis.

Parallel-Axis Theorem. The moment of inertia of each part should be determined about its centroidal axis, which is parallel to the reference axis. For the calculation use the table given in Appendix *B*. If the centroidal axis does not coincide with the reference axis, the parallel-axis theorem, $I = \bar{I} + Ad^2$, should be used to determine the moment of inertia of the part about the reference axis.

Summation. The moment of inertia of the entire area about the reference axis is determined by summing the results of its composite parts. In particular, if a composite part has a "hole," its moment of inertia is found by "subtracting" the moment of inertia for the hole from the moment of inertia of the entire part including the hole.

Example 6–17

Determine the moment of inertia of the cross-sectional area of the T-beam shown in Fig. 6–32a about the centroidal x' axis.

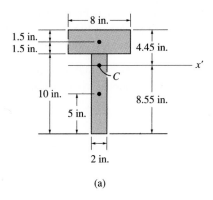

(a)

Fig. 6–32

SOLUTION I

The area is segmented into two rectangles as shown in Fig. 6–32a, and the distance from the x' axis and each centroidal axis is determined. Using the table in Appendix B, the moment of inertia of a rectangle about its centroidal axis is $I = \frac{1}{12}bh^3$. Applying the parallel-axis theorem, to each rectangle and adding the results, we therefore have

$$I = \Sigma \bar{I}_{x'} + Ad_y^2$$

$$= \left[\frac{1}{12} (2 \text{ in.})(10 \text{ in.})^3 + (2 \text{ in.})(10 \text{ in.})(8.55 \text{ in.} - 5 \text{ in.})^2 \right]$$

$$+ \left[\frac{1}{12} (8 \text{ in.})(3 \text{ in.})^3 + (8 \text{ in.})(3 \text{ in.})(4.45 \text{ in.} - 1.5 \text{ in.})^2 \right]$$

$$I = 645.6 \text{ in}^4 \qquad\qquad Ans.$$

SOLUTION II

The area can be considered as one large rectangle less two small rectangles, shown dashed in Fig. 6–32b. We have

$$I = \Sigma \bar{I}_{x'} + Ad_y^2$$

$$= \left[\frac{1}{12} (8 \text{ in.})(13 \text{ in.})^3 + (8 \text{ in.})(13 \text{ in.})(8.55 \text{ in.} - 6.5 \text{ in.})^2 \right]$$

$$- 2 \left[\frac{1}{12} (3 \text{ in.})(10 \text{ in.})^3 + (3 \text{ in.})(10 \text{ in.})(8.55 \text{ in.} - 5 \text{ in.})^2 \right]$$

$$I = 645.6 \text{ in}^4 \qquad\qquad Ans.$$

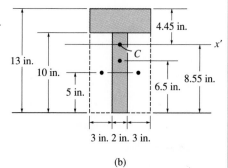

(b)

Example 6–18

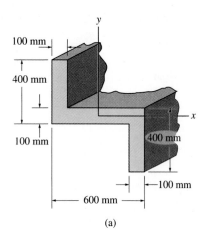

100 mm

400 mm

100 mm

400 mm

100 mm

600 mm

(a)

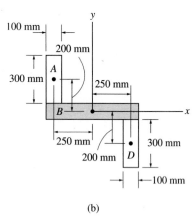

100 mm

200 mm

300 mm

A

250 mm

B ————— x

250 mm

300 mm

200 mm D

100 mm

(b)

Fig. 6–33

Determine the moments of inertia of the beam's cross-sectional area shown in Fig. 6–33a about the x and y centroidal axes.

SOLUTION

Composite Parts. The cross section can be considered as three composite rectangular areas A, B, and D shown in Fig. 6–33b. For the calculation, the centroid of each of these rectangles is located in the figure.

Parallel-Axis Theorem. Using the parallel-axis theorem for rectangles A and D, and $\bar{I} = \frac{1}{12}bh^2$, the calculations are as follows:

Rectangle A

$$I_x = \bar{I}_{x'} + Ad_y^2 = \frac{1}{12}(100)(300)^3 + (100)(300)(200)^2$$

$$= 1.425(10^9) \text{ mm}^4$$

$$I_y = \bar{I}_{y'} + Ad_x^2 = \frac{1}{12}(300)(100)^3 + (100)(300)(250)^2$$

$$= 1.90(10^9) \text{ mm}^4$$

Rectangle B

$$I_x = \frac{1}{12}(600)(100)^3 = 0.05(10^9) \text{ mm}^4$$

$$I_y = \frac{1}{12}(100)(600)^3 = 1.80(10^9) \text{ mm}^4$$

Rectangle D

$$I_x = \bar{I}_{x'} + Ad_y^2 = \frac{1}{12}(100)(300)^3 + (100)(300)(200)^2$$

$$= 1.425(10^9) \text{ mm}^4$$

$$I_y = \bar{I}_{y'} + Ad_x^2 = \frac{1}{12}(300)(100)^3 + (100)(300)(250)^2$$

$$= 1.90(10^9) \text{ mm}^4$$

Summation. The moments of inertia for the entire cross section are thus

$$I_x = 1.425(10^9) + 0.05(10^9) + 1.425(10^9)$$

$$= 2.90(10^9) \text{ mm}^4 \qquad \qquad Ans.$$

$$I_y = 1.90(10^9) + 1.80(10^9) + 1.90(10^9)$$

$$= 5.60(10^9) \text{ mm}^4 \qquad \qquad Ans.$$

PROBLEMS

6–81. Determine the moments of inertia of the shaded area with respect to the *x* and *y* centroidal axes.

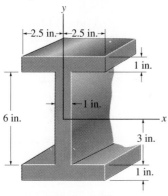

Prob. 6–81

6–82. The composite beam consists of a wide-flange beam and cover plates welded together as shown. Determine the moment of inertia of the cross-sectional area with respect to a horizontal axis passing through the beam's centroid.

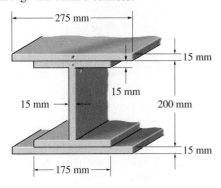

Prob. 6–82

6–83. Determine the distance $\bar{y}$ to the centroid of the channel's cross-sectional area, and then determine the moment of inertia with respect to the $x'-x'$ axis passing through the centroid.

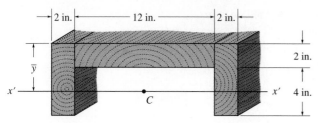

Prob. 6–83

*6–84. The beam is constructed from the two channels and two cover plates. If each channel has a cross-sectional area of $A_c = 11.8$ in^2 and a moment of inertia about a vertical axis passing through its own centroid, C_c, of $(\bar{I}_{\bar{y}})_{C_c} = 349$ in^4, determine the moment of inertia of the beam about the *y* axis.

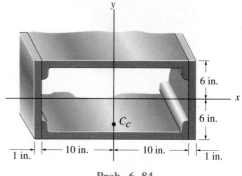

Prob. 6–84

6–85. Determine $\bar{y}$, which locates the centroid *C* of the *wing channel,* and then determine the moment of inertia $\bar{I}_{x'}$ about the centroidal x' axis. Neglect the effect of rounded corners. The material has a uniform thickness of 0.5 in.

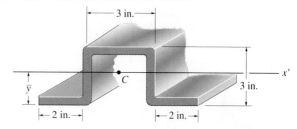

Prob. 6–85

6–86. Determine the distance $\bar{y}$ to the centroid of the beam constructed from the two channels and the cover plate. If each channel has a cross-sectional area of $A_c = 11.8$ in^2 and a moment of inertia about a horizontal axis passing through its own centroid, C_C, of $(\bar{I}_{\bar{x}})_{C_C} = 349$ in^4, determine the moment of inertia of the beam about the $x'-x'$ axis.

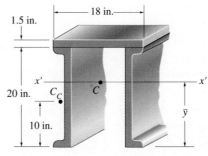

Prob. 6–86

6–87. Determine the moment of inertia of the shaded area with respect to a horizontal axis passing through the centroid of the section.

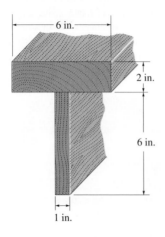

6 in.

2 in.

6 in.

1 in.

Prob. 6–87

*6–88.** Determine the moments of inertia I_x and I_y of the "Z" section about the x and y axes, which pass through the centroid C.

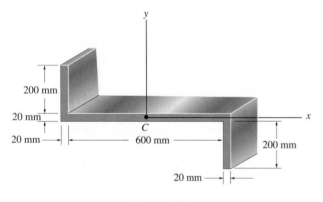

200 mm

20 mm

20 mm

600 mm

C

200 mm

20 mm

Prob. 6–88

6–89. Determine the distance $\bar{y}$ to the centroid of the beam's cross-sectional area; then find the moment of inertia about the x' axis.

6–90. Determine the moment of inertia of the beam's cross-sectional area about the y axis.

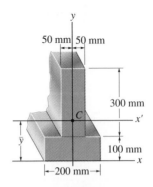

50 mm 50 mm

300 mm

C x'

$\bar{y}$

100 mm

x

200 mm

Probs. 6–89/6–90

6–91. Determine the distance $\bar{y}$ to the centroid of the beam's cross-sectional area; then find the moment of inertia about the x' axis.

*6–92.** Determine the moment of inertia of the beam's cross-sectional area about the y axis.

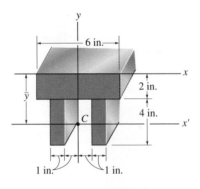

6 in.

x

2 in.

$\bar{y}$

4 in.

C x'

1 in. 1 in.

Probs. 6–91/6–92

6–93. Determine the location $(\bar{x}, \bar{y})$ of the centroid C of the cross-sectional area, and then compute the moments of inertia with respect to the x' and y' axes.

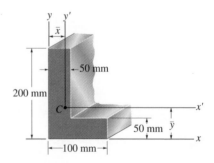

y y'

$\bar{x}$

50 mm

200 mm

C x'

50 mm $\bar{y}$

x

100 mm

Prob. 6–93

REVIEW PROBLEMS

6–94. Determine the distance $\bar{y}$ to the centroidal axis $\bar{x}$–$\bar{x}$ of the beam's cross-sectional area. Neglect the size of the corner welds at A and B for the calculation.

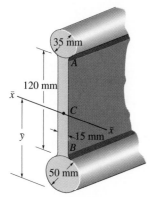

Prob. 6–94

6–95. Determine the distance $\bar{y}$ to the centroid of the plate area.

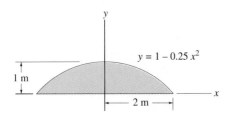

$y = 1 - 0.25\,x^2$

Prob. 6–95

*6–96. Determine the location $(\bar{x}, \bar{y})$ of the centroid of the parabolic area.

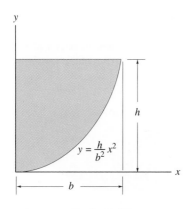

$y = \dfrac{h}{b^2}\,x^2$

Prob. 6–96

6–97. Determine the distance $\bar{z}$ to the center of gravity for the frustum of the paraboloid. The material is homogeneous.

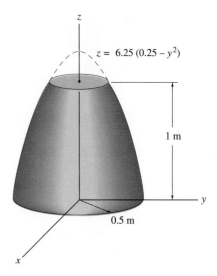

$z = 6.25\,(0.25 - y^2)$

Prob. 6–97

6–98. Determine the moments of inertia of the "Z" section with respect to the x and y centroidal axes.

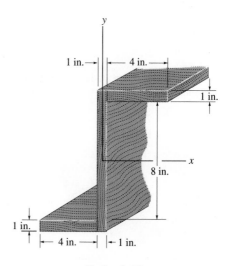

Prob. 6–98

6–99. The form is used to cast concrete columns. Determine the resultant force that wet concrete exerts along the plate A, 0.5 m $\leq$ $z \leq 3$ m, if the pressure due to the concrete varies as shown. Specify the location of the resultant force, measured from the top of the column.

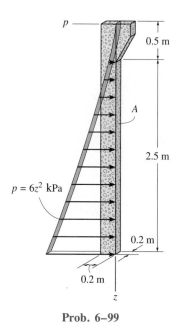

Prob. 6–99

*6–100. Determine the moment of inertia of the beam's cross-sectional area about the x axis.

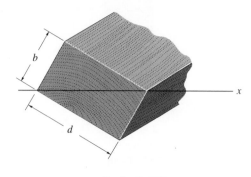

Prob. 6–100

6–101. Determine the location $(\bar{x}, \bar{y})$ of the centroid of the area.

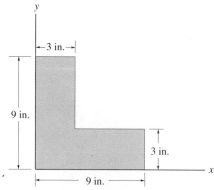

Prob. 6–101

6–102. Determine the location $(\bar{x}, \bar{y})$ of the centroid of the homogeneous plate.

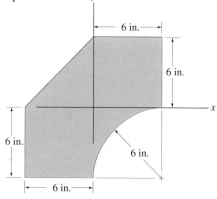

Prob. 6–102′

6–103. Determine the moment of inertia of the beam's cross-sectional area about the x axis which passes through the centroid C.

*6–104. Determine the moment of inertia of the beam's cross-sectional area about the y axis, which passes through the centroid C.

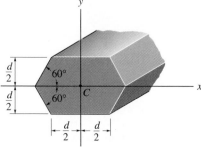

Probs. 6–103/6–104

7 Internal Loadings

One of the most important applications of statics in the analysis of problems involving mechanics of materials is to be able to determine the resultant force and movement acting within a body, which are necessary to hold the body together when the body is subjected to external loads. In this chapter we will first develop a technique for finding these internal resultant loads at a specific *point* within a member, and then we will generalize this method to find the point-to-point variation of loading along the axis of the member. A graph showing this variation of internal load will then allow us to find the critical points where the *maximum* internal loading occurs.

7.1 Resultant Internal Loadings Developed in a Body

As we shall see in our study of mechanics of materials, the design of a structural member or a mechanical element requires an investigation of the loadings acting *within* the member which are necessary to balance the loadings acting external to it. To illustrate how this is done, we will consider the body shown in Fig. 7–1a, which is held in equilibrium by the four external forces.* In order to obtain the *internal loadings* acting on a specific region within the body, it is necessary to use the *method of sections*. This requires

---

*The body's weight is not shown, since it is assumed to be quite small, and therefore negligible compared with the other loads.

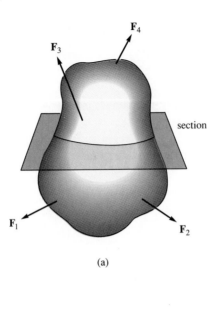

(a)

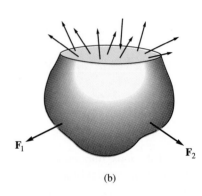

(b)

Fig. 7–1(a, b)

that an imaginary section or "cut" be made through the region where the internal loadings are to be determined. The two parts of the body are then separated, and a free-body diagram of one of the parts is drawn. If we consider the section shown in Fig. 7–1a, then the resulting free-body diagram of the bottom part of the body is shown in Fig. 7–1b. Here it can be seen that there is actually a *distribution* of internal force acting on the "exposed" area of the section. These forces represent the effects of the material of the top part of the body acting on the adjacent material of the bottom part. Although this exact distribution may be *unknown*, we can use statics to determine the *resultant internal force and moment*, $\mathbf{F}_R$ and $\mathbf{M}_{R_o}$, *this distribution exerts at a specific point O* on the sectioned area, Fig. 7–1c.

Since the entire body is in equilibrium, then each part of the body is also in equilibrium. Consequently, $\mathbf{F}_R$ and $\mathbf{M}_{R_o}$ can be determined by applying the equilibrium equations to any one of the two parts of the sectioned body. When doing so, note that $\mathbf{F}_R$ is at point O, although its computed value will *not* depend on the location of this point. On the other hand, $\mathbf{M}_{R_o}$ does depend on this location, since the moment arms must extend from O to the line of action of each force on the free-body diagram. It will be shown in later portions of the text that point O is most often chosen at the *centroid* of the sectioned area, and so we will always choose this location for O, unless otherwise stated. Also, if a member is long and slender, as in the case of a rod or beam, the section to be considered is generally taken *perpendicular* to the longitudinal axis of the member. This section is referred to as the *cross section*.

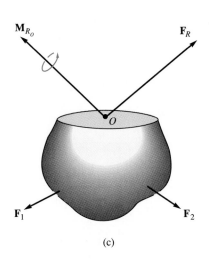

(c)

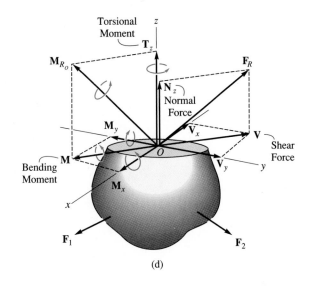

(d)

Later in this text we will show how to relate the resultant internal force and moment to the *distribution of force* on the sectioned area, Fig. 7–1*b*, and thereby develop equations that can be used for analysis and design of the body. To do this, however, the components of $\mathbf{F}_R$ and $\mathbf{M}_{R_O}$, acting both normal or perpendicular to the sectioned area and within the plane of the area, must be considered. If we establish x, y, z axes with origin at point O, as shown in Fig. 7–1*d*, then $\mathbf{F}_R$ and $\mathbf{M}_{R_O}$ can each be resolved into three components. Four different types of loadings can then be defined as follows:

$\mathbf{N}_z$ is called the *normal force*, since it acts perpendicular to the area. This force is developed when the external loads tend to push or pull on the two segments of the body.

$\mathbf{V}$ is called the *shear force*, and it can be determined from its two components using vector addition, $\mathbf{V} = \mathbf{V}_x + \mathbf{V}_y$. The shear force lies in the plane of the area and is developed when the external loads tend to cause the two segments of the body to slide over one another.

$\mathbf{T}_z$ is called the *torsional moment* or *torque*. It is developed when the external loads tend to twist one segment of the body with respect to the other.

$\mathbf{M}$ is called the *bending moment*. It is determined from the vector addition of its two components, $\mathbf{M} = \mathbf{M}_x + \mathbf{M}_y$. The bending moment is caused by the external loads that tend to bend the body about an axis lying within the plane of the area.

In this text, note that representation of a moment or torque is shown in three dimensions as a vector with an associated curl, Fig. 7–1*d*. By the *right-hand rule*, the thumb gives the arrowhead sense of the vector and the fingers or curl indicate the tendency for rotation (twist or bending).

Fig. 7–1(c, d)

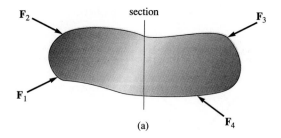

(a)

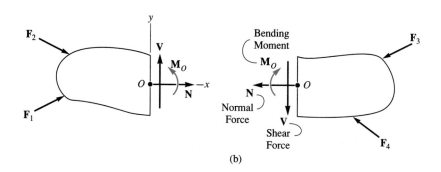

(b)

Fig. 7–2

Each of the six unknown x, y, z components of force and moment shown in Fig. 7–1d can be determined directly from the six equations of equilibrium, that is, Eqs. 4–7, applied to either segment of the body. If, however, the body is subjected to a *coplanar system of forces*, then only normal-force, shear, and bending-moment components will exist at the section. To show this, consider the body in Fig. 7–2a. If it is in equilibrium, then the internal resultant components, acting at the indicated section, can be determined by first "cutting" the body into two parts, as shown in Fig. 7–2b, and then applying the equations of equilibrium to one of the parts. Here the internal resultants consist of the normal force $\mathbf{N}$, shear force $\mathbf{V}$, and bending moment $\mathbf{M}_O$. These loadings must be equal in magnitude and opposite in direction on each of the sectioned parts (Newton's third law). Furthermore, the magnitude of each unknown is determined by applying the three equations of equilibrium to either one of these parts, Eqs. 4–3. If we use the x, y, z coordinate axes, with origin established at point O, as shown on the left segment, then a direct solution for $\mathbf{N}$ can be obtained by applying $\Sigma F_x = 0$, and $\mathbf{V}$ can be obtained directly from $\Sigma F_y = 0$. Finally, the bending moment $\mathbf{M}_O$ can be determined by summing moments about point O (the z axis), $\Sigma M_O = 0$, in order to eliminate the moments caused by the unknowns $\mathbf{N}$ and $\mathbf{V}$.

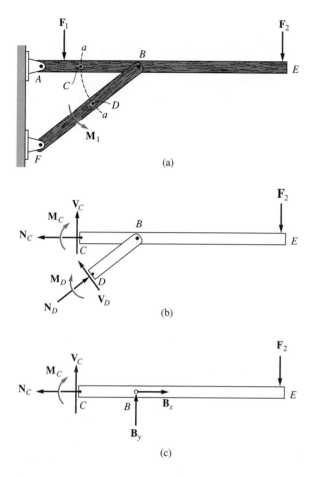

Fig. 7–3

Free-Body Diagrams. If the body represents one of the members or parts of a structure or machine, then the structure or machine may have to be disassembled in order to determine the support reactions on the member *before* computing the internal loadings. For example, consider the frame shown in Fig. 7–3*a*. If section *a–a* passes through the frame at *C* and *D*, the resulting free-body diagram of the right portion of the section is shown in Fig. 7–3*b*. At each cross-sectioned member there are *three* unknown internal resultants. As a result, we cannot apply the *three* available equations of equilibrium to obtain these *six unknowns*. Instead, to solve this problem we must *first* disassemble the frame and determine the reactions at the pin connections by applying the equations of equilibrium to each member. Then each member may be sectioned at its appropriate point and the equilibrium equations applied to determine the internal resultants. In this regard, the free-body diagram of segment *CBE*, Fig. 7–3*c*, can be used to determine the three internal loadings at *C*, provided the pin reactions $\mathbf{B}_x$ and $\mathbf{B}_y$ are known.

PROCEDURE FOR ANALYSIS

The method of sections is used to determine the *internal* loadings at a point located on the section of a body. These resultants are statically equivalent to the forces that are *distributed* over the material at the sectioned area. If the body is static, that is, at rest or moving with constant velocity, the resultants must be in equilibrium with the external loadings acting on either one of the sectioned segments of the body.

We will now present a procedure that can be used for applying the method of sections to determine the internal resultant normal force, shear force, bending moment, and torsional moment at a specific location in a body.

Support Reactions.   Before the body is sectioned, it may first be necessary to determine either its support reactions or the reactions at its connections. This is done by drawing the body's free-body diagram, establishing a coordinate system, and then applying the equations of equilibrium to the body.

Free-Body Diagram.   Keep all external distributed loadings, couple moments, torques, and forces acting on the body in their *exact locations*, then pass an imaginary section through the body at the point where the internal resultant loadings are to be determined. If the body represents a member of a structure or mechanical device, this section is often taken *perpendicular* to the axis of the member. Draw a free-body diagram of one of the "cut" segments and indicate the unknown resultants **N, V, M,** and **T** at the section. In most cases, these resultants are placed at the point representing the geometric center or *centroid* of the sectioned area. In particular, if the member is subjected to a *coplanar* system of forces, only **N, V,** and **M** act at the centroid.

Establish the *x, y, z* coordinate axes at the centroid and show the resultant components acting along the axes.

Equations of Equilibrium.   Apply the equations of equilibrium to obtain the unknown resultants. Moments should be summed at the section, about the axes where the resultants act. Doing this eliminates the unknown forces **N** and **V** and allows a direct solution for **M** (and **T**). If the solution of the equilibrium equations yields a negative value for a resultant, the assumed *directional sense* of the resultant is *opposite* to that shown on the free-body diagram.

The following examples illustrate this procedure numerically.

Example 7–1

Determine the resultant internal loadings acting on the cross section at *C* of the machine shaft shown in Fig. 7–4a. The shaft is supported by bearings at *A* and *B*, which exert only vertical forces on the shaft.

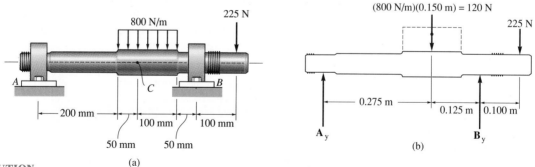

(a)

(b)

SOLUTION

Support Reactions. A free-body diagram of the entire shaft is shown in Fig. 7–4b. Note that the distributed loading has been replaced by its resultant, which has a magnitude equal to the area under the loading curve (rectangle) and acts through the centroid of this area.

$$\zeta + \ \Sigma M_A = 0; \quad -(120 \text{ N})(0.275 \text{ m}) + B_y(0.400 \text{ m})$$
$$- (225 \text{ N})(0.500 \text{ m}) = 0 \qquad B_y = 363.75 \text{ N}$$
$$+\uparrow \ \Sigma F_y = 0; \qquad A_y - 120 \text{ N} + 363.75 \text{ N} - 225 \text{ N} = 0$$
$$A_y = -18.75 \text{ N}$$

The negative sign for A_y indicates that $\mathbf{A}_y$ acts in the *opposite sense* to that shown in Fig. 7–4b.

Free-Body Diagrams. Passing an imaginary section perpendicular to the axis of the shaft through *C* yields the free-body diagrams of segments *AC* and *CB* shown in Fig. 7–4c. It is important when constructing these diagrams to keep the distributed loading *exactly* where it is until *after* the section is made. Only then should this loading be replaced by a single resultant force. Also, notice that $\mathbf{N}_C$, $\mathbf{V}_C$, and $\mathbf{M}_C$ have an equal magnitude but opposite direction on each segment—Newton's third law.

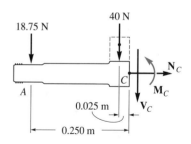

Equations of Equilibrium. Applying the equations of equilibrium to segment *AC*, we have

$$\xrightarrow{+} \ \Sigma F_x = 0; \qquad\qquad N_C = 0 \qquad\qquad Ans.$$
$$+\uparrow \ \Sigma F_y = 0; \qquad -18.75 \text{ N} - 40 \text{ N} - V_C = 0$$
$$V_C = -58.8 \text{ N} \qquad Ans.$$
$$\zeta + \ \Sigma M_C = 0; \quad M_C + 40 \text{ N}(0.025 \text{ m}) + 18.75 \text{ N}(0.250 \text{ m}) = 0$$
$$M_C = -5.69 \text{ N} \cdot \text{m} \qquad Ans.$$

What do the negative signs for V_C and M_C indicate? As an exercise, try to obtain the same results using segment *CB* of the shaft.

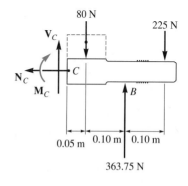

(c)

Fig. 7–4

Example 7–2

The hoist in Fig. 7–5a consists of the beam AB and attached pulleys, the cable, and the motor. Determine the resultant internal loadings acting on the cross section at C if the motor is lifting the 500-lb load W with constant velocity. Neglect the weight of the pulleys and beam.

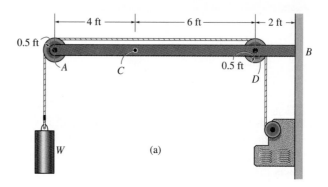

(a)

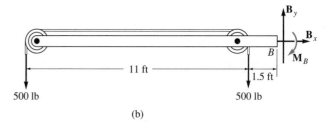

(b)

Fig. 7–5(a, b)

SOLUTION I

Since the load is hoisted at constant velocity, the equations of equilibrium apply.

Support Reactions. The free-body diagram of the beam *including* the pulleys and a portion of the cable is shown in Fig. 7–5b. The fixed wall at B exerts three reactions on the beam. Also, the cable is subjected to a constant tension force of 500 lb at its ends. Applying the equations of equilibrium yields

$\xrightarrow{+} \Sigma F_x = 0;$ $\qquad\qquad\qquad B_x = 0$

$+\uparrow \Sigma F_y = 0;$ $\qquad B_y - 500 \text{ lb} - 500 \text{ lb} = 0 \qquad B_y = 1000 \text{ lb}$

$\measuredangle + \Sigma M_B = 0;$ $\qquad -M_B + 500 \text{ lb}(1.5 \text{ ft}) + 500(12.5 \text{ ft}) = 0$

$\qquad\qquad\qquad\qquad M_B = 7000 \text{ lb} \cdot \text{ft}$

Free-Body Diagram. Using these results, the free-body diagrams of the pulley at D and beam segment CB are shown in Fig. 7–5c. Note that the 500-lb force components of the beam on the center of the pulley must be equal and opposite to the 500-lb force components of the pulley on the beam—Newton's third law.

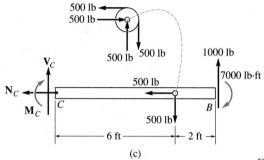

(c)

Fig. 7–5(c, d)

Equations of Equilibrium. For segment CB we have

$\xrightarrow{+} \Sigma F_x = 0; \quad -N_C - 500 \text{ lb} = 0 \qquad N_C = -500 \text{ lb} \qquad$ *Ans.*

$+\uparrow \Sigma F_y = 0; \quad V_C - 500 \text{ lb} + 1000 \text{ lb} = 0 \qquad V_C = -500 \text{ lb} \qquad$ *Ans.*

$\zeta+ \Sigma M_C = 0; \quad -M_C - 500 \text{ lb}(6 \text{ ft}) + 1000 \text{ lb}(8 \text{ ft}) - 7000 \text{ lb} \cdot \text{ft} = 0$

$$M_C = -2000 \text{ lb} \cdot \text{ft} \qquad \textit{Ans.}$$

The negative signs indicate that each of these loadings acts in the opposite direction to that shown in Fig. 7–5c.

SOLUTION II

This problem can be solved in a more *direct manner* by sectioning the cable and the beam at C and then considering the entire left segment.

Free-Body Diagram. See Fig. 7–5d.

Equations of Equilibrium

$\xrightarrow{+} \Sigma F_x = 0; \quad 500 \text{ lb} + N_C = 0 \qquad N_C = -500 \text{ lb} \qquad$ *Ans.*

$+\uparrow \Sigma F_y = 0; \quad -500 \text{ lb} - V_C = 0 \qquad V_C = -500 \text{ lb} \qquad$ *Ans.*

$\zeta+ \Sigma M_C = 0; \quad 500 \text{ lb}(4.5 \text{ ft}) - 500 \text{ .lb}(0.5 \text{ ft}) + M_C = 0$

$$M_C = -2000 \text{ lb} \cdot \text{ft} \qquad \textit{Ans.}$$

Try obtaining these same results by removing the pulley at A from the beam and showing the 500-lb force components of the pulley acting on the beam segment AC.

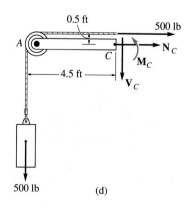

(d)

Example 7–3

Determine the resultant internal loadings acting on the cross section at G of the wooden beam shown in Fig. 7–6a. Assume the joints at A, B, C, D, and E are pin-connected.

Fig. 7–6

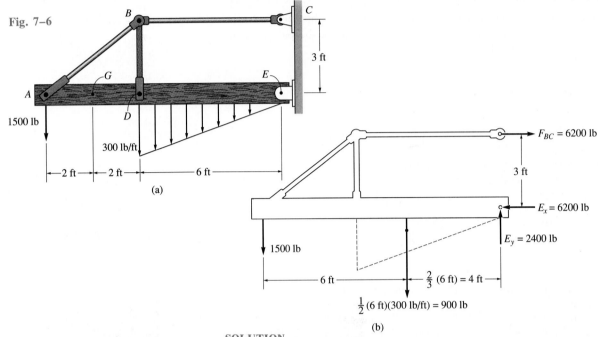

(a)

(b)

SOLUTION

Support Reactions. A free-body diagram of the *entire* structure is shown in Fig. 7–6b. Verify the computed reactions at E and C. In particular, note that BC is a *two-force member*, so the reaction at C must be horizontal. Why?

Also, since BA and BD are two-force members, the free-body diagram of joint B is shown in Fig. 7–6c. Again, verify the magnitudes of the computed forces $\mathbf{F}_{BA}$ and $\mathbf{F}_{BD}$.

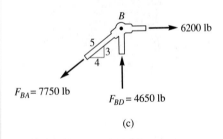

(c)

Free-Body Diagrams. Using the computed reaction at A, the left section AG of the beam is shown in Fig. 7–6d.

Equations of Equilibrium. Applying the equations of equilibrium to segment AG, since it involves the least number of forces, we have

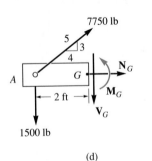

(d)

$$\xrightarrow{+} \Sigma F_x = 0; \quad 7750 \text{ lb}(\tfrac{4}{5}) + N_G = 0 \qquad N_G = -6200 \text{ lb} \qquad \textit{Ans.}$$

$$+\uparrow \Sigma F_y = 0; \quad -1500 \text{ lb} + 7750 \text{ lb}(\tfrac{3}{5}) - V_G = 0$$

$$V_G = 3150 \text{ lb} \qquad \textit{Ans.}$$

$$\downarrow^+ \Sigma M_G = 0; \quad M_G - (7750 \text{ lb})(\tfrac{3}{5})(2 \text{ ft}) + 1500 \text{ lb}(2 \text{ ft}) = 0$$

$$M_G = 6300 \text{ lb} \cdot \text{ft} \qquad \textit{Ans.}$$

As an exercise, compute these same results using segment GE.

Example 7–4

Determine the resultant internal loadings acting on the cross section at B of the pipe shown in Fig. 7–7a. The pipe has a mass of 2 kg/m and is subjected to both a vertical force of 50 N and a couple moment of 70 N · m at its end A. It is fixed-connected to the wall at C.

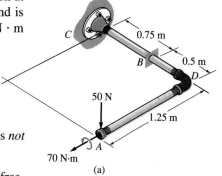

(a)

SOLUTION

The problem can be solved by considering segment AB, which does *not* involve the support reactions at C.

Free-Body Diagram. The x, y, z axes are established at B and the free-body diagram of segment AB is shown in Fig. 7–7b. The resultant force and moment components at the section are assumed to act in the positive coordinate directions and to pass through the *centroid* of the cross-sectional area at B. The weight of each segment of pipe is calculated as follows:

$$W_{BD} = (2 \text{ kg/m})(0.5 \text{ m})(9.81 \text{ N/kg}) = 9.81 \text{ N}$$
$$W_{AD} = (2 \text{ kg/m})(1.25 \text{ m})(9.81 \text{ N/kg}) = 24.52 \text{ N}$$

These forces act through the center of gravity of each segment.

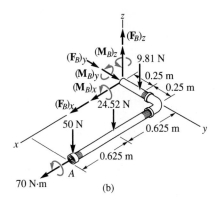

(b)

Fig. 7–7

Equations of Equilibrium. Applying the six scalar equations of equilibrium, we have*

$\Sigma F_x = 0$; $(F_B)_x = 0$ *Ans.*

$\Sigma F_y = 0$; $(F_B)_y = 0$ *Ans.*

$\Sigma F_z = 0$; $(F_B)_z - 9.81 \text{ N} - 24.52 \text{ N} - 50 \text{ N} = 0$

$(F_B)_z = 84.3 \text{ N}$ *Ans.*

$\Sigma (M_B)_x = 0$; $(M_B)_x + 70 \text{ N} \cdot \text{m} - 50 \text{ N}(0.5 \text{ m}) - 24.52 \text{ N}(0.5 \text{ m})$

$- 9.81 \text{ N}(0.25 \text{ m}) = 0$

$(M_B)_x = -30.3 \text{ N} \cdot \text{m}$ *Ans.*

$\Sigma (M_B)_y = 0$; $(M_B)_y + 24.52 \text{ N}(0.625 \text{ m}) + 50 \text{ N}(1.25 \text{ m}) = 0$

$(M_B)_y = -77.8 \text{ N} \cdot \text{m}$ *Ans.*

$\Sigma (M_B)_z = 0$; $(M_B)_z = 0$ *Ans.*

What do the negative signs for $(M_B)_x$ and $(M_B)_y$ indicate? Note that the normal force $N_B = (F_B)_y = 0$, whereas the shear force is $V_B = \sqrt{(0)^2 + (84.3)^2} = 84.3 \text{ N}$. Also, the torsional moment is $T_B = (M_B)_y = 77.8 \text{ N} \cdot \text{m}$ and the bending moment is $M_B = \sqrt{(30.3)^2 + (0)^2} = 30.3 \text{ N} \cdot \text{m}$.

---

*The *magnitude* of each moment about an axis is equal to the magnitude of each force times the perpendicular distance from the axis to the line of action of the force. The *direction* of each moment is determined using the right-hand rule, with positive moments (thumb) directed along the positive coordinate axes.

PROBLEMS

7–1. Determine the resultant internal normal force acting on the cross section through point A in each column. In (a), segment BC weighs 180 lb/ft and segment CD weighs 250 lb/ft. In (b), the column has a mass of 200 kg/m.

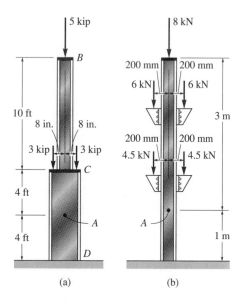

(a)   (b)

Prob. 7–1

7–2. Determine the resultant internal torque acting on the cross sections through points C and D on each shaft. The support bearings at A and B in (a) allow free turning of the shaft.

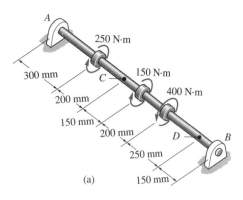

(a)

Prob. 7–2a

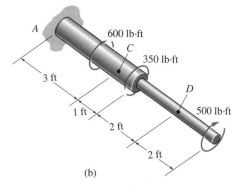

(b)

Prob. 7–2b

7–3. Determine the resultant internal normal and shear force in the member at (a) section a–a and (b) section b–b, each of which passes through point A. The 500-lb load is applied along the centroidal axis of the member.

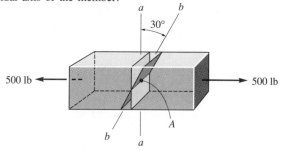

Prob. 7–3

*7–4. The beam supports the distributed load shown. Determine the resultant internal loadings on the cross section through point C. Assume the reactions at the supports A and B are vertical.

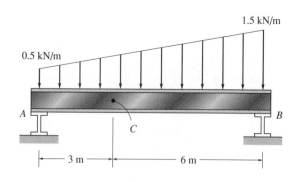

Prob. 7–4

7–5. Determine the resultant internal loadings on the cross section through point D of member AB.

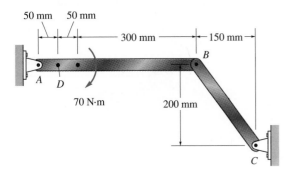

Prob. 7–5

7–6. Determine the resultant internal loadings on the cross sections located through points D and E of the frame.

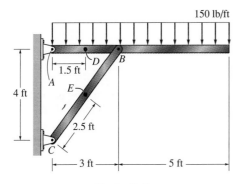

Prob. 7–6

7–7. Determine the resultant internal loadings on the cross sections located through points D and E of the frame.

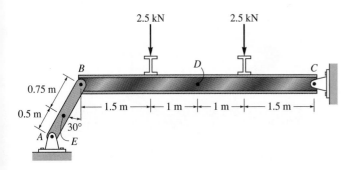

Prob. 7–7

*7–8. Determine the resultant internal loadings on the sections through points C and D of the pliers. There is a pin at A, and the jaws at B are smooth.

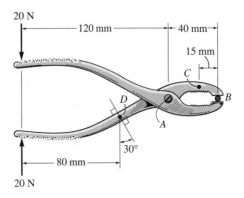

Prob. 7–8

7–9. Determine the resultant internal loadings acting on (a) section a–a and (b) section b–b. Each section is located through the centroid, point C.

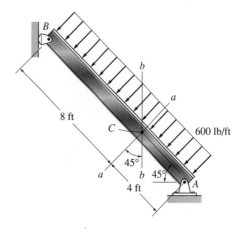

Prob. 7–9

7–10. The serving tray T used on an airplane is supported on *each side* by an arm. The tray is pin-connected to the arm at A and at B there is a smooth pin. (The pin can move within the slot in the arms to permit folding the tray against the front passenger seat when not in use.) Determine the resultant internal loadings in the arm on the cross section through point C when the tray arm supports the loads shown.

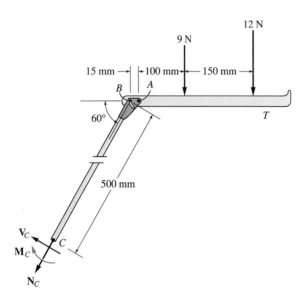

Prob. 7–10

7–11. Determine the resultant internal loadings acting on the cross section through point C in the beam. The load D has a mass of 300 kg and is being hoisted by the motor M with constant velocity.

*7–12. The 800-lb load is being hoisted at a constant speed using the motor M, which has a weight of 90 lb. Determine the resultant internal loadings acting on the cross section through point B in the beam. The beam has a weight of 40 lb/ft and is fixed to the wall at A.

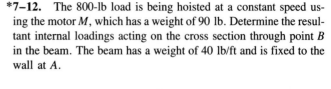

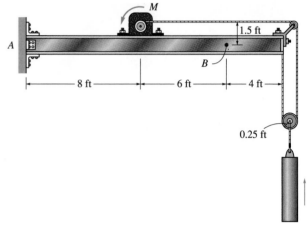

Prob. 7–12

7–13. The pipe has a mass of 12 kg/m. If it is fixed to the wall at A, determine the resultant internal loadings acting on the cross section located at B. Neglect the weight of the wrench CD.

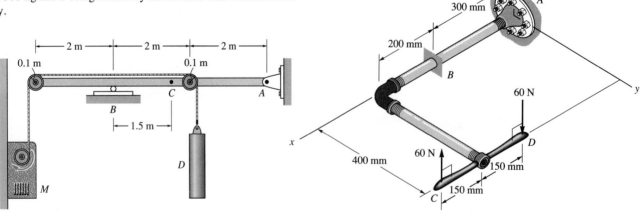

Prob. 7–11   Prob. 7–13

7–14. Determine the resultant internal loadings acting on the cross section through point B of the signpost. The post is fixed to the ground and a uniform pressure of $p = 7$ lb/ft² acts perpendicular to the face of the sign.

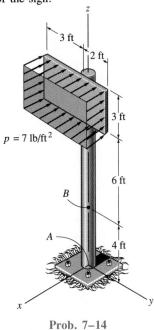

Prob. 7–14

7–15. The shaft is supported at its ends by two bearings A and B and is subjected to the forces applied to the pulleys fixed to the shaft. Determine the resultant internal loadings acting on the cross section located at point C. The 300-N forces act in the $-z$ direction and the 500-N forces act in the $+x$ direction. The journal bearings at A and B exert only x and z components of force on the shaft.

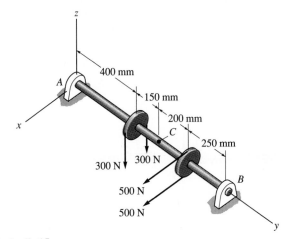

Prob. 7–15

*7–16. A hand crank that is used in a press has the dimensions shown. Determine the resultant internal loadings acting on the cross section at A if a vertical force of 50 lb is applied to the handle as shown. Assume the crank is fixed to the shaft at B.

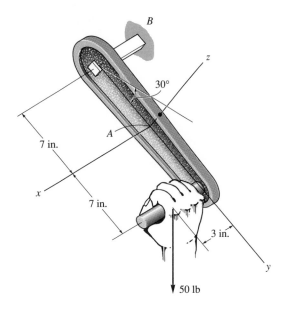

Prob. 7–16

7–17. The curved rod AD of radius r has a weight per length of w. If it lies in the horizontal plane, determine the resultant internal loadings acting on the cross section through point B. *Hint*: The distance from the centroid C of segment AB to point O is $CO = 0.9745r$.

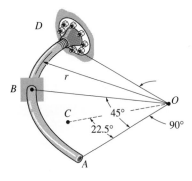

Prob. 7–17

7.2 Shear and Moment Equations and Diagrams

Beams are structural members which are designed to support loadings applied perpendicular to their axes. In general, they are long, straight bars having a constant cross-sectional area. The actual design of a beam requires a detailed knowledge of the *variation* of the internal shear force V and bending moment M acting at *each point* along the axis of the beam. After this force and bending-moment analysis is complete, we can then use the theory of mechanics of materials and an appropriate engineering design code to determine the beam's required cross-sectional area.

The *variations* of V and M as a function of the position x along the beam's axis can be obtained by using the method of sections discussed in Sec. 7.1. Here, however, it is necessary to section the beam at an *arbitrary distance x* from one end and compute V and M in terms of x. If the results are plotted, the graphical variations of V and M as a function of x are termed the *shear diagram* and *bending-moment diagram,* respectively.

In general, the internal shear and bending-moment functions obtained as a function of x will be *discontinuous,* or their slope will be discontinuous, at points where a distributed load changes or where concentrated forces or couples are applied. Because of this, shear and bending-moment functions must be determined for *each region* of the beam located *between* any two discontinuities of loading. For example, coordinates x_1, x_2, and x_3 will have to be used to describe the variation of V and M throughout the length of the beam in Fig. 7–8a. These coordinates will be valid *only* within the regions from A to B for x_1, from B to C for x_2, and from C to D to x_3. Although case of these coordinates has the *same* origin, this does not have to be the case. Instead, it may be easier to express V and M as functions of x_1, x_2, and x_3 having origins at A, C, and D as shown in Fig. 7–8b. Here x_1 is positive to the right and x_2 and x_3 are positive to the left.

The internal normal force will not be considered in the following discussion for two reasons. In most cases, the loads applied to a beam act perpendicular to the beam's axis and hence produce only an internal shear force and bending moment. For design purposes, the beam's resistance to shear, and particularly to bending, is more important than its ability to resist a normal force.

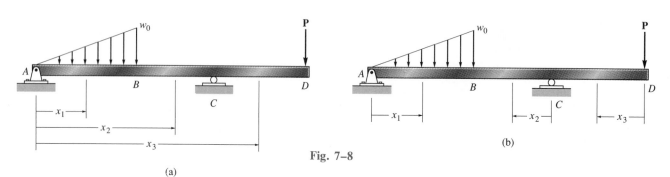

Fig. 7–8

Beam Sign Convention. Before presenting a method for determining the shear and moment as functions of x and later plotting these functions (shear and moment diagrams), it is first necessary to establish a *sign convention* so as to define "positive" and "negative" internal shear force and bending moment. [This is analogous to assigning coordinate directions x positive to the right and y positive upward when plotting a function $y = f(x)$.] Although the choice of a sign convention is arbitrary, here we will use the one used in the majority of books on mechanics. The convention is illustrated in Fig. 7–9. The *positive directions* require the *distributed load* to act *downward* on the beam, the internal *shear force* to cause a *clockwise* rotation of the beam segment on which it acts, and the internal *moment* to cause *compression* in the *top fibers* of the segment. Loadings that are opposite to these are considered negative.

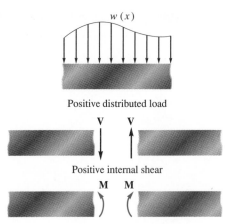

Positive distributed load

Positive internal shear

Positive internal moment

Beam sign convention

Fig. 7–9

PROCEDURE FOR ANALYSIS

The following procedure provides a method for determining the shear and moment functions and constructing the shear and moment diagrams for a beam.

Support Reactions. Draw a free-body diagram of the beam and determine all the support reactions. Resolve the forces into components acting perpendicular and parallel to the beam's axis.

Shear and Moment Functions. Select position coordinates x such that each coordinate extends into a region of the beam located *between* concentrated forces, couples, or discontinuities of distributed loading. The *origin* for each coordinate can be established at any suitable point, but usually it is at the beam's left end. Section the beam perpendicular to its axis at each position x and draw the free-body diagram of one of the two segments. Make sure that **V** and **M** are shown acting in their *positive sense,* in accordance with the sign conventions given in Fig. 7–9. Use the equilibrium equation $\Sigma F_y = 0$ to determine V as a function of x. The internal moment M as a function of x is obtained by summing moments about the axis of the beam that lies on the cross section and passes through the centroid of the cross-sectional area.

Once established, the results for V and M can be *checked* since they are related by Eq. 7–2, $V = dM/dx$. (See Sec. 7–3.) Also, the equation for shear can be checked, since by Eq. 7–1 its derivative must yield the negative of the loading, that is, $-w = dV/dx$.

Shear and Moment Diagrams. Plot the shear function (V versus x) and the moment function (M versus x). If numerical values of the functions describing V and M are *positive,* the values are plotted above the x axis, whereas negative values are plotted below the axis. Generally it is convenient to show the shear and moment diagrams directly below the free-body diagram of the beam.

Example 7–5

Draw the shear and bending-moment diagrams for the beam shown in Fig. 7–10a.

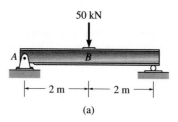

(a)

SOLUTION

Support Reactions. The support reactions have been computed, as shown on the beam's free-body diagram, Fig. 7–10d.

Shear and Moment Functions. The beam is sectioned at an arbitrary distance x from point A, extending within the region AB, and the free-body diagram of the left segment is shown in Fig. 7–10b. The unknowns **V** and **M** are assumed to act in the *positive sense* on the right-hand face of the segment according to the established sign convention. Why? Applying the equilibrium equations yields

$$+ \uparrow \Sigma F_y = 0; \qquad\qquad V = 25 \text{ kN} \qquad\qquad (1)$$

$$\zeta + \Sigma M = 0; \qquad\qquad M = 25x \text{ kN} \cdot \text{m} \qquad\qquad (2)$$

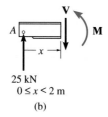

$0 \le x < 2 \text{ m}$

(b)

Fig. 7–10(a, b)

A free-body diagram for a left segment of the beam extending a distance x within the region BC is shown in Fig. 7–10c. As always, **V** and **M** are shown acting in the positive sense. Hence,

$$+ \uparrow \Sigma F_y = 0; \qquad\qquad 25 - 50 - V = 0$$

$$V = -25 \text{ kN} \qquad\qquad (3)$$

$$\zeta + \Sigma M = 0; \qquad\qquad M + 50(x - 2) - 25(x) = 0$$

$$M = (100 - 25x) \text{ kN} \cdot \text{m} \qquad\qquad (4)$$

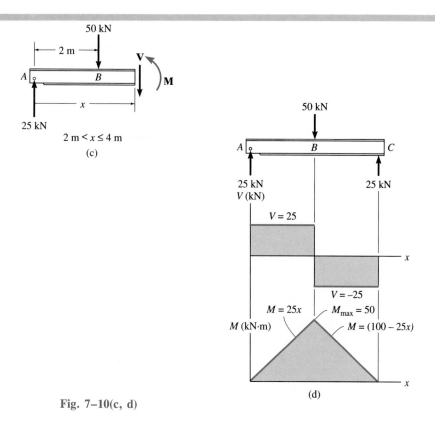

Fig. 7–10(c, d)

Shear and Moment Diagrams.   When Eqs. 1 through 4 are plotted within the regions in which they are valid, the shear and bending-moment diagrams shown in Fig. 7–10d are obtained. The shear diagram indicates that the internal shear force is always 25 kN (positive) within beam segment AB. Just to the right of point B, the shear force changes sign and remains at a constant value of −25 kN for segment BC. The moment diagram starts at zero, increases linearly to point C at x = 2 m, where M_{max} = 25 kN(2 m) = 50 kN · m, and thereafter decreases back to zero.

It is seen in Fig. 7–10d that the graph of the shear and moment diagrams is discontinuous at points of concentrated force, i.e., points A, B, and C. For this reason, as stated earlier, it is necessary to express both the shear and bending-moment functions separately for regions between concentrated loads. It should be realized, however, that all loading discontinuities are mathematical, arising from the *idealization of a concentrated force and couple moment.* Physically, loads are always applied over a finite area, and if this load variation could be accounted for, the shear and bending-moment diagrams would actually be continuous over the beam's entire length.

Example 7–6

Draw the shear and moment diagrams for the beam shown in Fig. 7–11a.

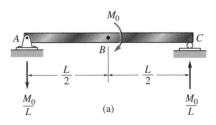

$$\frac{M_0}{L} \qquad (a) \qquad \frac{M_0}{L}$$

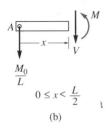

(b)

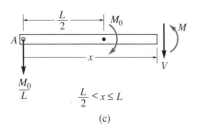

(c)

SOLUTION

Support Reactions. The support reactions have been computed in Fig. 7–11d.

Shear and Moment Functions. This problem is similar to the previous example, where two x coordinates must be used to express the shear and moment in the beam throughout its length. For the segment within region AB, Fig. 7–11b, we have

$$+\uparrow\ \Sigma F_y = 0; \qquad\qquad V = -\frac{M_0}{L}$$

$$\overset{\curvearrowleft}{+}\ \Sigma M_{NA} = 0; \qquad\qquad M = -\frac{M_0}{L}x$$

And for the segment within region BC, Fig. 11–5c,

$$+\uparrow\ \Sigma F_y = 0; \qquad\qquad V = -\frac{M_0}{L}$$

$$\overset{\curvearrowleft}{+}\ \Sigma M_{NA} = 0; \qquad\qquad M = M_0 - \frac{M_0}{L}x$$

$$M = M_0\left(1 - \frac{x}{L}\right)$$

Shear and Moment Diagrams. When the above functions are plotted, the shear and moment diagrams shown in Fig. 7–11d are obtained. In this case, notice that the shear is constant over the entire length of the beam; i.e., it is not affected by the couple moment $\mathbf{M_0}$ acting at the center of the beam. Just as a force creates a jump in the shear diagram, Example 7–5, a couple moment creates a jump in the moment diagram.

(d)

Fig. 7–11

Example 7–7

Draw the shear and moment diagrams for the beam shown in Fig. 7–12a.

SOLUTION

Support Reactions. The support reactions have been computed in Fig. 7–12c.

Shear and Moment Functions. A free-body diagram of the left segment of the beam is shown in Fig. 7–12b. The distributed loading on this segment is represented by its resultant force only *after* the segment is isolated as a free-body diagram. Since the segment has a length x, the *magnitude of the resultant force* is wx. This force acts through the centroid of the area comprising the distributed loading, a distance of $x/2$ from the right end. Applying the two equations of equilibrium yields

$$+\uparrow \ \Sigma F_y = 0; \qquad \frac{wL}{2} - wx - V = 0$$

$$V = w\left(\frac{L}{2} - x\right) \qquad (1)$$

$$\downarrow^+ \ \Sigma M_{NA} = 0; \qquad -\left(\frac{wL}{2}\right)x + (wx)\left(\frac{x}{2}\right) + M = 0$$

$$M = \frac{w}{2}(Lx - x^2) \qquad (2)$$

These results for V and M can be checked by noting that $dV/dx = -w$. This is indeed correct, since positive w acts downward. Also, notice that $dM/dx = V$, as expected.

Shear and Moment Diagrams. The shear and moment diagrams shown in Fig. 7–12c are obtained by plotting Eqs. 1 and 2. The point of *zero shear* can be found from Eq. 1:

$$V = w\left(\frac{L}{2} - x\right) = 0$$

$$x = \frac{L}{2}$$

From the moment diagram, this value of x happens to represent the point on the beam where the *maximum moment* occurs, since by Eq. 7–2, the slope $V = dM/dx = 0$. From Eq. 2, we have

$$M_{max} = \frac{w}{2}\left[L\left(\frac{L}{2}\right) - \left(\frac{L}{2}\right)^2\right]$$

$$= \frac{wL^2}{8}$$

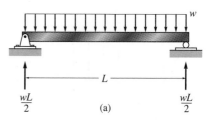

(a)

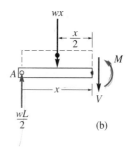

(b)

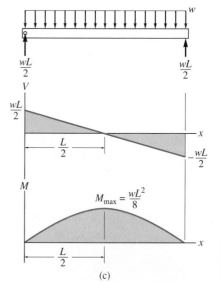

(c)

Fig. 7–12

Example 7–8

Draw the shear and bending-moment diagrams for the beam shown in Fig. 7–13a.

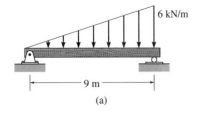

(a)

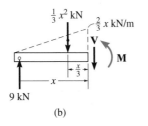

(b)

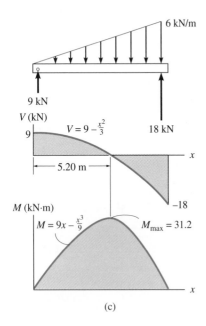

(c)

Fig. 7–13

SOLUTION

Support Reactions.   The support reactions have been computed as shown on the beam's free-body diagram, Fig. 7–13c.

Shear and Moment Functions.   A free-body diagram for a left segment of the beam having a length x is shown in Fig. 7–13b. The distributed loading acting on this segment has an intensity of $\frac{2}{3}x$ at its end and is replaced by a resultant force *after* the segment is isolated as a free-body diagram. The *magnitude* of the resultant force is equal to $\frac{1}{2}(x)(\frac{2}{3}x) = \frac{1}{3}x^2$. This force *acts through the centroid* of the distributed loading area, a distance $\frac{1}{3}x$ from the right end. Applying the two equations of equilibrium yields

$$+\uparrow \Sigma F_y = 0; \qquad 9 - \frac{1}{3}x^2 - V = 0$$

$$V = \left(9 - \frac{x^2}{3}\right) \text{ kN} \qquad (1)$$

$$\gtrdot +\Sigma M = 0; \qquad M + \frac{1}{3}x^2\left(\frac{x}{3}\right) - 9x = 0$$

$$M = \left(9x - \frac{x^3}{9}\right) \text{ kN} \cdot \text{m} \qquad (2)$$

Shear and Moment Diagrams.   The shear and bending-moment diagrams shown in Fig. 7–13c are obtained by plotting Eqs. 1 and 2.

The point of *zero shear* can be found using Eq. 1:

$$V = 9 - \frac{x^2}{3} = 0$$

$$x = 5.20 \text{ m}$$

This value of x happens to represent the point on the beam where the *maximum moment* occurs (see Sec. 7.3). Using Eq. (2), we have

$$M_{\max} = \left(9(5.20) - \frac{(5.20)^3}{9}\right) \text{ kN} \cdot \text{m}$$

$$= 31.2 \text{ kN} \cdot \text{m}$$

Example 7–9

Draw the shear and moment diagrams for the beam shown in Fig. 7–14a.

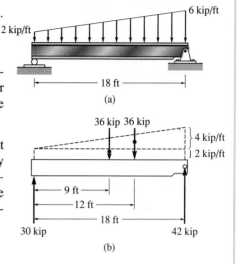

(a)

SOLUTION

Support Reactions. The distributed load is divided into component triangular and rectangular loadings and these loadings are then replaced by their resultant forces. The reactions have been computed as shown on the beam's free-body diagram, Fig. 7–14b.

Shear and Moment Functions. A free-body diagram of the left segment is shown in Fig. 7–14c. As above, the trapezoidal loading is replaced by rectangular and triangular distributions. Note that the intensity of the triangular load at the section is found by proportion. The resultant force and the location of each distributed loading are also shown. Applying the equilibrium equations, we have

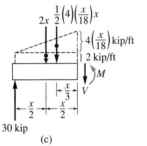

(b)

$$+\uparrow\ \Sigma F_y = 0; \qquad 30 - 2x - \frac{1}{2}(4)\left(\frac{x}{18}\right)x - V = 0$$

$$V = 30 - 2x - \frac{x^2}{9} \qquad (1)$$

$$\curvearrowleft+\ \Sigma M_{NA} = 0; \quad -30(x) + 2x\left(\frac{x}{2}\right) + \frac{1}{2}(4)\left(\frac{x}{18}\right)x\left(\frac{x}{3}\right) + M = 0$$

$$M = 30x - x^2 - \frac{x^3}{27} \qquad (2)$$

(c)

Equation 2 may be checked by noting that $dM/dx = V$, that is, Eq. 1. Also, $w = -dV/dx = 2 + \frac{2}{9}x$. This equation checks, since when $x = 0$, $w = 2$ kip/ft, and when $x = 18$ ft, $w = 6$ kip/ft, Fig. 7–14a.

Shear and Moment Diagrams. Equations 1 and 2 are plotted in Fig. 7–14d. Since the point of maximum moment occurs when $dM/dx = V = 0$, then, from Eq. 1,

$$0 = 30 - 2x - \frac{x^2}{9}$$

Choosing the positive root,

$$x = 9.73\ \text{ft}$$

Thus, from Eq. 2,

$$M_{\text{max}} = 30(9.73) - (9.73)^2 - \frac{(9.73)^3}{27}$$

$$= 163\ \text{kip}\cdot\text{ft}$$

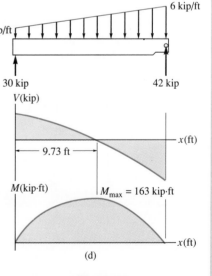

(d)

Fig. 7–14

PROBLEMS

For each of the following problems, establish the x axis with the origin at the left side of the member, and obtain the internal shear and moment as a function of x. Use these results to plot the shear and moment diagrams.

7–18. Draw the shear and moment diagrams for the shaft. The shaft is supported by journal bearings.

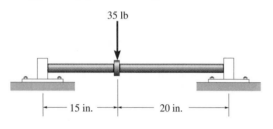

Prob. 7–18

7–19. The suspender bar supports the 600-lb engine. Draw the shear and moment diagrams for the bar.

Prob. 7–19

*7–20. Draw the shear and moment diagrams for the beam.

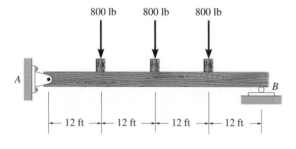

Prob. 7–20

7–21. Draw the shear and moment diagrams for the beam in terms of the parameters shown.

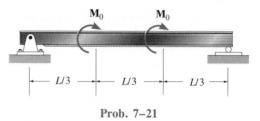

Prob. 7–21

7–22. Draw the shear and moment diagrams for the beam.

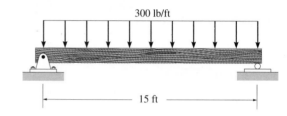

Prob. 7–22

7–23. Draw the shear and moment diagrams for the shaft. The supports at A and C are journal bearings.

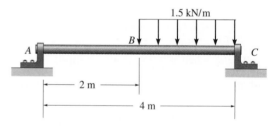

Prob. 7–23

*7–24. Draw the shear and moment diagrams for the beam.

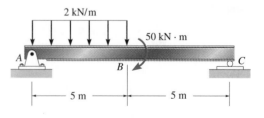

Prob. 7–24

7–25. Draw the shear and moment diagrams for the shaft. The supports at *A* and *B* are journal bearings.

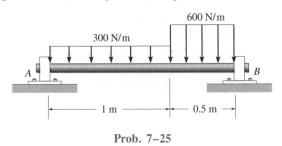

Prob. 7–25

7–26. The jib crane supports a load of 750 lb. If the boom *AB* has a uniform weight of 60 lb/ft, draw the shear and moment diagrams for the boom.

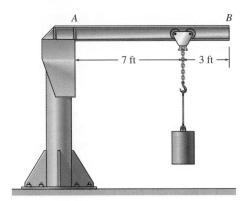

Prob. 7–26

7–27. Draw the shear and moment diagrams for beam *ABC*. Note that *A* is a fixed support and there is a pin at *B*. Solve the problem using the parameters shown.

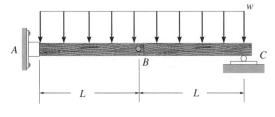

Prob. 7–27

*7–28. Draw the shear and moment diagrams for the beam.

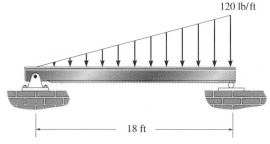

Prob. 7–28

7–29. Draw the shear and moment diagrams for the beam.

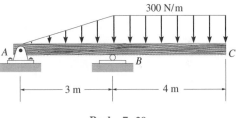

Prob. 7–29

7–30. Draw the shear and moment diagrams for the beam.

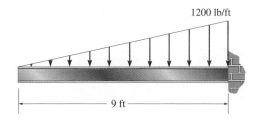

Prob. 7–30

7–31. Draw the shear and moment diagrams for the beam.

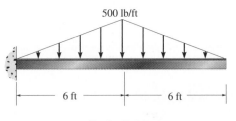

Prob. 7–31

7.3 Relations Between Distributed Load, Shear, and Moment

In cases where a beam is subjected to several concentrated forces, couple moments, and distributed loads, the method of constructing the shear and bending-moment diagrams discussed in Sec. 7.2 may become quite tedious. In this section a simpler method for constructing these diagrams is discussed—a method based on differential relations that exist between the load, shear, and bending moment.

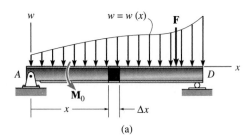

(a)

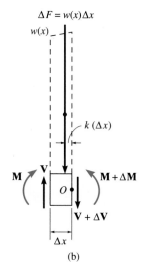

(b)

Fig. 7–15

Regions of Distributed Load. Consider the beam AD shown in Fig. 7–15a, which is subjected to an arbitrary distributed load $w = w(x)$ and a series of concentrated forces and couple moments. In the following discussion, the *distributed load* will be considered *positive* when the *loading acts downward* as shown. The free-body diagram for a small segment of the beam having a length Δx is shown in Fig. 7–15b. Since this segment has been chosen at a point x along the beam which is *not* subjected to a concentrated force or couple moment, any results obtained will not apply at points of concentrated loading. The internal shear force and bending moment shown on the free-body diagram are assumed to act in the *positive sense* according to the established sign convention, Fig. 7–15b. Note that both the shear force and moment acting on the right-hand face must be increased by a small, finite amount in order to keep the segment in equilibrium. The distributed loading has been replaced by a resultant force $\Delta F = w(x) \Delta x$ that acts at a fractional distance $k (\Delta x)$ from the right end, where $0 < k < 1$ [for example, if $w(x)$ is *uniform*, $k = \frac{1}{2}$]. Applying the equations of equilibrium, we have

$$+ \uparrow \Sigma F_y = 0; \qquad V - w(x) \, \Delta x - (V + \Delta V) = 0$$
$$\Delta V = -w(x) \, \Delta x$$

$$\underset{}{\downarrow} + \Sigma M_O = 0; \quad -V \, \Delta x - M + w(x) \, \Delta x[k(\Delta x)] + (M + \Delta M) = 0$$
$$\Delta M = V \, \Delta x - w(x)k(\Delta x)^2$$

Dividing by Δx and taking the limit as $\Delta x \to 0$, these two equations become

$$\frac{dV}{dx} = -w(x)$$

Slope of shear diagram	=	Negative of distributed load intensity

(7–1)

and

$$\frac{dM}{dx} = V$$

$$\frac{\text{Slope of}}{\text{moment diagram}} = \text{Shear}$$

(7–2)

These two equations provide a convenient means for quickly plotting the shear and moment diagrams for a beam. The shear diagram can be constructed by realizing that at each point along the beam the *slope* of the *shear diagram* equals the (negative) intensity of the *distributed loading* at the point (Eq. 7–1). In other words, if w is positive, i.e., acting downward, then the slope of the shear diagram will be negative, Fig. 7–16a and 7–16b. In a similar manner, the moment diagram can be constructed using data from the shear diagram, since the *slope* of the *moment diagram* at each point along the beam is equal to the *shear* at the point (Eq. 7–2), Fig. 7–16b and 7–16c. In particular, if the shear is equal to zero, then $dM/dx = 0$, and therefore a point of zero shear corresponds to a point of maximum (or possibly minimum) moment, Fig. 7–16b and 7–16c.

Equations 7–1 and 7–2 may also be rewritten in the form $dV = -w(x)\ dx$ and $dM = V\ dx$. Noting that $w(x)\ dx$ and $V\ dx$ represent differential areas under the distributed loading and shear diagram, respectively, we can integrate these areas between two points A and B along the beam, Fig. 7–16a, and write

$$\Delta V = -\int w(x)\ dx$$

$$\frac{\text{Change in}}{\text{shear}} = \frac{-\text{Area under}}{\text{distributed loading}}$$

(7–3)

$$\Delta M = \int V(x)\ dx$$

$$\frac{\text{Change in}}{\text{moment}} = \frac{\text{Area under}}{\text{shear diagram}}$$

(7–4)

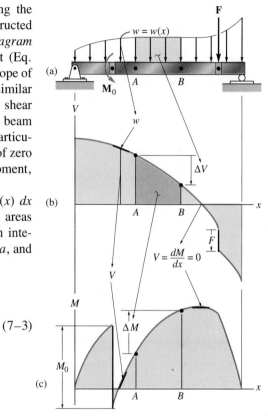

Fig. 7–16

Equation 7–3 states that the (negative) *area* under the distributed-loading curve between points A and B, Fig. 7–16a, is equal to the *change in shear* between these two points. Or stated another way, if the area under the loading curve is positive, i.e., due to a positive w, then the change in shear will be negative, Fig. 7–16b. Similarly, from Eq. 7–4, the area in shear diagram within the region from A to B is equal to the change in moment between A and B, Fig. 7–16c.

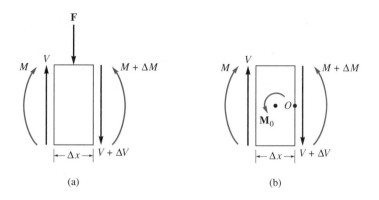

Fig. 7–17

Regions of Concentrated Force and Moment.

From the derivations given in Sec. 7.2, it should be apparent that Eqs. 7–1 and 7–3 cannot be used at points where an external force **F** acts, since these equations do not account for the sudden change in shear that occurs at these points. Similarly, Eq. 7–2 and Eq. 7–4 cannot be used at points where an external couple moment $\mathbf{M}_O$ is applied because of a discontinuity of moment. In order to account for these two cases, we must consider the free-body diagrams of differential segments of the beam that are located at a concentrated force and couple moment, Fig. 7–16a. These diagrams are shown in Fig. 7–17a and Fig. 7–17b, respectively.

From Fig. 7–17a it is seen that force equilibrium requires the change in shear to be

$$+\uparrow \ \Sigma F_y = 0; \qquad\qquad V - F - (V + \Delta V) = 0$$
$$\Delta V = -F \qquad\qquad (7\text{–}5)$$

Thus, when **F** acts *downward* on the beam, as in Fig. 7–16a, ΔV is *negative* so the shear "jumps" *downward*, Fig. 7–16b. Likewise, if **F** acts *upward*, the jump (ΔV) is *upward*.

From Fig. 7–17b, moment equilibrium requires the change in moment to be

$$\zeta + \ \Sigma M_O = 0; \qquad M + \Delta M + M_0 - V \Delta x - M = 0$$

Letting $\Delta x \rightarrow 0$, we get

$$\Delta M = -M_0 \qquad\qquad (7\text{–}6)$$

In this case, if $\mathbf{M}_0$ is applied *counterclockwise*, as in Fig. 7–16a, ΔM is *negative* so the moment diagram jumps *downward*, Fig. 7–16c. Likewise, when $\mathbf{M}_0$ acts *clockwise*, the jump (ΔM) must be *upward*.

Table 7–1 illustrates application of Eqs. 7–1, 7–2, 7–5, and 7–6 to some common loading cases. None of these results should be memorized; rather, each should be *carefully studied* so that you become fully aware of how the shear and moment diagrams can be constructed on the basis of knowing the *variation of the slope* from the load and shear diagrams, respectively. It would be well worth the time and effort to self-test your understanding of these concepts by covering over the shear and moment diagram columns in the table and then trying to reconstruct these diagrams on the basis of knowing the loading.

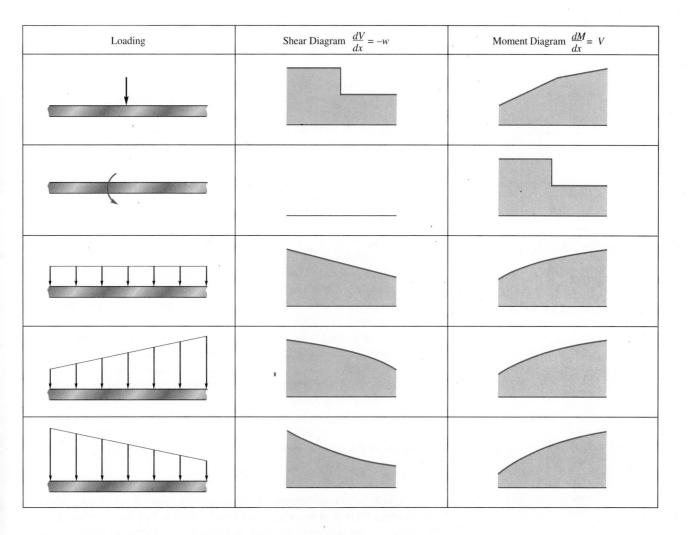

Loading	Shear Diagram $\dfrac{dV}{dx} = -w$	Moment Diagram $\dfrac{dM}{dx} = V$

Table 7–1

PROCEDURE FOR ANALYSIS

The following procedure provides a method for constructing the shear and moment diagrams for a beam based on the relations among distributed load, shear, and moment.

Support Reactions. Draw the free-body diagram of the beam and determine the support reactions. Resolve the forces acting on the beam into components that are perpendicular and parallel to the beam's axis.

Shear Diagram. Establish the V and x axes and plot the known values of the shear at the two *ends* of the beam.

Since $dV/dx = -w$, the *slope* of the *shear diagram* at any point is equal to the (negative) intensity of the *distributed loading* at the point; for example, if w acts downward, the slope of the shear diagram will be negative. See Table 7–1.

If a numerical value of the shear is to be determined at a point, one can find this value either by using the method of sections and the equation of force equilibrium, or by using $\Delta V = -\int w(x)\,dx$, which states that the *change in the shear* between any two points is equal to the (negative) *area under the load diagram* between the two points; for example, if w acts downward, the area under the curve gives a negative change in shear.

Since $w(x)$ must be *integrated* to obtain ΔV, then if $w(x)$ is a curve of degree n, $V(x)$ will be a curve of degree $n + 1$; for example, if $w(x)$ is uniform, $V(x)$ will be linear.

Moment Diagram. Establish the M and x axes and plot the known values of the moment at the *ends* of the beam.

Since $dM/dx = V$, the *slope* of the moment diagram at any point is equal to the *shear* at the point. See Table 7–1. In particular, note that at the point where the shear is zero, $dM/dx = 0$, and therefore this may be a point of maximum or minimum moment.

If a numerical value of the moment is to be determined at the point, one can find this value either by using the method of sections and the equation of moment equilibrium, or by using $\Delta M = \int V(x)\,dx$, which states that the *change in moment* between any two points is equal to the *area under the shear diagram* between the two points.

Since $V(x)$ must be *integrated* to obtain ΔM, then if $V(x)$ is a curve of degree n, $M(x)$ will be a curve of degree $n + 1$; for example, if $V(x)$ is linear, $M(x)$ will be parabolic.

The following examples illustrate this procedure. After working through them it is recommended that Examples 7–1 through 7–6 also be solved using this method. Also, it would be instructive to verify the solutions given in Figs. 7–10, 7–11, 7–12, 7–13, and 7–14.

Example 7-10

Draw the shear and moment diagrams for the beam in Fig. 7–18a.

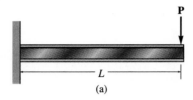

(a)

SOLUTION

Support Reactions.   The reactions are calculated and shown on a free-body diagram, Fig. 7–18b.

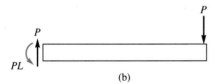

(b)

Shear Diagram.   The shear at each end of the beam is plotted first, i.e., at $x = 0$, $V = +P$ and at $x = L$, $V = +P$, Fig. 7–18c. Since $w = 0$ for $0 < x < L$, Fig. 7–18b, the slope of the shear diagram will be zero ($dV/dx = -w = 0$), and therefore a horizontal straight line connects the end points.

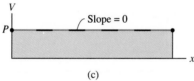

(c)

Moment Diagram.   The moment at each end of the beam is plotted first, i.e., at $x = 0$, $M = -PL$ and at $x = L$, $M = 0$, Fig. 7–18d. The shear diagram, Fig. 7–18c, indicates the slope of the moment diagram will be *constant-positive* for $0 < x < L$, such that $dM/dx = V = +P$. Hence, the end points are connected by a straight positive-sloped line as shown in the figure.

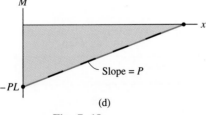

(d)

Fig. 7–18

■ Example 7–11

Draw the shear and moment diagrams for the beam shown in Fig. 7–19a.

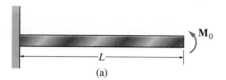

(a)

SOLUTION

Support Reactions. The reaction at the fixed support has been calculated and is shown on the free-body diagram, Fig. 7–19b.

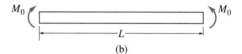

(b)

Shear Diagram. The shear at each end point, $x = 0$ and $x = L$, is plotted first, Fig. 7–19c. Since no load exists on the beam for $0 < x < L$, Fig. 7–19b, the shear diagram will have zero slope, $dV/dx = 0$. Therefore, a horizontal line connects the end points, which indicates that the shear is zero throughout the beam.

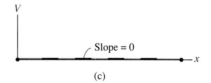

(c)

Moment Diagram. The moment M_0 at the beam's end points $x = 0$ and $x = L$ is plotted first, Fig. 7–19d. The shear diagram, Fig. 7–19c, indicates that the slope of the moment diagram will be zero for $0 < x < L$ since $dM/dx = V = 0$. Therefore, a horizontal line connects the end points as shown.

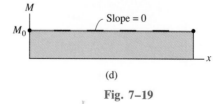

(d)

Fig. 7–19

Example 7–12

Draw the shear and moment diagrams for the beam shown in Fig. 7–20a.

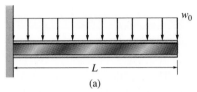

(a)

SOLUTION

Support Reactions. The reactions at the fixed support have been calculated and are shown on the free-body diagram, Fig. 7–20b.

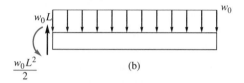

(b)

Shear Diagram. The shear at each end point, $x = 0$ and $x = L$, is plotted first, Fig. 7–20c. The distributed loading on the beam is constant-positive, and since $dV/dx = -w_0$, the slope of the shear diagram will be constant-negative. Thus, a straight negative-sloped line connects the end points.

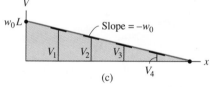

(c)

Moment Diagram. The moment at each end point, $x = 0$ and $x = L$, is plotted first, Fig. 7–20d. Successive values of shear on the shear diagram indicate that the slope of the moment diagram will always be positive, yet it decreases linearly, from $dM/dx = w_0L$ at $x = 0$ to $dM/dx = 0$ at $x = L$. Since the shear diagram is *linear,* the moment diagram will be *parabolic,* having a linear decreasing slope as shown in the figure.

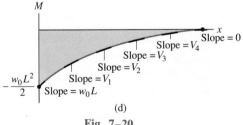

(d)

Fig. 7–20

Example 7–13

Draw the shear and moment diagrams for the beam shown in Fig. 7–21a.

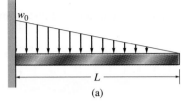

(a)

SOLUTION

Support Reactions. The reactions at the fixed support have been calculated and are shown on the free-body diagram, Fig. 7–21b.

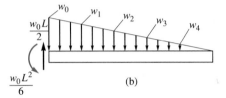

(b)

Shear Diagram. The shear at each end point, $x = 0$ and $x = L$, is plotted first, Fig. 7–21c. The distributed loading on the beam is positive yet linearly decreasing. Therefore the slope of the shear diagram will be *negatively decreasing* from $dV/dx = -w_0$ at $x = 0$ to $dV/dx = 0$ at $x = L$. Since the loading has a *linear distribution,* the shear diagram is a *parabola* having a negatively decreasing slope.

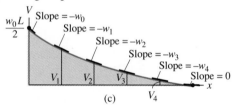

(c)

Moment Diagram. The moment at each end point, $x = 0$ and $x = L$, is plotted first, Fig. 7–21d. From the shear diagram, the slope of the moment diagram will vary parabolically; i.e., it will always be positive yet decreasing, from $dM/dx = +w_0 L/2$ at $x = 0$ to $dM/dx = 0$ at $x = L$. The curve connecting the plotted end points that has this characteristic is a *cubic* function of x, as shown in the figure.

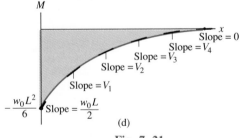

(d)

Fig. 7–21

Example 7–14

Draw the shear and moment diagrams for the beam in Fig. 7–22a.

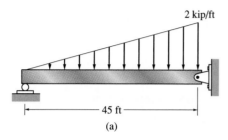

2 kip/ft

45 ft

(a)

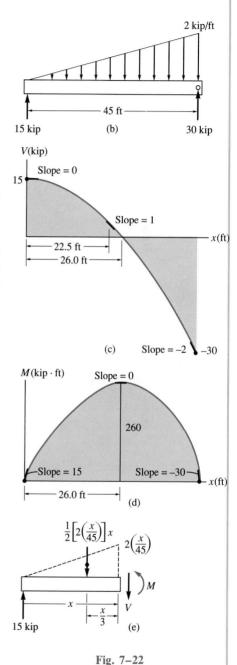

SOLUTION

Support Reactions. The reactions have been calculated and are shown on the free-body diagram of the beam, Fig. 7–22b.

Shear Diagram. The end points $x = 0$, $V = +15$ and $x = 45$, $V = -30$ are plotted first. As shown from the beam's distributed loading, the slope of the shear diagram will vary from $dV/dx = 0$ at $x = 0$ to $dV/dx = -2$ at $x = 45$. In general, for $0 \le x \le 45$, the slope of the shear diagram will be *increasingly negative* since the distributed load is increasingly positive $(dV/dx = -w)$, Fig. 7–22c. As a result, the shear diagram is a parabola having the slope shown.

The point of zero shear can be found by using the method of sections for a beam segment of length x, Fig. 7–22e. We require that $V = 0$, so

$$+\uparrow \ \Sigma F_y = 0; \quad 15 - \frac{1}{2}\left[2\left(\frac{x}{45}\right)\right]x = 0; \quad x = 26.0 \text{ ft}$$

Moment Diagram. The end points $x = 0$, $M = 0$ and $x = 45$, $M = 0$ are plotted first, Fig. 7–22d. From the shear diagram, the slope of the moment diagram will be $dM/dx = 15$ at $x = 0$ and $dM/dx = -30$ at $x = 45$. In general, for $0 \le x < 26.0$ the slope will be *decreasingly positive,* since the shear is decreasingly positive. Likewise, for $26.0 < x \le 45$ the slope will be *increasingly negative,* Fig. 7–22d. Here the moment diagram is a cubic function of x. Why?

Notice that the maximum moment is at $x = 26.0$, since $dM/dx = V = 0$ at this point. From the free-body diagram in Fig. 7–22e we have

$$\zeta + \ \Sigma M_{NA} = 0; \quad -15(26.0) + \frac{1}{2}\left[2\left(\frac{26.0}{45}\right)\right](26.0)\left(\frac{26.0}{3}\right) + M = 0$$

$$M = 260 \text{ kip} \cdot \text{ft}$$

Fig. 7–22

Example 7–15

Draw the shear and moment diagrams for the beam shown in Fig. 7–23a.

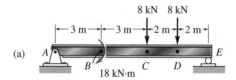

(a)

Fig. 7–23(a)

SOLUTION

Support Reactions. The reactions are calculated and indicated on the free-body diagram, Fig. 7–23b.

Shear Diagram. The values of the shear at the end points A and E are plotted first. At $x = 0$, $V_A = +3$ kN, and at $x = 10$, $V_E = -13$ kN, Fig. 7–23c. At an intermediate point between A and C, $w(x) = 0$, so the slope of the shear diagram will be zero, $dV/dx = -w(x) = 0$. Hence the shear retains its value of $+3$ kN within this region up to point C. At C the shear is *discontinuous,* since there is a *concentrated force* of 8 kN there. The value of the shear just to the right of C (-5 kN) can be found by sectioning the beam at this point. This yields the free-body diagram shown in equilibrium in Fig. 7–23e. This point ($V = -5$ kN) is plotted on the shear diagram. As before, $w(x) = 0$ from C to D and from D to E, Fig. 7–23b, so the slope of the shear diagram will be zero in these regions. The diagram "jumps" again at D, as shown, then closes to the value of -13 kN at E. Notice that no "jump" or discontinuity in shear occurs at B, the point where the 18-kN · m couple moment is applied, Fig. 7–23b. The reason for this can be shown by considering the equilibrium of the free-body diagram in Fig. 7–23f.

It should be noted that based on Eq. 7–5, the shear diagram can also be constructed by "following the load" on the free-body diagram, Fig. 7–23b. In this regard, beginning at A the 3-kN force acts upward, so $V_A = +3$ kN. No distributed load acts between A and C, so the shear remains constant ($dV/dx = 0$). At C the 8-kN force is down, so the shear jumps down 8 kN, from $+3$ kN to -5 kN. Again the shear is constant from C to D (no distributed load), then at D it jumps down another 8 kN to -13 kN. Finally, with no distributed load between D and E, it ends at -13 kN at point E.

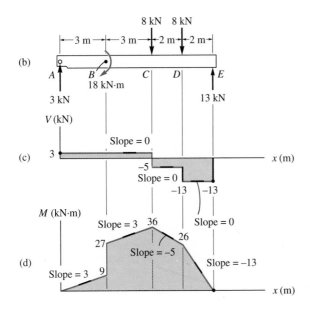

Fig. 7–23(b–f)

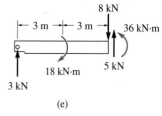

Moment Diagram. The moment at each end of the beam is zero. These two points are plotted first, Fig. 7–23d. From the shear diagram, the slope of the moment diagram from A to C is constant yet positively increasing at $+3$. The value of the moment just to the *left* of B can be determined by using the method of sections and statics or by computing the area under the shear diagram between A and B, that is, $\Delta M_{AB} = M_B - M_A = (3\ kN)(3\ m) = 9\ kN \cdot m$. Since $M_A = 0$, then $M_B = 0 + 9\ kN \cdot m = 9\ kN \cdot m$. A jump occurs at point B due to the concentrated couple moment of 18 kN · m. Using Eq. 7–6, this jump is 18 kN · m *upward*, since the couple moment is *clockwise*. Also, the method of sections, Fig. 7–23f, gives the same value of $M_{B^+} = +27\ kN \cdot m$ just to the right of B. From this point, the slope of $dM/dx = +3$ is maintained until the diagram reaches a peak of 36 kN · m. Again, this value can be obtained by using the method of sections or by finding the area under the shear diagram from B to C, that is, $\Delta M_{BC} = (3\ kN)(3\ m) = 9\ kN \cdot m$, so that $M_C = 27\ kN \cdot m + 9\ kN \cdot m = 36\ kN \cdot m$. Continuing in this manner, verify the value of 26 kN · m at D and closure to zero at E.

Example 7-16

Draw the shear and moment diagrams for the overhanging beam shown in Fig. 7–24a.

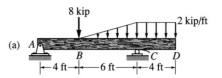

(a)

SOLUTION

Support Reactions. The free-body diagram with the calculated support reactions is shown in Fig. 7–24b.

Shear Diagram. As usual we start by plotting the end shears $V_A =$ +4.40 kip and $V_D = 0$, Fig. 7–24c. Since there is no load from A to B, the shear diagram will have zero slope. Furthermore, since w is increasingly positive from B to C, the slope of the shear diagram will be *increasingly negative*. And, lastly, from C to D, w is constant, so the slope of the shear diagram will be constant yet negative, Fig. 7–24c. Try to "follow the load" on the free-body diagram and establish the peak values on the shear diagram. Use the appropriate areas under the load diagram (w curve) to find the change in shear. For example, $\Delta V_{BC} =$ $-(1/2)(6 \text{ ft})(2 \text{ kip/ft}) = -6 \text{ kip}$, so that $V_C = -3.60 \text{ kip} - 6 \text{ kip} =$ -9.60 kip just to the left of point C.

Moment Diagram. The end moments $M_A = 0$ and $M_D = 0$ are plotted first. Study the diagram and note how the slopes and therefore the various curves are established from the shear diagram using $dM/dx = V$. Verify the numerical values for the peaks using the method of sections and statics or by computing the appropriate areas under the shear diagram to find the change in moment. Note that the point of zero moment can be determined by establishing M as a function of x, where, for convenience, x extends *from* point B into region BC, Fig. 7–24e. Hence,

$$\zeta+ \ \Sigma M_{NA} = 0; \ -4.40(4+x) + 8(x) + \frac{1}{2}\left(\frac{2}{6}\right)x(x)\left(\frac{x}{3}\right) + M = 0$$

$$M = -\frac{1}{18}x^3 - 3.60x + 17.6 = 0$$

$$x = 3.94 \text{ ft}$$

Reviewing these diagrams, we see that for region AB the load is zero, shear is constant, and moment is linear; for region BC the load is linear, shear is parabolic, and moment is cubic; and for region CD the load is constant, the shear is linear, and the moment is parabolic.

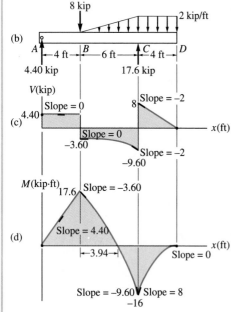

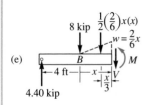

Fig. 7–24

PROBLEMS

*7–32. Draw the shear and moment diagrams for the beam in Prob. 7–20.

7–33. Draw the shear and moment diagrams for the beam in Prob. 7–23.

7–34. Draw the shear and moment diagrams for the beam in Prob. 7–24.

7–35. Draw the shear and moment diagrams for the beam in Prob. 7–30.

*7–36. Draw the shear and moment diagrams for the cantilever beam.

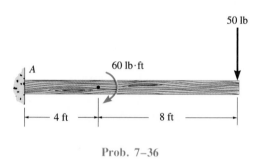

Prob. 7–36

7–37. Draw the shear and moment diagrams for the beam. There is a pin at *C*. And *A* is a fix support.

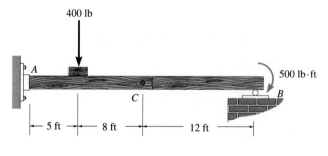

Prob. 7–37

7–38. Draw the shear and moment diagrams for the shaft. The supports at *A* and *B* are journal bearings.

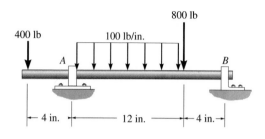

Prob. 7–38

7–39. The beam has a weight of 150 lb/ft. Draw the shear and moment diagrams for the beam.

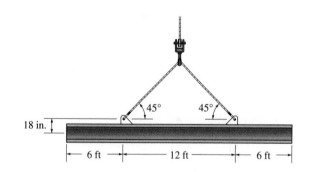

Prob. 7–39

*7–40. Draw the shear and moment diagrams for the beam.

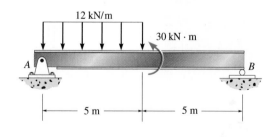

Prob. 7–40

7–41. Draw the shear and moment diagrams for the shaft. The supports at A and B are journal bearings.

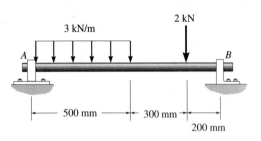

Prob. 7–41

7–42. Draw the shear and moment diagrams for the beam.

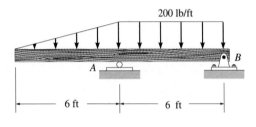

Prob. 7–42

7–43. The smooth pin is supported by two leaves A and B and subjected to a compressive load of 0.4 kN/m caused by bar C. Determine the intensity of the distributed load w_0 of the leaves on the pin and draw the shear and moment diagrams for the pin.

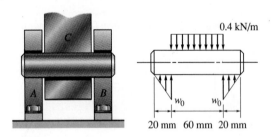

Prob. 7–43

*7–44. Draw the shear and moment diagrams for the beam.

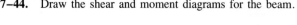
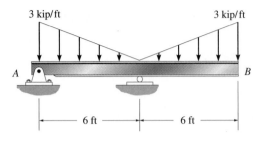

Prob. 7–44

7–45. The beam consists of two segments pin-connected at B. Draw the shear and moment diagrams for the beam.

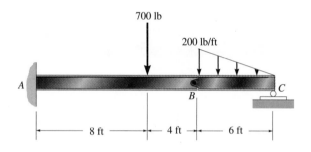

Prob. 7–45

7–46. Draw the shear and moment diagrams for the beam in terms of the parameters shown.

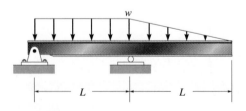

Prob. 7–46

REVIEW PROBLEMS

7–47. Draw the shear and moment diagrams for the beam.

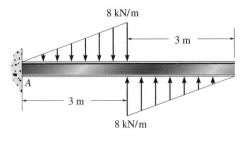

8 kN/m

3 m

A

3 m

8 kN/m

Prob. 7–47

7–49. Draw the shear and moment diagrams for the beam.

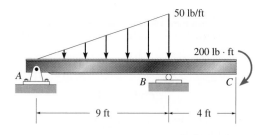

50 lb/ft

200 lb · ft

A

B

C

9 ft

4 ft

Prob. 7–49

7–50. Draw the shear and moment diagrams for the beam.

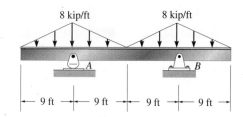

8 kip/ft

8 kip/ft

A

B

9 ft

9 ft

9 ft

9 ft

Prob. 7–50

7–51. The beam is supported by a pin at C and a rod AB. Determine the normal force, shear force, and moment at point D.

*7–48. Determine the normal force, shear force, and moment at point C. Assume the support at A is a roller and the support at B is a pin.

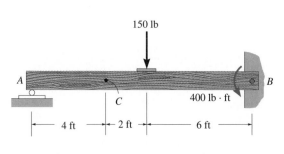

150 lb

A

C

B

400 lb · ft

4 ft

2 ft

6 ft

Prob. 7–48

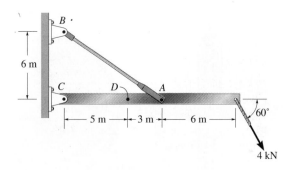

B

6 m

C

D

A

5 m

3 m

6 m

60°

4 kN

Prob. 7–51

*7–52. Draw the shear and moment diagrams for the shaft. The supports at *A* and *B* are journal bearings.

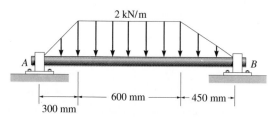

Prob. 7–52

7–53. Determine the normal force, shear force, and moment at points *D* and *E* of the frame.

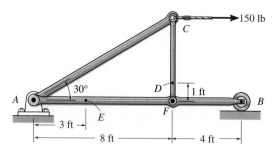

Prob. 7–53

7–54. The sheave on a traveling block supports a load of 15 kip. Draw the moment diagram for the supporting pin *AB* if the force is (*a*) assumed concentrated as a single force acting at the *center E* of the pin, and (*b*) uniformly distributed over the sheave contact area along *CD*.

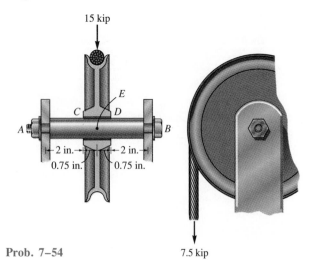

Prob. 7–54

7–55. Draw the shear and moment diagrams for the beam.

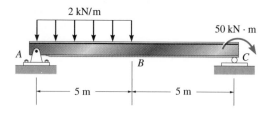

Prob. 7–55

*7–56. Each of the street lights has a weight of 20 lb, and the supporting arm weighs 5 lb/ft. Draw the shear and moment diagrams for the arm.

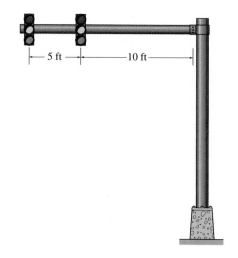

Prob. 7–56

MECHANICS OF MATERIALS

8 Stress and Strain

In this chapter we will first introduce the concepts of normal and shear stress, and then give specific applications for the analysis and design of members subjected to an axial load and direct shear. Afterwards we will define normal and shear strain, and show how they can be determined for various types of problems.

8.1 Introduction

As stated in Chapter 1, *mechanics of materials* is a branch of mechanics that develops relationships between the *external* loads applied to a deformable body and the intensity of *internal* forces acting within the body. This subject is also concerned with computing the *deformations* of the body, and it provides a study of the body's *stability* when the body is subjected to external forces.

In the design of any structure or machine, it is *first* necessary to use the principles of statics to determine the forces acting both on and within its various members. Furthermore, the size of the members, their deflection, and their stability depend not only on these internal loadings, but also on the nature of the material from which the members are made. As a result, an accurate determination and fundamental understanding of *material behavior* is of vital importance to the development of the necessary equations used in mechanics of materials. Here we will not be interested in the specific details of experimental methods. Instead, we will state the experimental results and explain how they are used. Realize that many formulas and rules for design, as defined in engineering codes, are based on the fundamentals of mechanics of materials, and for this reason an understanding of the principles of this subject is very important.

8.2 Stress

In Sec. 7.1 we showed how to determine the internal resultant force and moment acting at a specified point on the sectioned area of a body, Fig. 8–1a. It was stated that these two loadings represent the resultant effects of the actual *distribution of force* acting over the sectioned area, Fig. 8–1b. Obtaining this *distribution* of internal loading, however, is one of the major problems in mechanics of materials.

To solve this problem it will be necessary to study how the body deforms under load, since each internal force distribution will *deform* the material in a *unique* way. In later chapters we will show how measurements of material deformation can be related to the force distribution. Before this can be done, however, it is first necessary to develop a means for describing each of the internal forces at all *points* on the sectioned area. To do this we will establish the concept of stress.

Notice that if the sectioned area is subdivided into small areas, such as the one ΔA shown shaded in Fig. 8–1b, then, on each of these areas, the force distribution will become more *uniform* as the area gets smaller and smaller. As we reduce the area in this manner, however, we must make two assumptions regarding the properties of the material. We will consider the material to be *continuous*, that is, to consist of a continuum or uniform distribution of matter having no voids, rather than being composed of a finite number of distinct atoms or molecules. Furthermore, the material must be *cohesive*, meaning that all portions of it are connected together, rather than having breaks, cracks, or separations. Now, as the subdivided area ΔA of this continuous-cohesive material is reduced to one of *infinitesimal size*, the distribution of force acting over the entire sectioned area will consist of an *infinite number* of forces, each acting at a specific point on the area. A typical finite yet very small force $\Delta \mathbf{F}$, acting on its associated area ΔA, is shown in Fig. 8–1c. This force, like all the others, will have a unique direction, but for further discussion we will replace it by two of its *components*, namely, $\Delta \mathbf{F}_n$ and $\Delta \mathbf{F}_t$, which are taken normal and tangent to the area, respectively. As the area ΔA becomes smaller and smaller, and approaches zero, so do the force $\Delta \mathbf{F}$ and its components; however, the quotient of the force and area will, in general, approach a finite limit. This quotient is called *stress*, and it describes the *intensity of the internal force* on a *specific plane* passing through a point.

Fig. 8–1

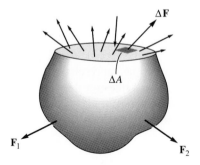

(a)

(b)

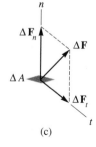

(c)

Normal Stress. The *intensity* of force, or force per unit area, acting normal to ΔA is defined as the *normal stress*, σ (sigma). Mathematically it can be expressed as

$$\sigma = \lim_{\Delta A \to 0} \frac{\Delta F_n}{\Delta A} \qquad (8\text{--}1)$$

If the normal force or stress "pulls" on the area element ΔA as shown in Fig. 8–1c, it is referred to as *tensile stress*, whereas if it "pushes" on ΔA it is called *compressive stress*.

Shear Stress. Likewise, the intensity of force, or force per unit area, acting tangent to ΔA is called the *shear stress*, τ (tau). This component is expressed mathematically as

$$\tau = \lim_{\Delta A \to 0} \frac{\Delta F_t}{\Delta A} \qquad (8\text{--}2)$$

In Fig. 8–1c, note that the orientation of the area ΔA completely specifies the direction of $\Delta \mathbf{F}_n$, which is always perpendicular to the area. On the other hand, each shear force $\Delta \mathbf{F}_t$ can act in an infinite number of directions within the plane of the area. Provided, however, the direction of $\Delta \mathbf{F}$ is known, then the direction of $\Delta \mathbf{F}_t$ can be established as shown in the figure.

Cartesian Stress Components. To specify further the direction of the shear stress, we will resolve it into rectangular components, and to do this we will make reference to x, y, z coordinate axes, oriented as shown in Fig. 8–2a. Here the element of area $\Delta A = \Delta x\, \Delta y$ and the three Cartesian components of $\Delta \mathbf{F}$ are shown in Fig. 8–2b. We can now express the normal-stress component as

$$\sigma_z = \lim_{\Delta A \to 0} \frac{\Delta F_z}{\Delta A}$$

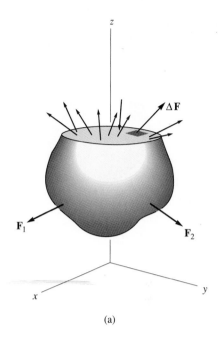

(a)

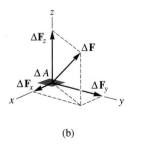

(b)

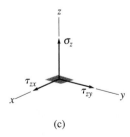

(c)

Fig. 8–2

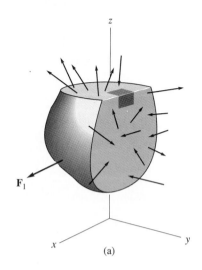

(a)

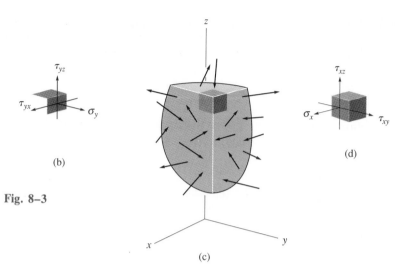

(b)

Fig. 8–3

(d)

(c)

and the two shear-stress components as

$$\tau_{zx} = \lim_{\Delta A \to 0} \frac{\Delta F_x}{\Delta A}$$

$$\tau_{zy} = \lim_{\Delta A \to 0} \frac{\Delta F_y}{\Delta A}$$

The subscript notation z in σ_z is used to reference the *direction* of the outward normal line, which specifies the orientation of the area ΔA. Two subscripts are used for the shear-stress components, τ_{zx} and τ_{zy}. The z specifies the orientation of the area, and x and y refer to the direction lines for the shear stresses.

To summarize these concepts, the intensity of the internal force at a *point* in a body *must* be described on an *area having a specified orientation*. This intensity can then be measured using three components of stress acting on the area. The normal component acts normal or perpendicular to the area, and the shear components act within the plane of the area. These three stress components are shown graphically in Fig. 8–2c.

Now consider passing another imaginary section through the body parallel to the x–z plane and intersecting the front side of the element shown in Fig. 8–2a. The resulting free-body diagram is shown in Fig. 8–3a. Resolving the force acting on the area $\Delta A = \Delta x\, \Delta z$ into its rectangular components, and then determining the intensity of these force components, leads to the normal-stress and shear-stress components shown in Fig. 8–3b. Using the same notation as before, the subscript y in σ_y, τ_{yx}, and τ_{yz} refers to the direction of the normal line associated with the orientation of the area, and x and z in τ_{yx} and τ_{yz} refer to the corresponding direction lines for the shear stress. Lastly, one more section of the body parallel to the y–z plane, as shown in Fig. 8–3c, gives rise to normal stress σ_x and shear stresses τ_{xy} and τ_{xz}, Fig. 8–3d. If we continue in this manner, using corresponding parallel planes, we can "cut out" a cubic volume element of material that represents the *state of stress* acting at the point of its location in the body, Fig. 8–4.

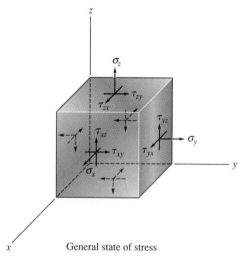

General state of stress

Fig. 8–4

Equilibrium Requirements. Although each of the six faces of the element in Fig. 8–4 will have three components of stress acting on it, if the stress at the point is *constant*, some of these stress components can be related by satisfying both force and moment equilibrium for the element. To show the relationships between the components we will consider a free-body diagram of the element, Fig. 8–5a. This element has a volume of $\Delta V = \Delta x\,\Delta y\,\Delta z$, and in accordance with Eqs. 8–1 and 8–2, the forces acting on each face are determined from the product of the average stress times the area of the face. For simplicity, we have not labeled the "dashed" forces acting on the "hidden" sides of the element. Instead, to view, and thereby label, some of these forces, the element is shown from a front view in Fig. 8–5b. Here it should be noted that the force components on the "hidden" sides of the element are designated with stresses having primes, and these forces are shown in the opposite direction to their counterparts acting on the opposite faces of the element.

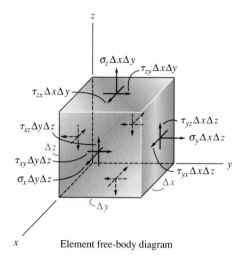

Element free-body diagram

(a)

If we now consider force equilibrium in the y direction, we have

$$\xrightarrow{+}\ \Sigma F_y = 0;\quad \underbrace{\underbrace{\sigma_y}_{\text{stress}}\ \underbrace{\Delta x\,\Delta z}_{\text{area}}}_{\text{force}} - \sigma_y'\,\Delta x\,\Delta z + \tau_{zy}\,\Delta x\,\Delta y - \tau_{zy}'\,\Delta x\,\Delta y + \tau_{xy}\,\Delta y\,\Delta z$$
$$-\tau_{xy}'\,\Delta y\,\Delta z = 0$$

Letting the side $\Delta x \to 0$, we get $\tau_{xy}\,\Delta y\,\Delta z - \tau_{xy}'\,\Delta y\,\Delta z = 0$, which requires $\tau_{xy} = \tau_{xy}'$. In a similar manner, if instead $\Delta y \to 0$, then it is necessary that $\sigma_y = \sigma_y'$; and lastly, if only $\Delta z \to 0$, then $\tau_{zy} = \tau_{zy}'$. In other words, as Δx, Δy, Δz approach zero each stress component acting in the y direction must be equal in magnitude but opposite in direction to its counterpart acting on the opposite face of the element. We must also satisfy force equilibrium in the x and z directions using the same type of analysis. When we are finished, we may again conclude that corresponding normal and shear stress components acting on opposite sides of the element must be equal in magnitude but opposite in direction.

In order to satisfy moment equilibrium, a further restriction has to be placed on the shear-stress components. To show this, consider summing moments about the x axis. From Fig. 8–5b, since $\sigma_y = \sigma_y'$, $\sigma_z = \sigma_z'$, $\tau_{xy} = \tau_{xy}'$, $\tau_{xz} = \tau_{xz}'$, the forces these components create are in equal but oppo-

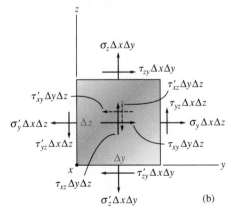

(b)   Fig. 8–5

site pairs, and so their moments about the x axis will also cancel. Therefore, excluding these components, we have

$$\zeta^+ \ \Sigma M_x = 0; \qquad (\tau_{yz} \ \Delta x \ \Delta z) \ \Delta y - (\tau_{zy} \ \Delta x \ \Delta y) \ \Delta z = 0$$

If we divide through by $\Delta x \ \Delta y \ \Delta z$, then it is necessary that

$$\tau_{yz} = \tau_{zy}$$

In a similar manner, we can show that moment equilibrium about the y and z axes requires

$$\tau_{xz} = \tau_{zx}$$

and

$$\tau_{xy} = \tau_{yx}$$

Thus, we may conclude that the above pairs of shear stresses on adjacent faces of the element must have equal magnitude and be directed either toward or away from the corners of the element, as in the manner shown in Fig. 8–5. This is sometimes referred to as the *complementary property* of shear.

To summarize the above analysis, the *state of stress* at a point is characterized by *six* independent stress components, namely, three normal stresses σ_x, σ_y, σ_z and three shear stresses τ_{xy}, τ_{yz}, τ_{xz}. Realize that these six components depend only on the *orientation* of the element, since the distribution of force acting on a sectioned plane through the body depends on the orientation of the sectioned plane. In other words, there is a unique set of six stress components describing the state of stress for each particular orientation of the element at the point. In Chapter 15 we will show that if these stress components are known on an element having a specified orientation, it will then be possible to determine the stress components on an element having some other orientation at the point.

Units.

In the International System, abbreviated SI, the magnitudes of both normal and shear stress components are specified in the basic units of newtons per square meter (N/m^2). This unit, called a *pascal* ($1 \ Pa = 1 \ N/m^2$) is rather small, and in engineering work prefixes such as kilo- (10^3), symbolized by k, mega- (10^6), symbolized by M, or giga- (10^9), symbolized by G, are used to represent larger, more realistic values of stress.* Likewise, in the U.S. Customary or Foot-Pound-Second system of units, engineers usually express stress in pounds per square inch (psi) or kilopounds per square inch (ksi), where 1 kilopound (kip) = 1000 lb.

---

*Sometimes stress is expressed in units of N/mm^2, where $1 \ mm = 10^{-3} \ m$. However, in the SI system, prefixes are not allowed in the denominator of a fraction and therefore it is better to use the equivalent $1 \ N/mm^2 = 1 \ MN/m^2 = 1 \ MPa$.

8.3 Average Normal Stress in an Axially Loaded Bar

Frequently structural or mechanical members are made long and slender. Also, they are subjected to axial loads that are usually applied to the ends of the member. Truss members, hangers, and bolts are typical examples. In this section we will determine the average stress distribution in an axially loaded bar, such as the one having the general form shown in Fig. 8–6a. For the analysis the internal loading on the cross section will be considered. This section defines the *cross-sectional area* of the bar, and since all such cross sections are the same, the bar is referred to as being *prismatic*. If we neglect the weight of the bar and section it as indicated, Fig. 8–6b, then, for equilibrium of the segment, the internal resultant force acting on the cross-sectional area must be equal in magnitude, opposite in direction, and collinear to the force acting at the end of the bar.

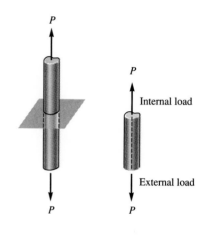

(a)   (b)

Assumptions. Before we determine the average stress distribution acting over the bar's cross section, it is first necessary to make some simplifying assumptions concerning the material description and the specific application of the load. In this regard, the analysis will depend on the bar remaining straight both before and after the load is applied, and also, the cross section should remain flat or plane during the deformation, that is, during the time the bar changes its volume and shape. If this occurs, then horizontal and vertical grid lines inscribed on the bar will *deform uniformly* when the bar is subjected to the load, Fig. 8–6c. Here we will not consider regions of the bar near its ends, where application of the external loads can cause localized distortions. Instead we will focus only on the stress distribution within the bar's midsection.

As will be shown below, uniform deformation occurs provided **P** is applied along the *centroidal axis* of the cross section. Also, uniform deformation of the bar will occur if the material is assumed to be homogeneous and isotropic. *Homogeneous material* has the same physical and mechanical properties throughout its volume, and *isotropic material* has these same properties in all directions. Many engineering materials may be approximated as being both homogeneous and isotropic as assumed here. Steel, for example, contains thousands of randomly oriented crystals in each cubic millimeter of its volume, and since most problems involving this material have a physical size that is much larger than a single crystal, the above assumption regarding its material composition is quite realistic. It should be mentioned, however, that steel can be made anisotropic by cold-rolling, i.e., rolling or forging it at subcritical temperatures. *Anisotropic materials* have different properties in different directions, and although this is the case, if the anisotropy is in a specified direction throughout the material, it will also deform uniformly when subjected to an axial load. For example, timber, due to its grains or fibers of wood, is an example of an engineering material that is homogeneous and anisotropic in a specified direction and is therefore suited for the foregoing analysis.

Region of uniform deformation of bar

(c)

Fig. 8–6(a–c)

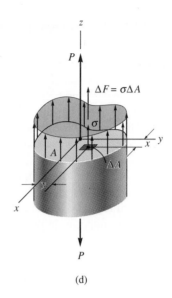

(d)

Fig. 8–6(d)

Average Normal Stress Distribution. Provided the bar is subjected to a constant uniform deformation as noted, it is reasonable to assume further that this deformation is caused by a *constant* normal stress σ, which is then uniformly distributed over the bar's cross-sectional area, Fig. 8–6*d*. Since each area ΔA on the cross section is subjected to a force $\Delta F = \sigma \Delta A$, then the sum of these forces acting over the entire cross-sectional area must be equivalent to the internal force resultant **P** at the section. If we let $\Delta A \rightarrow dA$ and therefore $\Delta F \rightarrow dF$, then, since σ is *constant,* we have

$$F_{R_z} = \Sigma F_z; \qquad\qquad \int dF = \int_A \sigma \, dA$$

$$P = \sigma A$$

or

$$\boxed{\sigma = \frac{P}{A}} \tag{8–3}$$

Here

σ = average normal stress at any point on the cross-sectional area

P = internal resultant normal force, which is applied through the *centroid* of the cross-sectional area. P is determined using the method of sections and the equations of equilibrium

A = cross-sectional area of the bar

In order to maintain the uniform deformation of the bar, the stress distribution cannot produce a resultant internal moment on the cross section. To show that this is actually the case, the sum of the moments about the centroidal x and y axes must be *equal to zero*, since P creates zero moment about these axes, Fig. 8–6d. We have

$$(M_R)_x = \Sigma\, M_x; \qquad 0 = \int_A y\, dF = \int_A y\sigma\, dA = \sigma\!\int_A y\, dA$$

$$(M_R)_y = \Sigma\, M_y; \qquad 0 = \int_A x\, dF = \int_A x\sigma\, dA = \sigma\!\int_A x\, dA$$

These equations are satisfied, since by definition of the centroid, $\int y\, dA = 0$ and $\int x\, dA = 0$.

In summary, then, an axial force **P,** acting on the ends of a homogeneous straight prismatic bar and passing through the centroid of the bar's cross-sectional area, will cause a uniform normal stress distribution over the cross-sectional area, Fig. 8–6e. This stress has a *magnitude* of $\sigma = P/A$ and an arrowhead *sense* of direction that is the *same* as that of the internal resultant force **P,** since all the normal stresses on the cross section develop this resultant. At a specific point in the bar it should therefore be evident that only a normal stress exists on a volume element of material located at the point. In other words, the stress components in Fig. 8–4 are all zero except σ_z, Fig. 8–6f. Under these conditions, the material is said to be subjected to a state of *uniaxial stress*.

Returning to the stress distribution shown in Fig. 8–6e, notice that *graphically* the *magnitude* of the internal resultant force **P** is *equivalent* to the *volume* under the stress diagram; that is, $P = \sigma A$ (volume = height × base). Furthermore, as a consequence of the balance of moments, this resultant *passes through the centroid of this volume*.

The above analysis applies as well to *short members* subjected to a compressive force, Fig. 8–7a. Here the cross section is subjected to a uniaxial compressive stress, Fig. 8–7b, and so is a volume element of material located at a point within the member, Fig. 8–7c. Note that a restriction on the length of the member is necessary, since if the member is long and slender and **P** is large, the member may become unstable and buckle.

Although we have developed this analysis for *prismatic* bars, this assumption can be relaxed somewhat to include bars that have a *slight taper*. For example, it can be shown, using the more exact analysis of the theory of elasticity, that for a tapered bar of rectangular cross section, for which the angle between two adjacent sides is 15°, the average normal stress, as calculated by Eq. 8–3, is only 2.2% less than its value found from the theory of elasticity.

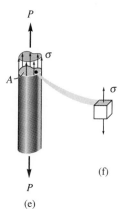

(f)

(e)

Fig. 8–6(e, f)

PROCEDURE FOR ANALYSIS

It is important to be aware that the equation $\sigma = P/A$ gives the *average* normal stress on the sectional area of a member when the section is subjected to a resultant normal force. For axially loaded members, application of the equation requires the following steps.

Internal Loading. Section the member *perpendicular* to its longitudinal axis at the point where the stress is to be determined and use the necessary free-body diagram and equation of force equilibrium to obtain the internal axial force **P** at the section.

Average Normal Stress. Determine the member's cross-sectional area at the section and compute the average normal stress $\sigma = P/A$. It is suggested that σ be shown acting on a small volume element of the material located at a point on the section where the stress is calculated. To do this, first draw σ on the face of the element coincident with the sectioned area A. This is indicated as the shaded face in Figs. 8–6*f* and 8–7*c*. Here σ acts in the *same direction* as **P**. For force equilibrium, the normal stress σ acting on the opposite face of the element can then be drawn in its appropriate direction.

The following examples numerically illustrate applications of this procedure.

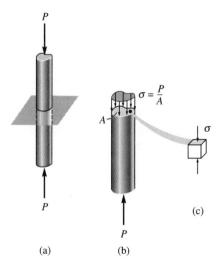

Fig. 8–7

Example 8–1

The bar in Fig. 8–8a has a constant width of 35 mm and a thickness of 10 mm. Determine the largest normal stress in the bar when it is subjected to the loading shown.

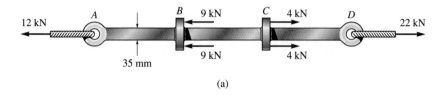

(a)

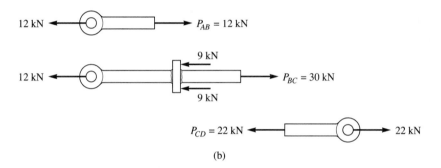

(b)

SOLUTION

Internal Loading.   By inspection, the internal axial forces in regions *AB*, *BC*, and *CD* are all constant yet have different magnitudes. Using the method of sections, these loadings are computed in Fig. 8–8b. By comparison, the largest loading is in region *BC*, where $P_{BC} = 30$ kN. Since the cross-sectional area of the bar is constant, the largest normal stress also occurs within this region of the bar.

Average Normal Stress.   Applying Eq. 8–3, we have

$$\sigma_{BC} = \frac{P_{BC}}{A} = \frac{30(10^3) \text{ N}}{(0.035 \text{ m})(0.010 \text{ m})} = 85.7 \text{ MPa} \qquad Ans.$$

The stress distribution acting on an arbitrary cross section of the bar within region *BC* is shown in Fig. 8–8c. Graphically the *volume* (or "block") represented by this distribution of stress is equivalent to the load of 30 kN; that is, 30 kN = (85.7 MPa)(35 mm)(10 mm).

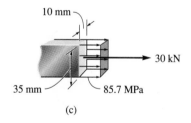

(c)

Fig. 8–8

Example 8–2

The 80-kg lamp is supported by two rods AB and BC as shown in Fig. 8–9a. If AB has a diameter of 10 mm and BC has a diameter of 8 mm, determine which rod is subjected to the greater normal stress.

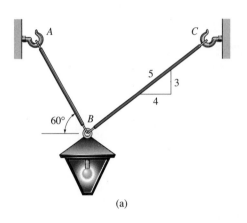

(a)

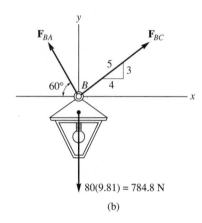

(b)

SOLUTION

Internal Loading. We must first determine the axial force in each rod. A free-body diagram of the lamp is shown in Fig. 8–9b. Applying the equations of force equilibrium yields

$$\xrightarrow{+} \Sigma F_x = 0; \qquad\qquad F_{BC} \left(\tfrac{4}{5}\right) - F_{BA} \cos 60° = 0$$

$$+\uparrow \Sigma F_y = 0; \qquad F_{BC} \left(\tfrac{3}{5}\right) + F_{BA} \sin 60° - 784.8 \text{ N} = 0$$

$$F_{BC} = 395.2 \text{ N}, \qquad F_{BA} = 632.4 \text{ N}$$

By Newton's third law of action, equal but opposite reaction, these forces also represent the internal axial forces developed within the rods.

Average Tensile Stress. Applying Eq. 8–3, we have

$$\sigma_{BC} = \frac{F_{BC}}{A_{BC}} = \frac{395.2 \text{ N}}{\pi (0.004 \text{ m})^2} = 7.86 \text{ MPa}$$

$$\sigma_{BA} = \frac{F_{BA}}{A_{BA}} = \frac{632.4 \text{ N}}{\pi (0.005 \text{ m})^2} = 8.05 \text{ MPa} \qquad\qquad \textit{Ans.}$$

The average normal stress distribution acting over a cross section of rod AB is shown in Fig. 8–9c, and at a point on this cross section, an element of material is stressed as shown in Fig. 8–9d.

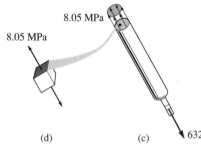

8.05 MPa

8.05 MPa

632.4 N

(d)   (c)

Fig. 8–9

Example 8–3

The casting shown in Fig. 8–10a is made of steel having a specific weight of $\gamma_{st} = 490$ lb/ft³. Determine the average compressive stress acting at points A and B.

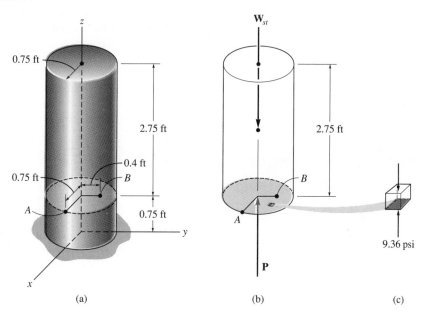

(a)   (b)   (c)

Fig. 8–10

SOLUTION

Internal Loading. A free-body diagram of the top segment of the casting where the section passes through points A and B is shown in Fig. 8–10b. The weight of this segment is determined from $W_{st} = \gamma_{st}V_{st}$. Thus the internal axial force P at the section is

$$+\uparrow \ \Sigma F_z = 0; \qquad\qquad P - W_{st} = 0$$
$$P - (490 \text{ lb/ft}^3)(2.75 \text{ ft})\pi(0.75 \text{ ft})^2 = 0$$
$$P = 2381 \text{ lb}$$

Average Compressive Stress. The cross-sectional area at the section is $A = \pi(0.75 \text{ ft})^2$, and so the compressive stress becomes

$$\sigma = \frac{P}{A} = \frac{2381 \text{ lb}}{\pi(0.75 \text{ ft})^2}$$
$$= 1347.5 \text{ lb/ft}^2 = 1347.5 \text{ lb/ft}^2(1 \text{ ft}^2/144 \text{ in}^2)$$
$$= 9.36 \text{ psi} \qquad\qquad\qquad\qquad\qquad Ans.$$

The stress shown on the volume element of material in Fig. 8–10c is representative of the conditions at either point A or B. Notice that this stress acts *upward* on the bottom or shaded face of the element since this face forms part of the bottom surface area of the cut section, and furthermore, the resultant internal force **P** is pushing upward on this section.

Example 8–4

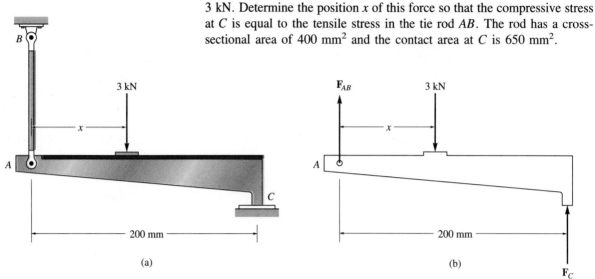

Member AC shown in Fig. 8–11a is subjected to a vertical force of 3 kN. Determine the position x of this force so that the compressive stress at C is equal to the tensile stress in the tie rod AB. The rod has a cross-sectional area of 400 mm² and the contact area at C is 650 mm².

Fig. 8–11

SOLUTION

Internal Loading. The free-body diagram for member AC is shown in Fig. 8–11b. There are three unknowns, namely, F_{AB}, F_C, and x. To solve this problem we will work in units of newtons and millimeters.

$$+\uparrow \ \Sigma F_y = 0; \qquad\qquad F_{AB} + F_C - 3000 \text{ N} = 0 \qquad (1)$$

$$\tikz \ \Sigma M_A = 0; \qquad -3000 \text{ N}(x) + F_C (200 \text{ mm}) = 0 \qquad (2)$$

Average Normal Stress. A necessary third equation can be written that requires the tensile stress in the bar AB and the compressive stress at C to be equivalent; i.e.,

$$\sigma = \frac{F_{AB}}{400 \text{ mm}^2} = \frac{F_C}{650 \text{ mm}^2}$$

$$F_C = 1.625 F_{AB}$$

Substituting this into Eq. 1, solving for F_{AB}, then solving for F_C, we obtain

$$F_{AB} = 1143 \text{ N}$$

$$F_C = 1857 \text{ N}$$

The position of the applied load is determined from Eq. 2.

$$x = 124 \text{ mm} \qquad\qquad Ans.$$

Note that $0 < x < 200$ mm, as required.

8.4 Average Shear Stress

As discussed in Sec. 8.2, shear stress has been defined as the stress component that acts *in the plane* of the sectioned area. In order to show how this stress can develop, we will consider the effect of applying a force **P** to the bar in Fig. 8–12*a*. If the supports are considered rigid, and **P** is large enough, it will cause the material of the bar to deform and fail along the planes identified by *AB* and *CD*. A free-body diagram of the unsupported center segment of the bar, Fig. 8–12*b*, indicates that the shear force $V = P/2$ must be applied at each section to hold the segment in equilibrium. The *average shear stress* distributed over each sectioned area that develops this shear force is defined by

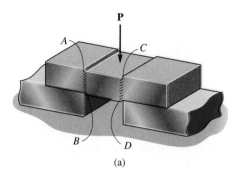

(a)

$$\tau_{avg} = \frac{V}{A} \qquad (8-4)$$

Here

τ_{avg} = average shear stress at the section, which is assumed to be the *same* at each point located on the section

V = internal resultant shear force at the section determined from the equations of equilibrium

A = area at the section

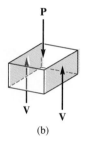

(b)

The distribution of average shear stress is shown acting over the right-hand section in Fig. 8–12*c*. Notice that τ_{avg} is in the *same direction* as **V**, since the shear stress must create forces that contribute to the internal resultant force **V** at the section.

The loading case discussed in Fig. 8–12 is an example of *simple* or *direct shear*, since the shear is caused by the *direct action* of the applied load **P**. This type of shear often occurs in various types of simple connections that use bolts, pins, welding material, etc. In all these cases, however, application of Eq. 8–4 is *only approximate*. A more precise investigation of the shear-stress distribution over the critical section often reveals that much larger shear stresses occur in the material than those predicted by Eq. 8–4. Although this may be the case, application of Eq. 8–4 may be acceptable for many problems in engineering design and analysis. For example, engineering codes allow the use of Eq. 8–4 when considering design sizes for fasteners such as bolts and for obtaining the bonding strength of joints subjected to shear loadings. In this regard, two types of shear frequently occur in practice, which deserve separate treatment.

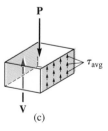

(c)

Fig. 8–12

Single Shear. The steel and wood joints shown in Fig. 8–13*a* and 8–13*b*, respectively, are examples of *single-shear connections* and are often referred to as *lap joints*. Here we will assume that the members are thin and that the nut *A* in Fig. 8–13*a* is not tightened to any great extent so friction between the members can be neglected. Passing a section between the members yields the free-body diagrams shown in Fig. 8–13*c* and 8–13*d*. Since the members are thin, we can neglect the moment created by the force **P**. Hence the cross-sectional area of the bolt in Fig. 8–13*c* and the bonding surface between the members in Fig. 8–13*d* are subjected only to a *single shear force* $V = P$. This force is used in Eq. 8–4 to determine the average shear stress acting on the colored section.

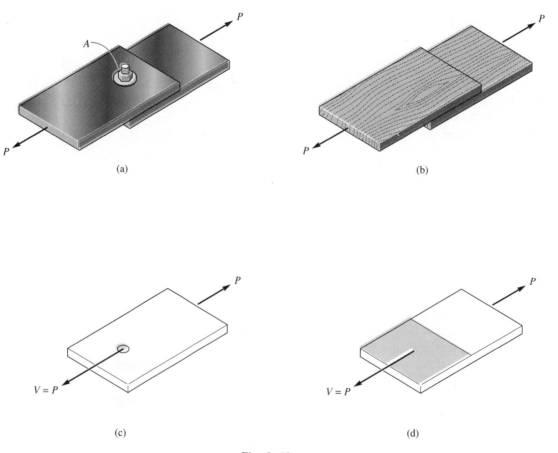

(a)

(b)

(c)

(d)

Fig. 8–13

Double Shear. When the joint is constructed as shown in Fig. 8–14*a* or 8–14*b*, *two shear surfaces* must be considered. These types of connections are often called *double lap joints*. If we pass a section between each of the members, the free-body diagrams of the center member are shown in Fig. 8–14*c* and 8–14*d*. Here we have a condition of *double shear*. Consequently, $V = P/2$ acts on *each* sectioned area and this shear, like that in Fig. 8–12*b*, must be considered when applying Eq. 8–4.

Consider now a volume element of material taken at a point located on the surface of any sectioned area on which the average shear stress acts, Fig. 8–15. It was shown in Sec. 8.2 that force and moment equilibrium requires τ_{avg}, acting on the top face of the element, to be accompanied by shear stress acting on three other faces. As shown, all four shear stresses must have equal magnitude and be directed either toward or away from each other at opposite edges of the element. Under these conditions, the material is subjected to *pure shear*.

Although we have considered here a case of simple shear as caused by the *direct* action of a load, in later chapters we will show that shear stress can also arise *indirectly* due to the action of a bending or torsional moment.

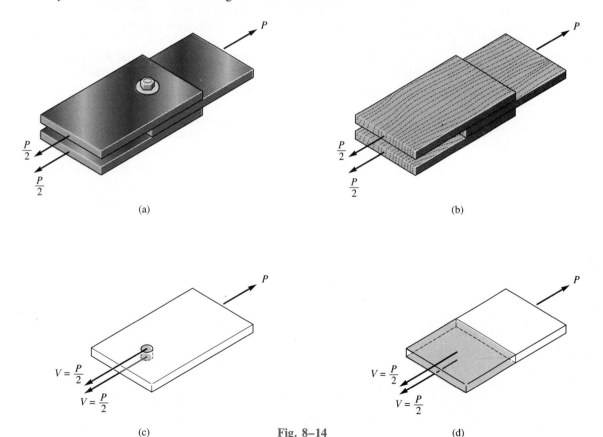

Fig. 8–14

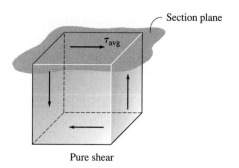

Pure shear

Fig. 8–15

PROCEDURE FOR ANALYSIS

Equation 8–4, $\tau_{avg} = V/A$, is used to compute only the *average shear stress* in the material. Its use in engineering is generally restricted to the design and analysis of members or parts that are thin or small, so that bending can be neglected. Application requires the following steps.

Internal Shear.   Section the member at the point where the average shear stress is to be determined. Draw the necessary free-body diagram and compute the internal shear force **V** acting at the section.

Average Shear Stress.   Determine the sectioned area A, and compute the average shear stress $\tau_{avg} = V/A$. It is suggested that τ_{avg} be shown on a small volume element of material located at a point on the section where it is computed. To do this, first draw τ_{avg} on the face of the element, coincident with the sectioned area A. This shear stress acts in the same direction as **V**. The shear stresses acting on the three adjacent planes can then be drawn in their appropriate directions following the scheme shown in Fig. 8–15.

Example 8–5

The bar shown in Fig. 8–16a has a square cross section for which the depth and thickness are 40 mm. If an axial force of 800 N is applied along the centroidal axis of the bar's cross-sectional area, determine the average normal stress and average shear stress acting on the material along (a) section plane a–a and (b) section plane b–b.

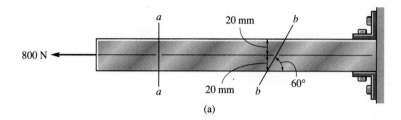

(a)

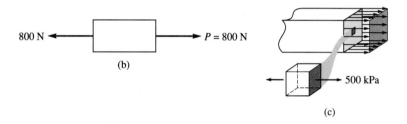

(b)

(c)

Fig. 8–16(a–c)

SOLUTION

Part (a)

Internal Loading.   The bar is sectioned, Fig. 8–16b, and the internal resultant loading consists only of an axial force for which $P = 800$ N.

Average Stress.   The average normal stress is determined from Eq. 8–3.

$$\sigma = \frac{800 \text{ N}}{(0.04 \text{ m})(0.04 \text{ m})} = 500 \text{ kPa} \qquad Ans.$$

No shear stress exists on the section, since the shear force at the section is zero.

$$\tau_{avg} = 0 \qquad Ans.$$

The distribution of average normal stress over the cross section and the stress on a shaded element located on the cross section are shown in Fig. 8–16c.

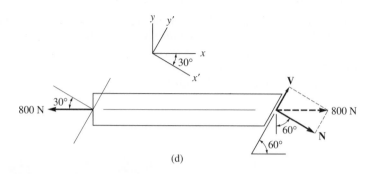

(d)

Part (b)

Internal Loading. If the bar is sectioned along *b–b*, the free-body diagram shown in Fig. 8–16*d* results. Here both a normal force (**N**) and shear force (**V**) act on the sectioned area. Using *x, y* axes, we require

$$\xrightarrow{+} \Sigma F_x = 0; \quad -800 \text{ N} + N \sin 60° + V \cos 60° = 0$$

$$+\uparrow \Sigma F_y = 0; \quad \quad \quad V \sin 60° - N \cos 60° = 0$$

or, more directly, using *x', y'* axes,

$$+\searrow \Sigma F_{x'} = 0; \quad \quad N - 800 \text{ N} \cos 30° = 0$$

$$+\nearrow \Sigma F_{y'} = 0; \quad \quad V - 800 \text{ N} \sin 30° = 0$$

Solving either set of equations,

$$N = 692.8 \text{ N}$$

$$V = 400 \text{ N}$$

Average Stresses. In this case the sectioned area has a thickness and depth of 40 mm and 40 mm/sin 60° = 46.19 mm, respectively, Fig. 8–16*a.* Thus the average normal stress is

$$\sigma = \frac{N}{A} = \frac{692.8 \text{ N}}{(0.04 \text{ m})(0.04619 \text{ m})} = 375 \text{ kPa} \qquad \textit{Ans.}$$

and the average shear stress is

$$\tau_{avg} = \frac{V}{A} = \frac{400 \text{ N}}{(0.04 \text{ m})(0.04619 \text{ m})} = 217 \text{ kPa} \qquad \textit{Ans.}$$

The stress distribution and a shaded element of material located at a point on the sectioned area are shown in Fig. 8–16*e.* Comparing this element with that of Fig. 8–16*c,* it can be seen that the stress components at the point depend on the *orientation of the element.* As discussed in Sec. 8.2, any one of these elements can be used to describe the state of stress at the point.

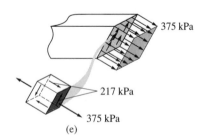

(e)

Fig. 8–16(d–e)

Example 8–6

The wooden strut shown in Fig. 8–17a is suspended from a 10-mm-diameter steel rod, which is fastened to the wall. If the strut supports a vertical load of 5 kN, compute the average shear stress in the rod at the wall and along the two shaded planes of the strut, one of which is indicated as *abcd*.

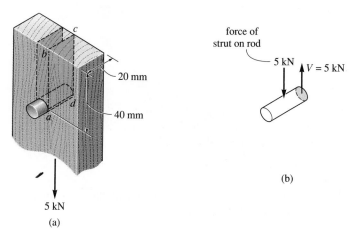

(a)

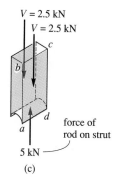

(b)

force of strut on rod

5 kN

$V = 5$ kN

$V = 2.5$ kN

$V = 2.5$ kN

force of rod on strut

5 kN

(c)

Fig. 8–17

SOLUTION

Internal Shear.   As shown on the free-body diagram in Fig. 8–17b, the rod resists a shear force of 5 kN where it is fastened to the wall. A free-body diagram of the sectioned segment of the strut that is in contact with the rod is shown in Fig. 8–17c. Here the shear force acting along each shaded plane is 2.5 kN.

Average Shear Stress.   For the rod,

$$\tau_{avg} = \frac{V}{A} = \frac{5000 \text{ N}}{\pi (0.005 \text{ m})^2} = 63.7 \text{ MPa} \qquad \textit{Ans.}$$

For the strut,

$$\tau_{avg} = \frac{V}{A} = \frac{2500 \text{ N}}{(0.04 \text{ m})(0.02 \text{ m})} = 3.12 \text{ MPa} \qquad \textit{Ans.}$$

The average-shear-stress distribution on the sectioned rod and strut segment is shown in Fig. 8–17d and 8–17e. Also shown with these figures is a typical volume element of the material taken at a point located on the surface of each section. Note carefully how the shear stress must act on each shaded face of these elements and then on the other faces of the elements.

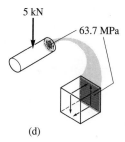

5 kN

63.7 MPa

(d)

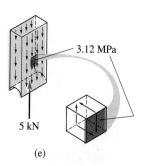

5 kN

3.12 MPa

5 kN

(e)

Example 8–7

The inclined member in Fig. 8–18a is subjected to a compressive force of 600 lb. Determine the average compressive stress along the areas of contact defined by *AB* and *BC*, and the average shear stress along the horizontal plane defined by *EDB*.

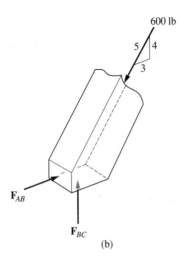

$\mathbf{F}_{AB}$

$\mathbf{F}_{BC}$

(b)

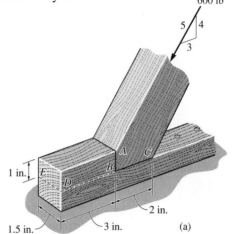

Fig. 8–18

1 in.

1.5 in.

3 in.

2 in.

(a)

SOLUTION

Internal Loadings. The free-body diagram of the inclined member is shown in Fig. 8–18b. The compressive forces acting on the areas of contact are

$\xrightarrow{+} \Sigma F_x = 0;$ $F_{AB} - 600 \text{ lb}(\frac{3}{5}) = 0$ $F_{AB} = 360 \text{ lb}$

$+\uparrow \Sigma F_y = 0;$ $F_{BC} - 600 \text{ lb}(\frac{4}{5}) = 0$ $F_{BC} = 480 \text{ lb}$

Also, from the free-body diagram of the top segment of the bottom member, Fig. 8–18c, the shear force acting on the sectioned horizontal plane *EDB* is

$\xrightarrow{+} \Sigma F_x = 0;$ $V = 360 \text{ lb}$

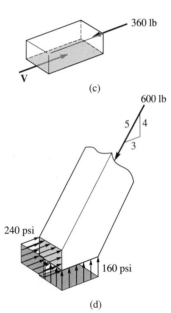

360 lb

V

(c)

600 lb

240 psi

160 psi

(d)

Average Stress. The average compressive stresses along the horizontal and vertical planes of the inclined member are

$$\sigma_{AB} = \frac{360 \text{ lb}}{(1 \text{ in.})(1.5 \text{ in.})} = 240 \text{ psi} \qquad \textit{Ans.}$$

$$\sigma_{BC} = \frac{480 \text{ lb}}{(2 \text{ in.})(1.5 \text{ in.})} = 160 \text{ psi} \qquad \textit{Ans.}$$

These stress distributions are shown in Fig. 8–18d.

The average shear stress acting on the horizontal plane defined by *EDB* is

$$\tau_{\text{avg}} = \frac{360 \text{ lb}}{(3 \text{ in.})(1.5 \text{ in.})} = 80 \text{ psi} \qquad \textit{Ans.}$$

This stress is shown distributed over the sectioned area in Fig. 8–18e.

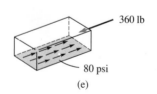

360 lb

80 psi

(e)

PROBLEMS

8–1. The column is subjected to an axial force of 8 kN at its top. If the cross-sectional area has the dimensions shown in the figure, determine the normal stress acting at section a–a. Show this distribution of stress acting over the area's cross section.

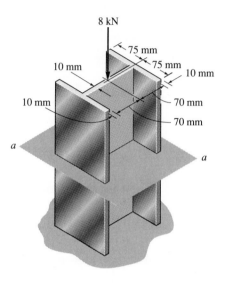

Prob. 8–1

8–2. The yoke-and-rod connection is subjected to a tensile force of 5 kN. Determine the normal stress in each rod and the average shear stress in the pin A between the members.

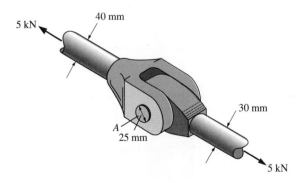

Prob. 8–2

8–3. The beam is supported by a pin at A and a short link BC. Determine the average shear stress developed in the pins at A, B, and C. All pins are in double shear as shown, and each has a diameter of 18 mm.

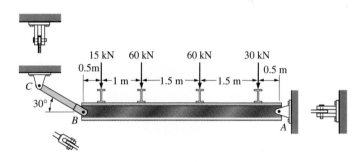

Prob 8–3

*8–4. The 50-lb lamp is supported by three steel rods connected together by a ring at A. Determine which rod is subjected to the greatest normal stress and compute its value. The diameter of each rod is given in the figure.

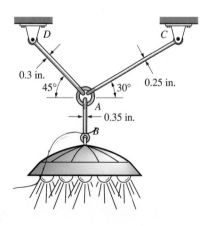

Prob. 8–4

8–5. The pedestal has a triangular cross section as shown. If it is subjected to a compressive force of 500 lb, specify the x and y coordinates for the location of point $P(x, y)$, where the load must be applied on the cross section, so that the bearing stress is uniform. Compute the bearing stress and sketch the distribution of the bearing stress acting on the cross section at a location removed from the point of load application.

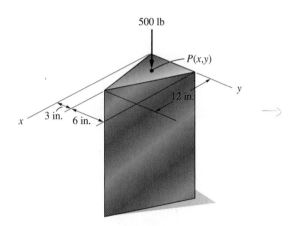

Prob. 8–5

8–6. The two steel members are joined together using a 60° scarf weld. Determine the normal stress and average shear stress resisted in the plane of the weld.

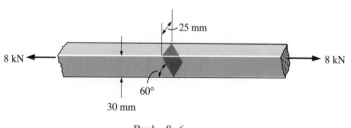

Prob. 8–6

8–7. The board is subjected to a tensile force of 85 lb. Determine the average normal and average shear stress developed in the wood fibers that are orientated along section a–a at 15° with the axis of the board.

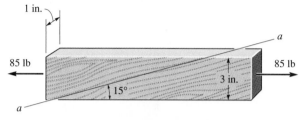

Prob. 8–7

*8–8. The bars of the truss each have a cross-sectional area of 1.25 in². Determine the normal stress in each member due to the loading shown. State whether the stress is tensile or compressive.

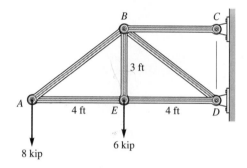

Prob. 8–8

8–9. The thrust bearing is subjected to the loads shown. Determine the average normal stress developed on the cross sections at B, C, and D. Sketch the results on a volume element located at each section.

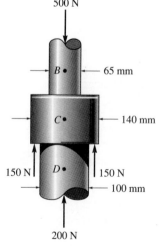

Prob. 8–9

8–10. The column is made of concrete having a density of 2.30 Mg/m³. At its top B it is subjected to an axial compressive force of 15 kN. Determine the compressive stress in the column as a function of the distance z measured from its base. *Note:* The result will be useful only for finding the average compressive stress at a section removed from the ends of the column, because of localized deformation at the ends.

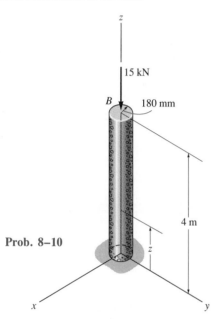

Prob. 8–10

8–11. The block in the shape of a frustum of a cone is made of concrete having a specific weight of 150 lb/ft³. Determine the bearing stress acting in the column at its midheight, $z = 4$ ft. *Hint:* The volume of a cone of radius r and height h is $V = \frac{1}{3}\pi r^2 h$.

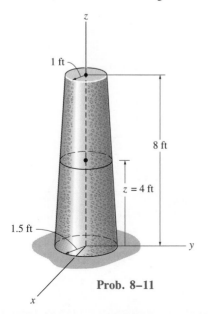

Prob. 8–11

*8–12. Rods AB and BC have diameters of 4 mm and 6 mm, respectively. If a vertical load of 8 kN is applied to the ring at B, determine the angle θ of rod BC so that the normal stress in each rod is equivalent.

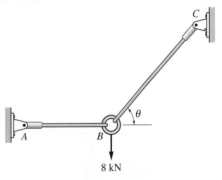

Prob. 8–12

8–13. The long bolt passes through the 30-mm-thick plate. If the force in the bolt shank is 8 kN, determine the normal stress in the shank, the average shear stress along the cylindrical area of the plate defined by the section lines a, and the average shear stress in the bolt head along the cylindrical area defined by the section lines b.

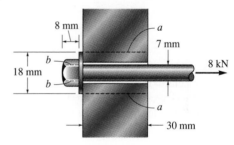

Prob. 8–13

8–14. The uniform beam is supported by two rods AB and CD that have cross-sectional areas of 12 mm² and 8 mm², respectively. Determine the position d of the 6-kN load so that the normal stress in each rod is the same.

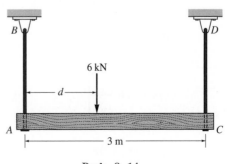

Prob. 8–14

8–15. The pulley is held fixed to the 20-mm-diameter shaft using a key that fits within a groove cut into the pulley and shaft. If the suspended load has a mass of 50 kg, determine the average shear stress in the key along section a–a. The key is 5 mm by 5 mm square and 12 mm long.

8–17. The two-member frame is subjected to the distributed loading shown. Determine the intensity w of the largest uniform loading that can be applied to the frame without causing either the normal stress or the average shear stress at section b–b to exceed $\sigma = 15$ MPa and $\tau = 16$ MPa, respectively. Member CB has a square cross-section of 35 mm on each side.

75 mm

Prob. 8–15

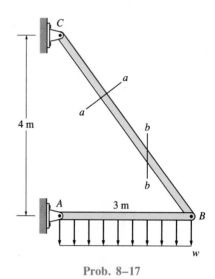

4 m

3 m

w

Prob. 8–17

8–18. The pier is made of material having a specific weight γ. If it has a square cross section, determine its width w as a function of z so that the normal stress in the pier remains constant. The pier supports a constant load P at its top where its width is w_1.

*8–16. The two-member frame is subjected to the loading shown. Determine the normal stress and the average shear stress acting at sections a–a and b–b. Member CB has a square cross section of 2 in. on each side.

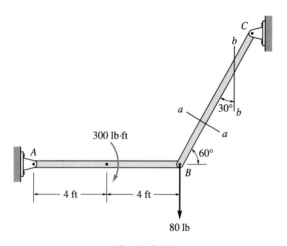

C

b

$30°$ b

a

300 lb·ft

a

A

$60°$

B

4 ft

4 ft

80 lb

Prob. 8–16

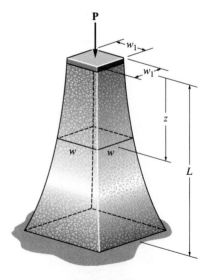

P

w_1

w_1

z

w

w

L

Prob. 8–18

8.5 Allowable Stress

An engineer in charge of the *design* of a structural member or mechanical element must restrict the stress in the material to a level that will be safe. Furthermore, a structure or machine that is currently in use may, on occasion, have to be *analyzed* to see what additional loadings its members or parts can support. So again it becomes necessary to perform the calculations using a safe or allowable stress.

To ensure safety, it is necessary to choose an allowable stress that restricts the applied load to one that is *less* than the load the member can fully support. There are several reasons for this. For example, the load for which the member is designed may be different from actual loadings placed on it. The intended measurements of a structure or machine may not be exact due to errors in fabrication or in the assembly of its component parts. Unknown vibrations, impact, or accidental loadings can occur that may not be accounted for in the design. Atmospheric corrosion, decay, or weathering tend to cause materials to deteriorate during service. And lastly, some materials, such as wood, concrete, or fiber-reinforced composites, can show high variability in mechanical properties.

One method of specifying the allowable load for the design or analysis of a member is to use a number called the factor of safety. The *factor of safety* (F.S.) is a ratio of a theoretical maximum load that can be carried by the member until it fails in a particular manner (P_{fail}) divided by an allowable load (P_{allow}), which has been determined from experience or experiments to be safe under similar conditions of loading and geometry. Stated mathematically,

$$\text{F.S.} = \frac{P_{fail}}{P_{allow}} \qquad (8\text{--}5)$$

If the load applied to the member is *linearly related* to the stress developed within the member, as in the case of using $\sigma = P/A$ and $\tau_{avg} = V/A$, then we can express the factor of safety as a ratio of the failure stress σ_{fail} (or τ_{fail}) to the allowable stress σ_{allow} (or τ_{allow});* that is,

$$\text{F.S.} = \frac{\sigma_{fail}}{\sigma_{allow}} \qquad (8\text{--}6)$$

or

$$\text{F.S.} = \frac{\tau_{fail}}{\tau_{allow}} \qquad (8\text{--}7)$$

---

*In some cases, such as columns, the applied load is *not* linearly related to stress and therefore only Eq. 8–5 can be used to compute the factor of safety. See Chapter 17.

In any of these equations, the factor of safety is generally chosen to be *greater* than 1 in order to avoid the potential for failure. Specific values depend on the types of materials to be used and the intended purpose of the structure or machine. For example, the F.S. used in the design of aircraft or space-vehicle components may be as low as 1 in order to reduce the weight of the vehicle. On the other hand, in the case of a nuclear power plant, the factor of safety for some of its components may be as high as 3 since there may be uncertainties in loading or material behavior. In general, however, factors of safety and therefore the allowable loads or stresses for both structural and mechanical elements have become well standardized, since their design uncertainties have been reasonably evaluated. Their values, which can be found in design codes and engineering handbooks, are intended to form a balance of ensuring public and environmental safety and providing for a reasonable economic solution to design.

8.6 Design of Simple Connections

By making simplifying assumptions regarding the behavior of the material, the equations $\sigma = P/A$ and $\tau_{avg} = V/A$ can often be used to analyze or design a simple connection or a mechanical element. In particular, if a member is subjected to a *normal force* at a section, its required area at the section is determined from

$$A = \frac{P}{\sigma_{allow}} \qquad (8\text{--}8)$$

On the other hand, if the section is subjected to a *shear force,* then the required area at the section is

$$A = \frac{V}{\tau_{allow}} \qquad (8\text{--}9)$$

As discussed in Sec. 8.5, the allowable stress used in each of these equations is determined either by applying a factor of safety to a specified normal or shear stress or by finding these stresses directly from an appropriate design code.

We will now discuss four common types of problems for which the above equations can be used for design.

Cross-Sectional Area of a Tension Member. The cross-sectional area of a prismatic member subjected to a tension force can be determined using Eq. 8–8, *provided* the force has a line of action that passes through the centroid of the cross section. For example, consider the "eye bar" shown in Fig. 8–19a. At the intermediate section a–a, the stress distribution is uniform over the cross section and the shaded area A is determined from Eq. 8–8, as shown in Fig. 8–19b.

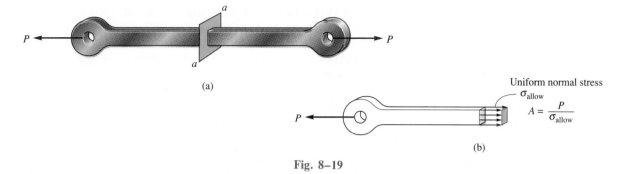

(a)

Uniform normal stress

σ_{allow}

$A = \dfrac{P}{\sigma_{allow}}$

(b)

Fig. 8–19

Cross-Sectional Area of a Connector Subjected to Shear.

Often bolts or pins are used to connect plates, boards, or several members together. When these connectors are subjected to shear, Eq. 8–9 can be used to determine their cross-sectional area. For example, consider the lap joint shown in Fig. 8–20a. If the bolt is loose or the clamping force of the bolt is unknown, it is safe to assume that any frictional force between the plates is negligible. As a result, the free-body diagram for a section passing *between* the plates and through the bolt is shown in Fig. 8–20b. The bolt is subjected to a resultant internal shear force of $V = P$ at its cross section. Assuming that the shear stress causing this force is *uniformly distributed* over the cross section, the bolt's cross-sectional area A is determined as shown in Fig. 8–20c.

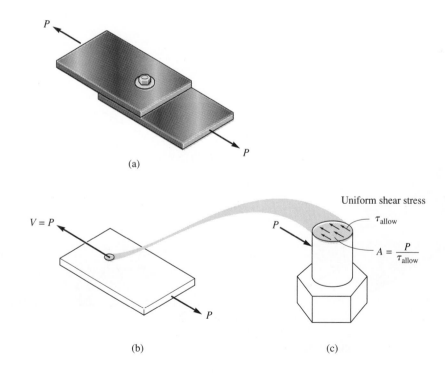

(a)

$V = P$

Uniform shear stress

τ_{allow}

$A = \dfrac{P}{\tau_{allow}}$

Fig. 8–20   (b)   (c)

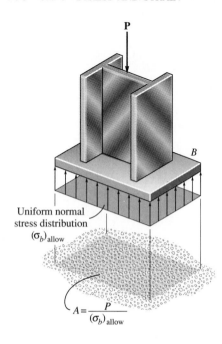

Required Area to Resist Bearing. A normal stress that is produced by the compression of one surface against another is called a bearing stress. If this stress becomes large enough, it may crush or locally deform one or both of the surfaces. Hence, in order to prevent failure it is necessary to determine the proper bearing area for the material using an allowable bearing stress. For example, the area A of the column base plate B shown in Fig. 8–21 is determined from the allowable bearing stress of the concrete, $(\sigma_b)_{allow}$, using Eq. 8–8. This assumes, of course, that the bearing stress is uniformly distributed between the plate and the concrete as shown in the figure.

Engineers often use Eq. 8–8 to determine the thickness of plates connected together by bolts or pins. The lap joint in Fig. 8–22a is an example of such a situation. Friction between the plates caused by the unknown clamping force of the bolt can be neglected, so the load $\mathbf{P}$ is transmitted from one plate to the next only by the *bearing* of the bolt shank against the edge of the hole in each plate. The material strength of the bolt will generally be greater than that of the plate, and therefore crushing of the plate, not the bolt, can occur. The actual stress distribution that the curved surface of the bolt shank exerts on the plate is very difficult to determine. To simplify the design, however, engineering codes often specify the use of the *projected area of contact* between the bolt and plate. As shown in Fig. 8–22b, this area is equal to the product of the plate thickness t and the diameter of the bolt, d_b, not the diameter of the hole. By assuming that the bearing stress is uniformly distributed over this area, it is possible to calculate t using Eq. 8–8, Fig. 8–22b.

Fig. 8–21

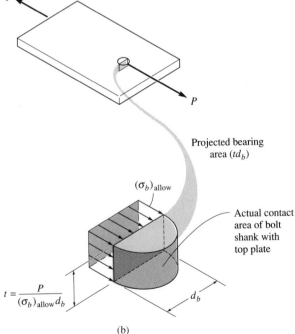

Projected bearing area $(t d_b)$

$(\sigma_b)_{allow}$

Actual contact area of bolt shank with top plate

d_b

$$t = \frac{P}{(\sigma_b)_{allow}\, d_b}$$

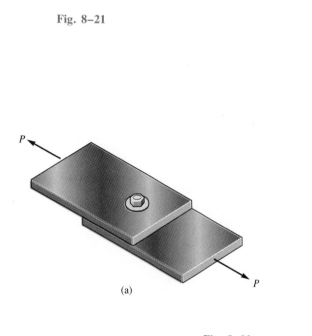

(a)

P

Fig. 8–22

(b)

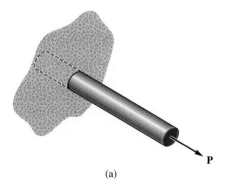

(a)

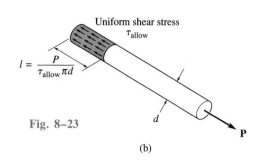

Uniform shear stress
τ_{allow}

$l = \dfrac{P}{\tau_{allow}\,\pi d}$

d

Fig. 8–23

(b)

Required Area to Resist Shear Caused by Axial Load. Occasionally rods or other members will be supported in such a way that shear stress can be developed in the member even though the member may be subjected to an axial load. An example of this situation would be a steel rod whose end is encased in concrete and loaded as shown in Fig. 8–23a. A free-body diagram of the rod, Fig. 8–23b, shows that *shear stress* acts over the area of contact of the rod with the concrete. This area is $(\pi d)l$, where d is the rod's diameter and l is the length of embedment. Although the actual shear-stress distribution along the rod would be difficult to determine, if we assume it is *uniform*, we can use Eq. 8–9 to calculate l, provided we know d and τ_{allow}, Fig. 8–23b.

PROCEDURE FOR ANALYSIS

The above four cases represent just a few of the many applications of Eqs. 8–8 and 8–9 in engineering design or analysis. Whenever these equations are applied, however, it is important to be aware that the stress distribution is assumed to be *uniformly distributed* over the section. When solving problems, a careful consideration should first be made as to the section over which the stress is to be determined. Once this section is determined, the member must then be designed to have a sufficient area at the section to resist the stress that acts on it. To determine this area, application requires the following steps.

Internal Loading. Section the member through the area and draw a free-body diagram of a segment of the member. The internal resultant force at the section is then determined using the equations of equilibrium.

Required Area. Provided the allowable stress is known or can be determined, the required area needed to sustain the load at the section is then computed from Eq. 8–8 or Eq. 8–9.

The following examples numerically illustrate the above concepts.

Example 8–8

The two members are pinned together at *B* as shown in Fig. 8–24*a*. Top views of the pin connections at *A* and *B* are also given in the figure. If the pins have an allowable shear stress of $\tau_{\text{allow}} = 12.5$ ksi and the allowable tensile stress of rod *CB* is $(\sigma_t)_{\text{allow}} = 16.2$ ksi, determine to the nearest $\frac{1}{16}$ in. the smallest diameter of pins *A* and *B* and the diameter of rod *CB* necessary to support the load.

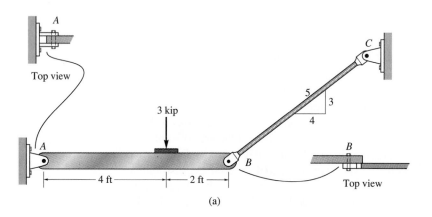

(a)

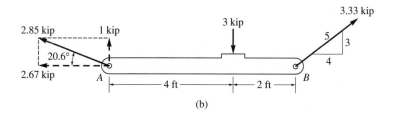

(b)

Fig. 8–24

SOLUTION

Recognizing *CB* to be a two-force member, the free-body diagram of member *AB* along with the computed reactions at *A* and *B* is shown in Fig. 8–24*b*. As an exercise, verify the computations and notice that the *resultant force* at *A* must be used for the design of pin *A*, since this is the force the pin resists.

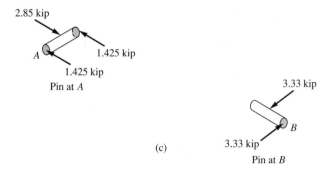

(c)

Diameter of the Pins. From Fig. 8–24a and the free-body diagrams shown in Fig. 8–24c, pin A is subjected to *double shear,* whereas pin B is subjected to *single shear.* We can compute the required diameter of each pin using Eq. 8–9, that is,

$$A_A = \frac{V_A}{\tau_{\text{allow}}} = \frac{1.425 \text{ kip}}{12.5 \text{ kip/in}^2} = 0.114 \text{ in}^2 = \pi\left(\frac{d_A^2}{4}\right) \qquad d_A = 0.381 \text{ in.}$$

$$A_B = \frac{V_B}{\tau_{\text{allow}}} = \frac{3.33 \text{ kip}}{12.5 \text{ kip/in}^2} = 0.267 \text{ in}^2 = \pi\left(\frac{d_B^2}{4}\right) \qquad d_B = 0.583 \text{ in.}$$

Although these values represent the *smallest* allowable pin diameters, a *fabricated* or available pin size will have to be chosen. We will choose a size larger to the nearest $\frac{1}{16}$ in. as required.

$$d_A = \tfrac{7}{16} \text{ in.} = 0.438 \text{ in.} \qquad\qquad Ans.$$
$$d_B = \tfrac{5}{8} \text{ in.} = 0.625 \text{ in.} \qquad\qquad Ans.$$

Diameter of Rod. The required diameter of the rod throughout its midsection is determined from Eq. 8–8; that is,

$$A_{BC} = \frac{P}{(\sigma_t)_{\text{allow}}} = \frac{3.33 \text{ kip}}{16.2 \text{ kip/in}^2} = 0.206 \text{ in}^2 = \pi\left(\frac{d_{BC}^2}{4}\right)$$

$$d_{BC} = 0.512 \text{ in.}$$

We will choose

$$d_{BC} = \tfrac{9}{16} \text{ in.} = 0.563 \text{ in.} \qquad\qquad Ans.$$

Example 8–9

The control arm is subjected to the loading shown in Fig. 8–25a. Determine to the nearest $\frac{1}{4}$ in. the required diameter of the steel pin at C if the allowable shear stress for the steel is $\tau_{allow} = 8$ ksi. Note in the figure that the pin is subjected to double shear.

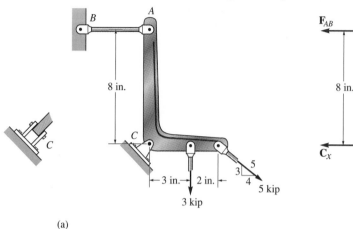

(a)

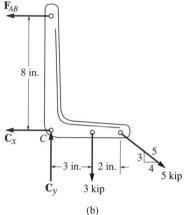

(b)

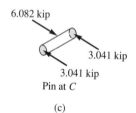

6.082 kip

3.041 kip

3.041 kip

Pin at C

(c)

Fig. 8–25

SOLUTION

Internal Shear Force. A free-body diagram of the arm is shown in Fig. 8–25b. For equilibrium we have

$\downarrow + \Sigma M_C = 0;$ $F_{AB}(8 \text{ in.}) - 3 \text{ kip}(3 \text{ in.}) - 5 \text{ kip}(\frac{3}{5})(5 \text{ in.}) = 0$

$$F_{AB} = 3 \text{ kip}$$

$\xrightarrow{+} \Sigma F_x = 0;$ $-3 \text{ kip} - C_x + 5 \text{ kip}(\frac{4}{5}) = 0$ $C_x = 1 \text{ kip}$

$+\uparrow \Sigma F_y = 0;$ $C_y - 3 \text{ kip} - 5 \text{ kip}(\frac{3}{5}) = 0$ $C_y = 6 \text{ kip}$

The resultant force at C is therefore

$$F_C = \sqrt{(1 \text{ kip})^2 + (6 \text{ kip})^2} = 6.082 \text{ kip}$$

Since the pin is subjected to double shear, a shear force of 3.041 kip is resisted over its cross-sectional area between the arm and each supporting leaf for the pin, Fig. 8–25c.

Required Area. Applying Eq. 8–9, we have

$$A = \frac{V}{\tau_{allow}} = \frac{3.041 \text{ kip}}{8 \text{ kip/in}^2} = 0.3802 \text{ in}^2$$

$$\pi\left(\frac{d}{2}\right)^2 = 0.3802 \text{ in}^2$$

$$d = 0.696 \text{ in.}$$

Use a pin having a diameter of

$$d = \frac{3}{4} \text{ in.} = 0.750 \text{ in.} \qquad \textit{Ans.}$$

Example 8–10

The suspender rod is supported at its end by a fixed-connected circular disk as shown in Fig. 8–26a. If the rod passes through a 40-mm-diameter hole, determine the minimum required diameter of the rod and the minimum thickness of the disk needed to support the 20-kN load. The allowable normal stress for the rod is $\sigma_{allow} = 60$ MPa, and the allowable shear stress for the disk is $\tau_{allow} = 35$ MPa.

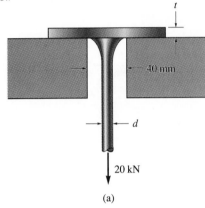

(a)

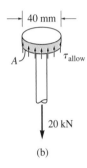

(b)

Fig. 8–26

SOLUTION

Diameter of Rod.   By inspection, the axial force in the rod is 20 kN. Thus the required cross-sectional area of the rod is

$$A = \frac{P}{\sigma_{allow}} = \frac{20(10^3) \text{ N}}{60(10^6) \text{ N/m}^2} = 0.333(10^{-3}) \text{ m}^2$$

So that

$$A = \pi \frac{d^2}{4} = 0.333(10^{-3}) \text{ m}^2$$

$$d = 0.0206 \text{ m} = 20.6 \text{ mm} \qquad\qquad Ans.$$

Thickness of Disk.   As shown on the free-body diagram of the core section of the disk, Fig. 8–26b, the material at the sectioned area must resist *shear stress* to prevent movement of the disk through the hole. If this shear stress is *assumed* to be distributed uniformly over the sectioned area, then, since $V = 20$ kN, we have

$$A = \frac{V}{\tau_{allow}} = \frac{20(10^3) \text{ N}}{35(10^6) \text{ N/m}^2} = 0.571(10^{-3}) \text{ m}^2$$

Since the sectioned area $A = 2\pi(0.02 \text{ m})(t)$, the required thickness of the disk is

$$t = \frac{0.571(10^{-3}) \text{ m}^2}{2\pi(0.02 \text{ m})} = 4.55(10^{-3}) \text{ m} = 4.55 \text{ mm} \qquad Ans.$$

Example 8–11

An axial load on the shaft shown in Fig. 8–27a is resisted by the collar at C, which is attached to the shaft and located on the right side of the bearing at B. Determine the largest value of P for the two axial forces so that the stress in the collar does not exceed an allowable bearing stress of $(\sigma_b)_{\text{allow}} = 75$ MPa and the normal stress in the shaft does not exceed an allowable tensile stress of $(\sigma_t)_{\text{allow}} = 55$ MPa.

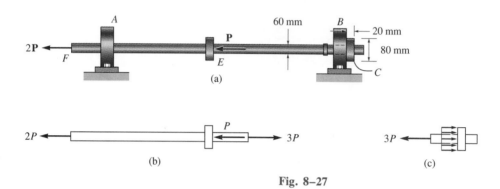

Fig. 8–27

SOLUTION

To solve the problem we will determine P for each possible failure condition. Then we will choose the *smallest* value. Why?

Axial Stress. Using the method of sections, the axial load within region FE of the shaft is $2P$, whereas the *largest* axial load occurs in region EC, Fig. 8–27b. Since the cross-sectional area of the entire shaft is constant, region EC will be subjected to the maximum normal stress. We have

$$55(10^6) \text{ N/m}^2 = \frac{3P}{\pi(0.03 \text{ m})^2}$$

$$P = 51.8 \text{ kN}$$

Bearing Stress. As shown on the free-body diagram in Fig. 8–27c, the collar at C must resist the load of $3P$, which acts over a bearing area of $A_b = [\pi(0.04 \text{ m})^2 - \pi(0.03 \text{ m})^2] = 2.20(10^{-3}) \text{ m}^2$. Thus,

$$75(10^6) \text{ N/m}^2 = \frac{3P}{2.20(10^{-3}) \text{ m}^2}$$

$$P = 55.0 \text{ kN}$$

By comparison, the largest load that can be applied to the shaft is $P = 51.8$ kN, since any load larger than this will cause the allowable tensile stress in the shaft to be exceeded.

Example 8-12

The rigid bar AB shown in Fig. 8–28a is supported by a steel rod AC having a diameter of 20 mm and an aluminum block having a cross-sectional area of 1800 mm². The 18-mm-diameter pins at A and C are subjected to *single shear*. If the failure stress for the steel and aluminum is $(\sigma_{st})_{fail} = 680$ MPa and $(\sigma_{al})_{fail} = 70$ MPa, respectively, and the failure shear stress for each pin is $\tau_{fail} = 900$ MPa, determine the largest load P that can be applied to the bar. Apply a factor of safety of F.S. = 2.0.

SOLUTION

Using Eqs. 8–6 and 8–7, the allowable stresses are

$$(\sigma_{st})_{allow} = \frac{(\sigma_{st})_{fail}}{\text{F.S.}} = \frac{680 \text{ MPa}}{2} = 340 \text{ MPa}$$

$$(\sigma_{al})_{allow} = \frac{(\sigma_{al})_{fail}}{\text{F.S.}} = \frac{70 \text{ MPa}}{2} = 35 \text{ MPa}$$

$$\tau_{allow} = \frac{\tau_{fail}}{\text{F.S.}} = \frac{900 \text{ MPa}}{2} = 450 \text{ MPa}$$

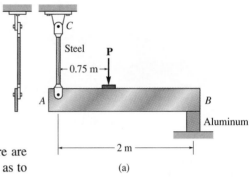

(a)

The free-body diagram for the bar is shown in Fig. 8–28b. There are three unknowns. Here we will apply the equations of equilibrium so as to express F_{AC} and F_B in terms of the applied load P. We have

$$\zeta^+ \ \Sigma M_B = 0; \qquad P(1.25 \text{ m}) - F_{AC}(2 \text{ m}) = 0 \qquad (1)$$
$$\zeta^+ \ \Sigma M_A = 0; \qquad F_B(2 \text{ m}) - P(0.75 \text{ m}) = 0 \qquad (2)$$

We will now determine each value of P that creates the allowable stress in the rod, block, and pins, respectively. In particular, note that the pins at A and C are each subjected to a shear F_{AC}, since single shear exists, Fig. 8–28a.

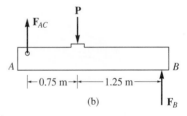

(b)

Fig. 8–28

Rod AC. This requires

$$F_{AC} = (\sigma_{st})_{allow}(A_{AC}) = 340(10^6) \text{ N/m}^2[\pi(0.01 \text{ m})^2] = 106.8 \text{ kN}$$

Using Eq. 1,

$$P = \frac{(106.8 \text{ kN})(2 \text{ m})}{1.25 \text{ m}} = 171 \text{ kN}$$

Block B. In this case,

$$F_B = (\sigma_{al})_{allow} \ A_B = 35(10^6) \text{ N/m}^2[1800 \text{ mm}^2(10^{-6}) \text{ m}^2/\text{mm}^2] = 63.0 \text{ kN}$$

Using Eq. 2,

$$P = \frac{(63.0 \text{ kN})(2 \text{ m})}{0.75 \text{ m}} = 168 \text{ kN}$$

Pin A or C. Here

$$V = F_{AC} = \tau_{allow} \ A = 450(10^6) \text{ N/m}^2[\pi(0.009 \text{ m})^2] = 114.5 \text{ kN}$$

From Eq. 1,

$$P = \frac{114.5 \text{ kN}(2 \text{ m})}{1.25 \text{ m}} = 183 \text{ kN}$$

By comparison, when P reaches its *smallest value* (168 kN), it develops the allowable normal stress in the aluminum block. Hence,

$$P = 168 \text{ kN} \qquad \qquad Ans.$$

PROBLEMS

8–19.   The steel lap joint is held together by two bolts. If the allowable shear stress for the bolts is $(\tau_{allow})_b = 12$ ksi, determine to the nearest $\frac{1}{4}$ in. the required diameter for each bolt.

18 kip

18 kip

Prob. 8–19

*8–20.   The joint is fastened together using two bolts. Determine the required diameter of the bolts if the allowable shear stress for the bolts is $\tau_{allow} = 110$ MPa.

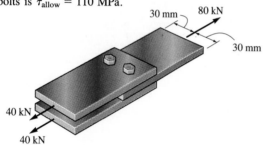

30 mm   80 kN

30 mm

40 kN

40 kN

Prob. 8–20

8–21.   The two aluminum rods support the horizontal force of 18 kN. Determine their required diameters if the allowable tensile stress for the aluminum is $\sigma_{allow} = 150$ MPa.

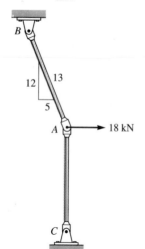

B

12   13

5

A   18 kN

C

Prob. 8–21

8–22.   Member B is subjected to a compressive force of 800 lb. If A and B are both made of wood and are $\frac{3}{8}$ in. thick, determine to the nearest $\frac{1}{4}$ in. the smallest dimension h of the support so that the shear stress does not exceed $\tau_{allow} = 600$ psi.

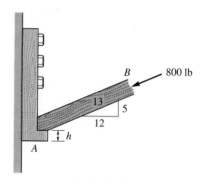

B   800 lb

13

5

12

h

A

Prob. 8–22

8–23.   If the allowable bearing stress for the material under the supports at A and B is $(\sigma_b)_{allow} = 400$ psi, determine the size of *square* bearing plates A' and B' required to support the load. Dimension the plates to the nearest $\frac{1}{2}$ in. The reactions at the supports are vertical.

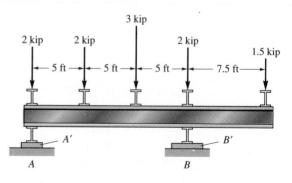

3 kip

2 kip   2 kip   2 kip

1.5 kip

5 ft   5 ft   5 ft   7.5 ft

A'   B'

A   B

Prob. 8–23

*8–24. The clevis C and pin are made of steel having an allowable normal stress of $\sigma_{\text{allow}} = 21$ ksi and an allowable shear stress of $\tau_{\text{allow}} = 12$ ksi. Determine to the nearest $\frac{1}{4}$ in. the diameters of the rod, d_r, and the pin, d_p needed to support the load. Notice that the screw at the end of the rod has "upset" threads, so the rod will *not* fail in this region.

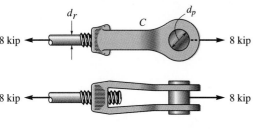

Prob. 8–24

8–26. Determine the required cross-sectional area of member BC and the diameter of the pins at A and B if the allowable normal stress is $\sigma_{\text{allow}} = 3$ ksi and the allowable shear stress is $\tau_{\text{allow}} = 4$ ksi.

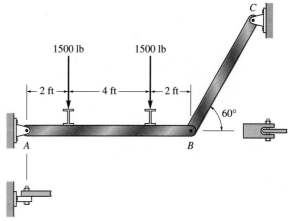

Prob. 8–26

8–25. The column has a cross-sectional area of $12(10^3)$ mm^2. It is subjected to an axial force of 50 kN. If the base plate to which the column is attached has a length of 250 mm, determine its width d so that the average bearing stress under the plate at the ground is one-third of the average compressive stress in the column. Sketch the stress distributions acting over the column's cross-sectional area and at the bottom of the base plate.

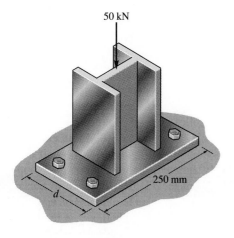

Prob. 8–25

8–27. The two steel cables AB and AC are used to support the chain. If both cables have an allowable tensile stress of $\sigma_{\text{allow}} = 170$ MPa, and cable AB has a diameter of 8 mm and AC has a diameter of 5 mm, determine the greatest force P that can be applied to the chain before one of the cables exceeds the allowable stress.

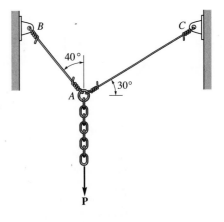

Prob. 8–27

*8–28. The rod BC is made of steel having an allowable tensile stress of $\sigma_{allow} = 155$ MPa. Determine its smallest diameter so that it can support the load shown. The beam is assumed to be pin-connected at A.

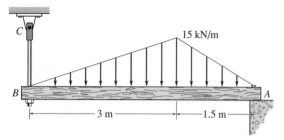

15 kN/m

B

3 m

1.5 m

A

Prob. 8–28

8–29. The tension member is fastened together using *two* bolts, one acting on each side of the member as shown. Each bolt has a diameter of 0.3 in. Determine the maximum load P that can be applied to the member if the allowable average shear stress for the bolts is $\tau_{allow} = 12$ ksi and the allowable normal stress is $\sigma_{allow} = 20$ ksi.

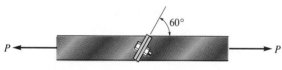

60°

P

P

Prob. 8–29

8–30. The compound wooden beam is connected together by a bolt at B. Assuming that the connections at A, B, C, and D exert only vertical forces on the beam, determine the required diameter of the bolt and the required outer diameter of the washers if the allowable tensile stress for the bolt is $(\sigma_t)_{allow} = 150$ MPa and the allowable bearing stress for the wood is $(\sigma_b)_{allow} = 28$ MPa. Assume that the hole in the washers is the same size as the bolt diameter.

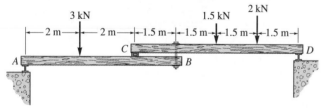

3 kN   1.5 kN   2 kN

2 m   2 m   1.5 m 1.5 m 1.5 m 1.5 m

C   B   D

A

Prob. 8–30

8–31. The assembly is used to support the distributed loading of $w = 500$ lb/ft. Determine the factor of safety with respect to yielding for the steel rod BC and the pins at B and C if the yield stress for the steel in tension is $\sigma_y = 36$ ksi and in shear $\tau_y = 18$ ksi. The rod has a diameter of 0.4 in., and the pins each have a diameter of 0.30 in.

*8–32. If the allowable shear stress for each of the 0.3-in.-diameter steel pins at A, B, and C is $\tau_{allow} = 12.5$ ksi, and the allowable normal stress for the 0.40-in.-diameter rod is $\sigma_{allow} = 22$ ksi, determine the largest intensity w of the uniform distributed load that can be suspended from the beam.

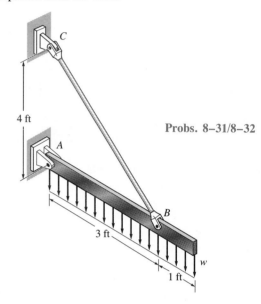

C

4 ft

Probs. 8–31/8–32

A

3 ft

B

1 ft

w

8–33. The hanger assembly is used to support a distributed loading of $w = 0.8$ kip/ft. Determine the average shear stress in the 0.40-in.-diameter bolt at A and the average tensile stress in rod AB, which has a diameter of 0.5 in. If the failure shear stress for the bolt is $\tau_{fail} = 25$ ksi, and the failure tensile stress for the rod is $\sigma_{fail} = 38$ ksi, determine the factor of safety in each case.

8–34. Determine the intensity w of the maximum distributed load that can be supported by the hanger assembly so that an allowable shear stress of $\tau_{allow} = 13.5$ ksi is not exceeded in the 0.40-in.-diameter bolts at A and B, and an allowable tensile stress of $\sigma_{allow} = 22$ ksi is not exceeded in the 0.5-in.-diameter rod AB.

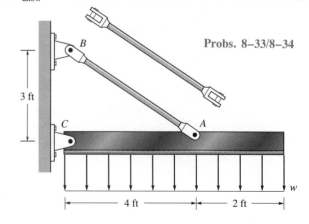

B

Probs. 8–33/8–34

3 ft

C

A

4 ft

2 ft

w

8.7 Deformation

Whenever a force is applied to a body, it will tend to change the body's shape and size. These changes are referred to as *deformation,* and they may be either highly visible or practically unnoticable, without the use of equipment to make precise measurements. For example, a rubber band will undergo a very large deformation when stretched. On the other hand, only slight deformations of structural members occur when a building is occupied with people walking about. Deformation of a body can also occur when the temperature of the body is changed. A typical example is the thermal expansion or contraction of a roof caused by the weather.

Displacement is a vector quantity that is used to measure the movement of a particle or point from one position to the next. Hence, if a body is classified as being deformable, then its adjacent particles can be displaced relative to one another when forces are applied to the body. On the other hand, if the body is rigid, then no relative displacement can occur between the particles.

Consider the body shown in Fig. 8–29, which is made from a continuous and cohesive material and is shown in the initial undeformed state. The three particles *A, B,* and *C* are located in the body at points measured from a fixed coordinate system. When a loading causes the body to deform, and thus move into its final position, the particles are displaced to points *A', B',* and *C',* Fig. 8–29. For example, the displacement of point *A* is indicated by the vector **u**(*A*). Because of the deformation, the once straight lines *AB* and *AC* become curves *A'B'* and *A'C';* and as a result, the length of *AB* and *AC* and the angle θ will be different from the curved length of *A'B'* and *A'C'* and the angle θ'. In other words, the difference between the lengths and relative orientations of the two lines in the body is a consequence of the displacements caused by the *deformation.* Measurements of the deformation must therefore account for *both* the changes made in the length of line segments inscribed in the body *and* the changes made in the angle between them.

In the general sense, the deformation of a body will not be uniform throughout its volume, and so the change in geometry of a line segment within the body may vary along its length. For example, one portion of the line may elongate, whereas another portion may contract. As shorter and shorter line segments are considered, however, they remain straighter after the deformation, and so to study deformational changes in a more uniform manner, we will consider the lines to be very short and located in the neighborhood of a point. In doing so, realize that any line segment located at one point in the body will change by a different amount from one located at some other point. Furthermore, these changes will also depend on the orientation of the line segment at the point. For example, a line segment may elongate if it is orientated in one direction, whereas it may contract if it is orientated in another direction.

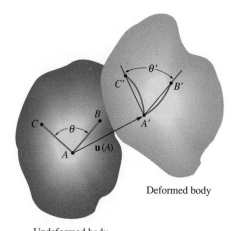

Deformed body

Undeformed body

Fig. 8–29

8.8 Strain

In order to describe the deformation by changes in length of line segments and the changes in the angles between them, we will develop the concept of strain. Measurements of strain are actually made by experiments, and once the strains are obtained, it will be shown later in the text how they may be related to the applied loads, or stresses, acting within the body.

Normal Strain.   The elongation or contraction of a line segment per unit of length is referred to as *normal strain*. To develop a formalized definition of normal strain, consider the line *AB*, which is contained within the undeformed body shown in Fig. 8–30a. This line lies along the *n* axis and has an original length of Δs. During deformation, points *A* and *B* are displaced to *A'* and *B'*, and the line becomes a curve having a length of $\Delta s'$, Fig. 8–30b. The change in length of the line is therefore $\Delta s' - \Delta s$. If we define the *average normal strain* using the symbol ϵ_{avg} (epsilon), then

$$\epsilon_{avg} = \frac{\Delta s' - \Delta s}{\Delta s} \qquad (8-10)$$

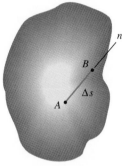

Undeformed body

(a)

As point *B* is chosen closer and closer to point *A*, the length of the line becomes shorter and shorter, such that $\Delta s \to 0$. Also, this causes *B'* to approach *A'*, such that $\Delta s' \to 0$. Consequently, in the limit the normal strain at *point A* and in the *direction* of *n* is

$$\epsilon = \lim_{B \to A \text{ along } n} \frac{\Delta s' - \Delta s}{\Delta s} \qquad (8-11)$$

If the normal strain is known, we can use this equation to obtain the approximate final length of a *short* line segment after it is deformed. We have

$$\Delta s' \approx (1 + \epsilon)\, \Delta s \qquad (8-12)$$

Hence, when ϵ is positive the initial line Δs will elongate, whereas if ϵ is negative the line contracts.

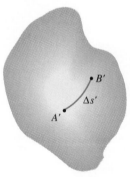

Deformed body

(b)

Fig. 8–30

Units.   Notice that normal strain is a *dimensionless quantity,* since it is a ratio of two lengths. Although this is the case, it is common practice to state it in terms of a ratio of length units. If the SI system is used, then the basic units will be meters/meter (m/m). Ordinarily, for most engineering applications ϵ will be very small, so measurements of strain are in micrometers per meter (μm/m), where $1\ \mu m = 10^{-6}$ m. In the Foot-Pound-Second system, strain can be stated in units of inches per inch (in./in.). Sometimes for experimental work, strain is expressed as a percent, e.g., 0.001 m/m = 0.1%. As an example, a normal strain of $480(10^{-6})$ can be reported as $480(10^{-6})$ in./in., 480 μm/m, or 0.0480%. Also, one can state this answer as simply 480 μ (480 ''micros'').

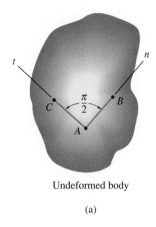

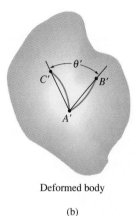

Undeformed body   Deformed body

(a)   (b)

Fig. 8–31

Shear Strain. The change in angle that occurs between two line segments that were originally *perpendicular* to one another is referred to as *shear strain*. This angle is denoted by γ (gamma) and is measured in radians (rad). To show how it is developed, consider the line segments AB and AC originating from the same point A in a body, and directed along the perpendicular n and t axes, Fig. 8–31a. After deformation, the ends of the lines are displaced, and the lines themselves become curves, such that the angle between them at A is θ', Fig. 8–31b. Hence we define the shear strain at point A that is associated with the n and t axes as

$$\gamma_{nt} = \frac{\pi}{2} - \lim_{\substack{B \to A \text{ along } n \\ C \to A \text{ along } t}} \theta' \qquad (8\text{–}13)$$

Notice that if θ' is smaller than $\pi/2$ the shear strain is positive, whereas if θ' is larger than $\pi/2$ the shear strain is negative.

Cartesian Strain Components. Using the above definitions of normal and shear strain, we will now show how they can be used to describe the deformation of the body shown in Fig. 8–32a. To do so, imagine the body to be subdivided into small elements such as the one shown in Fig. 8–32b. This element is rectangular, has undeformed dimensions Δx, Δy, and Δz, and is located in the neighborhood of a point in the body, Fig. 8–32a. Assuming that the element's dimensions are very small, the deformed shape of the element, shown in Fig. 8–32c, will be a parallelepiped, since very small line segments will remain approximately straight after the body is deformed. In order to

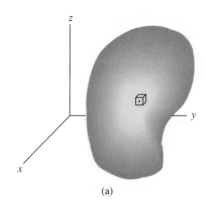

(a)

Undeformed
element

(b)

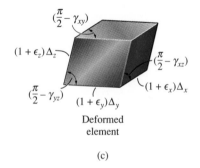

Deformed
element

(c)

Fig. 8–32

achieve this deformed shape, we may first consider how the normal strain changes the lengths of the sides of the rectangular element, and then how the shear strain changes the angles of each side. Hence, using Eq. 8–12, in reference to the lines Δx, Δy, and Δz, the approximate lengths of the sides of the parallelepiped are

$$(1 + \epsilon_x)\,\Delta x \qquad (1 + \epsilon_y)\,\Delta y \qquad (1 + \epsilon_z)\,\Delta z$$

And the approximate angles between the sides, again originally defined by the sides Δx, Δy, and Δz, are

$$\frac{\pi}{2} - \gamma_{xy} \qquad \frac{\pi}{2} - \gamma_{yz} \qquad \frac{\pi}{2} - \gamma_{zx}$$

In particular, notice that the *normal strains* cause a change in *volume* of the rectangular element, whereas the *shear strains* cause a change in its *shape*. Of course, both of these effects occur simultaneously during the deformation. It might also be added that because the other elements of the body deform, the element under consideration will not only deform but will also both translate and rotate, since it is being displaced by the deformations of all the other elements. For example, imagine the element located near the end of a fishing pole while the pole deflects when supporting a fish. These *rigid-body displacements,* however, will be neglected here, since we are interested only in the relative displacements of the element due to its deformation. Specifically, it is these *deformation displacements* that will allow us later to determine the stress in the body.

In summary, then, the *state of strain* at a point in a body requires specifying three normal strains, ϵ_x, ϵ_y, ϵ_z, and three shear strains, γ_{xy}, γ_{yz}, γ_{zx}. These strains completely describe the deformation of a rectangular volume element of material located at the point and orientated so that its sides are originally parallel to the x, y, z axes. Once these strains are defined at all points in the body, the deformed shape of the body can then be described. It should also be added that by knowing the state of strain at a point, defined by its six components, it is possible to determine the strain components on an element orientated at the point in any other direction. This is discussed in Chapter 15.

Small Strain Analysis. Most engineering design involves applications for which only small deformations are allowed. For example, almost all structures and machines appear to be rigid, and the deformations that occur during use are hardly noticeable. Furthermore, even if the deflection of a member such as a thin plate or slender rod may appear to be large, the material from which it is made may only be subjected to very small deformations. In this text, therefore, we will assume that the deformations that take place within a body are almost infinitesimal, so that the *normal strains* occurring within the material are *very small* compared to 1, that is, $\epsilon \ll 1$. This assumption based on the magnitude of the strain has wide practical application in engineering, and its application is often referred to as a *small strain analysis.*

Whenever such an analysis is used, since $\epsilon \ll 1$, it is permissible to *neglect* the product of either two strains or deformation displacements, or any term involving strain or deformation displacement that is raised to a power greater than 1. Such "first-order" approximations may be used to simplify calculations involving strain and deformation. The following examples illustrate numerical applications of some of these approximations in the calculation of normal and shear strain.

Example 8–13

A force acting on the grip of the lever arm shown in Fig. 8–33a causes the arm to rotate clockwise through an angle of $\theta = 0.002$ rad. Determine the average normal strain developed in the wire BC.

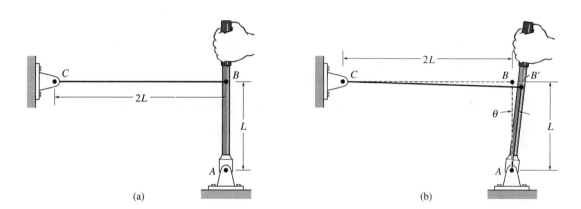

Fig. 8–33

SOLUTION

After deformation point B moves to point B', Fig. 8–33b. The approximate stretch of the wire Δ can then be determined by considering point B to move along the arc of a circle having a radius L. This arc can be determined from the formula $s = \theta r$, where in this case, $\Delta = \theta L$. Since the original length of the wire is $2L$, applying Eq. 8–10, the average normal strain in the wire is therefore

$$\epsilon_{avg} = \frac{\Delta s' - \Delta s}{\Delta s} = \frac{\theta L}{2L} = \frac{0.002}{2} = 0.001 \qquad Ans.$$

Example 8–14

The slender rod shown in Fig. 8–34 is subjected to an increase of temperature along its axis, which creates a normal strain in the rod of $\epsilon_z = 40(10^{-3})z^{1/2}$, where z is given in meters. Determine (a) the displacement of the end B of the rod due to the temperature increase, and (b) the average normal strain in the rod.

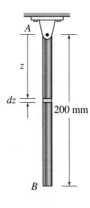

Fig. 8–34

SOLUTION

Part (a). Since the normal strain is reported at each point along the rod, a differential segment dz, located at position z, Fig. 8–34, has a deformed length that can be determined from Eq. 8–12; that is,

$$dz' = [1 + 40(10^{-3})z^{1/2}] \, dz$$

The sum total of these segments along the axis yields the *deformed length* of the rod, i.e.,

$$L' = \int_0^{0.2 \text{ m}} [1 + 40(10^{-3})z^{1/2}] \, dz$$

$$= z + 40(10^{-3})(\tfrac{2}{3} z^{3/2}) \Big|_0^{0.2 \text{ m}}$$

$$= 0.20239 \text{ m}$$

The displacement of the end of the rod is therefore

$$\Delta_B = 0.20239 \text{ m} - 0.2 \text{ m} = 0.00239 \text{ m} = 2.39 \text{ mm} \downarrow \qquad Ans.$$

Part (b). The average normal strain in the rod is determined from Eq. 8–10, which assumes that the rod or "line segment" has an original length of 200 mm and a change in length of 2.39 mm. Hence,

$$\epsilon_{\text{avg}} = \frac{2.39 \text{ mm}}{200 \text{ mm}} = 0.0119 \text{ mm/mm} \qquad Ans.$$

Example 8–15

The plate is deformed into the dashed shape shown in Fig. 8–35a. If in this deformed shape horizontal lines on the plate remain horizontal and do not change their length, determine (a) the average normal strain along the side AB, and (b) the average shear strain in the plate relative to the x and y axes.

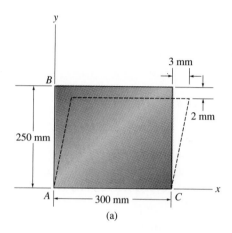

(a)

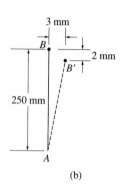

(b)

Fig. 8–35

SOLUTION

Part (a). Line AB, coincident with the y axis, becomes line AB' after deformation, as shown in Fig. 8–35b. The length of this line is

$$AB' = \sqrt{(250 - 2)^2 + (3)^2} = 248.018 \text{ mm}$$

The average normal strain for AB is therefore

$$(\epsilon_{AB})_{avg} = \frac{AB' - AB}{AB} = \frac{248.018 \text{ mm} - 250 \text{ mm}}{250 \text{ mm}}$$

$$= -7.93(10^{-3}) \text{ mm/mm} \qquad Ans.$$

The negative sign indicates the strain causes a contraction of AB.

Part (b). As noted in Fig. 8–35c, the once 90° angle BAC between the sides of the plate, referenced from the x, y axes, changes to θ' due to the displacement of B to B'. Since $\gamma_{xy} = \pi/2 - \theta'$, then γ_{xy} is the angle shown in the figure. Thus,

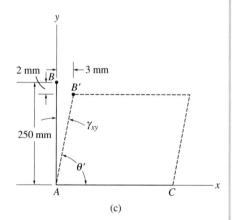

(c)

$$\gamma_{xy} = \tan^{-1}\left(\frac{3 \text{ mm}}{250 \text{ mm} - 2 \text{ mm}}\right) = 0.0121 \text{ rad} \qquad Ans.$$

Example 8–16

The plate shown in Fig. 8–36a is held in the rigid horizontal guides at its top and bottom, AD and BC. If its right side CD is given a uniform horizontal displacement of 2 mm, determine (a) the average normal strain along the diagonal AC, and (b) the shear strain at E relative to the x, y axes.

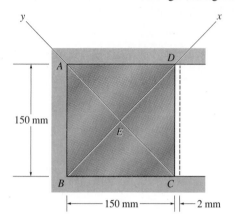

Fig. 8–36

(a)

(b)

SOLUTION

Part (a). When the plate is deformed, the diagonal AC becomes AC′, Fig. 8–36b. The length of diagonals AC and AC′ can be computed from the Pythagorean theorem. We have

$$AC = \sqrt{(0.150)^2 + (0.150)^2} = 0.21213 \text{ m}$$
$$AC' = \sqrt{(0.150)^2 + (0.152)^2} = 0.21355 \text{ m}$$

Therefore the average normal strain along the diagonal is

$$(\epsilon_{AC})_{\text{avg}} = \frac{AC' - AC}{AC} = \frac{0.21355 \text{ m} - 0.21213 \text{ m}}{0.21213 \text{ m}}$$

$$= 0.00669 \text{ mm/mm} \qquad \qquad \textit{Ans.}$$

Part (b). To find the shear strain at E relative to the x and y axes, it is first necessary to find the angle θ′, which specifies the angle between these axes after deformation, Fig. 8–36b. We have

$$\tan\left(\frac{\theta'}{2}\right) = \frac{76 \text{ mm}}{75 \text{ mm}}$$

$$\theta' = 90.759° = \frac{\pi}{180°}(90.759°) = 1.58404 \text{ rad}$$

Applying Eq. 8–13, the shear strain at E is therefore

$$\gamma_{xy} = \frac{\pi}{2} - 1.58404 \text{ rad} = -0.0132 \text{ rad} \qquad \textit{Ans.}$$

According to the sign convention, the *negative sign* indicates that the angle θ′ is *greater than* 90°.

PROBLEMS

8–35. A thin strip of rubber has an unstretched length of 15 in. If it is stretched around a pipe having an outer diameter of 5 in., determine the average normal strain in the strip.

*8–36. An air filled rubber ball has a diameter of 6 in. If the air pressure within it is increased until the ball's diameter becomes 7 in., determine the average normal strain in the rubber.

8–37. The rigid beam is supported by a pin at A and wires BD and CE. If the load $\mathbf{P}$ on the beam causes the end C to be displaced 10 mm downward, determine the strain developed in wires CE and BD.

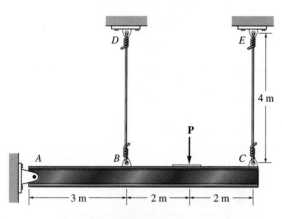

Prob. 8–37

8–38. Due to its weight, the rod is subjected to a normal strain that varies along its length such that $\epsilon = kz$, where k is a constant. Determine the displacement ΔL of its end B when it is suspended as shown.

Prob. 8–38

8–39. The wire AB is unstretched when $\theta = 45°$. If a load is applied to the bar AC, which causes $\theta = 47°$, determine the normal strain in the wire.

*8–40. If a load applied to bar AC causes point A to be displaced to the right by an amount ΔL, determine the normal strain in the wire AB. Originally, $\theta = 45°$.

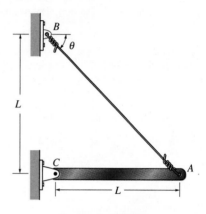

Probs. 8–39/8–40

8–41. The two wires are connected together at A. If the force $\mathbf{P}$ causes point A to be displaced horizontally 2 mm, determine the normal strain developed in each wire.

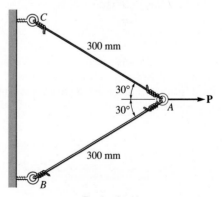

Prob. 8–41

8–42. A thin wire, lying along the x axis, is strained such that each point on the wire is displaced $\Delta x = kx^2$ along the x axis. What is the normal strain at any point P along the wire?

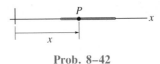

Prob. 8–42

8–43. The triangular plate is fixed at its base and its apex A is given a horizontal displacement of 5 mm. Determine the shear strain γ_{xy} at A.

*8–44. The triangular plate is fixed at its base, and its apex A is given a horizontal displacement of 5 mm. Determine the average normal strain ϵ_x along the x axis.

8–45. The triangular plate is fixed at its base, and its apex A is given a horizontal displacement of 5 mm. Determine the average normal strain $\epsilon_{x'}$ along the x' axis.

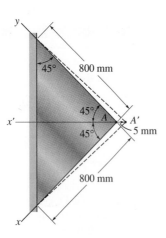

Probs. 8–43/8–44/8–45

8–46. The nonuniform loading causes a normal strain in the shaft that can be expressed as $\epsilon_x = kx^2$, where k is a constant. Determine the displacement of the end B. Also, what is the average strain in the rod?

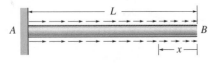

Prob. 8–46

8–47. The corners of the square plate are given the displacements indicated. Determine the average normal strains ϵ_x and ϵ_y along the x and y axes, and the shear strain γ_{xy} at point A.

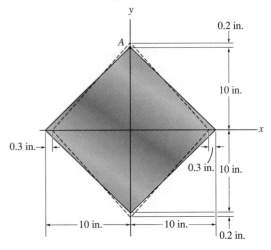

Prob. 8–47

*8–48. The rectangular plate is subjected to the deformation shown by the dashed line. Determine the shear strains γ_{xy} and $\gamma_{x'y'}$ developed at point A.

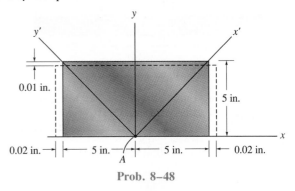

Prob. 8–48

8–49. The rectangular plate is subjected to the deformation shown by the dashed line. Determine the average shear strain γ_{xy} of the plate.

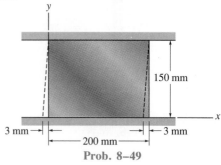

Prob. 8–49

REVIEW PROBLEMS

8–50. The truss is made from three pin-connected members having the cross-sectional areas shown in the figure. Determine the normal stress developed in each member when the truss is subjected to the load shown. State whether the stress is tensile or compressive.

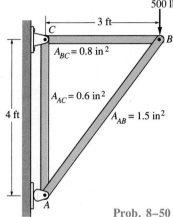

Prob. 8–50

8–51. A force of 8 kN is applied at the *center* of the wooden post. If the post is mounted at the corner of its base plate B, can the bearing stress that the base plate exerts on the slab S be assumed uniformly distributed? Why or why not? What is the average compressive stress in the wooden post?

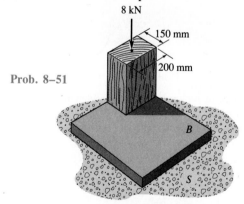

Prob. 8–51

*8–52. The built-up shaft consists of a pipe AB and solid rod BC. The pipe has an inner diameter of 20 mm and outer diameter of 28 mm. The rod has a diameter of 12 mm. Determine the normal stress at points D and E and represent the stress on a volume element located at each of these points.

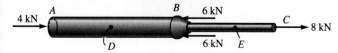

Prob. 8–52

8–53. The member ABC is supported by a pin at A and a rocker at C. Determine the resultant internal loadings acting on the cross sections located through points D and E.

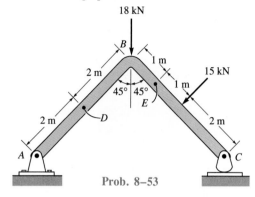

Prob. 8–53

8–54. The beam AB is fixed to the wall and has a uniform weight of 80 lb/ft. If the trolley supports a load of 1500 lb, determine the resultant internal loadings on the cross sections through points C and D.

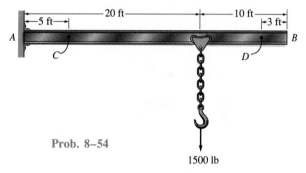

Prob. 8–54

8–55. The beam supports a distributed load and a couple moment. Determine the resultant internal loadings on the cross sections located just to the left and just to the right of the roller support at B.

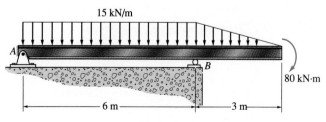

Prob. 8–55

*8–56. The plastic block is subjected to an axial compressive force of 600 N. Assuming that the caps at the top and bottom distribute the load uniformly throughout the block, determine the normal and average shear stress acting along section a–a.

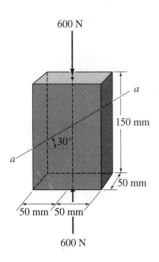

Prob. 8–56

8–57. The lever is attached to the shaft A using a key that has a width of 8 mm and length of 25 mm. If the shaft is fixed and a vertical force of 50 N is applied perpendicular to the handle, determine the average shear stress developed along section a–a of the key.

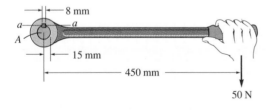

Prob. 8–57

8–58. Bar ABC is originally in a horizontal position. If loads cause the end A to be displaced downwards $\Delta_A = 0.002$ in. and the bar rotates $\theta = 0.2°$, determine the strains in the rods AD, BE, and CF.

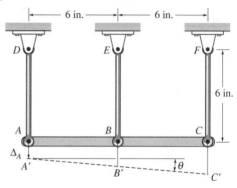

Prob. 8–58

8–59. The rectangular plate is distorted into a parallelogram as shown by the dashed lines. If the diagonal AB' is 50.03 mm, determine the shear strain γ_{xy}.

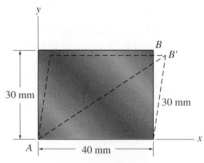

Prob. 8–59

*8–60. The piece of rubber is originally rectangular. Determine the average shear strain γ_{xy} if the corners B and D are subjected to the displacements that cause the rubber to distort as shown by the dashed lines.

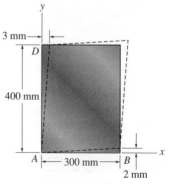

Prob. 8–60

9 Mechanical Properties of Materials

Having discussed the basic concepts of stress and strain, we will in this chapter show how stress can be related to strain by using experimental methods to determine the stress–strain diagram for a specific material. The properties of this diagram will then be discussed for materials that are commonly used in engineering. Also, other mechanical properties and tests that are related to the development of mechanics of materials will be discussed.

9.1 The Tension and Compression Test

The strength of a material depends on its ability to sustain a load without undue deformation or failure. This property is inherent in the material itself and must be determined by *experiment*. As a result, several types of tests have been developed to evaluate a material's strength under loads that are static, cyclic, extended in duration, or impulsive. Over the years, each of these tests has become standardized so that the results obtained from different laboratories can be compared. In the United States the American Society for Testing and Materials (ASTM) has published guidelines for performing such tests and provides limits for which the use of a particular material is considered acceptable.

One of the most important tests to perform is the *tension or compression test*. Although many important mechanical properties of a material can be determined from this test, it is used primarily to determine the relationship between the average normal stress and normal strain in many engineering materials consisting of metal, ceramics, polymers, and composites. To perform this test a specimen of the material is made into a "standard" shape and

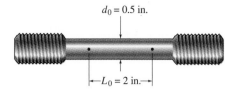

$d_0 = 0.5$ in.

$L_0 = 2$ in.

Fig. 9–1

size. Before testing, two small punch marks are made along the specimen's length. These marks are located away from both ends of the specimen because the stress distribution at the ends is somewhat complex due to gripping at the connections where the load is applied. Measurements are taken of both the specimen's initial cross-sectional area, A_0, and the *gauge-length* distance L_0 between the punch marks. For example, when a steel specimen is used in a tension test it generally has an initial diameter of $d_0 = 0.5$ in. and a gauge length of $L_0 = 2$ in., Fig. 9–1. In order to apply an axial load with no bending of the specimen, the ends are usually seated into ball-and-socket joints. A testing machine like the one shown in Fig. 9–2 is then used to stretch the specimen at a very slow, constant rate until it reaches the breaking point.

At frequent intervals during the test, data is recorded of the applied load P, as read on the dial of the machine. Also, the elongation $\Delta L = L - L_0$ between the punch marks on the specimen may be measured using either a caliper or a mechanical or optical device called an *extensometer*. This value of ΔL is then used to determine the average normal strain in the specimen. Sometimes, however, this measurement is not taken, since it is also possible to read the strain *directly* by using an *electrical-resistance strain gauge*, which looks like the one shown in Fig. 9–3. The operation of this gauge is based on the change in electrical resistance of a very thin wire or piece of metal foil under strain. Essentially the gauge is cemented to the specimen in a specified direction. If the cement is very strong in comparison to the gauge, then the gauge is in effect an integral part of the specimen, so that when the specimen is strained in the direction of the gauge, the wire and specimen will experience the same strain. By measuring the electrical resistance of the wire, the gauge may be calibrated to read values of normal strain directly.

Fig. 9–2

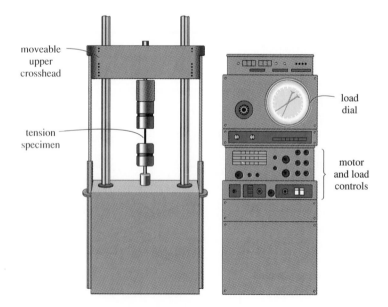

moveable upper crosshead

tension specimen

load dial

motor and load controls

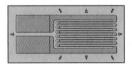

Electrical–resistance strain gauge

Fig. 9–3

9.2 The Stress–Strain Diagram

From the data of a tension or compression test, it is possible to compute the stress and corresponding strain in the specimen and then plot the results. The resulting curve is called the *stress–strain diagram,* and there are two ways in which it is normally described.

Conventional Stress and Strain Diagram.   Using the recorded data, we can determine the *nominal* or *engineering stress* by dividing the applied load P by the specimen's *original* cross-sectional area A_0. This calculation assumes that the stress is *constant* over the cross section and throughout the region between the gauge points. We have

$$\sigma = \frac{P}{A_0} \tag{9–1}$$

Likewise, the *nominal* or *engineering strain* is found directly from the strain gauge reading, or by dividing the change in the specimen's gauge length, ΔL, by the specimen's original gauge length L_0. Here the strain is assumed to be constant throughout the region between the gauge points. Thus,

$$\epsilon = \frac{\Delta L}{L_0} \tag{9–2}$$

If the corresponding values of σ and ϵ are plotted as a graph, for which the ordinate is the stress and the abscissa is the strain, the resulting curve is called a *conventional stress–strain diagram*. This diagram is very important in engineering since it provides the means for obtaining data about a material's tensile (or compressive) strength *without* regard for the material's physical size or shape. Realize, however, that no two stress–strain diagrams for a particular material will be *exactly* the same, since the results depend on such variables as the material's composition, microscopic imperfections, the way it is manufactured, the rate of loading, and the temperature during the time of the test.

We will now discuss the characteristics of the conventional stress–strain curve as it pertains to *steel,* a commonly used material for making both structural members and mechanical elements. Using the method described above, the characteristic stress–strain diagram for a steel specimen is shown in Fig. 9–4. From this curve we can identify four different ways in which the material behaves, depending on the amount of strain induced in the material.

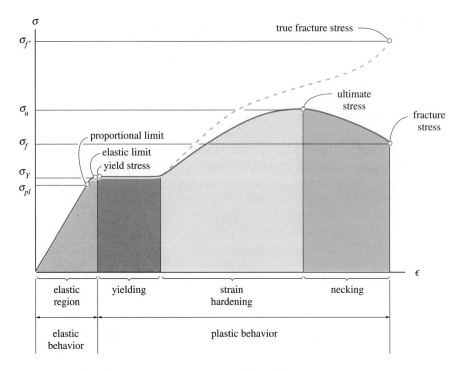

Conventional and true stress-strain diagrams for a ductile material (steel)
(not to scale)

Fig. 9–4

Elastic Behavior. The specimen is said to respond *elastically* if it returns to its *original shape* or length when the load acting on it is removed. This elastic behavior occurs when the strains in the specimen are within the lightly shaded region shown in Fig. 9–4. It can be seen that the curve is actually a *straight line* throughout most of this region, so the stress is *proportional* to the strain. In other words, the material is said to be *linear elastic*. The upper stress limit to this linear relationship is called the *proportional limit*, σ_{pl}. If the stress slightly exceeds the proportional limit, the material may still respond elastically; however, the curve tends to flatten out causing a greater increment of strain for a corresponding increment of stress. This continues until the stress reaches the *elastic limit*. To determine this point for any specimen, one must apply and then release an increased loading until a permanent deformation is detected in the specimen. Normally for steel, however, the elastic limit is seldom determined, since it is very close to the proportional limit and therefore rather difficult to detect.

Yielding. A slight increase in stress above the elastic limit will result in a breakdown of the material and cause it to *deform permanently*. This behavior is called *yielding,* and it is indicated by the dark-shaded region of the curve, Fig. 9–4. The stress that causes yielding is called the *yield stress* or *yield point,* σ_Y, and the deformation that occurs is called *plastic deformation*. Unlike elastic loading, a load that causes yielding of the material will permanently change the properties of the material. Although not shown in Fig. 9–4, for low-carbon steels or those that are hot-rolled, the yield point is often distinguished by two values. The *upper yield point* occurs first, followed by a sudden decrease in load-carrying capacity to a *lower yield point*. Once the lower yield point is reached, however, then as shown in Fig. 9–4, the specimen will continue to elongate with *no increase in load*. Realize that Fig. 9–4 is not drawn to scale. If it was, the induced strains due to yielding would be about 10 to 40 times greater than those produced up to the elastic limit. When the material is in this state, it is often referred to as being *perfectly plastic*.

Strain Hardening. When yielding has ended, a further load can be applied to the specimen, resulting in a curve that rises continuously but becomes flatter until it reaches a maximum stress referred to as the *ultimate stress, σ_u.* The rise in the curve in this manner is called *strain hardening,* and it is identified in Fig. 9–4 as the region in light-shaded color. Throughout the test, while the specimen is elongating, its cross-sectional area will decrease. This decrease in area is fairly *uniform* over the specimen's entire gauge length, even up to the strain corresponding to the ultimate stress.

Necking. At the ultimate stress, the cross-sectional area begins to decrease in a *localized* region of the specimen, instead of over its entire length. As a result, a constriction or "neck" gradually tends to form in this region as the specimen elongates further, Fig. 9–5a. Since the cross-sectional area within the region is continually decreasing, the smaller area can only carry an ever-decreasing load. Hence the stress–strain diagram tends to curve downward until the specimen breaks at the *fracture stress σ_f*. This region of the curve due to necking is indicated in dark color in Fig. 9–4. The result due to failure of a specimen is shown in Fig. 9–5b.

True Stress–Strain Diagram.

Instead of always using the *original* cross-sectional area and specimen length to calculate the (engineering) stress and strain, we could have used the *actual* cross-sectional area and specimen length at the *instant* the load is measured. The values of stress and strain computed from these measurements are called *true stress* and *true strain,* and a plot of their values is called the *true stress–strain diagram*. When this diagram is plotted it has a form shown by the dashed curve in Fig. 9–4. Note that both the conventional and true σ–ϵ diagrams are practically coincident when the strain is small. The differences between the diagrams begin to appear in the strain-hardening range, where the magnitude of strain becomes more significant. In particular, notice the large divergence within the necking

Failure of a ductile material

(a)   (b)

Fig. 9–5

region. Here it can be seen from the conventional σ–ϵ diagram that the specimen *actually* supports a *decreasing load,* since A_0 is constant when calculating engineering stress, $\sigma = P/A_0$. However, from the true σ–ϵ diagram, the actual area A within the necking region is always decreasing until fracture, $\sigma_{f'}$, and so the material actually sustains *increasing stress,* since $\sigma = P/A$.

Although the true and conventional stress–strain diagrams are different, most engineering design is done within the elastic range. Provided the material is "stiff," like most metals, the strain up to the elastic limit will remain small and the error in using the engineering values of σ and ϵ is very small (about 0.1%) compared with their true values. This is one of the primary reasons for using conventional stress–strain diagrams.

The above concepts can be summarized with reference to Fig. 9–6, which shows an actual conventional stress–strain diagram for a mild steel specimen. In order to enhance the details, the elastic region of the curve has been shown in color to an exaggerated strain scale, also shown in color. Tracing the behavior, the proportional limit is reached at $\sigma_{pl} = 35$ ksi (241 MPa), where $\epsilon_{pl} = 0.0012$ in./in. This is followed by an upper yield point of $(\sigma_Y)_u = 38$ ksi (262 MPa), then suddenly a lower yield point of $(\sigma_Y)_l = 36$ ksi (248 MPa). The end of yielding occurs at a strain of $\epsilon_1 = 0.03$ in./in., which is 25 times greater than the strain at the proportional limit! Continuing, the specimen is strain-hardened until it reaches the ultimate stress of $\sigma_u = 63$ ksi (434 MPa), then it begins to neck down until failure occurs, $\sigma_f = 47$ ksi (324 MPa). By comparison, the strain at failure, $\epsilon_f = 0.380$ in./in., is 317 times greater than ϵ_{pl}!

Fig. 9–6

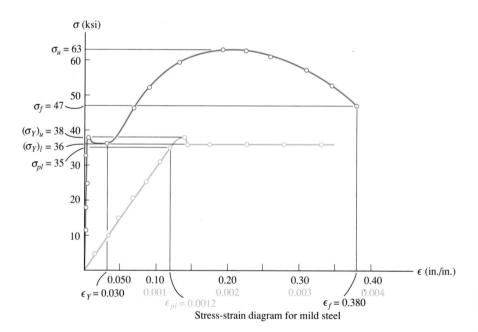

Stress-strain diagram for mild steel

9.3 Stress–Strain Behavior of Ductile and Brittle Materials

Materials can be classified as either being ductile or brittle, depending on their stress–strain characteristics. Each will now be given separate treatment.

Ductile Materials. Any material that can be subjected to large strains before it ruptures is called a *ductile material*. Mild steel, as discussed above, is a typical example. Engineers often choose ductile materials for design because these materials are capable of absorbing shock or energy, and if they become overloaded, they will usually exhibit large deformation before failing.

One way to specify the ductility of a material is to report its percent elongation or percent reduction in area at the time of fracture. The *percent elongation* is the specimen's fracture strain expressed as a percent. Thus, if the specimen's original gauge-mark length is L_0 and its length at fracture is L_f, then

$$\text{Percent elongation} = \frac{L_f - L_0}{L_0}\,(100\%) \qquad (9\text{--}3)$$

As seen in Fig. 9–6, since $\epsilon_f = 0.380$, this value would be 38% for a mild steel specimen.

The *percent reduction in area* is another way to specify ductility. It is defined within the region of necking as follows:

$$\text{Percent reduction of area} = \frac{A_0 - A_f}{A_0}\,(100\%) \qquad (9\text{--}4)$$

Here A_0 is the specimen's original cross-sectional area and A_f is the area at fracture. Mild steel has a typical value of 60%.

Besides steel, other metals such as brass, molybdenum, and zinc may also exhibit ductile stress–strain characteristics similar to steel, whereby they undergo elastic stress–strain behavior, yielding at constant stress, strain hardening, and finally necking until rupture. In most metals, however, constant yielding will *not occur* beyond the elastic range. One metal for which this is the case is aluminum. Actually, this metal often does not have a well-defined *yield point,* and consequently it is standard practice to define a *yield strength* for aluminum using a graphical procedure called the *offset method.* Normally a 0.2% strain (0.002 in./in.) is chosen, and from this point on the ϵ axis, a line parallel to the initial straight-line portion of the stress–strain diagram is drawn. The point where this line intersects the curve defines the yield strength. An example of the construction for determining the yield strength for an aluminum alloy is shown in Fig. 9–7. From the graph, the yield strength is $\sigma_{YS} = 51$ ksi (352 MPa).

Realize that the yield strength is not a physical property of the material, since it is a stress that caused a *specified* permanent strain in the material. In this text, however, we will assume that the yield strength, yield point, elastic

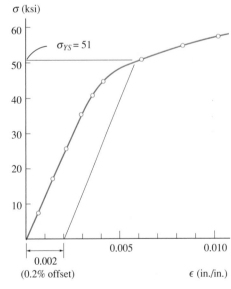

Yield strength for aluminum alloy

Fig. 9–7

limit, and proportional limit all *coincide* unless otherwise stated. For example, an exception would be natural rubber, which in fact does not even have a proportional limit, since stress and strain are *not* linearly related, Fig. 9–8. Instead, this material, which is known as a polymer, exhibits *nonlinear elastic behavior*.

Wood is a material that is often moderately ductile, and as a result it is usually designed to respond only to elastic loadings. The strength characteristics of wood vary greatly from one species to another, and for each species they depend on the moisture content, age, and the size and arrangement of knots in the wood. Since wood is a fibrous material, its tensile or compressive characteristics will differ greatly when it is loaded either parallel or perpendicular to its grain. Specifically, wood splits easily when it is loaded in tension perpendicular to its grain, and consequently tensile loads are almost always intended to be applied parallel to the grain of wood members.

Brittle Materials.

Materials that exhibit little or no yielding before failure are referred to as *brittle materials*. Gray cast iron is an example, having a stress–strain diagram in tension as shown by portion *AB* of the curve in Fig. 9–9. Here fracture at $\sigma_f = 22$ ksi (152 MPa) takes place initially at an imperfection or microscopic crack and then spreads rapidly across the specimen, causing complete fracture. As a result of this type of failure, brittle materials do not have a well-defined tensile fracture stress, since the appearance of cracks in a specimen is quite random. Instead the average fracture stress from a set of observed tests is generally reported. A typical failed specimen is shown in Fig. 9–10a.

Fig. 9–8

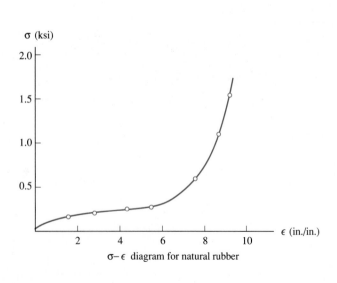

σ–ε diagram for natural rubber

Fig. 9–9

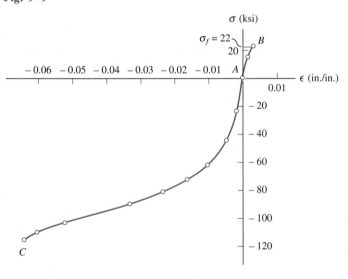

σ–ε diagram for gray cast iron

Compared with their behavior in tension, brittle materials, such as gray cast iron, exhibit a much higher resistance to axial compression, as evidenced by portion *AC* of the curve in Fig. 9–9. For this case any cracks or imperfections in the specimen tend to close-up, and as the load increases the material will generally bulge out or become barrel-shaped as the strains become larger, Fig. 9–10*b*.

Like gray cast iron, concrete is also classified as a brittle material, and it has a low strength capacity in tension. The characteristics of its stress–strain diagram depend primarily on the mix of concrete (water, sand, gravel, and cement) and the time and temperature of curing. A typical example of a ''complete'' stress–strain diagram for concrete is given in Fig. 9–11. By inspection, its maximum compressive strength is almost 12.5 times greater than its tensile strength, $(\sigma_c)_{max} = 5$ ksi (34.5 MPa) versus $(\sigma_t)_{max} = 0.40$ ksi (2.76 MPa). For this reason, concrete is almost always reinforced with steel bars or rods whenever it is designed to support tensile loads.

It can generally be stated that most materials exhibit both ductile and brittle behavior. For example, steel has brittle behavior when it contains a high carbon content, and it is ductile when the carbon content is reduced. Also, at low temperatures materials become harder and more brittle, whereas when the temperature rises they become softer and more ductile.

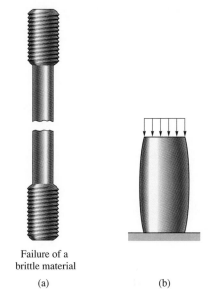

Failure of a
brittle material

(a)   (b)

Fig. 9–10

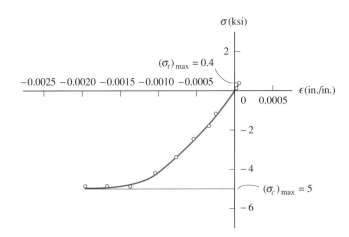

σ-ϵ diagram for typical concrete mix

Fig. 9–11

9.4 Hooke's Law

As noted in the previous section, the stress–strain diagrams for most engineering materials exhibit a *linear relationship* between stress and strain within the elastic region. Consequently, an increase in stress causes a proportionate increase in strain. This fact was discovered by Robert Hooke in 1676 using springs and is known as *Hooke's law*. It may be expressed mathematically as

$$\sigma = E\epsilon \qquad (9\text{–}5)$$

Here E represents the constant of proportionality, which is called the *modulus of elasticity* or *Young's modulus,* after Thomas Young, who published an account of it in 1807.

Equation 9–5 actually represents the equation of the *initial straight-lined portion* of the stress–strain diagram up to the proportional limit. Furthermore, the modulus of elasticity represents the *slope* of this line. Since strain is dimensionless, from Eq. 9–5, E will have units of stress, such as psi, ksi, or pascals. As an example of its calculation, consider the stress–strain diagram for steel shown in Fig. 9–6. Here $\sigma_{pl} = 35$ ksi and $\epsilon_{pl} = 0.0012$ in./in., so that

$$E = \frac{\sigma_{pl}}{\epsilon_{pl}} = \frac{35 \text{ ksi}}{0.0012 \text{ in./in.}} = 29(10^3) \text{ ksi}$$

As shown in Fig. 9–12, the proportional limit for a particular type of steel depends on its alloy content; however, most grades of steel, from the softest rolled steel to the hardest tool steel, have about the same modulus of elasticity, generally accepted to be $E_{st} = 29(10^3)$ ksi or 200 GPa. Common values of E for other engineering materials are often tabulated in engineering code and reference books. Also, see Appendix C. It should be noted that the modulus of elasticity is a mechanical property that indicates the *stiffness* of a material. Materials that are very stiff, such as steel, have large values of E [$E_{st} = 29(10^3)$ ksi or 200 GPa], whereas spongy materials such as vulcanized rubber may have low values [$E_r = 0.10(10^3)$ ksi or 0.70 MPa].

The modulus of elasticity is one of the most important mechanical properties used in the development of equations presented in this text. It must always be remembered, though, that E can be used only if a material has *linear-elastic* behavior. Also, if the stress in the material is *greater* than the proportional limit, the stress–strain diagram ceases to be a straight line and Eq. 9–5 is no longer valid.

Strain Hardening. If a specimen of ductile material, such as steel, is loaded into the *plastic range* and then unloaded, *elastic strain is recovered* as the material returns to its equilibrium state. The *plastic strain remains,* how-

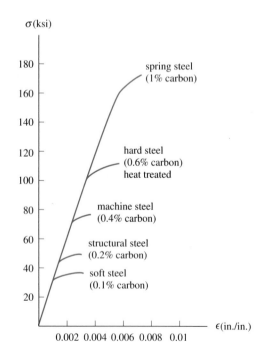

σ(ksi)

180 — spring steel
(1% carbon)

160 —

140 —

120 — hard steel
(0.6% carbon)
heat treated

100 —

80 — machine steel
(0.4% carbon)

60 — structural steel
(0.2% carbon)

40 — soft steel
(0.1% carbon)

20 —

ϵ(in./in.)

0.002 0.004 0.006 0.008 0.01

Fig. 9–12

ever, and as a result the material is subjected to a *permanent set*. For example, a wire when bent (plastically) will spring back a bit (elastically) when the load is removed; however, it will not fully return to its original position. This behavior can be illustrated on the stress–strain diagram shown in Fig. 9–13a. Here the specimen is first loaded beyond its yield point A to the point A'. Since interatomic forces have to be overcome to elongate the specimen *elastically,* then these same forces pull the atoms back together when the load is removed, Fig. 9–13a. Consequently, the modulus of elasticity, E, is the same, and therefore the slope of line O'A' has the same slope as line OA.

If the load is reapplied, the atoms in the material will again be displaced until yielding occurs at or near the stress A', and the stress–strain diagram continues along the same path as before, Fig. 9–13b. It should be noted, however, that this new stress–strain diagram, defined by O'A'B, now has a *higher* yield point (A'), a consequence of strain-hardening. In other words, the material now has a *greater elastic region;* however, it has *less ductility* than when it was in its original state.

It should be mentioned, however, that in the true sense some heat or *energy* may be *lost* as the specimen is unloaded from A' and then again loaded to this same stress. As a result, slight curves in the paths A' to O' and O' to A' will occur during a carefully measured cycle of loading. This is shown by the dashed curves in Fig. 9–13b. The area between these curves represents lost energy and is called *mechanical hysteresis*. It becomes an important consideration when selecting materials to serve as dampers for vibrating structural or mechanical equipment, although its effects will not be considered in this text.

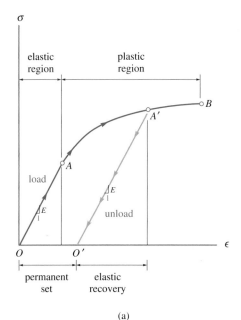

(a)

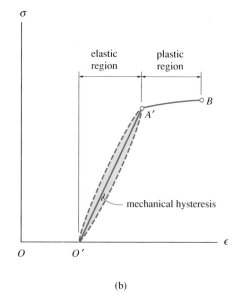

(b)

Fig. 9–13

9.5 Strain Energy

As a material is deformed by an external loading, the material tends to store energy *internally* throughout its volume. Since this energy is related to the strains in the material, it is referred to as *strain energy*. In this section we will formulate the strain energy stored in a volume element of material when the energy is caused by a uniaxial stress. The results will then be used to identify two important material properties that are often used to specify a material's ability to *absorb* energy.

When a tension-test specimen is subjected to an axial load, a volume element of the material is subjected to uniaxial stress as shown in Fig. 9–14. This stress develops a force $\Delta F = \sigma \, \Delta A = \sigma(\Delta x \, \Delta y)$ on the top and bottom faces of the element *after* the element undergoes a vertical displacement $\epsilon \, \Delta z$. The work of this force will now be formulated.

By definition, *work* is determined by the product of the force and displacement in the direction of the force. Since the force ΔF is increased uniformly from zero to its final magnitude ΔF when the displacement $\epsilon \, \Delta z$ is attained, the work done on the element by the force is equal to the *average* force magnitude $(\Delta F/2)$ times the displacement $\epsilon \, \Delta z$. This "external work" is equivalent to the internal work or strain energy stored in the element—assuming that no energy is lost in the form of heat. Consequently, the strain energy is

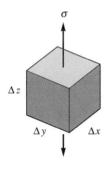

Fig. 9–14

$$\Delta U = \left(\frac{1}{2} \, \Delta F\right) \epsilon \, \Delta z = \left(\frac{1}{2}\sigma \, \Delta x \, \Delta y\right) \epsilon \, \Delta z$$

Since the volume of the element is $\Delta V = \Delta x \, \Delta y \, \Delta z$, then

$$\Delta U = \frac{1}{2}\sigma\epsilon \, \Delta V$$

It is sometimes convenient to formulate the strain energy per unit volume of material. This is called the *strain-energy density,* and from the above equation it can be expressed as

$$u = \frac{\Delta U}{\Delta V} = \frac{1}{2}\sigma\epsilon \tag{9–6}$$

If the material behavior is linear-elastic, then Hooke's law applies, $\sigma = E\epsilon$, and therefore we can express the strain-energy density in terms of the uniaxial stress as

$$u = \frac{1}{2} \frac{\sigma^2}{E} \tag{9–7}$$

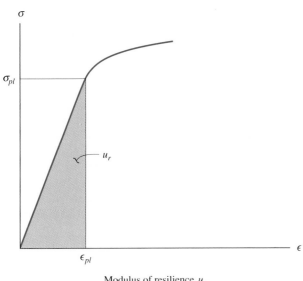

Modulus of resilience u_r

(a)

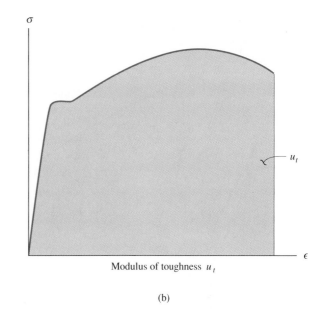

Modulus of toughness u_t

(b)

Fig. 9–15

Modulus of Resilience.

In particular, when the stress σ reaches the proportional limit, the strain-energy density, as calculated by Eq. 9–6 or 9–7, is referred to as the *modulus of resilience*, i.e.,

$$u_r = \frac{1}{2}\sigma_{pl}\epsilon_{pl} = \frac{1}{2}\frac{\sigma_{pl}^2}{E} \qquad (9\text{–}8)$$

From the elastic region of the stress–strain diagram, Fig. 9–15a, we can see that u_r is equivalent to the shaded *triangular area* under the diagram. Physically a material's resilience represents the ability of the material to absorb energy without any permanent damage to the material.

Modulus of Toughness.

Another important property of a material is the *modulus of toughness, u_t*. This quantity represents the *entire area* under the stress–strain diagram, Fig. 9–15b, and therefore it indicates the strain-energy density of the material just before it fractures. By comparing various values of u_t, one can determine the toughness of a particular material. This consideration becomes important when designing members that may be accidentally overloaded. Materials with a high modulus of toughness will distort greatly due to an overloading; however, they may be preferable to those with a low value, since materials having a low u_t may suddenly fracture without warning of an approaching failure. Alloying metals can also change their resilience and toughness. For example, by changing the percentage of carbon in steel, the resulting stress–strain diagrams in Fig. 9–16 indicate how the degrees of resilience and toughness can be changed among the three alloys.

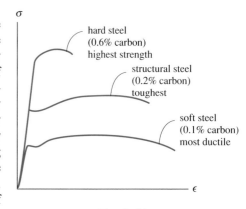

hard steel
(0.6% carbon)
highest strength

structural steel
(0.2% carbon)
toughest

soft steel
(0.1% carbon)
most ductile

Fig. 9–16

A tension test for a steel alloy results in the stress–strain diagram shown in Fig. 9–17. Calculate the modulus of elasticity and the yield strength based on a 0.2% offset. Identify on the graph the ultimate stress, and the fracture stress.

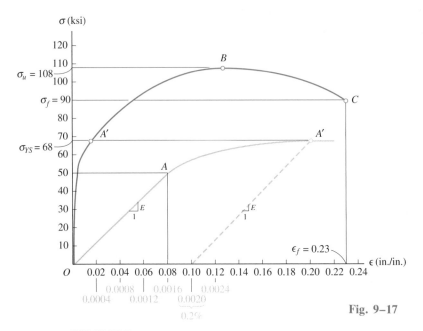

Fig. 9–17

SOLUTION

Modulus of Elasticity. We must calculate the *slope* of the initial straight-line portion of the graph. Using the magnified scale shown in color in Fig. 9–17, this line extends from point O to point A. A point on this line has coordinates of approximately (50 ksi, 0.0016 in./in.). Therefore,

$$E = \frac{50 \text{ ksi}}{0.0016 \text{ in./in.}} = 31.25(10^3) \text{ ksi} \qquad \textit{Ans.}$$

Note that the equation of the line OA is thus $\sigma = 31.25(10^3)\epsilon$.

Yield Strength. For a 0.2% offset, we begin at a strain of 0.2% or 0.002 in./in. and graphically extend a (dashed) line parallel to OA until it intersects the σ–ϵ curve at A', Fig. 9–17. The yield strength is approximately

$$\sigma_{YS} = 68 \text{ ksi} \qquad \textit{Ans.}$$

Ultimate Stress. This is defined by the peak of the σ–ϵ graph, point B in Fig. 9–17.

$$\sigma_u = 108 \text{ ksi} \qquad \textit{Ans.}$$

Fracture Stress. When the specimen is strained to its maximum of $\epsilon_f = 0.23$ in./in., it fractures at point C, Fig. 9–17. Thus,

$$\sigma_f = 90 \text{ ksi} \qquad \textit{Ans.}$$

Example 9–2

The stress–strain diagram for an aluminum alloy that is used to make parts for an aircraft is shown in Fig. 9–18. If a specimen of this material is stressed to 600 MPa, determine the permanent strain that remains in the specimen when the load is released. Also, compute the modulus of resilience both before and after the load application.

SOLUTION

Permanent Strain. When the specimen is subjected to a stress of 600 MPa, it strain-hardens until point B is reached on the σ–ϵ diagram, Fig. 9–18. The strain at this point is approximately 0.023 mm/mm. When the load is released, the material behaves by following the straight line BC, which is parallel to line OA. Since both lines have the same slope, the strain at point C can be determined analytically. The slope of line OA is the modulus of elasticity, i.e.,

$$E = \frac{\sigma_{pl}}{\epsilon_{pl}} = \frac{450 \text{ MPa}}{0.006 \text{ mm/mm}} = 75.0 \text{ GPa}$$

From triangle CBD in Fig. 3–18, we require

$$E = \frac{BD}{CD} = \frac{600(10^6) \text{ Pa}}{CD} = 75.0(10^9) \text{ Pa}$$

$$CD = 0.008 \text{ mm/mm}$$

This strain represents the amount of *recovered elastic strain*. The permanent strain, ϵ_{OC}, Fig. 9–18, is thus

$$\epsilon_{OC} = 0.023 \text{ mm/mm} - 0.00800 \text{ mm/mm}$$

$$= 0.0150 \text{ mm/mm} \qquad Ans.$$

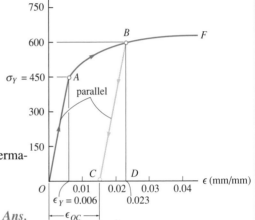

Fig. 9–18

Note: If gauge marks on the specimen were originally 50 mm apart, then after the load is *released* these marks will be 50 mm + (0.0150) (50 mm) = 50.75 mm apart.

Modulus of Resilience. Applying Eq. 9–8, we have[*]

$$(u_r)_{\text{initial}} = \frac{1}{2}\sigma_{pl}\epsilon_{pl} = \frac{1}{2}(450 \text{ MPa}) (0.006 \text{ mm/mm})$$

$$= 1.35 \text{ MJ/m}^3 \qquad Ans.$$

$$(u_r)_{\text{final}} = \frac{1}{2}\sigma_{pl}\epsilon_{pl} = \frac{1}{2}(600 \text{ MPa}) (0.008 \text{ mm/mm})$$

$$= 2.40 \text{ MJ/m}^3 \qquad Ans.$$

The effect of strain-hardening the material has caused an increase in the modulus of resilience as noted by comparing the answers; however, note that the modulus of toughness for the material has decreased since the area under the original curve, $OABF$, is larger than the area under curve CBF.

[*]Work in the SI system of units is measured in joules, where 1 J = 1 N $\cdot$ m.

Example 9–3

An aluminum rod shown in Fig. 9–19a has a circular cross section and is subjected to an axial load of 10 kN. If a portion of the stress–strain diagram for the material is shown in Fig. 9–19b, determine the approximate elongation of the rod when the load is applied. If the load is removed, does the rod return to its original length? Take $E_{al} = 70$ GPa.

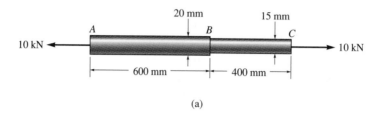

(a)

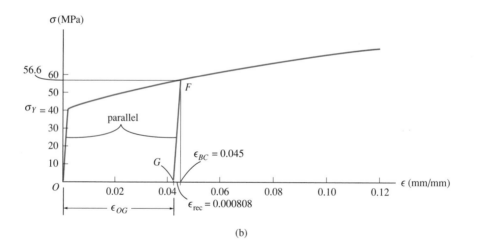

(b)

Fig. 9–19

SOLUTION

For the analysis we will neglect the *localized deformations* at the point of load application and where the rod's cross-sectional area suddenly changes. (These effects will be discussed in Sec. 10.1.) Throughout the midsection of each segment the normal stress and deformation are uniform.

In order to study the deformation of the rod, we must obtain the strain. This is done by first calculating the stress, then using the stress–strain diagram to obtain the strain. The normal stress within each segment is

$$\sigma_{AB} = \frac{P}{A} = \frac{10(10^3)\ N}{\pi\ (0.01\ m)^2} = 31.8\ MPa$$

$$\sigma_{BC} = \frac{P}{A} = \frac{10(10^3)\ N}{\pi\ (0.0075\ m)^2} = 56.6\ MPa$$

From the stress–strain diagram, the material in region AB is strained *elastically* since $\sigma_Y = 40\ MPa > 31.8\ MPa$. Using Hooke's law,

$$\epsilon_{AB} = \frac{\sigma_{AB}}{E_{al}} = \frac{31.8(10^6)\ Pa}{70(10^9)\ Pa} = 0.0004543\ mm/mm$$

The material within region BC is strained plastically, since $\sigma_Y = 40\ MPa < 56.6\ MPa$. From the graph, for $\sigma_{BC} = 56.6\ MPa$,

$$\epsilon_{BC} \approx 0.0450\ mm/mm$$

The approximate elongation of the rod is therefore

$$\Delta = \Sigma\epsilon L = 0.0004543(600\ mm) + 0.045(400\ mm)$$
$$= 18.3\ mm \qquad\qquad\qquad Ans.$$

When the 10-kN load is removed, segment AB of the rod will be restored to its original length. Why? On the other hand, the material in segment BC will recover elastically along line FG, Fig. 9–19b. Since the slope of FG is E_{al}, the elastic strain recovery is

$$\epsilon_{rec} = \frac{\sigma_{BC}}{E_{al}} = \frac{56.6(10^6)\ Pa}{70(10^9)\ Pa} = 0.000808\ mm/mm$$

The remaining plastic strain in segment BC is then

$$\epsilon_{OG} = 0.0450 - 0.000808 = 0.0442\ mm/mm$$

Therefore, when the load is removed the rod remains elongated by an amount

$$\Delta' = \epsilon_{OG}L_{BC} = 0.0442(400\ mm) = 17.7\ mm \qquad Ans.$$

PROBLEMS

9–1. Data taken from a stress–strain test for a ceramic is given in the table. The curve is linear between the origin and the first point. Plot the curve, and determine the modulus of elasticity and the modulus of resilience.

σ (MPa)	ϵ (mm/mm)
0	0
229	0.0008
314	0.0012
341	0.0016
355	0.0020
368	0.0024

Prob. 9–1

9–2. Data taken from a stress–strain test is given in the table. The curve is linear between the origin and the first point. Plot the curve, and determine the modulus of elasticity and the modulus of resilience.

σ (ksi)	ϵ (in./in.)
0	0
32.0	0.0016
33.5	0.0018
40.0	0.0030
41.2	0.0050

Prob. 9–2

9–3. A concrete cylinder having a diameter of 6.00 in. and gauge length of 12 in. is tested in compression. The results of the test are reported in the table as load versus contraction. Draw the stress–strain diagram using scales of 1 in. = 0.5 ksi and 1 in. = $0.2(10^{-3})$ in./in. From the diagram, determine approximately the modulus of elasticity.

Load (kip)	Contraction (in.)
0	0
5.0	0.0006
9.5	0.0012
16.5	0.0020
20.5	0.0026
25.5	0.0034
30.0	0.0040
34.5	0.0045
38.5	0.0050
46.5	0.0062
50.0	0.0070
53.0	0.0075

Prob. 9–3

*9–4. A tension test was performed on a steel specimen having an original diameter of 0.503 in. and a gauge length of 2.00 in. The data is listed in the table. Plot the stress–strain diagram and determine approximately the modulus of elasticity, the ultimate stress, and the rupture stress. Use a scale of 1 in. = 15 ksi and 1 in. = 0.05 in./in. Redraw the linear-elastic region, using the same stress scale but a strain scale of 1 in. = 0.001 in.

9–5. A tension test was performed on a steel specimen having an original diameter of 0.503 in. and gauge length of 2.00 in. Using the data listed in the table, plot the stress–strain diagram and determine approximately the modulus of toughness.

Load (kip)	Elongation (in.)
0	0
2.50	0.0009
6.50	0.0025
8.50	0.0040
9.20	0.0065
9.80	0.0098
12.0	0.0400
14.0	0.1200
14.5	0.2500
14.0	0.3500
13.2	0.4700

Probs. 9–4/9–5

9–6. A tension test was performed on a steel specimen having an original diameter of 0.503 in. and gauge length of 2.00 in. The data is listed in the table below. Plot the stress–strain diagram and determine approximately the modulus of elasticity, the yield stress, the ultimate stress, and the rupture stress. Use a scale of 1 in. = 20 ksi and 1 in. = 0.05 in./in. Redraw the elastic region, using the same stress scale but a strain scale of 1 in. = 0.001 in./in.

Load (kip)	Elongation (in.)
0	0
1.50	0.0005
4.60	0.0015
8.00	0.0025
11.00	0.0035
11.80	0.0050
11.80	0.0080
12.00	0.0200
16.60	0.0400
20.00	0.1000
21.50	0.2800
19.50	0.4000
18.50	0.4600

Prob. 9–6

9–7. The stress–strain diagram for a steel bar is shown in the figure. Determine approximately the modulus of elasticity, the proportional limit, the ultimate stress, and the modulus of resilience. If the bar is loaded until it is stressed to 72 ksi, determine the amount of elastic recovery and the permanent set or strain in the bar when it is unloaded.

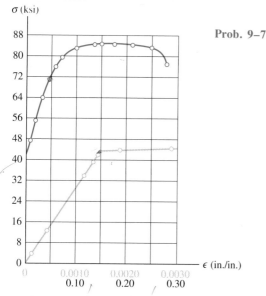

Prob. 9–7

9–9. The change in weight of an airplane is determined from reading the strain gauge A mounted in the plane's aluminum wheel strut. *Before* the plane is loaded, the strain-gauge reading in a strut is $\epsilon_1 = 0.00100$ in./in., whereas after loading $\epsilon_2 = 0.00243$ in./in. Determine the change in the force on the strut if the cross-sectional area of the strut is 3.5 in². $E_{al} = 10(10^3)$ ksi.

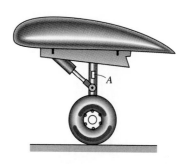

Prob. 9–9

*9–8. The stress–strain diagram for a steel specimen having an original diameter of 0.5 in. and a gauge length of 2 in. is given in the figure. Determine approximately the modulus of elasticity for the material, the load on the specimen that causes yielding, and the ultimate load the specimen will support.

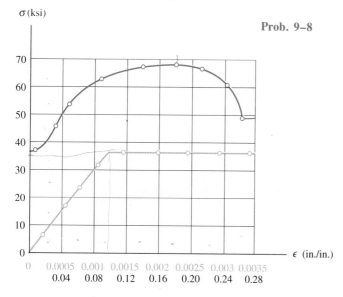

Prob. 9–8

9–10. The pole is supported by a pin at C and a steel guy wire AB. If the wire has a diameter of 0.2 in., determine how much it stretches when a horizontal force of 2.5 kip acts on the pole. $E_{st} = 29(10^3)$ ksi.

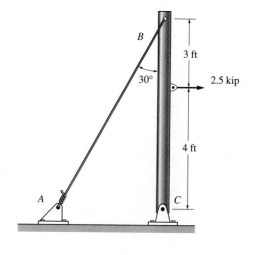

Prob. 9–10

9–11. The 8-mm-diameter bolt is made from an aluminum alloy. It fits through a magnesium sleeve that has an inner diameter of 12 mm and an outer diameter of 20 mm. If the original lengths of the bolt and sleeve are 80 mm and 50 mm, respectively, determine the strains in the sleeve and the bolt if the nut on the bolt is tightened so that the tension in the bolt is 8 kN. Assume the material at A is rigid. $E_{al} = 70$ GPa, $E_{mg} = 45$ GPa.

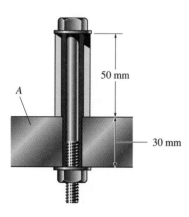

Prob. 9–11

9–13. The bar DB is rigid and is originally held in the horizontal position when the weight W is supported from C. If the weight causes the end B to be displaced downward 0.025 in., determine the strain in wires DE and BC. Also, if the wires are made of steel and have a cross-sectional area of 0.002 in^2, determine the weight W. $E_{st} = 29(10^3)$ ksi.

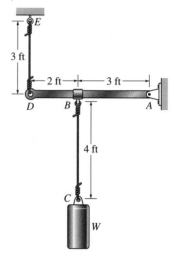

Prob. 9–13

9–14. A structural member in a nuclear reactor is made from a zirconium alloy. If an axial load of 4 kip is to be supported by the member, determine its required cross-sectional area. Use a factor of safety of 3 with respect to yielding. What is the load on the member if it is 3 ft long and its elongation is 0.02 in.? $E_{zr} = 14(10^3)$ ksi, $\sigma_Y = 57.5$ ksi. The material has elastic behavior.

9–15. The steel wire AB has a cross-sectional area of 10 mm^2 and is unstretched when $\theta = 45.0°$. Determine the applied load P needed to cause $\theta = 44.9°$. $E_{st} = 200$ GPa.

*9–12. The head H is connected to the cylinder of a compressor using six steel bolts. If the clamping force in each bolt is 800 lb, determine the normal strain in the bolts. Each bolt has a diameter of $\frac{3}{16}$ in. If $\sigma_Y = 40$ ksi and $E_{st} = 29(10^3)$ ksi, what is the strain in each bolt when the nut is unscrewed so that the clamping force is released?

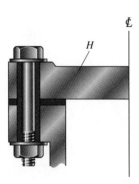

Prob. 9–12

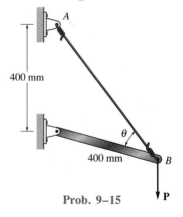

Prob. 9–15

9.6 Poisson's Ratio

When a deformable body is subjected to an axial tensile force, not only does it elongate but it also contracts laterally. For example, if a rubber band is stretched, it can be noted that both the thickness and width of the band are decreased. Likewise, a compressive force acting on a body causes it to contract in the direction of the force and yet its sides expand laterally. These two cases are illustrated in Fig. 9–20 for a bar having an original radius r and length L.

When a load **P** is applied to the bar, it changes the bar's length by an amount Δ and its radius by an amount δ (delta). Strains in the longitudinal or axial direction and in the lateral or radial direction are, respectively,

$$\epsilon_{\text{long}} = \frac{\Delta}{L} \quad \text{and} \quad \epsilon_{\text{lat}} = \frac{\delta}{r}$$

In the early 1800s, the French scientist S. D. Poisson realized that within the *elastic range* the *ratio* of these strains is a *constant,* since the deformations Δ and δ are proportional. This constant is referred to as *Poisson's ratio,* ν (nu), and it has a numerical value that is unique for a particular material that is both *homogeneous and isotropic.* Stated mathematically it is

$$\nu = -\frac{\epsilon_{\text{lat}}}{\epsilon_{\text{long}}} \qquad (9\text{–}9)$$

The negative sign is used here since *longitudinal elongation* (positive strain) causes *lateral contraction* (negative strain), and vice versa. Notice that this lateral strain is the *same* in all lateral (or radial) directions. Furthermore, this strain is caused only by the axial or longitudinal force; i.e., no force or stress acts in a lateral direction in order to strain the material in this direction.

Poisson's ratio is seen to be *dimensionless,* and for most nonporous solids it has a value that is generally between $\frac{1}{4}$ and $\frac{1}{3}$. Typical values for ν are given in Appendix C for various engineering materials. In particular, an ideal material having no lateral movement when it is stretched or compressed will have $\nu = 0$. Furthermore, it will be shown in Sec. 15.9 that the *maximum* possible value for Poisson's ratio is $\frac{1}{2}$. Therefore $0 \le \nu \le \frac{1}{2}$.

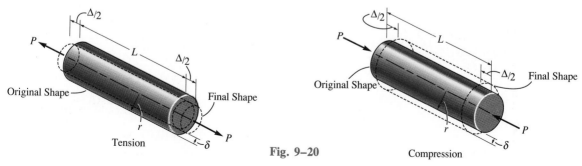

Tension

Compression

Fig. 9–20

Example 9–4

A steel bar has the dimensions shown in Fig. 9–21. If an axial force of $P = 80$ kN is applied to the bar, determine the change in its length and the change in the dimensions of its cross section after applying the load. Take $E_{st} = 200$ GPa and $\nu_{st} = 0.3$. The material behaves elastically.

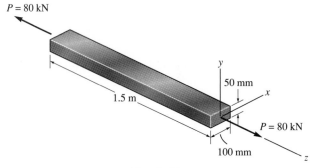

P = 80 kN

1.5 m

50 mm

100 mm

P = 80 kN

Fig. 9–21

SOLUTION

The normal stress in the bar is

$$\sigma_z = \frac{P}{A} = \frac{80(10^3)\ \text{N}}{(0.1\ \text{m})\ (0.05\ \text{m})} = 16.0(10^6)\ \text{Pa}$$

Thus the strain in the z direction is

$$\epsilon_z = \frac{\sigma_z}{E_{st}} = \frac{16.0(10^6)\ \text{Pa}}{200(10^9)\ \text{Pa}} = 80(10^{-6})\ \text{mm/mm}$$

The axial elongation of the bar is therefore

$$\Delta L_z = \epsilon_z L_z = [80(10^{-6})]\ (1.5\ \text{m}) = 120\ \mu\text{m} \qquad \textit{Ans.}$$

Using Eq. 9–9, the contraction strains in *both* the x and y directions are

$$\epsilon_x = \epsilon_y = -\nu_{st}\epsilon_z = -0.30[80(10^{-6})] = -24\ \mu\text{m/m}$$

Thus the changes in the dimensions of the cross section are

$$\Delta L_x = -\epsilon_x L_x = -[24(10^{-6})]\ (0.1\ \text{m}) = -2.40\ \mu\text{m} \qquad \textit{Ans.}$$

$$\Delta L_y = -\epsilon_y L_y = -[24(10^{-6})]\ (0.05\ \text{m}) = -1.20\ \mu\text{m} \qquad \textit{Ans.}$$

9.7 The Shear Stress–Strain Diagram

In Sec. 8.2 it was shown that when an element of material is subjected to *pure shear,* equilibrium requires that equal shear stresses must be developed on four faces of the element. These stresses must be directed toward or away from diagonally opposite corners of the element as shown on the side view in Fig. 9–22a. Furthermore, if the material is *homogeneous* and *isotropic,* then the shear stress will distort the element uniformly as shown in Fig. 9–22b. As mentioned in Sec. 8.8, the shear strain γ_{xy} measures the angular distortion of the element relative to the sides originally along the x and y axes.

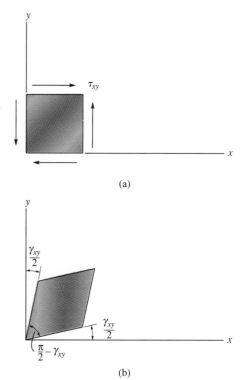

(a)

The behavior of a material subjected to pure shear can be studied in a laboratory by using specimens in the shape of thin tubes and subjecting them to a torsional loading. If measurements are made of the applied torque and the resulting angle of twist, then by the methods to be explained in Chapter 11, the data can be converted into shear stress and shear strain, and a shear stress–strain diagram plotted. An example of such a diagram for a ductile material is shown in Fig. 9–23. Like the tension test, this material when subjected to shear will exhibit linear-elastic behavior and it will have a defined *proportional limit* τ_{pl}. Also, strain hardening will occur until an *ultimate shear stress* τ_u is reached. And finally, the material will begin to lose its shear strength until it reaches a point where it fractures, τ_f.

For most engineering materials, like the one just described, the elastic behavior is *linear*, and so Hooke's law for shear can be written as

(b)

Fig. 9–22

$$\tau = G\gamma \qquad (9\text{–}10)$$

Here G is called the *shear modulus of elasticity* or the *modulus of rigidity*. Its value can be measured as the slope of the line on the τ–γ diagram, that is, $G = \tau_{pl}/\gamma_{pl}$, Fig. 9–23. Notice that the units of measurement for G will be the *same* as those for E (Pa or psi), since γ is measured in radians, a dimensionless quantity. Typical values for G are given in Appendix C.

It will be shown in Sec. 10.6 that the three material constants, E, ν, and G are actually *related* by the equation

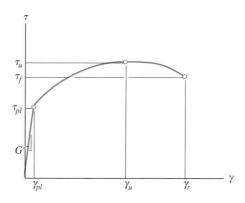

$$G = \frac{E}{2(1 + \nu)} \qquad (9\text{–}11)$$

Provided E and G are known, the value of ν can be determined from this equation rather than through experimental measurement. For example, for mild steel, $E_{st} = 29(10^3)$ ksi and $G_{st} = 11.2(10^3)$ ksi, so that, from Eq. 9–11, $\nu_{st} = 0.29$.

Fig. 9–23

Example 9–5

A specimen of titanium alloy is tested in torsion and the shear stress–strain diagram is shown in Fig. 9–24a. Determine the shear modulus G, the proportional limit, and the ultimate shear stress. Also, determine the maximum distance d that the top of a block of this material, shown in Fig. 9–24b, would be displaced horizontally if the material behaves elastically when acted upon by a shear force $\mathbf{V}$. What is the magnitude of $\mathbf{V}$ necessary to cause this displacement?

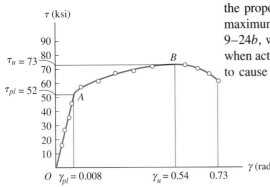

(a)

SOLUTION

Shear Modulus. This value represents the slope of the straight-line portion OA of the $\tau-\gamma$ diagram. The coordinates of point A are (0.008 rad, 52 ksi). Thus,

$$G = \frac{52 \text{ ksi}}{0.008 \text{ rad}} = 6500 \text{ ksi} \qquad \textit{Ans.}$$

The equation of line OA is therefore $\tau = 6500\gamma$, which is Hooke's law for shear.

Proportional Limit. By inspection, the graph ceases to be linear at point A. Thus,

$$\tau_{pl} = 52 \text{ ksi} \qquad \textit{Ans.}$$

Ultimate Stress. This value represents the maximum shear stress, point B. From the graph,

$$\tau_u = 73 \text{ ksi} \qquad \textit{Ans.}$$

Maximum Elastic Displacement and Shear Force. Since the maximum elastic shear strain is 0.008 rad, a very small angle, the top of the block in Fig. 9–24b will be displaced horizontally:

$$\tan(0.008 \text{ rad}) \approx 0.008 \text{ rad} = \frac{d}{2 \text{ in.}}$$

$$d = 0.016 \text{ in.} \qquad \textit{Ans.}$$

The corresponding *average* shear stress in the block is $\tau_{pl} = 52$ ksi. Thus, the shear force V needed to cause displacement is

$$\tau_{avg} = \frac{V}{A}; \qquad 52 \text{ ksi} = \frac{V}{(3 \text{ in.}) (4 \text{ in.})}$$

$$V = 624 \text{ kip} \qquad \textit{Ans.}$$

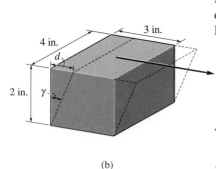

(b)

Fig. 9–24

Example 9–6

An aluminum specimen shown in Fig. 9–25 has a diameter of $d_0 = 25$ mm and a gauge length of $L_0 = 250$ mm. If a force of 165 kN elongates the gauge length 1.20 mm, determine the modulus of elasticity. Also, determine by how much the force causes the diameter of the specimen to contract. Take $G_{al} = 26$ GPa and $\sigma_Y = 440$ MPa.

SOLUTION

Modulus of Elasticity. The normal stress in the specimen is

$$\sigma = \frac{P}{A} = \frac{165(10^3)\ \text{N}}{(\pi/4)(0.025\ \text{m})^2} = 336.1\ \text{MPa}$$

and the normal strain is

$$\epsilon = \frac{\Delta L}{L} = \frac{1.20\ \text{mm}}{250\ \text{mm}} = 0.00480\ \text{mm/mm}$$

Since $\sigma < \sigma_y = 440$ MPa, the material behaves elastically. The modulus of elasticity is

$$E_{al} = \frac{\sigma}{\epsilon} = \frac{336.1(10^6)\ \text{Pa}}{0.00480} = 70.0\ \text{GPa} \qquad \textit{Ans.}$$

Contraction of Diameter. First we will compute Poisson's ratio for the material using Eq. 9–11.

$$G = \frac{E}{2(1 + \nu)}$$

$$26\ \text{GPa} = \frac{70\ \text{GPa}}{2(1 + \nu)}$$

$$\nu = 0.346$$

Since $\epsilon_{\text{long}} = 0.00480$, then by Eq. 9–9,

$$\nu = -\frac{\epsilon_{\text{lat}}}{\epsilon_{\text{long}}}$$

$$0.346 = -\frac{\epsilon_{\text{lat}}}{0.00480}$$

$$\epsilon_{\text{lat}} = -0.00166\ \text{mm/mm}$$

The contraction of the diameter is therefore

$$\Delta d = (0.00166)\ (25\ \text{mm})$$

$$= 0.0415\ \text{mm} \qquad \textit{Ans.}$$

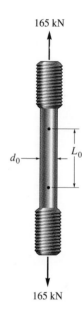

165 kN

165 kN

d_0 — L_0

Fig. 9–25

PROBLEMS

*9–16. The acrylic plastic rod is 200 mm long and 15 mm in diameter. If an axial load of 300 N is applied to it, determine the change in its length and the change in diameter. $E_p = 2.70$ GPa, $\nu_p = 0.4$.

300 N

200 mm

300 N

Prob. 9–16

9–17. The short cylindrical block of aluminum, having an original diameter of 0.5 in. and a length of 1.5 in., is placed in the smooth jaws of a vice and squeezed until the axial load applied is 800 lb. Determine (a) the decrease in its length and (b) its new diameter. $E_{al} = 10(10^3)$ ksi, $\nu_{al} = 0.33$.

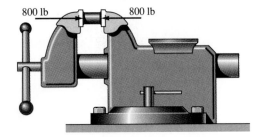

800 lb   800 lb

Prob. 9–17

9–18. The elastic portion of the stress–strain diagram for a steel alloy is shown in the figure. The specimen from which it was obtained had an original diameter of 13 mm and a gauge length of 50 mm. When the applied load on the specimen is 50 kN, the diameter is 12.99265 mm. Determine Poisson's ratio for the material.

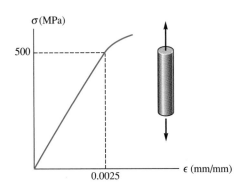

σ(MPa)

500

0.0025

ϵ (mm/mm)

Prob. 9–18

9–19. The rubber block is subjected to an elongation of 0.03 in. along the x axis, and its vertical faces are given a tilt so that $\theta = 89.3°$. Determine the strains ϵ_x, ϵ_y, and γ_{xy} at each point in the block. Take $\nu_r = 0.5$.

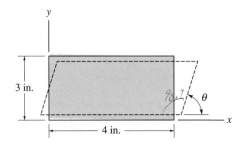

y

3 in.

θ

4 in.

x

Prob. 9–19

*9–20. The plug has a diameter of 30 mm and fits within a rigid sleeve having an inner diameter of 32 mm. Both the plug and the sleeve are 50 mm long. Determine the axial pressure p that must be applied to the top of the plug to cause it to contact the sides of the sleeve. Also, how far must the plug be compressed downward in order to do this? The plug is made from a material for which $E = 5$ MPa, $\nu = 0.45$.

Prob. 9–20

9–21. The support consists of three rigid plates, which are connected together using two symmetrically placed rubber pads. If a vertical force of 50 N is applied to plate A, determine the approximate vertical displacement of this plate due to shear strains in the rubber. Each pad has cross-sectional dimensions of 30 mm and 20 mm. $G_r = 0.20$ MPa.

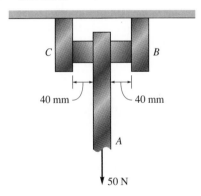

Prob. 9–21

9–22. The stone has a mass of 800 kg and center of gravity at G. It rests on a pad at A and a roller at B. The pad is fixed to the ground and has a compressed height of 30 mm, a width of 140 mm, and a length of 150 mm. If the coefficient of static friction between the pad and the stone is $\mu_s = 0.8$, determine the approximate horizontal displacement of the stone, caused by the shear strains in the pad, before the stone begins to slip. The pad is made from a material having $E = 4$ MPa and $\nu = 0.35$. Assume the normal force at A acts 1.5 m from G as shown.

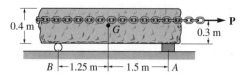

Prob. 9–22

9–23. A shear spring is made from two blocks of rubber, each having a height h, width b, and thickness a. The blocks are bonded to three plates as shown. If the plates are rigid and the shear modulus of the rubber is G, determine the displacement of point A if a vertical load P is applied at this point. Assume that the displacement is small so that $\delta = a \tan \gamma \approx a\gamma$.

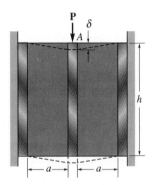

Prob. 9–23

*9–24. A shear spring is made by bonding the rubber annulus to a rigid fixed ring and a plug. When an axial load $\mathbf{P}$ is placed on the plug, show that the slope at point y in the rubber is $dy/dr = -\tan \gamma = -\tan(P/2\pi hGr)$. For small angles we can write $dy/dr = -P/(2\pi hGr)$. Integrate this expression and evaluate the constant of integration using the condition that $y = 0$ at $r = r_o$. From the result compute the deflection $y = \delta$ of the plug.

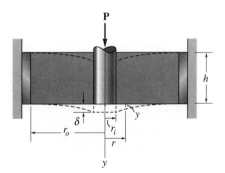

Prob. 9–24

REVIEW PROBLEMS

9–25. The aluminum block has a rectangular cross section and is subjected to an axial compressive force of 8 kip. If the 1.5-in. side changed its length to 1.500132 in., determine Poisson's ratio and the new length of the 2-in. side. $E_{al} = 10(10^3)$ ksi.

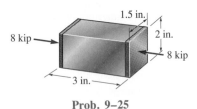

1.5 in.

2 in.

8 kip

8 kip

3 in.

Prob. 9–25

9–26. While undergoing a tension test, a copper-alloy specimen having a gauge length of 2 in. is subjected to a strain of 0.40 in./in. when the stress is 70 ksi. If $\sigma_Y = 45$ ksi when $\epsilon_y = 0.0025$ in./in., determine the distance between the gauge points when the load is released.

9–27. The steel wires AB and AC support the 200-kg mass. If the allowable axial stress for the wires is $\sigma_{allow} = 130$ MPa, determine the required diameter of each wire. Also, what is the new length of wire AB after the load is applied? Take the unstretched length of AB to be 750 mm. $E_{st} = 200$ GPa.

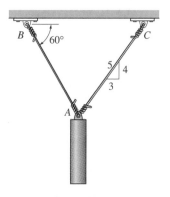

B   60°

C

5   4

3

A

Prob. 9–27

*9–28. An 8-mm-diameter brass rod has a modulus of elasticity of $E_{br} = 100$ GPa. If it is 3 m long and subjected to an axial load of 2 kN, determine its elongation. What is its elongation under the same load if its diameter is 6 mm?

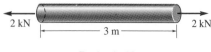

2 kN

3 m

2 kN

Prob. 9–28

9–29. The rigid beam rests in the horizontal position on two aluminum cylinders having the *unloaded* lengths shown. If each cylinder has a diameter of 30 mm, determine the placement x of the applied 80-kN load so that the beam remains horizontal. What is the new diameter of cylinder A after the load is applied? $E_{al} = 70$ GPa, $\nu_{al} = 0.33$.

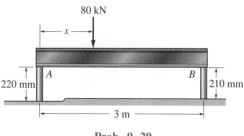

80 kN

x

A

B

220 mm

210 mm

3 m

Prob. 9–29

9–30. The shear stress–strain diagram for a steel alloy is shown in the figure. If a bolt having a diameter of 0.25 in. is made from this material and used in the lap joint, determine the force P required to cause the material to yield.

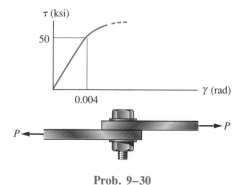

τ (ksi)

50

0.004

γ (rad)

P

P

Prob. 9–30

9–31. A bar having a length of 5 in. and cross-sectional area of 0.7 in.^{2} is subjected to an axial force of 8000 lb. If the bar stretches 0.002 in., determine the modulus of elasticity of the material. The material has linear-elastic behavior.

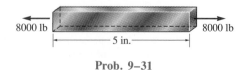

8000 lb

8000 lb

5 in.

Prob. 9–31

10 Axial Load

In Chapter 8 we developed the method for finding the normal stress in axially loaded members. In this chapter we will discuss how to determine the deformation of these members, and we will also develop a method for finding the support reactions when these reactions cannot be determined strictly from the equations of equilibrium. An analysis of the effects of thermal stress will also be discussed.

10.1 Saint-Venant's Principle

In the previous chapters we have developed the concept of *stress* as a means of measuring the force distribution within a body and *strain* as a means of measuring a body's deformation. We have also shown that the mathematical relationship between stress and strain depends on the type of material from which the body is made. In particular, if the stress creates a linear elastic response from the material, then Hooke's law applies and there is a proportional relationship between stress and strain. Furthermore, for this case, since stress can be related to the load and strain can be related to displacement, there must also be a proportional relationship between the applied load and the displacement of points in the body.

For example, consider the manner in which a rectangular bar will deform elastically when the bar is subjected to a force **P** applied along its centroidal axis, Fig. 10–1a. Here the bar is fixed-connected at one end, with the force applied through a hole at its other end. Due to the loading, the bar deforms as indicated by the distortions of the once horizontal and vertical grid lines

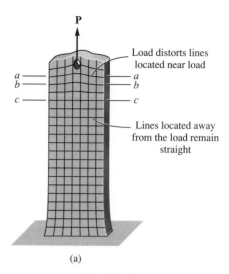

Load distorts lines located near load

Lines located away from the load remain straight

(a)

Fig. 10–1

drawn on the bar. Notice the *localized deformation* that occurs at each end. This effect tends to *decrease* as measurements are taken farther and farther away from the ends. Furthermore, the deformations even out and become uniform throughout the midsection of the bar.

Since the deformation is related to stress within the bar, we can state that stress will be distributed more uniformly throughout the cross-sectional area if the section is taken farther and farther from the point where the external load is applied. To show this, consider a profile of the variation of the stress distribution acting at sections *a–a*, *b–b*, and *c–c*, each of which is shown in Fig. 10–1*b*. In each of these three cases, force equilibrium requires the magnitude of the *resultant* force developed by the stress distribution to be equal to *P*. Furthermore, *moment equilibrium* requires each stress distribution to be *symmetrical* over the cross section, and it is for this reason that **P** is applied through the centroid of the cross section.

By comparison, the stress distribution *almost* reaches a uniform value at section *c–c*, which is sufficiently removed from the end. In other words, section *c–c* is far enough away from the application of **P** so that the localized deformation caused by **P** *vanishes*. The minimum distance from the bar's end where this occurs can be determined using a mathematical analysis based on the theory of elasticity. However, as a *general rule*, which applies as well to many other cases of loading and member geometry, we can consider this distance to be at least equal to the *largest dimension* of the loaded cross section. Hence, for the bar in Fig. 10–1*b*, section *c–c* should be located at a distance at least equal to the width (not the thickness) of the bar.* This rule is based on *experimental observation of material behavior,* and only in special

---

*When section *c–c* is so located, the theory of elasticity predicts the maximum stress to be $\sigma_{max} = 1.02\sigma_{avg}$.

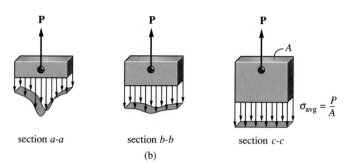

section *a-a*   section *b-b*   section *c-c*

(b)

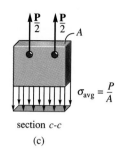

section *c-c*

(c)

cases, like the one discussed here, has it been validated mathematically. It should be noted, however, that this rule does not apply to every type of member and loading case. For example, members made from thin-walled elements, and subjected to loadings that cause large deflections, may create localized stresses and deformations that have an influence a considerable distance away from the point of application of loading.

At the support, in Fig. 10–1*a* notice how the bar is prevented from decreasing its width, which should occur due to the bar's lateral elongation—a consequence of the "Poisson effect," discussed in Sec. 9–6. By the same arguments given above, however, we could demonstrate that the stress distribution at the support will also even out and become uniform over the cross section at a short distance from the support; and furthermore, the magnitude of the resultant force created by this stress distribution must also equal *P*.

The fact that stress and deformation behave in this manner is referred to as *Saint-Venant's principle,* since it was first noticed by the French scientist Barré de Saint-Venant. Essentially it states that the stress and strain produced at points in a body sufficiently removed from the region of load application will be the *same* as the stress and strain produced by any applied loadings that have the same statically equivalent resultant and are applied to the body within the same region. In other words, by application of Saint-Venant's principle, we can assume that the *same* uniform stress distribution at section *c–c* in Fig. 10–1*b* will occur if the load **P** is replaced by *any other statically equivalent loading.* For example, if two symmetrically applied forces *P*/2 act on the bar, Fig. 10–1*c,* the stress distribution at section *c–c,* which is sufficiently removed from the localized effects of these loads, will be uniform and therefore equivalent to $\sigma_{\text{avg}} = P/A$ as before.

To summarize, then, we do not have to consider the somewhat complex stress distributions that may actually develop at points of load application, or at supports, when studying the stress distribution in a body at sections *sufficiently removed* from the points of load application. Saint-Venant's principle claims that the localized effects caused by any load acting on the body will dissipate or smooth out within regions that are sufficiently removed from the location of the load. Furthermore, the resulting stress distribution at these regions will be the *same* as that caused by any other statically equivalent load applied to the body within the same localized area.

10.2 Elastic Deformation of an Axially Loaded Member

Using Hooke's law and the definitions of stress and strain, we will now develop an equation that can be used to determine the elastic deformation of a member subjected to axial loads. To generalize the development, consider the bar shown in Fig. 10–2a, which has a cross-sectional area that *gradually* varies along its length *L*. The bar is subjected to concentrated loads at its ends and a variable external load distributed along its length. This distributed load could, for example, represent the weight of a vertical bar, or friction forces acting on the bar's surface. Here we wish to find the *relative displacement* Δ of one end of the bar with respect to the other end as caused by this loading. In the following analysis we will neglect the localized deformations that occur at points of concentrated loading and where the cross section suddenly changes. As noted in Sec. 10.1, these effects occur within small regions of the bar's length and will therefore have only a slight effect on the final result. For the most part, the bar will deform uniformly, so the normal stress will be uniformly distributed over the cross section.

Using the method of sections, a differential element (or wafer) of length *dx* and area *A(x)* is isolated from the bar at the arbitrary position *x* along the bar's length. The free-body diagram of this element is shown in Fig. 10–2b. The resultant internal axial force is represented as *P(x)*, since the external loading will cause it to vary along the length of the bar. This load, *P(x)*, will deform the element into the shape indicated by the dashed outline, and therefore the displacement of one end of the element with respect to the other end is δ*x*. The stress and strain in the element are

$$\sigma = \frac{P(x)}{A(x)} \quad \text{and} \quad \epsilon = \frac{\delta x}{dx}$$

Provided these quantities do not exceed the proportional limit, we can relate them using Hooke's law; i.e.,

$$\sigma = E\epsilon$$

$$\frac{P(x)}{A(x)} = E\left(\frac{\delta x}{dx}\right)$$

$$\delta x = \frac{P(x)\, dx}{A(x)\, E}$$

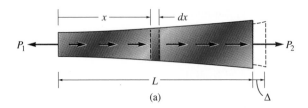

(a)

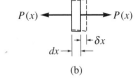

(b)

Fig. 10–2

For the entire length L of the bar, we must integrate this expression to find the required end displacement. This yields

$$\Delta = \int_0^L \frac{P(x)\, dx}{A(x)\, E}$$

(10–1)

where

Δ = displacement of one point on the bar relative to another point
L = distance between the points
$P(x)$ = internal axial force at the section, located a distance x from one end
$A(x)$ = cross-sectional area of the bar, expressed as a function of x
E = modulus of elasticity for the material

Constant Load and Cross-Sectional Area. In many cases the bar will have a constant cross-sectional area A, and the material will be homogeneous, so E is constant. Furthermore, if a constant external force is applied at each end, Fig. 10–3a, then the internal force P throughout the length of the bar is also constant, Fig. 10–3b. As a result, Eq. 10–1 can be integrated to yield

$$\Delta = \frac{PL}{AE}$$

(10–2)

If the bar is subjected to several different axial forces, or the cross-sectional area or modulus of elasticity changes abruptly from one region of the bar to the next, the above equation can be applied to each *segment* of the bar where these quantities are all *constant*. The displacement of one end of the bar with respect to the other is then found from the *vector addition* of the end displacements of each segment. For this general case,

$$\Delta = \sum \frac{PL}{AE}$$

(10–3)

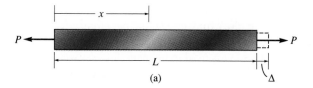

(a)

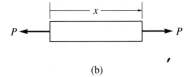

(b)

Fig. 10–3

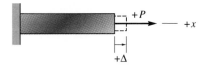

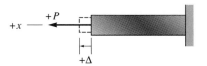

Positive sign convention for P and Δ

Fig. 10–4

Sign Convention.
In order to apply Eq. 10–3, we must develop a sign convention for the internal axial force and the displacement of one end of the bar with respect to the other end. To do so, we will consider both the force and displacement to be positive if they cause tension and elongation, respectively; whereas a negative force and displacement will cause compression and contraction, respectively, Fig. 10–4.

For example, consider the bar shown in Fig. 10–5a. The *internal axial forces* "P," computed by the method of sections for each segment, are $P_{AB} = +5$ kN, $P_{BC} = -3$ kN, $P_{CD} = -7$ kN, Fig. 10–5b. Applying Eq. 10–3 to obtain the displacement of end A relative to end D, we have

$$\Delta_{A/D} = \sum \frac{PL}{AE} = \frac{(5 \text{ kN}) L_{AB}}{AE} + \frac{(-3 \text{ kN}) L_{BC}}{AE} + \frac{(-7 \text{ kN}) L_{CD}}{AE}$$

If the other data are substituted and a positive answer is computed, it means that end A will move away from end D (the bar elongates), whereas a negative result would indicate that end A moves toward end D (the bar shortens). The double subscript notation is used to indicate this relative displacement ($\Delta_{A/D}$); however, if the displacement is to be determined relative to a *fixed point*, then only a single subscript will be used. For example, if D is located at a *fixed* support, then the computed displacement will be denoted as simply Δ_A.

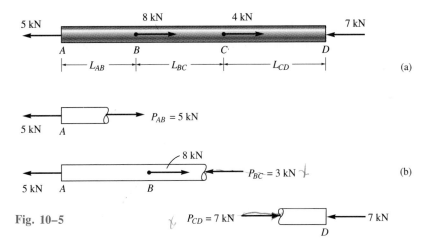

Fig. 10–5

PROCEDURE FOR ANALYSIS

The relative displacement between two points A and B on an axially loaded member can be determined by applying Eq. 10–1 (or Eq. 10–2). Since Hooke's law has been used in the development of these equations, it is important that the external loads do not cause yielding of the material and that the material is homogeneous and behaves in a linear-elastic manner. Application of the necessary equation requires the following steps.

Internal Force.   To obtain the internal axial force P, one should use the method of sections and the equation of force equilibrium applied along the axis of the member. If this force varies along the member's length, a section should be made at the arbitrary location x from one end of the member and the force represented as a function of x, $P(x)$. If several *constant external forces* act on the member, the internal force in each *segment* of the member, between any two external forces, must then be determined. For any segment, an internal *tensile force* is *positive* and an internal *compressive force* is *negative*.

Displacement.   When the member's cross-sectional area varies along its axis, the area must be expressed as a function of its position x, $A(x)$. On the other hand, if the cross-sectional area, the modulus of elasticity, or the internal loading suddenly changes between points A and B, then Eq. 10–2 should be applied to each segment for which these qualities are constant (Eq. 10–3). When substituting the data into Eqs. 10–1 through 10–3, be sure to account for the proper sign for P, as discussed above, and use a consistent set of units. For any segment, if the computed result is a *positive* numerical quantity, it indicates *elongation;* if it is *negative*, it indicates a *contraction*.

The following examples illustrate numerical applications of this method.

Example 10–1

The composite steel bar shown in Fig. 10–6a is made from two segments, AB and BD, having cross-sectional areas of $A_{AB} = 1$ in^2 and $A_{BD} = 2$ in^2, respectively. If it is subjected to the loads shown, determine the vertical displacement of end A and the displacement of B relative to C. Take $E_{st} = 29(10^3)$ ksi.

(a)

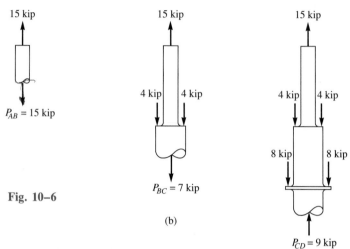

Fig. 10–6

(b)

SOLUTION

Internal Force. Due to the application of the external loadings, the *internal axial forces* in regions AB, BC, and CD will all be *different*. These forces are obtained by applying the method of sections and the equation of vertical force equilibrium as shown in Fig. 10–6b.

Displacement. Using the sign convention, i.e., the internal tensile forces are positive and the compressive forces are negative, the vertical displacement of A relative to the *fixed* support (D) is

$$\Delta_A = \sum \frac{PL}{AE} = \frac{[+15 \text{ kip}](2 \text{ ft})(12 \text{ in./ft})}{(1 \text{ in}^2)[29(10^3) \text{ kip/in}^2]} + \frac{[+7 \text{ kip}](1.5 \text{ ft})(12 \text{ in./ft})}{(2 \text{ in}^2)[29(10^3) \text{ kip/in}^2]}$$

$$+ \frac{[-9 \text{ kip}](1 \text{ ft})(12 \text{ in./ft})}{(2 \text{ in}^2)[29(10^3) \text{ kip/in}^2]}$$

$$= +0.0127 \text{ in.} \qquad\qquad Ans.$$

Since the result is *positive*, the bar *elongates* and so the displacement at A is upward.

Applying Eq. 10–2 between points B and C, we can obtain the relative displacement of B with respect to C; that is,

$$\Delta_{B/C} = \frac{P_{BC}L_{BC}}{A_{BC}E} = \frac{[+7 \text{ kip}](1.5 \text{ ft})(12 \text{ in./ft})}{(2 \text{ in}^2)[29(10^3) \text{ kip/in}^2]} = +0.00217 \text{ in.} \qquad Ans.$$

Here point B moves away from point C, since the segment elongates.

Example 10–2

The assembly shown in Fig. 10–7a consists of an aluminum tube AB having a cross-sectional area of 400 mm². A steel rod having a diameter of 10 mm is attached to a rigid collar and passes through the tube. If a tensile load of 80 kN is applied to the rod, determine the displacement of the end C of the rod. Take $E_{st} = 200$ GPa, $E_{al} = 70$ GPa.

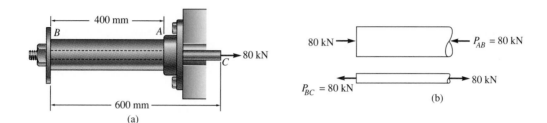

(a)

(b)

Fig. 10–7

SOLUTION

Internal Force.   The free-body diagram of the tube and rod, Fig. 10–7b, shows that the rod is subjected to a tension of 80 kN and the tube is subjected to a compression of 80 kN.

Displacement.   Using Eq. 10–2, we will first determine the displacement of end C with respect to end B. Working in units of newtons and meters, we have

$$\Delta_{C/B} = \frac{PL}{AE} = \frac{[+80(10^3) \text{ N}](0.6 \text{ m})}{\pi(0.005 \text{ m})^2[200(10^9) \text{ N/m}^2]} = +0.003056 \text{ m} \rightarrow$$

The positive sign indicates that end C moves *to the right* relative to end B, since the bar elongates.

Now, applying Eq. 10–2 to the tube, in order to determine the displacement of end B with respect to the *fixed* end A, yields

$$\Delta_B = \frac{PL}{AE} = \frac{[-80(10^3) \text{ N}](0.4 \text{ m})}{400 \text{ mm}^2((10^{-6}) \text{ m}^2/\text{mm}^2)[70(10^9) \text{ N/m}^2]}$$

$$= -0.001143 \text{ m} = 0.001143 \text{ m} \rightarrow$$

Here the negative sign indicates that the tube shortens and so B moves to the *right* relative to A.

Since both computed displacements are to the right, the resultant displacement of C relative to A is therefore

$$(\overset{+}{\rightarrow}) \qquad \Delta_C = \Delta_B + \Delta_{C/B} = 0.001143 \text{ m} + 0.003056 \text{ m}$$

$$= 0.00420 \text{ m} = 4.20 \text{ mm} \rightarrow \qquad\qquad \textit{Ans.}$$

■ Example 10–3

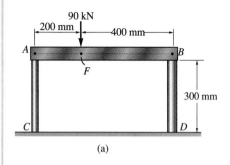

(a)

A *rigid* beam AB rests on the two short posts shown in Fig. 10–8a. AC is made of steel and has a diameter of 20 mm, and BD is made of aluminum and has a diameter of 40 mm. Determine the displacement of point F on AB if a vertical load of 90 kN is applied over this point. Take $E_{st} = 200$ GPa, $E_{al} = 70$ GPa.

SOLUTION

Internal Force. The compressive forces acting at the top of each post are determined from the equilibrium of member AB, Fig. 10–8b. These forces are equal to the internal forces in each post, Fig. 10–8c.

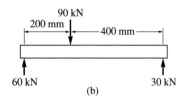

(b)

Displacement. Applying Eq. 10–2 to compute the displacement of the top of each post, we have

Post AC:

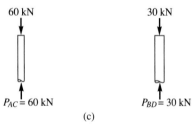

(c)

$$\Delta_A = \frac{P_{AC}L_{AC}}{A_{AC}E_{st}} = \frac{[-60(10^3)\text{ N}](0.300 \text{ m})}{\pi(0.010 \text{ m})^2[200(10^9) \text{ N/m}^2]} = -286(10^{-6}) \text{ m}$$

$$= 0.286 \text{ mm} \downarrow$$

Post BD:

$$\Delta_B = \frac{P_{BD}L_{BD}}{A_{BD}E_{al}} = \frac{[-30(10^3)\text{ N}](0.300 \text{ m})}{\pi(0.020 \text{ m})^2[70(10^9) \text{ N/m}^2]} = -102(10^{-6}) \text{ m}$$

$$= 0.102 \text{ mm} \downarrow$$

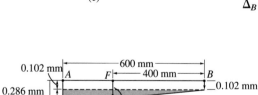

(d)

Fig. 10–8

A diagram showing the centerline displacements at points A, B, and F on the beam is shown in Fig. 10–8d. By proportion of the shaded triangle, the displacement of point F is therefore

$$\Delta_F = 0.102 \text{ mm} + (0.184 \text{ mm})\left(\frac{400 \text{ mm}}{600 \text{ mm}}\right) = 0.225 \text{ mm} \downarrow \quad Ans.$$

Example 10–4

A member is made from a material that has a specific weight γ and modulus of elasticity E. If it is formed into a *cone* having the dimensions shown in Fig. 10–9a, determine how far its end is displaced due to gravity when it is suspended in the vertical position.

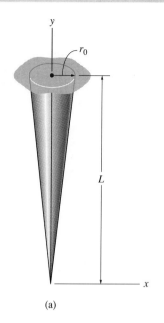

(a)

SOLUTION

Internal Force. The internal axial force varies along the member since it is dependent on the weight $W(y)$ of a segment of the member below any section, Fig. 10–9b. Hence, to calculate the displacement, we must use Eq. 10–1. At the section located a distance y from its bottom end, the radius x of the cone is determined by proportion; i.e.,

$$\frac{x}{y} = \frac{r_0}{L}; \qquad x = \frac{r_0}{L}y$$

The volume of a cone having a base of radius x and height y is

$$V = \frac{\pi}{3}yx^2 = \frac{\pi r_0^2}{3L^2}y^3$$

Since $W = \gamma V$, the internal force at the section becomes

$$+\uparrow \ \Sigma F_y = 0; \qquad P(y) = \frac{\gamma \pi r_0^2}{3L^2}y^3$$

Displacement. The area of the cross section is also a function of position y, Fig. 10–9b. We have

$$A(y) = \pi x^2 = \frac{\pi r_0^2}{L^2}y^2$$

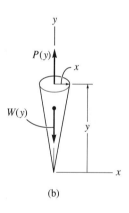

(b)

Applying Eq. 10–1 between the limits of $y = 0$ and $y = L$ yields

$$\Delta = \int_0^L \frac{P(y)\ dy}{A(y)\ E} = \int_0^L \frac{[(\gamma \pi r_0^2/3L^2)\ y^3]\ dy}{[(\pi r_0^2/L^2)\ y^2]\ E}$$

$$= \frac{\gamma}{3E}\int_0^L y\ dy$$

Fig. 10–9

$$= \frac{\gamma L^2}{6E} \qquad\qquad\qquad Ans.$$

As a partial check of this result, notice how the units of the terms, when canceled, give the deflection in units of length as expected.

PROBLEMS

10–1. The composite shaft, consisting of aluminum, copper, and steel sections, is subjected to the loading shown. Determine the displacement of end A with respect to end D and the normal stress in each section. The cross-sectional area and moduli of elasticity for each section are shown in the figure. Neglect the size of the collars at B and C.

10–2. Determine the displacement of B with respect to C of the composite shaft in Prob. 10–1.

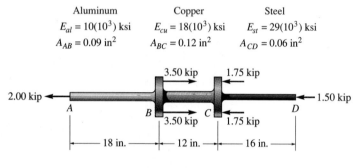

Aluminum	Copper	Steel
$E_{al} = 10(10^3)$ ksi	$E_{cu} = 18(10^3)$ ksi	$E_{st} = 29(10^3)$ ksi
$A_{AB} = 0.09$ in^2	$A_{BC} = 0.12$ in^2	$A_{CD} = 0.06$ in^2

Probs. 10–1/10–2

*10–4. The steel column is used to support the symmetric loads from the two floors of a building. Determine the vertical displacement of its top A if the column has a cross-sectional area of 23.4 in^2. $E_{st} = 29(10^3)$ ksi.

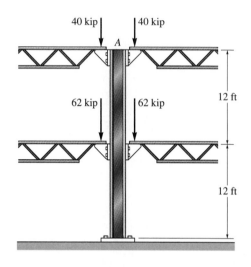

Prob. 10–4

10–3. The assembly consists of a steel rod CB and an aluminum rod BA, each having a diameter of 12 mm. If the rod is subjected to the axial loadings at A and at the coupling B, determine the displacement of the coupling B and the end A. The unstretched length of each segment is shown in the figure. Neglect the size of the connections at B and C, and assume that they are rigid. $E_{st} = 200$ GPa, $E_{al} = 70$ GPa.

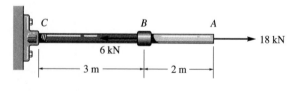

Prob. 10–3

10–5. The copper shaft is subjected to the axial loads shown. Determine the displacement of end A with respect to end D if the diameters of each segment are $d_{AB} = 0.75$ in., $d_{BC} = 1$ in., and $d_{CD} = 0.5$ in. Take $E_{cu} = 18(10^3)$ ksi.

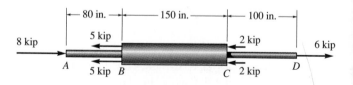

Prob. 10–5

10–6. The steel rod is subjected to the loading shown. If the cross-sectional area of the rod is 60 mm² and $E_{st} = 200$ GPa, determine the displacement of B and A. Neglect the size of the couplings at B, C, and D.

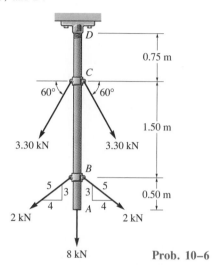

3.30 kN   3.30 kN

0.75 m

1.50 m

0.50 m

2 kN   2 kN

8 kN   **Prob. 10–6**

10–7. The assembly consists of a 30-mm-diameter aluminum bar ABC with fixed collar at B and a 10-mm-diameter steel rod CD. Determine the displacement of point D when the assembly is loaded as shown. Neglect the size of the collar at B and the connection at C. $E_{st} = 200$ GPa, $E_{al} = 70$ GPa.

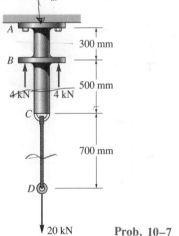

4 kN   4 kN

300 mm

500 mm

700 mm

20 kN   **Prob. 10–7**

*10–8.** The 15-mm-diameter steel shaft AC is supported by a rigid collar, which is fixed to the shaft at B. If it is subjected to an axial load of 80 kN at its end, determine the uniform pressure distribution p on the collar required for equilibrium. Also, what is the elongation of segment BC and segment BA? $E_{st} = 200$ GPa.

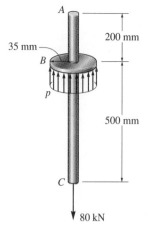

35 mm

200 mm

500 mm

80 kN

Prob. 10–8

10–9. The truss is made of three steel members, each having a cross-sectional area of 400 mm². If $E_{st} = 200$ GPa, determine the horizontal displacement of the roller at C when the truss supports the loads shown.

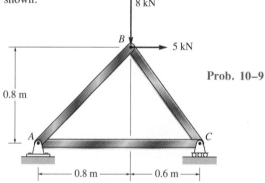

8 kN

5 kN

0.8 m

0.8 m   0.6 m

Prob. 10–9

10–10. The assembly consists of two steel suspender rods AC and BD attached to the 100-lb uniform rigid beam AB. Determine the position x for the 300-lb loading so that the beam remains in a horizontal position both before and after the load is applied. Each rod has a diameter of 0.5 in. $E_{st} = 29(10^3)$ ksi.

Prob. 10–10

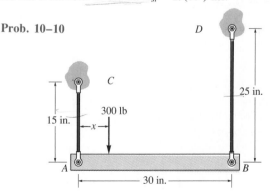

15 in.

25 in.

300 lb

30 in.

10–11. The linkage is made of three pin-connected steel members, each having a cross-sectional area of 0.730 in². If a vertical force of 50 kip is applied to the end B of member AB, determine the vertical displacement of point B. $E_{st} = 29(10^3)$ ksi.

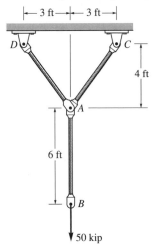

Prob. 10–11

*10–12. The segments of pipe and couplings used for drilling an oil well 12,000 ft deep are made of steel weighing 25 lb/ft. They have an outside diameter of 5.60 in. and an inside diameter of 4.70 in. In order to prevent buckling or sidesway of the pipe due to its own weight, it is partially supported at its top by the drawworks of the rig. If this force is $P = 298.7$ kip, determine the force $\mathbf{F}$ of the drill bit on the ground and the elongation of the pipe for this condition. $E_{st} = 29(10^3)$ ksi.

10–13. The bar has a length L and cross-sectional area A. Determine its elongation due to both the force $\mathbf{P}$ and its own weight. The material has a specific weight γ (weight/volume) and a modulus of elasticity E.

Prob. 10–12

Prob. 10–13

10–14. Show that the relative displacement of one end of the tapered plate with respect to the other end when it is subjected to an axial load P is $\Delta = \{Ph/[tE(d_2 - d_1)]\}\ln(d_2/d_1)$.

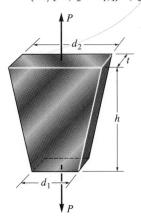

Prob. 10–14

10–15. The rod has a slight taper and length L. It is suspended from the ceiling and supports its own weight and a load $\mathbf{P}$ at its end. Determine the displacement of its end due to both of these loads. The material has a specific weight γ (weight/volume) and a modulus of elasticity E.

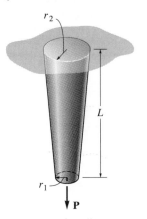

Prob. 10–15

*10–16. Determine the relative displacement of one end of the slightly tapered shaft with respect to the other end when it is subjected to the axial load P.

Prob. 10–16

10.3 Principle of Superposition

The principle of superposition is often used to determine the stress or displacement at a point in a member when the member is subjected to a complicated loading. By subdividing the loading into components, the *principle of superposition* states that the resultant stress or displacement at the point can be determined by first finding the stress or displacement caused by each component load acting *separately* on the member; then the resultant stress or displacement can be determined by adding the contributions caused by each of the components.

The following two conditions must be valid if the principle of superposition is to be applied.

1. *The loading must be linearly related to the stress or displacement that is to be determined.* For example, the equations $\sigma = P/A$ and $\Delta = PL/AE$ involve a linear relationship between P and σ or Δ.
2. *The loading must not significantly change the original geometry or configuration of the member.* If significant changes do occur, the direction and location of the applied forces and their moment arms will change, and consequently, application of the equilibrium equations will yield different results. For example, consider the slender rod shown in Fig. 10–10a, which is subjected to the load **P.** In Fig. 10–10b, **P** is replaced by two of its components, $\mathbf{P} = \mathbf{P}_1 + \mathbf{P}_2$. If **P** causes the rod to deflect a large amount, as shown, the moment of the load about its support, Pd, will *not* equal the sum of the moments of its component loads, $Pd \neq P_1 d_1 + P_2 d_2$, because $d_1 \neq d_2 \neq d$.

Most of the equations involving load, stress, and displacement that are developed in this text consist of linear relationships between these quantities. Also, members or bodies that are to be considered will be such that the loading will produce deformations that are so small that the change in position and direction of the loading will be insignificant and can be neglected. One exception to this rule, however, will be discussed in Chapter 17. It consists of a column that carries an axial load that is equivalent to the critical or buckling load. It will be shown that when this load increases only slightly, it will cause the column to have a large lateral deflection, even if the material remains linear-elastic. These deflections, associated with the components of any axial load, *cannot* be superimposed.

Fig. 10–10

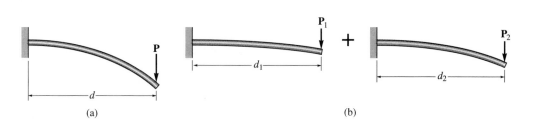

(a)   (b)

10.4 Statically Indeterminate Axially Loaded Member

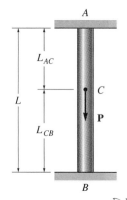

When a bar is fixed-supported at one end and is subjected to an axial force, the force equilibrium equation applied along the axis of the bar is *sufficient* to find the reaction at the fixed support. A problem such as this, where the reactions can be determined strictly from the equations of equilibrium, is called *statically determinate*. If the bar is fixed at *both ends,* however, as in Fig. 10–11a, then two unknown reactions occur, Fig. 10–11b, and the force equilibrium equation becomes

$$+\uparrow \ \Sigma F = 0; \qquad\qquad -F_B + F_A - P = 0$$

In this case the bar in Fig. 10–11a is called *statically indeterminate,* since the equilibrium equation(s) are not sufficient to determine the reactions.

In order to establish an additional equation needed for solution, it is necessary to consider the geometry of the deformation. Specifically, an equation that specifies the conditions for displacement is referred to as a *compatibility* or *kinematic condition.* For the bar in Fig. 10–11a, a suitable compatibility condition would require the relative displacement of one end of the bar with respect to the other end to be equal to zero, since the end supports are fixed. Hence, we can write

$$- \ \Delta_{A/B} = 0$$

This equation can be expressed in terms of the applied loads by using a *load-displacement relationship,* which depends on the material behavior. For example, if linear-elastic behavior occurs, Eq. 10–2 can be used. Realizing that the internal force in segment AC is $+F_A$, and in segment CB the internal force is $-F_B$, the compatibility equation can be written as

$$\frac{F_A L_{AC}}{AE} - \frac{F_B L_{CB}}{AE} = 0$$

Assuming that AE is constant, we can solve the above two equations for the reactions, which gives

$$F_A = P\left(\frac{L_{CB}}{L}\right) \quad \text{and} \quad F_B = P\left(\frac{L_{AC}}{L}\right)$$

Fig. 10–11

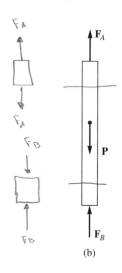

To summarize the above procedure, the unknown forces in statically indeterminate problems must be determined by satisfying both equilibrium and compatibility requirements for the bar. To do so it is necessary to transform the compatibility conditions, which involve displacement, into equations that involve force. This is done by using load-displacement relations that describe the material behavior.

Superposition of Forces. For some types of problems it may be easier to write the compatibility equation using the superposition of the forces acting on the free-body diagram. This method of solution is often referred to as the *flexibility* or *force method of analysis*. To show how it is applied, consider again the bar in Fig. 10–11*a*. In order to write the necessary equation of compatibility, we will first choose any one of the two supports as "redundant" and temporarily remove its effect on the bar. The word *redundant,* as used here, indicates that the support is not needed to hold the bar in stable equilibrium, so that when it is removed, the bar becomes statically determinate. Here we will choose the support at *B* as redundant. By using the principle of superposition, the bar having its original loading on it, Fig. 10–11*c*, is then equivalent to the bar subjected only to the external load **P,** Fig. 10–11*d*, plus the bar subjected only to the redundant load **F**$_B$, Fig. 10–11*e*. Although not shown, notice that the reaction at the support *A* satisfies force equilibrium, $F_A = P - F_B$, as determined either from Fig. 10–11*b* or from the *sum* of the reactions at *A* in Fig. 10–11*d* and 10–11*e*.

If the load **P** causes *B* to be displaced *downward* by an amount Δ_B, Fig. 10–11*d*, the reaction **F**$_B$ must be capable of displacing the end *B* of the bar *upward* by an amount δ_B, Fig. 10–11*e*, such that no displacement occurs at *B*, Fig. 10–11*c*, when the two loadings are superimposed.

$$(+\downarrow) \qquad\qquad 0 = \Delta_B - \delta_B$$

This equation represents the compatibility equation for displacements at point *B*, for which we have assumed that displacements are positive downward.

Applying the load–displacement relationship, we have $\Delta_B = PL_{AC}/AE$ and $\delta_B = F_B L/AE$. Consequently,

$$0 = \frac{PL_{AC}}{AE} - \frac{F_B L}{AE}$$

$$F_B = P\left(\frac{L_{AC}}{L}\right)$$

From the free-body diagram of the bar, Fig. 10–11*b*, the reaction at *A* can now be determined from the equation of equilibrium,

$$+\uparrow \ \Sigma F_y = 0; \qquad P\left(\frac{L_{AC}}{L}\right) + F_A - P = 0$$

Since $L_{CB} = L - L_{AC}$, then

$$F_A = P\left(\frac{L_{CB}}{L}\right)$$

These results are the same as those obtained previously, except that here we have applied the condition of compatibility and then the equilibrium condition to obtain the solution. Also note that the principle of superposition can be used here since the displacement and the load are linearly related ($\Delta = PL/AE$), which assumes, of course, that the material behaves in a linear-elastic manner.

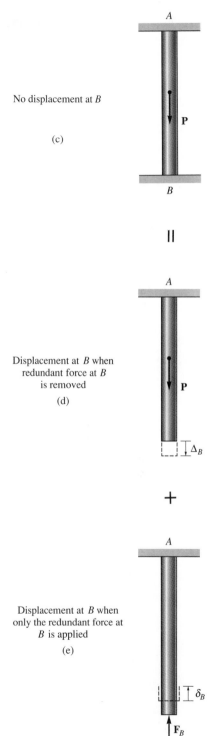

No displacement at *B*

(c)

||

Displacement at *B* when redundant force at *B* is removed

(d)

+

Displacement at *B* when only the redundant force at *B* is applied

(e)

PROCEDURE FOR ANALYSIS

Either one of the above two methods can be used to determine the unknown axial forces in a member that is statically indeterminate. Both methods require that the conditions of equilibrium and compatibility be satisfied.

To solve a particular problem it is important *first* to draw a *free-body diagram* of the member in order to identify all the forces that act on it. The sequence for applying the necessary equilibrium and compatibility equations is arbitrary. If it seems difficult to establish the compatibility equation, it is suggested that it be obtained using a superposition of forces. This method requires choosing one of the supports as *redundant*. The known displacement at the redundant support, which may be zero, is then equated to the displacement at this support caused by the external loads acting on the member, *excluding* the redundant support reaction, plus (vectorially) the displacement at the support caused *only* by the redundant reaction acting on the member. These two displacements are then expressed in terms of the loadings by using the load–displacement relationship $\Delta = PL/AE$. Once established, the compatibility equation can then be solved for the magnitude of the force at the redundant support. Any other unknown forces are then determined from the equilibrium equations.

If any of the unknown force magnitudes has a negative numerical value, it indicates that this force acts in the opposite sense of direction of that indicated on the free-body diagram.

The following examples numerically illustrate both methods of solution.

▪ Example 10–5

The steel rod shown in Fig. 10–12*a* has a diameter of 5 mm. It is attached to the fixed wall at *A*, and before it is loaded there is a gap between the wall at *B'* and the rod of 1 mm. Determine the reactions at *A* and *B'* if the rod is subjected to an axial force of *P* = 20 kN as shown. Neglect the size of the collar at *C*. Take E_{st} = 200 GPa.

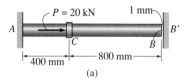

(a)

Fig. 10–12

SOLUTION I

Equilibrium. As shown on the free-body diagram, Fig. 10–12*b*, we will assume that the force *P* is large enough to cause the rod's end *B* to contact the wall at *B'*. The problem is statically indeterminate since there are two unknowns and only one equation of equilibrium.

Equilibrium of the rod requires

$$\xrightarrow{+} \Sigma F_x = 0; \qquad -F_A - F_B + 20(10^3) \text{ N} = 0 \qquad (1)$$

Compatibility. The compatibility condition for the rod is

$$\Delta_{B/A} = 0.001 \text{ m}$$

This displacement can be expressed in terms of the unknown reactions by using the load–displacement relationship, Eq. 10–2, applied to segments AC and CB, Fig. 10–12c. Working in units of newtons and meters, we have

$$\Delta_{B/A} = 0.001 \text{ m} = \frac{F_A L_{AC}}{AE} - \frac{F_B L_{CB}}{AE}$$

$$0.001 \text{ m} = \frac{F_A(0.4 \text{ m})}{\pi(0.0025 \text{ m})^2[200(10^9) \text{ N/m}^2]}$$

$$- \frac{F_B(0.8 \text{ m})}{\pi(0.0025 \text{ m})^2[200(10^9) \text{ N/m}^2]}$$

or

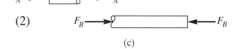

(b)

$$F_A(0.4 \text{ m}) - F_B(0.8 \text{ m}) = 3927.0 \text{ N} \cdot \text{m} \qquad (2)$$

(c)

Solving Eqs. 1 and 2 yields

$$F_A = 16.6 \text{ kN} \qquad F_B = 3.40 \text{ kN} \qquad \textit{Ans.}$$

Since the answer for F_B is *positive*, indeed the end B contacts the wall at B' as originally assumed.*

SOLUTION II
Compatibility. This problem can also be solved using the superposition of forces. Here we will consider the support at B' as redundant. Using the principle of superposition, Fig. 10–12d, we have

$$(\overset{+}{\rightarrow}) \qquad 0.001 \text{ m} = \Delta_B - \delta_B \qquad (3)$$

The deflections Δ_B and δ_B are determined from Eq. 10–2. Working in units of newtons and meters, we have

$$\Delta_B = \frac{PL_{AC}}{AE} = \frac{[20(10^3) \text{ N}](0.4 \text{ m})}{\pi(0.0025 \text{ m})^2[200(10^9) \text{ N/m}^2]} = 0.002037 \text{ m}$$

$$\delta_B = \frac{F_B L_{AB}}{AE} = \frac{F_B(1.20 \text{ m})}{\pi(0.0025 \text{ m})^2[200(10^9) \text{ N/m}^2]} = 0.3056(10^{-6}) F_B$$

Substituting into Eq. 3, we get

$$0.001 \text{ m} = 0.002037 \text{ m} - 0.3056(10^{-6}) F_B$$
$$F_B = 3.40(10^3) \text{ N} = 3.40 \text{ kN} \qquad \textit{Ans.}$$

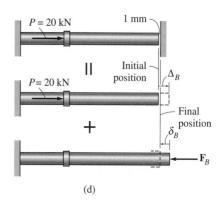

(d)

Equilibrium. From the free-body diagram, Fig. 10–12b, the reaction at A is thus

$$\overset{+}{\rightarrow} \Sigma F_x = 0; \qquad -F_A + 20 \text{ kN} - 3.40 \text{ kN} = 0$$
$$F_A = 16.6 \text{ kN} \qquad \textit{Ans.}$$

---

*If F_B were a negative quantity, the problem would be statically determinate, so that $F_B = 0$ and $F_A = 20$ kN.

■ **Example 10–6** ■

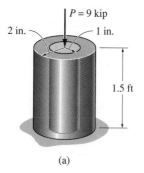

P = 9 kip

2 in.   1 in.

1.5 ft

(a)

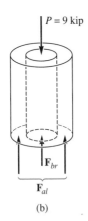

P = 9 kip

$\mathbf{F}_{br}$

$\mathbf{F}_{al}$

(b)

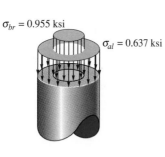

$\sigma_{br} = 0.955$ ksi

$\sigma_{al} = 0.637$ ksi

(c)

Fig. 10–13

The aluminum post shown in Fig. 10–13a is reinforced with a brass core. If it supports a resultant axial compressive load of $P = 9$ kip, determine the average normal stress in the aluminum and the brass. Take $E_{al} = 10(10^3)$ ksi and $E_{br} = 15(10^3)$ ksi.

SOLUTION

Equilibrium. The free-body diagram of the post is shown in Fig. 10–13b. Here the resultant axial force at the base is represented by the unknown components carried by the aluminum, $\mathbf{F}_{al}$, and brass, $\mathbf{F}_{br}$. The problem is statically indeterminate. Why?

Vertical force equilibrium requires

$$+\downarrow \Sigma F_y = 0; \qquad 9 \text{ kip} - F_{al} - F_{br} = 0 \qquad (1)$$

Compatibility. In order to satisfy compatibility requirements, the displacement at the top of the post for both the aluminum and brass must be the *same*; i.e.,

$$\Delta_{al} = \Delta_{br}$$

Using the load–displacement relationships,

$$\frac{F_{al}L}{A_{al}E_{al}} = \frac{F_{br}L}{A_{br}E_{br}}$$

$$F_{al} = F_{br}\left(\frac{A_{al}}{A_{br}}\right)\left(\frac{E_{al}}{E_{br}}\right)$$

$$F_{al} = F_{br}\left[\frac{\pi[(2 \text{ in.})^2 - (1 \text{ in.})^2]}{\pi(1 \text{ in.})^2}\right]\left[\frac{10(10^3) \text{ ksi}}{15(10^3) \text{ ksi}}\right]$$

$$F_{al} = 2F_{br} \qquad (2)$$

Solving Eqs. 1 and 2 simultaneously yields

$$F_{al} = 6 \text{ kip} \qquad F_{br} = 3 \text{ kip}$$

The average normal stress in the aluminum and brass is therefore

$$\sigma_{al} = \frac{6 \text{ kip}}{\pi[(2 \text{ in.})^2 - (1 \text{ in.})^2]} = 0.637 \text{ ksi} \qquad \textit{Ans.}$$

$$\sigma_{br} = \frac{3 \text{ kip}}{\pi(1 \text{ in.})^2} = 0.955 \text{ ksi} \qquad \textit{Ans.}$$

The stress distributions are shown on the top section of the post in Fig. 10–13c.

Example 10–7

The three steel bars shown in Fig. 10–14a are pin-connected to a *rigid* member. If the applied load on the member is 15 kN, determine the force developed in each bar. Bars *AB* and *EF* each have a cross-sectional area of 25 mm², and bar *CD* has a cross-sectional area of 15 mm². Take E_{st} = 200 GPa.

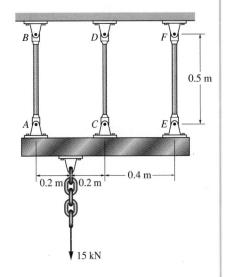

(a)

SOLUTION

Equilibrium. The free-body diagram of the rigid member is shown in Fig. 10–14b. This problem is statically indeterminate since there are three unknowns and only two available equilibrium equations. These equations are

$$+\uparrow \ \Sigma F_y = 0; \qquad F_A + F_C + F_E - 15 \text{ kN} = 0 \qquad (1)$$

$$\underset{+}{\curvearrowleft} \ \Sigma M_C = 0; \qquad -F_A(0.4 \text{ m}) + 15 \text{ kN}(0.2 \text{ m}) + F_E(0.4 \text{ m}) = 0 \qquad (2)$$

Compatibility. Due to the displacements at the ends of each bar, line *ACE* shown in Fig. 10–14c will move to the inclined position $A'C'E'$. From this position, the displacements of points *A*, *C*, and *E* can be related by proportional triangles. Thus the compatibility equation for these displacements is

$$\frac{\Delta_A - \Delta_E}{0.8 \text{ m}} = \frac{\Delta_C - \Delta_E}{0.4 \text{ m}}$$

$$\Delta_C = \frac{1}{2}\Delta_A + \frac{1}{2}\Delta_E$$

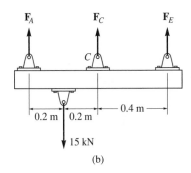

(b)

Using the load–displacement relationship, Eq. 10–2, we have

$$\frac{F_C L}{(15 \text{ mm}^2)E_{st}} = \frac{1}{2}\left[\frac{F_A L}{(25 \text{ mm}^2)E_{st}}\right] + \frac{1}{2}\left[\frac{F_E L}{(25 \text{ mm}^2)E_{st}}\right]$$

$$F_C = 0.3F_A + 0.3F_E \qquad (3)$$

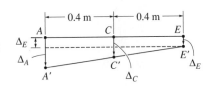

(c)

Solving Eqs. 1–3 simultaneously yields

$$F_A = 9.52 \text{ kN} \qquad \qquad Ans.$$

$$F_C = 3.46 \text{ kN} \qquad \qquad Ans.$$

$$F_E = 2.02 \text{ kN} \qquad \qquad Ans.$$

Fig. 10–14

Example 10–8

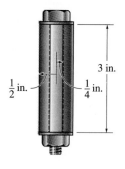

(a)

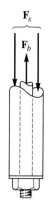

(b)

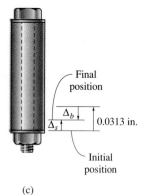

Final
position

Δ_b
Δ_s 0.0313 in.

Initial
position

(c)

Fig. 10–15

The bolt shown in Fig. 10–15a is made of an aluminum alloy and is tightened so it compresses a cylindrical spool made from a magnesium alloy. The spool has an outer radius of $\frac{1}{2}$ in. and it is assumed that both the inner radius of the spool and the radius of the bolt are $\frac{1}{4}$ in. The washers at the top and bottom of the spool are considered to be rigid and have a negligible thickness. Initially the nut is hand-tightened slightly; then, using a wrench, the nut is further tightened one-half turn. If the bolt has 16 threads per inch, determine the stress in the bolt. $E_{al} = 10(10^3)$ ksi, $E_{mg} = 6.5(10^3)$ ksi, $(\sigma_Y)_{al} = 70$ ksi, $(\sigma_Y)_{mg} = 40$ ksi.

SOLUTION

Equilibrium. The free-body diagram of a section of the bolt and the spool, Fig. 10–15b, is considered in order to relate the force in the bolt F_b to that in the spool, F_s. Equilibrium requires

$$+\uparrow \ \Sigma F_y = 0; \qquad\qquad F_b - F_s = 0 \qquad\qquad (1)$$

The problem is statically indeterminate since there are two unknowns in the equation.

Compatibility. Tightening the nut one-half turn advances it a distance of $(\frac{1}{2})(\frac{1}{16}$ in.$) = 0.0313$ in. along the bolt. This causes the bolt to *elongate* Δ_b and the spool to *shorten* Δ_s, Fig. 10–15c. As shown, compatibility requires

$$(+\uparrow) \qquad\qquad \Delta_s = 0.0313 \text{ in.} - \Delta_b$$

Assuming that the material has linear-elastic behavior, the load–displacement relationship is given by Eq. 10–2. Thus,

$$\frac{F_s(3 \text{ in.})}{\pi[(0.5 \text{ in.})^2 - (0.25 \text{ in.})^2][6.5(10^3) \text{ ksi}]} = 0.0313 \text{ in.} - \frac{F_b(3 \text{ in.})}{\pi(0.25 \text{ in.})^2[10(10^3) \text{ ksi}]}$$

$$0.8205 F_s = 32.725 - 1.60 F_b \qquad\qquad (2)$$

Solving Eqs. 1 and 2 simultaneously, we get

$$F_b = F_s = 13.52 \text{ kip}$$

The stresses in the bolt and spool are therefore

$$\sigma_b = \frac{F_b}{A_b} = \frac{13.52 \text{ kip}}{\pi(0.25 \text{ in.})^2} = 68.9 \text{ ksi} \qquad\qquad \textit{Ans.}$$

$$\sigma_s = \frac{F_s}{A_s} = \frac{13.52 \text{ kip}}{\pi[(0.5 \text{ in.})^2 - (0.25 \text{ in.})^2]} = 23.0 \text{ ksi}$$

These stresses are less than the reported yield stress for each material, $(\sigma_Y)_{al} = 70$ ksi and $(\sigma_Y)_{mg} = 40$ ksi, and therefore this "elastic" analysis is valid.

PROBLEMS

10–17. The steel pipe is filled with concrete and subjected to a compressive force of 80 kN. Determine the stress in the concrete and the steel due to this loading. The pipe has an outer diameter of 80 mm and an inner diameter of 70 mm. E_{st} = 200 GPa, E_c = 24 GPa.

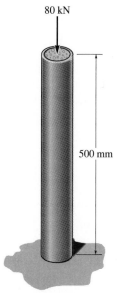

80 kN

500 mm

Prob. 10–17

10–18. The concrete column is reinforced using four steel reinforcing rods, each having a diameter of 18 mm. Determine the stress in the concrete and the steel if the column is subjected to an axial load of 800 kN. E_{st} = 200 GPa, E_c = 25 GPa.

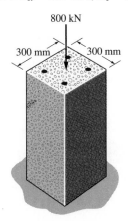

800 kN

300 mm   300 mm

Prob. 10–18

10–19. The column is constructed from concrete and six steel reinforcing rods. If it is subjected to an axial force of 18 kip, determine the stress in the concrete and in each rod. Each rod has a diameter of 0.5 in. E_{st} = 29(10^3) ksi, E_c = 3.5(10^3) ksi.

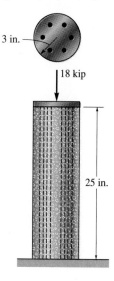

3 in.

18 kip

25 in.

Prob. 10–19

*10–20. The steel pipe has an outer radius of 20 mm and an inner radius of 15 mm. If it fits snugly between the fixed walls before it is loaded, determine the reaction at the walls when it is subjected to the load shown. E_{st} = 200 GPa.

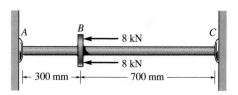

A   B   C
            8 kN
            8 kN
|— 300 mm —|—   700 mm   —|

Prob. 10–20

10–21. The composite bar consists of a 20-mm-diameter steel segment AB and 50-mm-diameter brass end segments DA and CB. Determine the normal stress in each segment due to the applied 250-kN load. $E_{br} = 100$ GPa, $E_{st} = 200$ GPa.

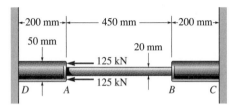

Prob. 10–21

10–22. The steel post A is surrounded by a brass tube B. Both rest on the rigid surface. If a force of 5 kip is applied to the rigid cap, determine the normal stress developed in the post and the tube. $E_{st} = 29(10^3)$ ksi, $E_{br} = 15(10^3)$ ksi.

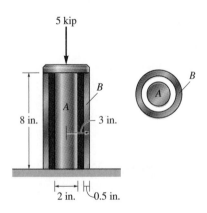

Prob. 10–22

10–23. The load of 2800 lb is to be supported by the two essentially vertical wires. If originally wire AB is 60 in. long and wire AC is 40 in. long, determine the force developed in each wire after the load is suspended. Each wire has a cross-sectional area of 0.02 in². $E_{st} = 29(10^3)$ ksi.

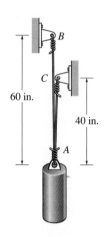

Prob. 10–23

*10–24. The steel bolt AB has a diameter of 20 mm and passes through a steel sleeve that has an inner diameter of 40 mm and an outer diameter of 50 mm. The bolt and sleeve are secured to the rigid brackets as shown. If the bolt length is 150 mm and the sleeve length is 120 mm, determine the tension in the bolt when a force of 20 kN is applied to the brackets. $E_{st} = 200$ GPa.

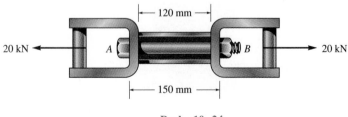

Prob. 10–24

10–25. The assembly consists of a steel bolt and a brass tube. If the nut is drawn up snug against the tube so that $L = 75$ mm, then turned an additional amount so that it advances 2 mm on the bolt, determine the force in the bolt and the tube. The bolt has a diameter of 7 mm and the tube has a cross-sectional area of 100 mm². $E_{st} = 200$ GPa, $E_{br} = 100$ GPa.

Prob. 10–25

10–26. Two steel pipes, each having a cross-sectional area of 0.32 in², are screwed together using a union at B as shown. Originally the assembly is adjusted so that no load is on the pipe. If the union is then tightened so that its screw, having a lead of 0.15 in., undergoes two full turns, determine the normal stress developed in the pipe. Assume that the union at B and couplings at A and C are rigid. Neglect the size of the union. $E_{st} = 29(10^3)$ ksi. *Note:* The lead would cause the pipe, when *unloaded,* to shorten 0.15 in. when the union is rotated one revolution.

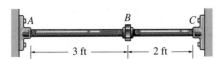

Prob. 10–26

10–27. The rigid member is held in the position shown by three steel tie rods. Assuming that each rod has an unstretched length of 0.75 m and a cross-sectional area of 125 mm², determine the forces in the rods if a turnbuckle on rod EF undergoes one full turn. The lead of the screw is 1.5 mm. Neglect the size of the turnbuckle and assume that it is rigid. *Note:* The lead would cause the rod, when *unloaded,* to shorten 1.5 mm when the turnbuckle is rotated one revolution. $E_{st} = 200$ GPa.

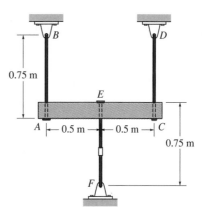

Prob. 10–27

*10–28. Two steel wires are used to hoist the 650-lb engine. Originally, AB is 32 in. long and $A'B'$ is 32.008 in. long. Determine the force supported by each wire when the engine is suspended from them. Each wire has a cross-sectional area of 0.01 in². $E_{st} = 29(10^3)$ ksi.

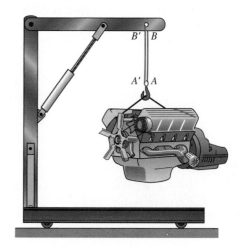

Prob. 10–28

10.5 Thermal Stress

A change in temperature can cause a material to change its dimensions. If the temperature increases, generally a material expands, whereas if the temperature decreases, the material will contract. Ordinarily this expansion or contraction is *linearly* related to the temperature increase or decrease that occurs. If this is the case, and the material is homogeneous and isotropic, it has been found from experiment that the deformation can be calculated using the formula

$$\Delta L = \alpha \Delta T L \qquad (10\text{--}4)$$

↳ coefficient of Thermal expansion

where

α = a property of the material, referred to as the *linear coefficient of thermal expansion*. The units measure strain per degree of temperature. They are 1/°F (Fahrenheit) in the Foot-Pound-Second or FPS system, and 1/°C (Celsius) or 1/°K (Kelvin) in the SI system (see Appendix C)

ΔT = the change in temperature $(T_2 - T_1)$

L = the original length of the member

ΔL = the change in length of the member

The change in length of a *statically determinate* member can readily be computed using Eq. 10–4, since the member is free to expand or contract when it undergoes a temperature change. However, in a *statically indeterminate* member these thermal displacements can be constrained by the supports, producing *thermal stresses* that must be considered in design.

Computations of these thermal stresses can be made using the procedure for analysis outlined in the previous section. The following examples illustrate some applications.

▪ Example 10–9

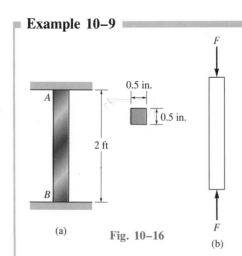

(a)   Fig. 10–16

(b)

The steel bar shown in Fig. 10–16 is constrained to just fit between two fixed supports when $T_1 = 60°F$. If the temperature is raised to $T_2 = 120°F$, determine the average normal thermal stress developed in the bar. Take $E_{st} = 29(10^3)$ ksi and $\alpha_{st} = 6.5(10^{-6})/°F$.

SOLUTION

Equilibrium.   The free-body diagram of the bar is shown in Fig. 4–16b. Since there is no external load, the force at A is equal but opposite to the force acting at B; that is,

$$+\uparrow \; \Sigma F_y = 0; \qquad\qquad F_A = F_B = F$$

The problem is statically indeterminate, since this force cannot be determined from equilibrium.

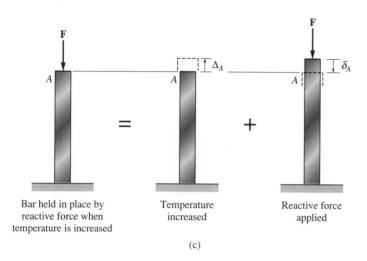

Bar held in place by
reactive force when
temperature is increased

Temperature
increased

Reactive force
applied

(c)

Compatibility.   Using the principle of superposition, Fig. 10–16c, the redundant support at A is removed, and the thermal displacement Δ_A at A occurs. The force **F**, developed at the redundant, A, then pushes the bar δ_A back to its original position; i.e., the compatibility condition at A becomes

$(+\uparrow)$ $0 = \Delta_A - \delta_A$

Applying the thermal and load–displacement relationships, Eq. 10–4 and Eq. 10–2, we have

$$0 = \alpha\Delta TL - \frac{FL}{AE}$$

Thus,

$$F = \alpha\Delta TAE$$
$$= [6.5(10^{-6})/°F](120°F - 60°F)(0.5 \text{ in.})^2[29(10^3)] \text{ kip/in}^2$$
$$= 2.83 \text{ kip}$$

From the magnitude of **F** it should be apparent that changes in temperature can cause large reaction forces in statically indeterminate members.

Since **F** also represents the internal axial force within the bar, the average normal compressive (thermal) stress is thus

$$\sigma = \frac{F}{A} = \frac{2.83 \text{ kip}}{(0.5 \text{ in.})^2} = 11.3 \text{ ksi} \qquad Ans.$$

PROBLEMS

10–29.   The three steel wires each have a diameter of 2 mm and unloaded lengths of $L_{CA} = 1.60$ m and $L_{AB} = L_{AD} = 2.00$ m. Determine the force in each wire after the 150-kg mass is suspended from the ring at A. $E_{st} = 200$ GPa.

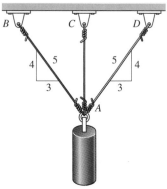

Prob. 10–29

10–31.   The rigid beam is supported by two posts, each having a width d and thickness d and a length L. If the modulus of elasticity for material A is E_A, and for material B it is E_B, determine the distance x for placement of the force **P** so that the beam remains horizontal.

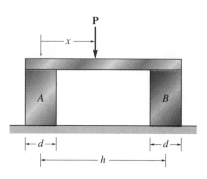

Prob. 10–31

10–30.   The three suspender bars are made of the same material and have equal cross-sectional areas A. Determine the stress in each bar if the rigid beam ACE is subjected to the force **P**.

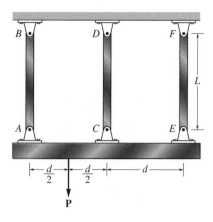

Prob. 10–30

*10–32.   The center post B of the assembly has an original length of 124.7 mm, whereas posts A and C have a length of 125 mm. If the caps on the top and bottom can be considered rigid, determine the axial stress in each post. The posts are made of aluminum and have a cross-sectional area of 400 mm². $E_{al} = 70$ GPa.

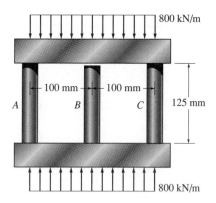

Prob. 10–32

10–33. The beam is pinned at A and supported by two aluminum rods, each having a diameter of 1 in. and a modulus of elasticity $E_{al} = 10(10^3)$ ksi. If the beam is assumed to be rigid and initially horizontal, determine the displacement of the end B when the force of 5 kip is applied.

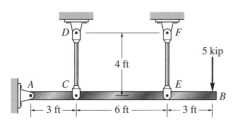

Prob. 10–33

10–34. The horizontal beam is assumed to be rigid and supports the distributed load shown. Determine the vertical reactions at the supports. Each support consists of a wooden post having a diameter of 120 mm and an unloaded (original) length of 1.40 m. Take $E_w = 12$ GPa.

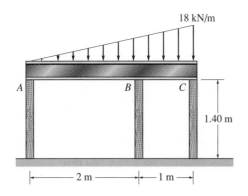

Prob. 10–34

10–35. The rigid link is supported by a pin at A and two steel wires, each having an unstretched length of 12 in. and cross-sectional area of 0.0125 in². Determine the force developed in the wires when the link supports the vertical load of 350 lb. $E_{st} = 29(10^3)$ ksi.

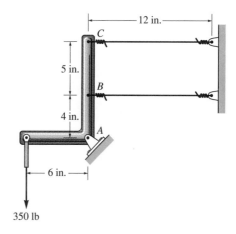

Prob. 10–35

*10–36. The rigid bar is supported by the two short wooden posts and a spring. If each of the posts has an unloaded length of 500 mm and a cross-sectional area of 800 mm², and the spring has a stiffness of $k = 1.8$ MN/m and an unstretched length of 520 mm, determine the force in each post after the load is applied to the bar. $E_w = 11$ GPa.

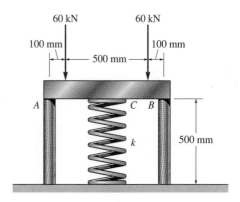

Prob. 10–36

Example 10–10

The rigid bar shown in Fig. 10–17a is fixed to the top of the three posts made of steel and aluminum. The posts each have a length of 250 mm when no load is applied to the bar and the temperature is $T_1 = 20°C$. Determine the force supported by each post if the bar is subjected to a uniform distributed load of 150 kN/m and the temperature is raised to $T_2 = 80°C$. The diameter of each post and its material properties are listed in the figure.

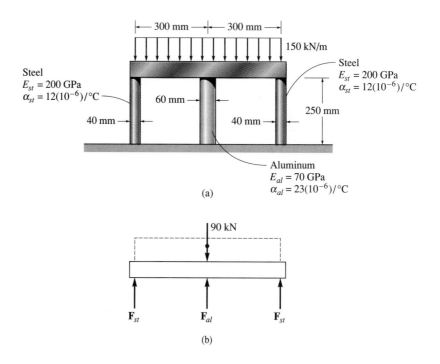

(a)

(b)

Fig. 10–17

SOLUTION

Equilibrium.   The free-body diagram of the bar is shown in Fig. 10–17b. Moment equilibrium about the bar's center requires the forces in the steel posts to be equal. Summing forces on the free-body diagram, we have

$$+\uparrow \ \Sigma F_y = 0; \qquad 2F_{st} + F_{al} - 90(10^3) \text{ N} = 0 \qquad (1)$$

Compatibility.   Due to load, geometry, and material symmetry, the top of each post is displaced by an equal amount. Hence, the compatibility equation is

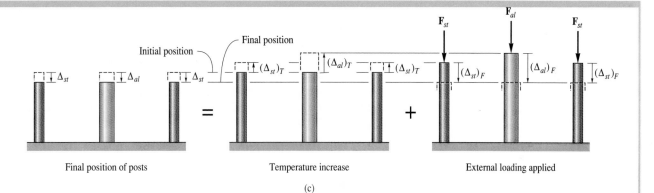

Final position of posts   Temperature increase   External loading applied

(c)

$(+\downarrow)$ $$\Delta_{st} = \Delta_{al} \qquad (2)$$

We will now consider the steel and aluminum posts separately and apply the principle of superposition, Fig. 10–17c. Here the rigid bar will be treated as the redundant. Hence, the actual displacement at the top of each post, or the displacement of the redundant, is equal to the displacement of the top of each post caused by temperature, plus the displacement of the top of each post caused by the axial force that the redundant creates in each post. Thus, for a steel and aluminum post we have

$(+\downarrow)$ $$\Delta_{st} = -(\Delta_{st})_T + (\Delta_{st})_F$$
$(+\downarrow)$ $$\Delta_{al} = -(\Delta_{al})_T + (\Delta_{al})_F$$

Applying Eq. 2 gives

$$-(\Delta_{st})_T + (\Delta_{st})_F = -(\Delta_{al})_T + (\Delta_{al})_F$$

Using Eqs. 10–2 and 10–4, we get

$$-[12(10^{-6})/°C](80°C - 20°C)(0.250 \text{ m}) + \frac{F_{st}(0.250 \text{ m})}{\pi(0.020 \text{ m})^2[200(10^9) \text{ N/m}^2]}$$

$$= -[23(10^{-6})/°C](80°C - 20°C)(0.250 \text{ m}) + \frac{F_{al}(0.250 \text{ m})}{\pi(0.03 \text{ m})^2[70(10^9) \text{ N/m}^2]}$$

$$F_{st} = 1.270F_{al} - 165.9(10^3) \qquad (3)$$

To be *consistent*, all numerical data has been expressed in terms of newtons, meters, and degrees Celsius. Solving Eqs. 1 and 3 simultaneously yields

$$F_{st} = -14.6 \text{ kN} \qquad \textit{Ans.}$$
$$F_{al} = 119 \text{ kN} \qquad \textit{Ans.}$$

The negative value for F_{st} indicates that this force acts opposite to that shown in Fig. 10–17b. In other words, the steel posts are in tension and the aluminum post is in compression.

Example 10–11

An aluminum tube having a cross-sectional area of 600 mm^2 is used as a sleeve for a steel bolt having a cross-sectional area of 400 mm^2, Fig. 10–18a. When the temperature is $T_1 = 15°C$, the nut holds the assembly in a snug position such that the axial force in the bolt is negligible. If the temperature increases to $T_2 = 80°C$, determine the average normal stress in the bolt and sleeve. Take $E_{st} = 200$ GPa, $\alpha_{st} = 12(10^{-6})/°C$, and $E_{al} = 70$ GPa, $\alpha_{al} = 23(10^{-6})/°C$.

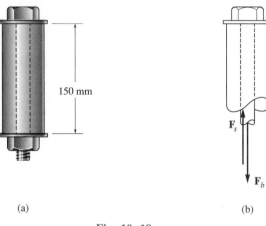

150 mm

F_s

F_b

(a)   (b)

Fig. 10–18

SOLUTION

Equilibrium. A free-body diagram of a sectioned segment of the assembly is shown in Fig. 10–18b. The forces F_b and F_s are produced since the bolt and sleeve have different coefficients of thermal expansion and therefore will expand by different amounts when the temperature is increased. The problem is statically indeterminate since these forces cannot be determined from equilibrium. However, it is required that

$$+\uparrow \ \Sigma F_y = 0; \qquad\qquad F_s = F_b \qquad\qquad (1)$$

Compatibility. Using the principle of superposition, Fig. 10–18c, the nut and washer (redundant) are removed, and the temperature increase causes the bolt and sleeve to expand Δ_b and Δ_s, respectively. The redundant forces F_b and F_s return these thermal displacements to the final position. Hence, the compatibility condition becomes

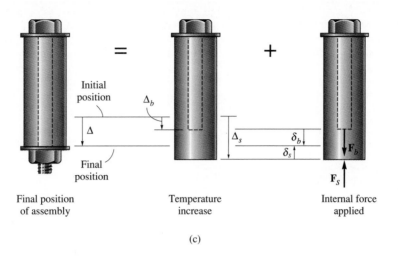

Final position   Temperature   Internal force
of assembly   increase   applied

(c)

$(+\downarrow)$ $\Delta = \Delta_b + \delta_b = \Delta_s - \delta_s$

Applying Eqs. 10–2 and 10–4, we have

$[12(10^{-6})/°C](80°C - 15°C)(0.150 \text{ m})$

$$+ \frac{F_b(0.150 \text{ m})}{(400 \text{ mm}^2)(10^{-6} \text{ m}^2/\text{mm}^2)[200(10^9) \text{ N/m}^2]}$$

$= [23(10^{-6})/°C](80°C - 15°C)(0.150 \text{ m})$

$$- \frac{F_s(0.150 \text{ m})}{600 \text{ mm}^2(10^{-6} \text{ m}^2/\text{mm}^2)[70(10^9) \text{ N/m}^2]}$$

Using Eq. 1 and solving gives

$$F_s = F_b = 19.67 \text{ kN}$$

The average normal stress in the bolt and sleeve is therefore

$$\sigma_b = \frac{19.67 \text{ kN}}{400 \text{ mm}^2(10^{-6} \text{ m}^2/\text{mm}^2)} = 49.2 \text{ MPa} \qquad \textit{Ans.}$$

$$\sigma_s = \frac{19.67 \text{ kN}}{600 \text{ mm}^2(10^{-6} \text{ m}^2/\text{mm}^2)} = 32.8 \text{ MPa} \qquad \textit{Ans.}$$

Since linear elastic material behavior was assumed in this analysis, the calculated stresses should be checked to make sure that they do not exceed the proportional limits for the material.

PROBLEMS

10–37. The 40-ft-long steel rails on a train track are laid with a small gap between them to allow for thermal expansion. Determine the required gap δ so that the rails just touch one another when the temperature is increased from $T_1 = -30°F$ to $T_2 = 70°F$. Using this gap, what would be the axial force in the rails if the temperature were to rise to $T_3 = 100°F$? The cross-sectional area of each rail is 4.25 in^2. $E_{st} = 29(10^3)$ ksi, $\alpha_{st} = 6.5(10^{-6})/°F$.

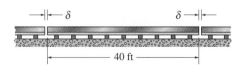

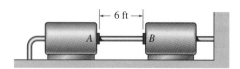

Prob. 10–37

10–38. A 6-ft-long steam pipe is connected directly to two turbines A and B as shown. The pipe has an outer radius of 1.5 in. and a wall thickness of 0.25 in. The connection was made at $T_1 = 85°F$. If the turbines' points of attachment are assumed rigid, determine the force the pipe exerts on the turbines when the steam and thus the pipe reach a temperature of $T_2 = 250°F$. $E_{st} = 29(10^3)$ ksi, $\alpha_{st} = 6.5(10^{-6})/°F$.

10–39. A 6-ft-long steam pipe is connected directly to two turbines A and B as shown. The pipe has an outer diameter of 3 in. and a wall thickness of 0.25 in. The connection was made at $T_1 = 85°F$. If the turbines' points of attachment are assumed to have a stiffness of $k = 80(10^3)$ kip/in., determine the force the pipe exerts on the turbines when the steam and thus the pipe reach a temperature of $T_2 = 250°F$. $E_{st} = 29(10^3)$ ksi, $\alpha_{st} = 6.5(10^{-6})/°F$.

Probs. 10–38/10–39

***10–40.** Three bars each made of different materials are connected together and placed between two walls when the temperature is $T_1 = 12°C$. Determine the force exerted on the (rigid) supports when the temperature becomes $T_2 = 18°C$. The material properties and cross-sectional area of each bar are given in the figure.

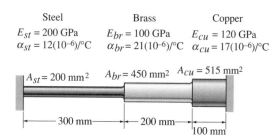

Prob. 10–40

10–41. The two circular rod segments, one of aluminum and the other of copper, are fixed to the rigid walls such that there is a gap of 0.008 in. between them when $T_1 = 60°F$. What larger temperature T_2 is required in order to just close the gap? Each rod has a diameter of 1.25 in. $\alpha_{al} = 13(10^{-6})/°F$, $E_{al} = 10(10^3)$ ksi, $\alpha_{cu} = 9.4(10^{-6})/°F$, $E_{cu} = 18(10^3)$ ksi. Determine the stress in each rod if $T_2 = 200°F$.

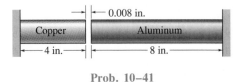

Prob. 10–41

10–42. A 0.25-in.-diameter steel rivet having a temperature of 1500°F is secured between two plates such that at this temperature it is 2 in. long and exerts a clamping force of 250 lb between the plates. Determine the approximate clamping force between the plates when the rivet cools to 70°F. For the calculation, assume that the heads of the rivet and the plates are rigid. Take $\alpha_{st} = 8(10^{-6})/°F$, $E_{st} = 29(10^3)$ ksi. Is the result a conservative estimate of the actual answer? Why or why not?

Prob. 10–42

10–43. The center rod CD of the assembly is heated from $T_1 = 30°C$ to $T_2 = 180°C$ using electrical resistance heating. At the lower temperature T_1 the gap between C and the rigid bar is 0.7 mm. Determine the force in rods AB and EF caused by the increase in temperature. Rods AB and EF are made of steel, and each has a cross-sectional area of 125 mm². CD is made of aluminum and has a cross-sectional area of 375 mm². $E_{st} = 200$ GPa, $E_{al} = 70$ GPa, and $\alpha_{al} = 23(10^{-6})/°C$.

*10–44.** The center rod CD of the assembly is heated from $T_1 = 30°C$ to $T_2 = 180°C$ using electrical resistance heating. Also, the two end rods AB and EF are heated from $T_1 = 30°C$ to $T_2 = 50°C$. At the lower temperature T_1 the gap between C and the rigid bar is 0.7 mm. Determine the force in rods AB and EF caused by the increase in temperature. Rods AB and EF are made of steel, and each has a cross-sectional area of 125 mm². CD is made of aluminum and has a cross-sectional area of 375 mm². $E_{st} = 200$ GPa, $E_{al} = 70$ GPa, $\alpha_{st} = 12(10^{-6})/°C$, and $\alpha_{al} = 23(10^{-6})/°C$.

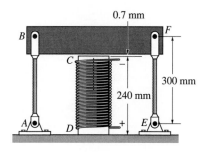

Probs. 10–43/10–44

10–45. The aluminum bar has a diameter of 0.6 in. and is attached to the rigid supports at A and B when $T_1 = 80°F$. If the temperature becomes $T_2 = -10°F$, and an axial force of 16 lb is applied to the rigid collar as shown, determine the reactions at A and B. $E_{al} = 10(10^3)$ ksi, $\alpha_{al} = 13(10^{-6})/°F$.

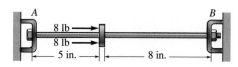

Prob. 10–45

10–46. The steel bolt has a diameter of 7 mm and fits through an aluminum sleeve as shown. The sleeve has an inner diameter of 8 mm and an outer diameter of 10 mm. The nut at A is adjusted so that it just presses up against the sleeve. If the assembly is originally at a temperature of $T_1 = 20°C$ and then is heated to a temperature of $T_2 = 100°C$, determine the stress developed in the bolt and the sleeve. $E_{st} = 200$ GPa, $E_{al} = 70$ GPa, $\alpha_{st} = 14(10^{-6})/°C$, $\alpha_{al} = 23(10^{-6})/°C$.

Prob. 10–46

10–47. The bar has a cross-sectional area A, length L, modulus of elasticity E, and coefficient of thermal expansion α. The temperature of the bar changes uniformly from an original temperature of T_A to T_B so that at any point x along the bar $T = T_A + x (T_B - T_A)/L$. Determine the force it exerts on the rigid walls. Initially no axial force is in the bar.

Prob. 10–47

10.6 Stress Concentrations

In Sec. 10.1 it was pointed out that when an axial force is applied to a member it creates a complex stress distribution within a localized region of the point of load application. Such typical stress distributions are shown in Fig. 10–1. Not only do complex stress distributions arise just under a concentrated loading, however, they also arise at sections where the member's cross-sectional area changes. For example, consider the bar in Fig. 10–19a, which is subjected to an axial force P. Here it can be seen that the once horizontal and vertical grid lines deflect into an irregular pattern around the hole centered in the bar. The maximum normal stress in the bar occurs on section a–a, which is taken through the bar's *smallest* cross-sectional area. Provided the material behaves in a linear-elastic manner, the stress distribution acting on this section can be determined either from a mathematical analysis using the theory of elasticity, or experimentally by measuring the strain normal to section a–a and then computing the stress using Hooke's law, $\sigma = E\epsilon$. Regardless of the method used, the general shape of the stress distribution will be like that shown in Fig. 10–19b. In a similar manner, if the bar has a reduction in its cross section, achieved using shoulder fillets, Fig. 10–20a, then again the maximum normal stress in the bar will occur at the *smallest* cross-sectional area, section a–a. Here the stress distribution is shown in Fig. 10–20b.

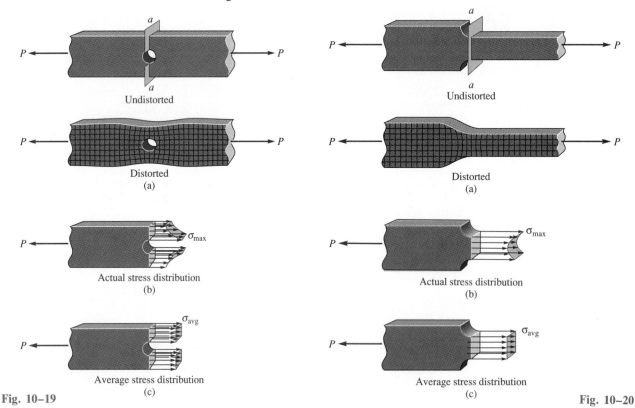

Undistorted

Distorted
(a)

Actual stress distribution
(b)

Average stress distribution
(c)

Fig. 10–19

Undistorted

Distorted
(a)

Actual stress distribution
(b)

Average stress distribution
(c)

Fig. 10–20

In both of the above cases, *force equilibrium* requires the magnitude of the *resultant force* developed by the stress distribution to be equal to P. In other words, from Eq. 8–3,

$$P = \int_A \sigma \, dA \qquad (10\text{–}5)$$

As stated in Sec. 8.3, this integral *graphically* represents the *volume* under each of the stress-distribution diagrams shown in Fig. 10–19*b* or 10–20*b*. Furthermore, moment equilibrium requires each stress distribution to be symmetrical over the cross section, so that **P** must pass through the *centroid* of each *volume*.

In engineering practice, though, the actual stress distribution does *not* have to be determined. Instead, only the *maximum stress* at these sections must be known, and the member is then designed to resist this stress when the axial load **P** is applied. In cases where a member's cross-sectional area changes, such as those discussed above, specific values of the maximum normal stress at the critical section can be determined by experimental methods or by advanced mathematical techniques using the theory of elasticity. The results of these investigations are usually reported in graphical form using a *stress-concentration factor K*. We define K as a ratio of the maximum stress to the average stress acting at the smallest cross section; i.e.,

$$K = \frac{\sigma_{\max}}{\sigma_{\text{avg}}} \qquad (10\text{–}6)$$

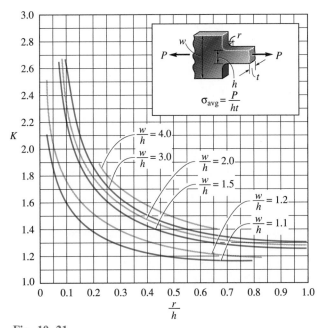

Fig. 10–21

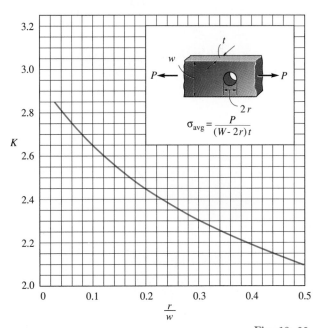

Fig. 10–22

Provided K is known, and the average normal stress has been calculated from $\sigma_{avg} = P/A$, where A is the *smallest* cross-sectional area, Figs. 10–19c and 10–20c, then from the above equation the maximum stress at the cross section is $\sigma_{max} = K(P/A)$.

Specific values of K are generally reported in graphical form in handbooks related to stress analysis. Examples of these graphs for the bars shown in Figs. 10–19a and 10–20a are given in Figs. 10–21 and 10–22, respectively.* In particular, note that K is independent of the bar's material properties; rather, it *depends* only on the bar's *geometry* and the type of discontinuity. As the size r of the discontinuity is *decreased*, the stress concentration is increased. For example, if a bar requires a change in cross section, theoretically it has been determined that a sharp corner, Fig. 10–23a, produces a stress-concentration factor greater than 3. In other words, the maximum stress will be three times greater than the average stress on the smallest cross section. However, this can be reduced to, say, 1.5 by introducing a fillet, Fig. 10–23b. And a still further reduction can be made by means of small grooves or holes placed at the transition, Fig. 10–23c and 10–23d. In all of these cases the designs help to reduce the rigidity of the material surrounding the corners, so that both the strain and the stress are more evenly spread throughout the bar.

The stress-concentration factors given in Figs. 10–21 and 10–22 were determined on the basis of a static loading, with the assumption that the stress in the material does not exceed the proportional limit. If the material is *very brittle*, the proportional limit may be at the rupture stress, and so for this material, failure will begin *at* the point of stress concentration when the proportional limit is reached. Essentially what happens is that a crack begins to form at this point and a higher stress concentration will develop at the *tip* of this crack. This, in turn, causes the crack to progress over the cross section, resulting in sudden fracture. For this reason, it is very important to use stress-concentration factors in design when using brittle materials. On the other hand, if the material is ductile and subjected to a static load, designers usually neglect using stress-concentration factors since any stress that exceeds the proportional limit will not result in a crack. Instead, the material will have reserve strength due to yielding and strain-hardening.

Stress concentrations are also responsible for many failures of structural members or mechanical elements subjected to *fatigue loadings*. For these cases, a stress concentration will cause the material to crack if the stress exceeds the material's endurance limit, whether or not the material is ductile or brittle. What happens is that the material *localized* at the tip of the crack remains in a *brittle state*, and so the crack continues to grow, leading to a progressive fracture. Consequently, engineers involved in the design of such members must seek ways to limit the amount of damage that can be caused by fatigue.

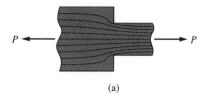

(a)

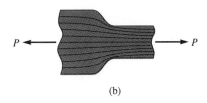

(b)

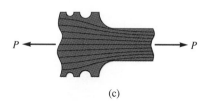

(c)

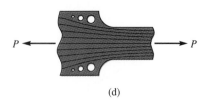

(d)

Fig. 10–23

*See Lipson, C. and Juvinall, R.C., *Handbook of Stress and Strength,* Macmillan, 1963.

Example 10–12

The steel strap shown in Fig. 10–24 is subjected to an axial load of 80 kN. Determine the maximum normal stress developed in the strap, and the displacement of one end of the strap with respect to the other end. The steel has a yield stress of $\sigma_Y = 700$ MPa, and $E_{st} = 200$ GPa.

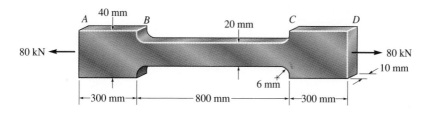

Fig. 10–24

SOLUTION

Maximum Normal Stress. By inspection, the maximum normal stress occurs at the smaller cross section, where the shoulder fillet begins at B or C. The stress-concentration factor is determined from Fig. 10–21. We require

$$\frac{r}{h} = \frac{6 \text{ mm}}{20 \text{ mm}} = 0.3; \qquad \frac{w}{h} = \frac{40 \text{ mm}}{20 \text{ mm}} = 2$$

Thus, $K = 1.6$.

The maximum stress is therefore

$$\sigma_{\max} = K\frac{P}{A} = 1.6 \, \frac{80(10^3) \text{ N}}{(0.02 \text{ m})(0.01 \text{ m})} = 640 \text{ MPa} \qquad Ans.$$

Notice that the material remains elastic, since 640 MPa $< \sigma_Y = 700$ MPa.

Displacement. Here we will neglect the localized deformations surrounding the applied load and at the sudden change in cross section of the shoulder fillet (Saint-Venant's principle). Applying Eq. 10–3, we have

$$\Delta_{A/D} = \Sigma\frac{PL}{AE} = 2\left\{\frac{80(10^3) \text{ N}(0.3 \text{ m})}{(0.04 \text{ m})(0.01 \text{ m})[200(10^9) \text{ N/m}^2]}\right\}$$

$$+\left\{\frac{80(10^3) \text{ N}(0.8 \text{ m})}{(0.02 \text{ m})(0.01 \text{ m})[200(10^9) \text{ N/m}^2]}\right\}$$

$$\Delta_{A/D} = 2.20 \text{ mm} \qquad Ans.$$

PROBLEMS

*10–48.   The steel bar has the dimensions shown in the figure. Determine the maximum axial force P that can be applied so as not to exceed an allowable tensile stress of $\sigma_{allow} = 150$ MPa.

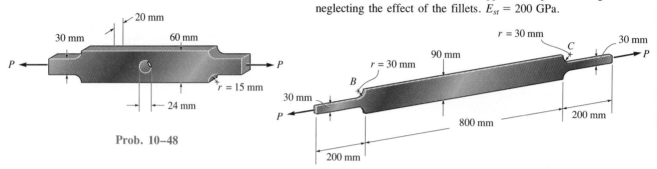

Prob. 10–48

10–49.   Determine the maximum normal stress developed in the bar when it is subjected to a tension of $P = 8$ kN.

10–50.   If the allowable normal stress for the bar is $\sigma_{allow} = 120$ MPa, determine the maximum axial force P that can be applied to the bar.

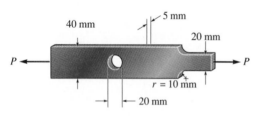

Probs. 10–49/10–50

10–51.   Determine the maximum axial force P that can be applied to the bar. The bar is made from steel and has an allowable stress of $\sigma_{allow} = 21$ ksi.

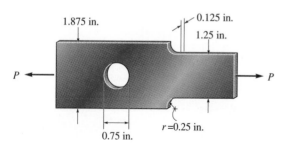

Prob. 10–51

*10–52.   The steel plate has a thickness of 12 mm. If there are shoulder fillets at B and C, and $\sigma_{allow} = 150$ MPa, determine the maximum axial load P that it can support. Compute its elongation neglecting the effect of the fillets. $E_{st} = 200$ GPa.

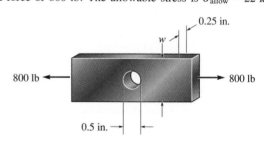

Prob. 10–52

10–53.   The member is to be made from a steel plate that is 0.25 in. thick. If a 0.5-in. hole is drilled through its center, determine the approximate width w of the plate so that it can support an axial force of 800 lb. The allowable stress is $\sigma_{allow} = 22$ ksi.

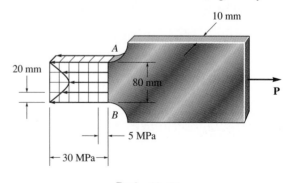

Prob. 10–53

10–54.   The resulting stress distribution along the section AB for the bar is shown in the figure. From this distribution, determine the approximate resultant axial force P applied to the bar. Also, what is the stress-concentration factor for this geometry?

Prob. 10–54

REVIEW PROBLEMS

10–55. A steel surveyor's tape is to be used to measure the length of a line. The tape has a rectangular cross section of 0.05 in. by 0.2 in. and a length of 100 ft when $T_1 = 60°F$ and the tension or pull on the tape is 20 lb. Determine the true length of the line if the tape shows the reading to be 463.25 ft when used with a pull of 35 lb at $T_2 = 90°F$. The ground on which it is placed is flat. $\alpha_{st} = 9.6(10^{-6})/°F$, $E_{st} = 29(10^3)$ ksi.

Prob. 10–55

10–57. The rigid link is supported by a pin at A, a steel wire BC having an unstretched length of 200 mm and cross-sectional area of 22.5 mm^2, and a short aluminum block having an unloaded length of 50 mm and cross-sectional area of 40 mm^2. If the link is subjected to the vertical load shown, determine the normal stress in the wire and the block. $E_{st} = 200$ GPa, $E_{al} = 70$ GPa.

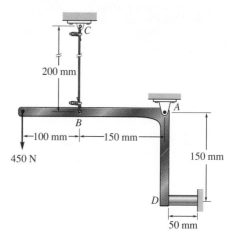

Prob. 10–57

10–58. The rigid beam is supported at its ends by steel tie rods. The rods have diameters $d_{AB} = 0.5$ in. and $d_{CD} = 0.3$ in. If the allowable stress for the steel is $\sigma_{\text{allow}} = 16.2$ ksi, determine the maximum value of P and the position x of this force on the beam so the beam remains in the horizontal position when it is loaded. $E_{st} = 29(10^3)$ ksi.

*10–56. The brass plug is force-fitted into the rigid casting. The uniform normal bearing pressure on the plug is estimated to be 15 MPa. If the coefficient of static friction between the plug and casting is $\mu_s = 0.3$, determine the axial force P needed to pull the plug out. Also, calculate the displacement of end B relative to end A just before the plug starts to slip out. $E_{br} = 98$ GPa.

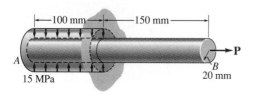

Prob. 10–56

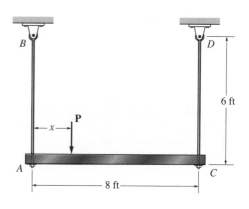

Prob. 10–58

10–59. The 0.4-in.-diameter steel bolt is used to hold the (rigid) assembly together. Determine the clamping force that must be provided by the bolt when $T_1 = 80°F$, so that the clamping force it exerts when $T_2 = 180°F$ is 200 lb. $E_{st} = 29(10^3)$ ksi, $\alpha_{st} = 6(10^{-6})/°F$.

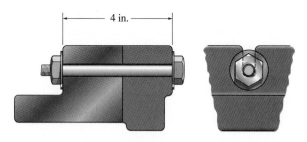

Prob. 10–59

*10–60. The joint is made from three steel plates that are bonded together at their seams. Determine the displacement of end A with respect to end B when the joint is subjected to the axial loads shown. Each plate has a thickness of 5 mm. $E_{st} = 200$ GPa.

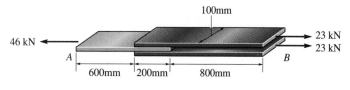

Prob. 10–60

10–61. The steel bar has the original dimensions shown in the figure. If it is subjected to an axial loading of 50 kN, determine the change in its length and its new cross-sectional dimensions at section a–a. $E_{st} = 200$ GPa, $\nu_{st} = 0.29$.

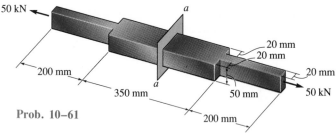

Prob. 10–61

10–62. The rigid beam is supported by three steel wires, each having a length of 4 ft. The cross-sectional area of AB and EF is 0.015 in^2, and the cross-sectional area of CD is 0.006 in^2. If $\sigma_Y = 36$ ksi, determine the largest distributed load w that can be supported by the beam before any of the wires begin to yield. $E_{st} = 29(10^3)$ ksi. If the steel is assumed to be elastic-perfectly plastic, determine how far the beam is displaced downward just before all the wires begin to yield.

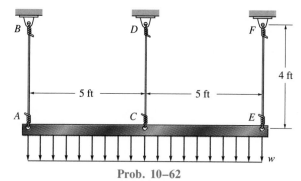

Prob. 10–62

10–63. The steel bolt has a diameter of 7 mm and fits through an aluminum sleeve as shown. The sleeve has an inner diameter of 8 mm and an outer diameter of 10 mm. The nut at A is adjusted so that it just presses up against the sleeve. If it is then tightened one-half turn, determine the force in the bolt and the sleeve. The single-threaded screw on the bolt has a lead of 1.5 mm. $E_{st} = 200$ GPa, $E_{al} = 70$ GPa. *Note:* The lead represents the distance the nut advances along the bolt for one complete turn of the nut.

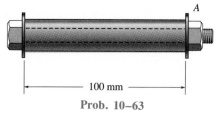

Prob. 10–63

*10–64. The observation cage C has a weight of 250 kip and, through a system of gears, travels upward at constant velocity along the steel column, which has a height of 200 ft. The column has an outer diameter of 3 ft and is made from steel plate having a thickness of 0.25 in. Neglect the weight of the column, and determine the normal stress in the column at its base, B, as a function of the cage's position y. Also, determine the relative displacement of end A with respect to end B as a function of y. $E_{st} = 29(10^3)$ ksi.

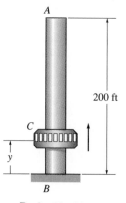

Prob. 10–64

11 Torsion

In this chapter we will discuss the effects of applying a torsional loading to a long straight member such as a shaft or tube. Initially we will consider the member to have a circular cross section. We will show how to determine both the stress distribution within the member and the angle of twist, when the material behaves in a linear-elastic manner.

11.1 Torsional Deformation of a Circular Shaft

Torque is a moment that tends to twist a member about its longitudinal axis. Its effect is of primary concern in the design of axles or drive shafts used in vehicles and machinery. In this section we will discuss the type of deformation that occurs when a torque is applied to a circular shaft or tube made of a homogeneous material. Later in the chapter we will then use these concepts to develop equations that give the stress distribution and rotation or twist that can be developed in shafts and tubes that transmit torsional loadings.

We can illustrate physically what happens when a torque is applied to a circular shaft by considering the shaft to be made of a highly deformable material such as rubber, Fig. 11–1a. When the torque is applied, the circles and longitudinal grid lines originally marked on the shaft tend to distort into the pattern shown in Fig. 11–1b. By inspection, twisting causes the circles to *remain circles,* and each longitudinal grid line deforms into a helix that inter-

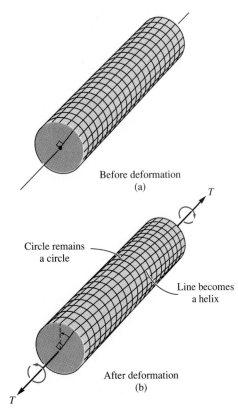

Before deformation
(a)

Circle remains
a circle

Line becomes
a helix

After deformation
(b)

Fig. 11–1

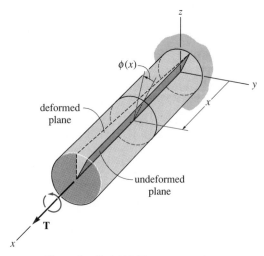

The angle of twist $\phi(x)$ increases as x increases.

Fig. 11–2

sects the circles at equal angles. Also, the cross sections at the *ends* of the shaft remain *flat*—that is, they do not warp or bulge in or out—and radial lines on these ends *remain straight* during the deformation. From these observations we will make the assumptions that *every cross section of the shaft remains plane,* and second radial lines on these cross sections *remain straight and rotate through the same angle* during deformation. Furthermore, we can assume that if the angle of rotation is *small,* the *length of the shaft* and *its radius* will *remain unchanged.* Thus, if the shaft is fixed at one end as shown in Fig. 11–2 and a torque is applied to its other end, the shaded plane will distort into a skewed form as shown. Here it is seen that a radial line located on the cross section at a distance x from the fixed end of the shaft will rotate through an angle $\phi(x)$. The angle $\phi(x)$, so defined, is called the *angle of twist.* It depends on the position x and will vary along the shaft as shown.

In order to understand how this distortion strains the material, we will now isolate a small element located at a radial distance ρ (rho) from the axis of the shaft, Fig. 11–3a. Notice that the front and back faces of the element have sides that are bounded by radial lines, located at x and $x + \Delta x$, respectively. Due to the deformation as noted in Fig. 11–2, each of these faces will undergo a rotation. The one at x by $\phi(x)$, and the one at $x + \Delta x$ by $\phi(x) + \Delta\phi$, Fig. 11–3b. As a result, the *difference* in these rotations, $\Delta\phi$, causes the element to be subjected to a *shear strain.* To show this, note that before deformation the angle between the edges AB and AC is 90°; after deformation, however, the edges of the element are AD and AC and the angle between them is θ'. From the definition of shear strain, Eq. 8–13, we have

$$\gamma = \frac{\pi}{2} - \lim_{\substack{C \to A \text{ along } CA \\ D \to A \text{ along } BA}} \theta'$$

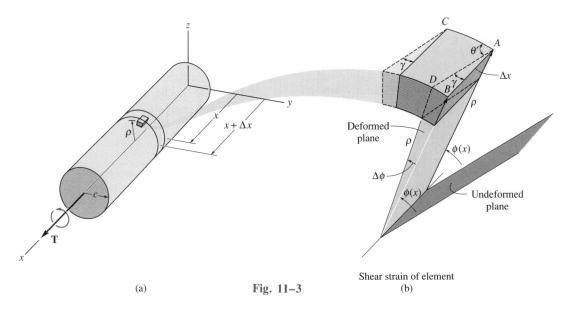

(a)   **Fig. 11–3**

Shear strain of element
(b)

This angle, γ, is indicated in Fig. 11–3b. It can be related to the length Δx of the element and the difference in the angle of rotation, $\Delta\phi$, between the shaded faces. If we let $\Delta x \to dx$ and $\Delta\phi \to d\phi$, then from Fig. 11–3b we have

$$BD = \rho \, d\phi = dx \, \gamma$$

Therefore,

$$\gamma = \rho \, \frac{d\phi}{dx} \qquad (11\text{–}1)$$

Since dx and $d\phi$ are the *same* for *all elements* located at points within the cross section at x, then $d\phi/dx$ is constant and Eq. 11–1 states that the magnitude of the shear strain for any of these elements varies only with its radial distance ρ from the axis of the shaft. In other words, *the shear strain within the shaft varies linearly along any radial line, from zero at the axis of the shaft to a maximum at its outer boundary.* Assuming that the outer radius of the shaft is c, the shear strain for typical elements located within the shaft at ρ, and at its surface $\rho = c$, is shown in Fig. 11–4. From Eq. 11–1, since $d\phi/dx = \gamma/\rho = \gamma_{max}/c$, then

$$\gamma = \left(\frac{\rho}{c}\right) \gamma_{max} \qquad (11\text{–}2)$$

The results obtained here are also valid for circular tubes. They depend only on the assumptions regarding the deformations mentioned above. Using a more complete analysis of the shaft, it is also possible to show that these assumptions require all other components of both normal and shear strain to be zero when the shaft is subjected to a torque.

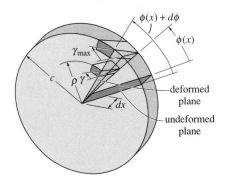

The shear strain for the material increases linearly with ρ, i.e., $\gamma = (\rho/c)\gamma_{max}$.

Fig. 11–4

11.2 The Torsion Formula

If a shaft is subjected to an external torque, then for equilibrium an internal torque **T** must also be developed within the shaft. At any given section along the shaft this internal torque is the resultant of the moment produced by the distribution of shear stress acting over the shaft's cross-sectional area, Fig. 11–5. In this section we will develop an equation that relates this shear-stress distribution to the resultant internal torque resisted at the section of a circular shaft or tube.

If the material is linear-elastic, then Hooke's law applies, $\tau = G\,\gamma$, and consequently a *linear variation in shear strain,* as noted in Sec. 11.1, leads to a corresponding *linear variation in shear stress* along any radial line on the cross section. Hence, like the shear-strain variation, for a solid shaft, τ will vary from zero at the shaft's longitudinal axis to a maximum value, τ_{max}, at its outer surface. This variation is shown in Fig. 11–5 on the front faces of a selected number of elements, located at an intermediate radial position ρ and at the outer radius c. Due to the proportionality of triangles, or by using Hooke's law ($\tau = G\,\gamma$) and Eq. 11–2 [$\gamma = (\rho/c)\gamma_{max}$], we can write

$$\tau = \left(\frac{\rho}{c}\right)\tau_{max} \qquad (11\text{–}3)$$

This equation expresses the shear-stress distribution as a *function* of the radial position ρ of the element; in other words, it defines the stress distribution in terms of the geometry of the shaft. Using it, we will now apply the condition that requires the torque produced by the stress distribution over the entire cross section to be equivalent to the resultant internal torque T at the section,

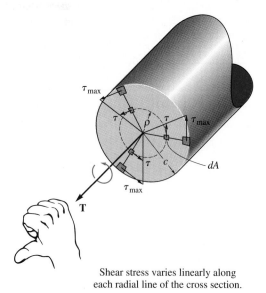

Shear stress varies linearly along each radial line of the cross section.

Fig. 11–5

which holds the shaft in equilibrium, Fig. 11–5. Specifically, each element of area dA, located at ρ, is subjected to a force of $dF = \tau\, dA$. The torque produced by this force is $dT = \rho(\tau\, dA)$. Using Eq. 11–3, we therefore have for the entire cross section

$$T = \int_A \rho(\tau\, dA) = \int_A \rho\left(\frac{\rho}{c}\right)\tau_{max}\, dA \qquad (11\text{–}4)$$

Since τ_{max}/c is constant,

$$T = \frac{\tau_{max}}{c} \int_A \rho^2\, dA \qquad (11\text{–}5)$$

The integral in this equation depends only on the geometry of the shaft. It represents the *polar moment of inertia* of the shaft's cross-sectional area computed about the shaft's longitudinal axis. We will symbolize its value as J, and therefore the above equation can be written in a more compact form, namely,

$$\boxed{\tau_{max} = \frac{Tc}{J}} \qquad (11\text{–}6)$$

where

τ_{max} = the maximum shear stress in the shaft, which occurs at the outer surface

T = the resultant internal torque acting at the cross section. Its value is determined from the method of sections and the equation of moment equilibrium applied about the shaft's longitudinal axis

J = the polar moment of inertia of the cross-sectional area

c = the outer radius of the shaft

Using Eqs. 11–3 and 11–6, the shear stress at the intermediate distance ρ can be determined from a similar equation:

$$\boxed{\tau = \frac{T\rho}{J}} \qquad (11\text{–}7)$$

Either of the above two equations is often referred to as the *torsion formula*. Recall that it is used only if the shaft is circular and the material is homogeneous and behaves in a linear-elastic manner, since the derivation is based on the fact that the shear stress is proportional to the shear strain, and thus both vary linearly along every radial line on the cross section.

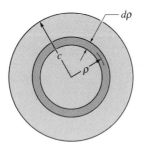

Fig. 11–6

Solid Shaft. If the shaft has a solid circular cross section, the polar moment of inertia J can be determined using an area element in the form of a *differential ring* or annulus having a thickness $d\rho$ and circumference $2\pi\rho$, Fig. 11–6. For this ring, $dA = 2\pi\rho \, d\rho$, so

$$J = \int_A \rho^2 \, dA = \int_0^c \rho^2 \, (2\pi\rho \, d\rho) = 2\pi \int_0^c \rho^3 \, d\rho = 2\pi \left(\frac{1}{4}\right)\rho^4 \Big|_0^c$$

$$\boxed{J = \frac{\pi}{2}c^4} \tag{11–8}$$

Note that J is always positive. Common units used for its measurement are mm^4 or in^4. It is a *geometric property* of the circular area to be used in *both* Eqs. 11–6 and 11–7.

As noted in the derivation, Eq. 11–7 gives specific values of the shear stress at any point located on the shaft's cross-sectional area. In particular, if the shear stress is plotted along radial line segments, it is represented by a linear distribution as shown in Fig. 11–7a. Notice that each value of τ is directed on the cross section so that the force it creates contributes a *counterclockwise torque* about the center of the shaft. This is necessary since the total torque developed by the entire stress distribution must yield the *counterclockwise* resultant torque **T** at the section, Eq. 11–4.

In Sec. 8.2 it was shown that any isolated *volume element* of material that is subjected to a shear stress on one of its faces must, by reason of both force and moment equilibrium, also develop equal shear stress on three of its adjacent faces. Consequently, typical elements, when isolated from the shaft in Fig. 11–7a, must be subjected to shear stresses directed as shown.

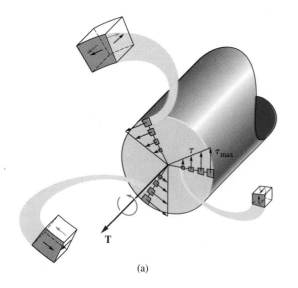

(a)

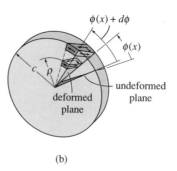

(b)

Fig. 11–7

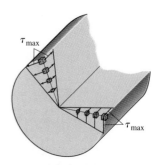

Shear stress varies linearly along each radial line of the cross section.

(c)

Perhaps another way of seeing this result is to isolate a differential disk of the shaft having a thickness dx, Fig. 11–7b. The deformation and associated shear stresses acting on two shaded elements, one located at the outer surface and the other at some intermediate radial point, ρ, are shown. Therefore, *not only does the internal torque* **T** *develop a linear distribution of shear stress along each radial line in the plane of the cross-sectional area, but an associated shear-stress distribution must also be developed along an axial plane as noted in Fig. 11–7c.* It is interesting to note that because of this axial distribution of shear stress, shafts made from wood tend to *split* along the axial plane when subjected to excessive torque, Fig. 11–8. This is because wood is an anisotropic material. Its shear resistance parallel to its grains or fibers, directed along the axis of the shaft, is much less than its resistance perpendicular to the fibers, directed in the plane of the cross section.

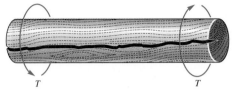

T T

Failure of a wooden shaft due to torsion

Fig. 11–8

Tubular Shaft. If a shaft has a tubular cross section, with inner radius c_i and outer radius c_o, then from Eq. 11–8 we can determine its polar moment of inertia by subtracting J for a shaft of radius c_i from that computed for a shaft of radius c_o. The result is

$$J = \frac{\pi}{2}(c_o^4 - c_i^4)$$ (11–9)

Like the solid shaft, the shear stress distributed over the tube's cross-sectional area varies linearly along any radial line, Fig. 11–9a. Furthermore, the shear stress varies along an axial plane in this same manner, Fig. 11–9b. Examples of the shear stress acting on typical volume elements are shown in Fig. 11–9a.

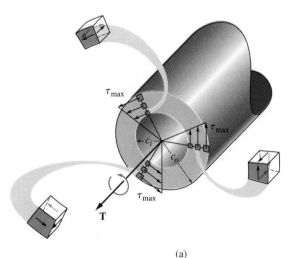

(a)

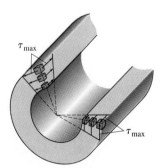

τ_{max}

τ_{max}

Shear stress varies linearly along each radial line of the cross section.

(b)

Fig. 11–9

PROCEDURE FOR ANALYSIS

The torsion formula can be used to find the shear-stress distribution in a solid shaft or tube having a *circular* cross section that is made of *homogeneous* material and has *linear-elastic* behavior. Saint-Venant's principle requires that this equation be applied at points located a sufficient distance away from any cross-section discontinuities or where an external torque acts. In order to apply the equation, the following procedure is suggested.

Internal Loading.   Section the shaft perpendicular to its axis at the point where the shear stress is to be determined, and use the necessary free-body diagram and equations of equilibrium to obtain the internal torque at the section.

Section Property.   Compute the polar moment of inertia of the cross-sectional area. For a solid section of radius c, $J = \pi c^4/2$, and for a tube of outer radius c_o and inner radius c_i, $J = \pi(c_o^4 - c_i^4)/2$.

Shear Stress.   Specify the radial distance ρ, measured from the center of the cross section to the point where the shear stress is to be computed. Then apply the torsion formula $\tau = T\rho/J$, or if the maximum shear stress is to be computed use $\tau_{max} = Tc/J$. When substituting the data, make sure to use a consistent set of units.

The shear stress acts on the cross section in a direction that is always perpendicular to ρ. The force it creates must contribute a torque about the axis of the shaft that is in the *same direction* as the internal resultant torque **T** acting on the section. Once this direction is established, a volume element of the material, located at the point where τ is computed, can be isolated, and the direction of τ acting on the remaining three faces of the element can be shown.

The following examples illustrate numerical applications of this procedure.

Example 11–1

The stress distribution in a solid shaft has been plotted along three arbitrary radial lines as shown in Fig. 11–10a. Determine the resultant internal torque at the section (a) using the torsion formula, (b) by finding the resultant of the stress distribution using basic principles.

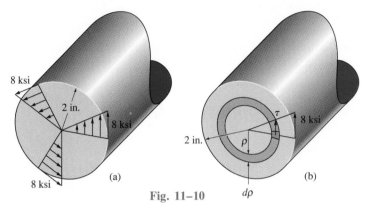

Fig. 11–10

SOLUTION

Part (a). The polar moment of inertia for the cross-sectional area is

$$J = \frac{\pi}{2}(2 \text{ in.})^4 = 25.13 \text{ in}^4$$

Applying the torsion formula, with $\tau_{max} = 8$ ksi, Fig. 11–10a, we have

$$\tau_{max} = \frac{Tc}{J}; \qquad 8 \text{ kip/in}^2 = \frac{T(2 \text{ in.})}{(25.13 \text{ in}^4)}$$

$$T = 101 \text{ kip} \cdot \text{in.} \qquad\qquad Ans.$$

Part (b). The shear stress τ acting at the arbitrary distance ρ from the center of the cross section is determined by proportional triangles, Fig. 11–10b. We have

$$\frac{\tau}{\rho} = \frac{8 \text{ ksi}}{2 \text{ in.}}$$

$$\tau = 4\rho$$

This stress acts on all portions of the differential ring element that has an area $dA = 2\pi\rho \, d\rho$. Since the force created by τ is $dF = \tau \, dA$, the torque is

$$dT = \rho \, dF = \rho(\tau dA) = \rho(4\rho)2\pi\rho \, d\rho = 8\pi\rho^3 \, d\rho$$

For the entire area over which τ acts, we require

$$T = \int_0^2 8\pi\rho^3 \, d\rho = 8\pi\left(\frac{1}{4}\rho^4\right)\Big|_0^2 = 101 \text{ kip} \cdot \text{in.} \qquad Ans.$$

■ Example 11–2

(a)

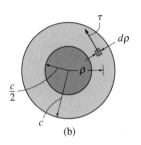

(b)

Fig. 11–11

The *solid* shaft of radius c is subjected to a torque $\mathbf{T}$, Fig. 11–11a. Determine the fraction of $\mathbf{T}$ that is resisted by the material contained within the lighter-shaded region of the shaft, which has an inner radius of $c/2$ and outer radius c.

SOLUTION

The stress in the shaft varies linearly, such that $\tau = (\rho/c)\tau_{max}$, Eq. 11–3. Therefore, the torque dT' on the ring (area) located within the lighter-shaded region, Fig. 11–11b, is

$$dT' = \rho(\tau \, dA) = \rho(\rho/c)\tau_{max}(2\pi\rho \, d\rho)$$

For the entire lighter-shaded area the torque is

$$T' = \frac{2\pi\tau_{max}}{c} \int_{c/2}^{c} \rho^3 \, d\rho$$

$$= \frac{2\pi\tau_{max}}{c} \frac{1}{4}\rho^4 \Big|_{c/2}^{c}$$

So that

$$T' = \frac{15\pi}{32}\tau_{max}c^3 \tag{1}$$

This torque T' can be expressed in terms of the applied torque T by first using the torsion formula to determine the maximum stress in the shaft. We have

$$\tau_{max} = \frac{Tc}{J} = \frac{Tc}{(\pi/2)c^4}$$

or

$$\tau_{max} = \frac{2T}{\pi c^3}$$

Substituting this into Eq. 1 yields

$$T' = \frac{15}{16}T$$

Here approximately 94% of the torque is resisted by the lighter-shaded region, and the remaining 6% of T (or $\frac{1}{16}$) is resisted by the inner "core" of the shaft, $\rho = 0$ to $\rho = c/2$. As a result, the material located at the *outer region* of the shaft is highly effective in resisting torque, which justifies the use of tubular shafts as an efficient means for transmitting torque and thereby saving material.

Example 11–3

The shaft shown in Fig. 11–12a is supported by two bearings and is subjected to three torques. Determine the shear stress developed at points A and B, located at section $a–a$ of the shaft, Fig. 11–12b.

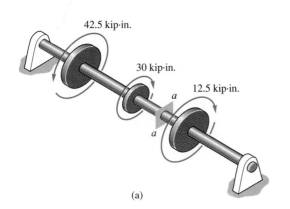

42.5 kip·in.

30 kip·in.

12.5 kip·in.

a

a

(a)

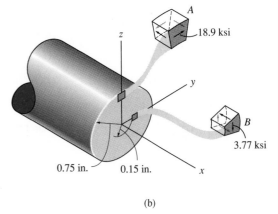

A

z

18.9 ksi

y

B

3.77 ksi

0.75 in.   0.15 in.   *x*

(b)

SOLUTION

Internal Torque.   The bearing reactions on the shaft are zero, provided the shaft's weight is neglected. Furthermore, the applied torques satisfy moment equilibrium about the shaft's axis.

The internal torque at section $a–a$ will be determined from the free-body diagram of the left segment, Fig. 11–12c. We have

$$\Sigma M_x = 0; \quad 42.5 \text{ kip} \cdot \text{in.} - 30 \text{ kip} \cdot \text{in.} - T = 0 \quad T = 12.5 \text{ kip} \cdot \text{in.}$$

Section Property.   The polar moment of inertia for the shaft is

$$J = \frac{\pi}{2}(0.75 \text{ in.})^4 = 0.497 \text{ in}^4$$

Shear Stress.   Since point A is at $\rho = c = 0.75$ in.,

$$\tau_A = \frac{Tc}{J} = \frac{(12.5 \text{ kip} \cdot \text{in.})(0.75 \text{ in.})}{(0.497 \text{ in}^4)} = 18.9 \text{ ksi} \qquad \textit{Ans.}$$

Likewise for point B, at $\rho = 0.15$ in., we have

$$\tau_B = \frac{T\rho}{J} = \frac{(12.5 \text{ kip} \cdot \text{in.})(0.15 \text{ in.})}{(0.497 \text{ in}^4)} = 3.77 \text{ ksi} \qquad \textit{Ans.}$$

The directions of these stresses on each element at A and B, Fig. 11–12b, are established from the direction of the resultant internal torque $\mathbf{T}$, shown in Fig. 11–12c. Note carefully how the shear stress acts on the planes of each of these elements.

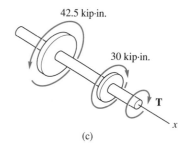

42.5 kip·in.

30 kip·in.

T

x

(c)

Fig. 11–12

Example 11-4

The pipe shown in Fig. 11–13a has an inner diameter of 80 mm and an outer diameter of 100 mm. If its end is tightened against the support at A using a torque wrench at B, determine the shear stress developed in the material at the inner and outer walls along the central portion of the pipe when the 80-N forces are applied to the wrench.

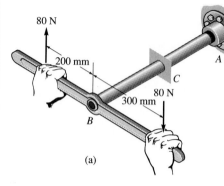

(a)

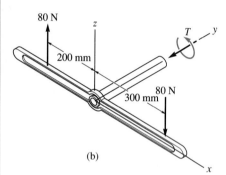

(b)

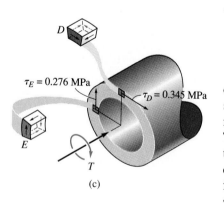

$\tau_E = 0.276$ MPa

$\tau_D = 0.345$ MPa

(c)

Fig. 11–13

SOLUTION

Internal Torque. A section is taken at an intermediate location C along the pipe's axis, Fig. 11–13b. The only unknown at the section is the internal torque **T**. Force equilibrium and moment equilibrium about the x and z axes are satisfied. We require

$$\Sigma M_y = 0; \qquad 80 \text{ N}(0.3 \text{ m}) + 80 \text{ N}(0.2 \text{ m}) - T = 0$$

$$T = 40 \text{ N} \cdot \text{m}$$

Section Property. The polar moment of inertia for the pipe's cross-sectional area is

$$J = \frac{\pi}{2}[(0.05 \text{ m})^4 - (0.04 \text{ m})^4] = 5.80(10^{-6}) \text{ m}^4$$

Shear Stress. For any point lying on the outside surface of the pipe, $\rho = c_o = 0.05$ m, we have

$$\tau_o = \frac{Tc_o}{J} = \frac{40 \text{ N} \cdot \text{m}(0.05 \text{ m})}{5.80(10^{-6}) \text{ m}^4} = 0.345 \text{ MPa} \qquad Ans.$$

And for any point located on the inside surface, $\rho = c_i = 0.04$ m, so

$$\tau_i = \frac{Tc_i}{J} = \frac{40 \text{ N} \cdot \text{m}(0.04 \text{ m})}{5.80(10^{-6}) \text{ m}^4} = 0.276 \text{ MPa} \qquad Ans.$$

To show how these stresses act at representative points D and E on the cross-sectional area, we will first view the cross section from the front of segment CA of the pipe, Fig. 11–13a. On this section, Fig. 11–13c, the resultant internal torque is equal but opposite to that shown in Fig. 11–13b. The shear stresses at D and E contribute to this torque and therefore act on the shaded faces of the elements in the directions shown. As a consequence, notice how the shear-stress components act on the other three faces. Furthermore, since the top face of D and the inner face of E are on stress-free regions taken from the pipe's inner and outer walls, no shear stress can exist on these faces or on the other corresponding faces of the elements.

11.3 Power Transmission

Shafts and tubes having circular cross sections are often used to transmit power developed by a machine. When used for this purpose they are subjected to torques that depend on the power generated by the machine and the angular speed of the shaft. *Power* is defined as the work performed per unit of time. The work transmitted by a rotating shaft equals the torque applied times the angle of rotation. Therefore, if during an instant of time dt an applied torque $\mathbf{T}$ will cause the shaft to rotate $d\theta$, then the instantaneous power is

$$P = \frac{T\, d\theta}{dt}$$

Since the shaft's angular velocity $\omega = d\theta/dt$, we can also express the power as

$$\boxed{P = T\omega} \qquad (11\text{–}10)$$

In the SI system, power is expressed in *watts* when torque is measured in newton-meters (N $\cdot$ m) and ω is in radians per second (rad/s) (1 W = 1 N $\cdot$ m/s). In the Foot-Pound-Second or FPS system, the basic units of power are foot-pounds per second (ft $\cdot$ lb/s); however, horsepower (hp) is often used in engineering practice, where

$$1 \text{ hp} = 550 \text{ ft} \cdot \text{lb/s}$$

For machinery, the *frequency* of a shaft's rotation, f, is often reported. This is a measure of the number of revolutions or cycles the shaft makes per second and is expressed in hertz (1 Hz = 1 cycle/s). Since 1 cycle = 2π rad, then $\omega = 2\pi f$ and the above equation for power becomes

$$\boxed{P = 2\pi f T} \qquad (11\text{–}11)$$

Shaft Design. When the power transmitted by a shaft and its frequency of rotation are known, the torque developed in the shaft can be determined from Eq. 11–11, that is, $T = P/2\pi f$. Knowing T and the allowable shear stress for the material, τ_{allow}, we can determine the size of the shaft's cross section using the torsion formula, provided the material behavior is linear-elastic. Specifically, the design or geometric parameter J/c becomes

$$\frac{J}{c} = \frac{T}{\tau_{\text{allow}}} \qquad (11\text{–}12)$$

For a *solid shaft,* $J = (\pi/2)c^4$, and thus, upon substitution, a *unique value* for the shaft's radius c can be determined. If the shaft is *tubular,* so that $J = (\pi/2)(c_o^4 - c_i^4)$, design permits a wide range of possibilities for the solution since an *arbitrary choice* can be made for either c_o or c_i and the other radius is determined from Eq. 11–12.

The following examples illustrate numerical applications of the above equations.

Example 11-5

A solid steel shaft AB shown in Fig. 11–14 is to be used to transmit 5 hp from the motor M to which it is attached. If the shaft rotates at $\omega = 175$ rpm and the steel has an allowable shear stress of $\tau_{allow} = 14.5$ ksi, determine the required diameter of the shaft to the nearest $\frac{1}{8}$ in.

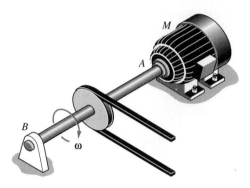

Fig. 11–14

SOLUTION

The torque on the shaft is determined from Eq. 11–10, that is, $P = T\omega$. Expressing P in foot-pounds per second and ω in radians/second, we have

$$P = 5 \text{ hp} \left(\frac{550 \text{ ft} \cdot \text{lb/s}}{1 \text{ hp}} \right) = 2750 \text{ ft} \cdot \text{lb/s}$$

$$\omega = \frac{175 \text{ rev}}{\text{min}} \left(\frac{2\pi \text{ rad}}{1 \text{ rev}} \right) \left(\frac{1 \text{ min}}{60 \text{ s}} \right) = 18.33 \text{ rad/s}$$

Thus,

$$P = T\omega; \qquad 2750 \text{ ft} \cdot \text{lb/s} = T(18.33 \text{ rad/s})$$
$$T = 150.1 \text{ ft} \cdot \text{lb}$$

Applying Eq. 11–12, we have

$$\frac{J}{c} = \frac{\pi}{2} \frac{c^4}{c} = \frac{T}{\tau_{allow}}$$

$$c = \left(\frac{2T}{\pi \tau_{allow}} \right)^{1/3} = \left(\frac{2(150.1 \text{ ft} \cdot \text{lb})(12 \text{ in./ft})}{\pi(14{,}500 \text{ lb/in}^2)} \right)^{1/3}$$

$$c = 0.429 \text{ in.}$$

Since $2c = 0.858$ in., select a shaft having a diameter of

$$d = \frac{7}{8} \text{ in.} = 0.875 \text{ in.} \qquad \qquad Ans.$$

Example 11–6

A tubular shaft, having an inner diameter of 30 mm and an outer diameter of 42 mm, is to be used to transmit 90 kW of power. Determine the frequency of rotation of the shaft so that the shear stress cannot exceed 50 MPa.

SOLUTION

The maximum torque that can be applied to the shaft is determined from the torsion formula.

$$\tau_{max} = \frac{Tc}{J}$$

$$50(10^6) \text{ N/m}^2 = \frac{T(0.021 \text{ m})}{(\pi/2)[(0.021 \text{ m})^4 - (0.015 \text{ m})^4]}$$

$$T = 538 \text{ N} \cdot \text{m}$$

Applying Eq. 11–11, the frequency of rotation is

$$P = 2\pi f T$$

$$90(10^3) \text{ N} \cdot \text{m/s} = 2\pi f(538 \text{ N} \cdot \text{m})$$

$$f = 26.6 \text{ Hz} \qquad\qquad Ans.$$

PROBLEMS

11–1. A shaft is made from a steel alloy having an allowable shear stress of $\tau_{allow} = 12$ ksi. If the diameter of the shaft is 1.5 in., determine the maximum torque **T** that can be transmitted. What would be the maximum torque **T′** if a 1-in.-diameter hole is bored through the shaft? Sketch the shear-stress distribution along a radial line in each case.

11–2. The copper pipe has an outer diameter of 40 mm and an inner diameter of 37 mm. If it is tightly secured to the wall at *A* and three torques are applied to it as shown, determine the maximum shear stress developed in the pipe.

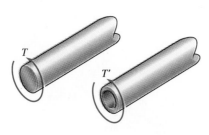

Prob. 11–1

Prob. 11–2

11–3. The solid shaft is fixed to the support at C and subjected to a torque of 950 N · m. Determine the shear stress at points A and B and sketch the shear stress on volume elements located at these points.

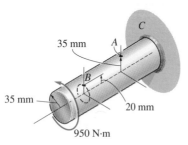

Prob. 11–3

*11–4.** The copper pipe has an outer diameter of 2.50 in. and an inner diameter of 2.30 in. If it is tightly secured to the wall at C and three torques are applied to it as shown, determine the shear stress developed at points A and B. These points lie on the pipe's outer surface. Sketch the shear stress on volume elements located at A and B.

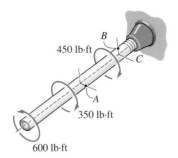

Prob. 11–4

11–5. The solid 1.25-in.-diameter shaft is used to transmit the torques applied to the gears. If it is supported by smooth bearings at A and B, which do not resist torque, determine the shear stress developed in the shaft at points C and D. Indicate the shear stress on volume elements located at these points.

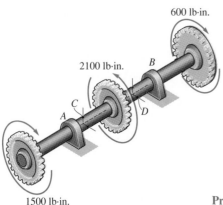

Prob. 11–5

11–6. The assembly consists of two sections of galvanized steel pipe connected together using a reducing coupling at B. The smaller pipe has an outer diameter of 0.75 in. and an inner diameter of 0.68 in., whereas the larger pipe has an outer diameter of 1 in. and an inner diameter of 0.86 in. If the pipe is tightly secured to the wall at C, determine the maximum shear stress developed in each section of the pipe when the couple shown is applied to the handles of the wrench.

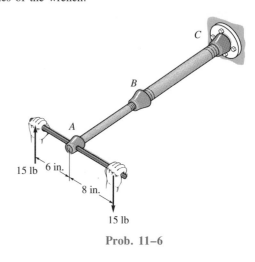

Prob. 11–6

11–7. The solid shaft has a diameter of 0.75 in. If it is subjected to the torques shown, determine the maximum shear stress developed in regions BC and DE of the shaft. The bearings at A and F allow free rotation of the shaft.

*11–8.** The solid shaft has a diameter of 0.75 in. If it is subjected to the torques shown, determine the maximum shear stress developed in regions CD and EF of the shaft. The bearings at A and F allow free rotation of the shaft.

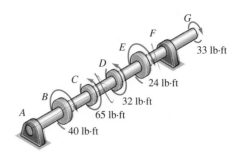

Probs. 11–7/11–8

11–9. The steel shaft is subjected to the torsional loading shown. Determine the shear stress developed at points A and B and sketch the shear stress on volume elements located at these points. The shaft where A and B are located has a radius of 60 mm.

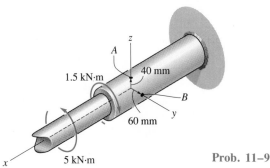

1.5 kN·m

40 mm

A

B

y

60 mm

5 kN·m

x

Prob. 11–9

11–10. If the solid shaft to which the valve handle is attached is made of brass and has a diameter of 10 mm, determine the maximum force F that can be applied to the handle just before the material starts to yield. Take $\tau_Y = 235$ MPa.

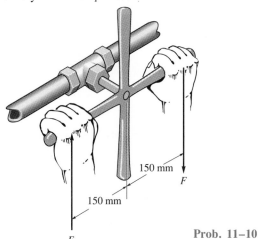

150 mm

F

150 mm

F

Prob. 11–10

11–11. The motor delivers a torque of 50 N · m to the shaft AB. This torque is transmitted to shaft CD using the gears at E and F. Determine the equilibrium torque $\mathbf{T}'$ on shaft CD and the maximum shear stress in each shaft. The bearings at B, C, and D allow free rotation of the shafts.

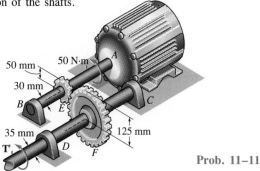

50 mm

50 N·m

A

30 mm

B

E

C

35 mm

T'

D

F

125 mm

Prob. 11–11

*11–12. The solid aluminum shaft has a diameter of 50 mm and an allowable shear stress of $\tau_{\text{allow}} = 6$ MPa. Determine the largest torques T_1 and T_2 that can be applied to the shaft if it is also subjected to the other torsional loadings. It is required that $\mathbf{T}_1$ and $\mathbf{T}_2$ act in the directions shown.

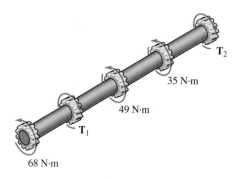

$\mathbf{T}_2$

35 N·m

49 N·m

$\mathbf{T}_1$

68 N·m

Prob. 11–12

11–13. The coupling consists of two disks fixed to separate shafts, each 25 mm in diameter. The shafts are supported on journal bearings that allow free rotation. In order to limit the torque $\mathbf{T}$ that can be transmitted, a "shear pin" P is used to connect the disks together. If this pin can sustain an *average* shear force of 550 N before it fails, determine the maximum constant torque T that can be transmitted from one shaft to the other. Also, what is the maximum shear stress in each shaft when the "shear pin" is about to fail?

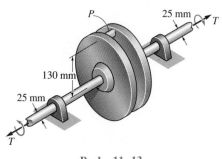

P

25 mm

T

130 mm

25 mm

T

Prob. 11–13

11–14. The wooden post, which is half buried in the ground, is subjected to a torsional moment of 50 N · m that causes the post to rotate at constant angular velocity. This moment is resisted by a *linear distribution* of torque developed by soil friction, which varies from zero at the ground to t_0 N · m/m at its base. Determine the equilibrium value for t_0 and then calculate the shear stress at points A and B, which lie on the outer surface of the post.

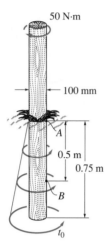

Prob. 11–14

11–15. When drilling a well at constant angular velocity, the bottom end of the drill pipe encounters a torsional resistance T_A. Also, soil along the sides of the pipe creates a distributed frictional torque along its length, varying uniformly from zero at the surface B to t_A at A. Determine the minimum torque T_B that must be supplied by the drive unit to overcome the resisting torques, and compute the maximum shear stress in the pipe. The pipe has an outer radius r_o and an inner radius r_i.

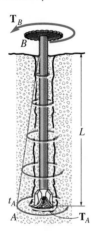

Prob. 11–15

11–16. The solid shaft has a linear taper from r_A at one end to r_B at the other. Derive an equation that gives the maximum shear stress in the shaft at a location x along the shaft's axis.

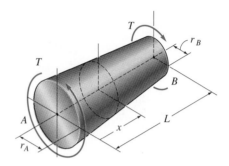

Prob. 11–16

11–17. Determine to the nearest $\frac{1}{8}$ in. the diameter of a shaft that is required to transmit 150 hp at 4000 rev/min. The material has an allowable shear stress of $\tau_{\text{allow}} = 8$ ksi.

11–18. The drilling pipe on an oil rig is made from steel pipe having an outside diameter of 4.5 in. and a thickness of 0.25 in. If the pipe is turning at 650 rev/min while being powered by a 15-hp motor, determine the maximum shear stress in the pipe.

11–19. The 0.75-in.-diameter shaft in the electric motor develops 0.5 hp and runs at 1740 rev/min. Determine the torque produced and compute the maximum shear stress in the shaft. The shaft is supported by ball bearings at A and B.

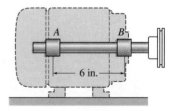

Prob. 11–19

*11–20. The 4-hp motor turns the shaft AB at a rate of 1300 rev/min. Determine the maximum shear stress developed in the shaft within region AB, where it has a diameter of 0.5 in., and within region BC, where it has a diameter of 0.65 in.

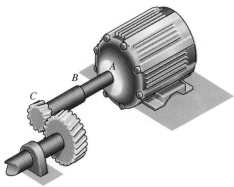

Prob. 11–20

11–21. The drive shaft AB of an automobile is made of a steel tube having an allowable shear stress of $\tau_{allow} = 8$ ksi. If the outer diameter is 2.5 in. and the engine delivers 200 hp to the shaft when it is turning at 1140 rev/min, determine the minimum required thickness of the shaft.

11–22. The drive shaft AB of an automobile is to be designed as a thin-walled tube. The engine delivers 150 hp when the shaft is turning at 1500 rev/min. Determine the minimum thickness if the outer diameter is 2.5 in. The material has an allowable shear stress of $\tau_{allow} = 7$ ksi.

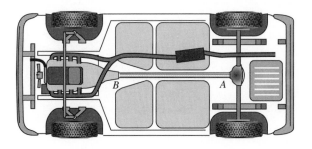

Probs. 11–21/11–22

11–23. The motor delivers 500 hp to the steel shaft AB, which is tubular and has an outer diameter of 2 in. and an inner diameter of 1.84 in. Determine the *smallest* angular velocity at which it can rotate if the allowable shear stress for the material is $\tau_{allow} = 25$ ksi.

Prob. 11–23

*11–24. The solid steel shaft AC has a diameter of 25 mm and is supported by smooth bearings at D and E. It is coupled to a motor at C, which delivers 3 kW of power to the shaft while it is turning at 50 rev/s. If gears A and B remove 1 kW and 2 kW, respectively, determine the maximum shear stress developed in the shaft within regions AB and BC. The shaft is free to turn in its support bearings D and E.

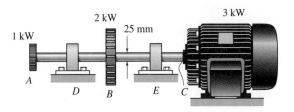

Prob. 11–24

11.4 Angle of Twist

Occasionally the design of a shaft depends on restricting the amount of rotation or twist that may occur when the shaft is subjected to a torque. Furthermore, being able to compute the angle of twist for a shaft is important when analyzing the reactions on statically indeterminate shafts.

In this section we will develop a formula for determining the *angle of twist* ϕ (phi) of one end of a shaft with respect to its other end. The shaft is assumed to have a circular cross section that can gradually vary along its length, Fig. 11–15a, and the material is assumed to be homogeneous and to behave in a linear-elastic manner when the torque is applied. As in the case of an axially loaded bar, we will neglect the localized deformations that occur at points of application of the torques and where the cross section changes abruptly. By Saint-Venant's principle, these effects occur within small regions of the shaft's length and generally have only a slight effect on the final result.

Using the method of sections, a differential disk of thickness dx, located at position x, is isolated from the shaft. Its free-body diagram is shown in Fig. 11–15b. The internal resultant torque is represented as $T(x)$, since the external loading may cause it to vary along the axis of the shaft. Due to $T(x)$ the disk will twist, such that the *relative rotation* of one of its faces with respect to the other face is $d\phi$, Fig. 11–15b. Furthermore, as explained in Sec. 11.1 and shown in Fig. 11–4, an element of material located at an arbitrary radius ρ within the disk will undergo a shear strain γ. The values of γ and $d\phi$ are related by Eq. 11–1, namely,

$$d\phi = \gamma \, \frac{dx}{\rho} \qquad (11\text{–}13)$$

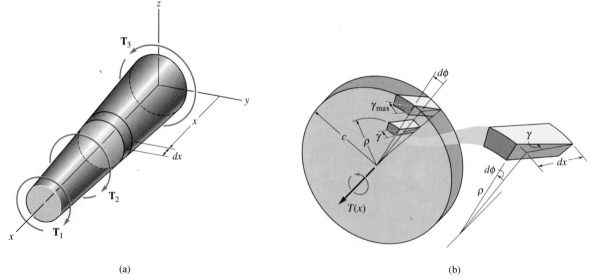

(a)   (b)

Fig. 11–15

Since Hooke's law, $\gamma = \tau/G$, applies, and the shear stress can be expressed in terms of the applied torque using the torsion formula $\tau = T(x)\rho/J(x)$, then $\gamma = T(x)\rho/J(x)G$. Substituting this result into Eq. 11–13, the angle of twist for the disk is

$$d\phi = \frac{T(x)}{J(x)G}\,dx$$

Integrating over the entire length L of the shaft, we obtain the angle of twist for the entire shaft, namely,

$$\phi = \int_0^L \frac{T(x)dx}{J(x)G} \tag{11-14}$$

Here

  $\phi =$ the angle of twist of one end of the shaft with respect to the other end, measured in radians
$T(x) =$ the internal torque at the arbitrary position x, found from the method of sections and the equation of moment-equilibrium applied about the shaft's axis
$J(x) =$ the shaft's polar moment of inertia expressed as a function of position x
  $G =$ the shear modulus for the material

Constant Torque and Cross-Sectional Area. Usually in engineering practice the shaft's cross-sectional area and the applied torque are constant along the length of the shaft, Fig. 11–16. If this is the case, the internal torque $T(x) = T$, the polar moment of inertia $J(x) = J$, and Eq. 11–14 can be integrated, which gives

$$\phi = \frac{TL}{JG} \tag{11-15}$$

The similarities between the above two equations and those for an axially loaded bar ($\Delta = \int P(x)\,dx/A(x)E$ and $\Delta = PL/AE$) should be noted.

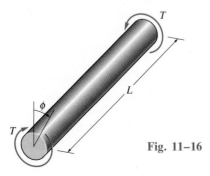

Fig. 11–16

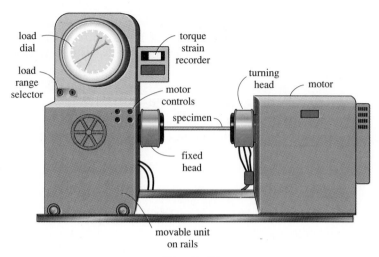

Fig. 11–17

In particular, Eq. 11–15 is often used to determine the shear modulus G. To do so a specimen of known length and diameter is placed in a torsion testing machine like the one shown in Fig. 11–17. The torque T and angle of twist ϕ are then measured between a gauge length L. Using Eq. 11–15, $G = TL/J\phi$. Usually, to obtain a more reliable value of G, several of these tests are performed and the average value used.

If the shaft is subjected to several different torques, or the cross-sectional area or shear modulus changes abruptly from one region of the shaft to the next, the above equation can be applied to each segment of the shaft where these quantities are all constant. The angle of twist of one end of the shaft with respect to the other is then found from the vector addition of the angles of twist of each segment. For this case,

$$\phi = \sum \frac{TL}{JG}$$

(11–16)

Sign Convention. In order to apply the above equations, we must develop a sign convention for the internal torque and the angle of twist of one end of the shaft with respect to the other end. To do this we will use the right-hand rule, whereby both the torque and angle will be *positive* at the location where they are to be calculated, provided that the *thumb* is directed *outward* from the shaft when the fingers curl to give the tendency for rotation, Fig. 11–18.

To illustrate the use of this sign convention, consider the shaft shown in Fig. 11–19a, which is subjected to four torques. The angle of twist of end A with respect to end D is to be determined. For this problem, three segments of the shaft must be considered, since the internal torque changes at B and C. Using the method of sections, the internal torques are computed for each

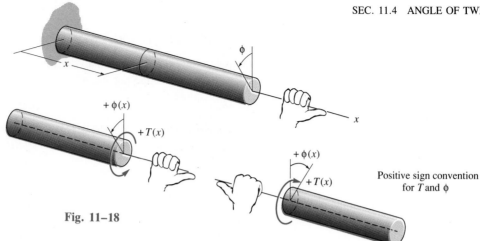

Fig. 11–18

Positive sign convention
for T and ϕ

segment, Fig. 11–19*b*. By the right-hand rule, with positive torques directed away from the *sectioned end* of the shaft, we have $T_{AB} = +80$ N · m, $T_{BC} = -70$ N · m, and $T_{CD} = -10$ N · m. Applying Eq. 11–16, we have

$$\phi_{A/D} = \frac{(+80 \text{ N} \cdot \text{m}) \, L_{AB}}{JG} + \frac{(-70 \text{ N} \cdot \text{m}) \, L_{BC}}{JG} + \frac{(-10 \text{ N} \cdot \text{m}) \, L_{CD}}{JG}$$

If the other data is substituted and the answer is computed as a *positive* quantity, it means that end *A* will rotate as indicated by the curl of the right-hand fingers when the thumb is directed *away* from the shaft, Fig. 11–19*a*. The double subscript notation is used to indicate this relative angle of twist ($\phi_{A/D}$); however, if the angle of twist is to be determined relative to a *fixed point*, then only a single subscript is used. For example, if *D* is located at a fixed support, then the computed angle of twist will be denoted as ϕ_A.

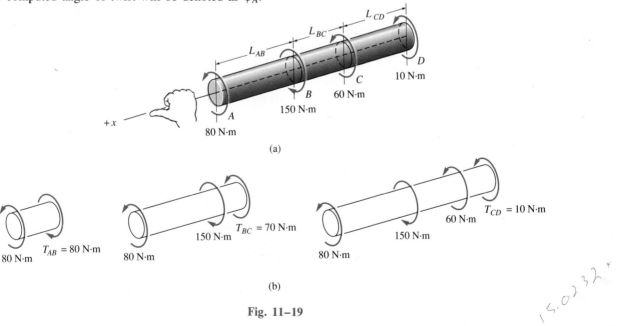

(a)

(b)

Fig. 11–19

PROCEDURE FOR ANALYSIS

The angle of twist of one end of a shaft or tube with respect to the other end can be determined by applying the above equations. The following is a suggested procedure for doing this.

Internal Torque. The internal torque $T(x)$ can be found at a point on the axis of the shaft by using the method of sections and the equation of moment equilibrium, applied along the shaft's axis. If the torque varies along the shaft's length, a section should be made at the arbitrary position x along the shaft and the torque represented as a function of x. If several constant external torques act on the shaft between its ends, the internal torque in each *segment* of the shaft, between any two external torques, must be determined.

Angle of Twist. When the circular cross-sectional area varies along the shaft's axis, the polar moment of inertia $J(x)$ must be expressed as a function of its position x along the axis. Furthermore, if the polar moment of inertia or the internal torque *suddenly changes* between the ends of the shaft, then Eq. 11–14, $\phi = \int (T(x)/(J(x)G))\,dx$, or Eq. 11–15, $\phi = TL/JG$, must be applied to *each segment* for which J, G, and T are continuous or constant. When substituting the data for the internal torque in each segment, be sure to use a consistent sign convention for the shaft, such as the one discussed above. Also make sure that a consistent set of units is used when substituting numerical data into the equations.

The following examples illustrate application of this procedure.

Example 11–7

The gears attached to the fixed-end steel shaft are subjected to the torques shown in Fig. 11–20a. If the shear modulus of elasticity is $G = 80$ GPa and the shaft has a diameter of 14 mm, determine the displacement of the tooth P on gear A. The shaft turns freely within the bearing at B.

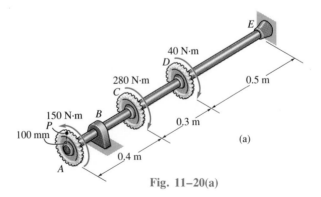

Fig. 11–20(a)

Example 11–7 (*Continued*)

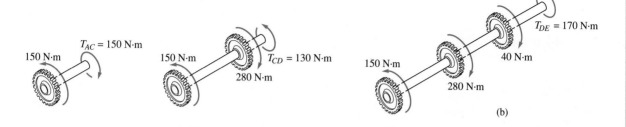

(b)

SOLUTION

Internal Torque. By inspection, the torques in segments AC, CD, and DE are different yet *constant* throughout each segment. Free-body diagrams of appropriate segments of the shaft along with the calculated internal torques are shown in Fig. 11–20*b*. Using the right-hand rule and the established sign convention that positive torque is directed away from the sectioned end of the shaft, we have

$$T_{AC} = +150 \text{ N} \cdot \text{m} \qquad T_{CD} = -130 \text{ N} \cdot \text{m} \qquad T_{DE} = -170 \text{ N} \cdot \text{m}$$

Angle of Twist. The polar moment of inertia for the shaft is

$$J = \frac{\pi}{2}(0.007 \text{ m})^4 = 3.77(10^{-9}) \text{ m}^4$$

Applying Eq. 11–16 to each segment and adding the results algebraically, we have

$$\phi_A = \sum \frac{TL}{JG} = \frac{(+150 \text{ N} \cdot \text{m})(0.4 \text{ m})}{3.77(10^{-9}) \text{ m}^4 [80(10^9) \text{ N/m}^2]}$$
$$+ \frac{(-130 \text{ N} \cdot \text{m})(0.3 \text{ m})}{3.77(10^{-9}) \text{ m}^4 [80(10^9) \text{ N/m}^2]}$$
$$+ \frac{(-170 \text{ N} \cdot \text{m})(0.5 \text{ m})}{3.77(10^{-9}) \text{ m}^4 [80(10^9) \text{ N/m}^2]} = -0.212 \text{ rad}$$

Since the answer is negative, by the right-hand rule the thumb is directed *toward* the end E of the shaft, and therefore gear A will rotate as shown in Fig. 11–20*c*.

The displacement of tooth P on gear A is

$$s_P = \phi_A r = (0.212 \text{ rad})(100 \text{ mm}) = 21.2 \text{ mm} \qquad \textit{Ans.}$$

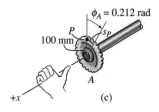

Fig. 11–20(b, c)

Remember that this analysis is valid only if the shear stress does not exceed the proportional limit of the material.

Example 11–8

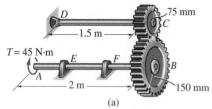

(a)

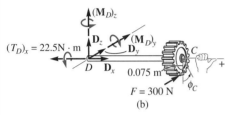

(b)

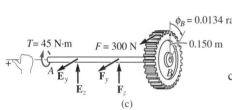

(c)

Fig. 11–21

The two solid steel shafts shown in Fig. 11–21a are coupled together using the meshed gears. Determine the angle of twist of end A of shaft AB when the torque $T = 45$ N · m is applied. Take $G = 80$ GPa. Shaft AB is free to rotate within bearings E and F, whereas shaft DC is fixed at D. Each shaft has a diameter of 20 mm.

SOLUTION

Since this problem involves two shafts, we must first calculate the angle of twist of each shaft separately, then add the results vectorially.

Internal Torque. Free-body diagrams for each shaft are shown in Fig. 11–21b and 11–21c. Summing moments along the x axis of shaft AB yields the tangential reaction between the gears of $F = 45$ N · m/0.15 m = 300 N. Summing moments about the x axis of shaft DC, this force then creates a torque of $(T_D)_x = 300$ N(0.075 m) = 22.5 N · m on shaft DC.

Angle of Twist. By the positive sign convention, the 22.5-N · m torque on shaft DC rotates gear C in the *positive direction;* i.e., by the right-hand rule the thumb is directed outward from the end of the shaft, Fig. 11–21b. This angle of twist is

$$\phi_C = \frac{TL_{DC}}{JG} = \frac{(+22.5 \text{ N} \cdot \text{m})(1.5 \text{ m})}{(\pi/2)(0.010 \text{ m})^4[80(10^9) \text{ N/m}^2]} = +0.0269 \text{ rad}$$

Since the gears at the end of the shaft are in *mesh,* the rotation ϕ_C causes gear B to rotate ϕ_B, Fig. 11–21c. Thus,

$$\phi_B(0.150 \text{ m}) = (0.0269 \text{ rad})(0.075 \text{ m})$$

$$\phi_B = 0.0134 \text{ rad}$$

Equation 11–15 may now be applied to determine the angle of twist of end A with respect to end B of shaft AB. The 45-N · m torque in the shaft is positive, since the thumb of the right hand is directed *away* from the end of the shaft at A. Therefore,

$$\phi_{A/B} = \frac{T_{AB}L_{AB}}{JG} = \frac{(+45 \text{ N} \cdot \text{m})(2 \text{ m})}{(\pi/2)(0.010 \text{ m})^4[80(10^9) \text{ N/m}^2]} = +0.0716 \text{ rad}$$

Here the positive sign indicates that the rotation, defined by the right-hand thumb at A, is directed to the left, Fig. 11–21c.

The rotation of end A is therefore determined by adding ϕ_B and $\phi_{A/B}$, since both angles are in the *same direction.* We have

$$\phi_A = \phi_B + \phi_{A/B} = 0.0134 \text{ rad} + 0.0716 \text{ rad} = +0.0850 \text{ rad} \quad \textbf{\textit{Ans.}}$$

This angle of twist is in the same direction as **T**, shown in Fig. 11–21c.

■ Example 11–9

The 2-in.-diameter cast-iron post shown in Fig. 11–22a is buried 24 in. in soil. If a torque is applied to its top using a rigid wrench, determine the maximum shear stress in the post and the angle of twist at its top. Assume that the torque is about to turn the post, and the soil exerts a uniform torsional resistance of t lb · in./in. along its 24-in. buried length. $G = 5.5(10^3)$ ksi.

SOLUTION

Internal Torque. The internal torque in segment AB of the post is constant. From the free-body diagram, Fig. 11–22b, we have

$$\Sigma M_z = 0; \qquad T_{AB} = 25 \text{ lb}(12 \text{ in.}) = 300 \text{ lb} \cdot \text{in.}$$

The magnitude of the uniform distribution of torque along the buried segment BC can be determined from equilibrium of the entire post, Fig. 11–22c. Here

$$\Sigma M_z = 0; \qquad 25 \text{ lb}(12 \text{ in.}) - t(24 \text{ in.}) = 0$$
$$t = 12.5 \text{ lb} \cdot \text{in./in.}$$

Hence, from a free-body diagram of a section of the post located at the position x within region BC, Fig. 11–22d, we have

$$\Sigma M_z = 0; \qquad T_{BC} - 12.5x = 0$$
$$T_{BC} = 12.5x$$

Maximum Shear Stress. The largest shear stress occurs in region AB, since the torque is largest there and J is constant for the post. Applying the torsion formula, we have

$$\tau_{max} = \frac{T_{AB}c}{J} = \frac{(300 \text{ lb} \cdot \text{in.})(1 \text{ in.})}{(\pi/2)(1 \text{ in.})^4} = 191 \text{ psi} \qquad Ans.$$

Angle of Twist. The angle of twist at the top can be determined relative to the bottom of the post, since it is fixed and yet is about to turn. Both segments AB and BC twist, and so in this case we have

$$\phi_A = \frac{T_{AB}L_{AB}}{JG} + \int_0^{L_{BC}} \frac{T_{BC} \, dx}{JG}$$

$$= \frac{(300 \text{ lb} \cdot \text{in.})(36 \text{ in.})}{JG} + \int_0^{24 \text{ in.}} \frac{12.5x \, dx}{JG}$$

$$= \frac{10{,}800 \text{ lb} \cdot \text{in}^2}{JG} + \frac{12.5[(24)^2/2]\text{lb} \cdot \text{in}^2}{JG}$$

$$= \frac{14{,}400 \text{ lb} \cdot \text{in}^2}{(\pi/2)(1 \text{ in.})^4 5500(10^3) \text{ lb/in}^2} = 0.00167 \text{ rad} \qquad Ans.$$

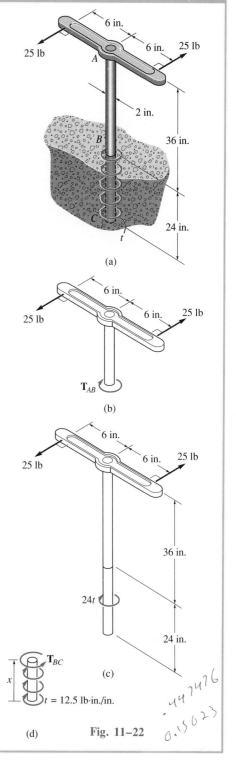

(a)

(b)

(c)

(d)

Fig. 11–22

489

Example 11–10

The tapered shaft shown in Fig. 11–23a is made of a material having a shear modulus G. Determine the angle of twist of its end B when subjected to the torque $\mathbf{T}$.

SOLUTION

Internal Torque. By inspection or from the free-body diagram of a section located at the arbitrary position x, Fig. 11–23b, the internal torque is T.

Angle of Twist. Here the polar moment of inertia varies along the shaft's axis and therefore we must express it in terms of the coordinate x. The radius c of the shaft at x can be determined in terms of x by proportion of the slope of line AB in Fig. 11–23c. We have

$$\frac{c_2 - c_1}{L} = \frac{c_2 - c}{x}$$

$$c = c_2 - x\left(\frac{c_2 - c_1}{L}\right)$$

Thus, at x,

$$J(x) = \frac{\pi}{2}\left[c_2 - x\left(\frac{c_2 - c_1}{L}\right)\right]^4$$

Applying Eq. 11–14, we have

$$\phi = \int_0^L \frac{T\,dx}{\left(\dfrac{\pi}{2}\right)\left[c_2 - x\left(\dfrac{c_2 - c_1}{L}\right)\right]^4 G} = \frac{2T}{\pi G}\int_0^L \frac{dx}{\left[c_2 - x\left(\dfrac{c_2 - c_1}{L}\right)\right]^4}$$

Performing the integration using an integral table, the result becomes

$$\phi = \left(\frac{2T}{\pi G}\right)\frac{1}{3\left(\dfrac{c_2 - c_1}{L}\right)\left[c_2 - x\left(\dfrac{c_2 - c_1}{L}\right)\right]^3}\Bigg|_0^L$$

$$= \frac{2T}{\pi G}\left(\frac{L}{3(c_2 - c_1)}\right)\left(\frac{1}{c_1^3} - \frac{1}{c_2^3}\right)$$

Rearranging terms yields

$$\phi = \frac{2TL}{3\pi G}\left(\frac{c_2^2 + c_1 c_2 + c_1^2}{c_1^3 c_2^3}\right) \qquad\qquad Ans.$$

To partially check this result, note that when $c_1 = c_2 = c$, then

$$\phi = \frac{TL}{[(\pi/2)c^4]G} = \frac{TL}{JG}$$

which is Eq. 11–15.

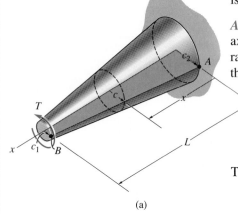

(a)

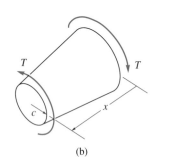

(b)

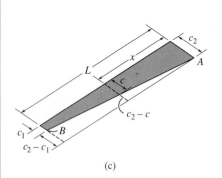

(c)

Fig. 11–23

PROBLEMS

11–25. The 0.75-in.-diameter steel shaft is supported by bearings at A and B. If it is subjected to torques of 35 lb · ft applied to each gear, determine in degrees the angle of twist of end C of the shaft with respect to end B. $G_{st} = 10.8(10^3)$ ksi.

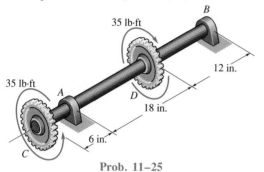

Prob. 11–25

11–26. The aluminum shaft consists of a solid section AB and a tube BC, which is fixed to the wall at C. Determine the angle of twist at its end A when it is subjected to the torsional loading shown. Take $G_{al} = 3.80(10^3)$ ksi. Section AB has a diameter of 1.5 in., and BC has an inner diameter of 1.5 in. and an outer diameter of 2.5 in.

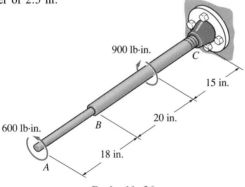

Prob. 11–26

11–27. The solid shaft of radius c is subjected to a torque T at its ends. Show that the maximum shear strain developed in the shaft is $\gamma_{max} = Tc/JG$. What is the shear strain on an element located at point A, $c/2$ from the center of the shaft? Sketch the strain distortion of this element.

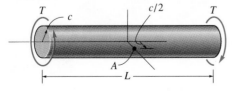

Prob. 11–27

*11–28.** The splined ends and gears attached to the steel shaft are subjected to the torques shown. Determine the angle of twist of end B with respect to end A. The shaft has a diameter of 40 mm and $G_{st} = 75$ GPa.

11–29. The splined ends and gears attached to the steel shaft are subjected to the torques shown. Determine the angle of twist of gear C with respect to gear D. The shaft has a diameter of 40 mm and $G_{st} = 75$ GPa.

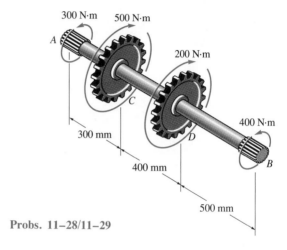

Probs. 11–28/11–29

11–30. The steel axle is made from tubes AB and CD and a solid section BC. It is supported on smooth bearings that allow it to rotate freely. If the gears, fixed to its ends, are subjected to 85-N · m torques, determine the angle of twist of gear A relative to gear D. The tubes have an outer diameter of 30 mm and an inner diameter of 20 mm. The solid section has a diameter of 40 mm. $G_{st} = 75$ GPa.

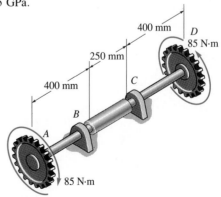

Prob. 11–30

11–31. The engine of the helicopter is delivering 600 hp to the rotor shaft AB when the blade is rotating at 1200 rev/min. Determine to the nearest $\frac{1}{8}$ in. the diameter of the shaft AB if the allowable shear stress is $\tau_{allow} = 8$ ksi and the vibrations limit the angle of twist of the shaft to 0.05 rad. The shaft is 2 ft long and made from steel. $G_{st} = 11.6(10^3)$ ksi.

*11–32. The engine of the helicopter is delivering 600 hp to the rotor shaft AB when the blade is rotating at 1200 rev/min. Determine to the nearest $\frac{1}{8}$ in. the diameter of the shaft AB if the allowable shear stress is $\tau_{allow} = 10.5$ ksi and the vibrations limit the angle of twist of the shaft to 0.05 rad. The shaft is 2 ft long and made from steel. $G_{st} = 11.6(10^3)$ ksi.

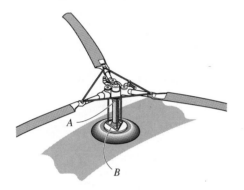

Probs. 11–31/11–32

11–33. The 8-mm-diameter steel bolt is screwed tightly into a block at A. Determine the couple forces F that should be applied to the wrench so that the maximum shear stress in the bolt becomes 180 MPa. Also, compute the corresponding displacement of each force F needed to cause this stress. Assume that the wrench is rigid. Take $G_{st} = 75$ GPa.

Prob. 11–33

11–34. The rotating flywheel-and-shaft, when brought to a sudden stop at D, begins to oscillate clockwise-counterclockwise such that a point A on the outer edge of the flywheel is displaced through a 6-mm arc. Determine the maximum shear stress developed in the tubular steel shaft due to this oscillation. The shaft has an inner diameter of 24 mm and an outer diameter of 32 mm. The bearings at B and C allow the shaft to rotate freely, whereas the support at D holds the shaft fixed. $G_{st} = 75$ GPa.

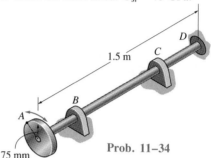

Prob. 11–34

11–35. A steel shaft is 2 m long and has an outer diameter of 40 mm. When it is rotating at 80 rad/s, it transmits 32 kW of power from the engine E to the generator G. Determine the smallest thickness of the shaft if the allowable shear stress is $\tau_{allow} = 140$ MPa and the shaft is restricted not to twist more than 0.05 rad. $G_{st} = 75$ GPa.

*11–36. A solid steel shaft is 3 m long and has a diameter of 50 mm. It is required to transmit 35 kW of power from the engine E to the generator G. Determine the smallest angular velocity the shaft can have if it is restricted not to twist more than 1°. $G_{st} = 75$ GPa.

11–37. The motor delivers 40 hp to the solid steel shaft while it rotates at 20 Hz. The shaft is supported on smooth bearings at A and B, which allow free rotation of the shaft. The gears C and D fixed to the shaft remove 25 hp and 15 hp, respectively. Determine the diameter of the shaft to the nearest $\frac{1}{8}$ in. if the allowable shear stress is $\tau_{allow} = 8$ ksi and the allowable angle of twist of C with respect to D is 0.20°. $G_{st} = 11.0(10^3)$ ksi.

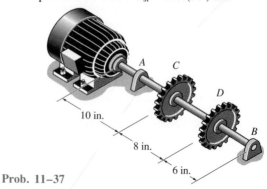

Prob. 11–37

11–38. The two steel shafts AC and FD are coupled together using the meshed-gear arrangement shown. If F is a fixed support, determine the angle of twist at end A due to the applied torsional loading. Each shaft has a diameter of 30 mm, and the bearings at A, B, and G resist no torque. $G_{st} = 80$ GPa.

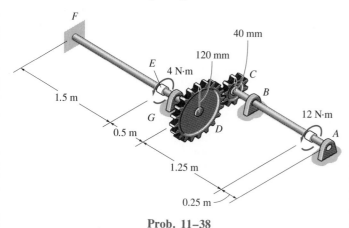

Prob. 11–38

11–39. The two steel shafts each have a diameter of 1 in. and are supported by bearings at A, B, and C, which allow free rotation. If the support at D is fixed, determine the angle of twist of end B when the torque of 60 ft · lb is applied to gear G. $G_{st} = 10.8(10^3)$ ksi.

*11–40. The two steel shafts each have a diameter of 1 in. and are supported by bearings at A, B, and C, which allow free rotation. If the support at D is fixed, determine the angle of twist of end A when the torque of 60 ft · lb is applied to gear G. $G_{st} = 10.8(10^3)$ ksi.

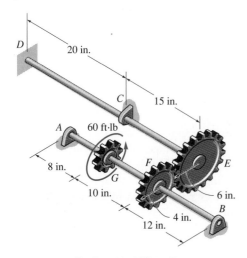

Probs. 11–39/11–40

11–41. The shaft has a radius c and is subjected to a torque per unit length of t_0, which is distributed uniformly over the shaft's entire length L. If it is fixed at its far end A, determine the angle of twist ϕ of end B.

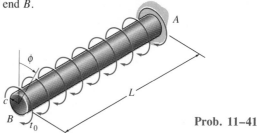

Prob. 11–41

11–42. When drilling a well, the deep end of the drill pipe is assumed to encounter a torsional resistance T_A. Furthermore, soil friction along the sides of the pipe creates a linear distribution of torque per unit length, varying from zero at the surface B to t_0 at A. Determine the necessary torque T_B that must be supplied by the drive unit to turn the pipe. Also, what is the relative angle of twist of one end of the pipe with respect to the other end at the instant the pipe is about to turn? The pipe has an outer radius r_o and an inner radius r_i. The shear modulus is G.

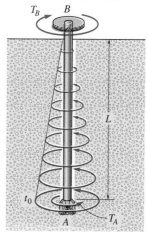

Prob. 11–42

11–43. The tapered shaft has a length L and a radius r at end A and $2r$ at end B. If it is fixed at end B and is subjected to a torque T, determine the angle of twist of end A. The shear modulus is G.

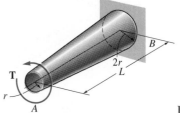

Prob. 11–43

11.5 Statically Indeterminate Torque-Loaded Members

A torsionally loaded shaft may be classified as statically indeterminate if the moment equation of equilibrium, applied about the axis of the shaft, is not adequate to determine the unknown torques acting on the shaft. An example of this situation is shown in Fig. 11–24a. As shown on the free-body diagram, Fig. 11–24b, the reactive torques at the supports A and B are unknown. We require

$$\Sigma M_x = 0; \qquad\qquad T - T_A - T_B = 0$$

Since only one equilibrium equation is relevant and there are two unknowns, this problem is statically indeterminate. In order to obtain a solution we will use the method of analysis discussed in Sec. 10.4.

The necessary condition of compatibility, or the kinematic condition, requires the angle of twist of one end of the shaft with respect to the other end to be equal to zero, since the end supports are fixed. Therefore,

$$\phi_{A/B} = 0$$

In order to write this equation in terms of the unknown torques, we will assume that the material behaves in a linear-elastic manner, so that the load–displacement relationship is expressed by Eq. 11–16. Realizing that the internal torque in segment AC is $+T_A$, and in segment CB the internal torque is $-T_B$, the above compatibility equation can be written as

$$\frac{T_A L_{AC}}{JG} - \frac{T_B L_{BC}}{JG} = 0$$

Here JG is assumed to be constant.

Solving the above two equations for the reactions, realizing that $L = L_{AC} + L_{BC}$, we get

$$T_A = T\left(\frac{L_{BC}}{L}\right)$$

and

$$T_B = T\left(\frac{L_{AC}}{L}\right)$$

To summarize the above procedure, the unknown torques in statically indeterminate problems must be determined by satisfying both equilibrium and compatibility for the shaft. To do so it is necessary to transform the compatibility equation, which involves rotation, into an equation that involves torque. This is done using the load–displacement relation that describes the material behavior.

Superposition of Torques. It is also possible to write the equation of compatibility using the superposition of torques. For example, we will choose the support at B to be redundant. When this support is temporarily removed, the shaft, being fixed at A, becomes statically determinate and stable. Using the principle of superposition, the external torque $\mathbf{T}$ is first applied to the shaft, Fig. 11–24d. This causes the end B of the shaft to undergo an angle of twist ϕ_B. The redundant torque $\mathbf{T}_B$ is then applied to the unloaded shaft, which causes the end B of the shaft to twist $-\phi'_B$, Fig. 11–24e. Since the redundant actually holds the shaft fixed at B, Fig. 11–24c, the compatibility equation at B becomes

$$0 = \phi_B - \phi'_B$$

Applying the load–displacement relationship, we find $\phi_B = TL_{AC}/JG$ and $\phi'_B = T_B L/JG$. Consequently,

$$0 = \frac{TL_{AC}}{JG} - \frac{T_B L}{JG}$$

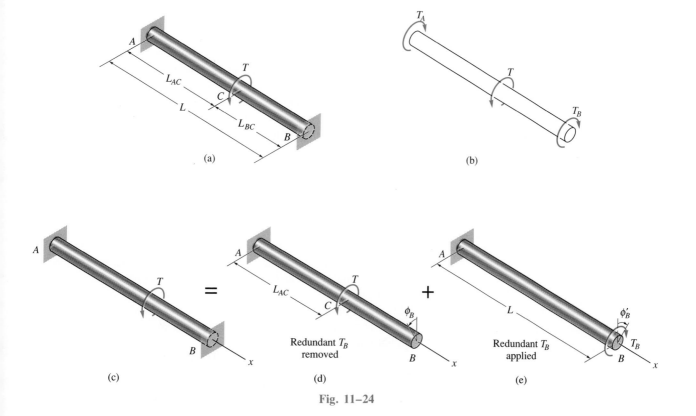

Fig. 11–24

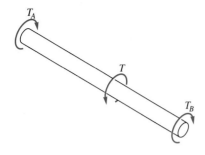

Fig. 11–24(b)

or

$$T_B = T\left(\frac{L_{AC}}{L}\right)$$

From the free-body diagram of the shaft, Fig. 11–24b, the torque at A can now be determined using the moment equation of equilibrium,

$$\Sigma M_x = 0; \qquad\qquad T - T_A - T_B = 0$$

$$T_A = T\left(\frac{L - L_{AC}}{L}\right)$$

Since $L_{BC} = L - L_{AC}$, we have

$$T_A = T\left(\frac{L_{BC}}{L}\right)$$

These results are the same as those obtained previously, except that here we first applied the compatibility condition and then the equilibrium condition to obtain the solution. Also note that the principle of superposition can indeed be used here, since the angles of twist and torque are linearly related, i.e., $\phi = TL/JG$.

PROCEDURE FOR ANALYSIS

Either of the above two methods can be used to determine the unknown torques in a shaft that is statically indeterminate. Each method requires that the conditions of equilibrium and compatibility be satisfied.

In order to solve a particular problem, it is important first to draw a free-body diagram of the shaft to identify all the torques. The sequence for applying the necessary equilibrium and compatibility equations is arbitrary. If it seems difficult to establish the compatibility equation, it is suggested that a superposition of torques be used. This method requires choosing one of the supports as a redundant. This support is then temporarily removed. The known angle of twist at the support (which may be zero) is then expressed as the algebraic addition of the angle of twist at the support caused by the external torques acting on the shaft and the angle of twist at the support caused only by the redundant torque acting on the shaft. These two angles are then expressed in terms of the torques using the load–displacement relationship $\phi = TL/JG$. Once established, the compatibility equation can then be solved for the magnitude of the redundant torque. The other unknown torque is determined from the conditions for equilibrium.

If any of the unknown torque magnitudes has a negative numerical value once it has been determined, then the torque acts in the opposite sense of direction of that indicated on the free-body diagram.

The following examples numerically illustrate both methods of solution.

Example 11–11

The solid steel shaft shown in Fig. 11–25a has a diameter of 20 mm. If it is subjected to the two torques, determine the reactions at the fixed supports A and B. Take $G = 75$ GPa.

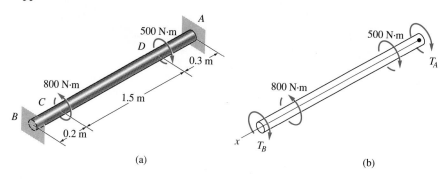

Fig. 11–25(a–c)

(a)

(b)

(c)

SOLUTION I

Equilibrium.   By inspection of the free-body diagram, Fig. 11–25b, it is seen that the problem is statically indeterminate since there is only *one* available equation of equilibrium, whereas $\mathbf{T}_A$ and $\mathbf{T}_B$ are unknown. We require

$$\Sigma M_x = 0; \qquad -T_B + 800 - 500 - T_A = 0 \qquad (1)$$

Compatibility.   Since the ends of the shaft are fixed, the angle of twist of one end of the shaft with respect to the other must be zero. Hence, the compatibility equation can be written as

$$\phi_{A/B} = 0$$

This condition can be expressed in terms of the unknown torques by using the load–displacement relationship, Eq. 11–15. Here there are three regions of the shaft where the internal torque is constant. On the free-body diagrams in Fig. 11–25c we have shown these internal torques acting on segments of the shaft.* By the sign conventions established in Sec. 11.4, we have

$$\frac{-T_B(0.2 \text{ m})}{JG} + \frac{(T_A + 500 \text{ N} \cdot \text{m})(1.5 \text{ m})}{JG} + \frac{T_A(0.3 \text{ m})}{JG} = 0$$

or

$$1.8T_A - 0.2T_B = -750 \qquad (2)$$

Solving Eqs. 1 and 2 yields

$$T_A = -345 \text{ N} \cdot \text{m} \qquad T_B = 645 \text{ N} \cdot \text{m} \qquad \textit{Ans.}$$

The negative sign indicates that $\mathbf{T}_A$ acts in the opposite direction of that shown in Fig. 11–25b.

*Alternatively, we can use internal loadings of $(T_A - 300)$, $(800 - T_B)$, and $(-T_B + 300)$.

Example 11–11 (Continued)

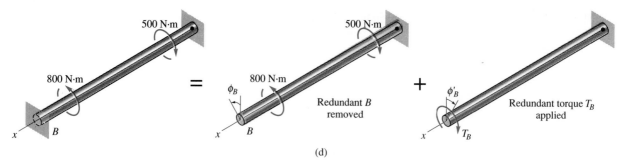

(d)

Fig. 11–25(d, e)

SOLUTION II

We can also solve this problem using the method of superposition, with the support at B considered as the redundant.

Compatibility. Figure 11–25d shows the shaft loaded with the external torques, causing end B to rotate ϕ_B, plus the shaft loaded only with the redundant torque $\mathbf{T}_B$, which causes a rotation of ϕ'_B. We require

$$0 = \phi_B + \phi'_B \tag{3}$$

The internal torque in each section of the shaft used for computing ϕ_B is shown on the segments in Fig. 11–25e. Thus,

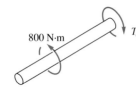

$$\phi_B = 0 + \frac{(+800 \text{ N} \cdot \text{m})(1.5 \text{ m})}{JG} + \frac{(+300 \text{ N} \cdot \text{m})(0.3 \text{ m})}{JG}$$

$$= \frac{1290}{JG}$$

For the shaft subjected only to $\mathbf{T}_B$, Fig. 11–25d, we have

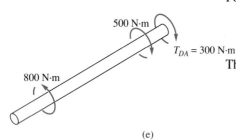

$$\phi'_B = \frac{-T_B \,(2.0 \text{ m})}{JG}$$

Therefore, Eq. 3 becomes

$$0 = \frac{1290}{JG} - \frac{T_B \,(2.0)}{JG}$$

$$T_B = 645 \text{ N} \cdot \text{m} \qquad\qquad Ans.$$

Equilibrium. Using this result and applying moment equilibrium to the shaft in Fig. 11–25b, we obtain the previous result for T_A, i.e.,

$$\Sigma M_x = 0; \quad -645 \text{ N} \cdot \text{m} + 800 \text{ N} \cdot \text{m} - 500 \text{ N} \cdot \text{m} - T_A = 0$$

$$T_A = -345 \text{ N} \cdot \text{m} \qquad\qquad Ans.$$

Example 11–12

The shaft shown in Fig. 11–26a is made from a steel tube and brass core. If a torque of $T = 250$ lb · ft is applied at its end, plot the shear-stress distribution along a radial line of its cross-sectional area. Take $G_{st} = 11.4(10^3)$ ksi, $G_{br} = 5.20(10^3)$ ksi.

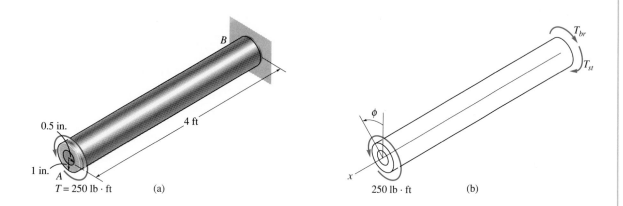

Fig. 11–26(a, b)

SOLUTION

Equilibrium. A free-body diagram of the shaft is shown in Fig. 11–26b. The reaction at the wall has been represented by the unknown amount of torque resisted by the steel, $\mathbf{T}_{st}$, and by the brass, $\mathbf{T}_{br}$. Working in units of pounds and inches, equilibrium requires

$$-T_{st} - T_{br} + 250 \text{ lb} \cdot \text{ft}(12 \text{ in./ft}) = 0 \qquad (1)$$

Compatibility. We require the angle of twist of end A to be the *same* for both the steel and brass. Thus,

$$\phi = \phi_{st} = \phi_{br}$$

Applying the load–displacement relationship, Eq. 11–15, we have

$$\frac{T_{st}L}{(\pi/2)[(1 \text{ in.})^4 - (0.5 \text{ in.})^4]11.4(10^3) \text{ kip/in}^2}$$

$$= \frac{T_{br}L}{(\pi/2)(0.5 \text{ in.})^4 5.20(10^3) \text{ kip/in}^2}$$

$$T_{st} = 32.88T_{br} \qquad (2)$$

Example 11–12 (*Continued*)

Solving Eqs. 1 and 2, we get

$$T_{st} = 2911.0 \text{ lb} \cdot \text{in.} = 242.6 \text{ lb} \cdot \text{ft}$$
$$T_{br} = 88.5 \text{ lb} \cdot \text{in.} = 7.38 \text{ lb} \cdot \text{ft}$$

These torques act throughout the entire length of the shaft, since no external torques act at intermediate points along the shaft's axis. The shear stress in the brass core varies from zero at its center to a maximum at the interface where it contacts the steel tube. Using the torsion formula,

$$(\tau_{br})_{\text{max}} = \frac{(88.5 \text{ lb} \cdot \text{in.})(0.5 \text{ in.})}{(\pi/2)(0.5 \text{ in.})^4} = 451 \text{ psi}$$

For the steel, the minimum shear stress is also at this interface,

$$(\tau_{st})_{\text{min}} = \frac{(2911.0 \text{ lb} \cdot \text{in.})(0.5 \text{ in.})}{(\pi/2)[(1 \text{ in.})^4 - (0.5 \text{ in.})^4]} = 988 \text{ psi}$$

and the maximum shear stress is at the outer surface,

$$(\tau_{st})_{\text{max}} = \frac{(2911.0 \text{ lb} \cdot \text{in.})(1 \text{ in.})}{(\pi/2)[(1 \text{ in.})^4 - (0.5 \text{ in.})^4]} = 1977 \text{ psi}$$

The results are plotted in Fig. 11–26c. Note the discontinuity of *shear stress* at the brass and steel interface. This is to be expected, since the materials have different moduli of rigidity; i.e., steel is stiffer than brass ($G_{st} > G_{br}$) and thus it carries more shear stress at the interface. Although the shear stress is discontinuous here, the *shear strain* is not. Rather, the shear strain is the *same* for both the brass and the steel. This can be shown by using Hooke's law, $\gamma = \tau/G$. At the interface, Fig. 11–26d, the shear strain is

$$\gamma = \frac{\tau}{G} = \frac{451 \text{ psi}}{5,200 \text{ psi}} = \frac{988 \text{ psi}}{11,400 \text{ psi}} = 0.0867 \text{ rad}$$

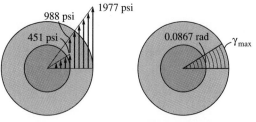

Shear stress distribution   Shear strain distribution

(c)   (d)

Fig. 11–26(c, d)

PROBLEMS

*11–44. The copper pipe has an outer diameter of 1.5 in. and a thickness of 0.125 in. The coupling on it at C is being tightened using a wrench. If the torque developed at A is 125 lb · in., determine the magnitude F of the couple forces. The pipe is fixed-supported at end B. $G_{cu} = 62(10^3)$ ksi.

11–45. The copper pipe has an outer diameter of 1.5 in. and a thickness of 0.125 in. The coupling on it at C is being tightened using a wrench. If the applied force $F = 30$ lb, determine the maximum shear stress developed in the pipe. The pipe is fixed-supported at end B. $G_{cu} = 62(10^3)$ ksi.

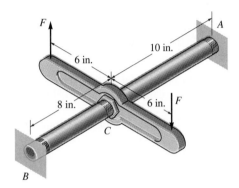

Probs. 11–44/11–45

11–46. A rod is made from two segments: AB is steel and BC is brass. It is fixed at its ends and subjected to a torque of $T = 680$ N · m. If the steel portion has a diameter of 30 mm, determine the required diameter of the brass portion so the reactions at the walls will be the same. $G_{st} = 75$ GPa, $G_{br} = 39$ GPa.

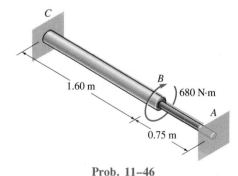

Prob. 11–46

11–47. The composite shaft is made from two segments, one having a diameter $2c$ and the other $4c$. Determine the position x of the applied torque T so that the reactive torques at A and B are equal. The material has a shear modulus G.

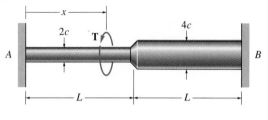

Prob. 11–47

*11–48. The steel shaft is made from two segments: Segment AC has a diameter of 0.5 in. and CB has a diameter of 1 in. If it is fixed at its ends A and B and subjected to a torque of 500 lb · ft, determine the maximum shear stress in the shaft. $G_{st} = 10.8(10^3)$ ksi.

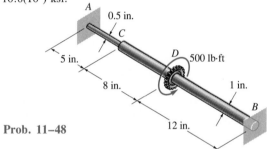

Prob. 11–48

11–49. The shaft is made from a solid steel section AB and a tubular portion made of steel and having a brass core. If it is fixed to a rigid support at A, and a torque of $T = 50$ lb · ft is applied to it at C, determine the angle of twist that occurs at C and compute the maximum shear stress and shear strain in the brass and steel. Take $G_{st} = 11.5(10^3)$ ksi, $G_{br} = 5.6(10^3)$ ksi.

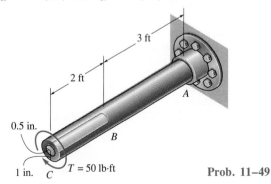

Prob. 11–49

11–50. The aluminum tube has an inner diameter of 35 mm and an outer diameter of 60 mm. A torque of $T = 850$ N · m is applied to its end B. Determine the depth d to which it has to be filled with a steel plug so that the rotation of end B is limited to 0.015 rad. The tube is fixed to the wall at A. $G_{st} = 80$ GPa, $G_{al} = 30$ GPa.

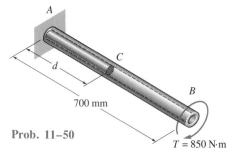

Prob. 11–50

11–51. The two steel shafts each have a diameter of 25 mm and are connected together using the gears fixed to their ends. Their other ends are attached to fixed supports at A and B. They are also supported by bearings at C and D, which allow free rotation of the shafts along their axes. If a torque of 500 N · m is applied to the top gear as shown, determine the reactions at A and B. $G_{st} = 80$ GPa.

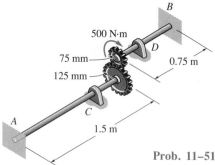

Prob. 11–51

*__11–52.__ The motor A develops a torque at gear B of 450 lb · ft, which is applied along the axis of the 2-in.-diameter steel shaft CD. This torque is to be transmitted to the pinion gears at E and F. If these gears are temporarily fixed, determine the maximum shear stress in segments CB and BD of the shaft. Also, what is the angle of twist of each of these segments? The bearings at C and D only exert force reactions on the shaft and do not resist torque. $G_{st} = 12(10^3)$ ksi.

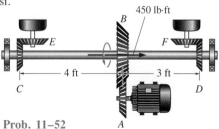

Prob. 11–52

11–53. The two 3-ft-long aluminum shafts each have a diameter of 1.5 in. and are connected together using the gears fixed to their ends. Their other ends are attached to fixed supports at A and B. They are also supported by bearings at C and D, which allow free rotation of the shafts along their axes. If a torque of 600 lb · ft is applied to the top gear as shown, determine the maximum shear stress in each shaft. $G_{al} = 10(10^3)$ ksi.

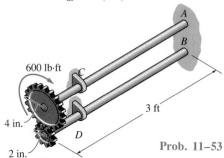

Prob. 11–53

11–54. A portion of the steel shaft is subjected to a constant distributed torsional loading of 200 lb · ft/ft. If the shaft has the dimensions shown, determine the reactions at the fixed supports A and C. Segment AB has a diameter of 1.5 in. and segment BC has a diameter of 0.75 in. $G_{st} = 11.3(10^3)$ ksi.

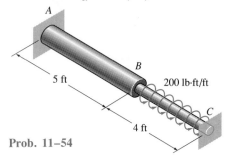

Prob. 11–54

11–55. The shaft of radius c is subjected to a distributed torque t, measured as torque/length of shaft. Determine the reactions at the fixed supports A and B.

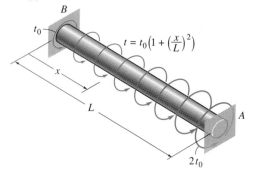

Prob. 11–55

11.6 Stress Concentration

The torsion formula, $\tau_{max} = Tc/J$, can be applied to regions of a shaft having a circular cross section that is constant or tapers slightly. When sudden changes arise in the cross section, both the shear-stress and shear-strain distributions in the shaft become complex and can be obtained only by using experimental methods or possibly by a mathematical analysis based on the theory of elasticity. Three common discontinuities of the cross section that occur in practice are shown in Fig. 11–27. They are at *couplings,* which are used to connect two collinear shafts together, Fig. 11–27a, *keyways,* used to connect gears or pulleys to a shaft, Fig. 11–27b, and *shoulder fillets,* used to fabricate a single collinear shaft from two shafts having different diameters, Fig. 11–27c. In each case the maximum shear stress will occur at the point (dot) indicated on the cross section.

In order to eliminate the necessity for the engineer to perform a complex stress analysis at a shaft discontinuity, the maximum shear stress can be determined for a specified geometry using a *torsional stress-concentration factor, K.* As in the case of axially loaded members, Sec. 10.6, K is usually taken from a graph. An example, for the shoulder-fillet shaft, is shown in Fig. 11–28.* To use this graph, one first computes the geometric ratio D/d to define the appropriate curve, and then once the abscissa r/d is calculated, the

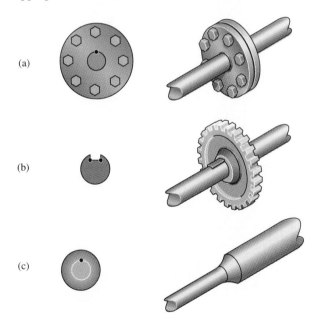

(a)

(b)

(c)

Fig. 11–27

*See Lipson, C. and Juvinall, R.C., *Handbook of Stress and Strength,* Macmillan, 1963.

value of K is found along the ordinate. The maximum shear stress is then determined from the equation

$$\tau_{max} = K\,\frac{Tc}{J} \qquad (11\text{--}17)$$

Here the torsion formula is applied to the *smaller* of the two connected shafts, since τ_{max} occurs at the base of the fillet, Fig. 11–27c.

It can be noted from the graph in Fig. 11–28 that an *increase* in fillet radius r causes a *decrease* in K. Hence the maximum shear stress in the shaft can be reduced by *increasing* the fillet radius. Also, if the diameter of the larger shaft is reduced, the D/d ratio will be lower and so the value of K and therefore τ_{max} will be lower.

Like the case of axially loaded members, torsional stress concentration factors should *always* be used when designing shafts made from *brittle materials,* or when designing shafts that will be subjected to *fatigue or cyclic torsional loadings*. These types of loadings give rise to the formation of cracks at the stress concentration, and this can often lead to a sudden failure of the shaft. Also realize that if a large *static* torsional loading is applied to a shaft made from *ductile material,* then *inelastic strains* may develop within the shaft. As a result of yielding, the stress distribution will become more *evenly distributed* throughout the shaft, so that the maximum stress that results will be limited at the stress concentration. This phenomenon will be discussed further in the next section.

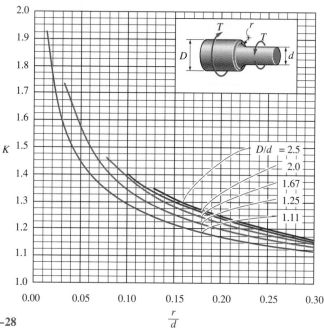

Fig. 11–28

11.7 Inelastic Torsion

The equations of stress and deformation developed thus far are valid only if the applied torque causes the material to behave in a linear-elastic manner. If the torsional loadings are excessive, however, the material may yield, and consequently a "plastic analysis" must then be used to determine the shear-stress distribution and the angle of twist. To perform this analysis, it is necessary to meet the conditions of both deformation and equilibrium for the shaft.

It was shown in Sec. 11.1 that the shear strains that develop in the material must vary *linearly* from zero at the center of the shaft to a maximum at its outer boundary, Fig. 11–29a. This conclusion was based entirely on geometric considerations and not the material's behavior. Also, the resultant torque at the section must be equivalent to the torque caused by the entire shear-stress distribution about the shaft's axis. This condition can be expressed mathematically by considering the shear stress τ acting on an element of area dA located a distance ρ from the center of the shaft, Fig. 11–29b. The force produced by this stress is $dF = \tau\, dA$, and the torque produced is $dT = \rho\, dF = \rho\tau\, dA$. For the entire shaft we require

$$T = \int_A \rho\tau\, dA \qquad (11\text{–}18)$$

If the area dA over which τ acts can be defined as a *differential ring* having an area of $dA = 2\pi\rho\, d\rho$, Fig. 11–29c, then the above equation can be written as

$$T = 2\pi \int_A \tau\rho^2\, d\rho \qquad (11\text{–}19)$$

These conditions of geometry and loading will now be used to determine the shear-stress distribution in a shaft when the shaft is subjected to three types of torque.

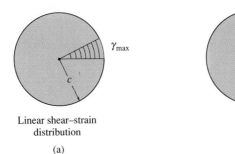

Linear shear–strain
distribution

(a)

(b)

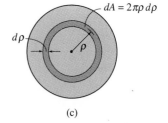

(c)

Fig. 11–29

Maximum Elastic Torque. If the torque produces the maximum *elastic* shear strain γ_Y, at the outer boundary of the shaft, then the shear-strain distribution along a radial line of the shaft will look like that shown in Fig. 11–30b. To establish the shear-stress distribution, we must either use Hooke's law or find the corresponding values of shear stress from the material's τ–γ diagram, Fig. 11–30a. For example, a shear strain γ_Y produces the shear stress τ_Y at $\rho = c$. Likewise, at $\rho = \rho_1$ in Fig. 11–30b, the shear strain is $\gamma_1 = (\rho_1/c)\gamma_Y$. From the τ–γ diagram γ_1 produces τ_1. When these stresses and others like them are plotted at $\rho = c$, $\rho = \rho_1$, etc., the expected *linear* shear-stress distribution in Fig. 11–30c results. Since this shear-stress distribution can be described mathematically as $\tau = \tau_Y(\rho/c)$, the maximum elastic torque can be determined from Eq. 11–19; i.e.,

$$T_Y = 2\pi \int_0^c \tau_Y \left(\frac{\rho}{c} \right) \rho^2 \, d\rho$$

or

$$T_Y = \frac{\pi}{2} \tau_Y c^3 \tag{11–20}$$

This same result can of course be obtained in a more direct manner using the torsion formula; i.e., $\tau_Y = T_Y c/[(\pi/2)c^4]$. Furthermore, the angle of twist can be determined from Eq. 11–13, namely,

$$d\phi = \gamma \frac{dx}{\rho} \tag{11–21}$$

As noted in Sec. 11.4, this equation results in $\phi = TL/JG$, when the shaft is subjected to a constant torque and has a constant cross-sectional area.

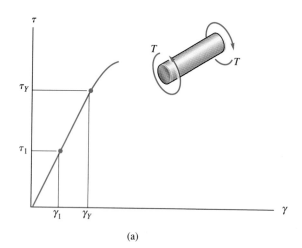

(a)

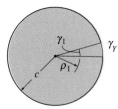

Shear–strain distribution

(b)

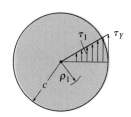

Shear–stress distribution

(c)

Fig. 11–30

Elastic-Plastic Torque.   Let us now consider the material in the shaft to exhibit an elastic-perfectly plastic behavior. As shown in Fig. 11–31a, this is characterized by a shear stress–strain diagram for which the material undergoes an increasing amount of shear strain when the shear stress in the material reaches the yield point τ_Y. Thus, as the applied torque increases in magnitude above T_Y, it will begin to cause yielding. First at the outer boundary of the shaft, $\rho = c$, and then, as the shear strain increases to, say, γ', the yielding boundary will progress inward toward the shaft's center, Fig. 11–31b. As shown, this produces an *elastic core,* where, by proportion, the outer radius of the core is $\rho_Y = (\gamma_Y/\gamma')c$. Also, the outer portion of the shaft forms a *plastic annulus* or ring, since the shear strains γ are greater than γ_Y within this region. The corresponding shear-stress distribution along a radial line of the shaft is shown in Fig. 11–31c. It was established by taking successive points on the shear-strain distribution, Fig. 11–31b, and finding the corresponding value of shear stress from the τ–γ diagram, Fig. 11–31a. For example, at $\rho = c$, γ' gives τ_Y, and at $\rho = \rho_Y$, γ_Y also gives τ_Y; etc.

Since τ can now be established as a function of ρ, we can apply Eq. 11–19 to determine the torque. As a general formula for elastic-plastic material behavior, we have

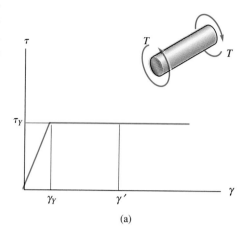

(a)

$$T = 2\pi \int_0^c \tau\rho^2 \, d\rho$$

$$= 2\pi \int_0^{\rho_Y} \left(\tau_Y \frac{\rho}{\rho_Y}\right)\rho^2 \, d\rho + 2\pi \int_{\rho_Y}^c \tau_Y \rho^2 \, d\rho$$

$$= \frac{2\pi}{\rho_Y}\tau_Y \int_0^{\rho_Y} \rho^3 \, d\rho + 2\pi\tau_Y \int_{\rho_Y}^c \rho^2 \, d\rho$$

$$= \frac{\pi}{2\rho_Y}\tau_Y\rho_Y^4 + \frac{2\pi}{3}\tau_Y(c^3 - \rho_Y^3)$$

$$= \frac{\pi\tau_Y}{6}(4c^3 - \rho_Y^3) \qquad\qquad (11\text{–}22)$$

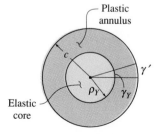

Shear–strain distribution

(b)

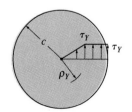

Shear–stress distribution

(c)

Fig. 11–31(a–c)

Plastic Torque. Further increases in T tend to shrink the radius of the elastic core until all the material will yield, i.e., $\rho_Y \to 0$. The material of the shaft is then subjected to *perfectly plastic behavior* and the shear-stress distribution is constant as shown in Fig. 11–31*d*. Since then $\tau = \tau_Y$, we can apply Eq. 11–19 to determine the *plastic torque* T_p, which represents the largest possible torque the shaft will support.

$$T_p = 2\pi \int_0^c \tau_Y \rho^2 \, d\rho$$

$$= \frac{2\pi}{3} \tau_Y c^3 \tag{11–23}$$

By comparison with the maximum elastic torque T_Y, Eq. 11–20, it can be seen that

$$T_p = \frac{4}{3} T_Y$$

In other words, the plastic torque is 33% greater than the maximum elastic torque.

The angle of twist ϕ for the shear-stress distribution in Fig. 11–31*d can-not* be uniquely defined. This is because $\tau = \tau_Y$ does not correspond to any unique value of shear strain $\gamma \geq \gamma_Y$. As a result, according to our "model" of behavior for the material, Fig. 11–31*a*, once $\mathbf{T}_p$ is applied, the shaft will continue to deform or twist with no corresponding increase in shear stress.

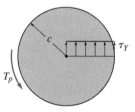

Fully plastic torque

(d)

Fig. 11–31(d)

Ultimate Torque. In the general case, most engineering materials will have a shear stress–strain diagram as shown in Fig. 11–32a. Consequently, if T is increased so that the maximum shear strain in the shaft becomes $\gamma = \gamma_u$, Fig. 11–32b, then by proportion γ_Y occurs at $\rho_Y = (\gamma_Y/\gamma_u)c$. Likewise, the shear strains at, say, $\rho = \rho_1$ and $\rho = \rho_2$ can be found by proportion, i.e., $\gamma_1 = (\rho_1/c)\gamma_u$ and $\gamma_2 = (\rho_2/c)\gamma_u$. If corresponding values of τ_1, τ_Y, τ_2, and τ_u are taken from the τ–γ diagram and plotted, we obtain the shear-stress distribution, which acts over a radial line on the cross section, Fig. 11–32c. The torque produced by this stress distribution is called the *ultimate torque, T_u,* since any further increase in shear strain would cause the maximum shear stress at the outer boundary of the shaft to be less than τ_u, and therefore the torque produced by the resulting shear-stress distribution would be *less* than T_u.

The magnitude of $\mathbf{T}_u$ can be determined by "graphically" integrating Eq. 11–19. To do this, the cross-sectional area of the shaft is segmented into a finite number of rings, such as the one shown shaded in Fig. 11–32d. The area of this ring, $\Delta A = 2\pi\rho\,\Delta\rho$, is multiplied by the shear stress τ that acts on it, so that the force $\Delta F = \tau\,\Delta A$ can be determined. The torque created by this force is then $\Delta T = \rho\,\Delta F = \rho(\tau\,\Delta A)$. The addition of all the torques for the entire cross section, as determined in this manner, gives the ultimate torque T_u; that is, Eq. 11–19 becomes $T_u \approx 2\pi\Sigma\tau\rho^2\,\Delta\rho$. On the other hand, if the stress distribution can be expressed as an analytical function, $\tau = f(\rho)$, as in the elastic and plastic torque cases, then the integration of Eq. 11–19 can be carried out directly.

The following examples further illustrate application of the above concepts.

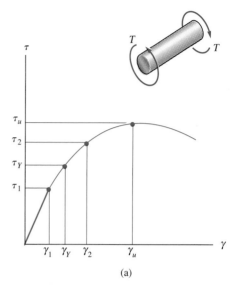

(a)

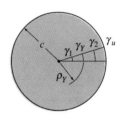

Ultimate shear–strain distribution

(b)

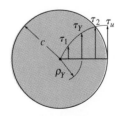

Ultimate shear–stress distribution

(c)

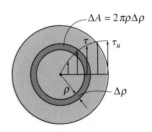

$\Delta A = 2\pi\rho\Delta\rho$

(d)

Fig. 11–32

Example 11–13

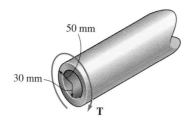

50 mm

30 mm

T

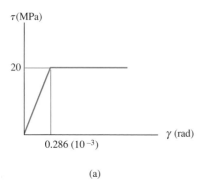

τ(MPa)

20

0.286 (10^{-3})

γ (rad)

(a)

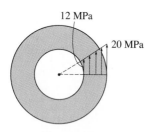

12 MPa

20 MPa

Elastic shear–stress distribution

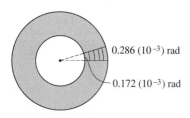

0.286 (10^{-3}) rad

0.172 (10^{-3}) rad

Elastic shear–strain distribution

(b)

Fig. 11–33

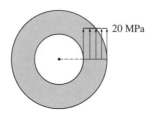

20 MPa

Plastic shear–stress distribution

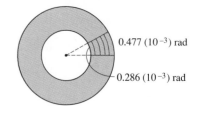

0.477 (10^{-3}) rad

0.286 (10^{-3}) rad

(c)

Initial plastic shear–strain distribution

The tubular shaft in Fig. 11–33a is made of an aluminum alloy that is assumed to have an elastic-plastic τ–γ diagram as shown. Determine (a) the maximum torque that can be applied to the shaft without causing material to yield, (b) the maximum torque or plastic torque that can be applied to the shaft. What should the minimum shear strain at the outer radius be in order to develop a plastic torque?

SOLUTION

Maximum Elastic Torque. We require the shear stress at the outer fiber to be 20 MPa. Using the torsion formula, we have

$$\tau_Y = \frac{T_Y c}{J}; \quad 20(10^6)\ \text{N/m}^2 = \frac{T_Y(0.05\ \text{m})}{(\pi/2)[(0.05\ \text{m})^4 - (0.03\ \text{m})^4]}$$

$$T_Y = 3.42\ \text{kN} \cdot \text{m} \qquad\qquad Ans.$$

The shear-stress and shear-strain distributions for this case are shown in Fig. 5–42b. The values at the tube's inner wall are obtained by proportion.

Plastic Torque. The shear-stress distribution in this case is shown in Fig. 11–33c. Application of Eq. 11–19 requires $\tau = \tau_Y$. We have

$$T_p = 2\pi \int_{0.03\ \text{m}}^{0.05\ \text{m}} [20(10^6)\ \text{N/m}^2]\rho^2\, d\rho = 125.66(10^6)\left.\frac{1}{3}\rho^3\right|_{0.03\ \text{m}}^{0.05\ \text{m}}$$

$$= 4.10\ \text{kN} \cdot \text{m} \qquad\qquad Ans.$$

For this tube T_p represents a 20% increase in torque capacity compared with the elastic torque T_Y.

Outer Radius Shear Strain. The tube becomes fully plastic when the shear strain at the *inner wall* becomes $0.286(10^{-3})$ rad as shown in Fig. 11–33c. Since the shear strain *remains linear* over the cross section, the plastic strain at the outer fibers of the tube in Fig. 11–33c is determined by proportion; i.e.,

$$\frac{\gamma_o}{50\ \text{mm}} = \frac{0.286(10^{-3})\ \text{rad}}{30\ \text{mm}}$$

$$\gamma_o = 0.477(10^{-3})\ \text{rad} \qquad\qquad Ans.$$

PROBLEMS

*11–56. The built-up shaft is to be designed to rotate at 650 rpm while transmitting 25 kW of power. If the allowable shear stress is $\tau_{\text{allow}} = 11.7$ MPa, determine the radius of the smallest fillet that can be used.

11–57. The built-up shaft is designed to rotate at 480 rpm. The diameter of the smaller shaft is $d = 60$ mm, the radius of the fillet weld connecting the shafts is $r = 7.20$ mm, and the allowable shear stress for the material is $\tau_{\text{allow}} = 55$ MPa. Determine the maximum power the shaft can transmit.

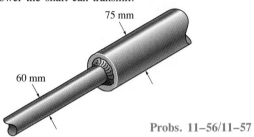

Probs. 11–56/11–57

11–58. The steel used for the shaft has an allowable shear stress of $\tau_{\text{allow}} = 8$ MPa. If the members are connected together with a fillet weld of radius $r = 2.25$ mm, determine the maximum torque T that can be applied.

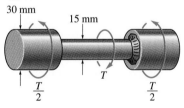

Prob. 11–58

11–59. A shaft of radius $c = 0.75$ in. is made from an elastic-plastic material as shown. Determine the torque T that must be applied to its ends so that it has an elastic core of radius $\rho = 0.6$ in. If the shaft is 30 in. long, determine the angle of twist.

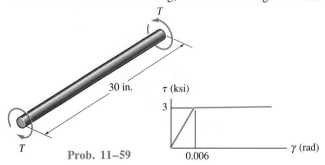

Prob. 11–59

*11–60. A bar having a circular cross section of 3 in. diameter is subjected to a torque of 100 in. · kip. If the material is elastic-plastic, with $\tau_Y = 16$ ksi, determine the radius of the elastic core.

11–61. The solid shaft is made from an elastic-plastic material as shown. Determine the torque T needed to form an elastic core in the shaft having a radius of $\rho_Y = 23$ mm. If the shaft is 2 m long, through what angle does one end of the shaft twist with respect to the other end? When the torque is removed, determine the residual stress distribution in the shaft and the permanent angle of twist.

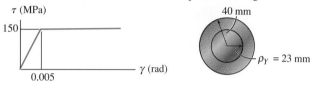

Prob. 11–61

11–62. A solid shaft is subjected to the torque T, which causes the material to yield. If the material is elastic-plastic, show that the torque can be expressed in terms of the angle of twist ϕ of the shaft as $T = \frac{4}{3}T_Y(1 - \phi_Y^3/4\phi^3)$, where T_Y and ϕ_Y are the torque and angle of twist when the material begins to yield.

11–63. A solid shaft has a diameter of 40 mm and length of 1 m. It is made from an elastic-plastic material having a yield stress of $\tau_Y = 100$ MPa. Determine the maximum elastic torque T_Y and the corresponding angle of twist. What is the angle of twist if the torque is increased to $T = 1.2T_Y$? $G = 80$ GPa.

*11–64. A tubular shaft has an inside diameter of 20 mm, an outside diameter of 40 mm, and a length of 1 m. It is made from an elastic-plastic material having a yield stress of $\tau_Y = 100$ MPa. Determine the maximum elastic torque T_Y and the corresponding angle of twist. What is the angle of twist if the torque is increased to $T = 1.2T_Y$? $G = 80$ GPa.

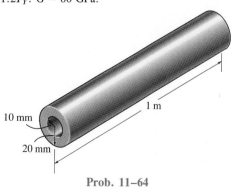

Prob. 11–64

11–65.   The 2-m-long tube is made from an elastic-plastic material as shown. Determine the applied torque T, which subjects the material at the tube's outer edge to a shearing strain of $\gamma_{max} = 0.008$ rad. What would be the permanent angle of twist of the tube when this torque is removed? Sketch the residual stress distribution in the tube.

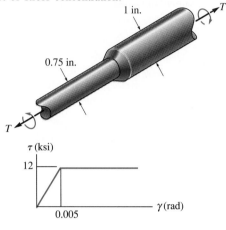

Prob. 11–65

11–66.   The tube has a length of 2 m and is made from an elastic-plastic material as shown. Determine the torque needed to just cause the material to become fully plastic. What is the permanent angle of twist of the tube when this torque is removed?

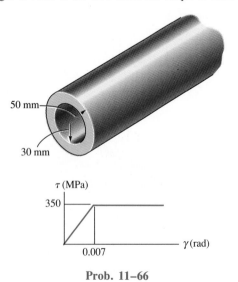

Prob. 11–66

11–67.   The shaft consists of two sections that are rigidly connected. If the material is elastic-plastic as shown, determine the largest torque T that can be applied to the shaft. Also, draw the shear-stress distribution over a radial line for each section. Neglect the effect of stress concentration.

Prob. 11–67

***11–68.**   The shear stress-strain diagram for a solid 50-mm-diameter shaft can be approximated as shown in the figure. Determine the torque required to cause a maximum shear stress in the shaft of 125 MPa. If the shaft is 3 m long, what is the corresponding angle of twist?

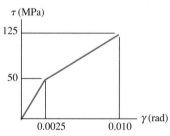

Prob. 11–68

11–69.   A torque is applied to the shaft of radius r. If the material obeys a shear stress–strain relation of $\tau = k\gamma^{1/6}$, where k is a constant, determine the maximum shear stress in the shaft.

Prob. 11–69

REVIEW PROBLEMS

11–70. The glass tube is confined within a rubber stopper, such that when the tube is twisted at constant angular velocity the stopper creates a *constant distribution* of frictional torque along the contacting length *AB* of the tube. If the tube has an inner diameter of 2 mm and an outer diameter of 4 mm, determine the shear stress developed at a point located at its inner and outer walls at a section through point *C*. Show the shear-stress distribution acting along a radial line segment at this section.

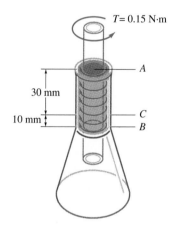

Prob. 11–70

11–71. The tube is subjected to a torque of 600 N · m. Determine the amount of this torque that is resisted by the shaded section. Solve the problem two ways: (*a*) by using the torsion formula; (*b*) by finding the resultant of the shear-stress distribution.

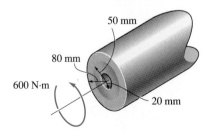

Prob. 11–71

*11–72. The motor of a fan delivers 150 W of power to the blade when it is turning at 18 rev/s. Determine the smallest diameter of shaft *A* that can be used to connect the fan blade to the motor if the allowable shear stress is $\tau_{allow} = 80$ MPa.

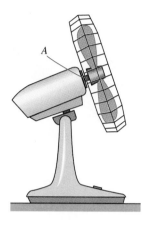

Prob. 11–72

11–73. A shaft is subjected to a torque **T**. Compare the effectiveness of using the tube shown in the figure with that of a solid section of radius *c*. To do this, compute the percent increase in torsional stress and angle of twist per unit length for the tube versus the solid section.

Prob. 11–73

11–74. The solid shaft of radius r is subjected to a torque T. Determine the radius r' of the inner core of the shaft that resists one-half of the applied torque $(T/2)$. Solve the problem two ways: (a) by using the torsion formula, (b) by finding the resultant of the shear-stress distribution.

11–75. The solid shaft of radius r is subjected to a torque T. Determine the radius r' of the inner core of the shaft that resists one-quarter of the applied torque $(T/4)$. Solve the problem two ways: (a) by using the torsion formula, (b) by finding the resultant of the shear-stress distribution.

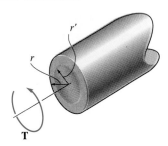

Probs. 11–74/11–75

*11–76.** The pipe is stuck in the ground so that when it is pulled upward the frictional force along its length varies linearly from zero at B to f_{max} (force/length) at C. Determine the initial force P required to pull the pipe out and the pipe's associated elongation just before it starts to slip. The pipe has a length L, cross-sectional area A, and the material from which it is made has a modulus of elasticity E.

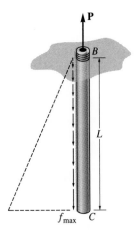

Prob. 11–76

11–77. The contour of the surface of the shaft is defined by the equation $y = e^{ax}$, where a is a constant. If the shaft is subjected to a torque T at its ends, determine the angle of twist of end A with respect to end B. The shear modulus is G.

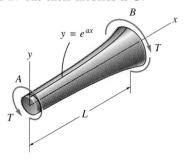

Prob. 11–77

11–78. A cylindrical spring consists of a rubber annulus bonded to a rigid ring and shaft. If the ring is held fixed and a torque T is applied to the shaft, determine the maximum shear stress in the rubber.

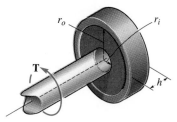

Pro. 11–78

11–79. The shaft is made from a strain-hardening material having a τ-γ diagram as shown. Determine the torque T that must be applied to the shaft in order to create an elastic core in the shaft having a radius of $\rho_c = 0.5$ in.

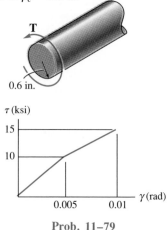

Prob. 11–79

12 Bending

In this chapter we will determine the stress in a member that is subjected to bending. We will begin by assuming the member is straight, has a symmetric cross section, and is made of homogeneous linear-elastic material. Afterwards, we will discuss special cases involving unsymmetric bending and inelastic bending.

12.1 Bending Deformation of a Straight Member

Members that are slender and support loadings that are applied perpendicular to their longitudinal axis are called *beams*. In this section we will discuss the deformations that occur when a straight prismatic beam, made of a homogeneous material, is subjected to bending. The discussion will be limited to beams having a cross-sectional area that is symmetrical with respect to an axis, and the bending moment is applied about an axis perpendicular to this axis of symmetry as shown in Fig. 12–1. The behavior of members that have unsymmetrical cross sections, or are made from several different materials, is based on similar observations and will be discussed separately in later sections of this chapter.

By using a highly deformable material such as rubber, we can physically illustrate what happens when a straight prismatic member is subjected to a bending moment. Consider, for example, the undeformed bar in Fig. 12–2a, which has a square cross section and is marked with longitudinal and transverse grid lines. When a bending moment is applied, it tends to distort these lines into the pattern shown in Fig. 12–2b. Here it can be seen that the

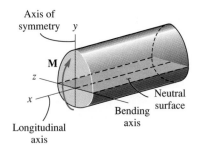

Axis of
symmetry *y*

M

z

x

Neutral
surface

Bending
axis

Longitudinal
axis

Fig. 12–1

longitudinal lines become *curved* and the vertical transverse lines *remain straight* and yet undergo a *rotation*.

The behavior of the deformable bar in Fig. 12–2 is the same as that exhibited by the member of arbitrary cross section shown in Fig. 12–1. In both cases, the bending moment causes the material within the bottom portion of the member to stretch and the material within the top portion to compress. Consequently, between these two regions there must be a surface, called the *neutral surface*, in which longitudinal fibers of the material will not undergo a change in length. This surface is located in the *x–z* plane as shown in Fig. 12–1.

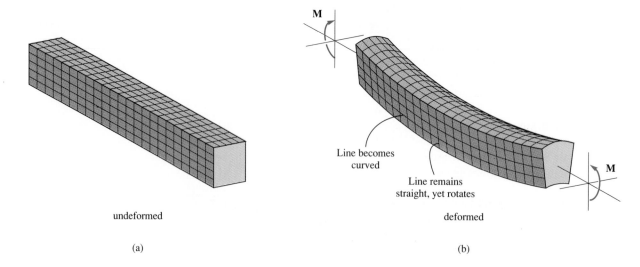

M

Line becomes
curved

Line remains
straight, yet rotates

M

undeformed

deformed

(a)

(b)

Fig. 12–2

From these observations we will make the following three assumptions regarding the way the stress deforms the material. First, the *longitudinal axis* x, which lies within the neutral surface, does *not* experience any *change in length*. Rather the moment will tend to deform the beam so that this line *becomes a curve* that lies in the x–y plane of symmetry, Fig. 12–3b. Second, all *cross sections* of the beam *remain plane* and perpendicular to the longitudinal axis during the deformation. And third, any *deformation* of the *cross section* within its own plane, as noticed in Fig. 12–2b, will be *neglected*. Thus, if the beam is fixed at one end as shown in Fig. 12–3a, and a moment is applied to its other end, the neutral surface and longitudinal x axis will deform into a curve, while the cross section remains plane, Fig. 12–3b. In particular, the z axis, lying in the plane of the cross section and about which the cross section rotates, is called the *neutral axis*. Its location will be determined in the next section.

In order to show how this distortion will strain the material in the longitudinal direction, we will isolate a segment of the beam that is located a distance x along the beam's length and has an undeformed thickness Δx, Fig. 12–3. This element, taken from the beam in Fig. 12–3, is shown in profile view in

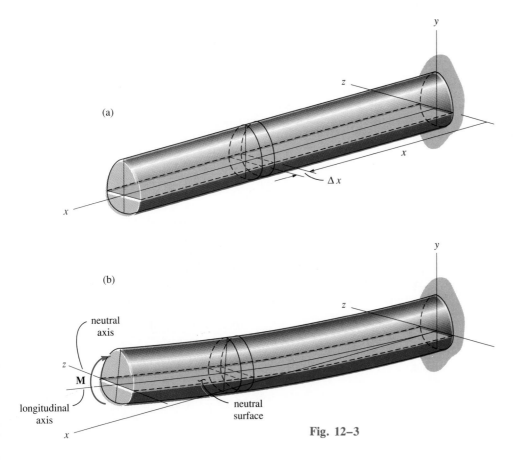

(a)

(b)

neutral axis

z

M

longitudinal axis

x

neutral surface

Fig. 12–3

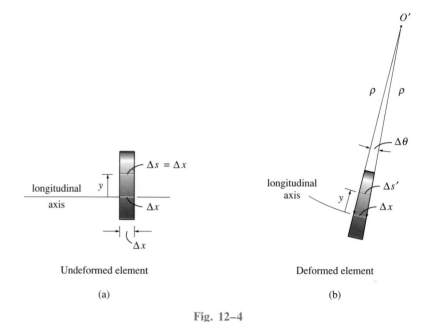

longitudinal axis

$\Delta s = \Delta x$

y

Δx

Δx

Undeformed element

(a)

O'

ρ ρ

$\Delta \theta$

longitudinal axis

$\Delta s'$

y

Δx

Deformed element

(b)

Fig. 12–4

the undeformed and deformed positions in Fig. 12–4. Notice that any line segment Δx, located on the neutral surface, does not change its length, whereas any line segment Δs, located at the arbitrary distance y above the neutral surface, Fig. 12–4a, will contract and become $\Delta s'$ after deformation, Fig. 12–4b. By definition, the normal strain along Δs is determined from Eq. 8–11, namely,

$$\epsilon = \lim_{\Delta s \to 0} \frac{\Delta s' - \Delta s}{\Delta s}$$

We will now represent this strain in terms of the location y of the segment and the radius of curvature ρ of the longitudinal axis of the element. Before deformation, $\Delta s = \Delta x$, Fig. 12–4a. After deformation Δx has a radius of curvature ρ, with center of curvature at point O', Fig. 12–4b. Since $\Delta \theta$ defines the angle between the cross-sectional sides of the element, $\Delta x = \Delta s = \rho \, \Delta \theta$. In the same manner, the deformed length of Δs becomes $\Delta s' = (\rho - y) \, \Delta \theta$. Substituting into the above equation, we get

$$\epsilon = \lim_{\Delta \theta \to 0} \frac{(\rho - y) \, \Delta \theta - \rho \, \Delta \theta}{\rho \, \Delta \theta}$$

or

$$\epsilon = \frac{-y}{\rho} \qquad\qquad (12\text{--}1)$$

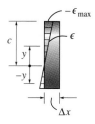

Normal strain distribution

Fig. 12–5

This important result indicates that the longitudinal normal strain of any element within the beam depends on its location y on the cross section and the radius of curvature of the beam's longitudinal axis at the point. In other words, for any specific cross section, the *longitudinal normal strain* will *vary linearly* with y from the neutral axis. Equation 12–1 states that a contraction $(-\epsilon)$ will occur in fibers located above the neutral axis $(+y)$, whereas elongation $(+\epsilon)$ will occur in fibers located below the axis $(-y)$. This variation in strain over the cross section is shown in Fig. 12–5. Here the maximum strain occurs at the outermost fiber, located a distance c from the neutral axis. Using Eq. 12–1, since $\epsilon_{max} = c/\rho$, then by division,

$$\frac{\epsilon}{\epsilon_{max}} = \frac{-y/\rho}{c/\rho}$$

So that

$$\epsilon = -\left(\frac{y}{c}\right)\epsilon_{max} \qquad (12-2)$$

This normal strain depends only on the assumptions made with regard to the *deformation*. Provided only a moment is applied to the beam, as shown in Fig. 12–1, then it is reasonable to further assume that this moment causes a *normal stress only* in the longitudinal or x direction. All the other components of normal and shear stress are zero, since the beam's surface is free of any other load. It is this uniaxial state of stress that causes the material to have the longitudinal normal strain component ϵ_x, $(\sigma_x = E\epsilon_x)$, defined by Eq. 12–1 or Eq. 12–2. Furthermore, by Poisson's ratio, there must also be associated strain components $\epsilon_y = -\nu\epsilon_x$ and $\epsilon_z = -\nu\epsilon_x$, which deform the plane of the cross-sectional area, although we have neglected these deformations in the development of Eqs. 12–1 and 12–2. Such deformations will, however, cause the *cross-sectional dimensions* to become smaller below the neutral axis and larger above the neutral axis. For example, if the beam has a square cross section, it will actually deform as shown in Fig. 12–2 or Fig. 12–6.

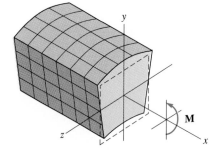

Fig. 12–6

12.2 The Flexure Formula

In this section we will develop an equation that relates the longitudinal stress distribution in a beam to the internal resultant bending moment acting on the beam's cross section. To do this we will assume that the material behaves in a linear-elastic manner so that Hooke's law applies, that is, $\sigma = E\epsilon$. Consequently, a linear variation of *normal strain,* Fig. 12–7a, must then be the result of a corresponding linear variation in *normal stress,* Fig. 12–7b. Hence, like the normal strain variation, σ will vary from zero at the member's neutral axis to a maximum value, σ_{max}, a distance c farthest from the neutral axis. A three-dimensional view of this stress distribution is shown in Fig. 12–7c. Because of the proportionality of triangles, or by using Hooke's law, $\sigma = E\epsilon$, and Eq. 12–2, we can write

$$\sigma = -\left(\frac{y}{c}\right)\sigma_{max} \tag{12–3}$$

Since σ_{max} and c are constant, this equation expresses the normal stress as a function of the y coordinate; in other words, it represents the stress distribution over the cross-sectional area. The sign convention established here is significant. By the right-hand rule, the moment **M** in Fig. 12–7c is considered positive since it is applied along the *positive z* axis. As a result, positive values of y give negative values for σ, that is, a compressive stress since it acts in the negative x direction. Similarly, negative y values will give positive or tensile values for σ. If a volume element of material is selected at a specific point on the cross section, only these tensile or compressive normal stresses will act on it. For example, the element located at $+y$ is shown in Fig. 12–7d.

We can locate the neutral axis on the cross section by satisfying the condition that the *resultant force* produced by the stress distribution over the cross-sectional area must be equal to *zero*. Noting that the force $dF = \sigma \, dA$ acts on the arbitrary element dA in Fig. 12–7c, then using Eq. 12–3 we require

$$F_R = \Sigma F_x; \quad 0 = \int_A dF = \int_A \sigma \, dA = \int_A -\left(\frac{y}{c}\right)\sigma_{max} \, dA = \frac{-\sigma_{max}}{c}\int_A y \, dA$$

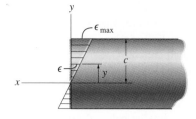

Normal strain variation
(profile view)

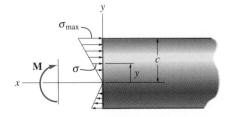

Bending stress variation
(profile view)

Fig. 12–7   (a)   (b)

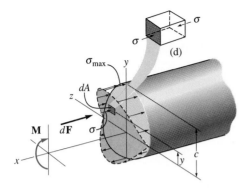

Bending stress variation

(c)

Since σ_{max}/c is not equal to zero, this equation is satisfied as long as

$$\int_A y \, dA = 0 \qquad (12\text{–}4)$$

In other words, the first moment of the member's cross-sectional area about the neutral axis must be zero. This condition can only be satisfied if the *neutral axis* is also the horizontal *centroidal axis* for the cross section.* Consequently, once the centroid for the member's cross-sectional area is determined, the location of the neutral axis is known.

In order to relate the stress in the beam to the resultant internal moment M acting at the cross section, we must require this moment to be equal to the moment produced by the stress distribution about the neutral axis. The moment of $d\mathbf{F}$ in Fig. 12–7c about the neutral axis is $dM = y \, dF$. This moment is *positive* since, by the right-hand rule, the thumb is directed along the positive z axis when the fingers are curled with the sense of rotation caused by $d\mathbf{M}$. Since $dF = \sigma \, dA$, using Eq. 12–3, and realizing that the moment is positive when y is positive and σ is negative, we have for the entire cross-section,

$$(M_R)_z = \Sigma M_z; \qquad M = \int_A y \, dF = \int_A y \, (-\sigma \, dA) = \int_A y \left(\frac{y}{c} \sigma_{max}\right) dA$$

or

$$M = \frac{\sigma_{max}}{c} \int_A y^2 \, dA \qquad (12\text{–}5)$$

*Recall that the location $\bar{y}$ for the centroid of the cross-sectional area is defined from the equation $\bar{y} = \int y \, dA/\int dA$. If $\int y \, dA = 0$, then $\bar{y} = 0$, and so the centroid lies on the reference (neutral) axis. See Sec. 6.1.

Here the integral represents the *moment of inertia* of the beam's cross-sectional area, computed about the neutral axis. We symbolize its value as *I*. Hence, Eq. 12–5 can be solved for σ_{max} and written in general form as

$$\sigma_{max} = \frac{Mc}{I}$$

(12–6)

Here

σ_{max} = the maximum normal stress in the member, which occurs at a point on the cross-sectional area *farthest away* from the neutral axis

M = the resultant internal moment, determined from the method of sections and the equations of equilibrium, and computed about the neutral axis of the cross section

I = the moment of inertia of the cross-sectional area computed about the neutral axis

c = the perpendicular distance from the neutral axis to a point farthest away from the neutral axis, where σ_{max} acts

Since $\sigma_{max}/c = -\sigma/y$, Eq. 12–3, the normal stress at the intermediate distance y can be determined from an equation similar to Eq. 12–6. We have

$$\sigma = -\frac{My}{I}$$

(12–7)

Note that the negative sign is necessary since it agrees with the established *x, y, z* axes. By the right-hand rule, *M* is positive along the +*z* axis, *y* is positive upward, and σ therefore must be negative since it acts in the negative *x* direction, Fig. 12–7*c*.

Either of the above two equations is often referred to as the *flexure formula*. Recall that it is used to determine the normal stress in a straight member, having a cross section that is symmetrical with respect to an axis, and the moment is applied perpendicular to this axis. Furthermore, the material is assumed to be homogeneous and to behave in a linear-elastic manner. Although we have also assumed that the member is prismatic, we can in most cases of engineering design use the flexure formula to determine as well the normal stress in members that have a *slight taper*. For example, using a mathematical analysis based on the theory of elasticity, a member having a rectangular cross section and a taper of 15° on both its top and bottom sides will have an actual maximum normal stress that is about 5.4% *less* than that calculated using the flexure formula.

PROCEDURE FOR ANALYSIS

The flexure formula can be used to find the normal-stress distribution in a *straight prismatic member,* which is made from homogeneous material and has linear-elastic behavior. Saint-Venant's principle requires that this formula be applied at points located a sufficient distance away from any supports, discontinuities in the cross section, or points where any concentrated loading acts. In order to apply the equation, the following procedure is suggested.

Internal Moment.   Section the member perpendicular to its longitudinal axis at the point where the bending or normal stress is to be determined, and use the necessary free-body diagrams and equations of equilibrium to obtain the internal moment M at the section. For this purpose, the centroidal or neutral axis for the cross section must be known, since M *must* be computed about this axis.

Section Property.   Compute the moment of inertia of the cross-sectional area about the neutral axis. Methods used for its computation are discussed in Sec. 6.8, and a table listing values of I for several common shapes is given in Appendix B.

Normal Stress.   Specify the distance y, measured perpendicular to the neutral axis to the point where the normal stress is to be determined. Then apply the equation $\sigma = My/I$, or if the maximum bending stress is to be computed, $\sigma_{max} = Mc/I$. When substituting the data, make sure to use a consistent set of units.

The stress acts in a direction such that the force $d\mathbf{F}$ it creates at the point contributes a moment about the neutral axis that is in the same direction as the internal moment $\mathbf{M}$, Fig. 12–7c. In this manner the stress distribution acting over the entire cross section can be graphed, or a volume element of the material can be isolated and used to represent graphically the normal stress acting at the point.

The following examples illustrate numerical applications of this procedure.

Example 12–1

A beam has a rectangular cross section and is subjected to a stress distribution as shown in Fig. 12–8a. Determine the internal moment **M** at the section caused by the stress distribution (a) using the flexure formula, (b) by finding the resultant of the stress distribution using basic principles.

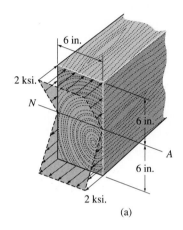

(a)

Fig. 12–8

SOLUTION

Part (a). The flexure formula is $\sigma_{max} = Mc/I$. From Fig. 12–8a, $c = 6$ in. and $\sigma_{max} = 2$ ksi. The neutral axis is defined as line NA, because the stress is zero along this line. Since the cross section has a rectangular shape, the moment of inertia for the area about NA is determined from the formula for a rectangle given in Appendix B; i.e.,

$$I = \frac{1}{12}bh^3 = \frac{1}{12}\,(6\text{ in.})(12\text{ in.})^3 = 864\text{ in}^4$$

Therefore,

$$\sigma_{max} = \frac{Mc}{I}; \qquad 2\text{ kip/in}^2 = \frac{M(6\text{ in.})}{864\text{ in}^4}$$

$$M = 288\text{ kip} \cdot \text{in.} = 24\text{ kip} \cdot \text{ft} \qquad\qquad Ans.$$

Part (b).   First we will show that the resultant force of the stress distribution is zero. As shown in Fig. 12–8b, the stress acting on the arbitrary element strip $dA = (6 \text{ in.}) \, dy$, located y from the neutral axis, is

$$\sigma = \left(\frac{-y}{6 \text{ in.}}\right)(2 \text{ kip/in}^2)$$

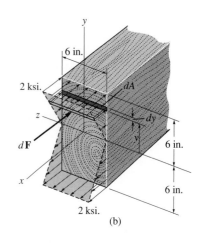

(b)

The force created by this stress is $dF = \sigma \, dA$, and thus, for the entire cross section,

$$F_R = \int_A \sigma \, dA = \int_{-6 \text{ in.}}^{6 \text{ in.}} \left[\left(\frac{-y}{6 \text{ in.}}\right)(2 \text{ kip/in}^2)\right](6 \text{ in.}) \, dy$$

$$= (-1 \text{ kip/in}^2) \, y^2 \, \Big|_{-6 \text{ in.}}^{+6 \text{ in.}} = 0$$

The resultant moment of the stress distribution about the neutral axis (z axis) must equal M. Since the magnitude of the moment of $d\mathbf{F}$ about this axis is $dM = y \, dF$, and $d\mathbf{M}$ is *always positive*, Fig. 12–8b, then for the entire area,

$$M = \int_A y \, dF = -\int_{-6 \text{ in.}}^{6 \text{ in.}} y\left[\left(\frac{-y}{6 \text{ in.}}\right)(2 \text{ kip/in}^2)\right](6 \text{ in.}) \, dy$$

$$= \left(\frac{2}{3} \text{ kip/in}^2\right) y^3 \, \Big|_{-6 \text{ in.}}^{+6 \text{ in.}}$$

$$= 288 \text{ kip} \cdot \text{in.} = 24 \text{ kip} \cdot \text{ft} \qquad\qquad \textit{Ans.}$$

The above result can *also* be determined without the need for integration. The resultant force for each of the two *triangular* stress distributions in Fig. 12–8c is graphically equivalent to the *volume* contained within each stress distribution. Thus, each volume is

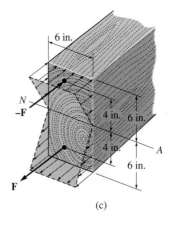

(c)

$$F = \frac{1}{2} \, (6 \text{ in.})(2 \text{ kip/in}^2)(6 \text{ in.}) = 36 \text{ kip}$$

These forces, which form a couple, act in the same direction as the stresses within each distribution, Fig. 12–8c. Furthermore, they act through the *centroid* of each volume, i.e., $\frac{1}{3}(6 \text{ in.}) = 2 \text{ in.}$ from the top and bottom of the beam. Hence the distance between them is 8 in. as shown. The moment of the couple is therefore

$$M = 36 \text{ kip} \, (8 \text{ in.}) = 288 \text{ kip} \cdot \text{in.} = 24 \text{ kip} \cdot \text{ft} \qquad \textit{Ans.}$$

Example 12–2

The simply-supported beam in Fig. 12–9a has the cross-sectional area shown in Fig. 12–9b. Determine the bending stress that acts at points B and D, located at section a–a.

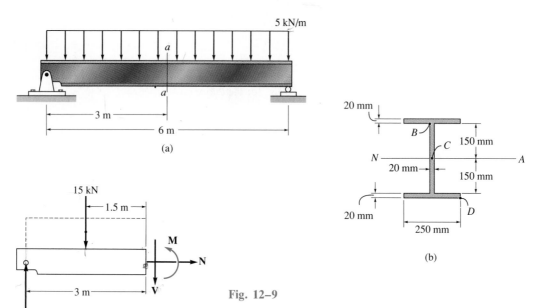

Fig. 12–9

SOLUTION

Internal Moment. Due to symmetry of both geometry and loading, the supports each exert a vertical reaction of 15 kN on the beam. Using the method of sections, the free-body diagram of the left segment of section a–a is shown in Fig. 12–9c. The moment is

$$M = 15 \text{ kN } (1.5 \text{ m}) = 22.5 \text{ kN} \cdot \text{m}$$

Section Property. By reasons of symmetry, the centroid C and thus the neutral axis pass through the midheight of the beam, Fig. 12–9b. The area is subdivided into the three parts shown, and the moment of inertia of each part is computed about the neutral axis using the parallel-axis theorem, Eq. 6–15. Choosing to work in meters, we have

$$I = \Sigma(\bar{I} + Ad^2)$$

$$= 2\left[\frac{1}{12}(0.25)(0.020)^3 + (0.25)(0.020)(0.160)^2\right]$$

$$+ \left[\frac{1}{12}(0.020)(0.300)^3\right]$$

$$= 301.3(10^{-6}) \text{ m}^4$$

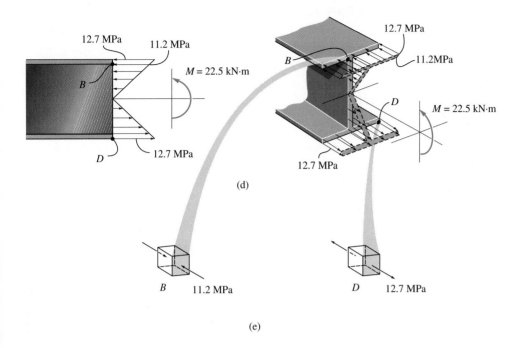

(d)

(e)

Normal Stress. Applying the flexure formula, with y_B = 150 mm for σ_B, we have

$$\sigma_B = \frac{My_B}{I}; \qquad \sigma_B = \frac{22.5 \text{ kN} \cdot \text{m}(0.150 \text{ m})}{301.3(10^{-6}) \text{ m}^4} = 11.2 \text{ MPa} \qquad \textit{Ans.}$$

For σ_D, $y_D = c$ = 170 mm, so

$$\sigma_D = \frac{Mc}{I}; \qquad \sigma_D = \frac{22.5 \text{ kN} \cdot \text{m}(0.170 \text{ m})}{301.3(10^{-6}) \text{ m}^4} = 12.7 \text{ MPa} \qquad \textit{Ans.}$$

Two-and-three-dimensional views of the stress distribution are shown in Fig. 12–9*d*. Here the stress at each point on the cross section develops a force that contributes a moment $d\mathbf{M}$ about the neutral axis such that it has the same direction as $\mathbf{M}$. Specifically, the normal stress acting on elements of material located at points *B* and *D* is shown in Fig. 12–9*e*.

Example 12–3

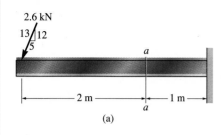

(a)

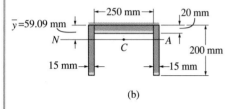

(b)

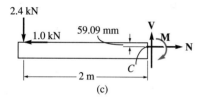

(c)

Fig. 12–10

The beam shown in Fig. 12–10a has a cross-sectional area in the shape of a channel, Fig. 12–10b. Determine the maximum bending stress that occurs in the beam at section a–a.

SOLUTION

Internal Moment. Here the beam's support reactions do not have to be determined. Instead, by the method of sections, the segment to the left of section a–a can be used, Fig. 12–10c. In particular, note that the resultant internal axial force **N** passes through the centroid of the cross section. Also, realize that the resultant internal moment must be computed about the beam's neutral axis at section a–a.

To find the location of the neutral axis, the cross-sectional area is subdivided into three composite parts as shown in Fig. 12–10b. Since the neutral axis passes through the centroid, then using Eq. 6–6, we have

$$\bar{y} = \frac{\Sigma \tilde{y} A}{\Sigma A} = \frac{2[100](200)(15) + [10](20)(250)}{2(200)(15) + 20(250)}$$

$$= 59.09 \text{ mm} = 0.05909 \text{ m}$$

This dimension is shown in Fig. 12–10c.

Applying the moment equation of equilibrium about the neutral axis, we have

$$\zeta + \Sigma M_{NA} = 0; \quad 2.4 \text{ kN}(2 \text{ m}) + 1 \text{ kN}(0.05909 \text{ m}) - M = 0$$

$$M = 4.859 \text{ kN} \cdot \text{m}$$

Section Property. The moment of inertia about the neutral axis is determined using the parallel-axis theorem applied to each of the three composite parts of the cross-sectional area. Working in meters, we have

$$I = \left[\frac{1}{12}(0.250)(0.020)^3 + (0.250)(0.020)(0.05909 - 0.010)^2 \right]$$

$$+ 2\left[\frac{1}{12}(0.015)(0.200)^3 + (0.015)(0.200)(0.100 - 0.05909)^2 \right]$$

$$= 42.26(10^{-6}) \text{ m}^4$$

Maximum Bending Stress. The maximum bending stress occurs at points farthest away from the neutral axis. This is at the bottom of the beam, $c = 200 \text{ mm} - \bar{y} = 140.9 \text{ mm}$. Thus,

$$\sigma_{\max} = \frac{Mc}{I} = \frac{4.859 \text{ kN} \cdot \text{m}(0.1409 \text{ m})}{42.26(10^{-6}) \text{ m}^4} = 16.2 \text{ MPa} \qquad \textit{Ans.}$$

By comparison, the bending stress at the *top* of the beam is

$$\sigma' = \frac{M\bar{y}}{I} = \frac{4.859 \text{ kN}(0.05909 \text{ m})}{42.26(10^{-6}) \text{ m}^4} = 6.79 \text{ MPa}$$

Example 12–4

The member having a rectangular cross section, Fig. 12–11a, is designed to resist a moment of 40 N · m. In order to increase its strength and rigidity, it is proposed that two small ribs be added at its bottom, Fig. 12–11b. Determine the maximum normal stress in the member for both cases.

SOLUTION

Without Ribs. Clearly the neutral axis is at the center of the cross section, so $\bar{y} = c = 15$ mm $= 0.015$ m. Thus,

$$I = \frac{1}{12} (0.06 \text{ m})(0.03 \text{ m})^3 = 0.135(10^{-6}) \text{ m}^4$$

Therefore the maximum normal stress is

$$\sigma_{max} = \frac{Mc}{I} = \frac{(40 \text{ N} \cdot \text{m})(0.015 \text{ m})}{0.135(10^{-6}) \text{ m}^4} = 4.44 \text{ MPa} \qquad Ans.$$

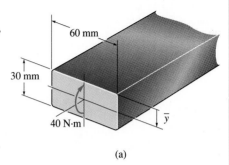

(a)

With Ribs. From Fig. 12–11b, segmenting the area into the large main rectangle and the bottom two rectangles (ribs), the location $\bar{y}$ of the centroid and the neutral axis is determined as follows:

$$\bar{y} = \frac{\Sigma \tilde{y} A}{\Sigma A}$$

$$= \frac{[0.015 \text{ m}](0.030 \text{ m})(0.060 \text{ m}) + 2[0.0325 \text{ m}](0.005 \text{ m})(0.010 \text{ m})}{(0.03 \text{ m})(0.060 \text{ m}) + 2(0.005 \text{ m})(0.010 \text{ m})}$$

$$= 0.01592 \text{ m}$$

This value does not represent c. Instead

$$c = 0.035 \text{ m} - 0.01592 \text{ m} = 0.01908 \text{ m}$$

Using the parallel-axis theorem, the moment of inertia about the neutral axis is

$$I = \left[\frac{1}{12} (0.06)(0.03)^3 + (0.06)(0.03)(0.01592 - 0.015)^2 \right]$$

$$+ 2 \left[\frac{1}{12} (0.01)(0.005)^3 + (0.01)(0.005)(0.0325 - 0.01592)^2 \right]$$

$$= 0.1642 (10^{-6}) \text{ m}^4$$

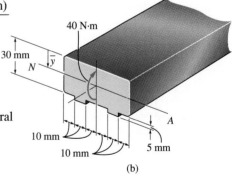

(b)

Fig. 12–11

Therefore, the maximum normal stress is

$$\sigma_{max} = \frac{Mc}{I} = \frac{40 \text{ N} \cdot \text{m}(0.01908 \text{ m})}{0.1642 (10^{-6}) \text{ m}^4} = 4.65 \text{ MPa} \qquad Ans.$$

The result indicates that the addition of the ribs to the cross section will *increase* the normal stress rather than decrease it, and for this reason they should be omitted.

PROBLEMS

12–1.   A member having the dimensions shown is to be used to resist an internal bending moment of $M = 17$ lb · ft. Determine the maximum stress in the member if the moment is applied (*a*) about the *z–z* axis, (*b*) about the *y–y* axis. Sketch the stress distribution for each case.

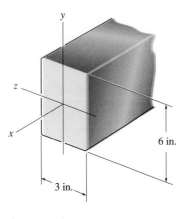

Prob. 12–1

12–3.   The tapered casting supports the loading shown. Determine the bending stress at points *A* and *B*. The cross section at section *a–a* is given in the figure.

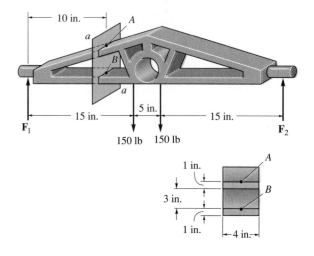

Prob. 12–3

12–2.   A beam is constructed from four pieces of wood, glued together as shown. If the internal bending moment is $M = 80$ kip · ft, determine the maximum bending stress in the beam. Sketch a three-dimensional view of the stress distribution acting over the cross section.

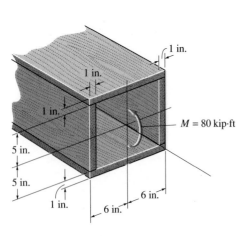

Prob. 12–2

*12–4.   The aluminum strut has a cross-sectional area in the form of a cross. If it is subjected to the moment $M = 5$ kN · m, determine the stress acting at points *A* and *B*. Also, sketch a three-dimensional view of the stress distribution acting over the entire cross-sectional area.

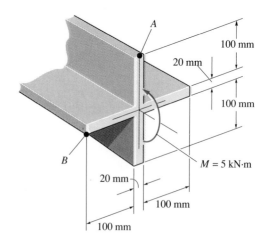

Prob. 12–4

12–5. The chair is supported by an arm that is hinged so it rotates about the vertical axis at A. If the load on the chair is 180 lb and the arm is a hollow tube section having the dimensions shown, determine the maximum bending stress at section a–a.

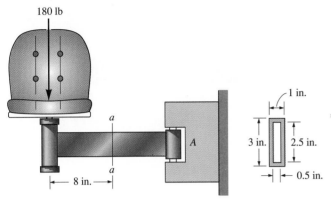

Prob. 12–5

12–6. The beam is made from three boards nailed together as shown. If the moment acting on the cross section is $M = 650$ N $\cdot$ m, determine the resultant force the stress produces on the top board, shown shaded in the figure.

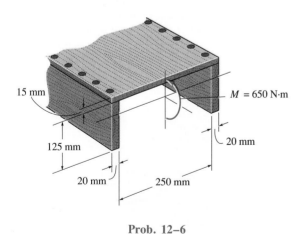

Prob. 12–6

12–7. The aluminum machine part is subjected to an internal bending moment of $M = 75$ N $\cdot$ m. Determine the stress created at points B and C on the cross section. Sketch the results on a volume element located at each of these points.

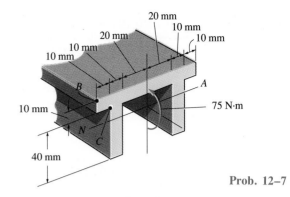

Prob. 12–7

*12–8. A beam has the cross section shown. If it is made of steel that has an allowable stress of $\sigma_{allow} = 24$ ksi, determine the largest internal moment the beam can resist if the moment is applied (a) about the z axis, (b) about the y axis.

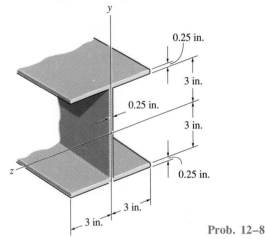

Prob. 12–8

12–9. A beam is constructed from four pieces of wood, glued together as shown. If the moment acting on the cross section is $M = 450$ N $\cdot$ m, determine the resultant force the stress produces on the top board A and on the side board B.

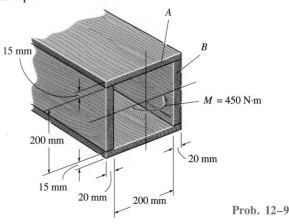

Prob. 12–9

12–10. The beam is subjected to a moment M. Determine the percent of this moment that is resisted by the stresses acting on both the top and bottom boards, A and B, of the beam.

12–11. Determine the moment M that should be applied to the beam in order to create a compressive stress at point D of $\sigma_D = 30$ MPa. Also sketch the stress distribution acting over the cross section and compute the maximum stress developed in the beam.

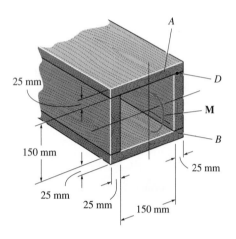

Probs. 12–10/12–11

*12–12. The simply-supported beam is subjected to the load of 1.5 kN. Draw the bending moment diagrams and determine the maximum bending stress in the beam.

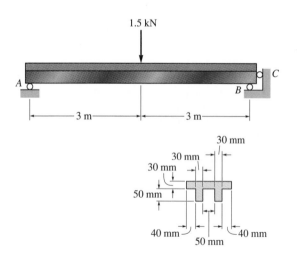

Prob. 12–12

12–13. Draw the bending moment diagrams, and then determine the bending stress distribution in the beam at the section where it is maximum. Sketch the distribution in three dimensions acting over the cross section.

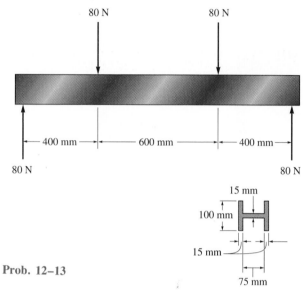

Prob. 12–13

12–14. The beam is constructed from four boards as shown. If it is subjected to an internal bending moment of $M_z = 16$ kip · ft, determine the stress at points A and B. Sketch a three-dimensional view of the stress distribution.

12–15. The beam is constructed from four boards as shown. If it is subjected to an internal bending moment of $M_z = 16$ kip · ft, determine the resultant force the stress produces on the top board C.

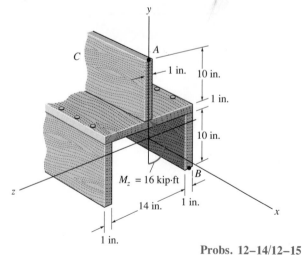

Probs. 12–14/12–15

*12–16. The bolster or main supporting girder of a truck body is subjected to the uniform distributed load. Determine the bending stress at points A and B.

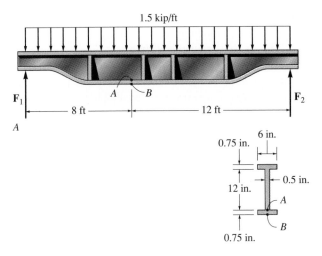

Prob. 12–16

12–19. The beam has a rectangular cross section as shown. Determine the largest load P that can be supported on its overhanging ends so that the maximum bending stress does not exceed $\sigma_{max} = 10$ MPa.

*12–20. The beam has the rectangular cross section shown. If $P = 12$ kN, determine the maximum bending stress acting in the beam. Sketch the stress distribution acting over the cross section.

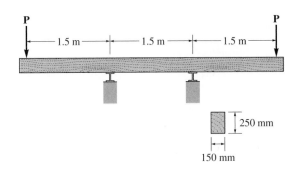

Probs. 12–19/12–20

12–17. Draw the bending moment diagram for the beam. Then determine the magnitude of the maximum loads **P** that can be applied to the beam if the beam is made of a material having an allowable bending stress of $(\sigma_{allow})_c = 12$ ksi in compression and $(\sigma_{allow})_t = 22$ ksi in tension.

12–18. Draw the bending moment diagram for the beam. Then determine the magnitude of the maximum loads **P** that can be applied to the beam if the beam is made of a material having an allowable bending stress of $(\sigma_{allow})_c = 16$ ksi in compression and $(\sigma_{allow})_t = 18$ ksi in tension. Maximum bending moment in the beam will occur at any section a–a between the loads.

12–21. The steel beam has the cross-sectional area shown. Determine the largest intensity of distributed load w that it can support so that the bending stress does not exceed $\sigma_{max} = 22$ ksi.

12–22. The steel beam has the cross-sectional area shown. If $w = 5$ kip/ft, determine the maximum bending stress in the beam.

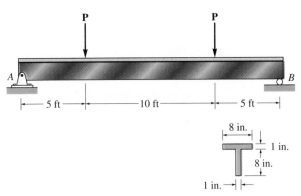

Probs. 12–17/12–18

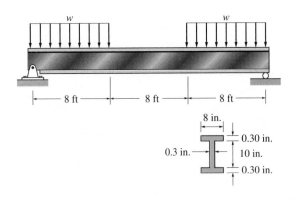

Probs. 12–21/12–22

12.3 Unsymmetric Bending

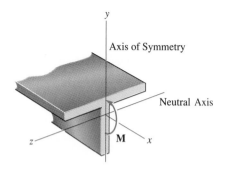

Axis of Symmetry

Neutral Axis

M

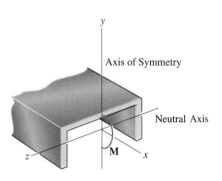

Axis of Symmetry

Neutral Axis

M

Fig. 12–12

Fig. 12–13

When developing the flexure formula we imposed a condition that the cross-sectional area be *symmetric* about an axis perpendicular to the neutral axis; furthermore, the resultant internal moment **M** acts along the neutral axis. Such is the case for the "T" or channel sections shown in Fig. 12–12. These conditions, however, are unnecessary, and in this section we will show that the flexure formula can also be applied either to a beam having a cross-sectional area of any shape or to a beam having a resultant internal moment that acts in any direction.

Moment Applied Along Principal Axis. Consider the beam's cross section to have the unsymmetrical shape shown in Fig. 12–13a. We will assume that the beam is both *straight* and *prismatic*. As in Sec. 12.2, the right-handed x, y, z coordinate system is established such that the origin is located at the centroid C of the cross section, and the resultant internal moment **M** acts along the $+z$ axis. We require the stress distribution acting over the entire cross-sectional area to have a zero force resultant, the resultant internal moment about the y axis to be zero, and the resultant internal moment about the z axis to equal **M**.* These three conditions can be expressed mathematically by considering the force acting on the differential element dA located at $(0, y, z)$, Fig. 12–13a. This force is $dF = \sigma \, dA$, and therefore we have

$$F_R = \Sigma F_x; \qquad\qquad 0 = \int_A \sigma \, dA \qquad\qquad (12\text{–}8)$$

$$(M_R)_y = \Sigma M_y; \qquad\qquad 0 = \int_A z\sigma \, dA \qquad\qquad (12\text{–}9)$$

$$(M_R)_z = \Sigma M_z; \qquad\qquad M = \int_A -y\sigma \, dA \qquad\qquad (12\text{–}10)$$

*The condition that moments about the y axis be equal to zero was not considered in Sec. 12.2, since the bending-stress distribution was *symmetric* with respect to the y axis and such a distribution of stress automatically produces zero moment about the y axis. See Fig. 12–7c.

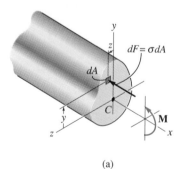

$dF = \sigma dA$

dA

C

M

(a)

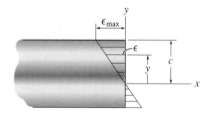

ϵ_{max}

ϵ

c

Normal Strain Distribution
(Profile View)
(b)

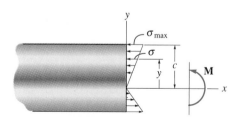

σ_{max}

σ

c

M

Bending Stress Distribution
(Profile View)
(c)

As shown in Sec. 12.2, Eq. 12–8 is satisfied since the z axis passes through the *centroid* of the cross-sectional area. Also, since the z axis represents the *neutral axis* for the cross section, the normal strain will vary linearly from zero at the neutral axis, to a maximum at a point located the largest y coordinate distance, $y = c$, from the neutral axis, Fig. 12–13*b*. Provided the material behaves in a linear-elastic manner, the normal-stress distribution over the cross-sectional area is *also* linear, so that $\sigma = -(y/c)\sigma_{max}$, Fig. 12–13*c*. When this equation is substituted into Eq. 12–10 and integrated, it leads to the flexure formula $\sigma_{max} = Mc/I$. When it is substituted into Eq. 12–9, we get

$$0 = -\frac{\sigma_{max}}{c} \int_A yz \, dA$$

which requires

$$\int_A yz \, dA = 0$$

This integral is called the *product of inertia* for the area. It will be zero provided the y and z axes are chosen as *principal axes of inertia* for the area. For an arbitrarily shaped area, like the one shown in Fig. 12–13*a*, the orientation of the principal axes can always be determined, using methods which are beyond the scope of this text. If the area has an axis of symmetry, however, the *principal axes* can easily be established since they will always be oriented *along the axis of symmetry* and *perpendicular* to it.

In summary, then, Eqs. 12–8 through 12–10 will *always* be satisfied *provided* the moment **M** is applied about one of the centroidal principal axes of inertia. For example, consider the members shown in Fig. 12–14, which are subjected to bending. In both of these cases, y and z define the principal axes of inertia for the cross section having the origin located at the area's centroid. Since **M** is applied about one of the principal axes (z axis), the stress distribution is determined from the flexure formula, $\sigma = My/I_z$, and is shown for each case.

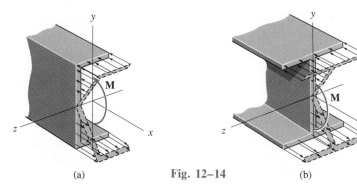

(a)   Fig. 12–14   (b)

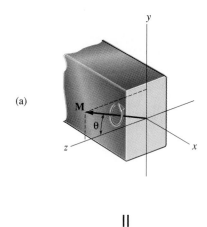

(a)

$\parallel$

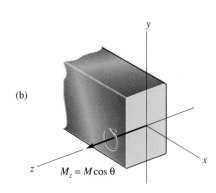

(b)

$+$

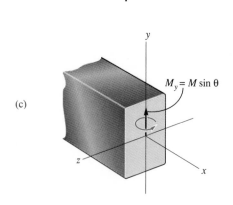

(c)

Fig. 12–15

Moment Arbitrarily Applied. Sometimes a member may be loaded such that the resultant internal moment does not act about one of the principal axes of the cross section. When this occurs, the moment should first be re-solved into components directed along the principal axes. The flexure formula can then be used to determine the normal stress caused by each moment component. Finally, using the principle of superposition, the resultant normal stress at the point can be determined.

For example, consider the beam to have a rectangular cross section and to be subjected to the moment **M**, Fig. 12–15a. By the right-hand rule, **M** is represented as a vector such that it makes an angle θ with the *principal z* axis. In particular, θ is positive when it is directed from the $+z$ axis towards the $+y$ axis, as shown. Resolving **M** into components along the z and y axes, we have $M_z = M \cos \theta$ and $M_y = M \sin \theta$, respectively. Each of these components is shown separately on the cross section in Fig. 12–15b and 12–15c. The normal-stress distributions that produce **M** and its components $\mathbf{M}_z$ and $\mathbf{M}_y$ are shown in Fig. 12–15d, 12–15e, and 12–15f, respectively. Here it is assumed that $(\sigma_x)_{max} > (\sigma'_x)_{max}$. By inspection, the maximum tensile and compressive stresses $[(\sigma_x)_{max} + (\sigma'_x)_{max}]$ occur at two of the beam's opposite corners, Fig. 12–15d.

Applying the flexure formula to each moment component in Fig. 12–15b and 12–15c, we can express the resultant normal stress at any point on the cross section, Fig. 12–15d, in general terms as

$$\sigma = -\frac{M_z y}{I_z} + \frac{M_y z}{I_y} \qquad (12\text{–}11)$$

Here

$\sigma =$ the normal stress at the point

$y, z =$ the coordinates of the point measured from x, y, z axes having their origin at the centroid of the cross-sectional area. The x axis is directed outward from the cross-section and the y and z axes repre-sent respectively the principal axes of minimum and maximum moment of inertia for the area

$M_y, M_z =$ the resultant internal moment components directed along the prin-cipal y and z axes. They are positive if directed along the $+y$ and $+z$ axes, otherwise they are negative. Or, stated another way, $M_y = M \sin \theta$ and $M_z = M \cos \theta$, where θ is measured positive from the $+z$ axis toward the $+y$ axis

$I_y, I_z =$ the *principal moments of inertia* computed about the z and y axes, respectively.

As noted above, it is *very important* that the x, y, z axes form a right-handed system and that the proper algebraic signs be assigned to the moment components and the coordinates when applying Eq. 12–11. The resulting stress will be *tensile* if it is *positive* and *compressive* if it is *negative*.

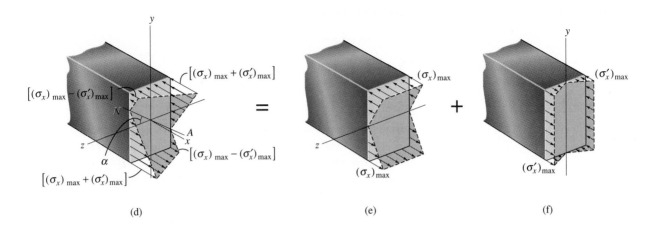

(d)　(e)　(f)

Orientation of the Neutral Axis.

The angle α of the neutral axis in Fig. 12–15d can be determined by applying Eq. 12–11 with $\sigma = 0$, since by definition no normal stress acts on the neutral axis. We have

$$y = \frac{M_y I_z}{M_z I_y} z$$

Since $M_z = M \cos \theta$ and $M_y = M \sin \theta$, Fig. 12–15a, then

$$y = \left(\frac{I_z}{I_y} \tan \theta\right) z \qquad (12\text{–}12)$$

This is the equation of the line that defines the neutral axis for the cross section, Fig. 12–15d. Since the term in parentheses represents the slope of this line (y/z),

$$\boxed{\tan \alpha = \frac{I_z}{I_y} \tan \theta} \qquad (12\text{–}13)$$

Here it can be seen that for *unsymmetrical bending* the angle θ, defining the direction of the moment **M**, Fig. 12–15a, is *not equal* to α, the angle defining the inclination of the neutral axis, Fig. 12–15d, unless $I_z = I_y$. Instead, if the *y axis* is chosen as the principal axis for the *minimum* moment of inertia, and the *z axis* is chosen as the principal axis for the *maximum* moment of inertia, so that $I_y < I_z$, Fig. 12–15a, then from Eq. 12–13 we can conclude that the angle α, which is measured positive from the $+z$ axis toward the $+y$ axis, will lie *between* the line of action of **M** and the y axis, Fig. 12–15d.

The following examples numerically illustrate these concepts.

Example 12-5

The rectangular cross section shown in Fig. 12–16a is subjected to a bending moment of $M = 12$ kN · m. Determine the normal stress developed at each corner of the section, and specify the orientation of the neutral axis.

SOLUTION

Internal Moment Components.   By inspection it is seen that the y and z axes represent the principal axes of inertia since they are axes of symmetry for the cross section. Here we have established the z axis as the principal axis for *maximum* moment of inertia. The moment is resolved into its y and z components, where

$$M_y = -\frac{4}{5}(12 \text{ kN} \cdot \text{m}) = -9.60 \text{ kN} \cdot \text{m}$$

$$M_z = \frac{3}{5}(12 \text{ kN} \cdot \text{m}) = 7.20 \text{ kN} \cdot \text{m}$$

Section Properties.   The moments of inertia about the y and z axes are

$$I_y = \frac{1}{12}(0.4 \text{ m})(0.2 \text{ m})^3 = 0.2667(10^{-3}) \text{ m}^4$$

$$I_z = \frac{1}{12}(0.2 \text{ m})(0.4 \text{ m})^3 = 1.067(10^{-3}) \text{ m}^4$$

Bending Stress.   Applying Eq. 12–11, that is,

$$\sigma = -\frac{M_z y}{I_z} + \frac{M_y z}{I_y}$$

we have

$$\sigma_B = -\frac{7.20(10^3)(0.2)}{1.067(10^{-3})} + \frac{-9.60(10^3)(-0.1)}{0.2667(10^{-3})} = 2.25 \text{ MPa} \qquad Ans.$$

$$\sigma_C = -\frac{7.20(10^3)(0.2)}{1.067(10^{-3})} + \frac{-9.60(10^3)(0.1)}{0.2667(10^{-3})} = -4.95 \text{ MPa} \qquad Ans.$$

$$\sigma_D = -\frac{7.20(10^3)(-0.2)}{1.067(10^{-3})} + \frac{-9.60(10^3)(0.1)}{0.2667(10^{-3})} = -2.25 \text{ MPa} \qquad Ans.$$

$$\sigma_E = -\frac{7.20(10^3)(-0.2)}{1.067(10^{-3})} + \frac{-9.60(10^3)(-0.1)}{0.2667(10^{-3})} = 4.95 \text{ MPa} \qquad Ans.$$

The resultant normal-stress distribution has been sketched using these values, Fig. 12–16b. Since superposition applies, the distribution is linear as shown.

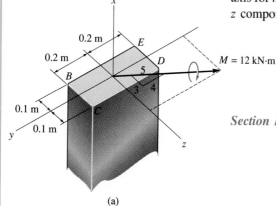

0.2 m
0.2 m
0.1 m
0.1 m
x
E
D
B
C
5
3   4
$M = 12$ kN·m
y
z

(a)

Fig. 12–16

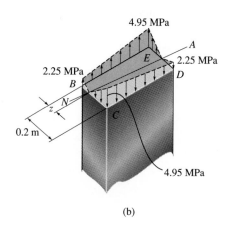

(b)

Orientation of Neutral Axis.

The location z of the neutral axis (NA), Fig. 12–16b, can be established by proportion. Along the edge BC, we require

$$\frac{2.25 \text{ MPa}}{z} = \frac{4.95 \text{ MPa}}{(0.2 \text{ m} - z)}$$

$$0.450 - 2.25z = 4.95z$$

$$z = 0.0625 \text{ m}$$

In the same manner this is also the distance from D to the neutral axis in Fig. 12–16b.

We can also establish the orientation of the NA using Eq. 12–13, which is used to specify the angle α that the axis makes with the z or *maximum* principal axis. According to our sign convention, θ must be measured from the $+z$ axis towards the $+y$ axis, Fig. 12–15a. By comparison, in Fig. 12–16a, $\theta = -\tan^{-1}\frac{4}{3} = -53.1°$ (or $\theta = +306.9°$). Thus,

$$\tan \alpha = \frac{I_z}{I_y} \tan \theta$$

$$\tan \alpha = \frac{1.067(10^{-3}) \text{ m}^4}{0.2667(10^{-3}) \text{ m}^4} \tan(-53.1°)$$

$$\alpha = -79.4° \qquad \textit{Ans.}$$

To verify the previous calculation for z, shown in Fig. 12–16c, we can recalculate the same angle α. From the geometry of the figure, we have

$$\tan \alpha = \frac{0.2 \text{ m}}{0.0375 \text{ m}}$$

$$\alpha = 79.4° \qquad \textit{Ans.}$$

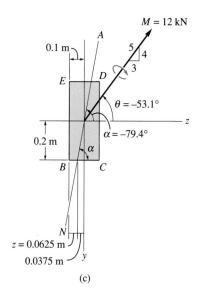

(c)

■ Example 12–6

A *T*-beam is subjected to the bending moment of 15 kN · m as shown in Fig. 12–17*b*. Determine the maximum normal stress in the beam and the orientation of the neutral axis.

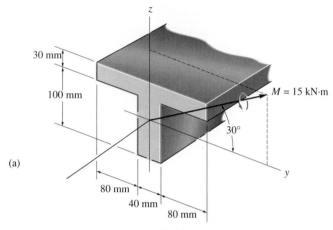

(a)

Fig. 12–17

SOLUTION

Internal Moment Components. The *y* and *z* axes are principal axes of inertia. Why? From Fig. 12–17*a*, both moment components are positive. We have

$$M_y = (15 \text{ kN} \cdot \text{m}) \cos 30° = 13.0 \text{ kN} \cdot \text{m}$$

$$M_z = (15 \text{ kN} \cdot \text{m}) \sin 30° = 7.50 \text{ kN} \cdot \text{m}$$

Section Properties. With reference to Fig. 12–17*b*, working in units of meters, we have

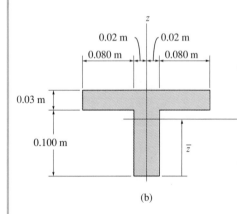

(b)

$$\bar{z} = \frac{\Sigma \bar{z} A}{\Sigma A} = \frac{[0.05](0.100)(0.04) + [0.115](0.03)(0.200)}{(0.100)(0.04) + (0.03)(0.200)}$$

$$= 0.0890 \text{ m}$$

Using the parallel-axis theorem, the principal moments of inertia are thus

$$I_z = \frac{1}{12}(0.100)(0.04)^3 + \frac{1}{12}(0.03)(0.200)^3 = 20.53(10^{-6}) \text{ m}^4$$

$$I_y = \left[\frac{1}{12}(0.04)(0.100)^3 + (0.100)(0.04)(0.0890 - 0.05)^2 \right]$$

$$+ \left[\frac{1}{12}(0.200)(0.03)^3 + (0.200)(0.03)(0.115 - 0.0890)^2 \right]$$

$$= 13.92(10^{-6}) \text{ m}^4$$

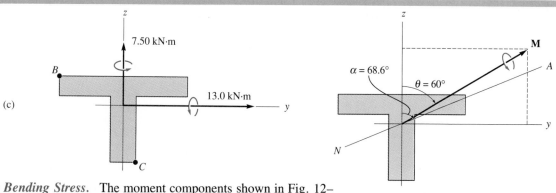

(c)

(d)

Maximum Bending Stress. The moment components shown in Fig. 12–17c are both positive. By inspection, the largest *tensile* stress occurs at point B, since by superposition both moment components create a tensile stress there. Likewise, the greatest *compressive* stress occurs at point C. Applying Eq. 12–11 to compute these stresses yields

$$\sigma = -\frac{M_z y}{I_z} + \frac{M_y z}{I_y}$$

$$\sigma_B = -\frac{7.50 \text{ kN} \cdot \text{m} \ (-0.100 \text{ m})}{20.53(10^{-6}) \text{ m}^4}$$
$$+ \frac{13.0 \text{ kN} \cdot \text{m} \ (0.130 \text{ m} - 0.0890 \text{ m})}{13.92 \ (10^{-6}) \text{ m}^4}$$

$$= 74.8 \text{ MPa}$$

$$\sigma_C = -\frac{7.50 \text{ kN} \cdot \text{m} \ (0.020 \text{ m})}{20.53(10^{-6}) \text{ m}^4} + \frac{13.0 \text{ kN} \cdot \text{m} \ (-0.0890 \text{ m})}{13.92(10^{-6}) \text{ m}^4}$$

$$= -90.4 \text{ MPa} \qquad\qquad\qquad\qquad \textit{Ans.}$$

The largest normal stress is therefore compressive and occurs at point C.

Orientation of Neutral Axis. When applying Eq. 12–13 it is important to be sure the angles α and θ are defined correctly. As previously stated, y must represent the axis for *minimum* principal moment of inertia, and z must represent the axis for *maximum* principal moment of inertia. These axes are properly positioned here since $I_y < I_z$. Using this setup, θ and α are measured positive from the $+z$ axis toward the $+y$ axis. Hence, from Fig. 12–17a, $\theta = +60°$. Thus,

$$\tan \alpha = \frac{20.53(10^{-6}) \text{ m}^4}{13.92(10^{-6}) \text{ m}^4} \tan 60°$$

$$\alpha = 68.6° \qquad\qquad\qquad\qquad \textit{Ans.}$$

The neutral axis is shown in Fig. 12–17d. As expected, it lies between the y axis and the line of action of **M**.

PROBLEMS

12–23. The member has a square cross section and is subjected to a resultant internal bending moment of $M = 850$ N · m as shown. Determine the stress at each corner and sketch the stress distribution produced by **M**. Set $\theta = 45°$.

*12–24. The member has a square cross section and is subjected to a resultant internal bending moment of $M = 850$ N · m as shown. Determine the stress at each corner and sketch the stress distribution produced by **M**. Set $\theta = 30°$.

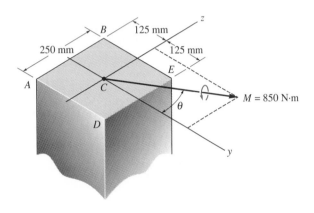

Probs. 12–23/12–24

12–25. The beam has a rectangular cross section. If it is subjected to a bending moment of $M = 3500$ N · m directed as shown, determine the maximum bending stress in the beam and the orientation of the neutral axis.

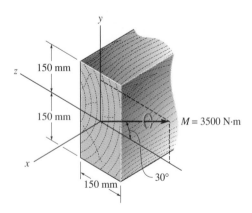

Prob. 12–25

12–26. The T-beam is subjected to a bending moment of $M = 150$ kip · in. directed as shown. Determine the maximum bending stress in the beam and the orientation of the neutral axis. The location $\bar{y}$ of the centroid, C, must be determined.

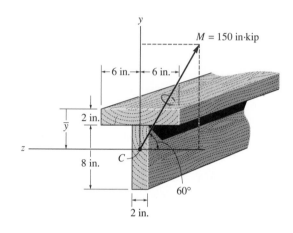

Prob. 12–26

12–27. The resultant internal moment acting on the cross section of the T-beam has a magnitude of $M = 15$ kip · ft and is directed as shown. Determine the bending stress at points A and B. The location $\bar{y}$ of the centroid C must be determined.

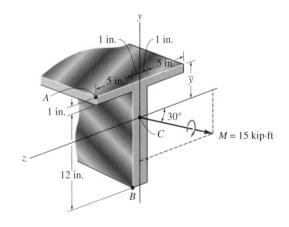

Prob. 12–27

*12–28. If the internal moment acting on the cross section of the strut has a magnitude of $M = 800$ N · m and is directed as shown, determine the bending stress at points A and B. The location $\bar{z}$ of the centroid C of the strut's cross-sectional area must be determined. Also, specify the orientation of the neutral axis.

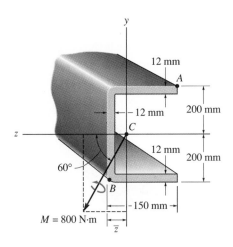

Prob. 12–28

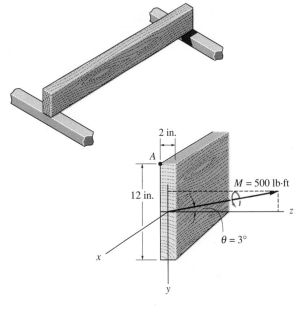

Prob. 12–29

12–29. The board is used as a simply-supported floor joist. If a bending moment of $M = 500$ lb · ft is applied 3° from the z axis, determine the stress developed in the board at the corner A. Compare this stress with that developed by the same moment applied along the z axis ($\theta = 0°$). What is the angle α for the neutral axis when $\theta = 3°$? *Comment:* Normally, floor boards would be nailed to the top of the beam so that $\theta \approx 0°$ and the high stress due to misalignment would not occur.

12–30. The cantilevered wide-flange steel beam is subjected to the concentrated force **P** at its end. Determine the largest magnitude of this force so that the bending stress developed at A does not exceed $\sigma_{\text{allow}} = 180$ MPa.

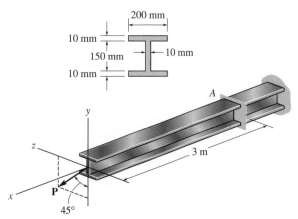

Prob. 12–30

12.4 Stress Concentrations

The flexure formula, $\sigma_{max} = Mc/I$, can be used to determine the stress distribution within regions of a member where the cross-sectional area is constant or tapers slightly. If the cross section suddenly changes, however, the normal stress and strain distributions at the section become *nonlinear* and can be obtained only through experiment or, in some cases, by a mathematical analysis using the theory of elasticity. Common discontinuities include members having notches on their surfaces, Fig. 12–18*a*, holes for passage of fasteners or other items, Fig. 12–18*b*, or abrupt changes in the outer dimensions of the member's cross section, Fig. 12–18*c*. The *maximum* normal stress at each of these discontinuities occurs at the section taken through the *smallest* cross-sectional area.

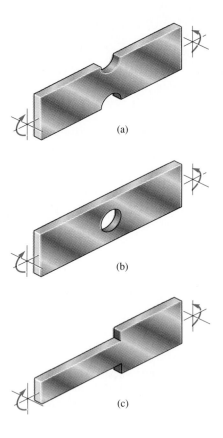

(a)

(b)

(c)

Fig. 12–18

For design, it is generally important to know the maximum normal stress developed at these sections, not the actual stress distribution itself. As in the previous cases of axially loaded bars and torsionally loaded shafts, we can obtain the maximum normal stress due to bending using a stress-concentration factor K. For example, Fig. 12–19 gives values of K for a flat bar that has a change in cross section using shoulder fillets.* To use this graph simply compute the geometric ratios w/h and r/h and then find the corresponding value of K for a particular geometry. Once K is obtained, the maximum bending stress is determined using

$$\sigma_{\text{max}} = K \frac{Mc}{I} \qquad (12\text{–}14)$$

Here the flexure formula is applied to the *smaller* cross-sectional area, since σ_{max} occurs at the base of the fillet, Fig. 12–20. In the same manner, Fig. 12–21 can be used if the discontinuity consists of circular grooves or notches.

Like axial load and torsion, stress concentration for bending should always be considered when designing members made of brittle materials or those that are subjected to fatigue or cyclic loadings. Realize also that stress-concentration factors apply only when the material is subjected to *elastic behavior*. If the applied moment causes yielding of the material, the stress becomes redistributed throughout the member, and the maximum stress that results will be *lower* than that computed using stress-concentration factors. This phenomenon is discussed further in the next section.

*See C. Lipson and R. C. Juvinall, *Handbook of Stress and Strength*, Macmillan, 1963.

Fig. 12–20

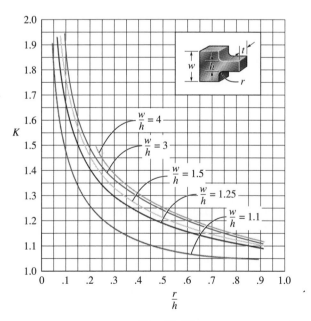

Fig. 12–19

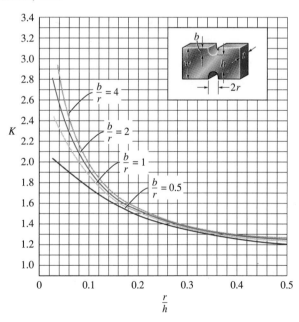

Fig. 12–21

12.5 Inelastic Bending

The equations for determining the normal stress due to bending that have previously been developed are valid only if the material behaves in a linear-elastic manner. If the applied moment causes the material to *yield*, a plastic analysis must then be used to determine the stress distribution. For both elastic and plastic cases, however, realize that for bending of straight members three conditions must be met.

Linear Normal-Strain Distribution. Based on geometric considerations, it was shown in Sec. 12.1 that the normal strains that develop in the material always vary *linearly* from zero at the neutral axis of the cross section to a maximum at the farthest point from the neutral axis.

Resultant Force Equals Zero. Since there is only a resultant internal moment acting on the cross section, the resultant force caused by the stress distribution must be equal to zero. This condition can be expressed mathematically by realizing that the normal stress σ, acting on an element of area dA, creates a force on the area of $dF = \sigma \, dA$, Fig. 12–22. Hence, for the entire cross-sectional area A, we have

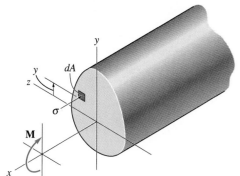

Fig. 12–22

$$F_R = \Sigma F_x; \qquad\qquad \int_A \sigma \, dA = 0 \qquad\qquad (12\text{–}15)$$

This equation provides a means for obtaining the *location of the neutral axis*.

Resultant Moment. The resultant moment at the section must be equivalent to the moment caused by the entire stress distribution about the neutral axis. This condition can be expressed mathematically by considering the moment of the force $dF = \sigma \, dA$ about the neutral axis and then summing the results over the entire cross section, Fig. 12–22. We have,

$$(M_R)_z = \Sigma M_z; \qquad\qquad M = \int_A y \, (\sigma \, dA) \qquad\qquad (12\text{–}16)$$

These conditions of geometry and loading will now be used to show how to determine the stress distribution in a beam when it is subjected to a resultant internal moment that causes yielding of the material. Throughout the discussion we will assume that the material has a stress–strain diagram that is the *same* in tension as it is in compression. For the sake of simplicity, we will begin by considering the beam to have a cross-sectional area with two axes of symmetry; for example, a rectangle of height h and width b, as shown in Fig. 12–23a. Three cases of loading that are of special interest will be considered.

Maximum Elastic Moment. Assume that the applied moment $M = M_Y$ is just sufficient to produce yielding strains in the top and bottom boundaries of the beam as shown in Fig. 12–23b. Since the strain distribution is linear, we can determine the corresponding stress distribution by using the stress–strain diagram, Fig. 12–23c. Here it is seen that the yield strain ϵ_Y causes the yield stress σ_Y, and the intermediate strains ϵ_1 and ϵ_2 cause stresses σ_1 and σ_2, respectively. When these stresses, and others like them, are plotted at the measured points $y = h/2$, $y = y_1$, $y = y_2$, etc., the stress distribution in Fig. 12–23d or 12–23e results. The linearity of the stress is, of course, a consequence of Hooke's law.

Now that the stress distribution has been established, we can check to see if Eq. 12–15 is satisfied. To do so we will first calculate the resultant force for each of the two portions of the stress distribution in Fig. 12–23e. Geometrically this is equivalent to finding the *volumes* under the two triangular blocks. As shown, the top cross section of the member is subjected to compression and the bottom portion is subjected to tension. We have

$$T = C = \frac{1}{2}\left(\frac{h}{2}\sigma_Y\right)b = \frac{1}{4}bh\sigma_Y$$

Since **T** is equal but opposite to **C**, Eq. 12–15 is satisfied and indeed the neutral axis passes through the centroid of the cross-sectional area.

The maximum elastic moment M_Y is computed from Eq. 12–16, which states that M_Y is equivalent to the moment of the stress distribution about the neutral axis. To apply this equation geometrically, we must compute the moments created by **T** and **C** in Fig. 12–23e about the neutral axis. Since each of the forces acts through the centroid of the volume of its associated triangular stress block, we have

$$M_Y = C\left(\frac{2}{3}\right)\frac{h}{2} + T\left(\frac{2}{3}\right)\frac{h}{2} = 2\left(\frac{1}{4}bh\,\sigma_Y\right)\left(\frac{2}{3}\right)\frac{h}{2}$$

$$= \frac{1}{6}bh^2\sigma_Y \tag{12–17}$$

This same result can of course be obtained in a more direct manner by using the flexure formula, that is, $\sigma_Y = M_Y(h/2)/[bh^3/12]$, or $M_Y = bh^2\sigma_Y/6$.

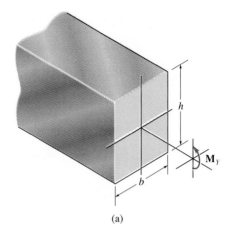

(a)

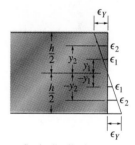

Strain distribution
(profile view)

(b)

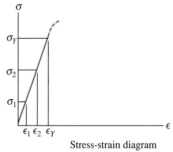

Stress-strain diagram
(elastic region)

(c)

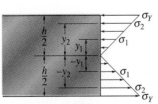

Stress distribution
(profile view)

(d)

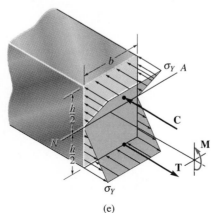

(e)

Fig. 12–23

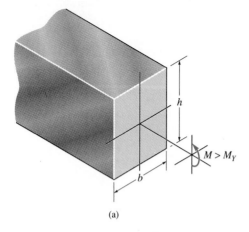

(a)

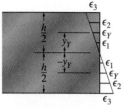

Strain distribution
(profile view)

(b)

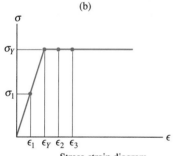

Stress-strain diagram
(elastic-plastic region)

(c)

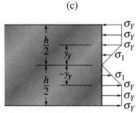

Stress distribution
(profile view)

(d)

Plastic Moment. Some materials, such as steel, tend to exhibit elastic-perfectly plastic behavior when the stress in the material exceeds σ_Y. Consider, for example, the member in Fig. 12–24a. If the internal moment $M > M_Y$, the material at the top and bottom of the beam will begin to yield, causing a redistribution of stress over the cross section until the required internal moment M is developed. If the normal-strain distribution so produced is as shown in Fig. 12–24b, the corresponding normal-stress distribution is determined from the stress–strain diagram in the same manner as in the elastic case. Using the stress–strain diagram for the material shown in Fig. 12–24c, the strains ϵ_1, ϵ_Y, ϵ_2, ϵ_3 correspond to stresses σ_1, σ_Y, σ_Y, σ_Y, respectively. When these and other stresses are plotted on the cross section, we obtain the stress distribution shown in Fig. 12–24d or 12–24e. Here the compression and tension stress "blocks" each consist of component rectangular and triangular blocks. Their volumes are

$$T_1 = C_1 = \frac{1}{2}\, y_Y \sigma_Y b$$

$$T_2 = C_2 = \left(\frac{h}{2} - y_Y\right)\sigma_Y b$$

Because of the symmetry, Eq. 12–15 is satisfied and the neutral axis passes through the centroid of the cross section as shown. The applied moment M can be related to the yield stress σ_Y using Eq. 12–16. From Fig. 12–24e, we require

$$
\begin{aligned}
M &= T_1\left(\frac{2}{3}y_Y\right) + C_1\left(\frac{2}{3}y_Y\right) + T_2\left[y_Y + \frac{1}{2}\left(\frac{h}{2} - y_Y\right)\right] \\
&\quad + C_2\left[y_Y + \frac{1}{2}\left(\frac{h}{2} - y_Y\right)\right] \\
&= 2\left(\frac{1}{2}y_Y\sigma_Y b\right)\left(\frac{2}{3}y_Y\right) + 2\left[\left(\frac{h}{2} - y_Y\right)\sigma_Y b\right]\left[\frac{1}{2}\left(\frac{h}{2} + y_Y\right)\right] \\
&= \frac{1}{4}\, bh^2\sigma_Y\left(1 - \frac{4}{3}\frac{y_Y^2}{h^2}\right)
\end{aligned}
$$

Or, using Eq. 12–17,

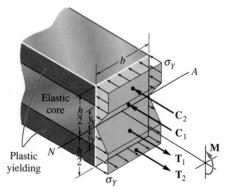

(e)

Fig. 12–24

$$M = \frac{3}{2} M_Y \left(1 - \frac{4}{3} \frac{y_Y^2}{h^2} \right) \qquad (12\text{–}18)$$

Inspection of Fig. 12–24e reveals that **M** produces two zones of plastic yielding (shown shaded) and an elastic core (shown colored) in the member. The boundary between them is located a distance $\pm y_Y$ from the neutral axis. As **M** increases in magnitude, y_Y approaches zero. This would render the material entirely plastic and the stress distribution would then look like that shown in Fig. 12–24f. From Eq. 12–18 with $y_Y = 0$, or by computing the moments of the stress "blocks" around the neutral axis, we can write this limiting value as

$$M_p = \frac{1}{4} bh^2 \sigma_Y \qquad (12\text{–}19)$$

or, using Eq. 12–17, we have

$$M_p = \frac{3}{2} M_Y \qquad (12\text{–}20)$$

This moment is referred to as the *plastic moment*. Its value is unique only for the rectangular section shown in Fig. 12–24f, since the analysis depends on the geometry of the cross section.

Beams used in steel buildings are sometimes designed to resist a plastic moment. When this is the case, codes usually list a design property for a beam called the shape factor. The *shape factor* is defined as a ratio,

$$\boxed{k = \frac{M_p}{M_Y}} \qquad (12\text{–}21)$$

This value specifies the additional moment capacity that a beam can support beyond its maximum elastic moment. For example, from Eq. 12–19, a beam having a rectangular cross section has a shape factor of $k = 1.5$. We may therefore conclude that this section will support 50% more bending moment than its maximum elastic moment when it becomes fully plastic.

Ultimate Moment.

Consider now the more general case of a beam having a cross section that is symmetrical only with respect to the vertical axis, while the moment is applied about the horizontal axis, Fig. 12–25a. We will assume that the material exhibits strain hardening and that its stress–strain diagrams for tension and compression are different, Fig. 12–25b.

If the moment **M** produces yielding of the beam, difficulty arises in finding *both* the location of the neutral axis and the maximum strain that is produced in the beam. This is because the cross section is unsymmetrical about the horizontal axis and the stress–strain behavior of the material is unsymmetrical in tension and compression. To solve this problem, a trial-and-error procedure requires the following steps:

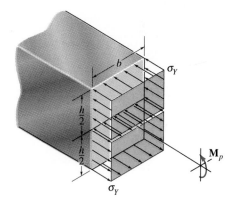

Plastic moment

(f)

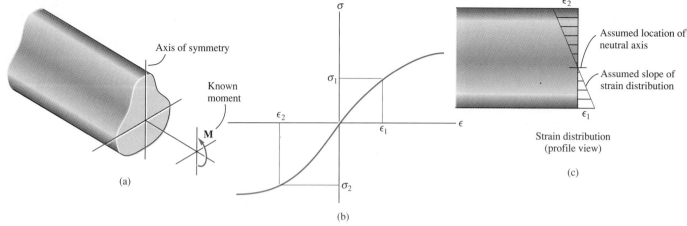

(a)

(b)

(c)

Strain distribution
(profile view)

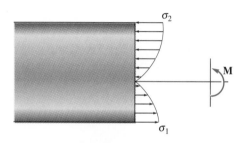

Stress distribution
(profile view)

(d)

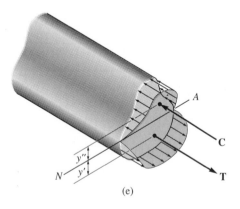

(e)

Fig. 12–25

1. For a given moment **M**, *assume* the location of the neutral axis and the slope of the "linear" strain distribution, Fig. 12–25c.

2. Graphically establish the stress distribution on the member's cross section using the σ–ϵ curve to plot values of stress corresponding to values of strain. The resulting stress distribution, Fig. 12–25d, will then have the same shape as the σ–ϵ curve.

3. Determine the volumes enclosed by the tensile and compressive stress "blocks." (As an approximation, this may require dividing each block into composite regions.) Equation 12–15 requires the volumes of these blocks to be *equal,* since they represent the resultant tensile force **T** and resultant compressive force **C** on the section, Fig. 12–25e. If these forces are unequal, an adjustment as to the *location* of the neutral axis must be made (point of *zero strain*) and the process repeated until Eq. 12–15 ($T = C$) is satisfied.

4. Once $T = C$, the moments produced by **T** and **C** can be computed about the neutral axis. Here the moment arms for **T** and **C** are measured from the neutral axis to the *centroids of the volumes* defined by the stress distributions, Fig. 12–25e. Equation 12–16 requires $M = Ty' + Cy''$. If this equation is not satisfied, the *slope* of the *strain distribution* must be adjusted and the computations for T and C and the moment must be repeated until close agreement is obtained.

This calculation process is obviously very tedious, and fortunately it does not occur very often in engineering practice. Most beams are symmetric about two axes, and they are constructed from materials that are assumed to have similar tension-and-compression stress–strain diagrams. Whenever this occurs, the neutral axis will pass through the centroid of the cross section, and the process of relating the stress distribution to the resultant moment is thereby simplified.

The following examples further illustrate application of the above concepts.

The steel wide-flange beam has the dimensions shown in Fig. 12–26a. If it is made of an elastic-plastic material having a tensile and compressive yield stress of $\sigma_Y = 36$ ksi, determine the shape factor for the beam.

SOLUTION

In order to determine the shape factor, it is first necessary to compute the maximum elastic moment M_Y and the plastic moment M_p.

Maximum Elastic Moment. The normal-stress distribution for the maximum elastic moment is shown in Fig. 12–26b. The moment of inertia about the neutral axis is

$$I = \left[\frac{1}{12}(0.5)(9)^3\right] + 2\left[\frac{1}{12}(8)(0.5)^3 + 8(0.5)(4.75)^2\right] = 211.0 \text{ in}^4$$

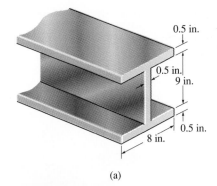

(a)

Applying the flexure formula, we have

$$\sigma_{max} = \frac{Mc}{I}; \qquad 36 \text{ kip/in}^2 = \frac{M_Y(5 \text{ in.})}{211.0 \text{ in}^4}$$

$$M_Y = 1519.5 \text{ kip} \cdot \text{in.}$$

Plastic Moment. The plastic moment causes the steel over the entire cross section of the beam to yield, so that the normal-stress distribution looks like that shown in Fig. 12–26c. Due to symmetry of the cross-sectional area and since the tension and compression stress–strain diagrams are the same, the neutral axis passes through the centroid of the cross section. In order to compute the plastic moment, the stress distribution is divided into four composite rectangular "blocks," and the force produced by each "block" is equal to the volume of the block. Therefore, we have

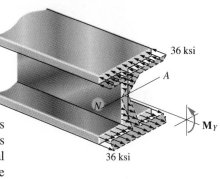

(b)

$$C_1 = T_1 = 36 \text{ kip/in}^2(0.5 \text{ in.})(4.5 \text{ in.}) = 81 \text{ kip}$$
$$C_2 = T_2 = 36 \text{ kip/in}^2(0.5 \text{ in.})(8 \text{ in.}) = 144 \text{ kip}$$

These forces act through the *centroid* of the volume for each block. Computing the moments of these forces about the neutral axis, we obtain the plastic moment:

$$M_p = 2[(2.25 \text{ in.})(81 \text{ kip})] + 2[(4.75 \text{ in.})(144 \text{ kip})] = 1732.5 \text{ kip} \cdot \text{in.}$$

Shape Factor. Applying Eq. 12–21 gives

$$k = \frac{M_p}{M_Y} = \frac{1732.5}{1519.5} = 1.14 \qquad \textit{Ans.}$$

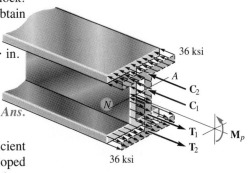

(c)

This value indicates that a wide-flange beam provides a very efficient section for resisting an *elastic moment*. Most of the moment is developed in the flanges, i.e., in the top and bottom segments, whereas the web or vertical segment contributes very little. In this particular case, only 14% additional moment can be supported by the beam beyond that which can be supported elastically.

Fig. 12–26

■ Example 12–8

A T-beam has the dimensions shown in Fig. 12–27a. If it is made of an elastic-plastic material having a tensile and compressive yield stress of $\sigma_Y = 250$ MPa, determine the plastic moment that can be resisted by the beam.

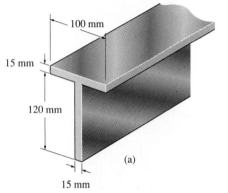

(a)

Fig. 12–27

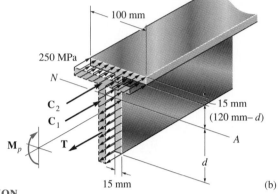

(b)

SOLUTION

The "plastic" stress distribution acting over the beam's cross-sectional area is shown in Fig. 12–27b. In this case the cross section is not symmetric with respect to a horizontal axis, and consequently, the neutral axis will *not* pass through the centroid of the cross section. To determine the *location* of the neutral axis, d, we require the stress distribution to produce a zero resultant force on the cross section. Assuming that $d \leq 120$ mm, we have

$$\int_A \sigma \, dA = 0; \qquad T - C_1 - C_2 = 0$$

$$250(15)(d) - 250(15)(120 - d) - 250(15)(100) = 0$$

$$2d = 220$$

$$d = 110 \text{ mm} < 120 \text{ mm} \qquad \text{OK}$$

Using this result, the forces acting on each segment are

$$T = 250 \text{ MN/m}^2 (0.015 \text{ m})(0.110 \text{ m}) = 412.5 \text{ kN}$$

$$C_1 = 250 \text{ MN/m}^2 (0.015 \text{ m})(0.010 \text{ m}) = 37.5 \text{ kN}$$

$$C_2 = 250 \text{ MN/m}^2 (0.015 \text{ m})(0.100 \text{ m}) = 375 \text{ kN}$$

Hence the resultant plastic moment about the neutral axis is

$$M_p = 412.5 \text{ kN} \left(\frac{0.110 \text{ m}}{2} \right) + 37.5 \text{ kN} \left(\frac{0.01 \text{ m}}{2} \right)$$

$$+ 375 \text{ kN} \left(0.01 \text{ m} + \frac{0.015 \text{ m}}{2} \right)$$

$$M_p = 29.4 \text{ kN} \cdot \text{m} \qquad \qquad \textit{Ans.}$$

PROBLEMS

12–31. The stepped bar has a thickness of 12 mm. Determine the maximum moment that can be applied to its ends if it is made of a material having an allowable bending stress of $\sigma_{\text{allow}} = 150$ MPa.

*12–32. The stepped bar has a thickness of 12 mm. If $M = 20$ N · m, determine the maximum bending stress in the plate. Specify where this stress occurs and sketch the stress distribution over the cross section.

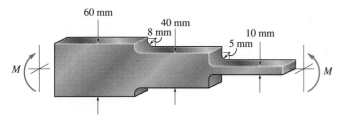

Probs. 12–31/12–32

12–33. Determine the maximum bending stress developed in the bar if its is subjected to the couples shown. The bar has a thickness of 0.25 in.

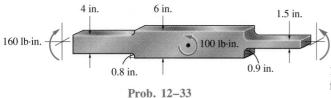

Prob. 12–33

12–34. Determine the plastic section modulus and the shape factor for the cross section of the beam.

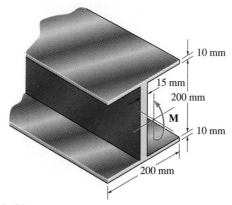

Prob. 12–34

12–35. The T-beam is made of an elastic-plastic material. Determine the maximum elastic moment and the plastic moment that can be applied to the cross section. $\sigma_Y = 36$ ksi.

*12–36. Determine the plastic section modulus and the shape factor for the beam.

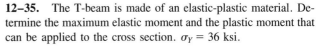
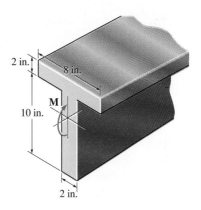

Probs. 12–35/12–36

12–37. The channel strut is made of an elastic-plastic material for which $\sigma_Y = 250$ MPa. Determine the maximum elastic moment and the plastic moment that can be applied to the cross section.

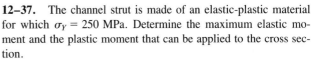
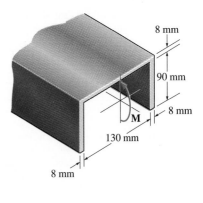

Prob. 12–37

12–38. The member has a square cross section. If it is made of an elastic-plastic material, determine the shape factor and the plastic section modulus Z.

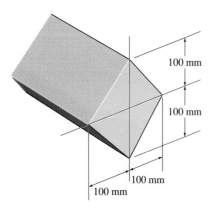

100 mm

100 mm

100 mm

100 mm

Prob. 12–38

12–39. The rod has a circular cross section. If it is made of an elastic-plastic material, determine the shape factor and the plastic section modulus Z.

3 in.

Prob. 12–39

*12–40. The beam is made of an elastic-plastic material for which $\sigma_Y = 30$ ksi. If the largest moment in the beam occurs at the center section $a-a$, determine the intensity of the distributed load w that causes this moment to be (a) the largest elastic moment and (b) the largest plastic moment.

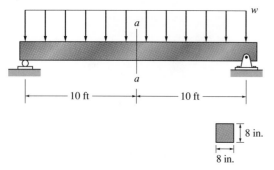

w

a

a

10 ft

10 ft

8 in.

8 in.

Prob. 12–40

12–41. The beam is made of an elastic-plastic material for which $\sigma_Y = 200$ MPa. If the largest moment in the beam occurs within the center section $a-a$, determine the magnitude of each force **P** that causes this moment to be (a) the largest elastic moment and (b) the largest plastic moment.

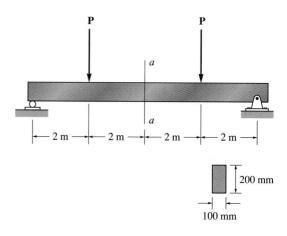

P   **P**

a

a

2 m   2 m   2 m   2 m

200 mm

100 mm

Prob. 12–41

12–42. The box beam is made of an elastic-plastic material for which $\sigma_Y = 25$ ksi. If the largest moment in the beam occurs at the center section $a-a$, determine the intensity of the distributed load w_0 that will cause this moment to be (a) the largest elastic moment and (b) the largest plastic moment.

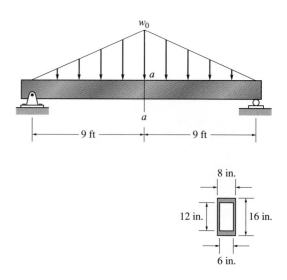

w_0

a

a

9 ft

9 ft

8 in.

12 in.   16 in.

6 in.

Prob. 12–42

REVIEW PROBLEMS

12–43.   The beam is subjected to a moment of 15 kip · ft. Determine the resultant force the stress produces on the top flange A and bottom flange B. Also compute the maximum stress developed in the beam.

*12–44.   The beam is subjected to a moment of 15 kip · ft. Determine the percent of this moment that is resisted by the web D of the beam.

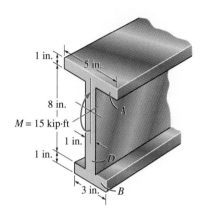

Probs. 12–43/12–44

12–45.   The beam is made from three boards nailed together as shown. If the moment acting on the cross section is $M = 600$ N · m, determine the maximum bending stress in the beam. Sketch a three-dimensional view of the stress distribution acting over the cross section.

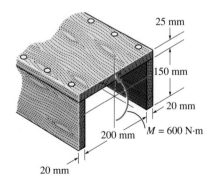

Prob. 12–45

12–46.   Determine the moment M that will produce a maximum stress of 12 ksi on the cross section.

12–47.   Determine the maximum tensile and compressive bending stress in the beam if it is subjected to a moment of $M = 2$ kip · ft.

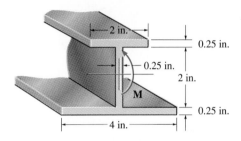

Probs. 12–46/12–47

*12–48.   The steel rod having a diameter of 1 in. is subjected to an internal moment of $M = 300$ lb · ft. Determine the stress created at points A and B. Also, sketch a three-dimensional view of the stress distribution acting over the cross-section.

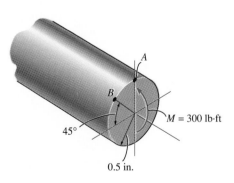

Prob. 12–48

12–49. Determine the plastic moment $\mathbf{M}_p$ that can be supported by a beam having the cross section shown. $\sigma_Y = 30$ ksi.

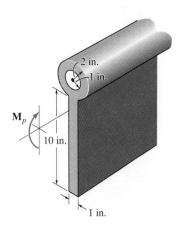

Prob. 12–49

12–50. The axle of the freight car is subjected to wheel loadings of 20 kip. If it is supported by two journal bearings at C and D, determine the maximum bending stress developed at the center of the axle, where the diameter is 5.5 in.

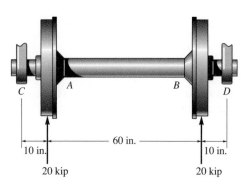

Prob. 12–50

12–51. The pin is used to connect the three links together. Due to wear, the load is distributed over the top and bottom of the pin as shown on the free-body diagram. If the diameter of the pin is 0.40 in., determine the maximum bending stress on the cross-sectional area. For the solution it is first necessary to determine the load intensities w_1 and w_2 and then draw the moment diagram.

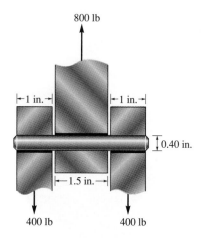

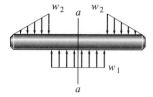

Prob. 12–51

13 Transverse Shear

In this chapter we will develop a method for finding the shear stress in a beam having a prismatic cross section and made from homogeneous material that behaves in a linear-elastic manner. The method of analysis to be developed will be somewhat limited to special cases of cross-sectional geometry. Although this is the case, it has many wide-range applications in engineering design and analysis.

13.1 Shear in Straight Members

Before we develop a relationship that describes the shear-stress distribution over the cross section of a beam, we will first make some preliminary remarks regarding the way shear stress acts within the beam. Realize that since beams are generally subjected to transverse loadings, these loadings not only cause an internal moment in the beam but *also* an internal shear force. This force **V**, shown in Fig. 13–1*a*, is necessary for translational equilibrium, and it is the result of a *transverse shear-stress* distribution that acts over the beam's cross section, Fig. 13–1*b*. As a result of this distribution, associated *longitudinal shear stresses* will *also* act along longitudinal planes of the beam. For example, a typical element removed from the interior point *A* on the cross section is subjected to both transverse and longitudinal shear stress as shown in Fig. 13–1*b*. In particular, note that the longitudinal shear stress at points *B* and *C*, located on the top and the bottom boundaries of the beam, must be zero since the top and bottom surfaces of the beam are free of any load. Consequently, the transverse shear stress at these points must also be zero.

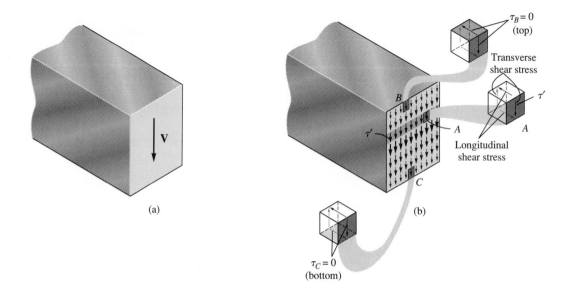

(a)

(b)

$\tau_B = 0$
(top)

Transverse
shear stress

τ'

A

A

Longitudinal
shear stress

τ'

B

C

$\tau_C = 0$
(bottom)

Fig. 13–1

It is also possible to physically illustrate why shear stress develops on the longitudinal planes of a beam that supports an internal shear loading by considering the beam to be made from three boards, Fig. 13–2a. If the top and bottom surfaces of each board are smooth, and the boards are not bonded together, then application of the load **P** will cause the boards to slide relative to one another, and so the beam will deflect as shown. On the other hand, if the boards are bonded together, then the longitudinal shear stresses between the boards will prevent the relative sliding of the boards, and consequently the beam will act as a single unit, Fig. 13–2b.

As a result of the internal shear-stress distribution, shear strains will be developed and these will tend to distort the cross section in a rather complex manner. To show how the shear strains vary, consider a bar made of a highly deformable material and marked with horizontal and vertical grid lines as shown in Fig. 13–3a. When the shear force is applied, it tends to deform these lines into the pattern shown in Fig. 13–3b. Notice that the squares near the top and bottom of the bar retain their shapes, since there it was shown in Fig. 13–1b that the shear stress and likewise the shear strain are zero (or very small). On the other hand, the shear strain on the square in the center of the bar will cause it to have the greatest deformation. In general then, the nonuniform shear-strain distribution over the cross section will cause the cross section to *warp,* that is, *not* to remain plane.

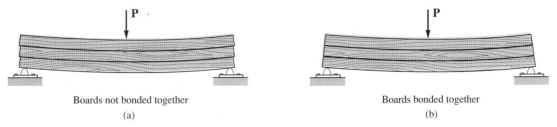

Boards not bonded together

(a)

Boards bonded together

(b)

Fig. 13–2

Recall that in the development of the flexure formula, we assumed that cross sections must *remain plane* and perpendicular to the longitudinal axis of the beam after deformation. Although these assumptions are *violated* when the beam is subjected to *both* bending and shear, we can generally assume the cross-sectional warping described above is small enough so that it can be neglected. This assumption is particularly true for the most common case of a *slender beam;* that is, one that has a small depth compared with its length.

In the previous chapters we developed the axial load, torsion, and flexure formulas by first determining the strain distribution, based on assumptions regarding the deformation of the cross section. Then using Hooke's law we related the strain to the stress; finally, we used the equilibrium requirements to relate the stress to the loading. Unlike these three cases, however, the shear-strain distribution throughout the depth of a beam *cannot* be easily expressed mathematically; for example, it is not uniform or linear for rectangular cross sections as we have shown. Therefore, the analysis of shear stress will be developed in a manner different from that used to study axial load, torsion, and bending moment.

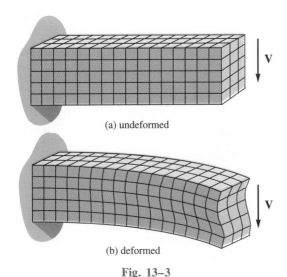

(a) undeformed

(b) deformed

Fig. 13–3

13.2 The Shear Formula

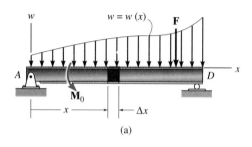

(a)

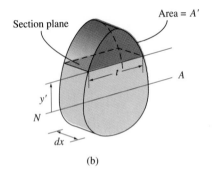

Section plane

Area = A'

(b)

Development of a relationship between the shear-stress distribution, acting over the cross section of a beam, and the resultant shear force at the section is based on a study of the *longitudinal shear stress* and the results of Eq. 7–2 ($V = dM/dx$). To show how this relationship is established, we will consider the differential element dx taken from the beam in Fig. 7–15a. This element is shown in Fig. 13–4a. A free-body diagram of the *entire element* that shows *only* the normal-stress distribution acting on the element is shown, Fig. 13–4b. This distribution is caused by the bending moments M and $M + dM$. We have excluded the effects of V, $V + dV$, and $w(x)$ on the free-body diagram since these loadings are vertical and are therefore not involved in a horizontal force summation. The element in Fig. 13–4b will indeed satisfy $\Sigma F_x = 0$ since the stress distribution on each side of the element forms only a couple moment and therefore a zero force resultant.

Now consider the shaded top *segment* of the element that has been sectioned at y' from the neutral axis, Fig. 13–4c. This segment has a width t at the section, and the cross-sectional sides each have an area A'. Because the resultant moments on each side of the element differ by dM, it can be seen that $\Sigma F_x = 0$ will not be satisfied *unless* a longitudinal shear stress τ acts over the bottom face of the segment. In the following analysis, we will assume this shear stress is *constant* across the width t of the bottom face. It acts on the area $t\,dx$. Applying the equation of horizontal force equilibrium, and using the flexure formula, Eq. 12–6, we have

$$\xrightarrow{+}\ \Sigma F_x = 0; \qquad \int_{A'} \sigma'\,dA' - \int_{A'} \sigma\,dA' - \tau(t\,dx) = 0$$

$$\int_{A'}\left(\frac{M + dM}{I}\right)y\,dA' - \int_{A'}\left(\frac{M}{I}\right)y\,dA' - \tau(t\,dx) = 0$$

$$\left(\frac{dM}{I}\right)\int_{A'} y\,dA' = \tau(t\,dx) \tag{13–1}$$

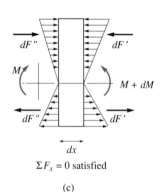

$\Sigma F_x = 0$ satisfied

(c)

Fig. 13–4

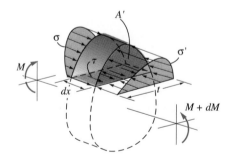

Three dimentional view

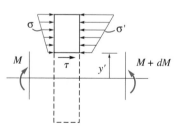

Profile view

(d)

Solving for τ, we get

$$\tau = \frac{1}{It}\left(\frac{dM}{dx}\right)\int_{A'} y\, dA'$$

This equation can be simplified by noting that $V = dM/dx$ (Eq. 13–2). Also, the integral represents the first moment of the area A' about the neutral axis. We will denote it by the symbol Q. Since the location of the centroid of the area A' is determined from $\bar{y}' = \int y\, dA'/A'$, we can also write

$$Q = \int_{A'} y\, dA' = \bar{y}'A' \qquad (13\text{--}2)$$

The final result is therefore

$$\boxed{\tau = \frac{VQ}{It}} \qquad (13\text{--}3)$$

Here

τ = the shear stress in the member at the point located a distance y' from the neutral axis, Fig. 13–4b. This stress is assumed to be constant and therefore *averaged* across the width t of the member.

V = the internal resultant shear force, determined from the method of sections and the equations of equilibrium.

I = the moment of inertia of the *entire* cross-sectional area computed about the neutral axis.

t = the width of the member's cross-sectional area, measured at the point where τ is to be determined.

Q = $\int_{A'} y\, dA' = \bar{y}'A'$, where A' is the top (or bottom) portion of the member's cross-sectional area, defined from the section where t is measured, and $\bar{y}'$ is the distance to the centroid of A', measured from the neutral axis.

The above equation is referred to as the *shear formula*. Although in the derivation we considered only the shear stresses acting on the beam's longitudinal plane, the formula applies as well for finding the transverse shear stress on the beam's cross-sectional area. This, of course, is because the transverse and longitudinal shear stresses are complementary and numerically equal.

Since Eq. 13–3 was derived from the flexure formula, it is necessary that the material behave in a linear-elastic manner and have a modulus of elasticity that is the *same* in tension as it is in compression.

13.3 Shear Stresses in Beams

In order to develop some insight as to the method of applying the shear formula and also discuss some of its limitations, we will now study the shear-stress distributions in a few common types of beam cross sections. Numerical applications of the shear formula will then be given in the examples that follow.

Rectangular Cross Section. Consider the beam to have a rectangular cross section of width b and height h as shown in Fig. 13–5a. The distribution of the shear stress throughout the cross section can be determined by computing the shear stress at an *arbitrary height y* from the neutral axis, Fig. 13–5b, and then plotting this function. Here the shaded area A' will be used for computing τ.* Hence

$$Q = \bar{y}'A' = \left[y + \frac{1}{2}\left(\frac{h}{2} - y\right)\right]\left(\frac{h}{2} - y\right)b$$

$$= \frac{1}{2}\left(\frac{h^2}{4} - y^2\right)b$$

Applying the shear formula, we have

$$\tau = \frac{VQ}{It} = \frac{V(\frac{1}{2})[(h^2/4) - y^2]b}{(\frac{1}{12}bh^3)b}$$

or

$$\tau = \frac{6V}{bh^3}\left(\frac{h^2}{4} - y^2\right) \tag{13–4}$$

This result indicates that the shear-stress distribution over the cross section is *parabolic*. As shown in Fig. 13–5c, the intensity varies from zero at the top and bottom, $y = \pm h/2$, to a maximum value at the neutral axis, $y = 0$. Specifically, since the area of the cross section is $A = bh$, then at $y = 0$ we have, from Eq. 13–4,

$$\tau_{max} = 1.5\frac{V}{A} \tag{13–5}$$

This same value for τ_{max} can be obtained directly from the shear formula, $\tau = VQ/It$, by realizing that τ_{max} occurs where Q is *largest*, since V, I, and t are *constant*. By inspection, Q will be a maximum when the area above (or below) the neutral axis is considered; that is, $A' = bh/2$ and $y' = h/4$. Thus,

---

*The area below y can also be used [$A' = b(h/2 + y)$], but doing so involves a bit more algebraic manipulation.

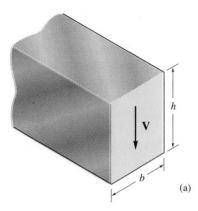

(a)

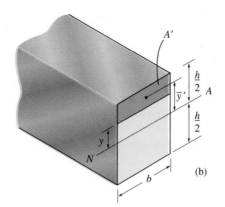

(b)

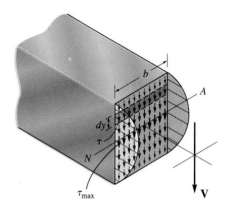

Shear stress distribution

(c)

$$\tau_{max} = \frac{VQ}{It} = \frac{V(h/4)(bh/2)}{[\frac{1}{12}bh^3]b} = 1.5\frac{V}{A}$$

By comparison, τ_{max} is 50% greater than the *average* shear stress computed from Eq. 8–4; that is, $\tau_{avg} = V/A$.

It is important to remember that for every τ acting on the cross-sectional area in Fig. 13–5c, there is a corresponding τ acting in the longitudinal direction along the beam. For example, if the beam is sectioned by a longitudinal plane through its neutral axis, then as noted above, the *maximum shear stress* acts on this plane, Fig. 13–5d. It is this stress that will cause a timber beam to fail as shown in Fig. 13–6. Here horizontal splitting of the wood starts to occur through the neutral axis at the beam's ends, since there the vertical reactions subject the beam to large shear stress and wood has a low resistance to shear along its grains, which are oriented in the longitudinal direction.

Since Eq. 13–4 expresses the shear-stress distribution as a function of position y, it is instructive to show that when integrated over the cross section it yields the resultant shear V. To do this, a differential strip of area $dA = b\ dy$ is chosen, Fig. 13–5c, and since τ acts uniformly over this strip, we have

$$\int_A \tau\ dA = \int_{-h/2}^{h/2} \frac{6V}{bh^3}\left(\frac{h^2}{4} - y^2\right) b\ dy$$

$$= \frac{6V}{h^3}\left[\frac{h^2}{4}y - \frac{1}{3}y^3\right]_{-h/2}^{h/2}$$

$$= \frac{6V}{h^3}\left[\frac{h^2}{4}(h) - \frac{1}{3}\left(\frac{h^3}{8} + \frac{h^3}{8}\right)\right] = V$$

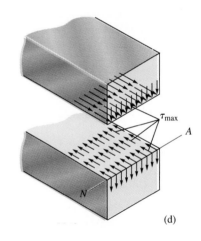

(d)

Fig. 13–5

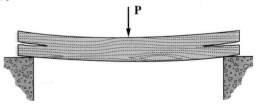

Fig. 13–6

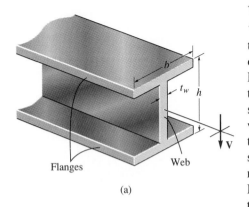

Flanges   Web

(a)

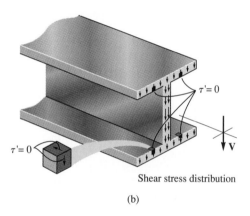

$\tau' = 0$ $\tau' = 0$

$\blacktriangledown V$

Shear stress distribution

(b)

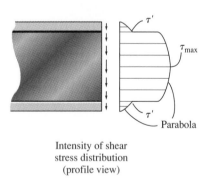

τ'

τ_{max}

τ'

Parabola

Intensity of shear
stress distribution
(profile view)

(c)

Fig. 13–7

Wide-Flange Beam. A *wide-flange beam* consists of two (wide) "flanges" and a "web" as shown in Fig. 13–7a. Using an analysis similar to that given before we can determine the shear-stress distribution acting over its cross section. The results are depicted graphically in Fig. 13–7b and 13–7c. Like the rectangular cross section, the shear stress varies *parabolically* over the beam's depth, since the cross section can be treated like the rectangular section, first having the width of the top flange, *b*, then the thickness of the web, t_w, and again the width of the bottom flange, *b*. In particular, notice that the shear stress will vary *only slightly* throughout the web, and also, a *jump* in shear stress occurs at the flange–web junction since the cross-sectional thickness changes at this point, or in other words, *t* in the shear formula changes. By comparison, the web will carry significantly more of the shear force than the flanges. This will be illustrated numerically in Example 13–2.

Limitations on the Use of the Shear Formula. One of the major assumptions used in the development of the shear formula is that the shear stress is *uniformly* distributed over the *width t* at the section where the shear stress is computed. In other words, the *average* shear stress is computed across the width. We can test the accuracy of this assumption by comparing it with a more exact mathematical analysis based on the theory of elasticity. In this regard, if the beam's cross section is rectangular, the actual shear-stress distribution across the neutral axis varies as shown in Fig. 13–8. The maximum value, τ'_{max}, occurs at the *edges* of the cross section, and its magnitude depends on the ratio *b/h* (width/depth). For sections having a *b/h* = 0.5, τ'_{max} is only about 3% greater than the shear stress calculated from the shear formula, Fig. 13–8a. However, for *flat sections,* say *b/h* = 2, τ'_{max} is about 40% greater than τ_{max}, Fig. 13–8b. The error becomes even greater as the section becomes flatter, or as the *b/h* ratio increases. Errors of this magnitude are certainly intolerable if one uses the shear formula to determine the shear stress in the *flange* of a wide-flange beam, as discussed above.

It should also be pointed out that the shear formula will not give accurate results when used to determine the shear stress at the flange–web junction of a wide-flange beam, since this is a point of sudden cross-sectional change and therefore a *stress concentration* occurs here. Furthermore, the inner regions of the flanges are free boundaries, Fig. 13–7b, and as a result the shear stress on these boundaries must be zero. If the shear formula is applied to determine the shear stress at these boundaries, however, one obtains a value of τ' that is *not* equal to zero, Fig. 13–7c. Fortunately, these limitations for applying the shear formula to the flanges of a wide-flange beam are not important in engineering practice. Most often engineers must only calculate the *average maximum shear stress* occurring at the neutral axis, where the *b/h* (width/depth) ratio is *very small,* and therefore the calculated result is very close to the *actual* maximum shear stress as explained above.

Another important limitation on the use of the shear formula can be pointed out with reference to Fig. 13–9a, which shows a beam having a cross section with an irregular or nonrectangular boundary. If we apply the shear formula to determine the (average) shear stress τ along the line AB, it will be directed as shown in Fig. 13–9b. Consider now an element of material taken from the boundary point B, such that one of its faces is located on the outer surface of the beam, Fig. 13–9c. Here the calculated shear stress τ on the front face of the element is resolved into components, τ' and τ''. By inspection, the component τ' must be equal to zero since its corresponding longitudinal component τ', acting on the stress-free boundary surface, must be zero. To satisfy this boundary condition therefore, the shear stress acting on the element at the boundary must be directed tangent to the boundary. The shear-stress distribution across line AB would then be directed as shown in Fig. 13–9d. Due to the inclination of the shear stresses at the boundaries, the maximum shear stress will occur at points A and B. Specific values for the shear stress must be obtained using the principles of the theory of elasticity. Note that we can, however, apply the shear formula to obtain the shear stress acting across each of the colored lines in Fig. 13–9a. These lines intersect the boundary of the cross section at *right angles,* and so the transverse shear stress at the boundary points will *not* have components normal to the boundary, Fig. 13–9e.

To summarize the above points, the shear formula should *not* be applied to members having cross sections that are *short or flat,* or at points where the cross section suddenly changes. Nor should it be applied across a section that intersects the boundary of the member at an angle other than 90°. Instead, for these cases the shear stress should be determined using more advanced methods based on the theory of elasticity.

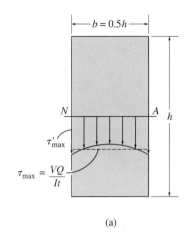

(a)

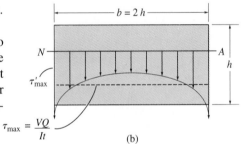

(b)

Fig. 13–8

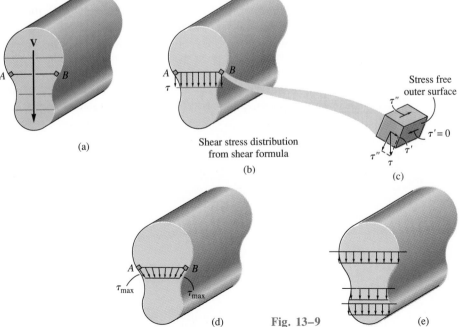

Shear stress distribution from shear formula

Stress free outer surface

Fig. 13–9

PROCEDURE FOR ANALYSIS

The shear formula can be used to find the shear-stress distribution acting over the cross-sectional area of a straight prismatic member made of homogeneous material that has linear-elastic behavior. It is required that the internal resultant shear force be directed along an axis of symmetry for the cross-sectional area. Also, Saint-Venant's principle requires that the shear formula be applied at points located away from any discontinuities in the cross section and away from points of concentrated loading. In order to apply the equation, the following procedure is suggested.

Internal Shear.   Section the member perpendicular to its axis at the point where the shear stress is to be determined, and use an appropriate free-body diagram and equation of equilibrium to obtain the internal shear **V** at the section.

Section Properties.   Determine the location of the centroid for the cross-sectional area in order to specify the position of the neutral axis. Then compute the moment of inertia I of the *entire area* about the neutral axis. Section the cross-sectional area through the point where the shear stress is to be determined, and measure the width t of the area at this section. The portion of the area lying either above or below this section is A'. Compute Q either by integration, $Q = \int_{A'} y \, dA'$, or by using $Q = \bar{y}'A'$. Here $\bar{y}'$ is the distance to the centroid of A', measured from the neutral axis. As noted in the derivation of the shear formula, it may be helped to realize that A' is the portion of the member's cross-sectional area that is being "held onto the member" by the longitudinal shear stresses, Fig. 13–4c.

Shear Stress.   Using a consistent set of units, substitute the data into the shear formula and compute the shear stress τ.

It is suggested that the proper direction of the transverse shear stress τ be established on a volume element of material located at the point where it is computed. This can be done by realizing that τ acts on the cross section in the same direction as **V**. From this, corresponding shear stresses acting on the other three planes of the element can then be established.

Example 13–1

The beam shown in Fig. 13–10a is made of wood and is subjected to a resultant internal vertical shear force of $V = 3$ kip. *(a)* Determine the shear stress in the beam at point P, and *(b)* compute the maximum shear stress in the beam.

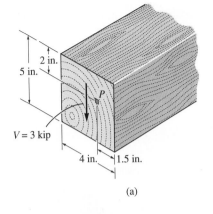

(a)

SOLUTION

Part (a)

Section Properties. The moment of inertia of the cross-sectional area computed about the neutral axis is

$$I = \frac{1}{12}bh^3 = \frac{1}{12}(4 \text{ in.})(5 \text{ in.})^3 = 41.7 \text{ in}^4$$

A horizontal section line is drawn through point P and the partial area A' is shown shaded in Fig. 13–10b. Hence

$$Q = \bar{y}'A' = \left[0.5 \text{ in.} + \frac{1}{2}(2 \text{ in.})\right](2 \text{ in.})(4 \text{ in.}) = 12 \text{ in}^3$$

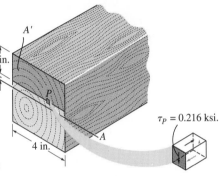

(b)   (c)

Shear Stress. The shear force at the section is $V = 3$ kip. Applying the shear formula, we have

$$\tau_P = \frac{VQ}{It} = \frac{(3 \text{ kip})(12 \text{ in}^3)}{(41.7 \text{ in}^4)(4 \text{ in.})} = 0.216 \text{ ksi} \qquad \textit{Ans.}$$

Since τ_P contributes to V, it acts downward at P on the cross section. Consequently, a volume element of the material at this point would have shear stresses acting on it as shown in Fig. 13–10c.

Part (b)

Section Properties. Maximum shear stress occurs at the neutral axis, since t is constant throughout the cross section and Q is largest for this case. For the shaded area A' in Fig. 13–10d, we have

$$Q = \bar{y}'A' = \left[\frac{2.5 \text{ in.}}{2}\right](4 \text{ in.})(2.5 \text{ in.}) = 12.5 \text{ in}^3$$

Shear Stress. Applying the shear formula yields

$$\tau_{\max} = \frac{VQ}{It} = \frac{(3 \text{ kip})(12.5 \text{ in}^3)}{(41.7 \text{ in}^4)(4 \text{ in.})} = 0.225 \text{ ksi} \qquad \textit{Ans.}$$

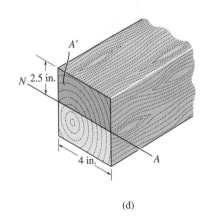

(d)

Fig. 13–10

▪ Example 13–2

A steel wide-flange beam has the dimensions shown in Fig. 13–11a. If it is subjected to a shear of $V = 80$ kN, *(a)* plot the shear-stress distribution acting over the beam's cross-sectional area, and *(b)* determine the shear force resisted by the web.

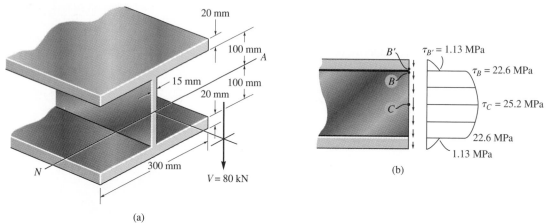

(a)

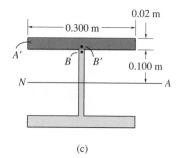

(c)

Fig. 13–11

SOLUTION

Part (a). The shear-stress distribution will be parabolic and varies in the manner shown in Fig. 13–11b. Due to symmetry, only the shear stresses at points B', B, and C have to be computed. To show how these values are obtained, we must first determine the moment of inertia of the cross-sectional area about the neutral axis. Working in meters, we have

$$I = \left[\frac{1}{12} (0.015)(0.200)^3 \right]$$

$$+ 2\left[\frac{1}{12} (0.300)(0.02)^3 + (0.300)(0.02)(0.110)^2 \right]$$

$$= 155.6(10^{-6}) \text{ m}^4$$

For point B', $t_{B'} = 0.300$ m, and A' is the shaded area shown in Fig. 13–11c. Thus,

$$Q_{B'} = \bar{y}'A' = [0.110](0.300)(0.02) = 0.660(10^{-3}) \text{ m}^3$$

so that

$$\tau_{B'} = \frac{VQ_{B'}}{It_{B'}} = \frac{80 \text{ kN}(0.660(10^{-3}) \text{ m}^3)}{155.6(10^{-6}) \text{ m}^4(0.300 \text{ m})} = 1.13 \text{ MPa}$$

For point B, $t_B = 0.015$ m and $Q_B = Q_{B'}$, Fig. 7–13c. Hence

$$\tau_B = \frac{VQ_B}{It_B} = \frac{80 \text{ kN}(0.660(10^{-3}) \text{ m}^3)}{155.6(10^{-6}) \text{ m}^4(0.015 \text{ m})} = 22.6 \text{ MPa}$$

Note from the discussion of "Limitations on the Use of the Shear Formula" that the calculated value for both $\tau_{B'}$ and τ_B will actually be very misleading. Why?

For point C, $t_C = 0.015$ m and A' is the shaded area shown in Fig. 13–11d. Considering this area to be composed of two rectangles, we have

$$Q_C = \Sigma \bar{y}'A' = [0.110](0.300)(0.02) + [0.05](0.015)(0.100)$$

$$= 0.735(10^{-3}) \text{ m}^3$$

Thus,

$$\tau_C = \tau_{max} = \frac{VQ_C}{It_C} = \frac{80 \text{ kN}[0.735(10^{-3}) \text{ m}^3]}{155.6(10^{-6}) \text{ m}^4(0.015 \text{ m})} = 25.2 \text{ MPa}$$

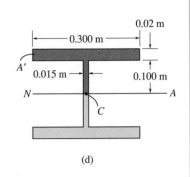

(d)

Part (b). The shear force in the web will be determined by first computing the shear force in each flange and then subtracting this result from $V = 80$ kN. To obtain the shear force in a flange, we must first determine the shear stress at the *arbitrary* location y, Fig. 13–11e. Using units of meters, we have

$$I = 155.6(10^{-6}) \text{ m}^4$$

$$t = 0.300 \text{ m}$$

$$A' = (0.300)(0.120 - y) \text{ m}^2$$

$$\bar{y}' = y + \frac{1}{2}(0.120 - y) = \frac{1}{2}(0.120 + y) \text{ m}$$

$$Q = \bar{y}'A' = [0.150]((0.120)^2 - y^2) \text{ m}^3$$

so that

$$\tau = \frac{VQ}{It} = \frac{80 \text{ kN}[0.150]((0.120)^2 - y^2) \text{ m}^3}{(155.6(10^{-6}) \text{ m}^4)(0.300 \text{ m})}$$

$$= 257((0.120)^2 - y^2) \text{ MPa}$$

This stress acts on the area strip $dA = 0.300 \, dy$ shown in Fig. 13–11e, and therefore the shear force resisted by the top flange is

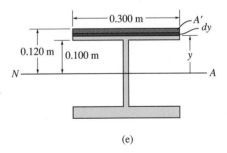

(e)

$$V_f = \int_{A_f} \tau \, dA = \int_{0.100}^{0.12} 257(10^6)((0.120)^2 - y^2)0.300 \, dy = 3.496 \text{ kN}$$

By symmetry, this force also acts in the bottom flange. Thus the shear force in the web is

$$V_w = V - 2V_f = 80 \text{ kN} - 2(3.496 \text{ kN}) = 73.0 \text{ kN} \qquad \textit{Ans.}$$

By comparison, the web supports 91% of the total shear (80 kN), whereas the flanges support the remaining 9%.

Example 13–3

The beam shown in Fig. 13–12*a* is made from two boards. Determine the shear stress in the glue necessary to hold the boards together at point *D*. The supports at *B* and *C* exert only vertical reactions on the beam.

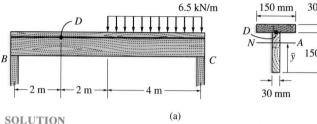

(a)

SOLUTION

Internal Shear. The support reactions on the beam are computed as shown in Fig. 13–12*b*. From the left segment it is seen that the resultant internal loading at a section through *D* consists of a shear of $V = 6.5$ kN and a moment of $M = 13$ kN · m.

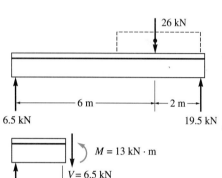

Section Properties. The centroid and therefore the neutral axis will be determined from the reference axis placed at the bottom of the cross-sectional area, Fig. 13–12*a*. Working in units of meters, we have

$$\bar{y} = \frac{\Sigma \tilde{y}A}{\Sigma A} = \frac{[0.075](0.150)(0.030) + [0.165](0.030)(0.150)}{(0.150)(0.030) + (0.030)(0.150)} = 0.120 \text{ m}$$

The moment of inertia, computed about the neutral axis, Fig. 13–12*a*, is therefore

$$I = \left[\frac{1}{12}(0.030)(0.150)^3 + (0.150)(0.030)(0.120 - 0.075)^2 \right]$$

$$+ \left[\frac{1}{12}(0.150)(0.030)^3 + (0.030)(0.150)(0.165 - 0.120)^2 \right]$$

$$= 27.0(10^{-6}) \text{ m}^4$$

The top board (flange) is being held onto the bottom board (web) by the glue, which is applied over the thickness $t = 0.03$ m. Consequently A' is defined as the area of the top board, Fig. 13–12*a*. We have

$$Q = \bar{y}'A' = [0.180 - 0.015 - 0.120](0.03)(0.150) = 0.2025(10^{-3}) \text{ m}^3$$

Shear Stress. Using the above data and applying the shear formula yields

$$\tau_D = \frac{VQ}{It} = \frac{6.5 \text{ kN}(0.2025(10^{-3}) \text{ m}^3)}{27.0(10^{-6}) \text{ m}^4(0.030 \text{ m})} = 1.62 \text{ MPa} \qquad Ans.$$

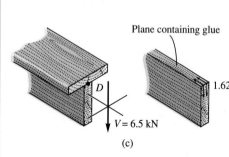

Fig. 13–12

The shear stress acting at the top of the bottom board is shown in Fig. 13–12*c*. Note that it is the glue's resistance to this lateral or *horizontal shear* stress that is necessary to hold the boards from slipping.

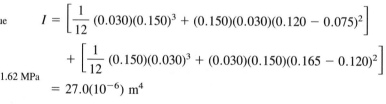

PROBLEMS

13–1. If the beam is subjected to a shear of $V = 15$ kN, determine the web's shear stress at A and B. Indicate the shear-stress components on a volume element located at these points.

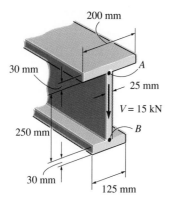

Prob. 13–1

13–3. If the wide-flange beam is subjected to a shear of $V = 25$ kip, determine (*a*) the maximum shear stress in the beam and (*b*) the average shear stress in the web.

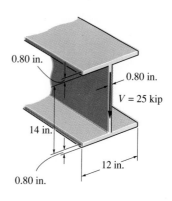

Prob. 13–3

13–2. The beam is made from three boards glued together at the seams A and B. If it is subjected to the loading shown, determine the shear stress developed in the glued joints at section $a–a$. The supports at C and D exert only vertical reactions on the beam.

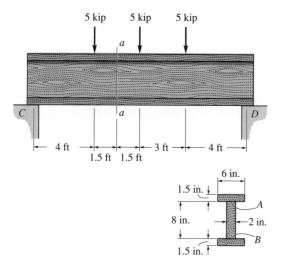

Prob. 13–2

*13–4. Railroad ties must be designed to resist large shear loadings. If the tie is subjected to the 34-kip rail loadings and an assumed uniformly distributed ground reaction, determine the intensity w for equilibrium, and compute the maximum shear stress in the tie at section $a–a$, which is located just to the left of the rail.

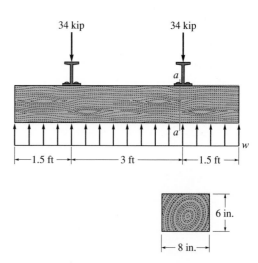

Prob. 13–4

13–5. If the T-beam is subjected to a vertical shear of $V = 10$ kip, determine the maximum shear stress in the beam. Also, compute the shear-stress jump at the flange–web junction AB. Sketch the variation of the shear-stress intensity over the entire cross section.

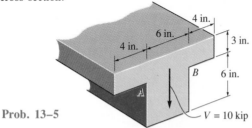

Prob. 13–5

13–6. The steel rod has a radius of 1.25 in. If it is subjected to a shear of $V = 5$ kip, determine the maximum shear stress.

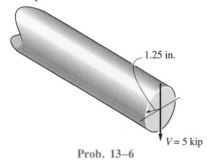

Prob. 13–6

13–7. Determine the maximum shear stress in the strut if it is subjected to a shear force of $V = 15$ kN.

*__13–8.__ Determine the maximum shear force V that the strut can support if the allowable shear stress for the material is $\tau_{allow} = 50$ MPa.

13–9. Determine the intensity of the shear stress distributed over the cross section of the strut if it is subjected to a shear force of $V = 12$ kN.

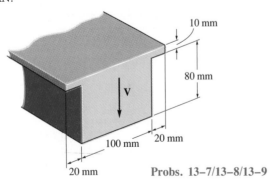

Probs. 13–7/13–8/13–9

13–10. Determine the shear stress at points B and C on the web of the beam located at section a–a.

13–11. Determine the maximum shear stress acting at section a–a in the beam.

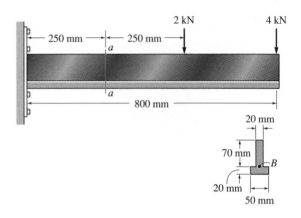

Probs. 13–10/13–11

*__13–12.__ Determine the shear stress at point B on the web of the cantilevered strut at section a–a.

13–13. Determine the maximum shear stress acting at section a–a of the cantilevered strut.

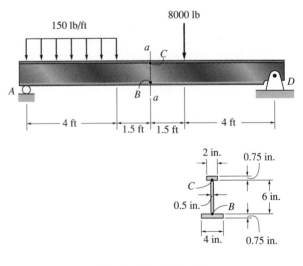

Probs. 13–12/13–13

13–14. Sketch the intensity of the shear-stress distribution acting over the beam's cross-sectional area, and determine the resultant shear force acting on the segment AB. The shear acting at the section is $V = 35$ kip.

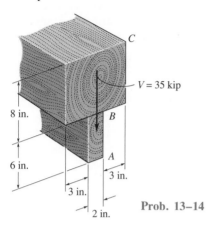

Prob. 13–14

13–15. The T-beam is subjected to a shear force of $V = 150$ kN. Determine the amount of this force that is supported by the web B.

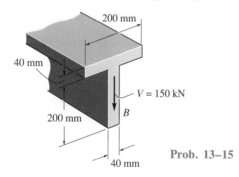

Prob. 13–15

*13–16.** The beam has a square cross section and is made of wood having an allowable shear stress of $\tau_{allow} = 1.4$ ksi. If it is subjected to a shear of $V = 1.5$ kip, determine the smallest dimension a of its sides.

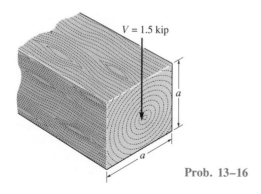

Prob. 13–16

13–17. The strut is subjected to a vertical shear of $V = 130$ kN. Plot the intensity of the shear-stress distribution acting over the cross-sectional area, and compute the resultant shear force developed in the vertical segment AB.

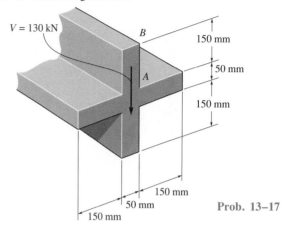

Prob. 13–17

13–18. Develop an expression for the average shear stress acting on the horizontal plane through the strut, located a distance y from the neutral axis.

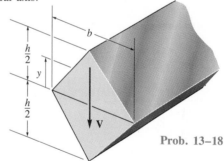

Prob. 13–18

13–19. The supports at A and B exert vertical reactions on the wood beam. If the distributed load $w = 4$ kip/ft, determine the maximum shear stress in the beam at section a–a.

*13–20.** The supports at A and B exert vertical reactions on the wood beam. If the allowable shear stress is $\tau_{allow} = 400$ psi, determine the intensity w of the largest distributed load that can be applied to the beam. Hint: Draw the shear diagram.

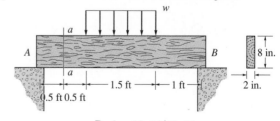

Probs. 13–19/13–20

13.4 Shear Flow in Built-up Members

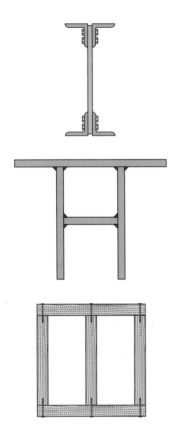

Fig. 13-13

Occasionally in engineering practice members are ''built up'' from several composite parts in order to achieve a greater resistance to loads. Some examples are shown in Fig. 13–13. If the loads cause the members to bend, fasteners such as nails, bolts, welding material, or glue, may be needed to keep the component parts from sliding relative to one another, Fig. 13-2. In order to design these fasteners it is necessary to know the shear force to be resisted by the fastener along the member's *length*. This loading, when measured as a force per unit length, is referred to as the *shear flow q*.

The magnitude of the shear flow along any longitudinal section of a beam can be obtained using a development similar to that for finding the shear stress in the beam. To show this, we will consider finding the shear flow along the juncture where the composite part in Fig. 13–14a is connected to the flange of the beam. As shown in Fig. 13–14b, three horizontal forces must act on this part. Two of these forces, F and $F + dF$, are developed by normal stresses caused by the moments M and $M + dM$, respectively. The third force, which for equilibrium equals dF, acts at the juncture and is to be supported by the fastener. Realizing that dF is the result of dM, then, as in the case of the shear formula, Eq. 13–1, we have

$$dF = \frac{dM}{I} \int_{A'} y \, dA'$$

The integral represents Q, that is, the moment of the colored area A' in Fig. 13–14b about the neutral axis for the cross section. Since the segment has a length dx, the shear flow, or force per unit length along the beam, is $q = dF/dx$. Hence dividing both sides by dx and noting that $V = dM/dx$, Eq. 7–2, we can write

$$q = \frac{VQ}{I} \tag{13–6}$$

Here

q = the shear flow, measured as a force per unit length along the beam

V = the internal resultant shear force, determined from the method of sections and the equations of equilibrium

I = the moment of inertia of the *entire* cross-sectional area computed about the neutral axis

$Q = \int_{A'} y \, dA = \bar{y}'A'$, where A' is the cross-sectional area of the segment that is connected to the beam at the juncture where the shear flow is to be calculated, and $\bar{y}'$ is the distance from the neutral axis to the centroid of A'

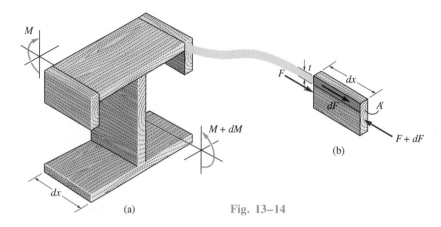

(a)

(b)

Fig. 13–14

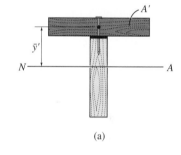

(a)

As previously stated, Eq 13–6 is often used in engineering practice to determine the fastener spacing, glue capacity, or weld size needed to hold the composite parts of a "built-up" member together when the member is subjected to shear caused by bending. Application follows the same "procedure for analysis" as outlined in Sec. 13.3 for the shear formula. In this regard it is very important to identify correctly the proper value for Q when computing the shear flow at a particular junction on the cross section. A proper understanding of this can always be realized if one recalls how the shear-flow equation was derived in reference to Fig. 13–14. A few examples should also serve to illustrate some of these points. Consider the beam cross sections shown in Fig. 13–15. The shaded composite parts are connected to the beam by fasteners such that the necessary shear flow q in Eq. 13–6 is determined by using a value of Q computed from A' and $\bar{y}'$ as indicated in each figure. Notice that this value of q will be resisted by a *single* fastener in Fig. 13–15a and 13–15b, by *two* fasteners in Fig. 13–15c, and by three fasteners in Fig. 13–15d. Hence, multiple fasteners will support an equal fraction of the shear flow as indicated in the figure; in other words, the fastener in Figs. 13–15a and 13–15b support the calculated value of q and in Figs. 13–15c and 13–15d the calculated value of q is divided by 2 and 3, respectively.

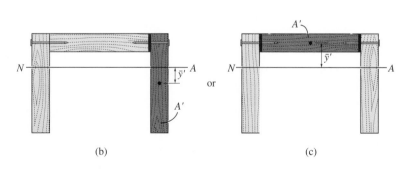

(b)   or   (c)

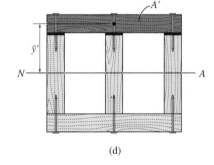

(d)

Fig. 13–15

Example 13–4

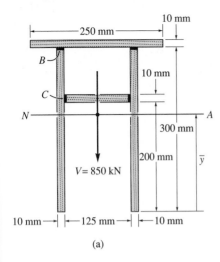

(a)

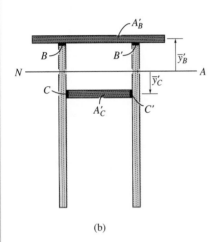

(b)

Fig. 13–16

The beam is constructed from four boards glued together as shown in Fig. 13–16a. If it is subjected to a shear of $V = 850$ kN, determine the shear flow at B and C that must be resisted by the glue.

SOLUTION

Section Properties. The neutral axis (centroid) will be located from the bottom of the beam, Fig. 13–16a. Working in units of millimeters, we have

$$\bar{y} = \frac{\Sigma \bar{y}A}{\Sigma A} = \frac{2[150](300)(10) + [205](125)(10) + [305](250)(10)}{2(300)(10) + 125(10) + 250(10)}$$

$$= 196.8 \text{ mm}$$

The moment of inertia computed about the neutral axis is thus

$$I = 2\left[\frac{1}{12}(10)(300)^3 + (10)(300)(196.8 - 150)^2 \right]$$

$$+ \left[\frac{1}{12}(125)(10)^3 + (125)(10)(205 - 196.8)^2 \right]$$

$$+ \left[\frac{1}{12}(250)(10)^3 + (250)(10)(305 - 196.8)^2 \right]$$

$$= 87.52(10^6) \text{ mm}^4 = 87.52(10^{-6}) \text{ m}^4$$

Since the glue at B and B' holds the top board to the beam, Fig. 13–16b, we have

$$Q_B = \bar{y}'_B A'_B = [305 - 196.8](250)(10) = 270.5(10^3) \text{ mm}^3$$

$$= 0.270(10^{-3}) \text{ m}^3$$

Likewise, the glue at C and C' holds the inner board to the beam, Fig. 13–16b, and so

$$Q_C = \bar{y}'_C A'_C = [205 - 196.8](125)(10) = 10.25(10^3) \text{ mm}^3$$

$$= 0.01025(10^{-3}) \text{ m}^3$$

Shear Flow. For B and B' we have

$$q'_B = \frac{VQ_B}{I} = \frac{850 \text{ kN}(0.270(10^{-3}) \text{ m}^3)}{87.52(10^{-6}) \text{ m}^4} = 2.62 \text{ MN/m}$$

And for C and C',

$$q'_C = \frac{VQ_C}{I} = \frac{850 \text{ kN}(0.01025(10^{-3}) \text{ m}^3)}{87.52(10^{-6}) \text{ m}^4} = 0.0995 \text{ MN/m}$$

Since *two seams* are used to secure each board, the glue per meter length of beam at each seam must be strong enough to resist *one-half* of each calculated value of q'. Thus,

$$q_B = 1.32 \text{ MN/m} \quad \text{and} \quad q_C = 0.0498 \text{ MN/m} \qquad Ans.$$

Example 13-5

A box beam is to be constructed from four boards nailed together as shown in Fig. 13–17a. If each nail can support a shear force of 30 lb, determine the maximum spacing s of nails at B and at C so that the beam will support the vertical force of 80 lb.

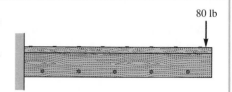

(a)

SOLUTION

Internal Shear. If the beam is sectioned at an *arbitrary point* along its length, Fig. 13–17b, the internal shear required for equilibrium is $V = 80$ lb.*

Section Properties. The moment of inertia of the cross-sectional area about the neutral axis can be determined by considering a 7.5-in. × 7.5-in. square minus a 4.5-in. × 4.5-in. square.

$$I = \frac{1}{12}(7.5)(7.5)^3 - \frac{1}{12}(4.5)(4.5)^3 = 229.5 \text{ in}^4$$

The shear flow at B is determined using a Q_B defined by the darker shaded area shown in Fig. 13–17c. It is this "symmetric" portion of the beam that is to be "held" onto the rest of the beam by nails on the left side and by the fibers of the board on the right side. Thus,

$$Q_B = \bar{y}'A' = [3](7.5)(1.5) = 33.75 \text{ in}^3$$

Likewise, the shear flow at C can be determined using the "symmetric" shaded area shown in Fig. 13–17d. We have

$$Q_C = \bar{y}'A' = [3](4.5)(1.5) = 20.25 \text{ in}^3$$

Shear Flow

$$q_B = \frac{VQ_B}{I} = \frac{80 \text{ lb}(33.75 \text{ in}^3)}{229.5 \text{ in}^4} = 11.76 \text{ lb/in.}$$

$$q_C = \frac{VQ_C}{I} = \frac{80 \text{ lb}(20.25 \text{ in}^3)}{229.5 \text{ in}^4} = 7.059 \text{ lb/in.}$$

These values represent the shear force per unit length of the beam that must be resisted by the nails at B and the fibers at B', Fig. 13–17c, and the nails at C and the fibers at C', Fig. 13–17d, respectively. Since in each case the shear flow is resisted at *two* surfaces and each nail can resist 30 lb, for B the spacing is

$$s_B = \frac{30 \text{ lb}}{(11.76/2) \text{ lb/in.}} = 5.10 \text{ in.} \qquad \text{Use } s_B = 5 \text{ in.} \qquad Ans.$$

And for C,

$$s_C = \frac{30 \text{ lb}}{(7.059/2) \text{ lb/in.}} = 8.50 \text{ in.} \qquad \text{Use } s_C = 8.5 \text{ in.} \qquad Ans.$$

*The magnitude of the moment **M** depends on the location x of the section. Its value, however, is not needed here.

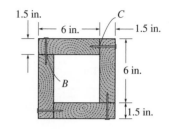

(b)

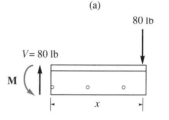

(c)

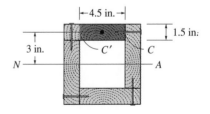

(d)

Fig. 13–17

Example 13–6

Nails having a total shear strength of 40 lb are used in a beam that can be constructed either as in Case I or as in Case II, Fig. 13–18. Determine the largest vertical shear in each case based on a nail spacing of 9 in.

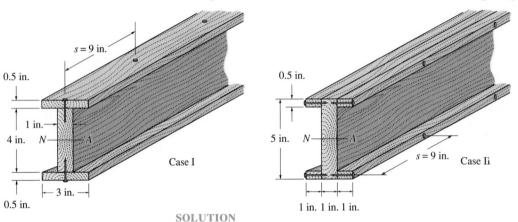

Fig. 13–18

SOLUTION

Since the geometry is the same in both cases, the moment of inertia about the neutral axis is

$$I = \frac{1}{12}(3 \text{ in.})(5 \text{ in.})^3 - 2\left[\frac{1}{12}(1 \text{ in.})(4 \text{ in.})^3\right] = 20.58 \text{ in}^4$$

Case I. For this design a single row of nails holds the top or bottom flange onto the web. For one of these flanges,

$$Q = \bar{y}'A' = [2.25 \text{ in.}](3 \text{ in.}(0.5 \text{ in.})) = 3.375 \text{ in}^3$$

so that

$$q = \frac{VQ}{I}$$

$$\frac{40 \text{ lb}}{9 \text{ in.}} = \frac{V(3.375 \text{ in}^3)}{20.58 \text{ in}^4}$$

$$V = 27.1 \text{ lb} \qquad\qquad Ans.$$

Case II. Here a single row of nails holds one of the side boards onto the web. Thus,

$$Q = \bar{y}'A' = [2.25 \text{ in.}](1 \text{ in.}(0.5 \text{ in.})) = 1.125 \text{ in}^3$$

$$q = \frac{VQ}{I}$$

$$\frac{40 \text{ lb}}{9 \text{ in.}} = \frac{V(1.125 \text{ in}^3)}{20.58 \text{ in}^4}$$

$$V = 81.3 \text{ lb} \qquad\qquad Ans.$$

PROBLEMS

13–21. The beam is subjected to a shear of $V = 800$ kN. Determine the average shear stress developed in the nails along the sides A and B if the nails are spaced $s = 100$ mm apart. Each nail has a diameter of 2 mm.

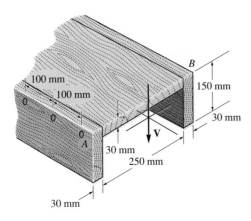

Prob. 13–21

13–22. The beam is constructed from three boards. If it is subjected to a shear of $V = 5$ kip, determine the spacing s of the nails used to hold the top and bottom flanges to the web. Each nail can support a shear force of 500 lb.

13–23. The beam is constructed from three boards. Determine the maximum shear V that it can sustain if the allowable shear stress for the wood is $\tau_{\text{allow}} = 400$ psi. What is the required spacing s of the nails if each nail can resist a shear force of 400 lb?

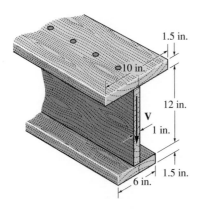

Probs. 13–22/13–23

*13–24. The beam is constructed from two boards fastened together at the top and bottom with two rows of nails spaced every 6 in. If each nail can support a 500-lb shear force, determine the maximum shear force V that can be applied to the beam.

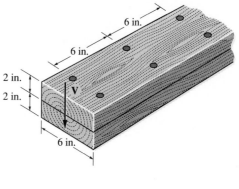

Prob. 13–24

13–25. The beam is fabricated from four boards nailed together as shown. Determine the shear force each nail along the sides C and the top D must resist if the nails are uniformly spaced at $s = 3$ in. Also, compute the maximum shear stress developed in the beam's web AB. The beam is subjected to a shear of $V = 4.5$ kip.

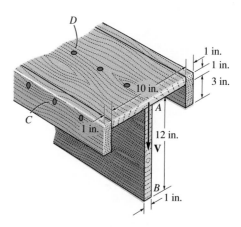

Prob. 13–25

13–26. The beam is made from four boards nailed together as shown. If the nails can each support a shear force of 100 lb, determine their required spacings s' and s if the beam is subjected to a shear of $V = 700$ lb.

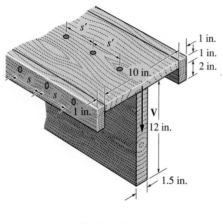

Prob. 13–26

13–27. The beam is fabricated from two equivalent structural tees and two plates; each plate has a width of 6 in. and a thickness of 0.5 in. If a shear of $V = 50$ kip is applied to the cross section, determine the maximum spacing of the bolts. Each bolt can resist a shear force of 15 kip.

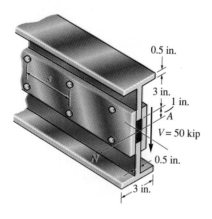

Prob. 13–27

*13–28. A beam is constructed from three boards bolted together as shown. Determine the shear force developed in each bolt if the bolts are spaced $s = 250$ mm apart and the applied shear is $V = 35$ kN.

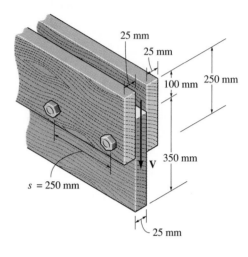

Prob. 13–28

13–29. The beam is constructed from four boards glued together at their seams. Based only on the allowable shear stress for the glue $\tau_{allow} = 150$ lb/in^2, what is the maximum vertical shear V that the beam can support (a) if it is used in the position shown, (b) if it is rotated 90°?

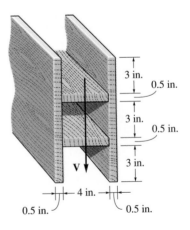

Prob. 13–29

REVIEW PROBLEMS

13–30. Show that the maximum shear stress in the shaft, which has a cross section of radius r and area $A = \pi r^2$, is $\tau_{max} = 4V/3A$.

Prob. 13–30

13–31. Develop an expression for the average vertical component of shear stress acting on the horizontal plane through the shaft, located a distance y from the neutral axis.

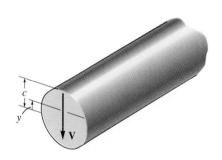

Prob. 13–31

*13–32. Determine the largest shear force V that the member can sustain if the allowable shear stress is $\tau_{allow} = 8$ ksi.

13–33. If the applied shear force $V = 18$ kip, determine the maximum shear stress in the member.

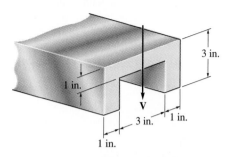

Probs. 13–32/13–33

13–34. The box beam is constructed from four boards that are fastened together using nails spaced along the beam every 2 in. If each nail can resist a shear force of 50 lb, determine the greatest shear force V that can be applied to the beam without causing failure of the nails.

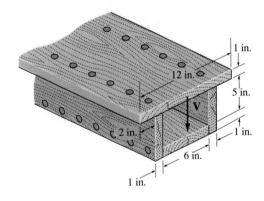

Prob. 13–34

13–35. The beam supports a vertical shear of $V = 7$ kip. Determine the resultant force this develops in segment AB of the beam.

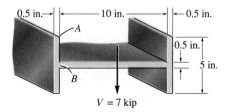

Prob. 13–35

*13–36.   The member consists of two triangular plastic strips bonded together along *AB*. If the glue can support an allowable shear stress of $\tau_{allow} = 600$ psi, determine the maximum vertical shear *V* that can be applied to the member based on the strength of the glue.

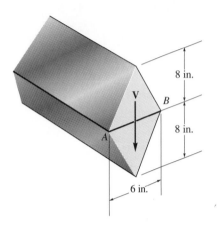

Prob. 13–36

13–37.   If the pipe is subjected to a shear of $V = 15$ kip, determine the maximum shear stress in the pipe.

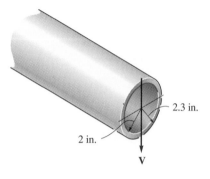

Prob. 13–37

14 Combined Loadings

This chapter serves as a review of the stress analysis that has been developed in the previous chapters regarding axial load, torsion, bending, and shear. We will discuss the solution of problems where several of these internal loads occur simultaneously on a member's cross section. Before doing this, however, the chapter begins with an analysis of stress developed in thin-walled pressure vessels.

14.1 Thin-Walled Pressure Vessels

Cylindrical or spherical vessels are commonly used in industry to serve as boilers or tanks. When under pressure, the material of which they are made is subjected to a loading from all directions. Although this is the case, the vessel can be analyzed in a simple manner provided it has a thin wall. In general, *"thin wall"* refers to a vessel having an inner-radius-to-wall-thickness ratio of 10 or more ($r/t \geq 10$). Specifically, when $r/t = 10$ the results of a thin-wall analysis will predict a stress that is approximately 4% less than the actual maximum stress in the vessel. For larger r/t ratios this error will be even smaller.

When the vessel wall is "thin," the stress distribution throughout its thickness will not vary significantly, and so we will assume that it is *uniform* or *constant*. Using this assumption, we will now analyze the state of stress in thin-walled cylindrical and spherical pressure vessels. In both cases the pressure in the vessel is understood to be the *gauge pressure,* since it measures the pressure *above* atmospheric pressure, which is assumed to exist both inside and outside the vessel's wall.

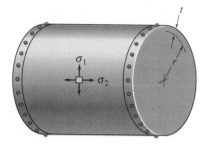

(a)

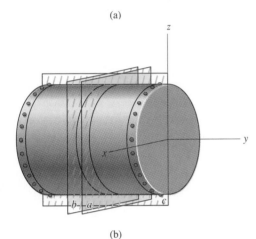

(b)

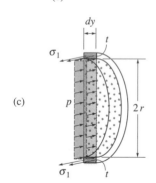

(c)

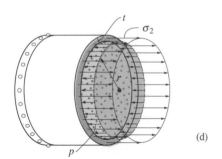

(d)

Fig. 14–1

Cylindrical Vessels. Consider the cylindrical vessel having a wall thickness t and inner radius r as shown in Fig. 14–1a. A gauge pressure p is developed within the vessel by a contained gas or fluid, which is assumed to have negligible weight. Due to the uniformity of this loading, an element of the vessel that is sufficiently removed from the end and oriented as shown in Fig. 14–1a is subjected to normal stresses σ_1 in the *circumferential or hoop direction* and σ_2 in the *longitudinal or axial direction*. Both of these stress components exert tension on the material. We wish to determine the magnitude of each of these components in terms of the vessel's geometry and the internal pressure. To do this requires using the method of sections and applying the equations of force equilibrium.

For the hoop stress, consider the vessel to be sectioned by planes a, b, and c, Fig. 14–1b. A free-body diagram of the back segment along with the contained gas or fluid is shown in Fig. 14–1c. Here only the loadings in the x direction are shown. These loadings are developed by the uniform hoop stress σ_1, acting throughout the vessel's wall, and the pressure acting on the vertical face of the sectioned gas or fluid. For equilibrium in the x direction, we require

$$\Sigma F_x = 0; \qquad 2[\sigma_1(t\,dy)] - p(2r\,dy) = 0$$

$$\boxed{\sigma_1 = \frac{pr}{t}} \qquad (14\text{–}1)$$

In order to obtain the longitudinal stress σ_2, we will consider the left portion of section b of the cylinder, Fig. 14–1b. As shown in Fig. 14–1d, σ_2 acts uniformly throughout the wall, and p acts on the section of gas or fluid. Since the mean radius is approximately equal to the vessel's inner radius, equilibrium in the y direction requires

$$\Sigma F_y = 0; \qquad \sigma_2(2\pi rt) - p(\pi r^2) = 0$$

$$\boxed{\sigma_2 = \frac{pr}{2t}} \qquad (14\text{–}2)$$

In the above equations,

σ_1, σ_2 = the normal stress in the hoop and longitudinal directions, respectively. Each is assumed to be *constant* throughout the wall of the cylinder, and each subjects the material to tension.

p = the internal gauge pressure developed by the contained gas or fluid.

r = the inner radius of the cylinder.

t = the thickness of the wall ($r/t \geq 10$).

Comparing Eqs. 14–1 and 14–2, it should be noted that the hoop or circumferential stress is twice as large as the longitudinal or axial stress. Consequently, when fabricating cylindrical pressure vessels from rolled-formed plates, the longitudinal joints must be designed to carry twice as much stress as the circumferential joints.

Spherical Vessels. We can analyze a spherical pressure vessel in a similar manner. For example, consider the vessel to have a wall thickness t and inner radius r and to be subjected to an internal gauge pressure p, Fig. 14–2a. If the vessel is sectioned in half using section a, the resulting free-body diagram is shown in Fig. 14–2b. Like the cylinder, equilibrium in the x direction requires

$$\Sigma F_x = 0; \qquad \sigma_2(2\pi rt) - p(\pi r^2) = 0$$

$$\boxed{\sigma_2 = \frac{pr}{2t}} \tag{14–3}$$

By comparison, this is the *same result* as that obtained for the longitudinal stress in the cylindrical pressure vessel. Furthermore, from the analysis, this stress will be the same *regardless* of the orientation of the hemispheric free-body diagram. Consequently, an element of the material is subjected to the state of stress shown in Fig. 14–2a.

The above analysis indicates that an element of material taken from either a cylindrical or a spherical pressure vessel is subjected to *biaxial stress,* i.e., normal stress existing in only two directions, Figs. 14–1a and 14–2a. Actually, the material of the vessel is also subjected to a *radial stress, σ_3,* which acts along a radial line. This stress has a maximum value equal to the pressure p at the interior wall and decreases through the wall to zero at the exterior surface of the vessel, since the gauge pressure there is zero. For thin-walled vessels, however, we will *ignore* the radial-stress component, since our limiting assumption of $r/t = 10$ results in σ_2 and σ_1 being, respectively, 5 and 10 times *higher* than the maximum radial stress, $(\sigma_3)_{\max} = p$. Lastly, realize that the above formulas apply only for vessels subjected to an internal gauge pressure. If the vessel is subjected to an external pressure, it may cause the vessel to become unstable, and collapse may occur by buckling.

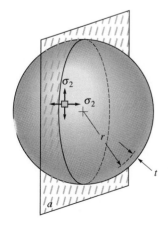

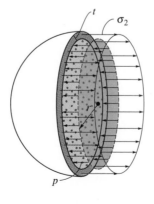

(a) **Fig. 14–2** (b)

Example 14–1

A cylindrical pressure vessel has an inner diameter of 4 ft and a thickness of $\frac{1}{2}$ in. Determine the maximum internal pressure it can sustain so that neither its circumferential nor its longitudinal stress component exceeds 20 ksi. Under the same conditions, what is the maximum internal pressure that a similar-size spherical vessel can sustain?

SOLUTION

Cylindrical Pressure Vessel.   The maximum stress occurs in the circumferential direction. From Eq. 14–1 we have

$$\sigma_1 = \frac{pr}{t}; \qquad\qquad 20 \text{ kip/in}^2 = \frac{p(24 \text{ in.})}{\frac{1}{2} \text{ in.}}$$

$$p = 417 \text{ psi} \qquad\qquad Ans.$$

Note that when this pressure is reached, from Eq. 14–2, the stress in the longitudinal direction will be $\sigma_2 = \frac{1}{2}(20 \text{ ksi}) = 10$ ksi. Furthermore, the *maximum stress* in the *radial direction* occurs on the material at the inner wall of the vessel and is $(\sigma_3)_{max} = p = 417$ psi. This value is 48 times smaller than the circumferential stress (20 ksi), and as stated earlier, its effects will be neglected.

Spherical Vessel.   Here the maximum stress occurs in any two perpendicular directions on an element of the vessel, Fig. 14–2a. From Eq. 14–3, we have

$$\sigma_2 = \frac{pr}{2t}; \qquad\qquad 20 \text{ kip/in}^2 = \frac{p(24 \text{ in.})}{2(\frac{1}{2} \text{ in.})}$$

$$p = 833 \text{ psi} \qquad\qquad Ans.$$

Although it is more difficult to fabricate, the spherical pressure vessel will carry twice as much internal pressure as a cylindrical vessel.

PROBLEMS

14–1.   The spherical gas tank has an inner radius of $r = 1.5$ m. (If it is subjected to an internal pressure of $p = 300$ kPa, determine its required thickness if the maximum normal stress is not to exceed 12 MPa.

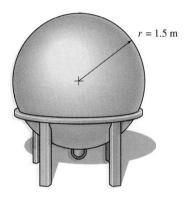

Prob. 14–1

14–2.   Air pressure in the cylinder is increased by exerting forces of 800 lb on the two pistons, each having a radius of 4.875 in. If the cylinder has a wall thickness of 0.25 in., determine the state of stress in the wall of the cylinder.

14–3.   The cap on the cylindrical tank is bolted to the tank along the flanges. The tank has an inner diameter of 1.5 m and a wall thickness of 18 mm. If the largest normal stress is not to exceed 150 MPa, determine the maximum pressure the tank can sustain. Also, compute the number of bolts required to attach the cap to the tank if each bolt has a diameter of 20 mm. The allowable stress for the bolts is $(\sigma_{allow})_b = 180$ MPa.

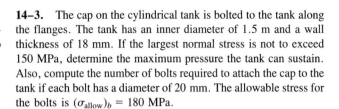

Prob. 14–3

*14–4.   The barrel is filled to the top with water. Determine the distance s that the top hoop should be placed from the bottom hoop so that the tensile force in each hoop is the same. Also, what is the force in each hoop? The barrel has an inner diameter of 4 ft. Neglect its wall thickness. Assume that only the hoops resist the water pressure. *Note:* Water develops pressure in the barrel according to Pascal's law, $p = (62.4z)$ lb/ft^2, where z is the depth from the surface of the water in feet.

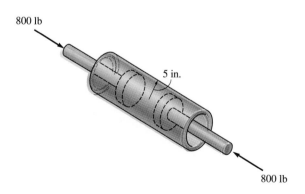

Prob. 14–2

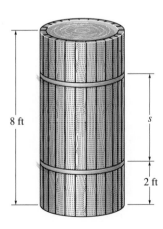

Prob. 14–4

14–5. A boiler is constructed of 8-mm steel plates that are fastened together at their ends using a butt joint consisting of two 8-mm cover plates and rivets having a diameter of 10 mm and spaced 50 mm apart as shown. If the steam pressure in the boiler is 1.35 MPa, determine (a) the circumferential stress in the boiler's plate apart from the seam, (b) the circumferential stress in the outer cover plate along the rivet line a–a, and (c) the shear stress in the rivets.

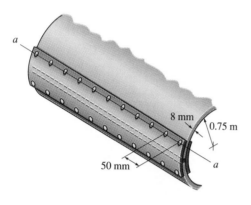

Prob. 14–5

14–6. The staves or vertical members of the wooden tank are held together using semicircular hoops having a thickness of 0.5 in. and a width of 2 in. Determine the normal stress in hoop AB if the tank is subjected to an internal gauge pressure of 2 psi and this loading is transmitted directly to the hoops. Also, if 0.25-in.-diameter bolts are used to connect each hoop together, determine the tensile stress in each bolt at A and B. Assume hoop AB supports the pressure loading within a 12-in. length of the tank as shown.

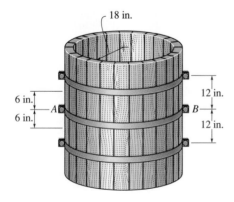

Prob. 14–6

14–7. A wood stave pipe having an inner diameter of 3 ft is bound together using steel hoops having a cross-sectional diameter of 0.5 in. If the allowable stress for the hoops is $\sigma_{allow} = 12$ ksi, determine their maximum spacing s along the section of pipe so that the pipe can resist an internal gauge pressure of 4 psi. Assume each hoop supports the pressure loading acting along the length s of the pipe.

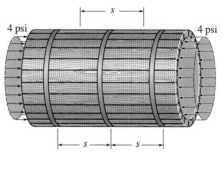

Prob. 14–7

14.2 State of Stress Caused by Combined Loading

In previous chapters we developed methods for determining the stress distributions in a member subjected to either an internal axial force, a shear force, a bending moment, or a torsional moment. Most often, however, the cross section of a member is subjected to several of these types of loadings *simultaneously*, and as a result, the method of superposition, if it applies, can be used to determine the *resultant* stress distribution caused by the loads. For application, the stress distribution due to *each loading* is first determined, and then these distributions are superimposed to determine the resultant stress distribution. As stated in Sec. 10.3, the principle of superposition can be used for this purpose provided a *linear relationship* exists between the *stress* and the *loads*. Also, the geometry of the member should *not* undergo *significant change* when the loads are applied. This is necessary in order to ensure that the stress produced by one load is not related to the stress produced by any other load. The discussion will be confined to meet these two criteria.

PROCEDURE FOR ANALYSIS

The following procedure provides a general means for establishing the normal and shear stress components at a point in a member when the member is subjected to several different types of loadings simultaneously. It will be assumed the material is homogeneous and behaves in a linear-elastic manner. Also, Saint-Venant's principle requires that the point where the stress is to be determined is far removed from any discontinuities in the cross section or points of applied load.

Internal Loadings.   Section the member perpendicular to its axis at the point where the stress is to be determined. Use the necessary free-body diagrams and equations of equilibrium to obtain the resultant internal normal and shear force components and the bending and torsional moment components. The force components should act through the *centroid* of the cross section, and the moment components should be computed about *centroidal axes,* which represent the principal axes of inertia for the cross section.

Stress Components.   Compute the stress component associated with *each* internal loading. For each case, represent the effect either as a distribution of stress acting over the entire cross-sectional area, or show the stress on an element of the material located at a specified point on the cross section.

Normal Force. The internal normal force is developed by a uniform normal-stress distribution determined from $\sigma = P/A$.

Shear Force. The internal shear force in a member which is also subjected to bending is developed by a shear-stress distribution determined from the shear formula, $\tau = VQ/It$. Special care, however, must be exercised when applying this equation, as noted in Sec. 13.3.

Bending Moment. For *straight members* the internal bending moment is developed by a normal-stress distribution that varies linearly from zero at the neutral axis to a maximum at the outer boundary of the member. The stress distribution is determined from the flexure formula, $\sigma = -My/I$.

Torsional Moment. For circular shafts and tubes the internal torsional moment is developed by a shear-stress distribution that varies linearly from the central axis of the shaft to a maximum at the shaft's outer boundary. The shear-stress distribution is determined from the torsion formula, $\tau = T\rho/J$.

Thin-Walled Pressure Vessels.   If the vessel is a thin-walled cylinder, the internal pressure p will cause a biaxial state of stress in the material such that the hoop or circumferential stress component is $\sigma_1 = pr/t$ and the longitudinal stress component is $\sigma_2 = pr/2t$. If the vessel is a thin-walled sphere, then the biaxial state of stress is represented by two equivalent components, each having a magnitude of $\sigma_2 = pr/2t$.

Superposition.   Once the normal and shear stress components for each loading case have been calculated, use the principle of superposition and determine the resultant normal and shear stress components. Represent the results on an element of material located at the point, or show the results as a distribution of stress acting over the member's cross-sectional area.

Problems in this section, which involve combined loadings, serve as a basic *review* of the application of many of the important stress equations mentioned above. A thorough understanding of how these equations are applied, as indicated in the previous chapters, is necessary if one is to successfully solve the problems at the end of this section. The following examples should be carefully studied before proceeding to solve the problems.

A force of 150 lb is applied to the edge of the member shown in Fig. 14–3a. Neglect the weight of the member and determine the state of stress at points B and C.

SOLUTION

Internal Loadings. The member is sectioned through B and C. For equilibrium at the section there must be an axial force of 150 lb acting through the centroid and a bending moment of 750 lb · in. about the centroidal or principal axis, Fig. 14–3b.

Stress Components

Normal Force. The uniform normal-stress distribution due to the normal force is shown in Fig. 14–3c. Here

$$\sigma = \frac{P}{A} = \frac{150 \text{ lb}}{(10 \text{ in.})(4 \text{ in.})} = 3.75 \text{ psi}$$

Bending Moment. The normal-stress distribution due to the bending moment is shown in Fig. 14–3d. The maximum stress is

$$\sigma_{\max} = \frac{Mc}{I} = \frac{750 \text{ lb} \cdot \text{in.}(5 \text{ in.})}{[\frac{1}{12} (4 \text{ in.})(10 \text{ in.})^3]} = 11.25 \text{ psi}$$

Superposition. If the above normal-stress distributions are added algebraically, the resultant stress distribution is shown in Fig. 14–3e. Although it is not needed here, the location of the line of zero stress can be determined by proportional triangles; i.e.,

$$\frac{7.5 \text{ psi}}{x} = \frac{15 \text{ psi}}{(10 \text{ in.} - x)}$$

$$x = 3.33 \text{ in.}$$

Elements of material at B and C are subjected only to normal or *uniaxial stress* as shown in Fig. 14–3f and 14–3g. Hence,

$$\sigma_B = 7.5 \text{ psi} \quad \text{(tension)} \qquad \textit{Ans.}$$
$$\sigma_C = 15 \text{ psi} \quad \text{(compression)} \qquad \textit{Ans.}$$

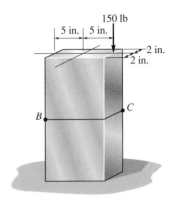

(a)

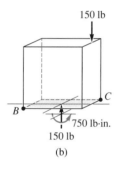

(b)

Fig. 14–3

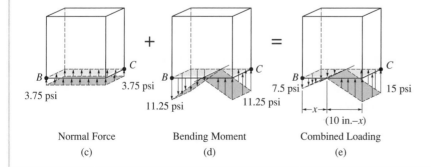

Normal Force	Bending Moment	Combined Loading
(c)	(d)	(e)

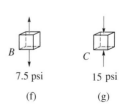

(f)   (g)

Example 14–3

The member shown in Fig. 14–4a has a rectangular cross section. Determine the state of stress that the loading produces at point C.

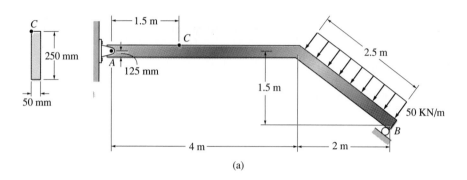

(a)

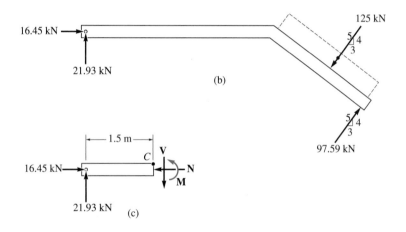

(b)

(c)

Fig. 14–4

SOLUTION

Internal Loadings. The support reactions on the member have been computed and are shown in Fig. 14–4b. If the left segment AC of the member is considered, Fig. 14–4c, the resultant internal loadings at the section consist of a normal force, a shear force, and a bending moment. Solving,

$$N = 16.45 \text{ kN} \qquad V = 21.93 \text{ kN} \qquad M = 32.89 \text{ kN} \cdot \text{m}$$

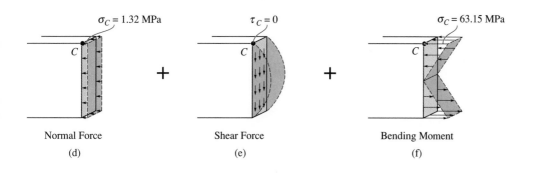

Normal Force
(d)

Shear Force
(e)

Bending Moment
(f)

64.5 MPa

(g)

Stress Components

Normal Force. The force is produced by a uniform normal-stress distribution over the cross section. At point C, Fig. 14–4d, it has a magnitude of

$$\sigma_C = \frac{16.45 \text{ kN}}{(0.050 \text{ m})(0.250 \text{ m})} = 1.32 \text{ MPa}$$

Shear Force. Here the area $A' = 0$, since point C is located at the top of the member. Thus $Q = \bar{y}'A' = 0$ and for C, Fig. 14–4e, the shear stress

$$\tau_C = 0$$

Bending Moment. Point C is located at $y = c = 125$ mm from the neutral axis, so the normal stress at C, Fig. 14–4f, is

$$\sigma_C = \frac{Mc}{I} = \frac{(32.89 \text{ kN} \cdot \text{m})(0.125 \text{ m})}{[\frac{1}{12}(0.050 \text{ m})(0.250)^3]} = 63.15 \text{ MPa}$$

Superposition. The shear stress is zero. Adding the normal stresses computed above gives a compressive stress at C having a value of

$$\sigma_C = 1.32 \text{ MPa} + 63.15 \text{ MPa} = 64.5 \text{ MPa} \qquad \textit{Ans.}$$

This result, acting on an element at C, is shown in Fig. 14–4g.

Example 14–4

The solid rod shown in Fig. 14–5a has a radius of 0.75 in. If it is subjected to the loading shown, determine the state of stress at point A.

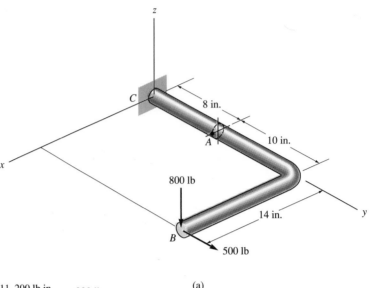

(a)

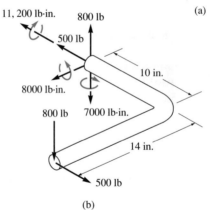

(b)

Fig. 14–5

SOLUTION

Internal Loadings.   The rod is sectioned through point A. Using the free-body diagram of segment AB, Fig. 14–5b, the resultant internal loadings can be determined for equilibrium. The normal force (500 lb) and shear force (800 lb) must act through the centroid of the cross section and the bending-moment components (8000 lb · in. and 7000 lb · in.) are applied about centroidal (principal) axes. In order to better "visualize" the stress distributions due to each of these loadings, we will consider the *equal but opposite resultants* acting on segment AC of the rod, Fig. 14–5c.

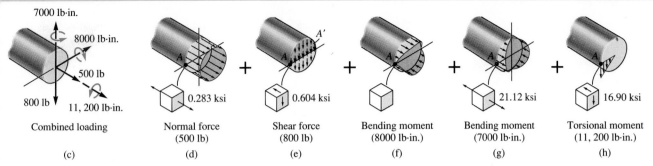

Combined loading
(c)

Normal force
(500 lb)
(d)

Shear force
(800 lb)
(e)

Bending moment
(8000 lb·in.)
(f)

Bending moment
(7000 lb·in.)
(g)

Torsional moment
(11, 200 lb·in.)
(h)

Stress Components

Normal Force. The normal-stress distribution is shown in Fig. 14–5d. For point A, we have

$$\sigma_A = \frac{P}{A} = \frac{500\ \text{lb}}{\pi(0.75\ \text{in.})^2} = 283\ \text{psi} = 0.283\ \text{ksi}$$

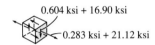

Shear Force. The shear-stress distribution is shown in Fig. 14–5e. For point A, Q is determined from the shaded *semicircular* area. Using the table on the inside front cover, we have

$$Q = \bar{y}'A' = \frac{4(0.75\ \text{in.})}{3\pi}\left[\frac{1}{2}\pi\left(0.75\ \text{in.}\right)^2\right] = 0.2813\ \text{in}^3$$

so that

$$\tau_A = \frac{VQ}{It} = \frac{800\ \text{lb}(0.2813\ \text{in}^3)}{[\frac{1}{4}\pi(0.75\ \text{in.})^4]2(0.75\ \text{in.})} = 604\ \text{psi} = 0.604\ \text{ksi}$$

or

(i)

Bending Moments. For the 8000-lb · in. component, point A lies on the neutral axis, Fig. 14–5f, so the normal stress is

$$\sigma_A = 0$$

For the 7000-lb · in. moment, c = 0.75 in., so the normal stress at point A, Fig. 14–5g, is

$$\sigma_A = \frac{Mc}{I} = \frac{7000\ \text{lb} \cdot \text{in.}(0.75\ \text{in.})}{[\frac{1}{4}\pi(0.75\ \text{in.})^4]} = 21{,}126\ \text{psi} = 21.12\ \text{ksi}$$

Torsional Moment. At point A, $\rho_A = c = 0.75$ in., Fig. 14–5h. Thus the shear stress is

$$\tau_A = \frac{Tc}{J} = \frac{11{,}200\ \text{lb} \cdot \text{in.}(0.75\ \text{in.})}{[\frac{1}{2}\pi(0.75\ \text{in.})^4]} = 16{,}901\ \text{psi} = 16.90\ \text{ksi}$$

Superposition. When the above results are superimposed, it is seen that an element of material at A is subjected to both normal and shear stress components, Fig. 14–5i.

Example 14–5

The rectangular block of negligible weight in Fig. 14–6a is subjected to a vertical force of 40 kN, which is applied to its corner. Determine the normal-stress distribution acting on a section through $ABCD$.

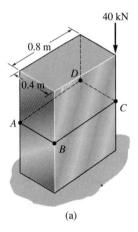
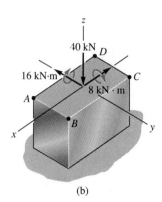

(a)   (b)

Fig. 14–6

SOLUTION

Internal Loadings. If we consider the equilibrium of the bottom segment of the block, Fig. 14–6b, it is seen that the 40-kN force must act through the centroid of the cross section and *two* bending-moment components must also act about the centroidal or principal axes of inertia for the section.

Stress Components

Normal Force. The uniform normal-stress distribution is shown in Fig. 14–6c. We have

$$\sigma = \frac{P}{A} = \frac{40 \text{ kN}}{(0.8 \text{ m})(0.4 \text{ m})} = 125 \text{ kPa}$$

Bending Moments. The normal-stress distribution for the 8-kN · m moment is shown in Fig. 14–6d. The maximum stress is

$$\sigma_{max} = \frac{M_x c_y}{I_x} = \frac{8 \text{ kN} \cdot \text{m}(0.2 \text{ m})}{[\frac{1}{12}(0.8 \text{ m})(0.4 \text{ m})^3]} = 375 \text{ kPa}$$

Likewise, for the 16-kN · m moment, Fig. 14–6e, the maximum normal stress is

$$\sigma_{max} = \frac{M_y c_x}{I_y} = \frac{16 \text{ kN} \cdot \text{m}(0.4 \text{ m})}{[\frac{1}{12}(0.4 \text{ m})(0.8 \text{ m})^3]} = 375 \text{ kPa}$$

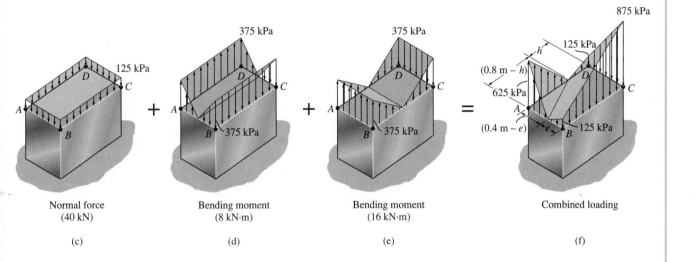

Normal force
(40 kN)

(c)

Bending moment
(8 kN·m)

(d)

Bending moment
(16 kN·m)

(e)

Combined loading

(f)

Superposition.   The normal stress at each corner point can be determined by algebraic addition. Assuming that tensile stress is positive, we have

$$\sigma_A = -125 + 375 + 375 = 625 \text{ kPa}$$
$$\sigma_B = -125 - 375 + 375 = -125 \text{ kPa}$$
$$\sigma_C = -125 - 375 - 375 = -875 \text{ kPa}$$
$$\sigma_D = -125 + 375 - 375 = -125 \text{ kPa}$$

Since the stress distributions due to bending moment are linear, the resultant stress distribution is also linear and therefore looks like that shown in Fig. 14–6f. The line of zero stress can be located along each side by proportional triangles. From the figure we require

$$\frac{(0.4 \text{ m} - e)}{625 \text{ kPa}} = \frac{e}{125 \text{ kPa}}$$
$$e = 0.0667 \text{ m}$$

and

$$\frac{(0.8 \text{ m} - h)}{625 \text{ kPa}} = \frac{h}{125 \text{ kPa}}$$
$$h = 0.133 \text{ m}$$

Example 14–6

A rectangular block has a negligible weight and is subjected to a vertical force **P**, Fig. 14–7a. (a) Determine the range of values for the eccentricity e_y of the load along the y axis so that it does not cause any tensile stress in the block. (b) Specify the region on the cross section where **P** may be applied without causing a tensile stress in the block.

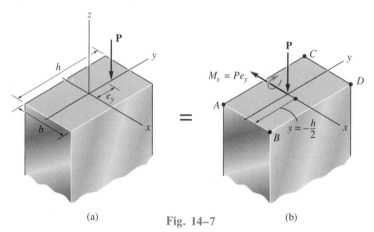

(a)

Fig. 14–7

(b)

SOLUTION

Part (a).   When **P** is moved to the centroid of the cross section, Fig. 14–7b, it is necessary to add a couple moment $M_x = Pe_y$ in order to maintain a statically equivalent loading. The combined normal stress at any coordinate location $\pm y$ on the cross section caused by these two loadings is

$$\sigma = -\frac{P}{A} - \frac{(Pe_y)y}{I_x} = -\frac{P}{A}\left(1 + \frac{Ae_y y}{I_x}\right)$$

Here the negative sign indicates compressive stress. For positive e_y, Fig. 14–7a, the smallest compressive stress will occur along edge AB at $y = -h/2$, Fig. 14–7b. Hence

$$\sigma_{min} = -\frac{P}{A}\left(1 - \frac{Ae_y h}{2I_x}\right)$$

This stress will remain negative, i.e., compressive, provided the term in parentheses is positive; i.e.,

$$1 > \frac{Ae_y h}{2I_x}$$

Since $A = bh$ and $I_x = \frac{1}{12} bh^3$, then

$$1 > \frac{6e_y}{h}$$

or

$$e_y < \frac{1}{6} h \qquad \qquad Ans.$$

In other words, if $-\frac{1}{6}h \leq e_y \leq \frac{1}{6}h$, the stress in the block along edge AB or CD will be zero or remain *compressive*. This is sometimes referred to as the "*middle-third rule.*" It is obviously important to keep this rule in mind when loading columns or arches having a rectangular cross section and made of materials such as stone or concrete, which can support little or no tensile stress.

Part (b). We can extend the above analysis in two directions by assuming that **P** acts in the positive quadrant of the x–y plane, Fig. 14–7c. The equivalent static loading when **P** acts at the centroid is shown in Fig. 14–7d. At any coordinate point x,y on the cross section, the combined normal stress due to both normal and bending loadings is

$$\sigma = -\frac{P}{A} - \frac{Pe_y y}{I_x} - \frac{Pe_x x}{I_y}$$

$$= -\frac{P}{A}\left(1 + \frac{Ae_y y}{I_x} + \frac{Ae_x x}{I_y}\right)$$

By inspection, Fig. 14–7d, the moments both create tensile stress at point A and the normal force creates a compressive stress there. Hence, the smallest compressive stress will occur at point A, for which $x = -b/2$ and $y = -h/2$. Thus,

$$\sigma_A = -\frac{P}{A}\left(1 - \frac{Ae_y h}{2I_x} - \frac{Ae_x b}{2I_y}\right)$$

As before, the normal stress remains negative or compressive at point A, provided the terms in the parentheses remain positive; i.e.,

$$0 < \left(1 - \frac{Ae_y h}{2I_x} - \frac{Ae_x b}{2I_y}\right)$$

Substituting $A = bh$, $I_x = \frac{1}{12}bh^3$, $I_y = \frac{1}{12}hb^3$ yields

$$0 < 1 - \frac{6e_x}{h} - \frac{6e_y}{b} \qquad \qquad Ans.$$

Hence, regardless of the magnitude of **P,** if it is applied at any point within the boundary of line GH shown in Fig. 14–7e, the normal stress at point A will remain compressive. In a similar manner, the normal stress at the other corners of the cross section will be compressive if **P** is confined within the boundaries of lines EG, FE, and HF. The shaded parallelogram so defined is referred to as the *core* or *kern* of the section. From the "middle-third rule" of part (*a*), the diagonals of the parallelogram will have lengths of $b/3$ and $h/3$.

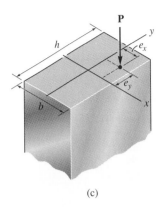

(c)

$\|$

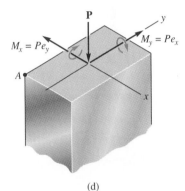

(d)

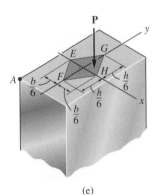

(e)

Example 14–7

$t = 0.5$ in.

$r = 24$ in.

3 ft

A

(a)

The tank in Fig. 14–8a has an inner radius of 24 in. and thickness of 0.5 in. It is filled to the top with water having a specific weight of $\gamma_w = 62.4$ lb/ft³. If it is made of steel having a specific weight of $\gamma_{st} = 490$ lb/ft³, determine the state of stress at point A. The tank is open at the top.

SOLUTION

Internal Loadings. The free-body diagram of the section of both the tank and the water above point A is shown in Fig. 14–8b. Notice that the weight of the water is supported by the water surface just *below* the section, *not* by the walls of the tank. In the vertical direction, the walls simply hold up the weight of the tank. This weight is

$$W_{st} = \gamma_{st} V_{st} = (490 \text{ lb/ft}^3)\left[\pi\left(\frac{24.5}{12} \text{ ft}\right)^2 - \pi\left(\frac{24}{12} \text{ ft}\right)^2\right](3 \text{ ft}) = 777.7 \text{ lb}$$

The stress in the circumferential direction is developed by the water pressure at level A. To obtain this pressure we must use *Pascal's law*, which states that the pressure at a point located a depth z in the water is $p = \gamma_w z$. Consequently, the pressure on the tank at level A is

$$p = \gamma_w z = (62.4 \text{ lb/ft}^3)(3 \text{ ft}) = 187.2 \text{ lb/ft}^2 = 1.30 \text{ psi}$$

Stress Components

Circumferential Stress. Applying Eq. 14–1, using the inner radius $r = 24$ in., we have

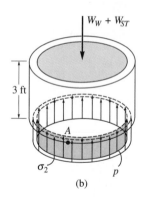

$W_W + W_{ST}$

3 ft

A

σ_2

p

(b)

$$\sigma_1 = \frac{pr}{t} = \frac{1.30 \text{ lb/in}^2 (24 \text{ in.})}{(0.5 \text{ in.})} = 62.4 \text{ psi} \qquad Ans.$$

Longitudinal Stress. Since the weight of the tank is supported uniformly by the walls, we have

$$\sigma_2 = \frac{W_{st}}{A_{st}} = \frac{777.7 \text{ lb}}{\pi[(24.5 \text{ in.})^2 - (24 \text{ in.})^2]} = 10.2 \text{ psi} \qquad Ans.$$

Note that Eq. 14–2, $\sigma_2 = pr/2t$, does *not apply* here, since the tank is open at the top and therefore, as stated previously, the water cannot develop a loading on the walls in the longitudinal direction.

Point A is subjected to the biaxial stress shown in Fig. 14–8c.

Fig. 14–8

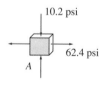

10.2 psi

62.4 psi

A

(c)

PROBLEMS

*14–8. The screw of the clamp exerts a compressive force of 500 lb on the wood blocks. Determine the maximum normal stress developed along section a–a. The cross section there is rectangular, 0.75 in. by 0.50 in.

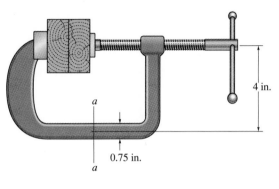

Prob. 14–8

14–10. The joint is subjected to a force of 80 lb as shown. Sketch the normal-stress distribution acting over section a–a if the member has a rectangular cross-sectional area of width 2 in. and thickness 0.5 in.

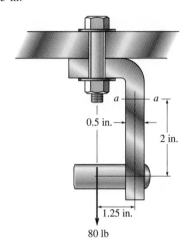

Prob. 14–10

14–9. The clamp is made from members AB and AC, which are pin-connected at A. If the compressive force at C and B is 180 N as shown, determine the maximum compressive stress in the clamp at section a–a. The screw EF is subjected only to a tensile force along its axis.

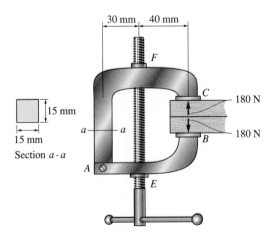

Prob. 14–9

14–11. The offset link supports the loading shown. Determine its required width w if the allowable normal stress is $\sigma_{allow} = 73$ MPa. The link has a thickness of 40 mm.

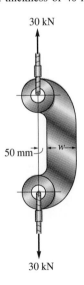

Prob. 14–11

*14–12.   The steel bracket is used to connect the ends of two cables. If the allowable normal stress is $\sigma_{allow} = 24$ ksi, determine the largest tensile force P that can be applied to the cables. The bracket has a thickness of 0.5 in. and a width of 0.75 in.

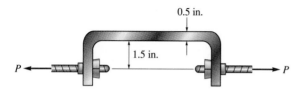

Prob. 14–12

14–13.   The gondola and passengers have a weight of 1500 lb and center of gravity at G. The suspender arm AE has a square cross-sectional area of 1.5 in. by 1.5 in., and is pin-connected at its ends A and E. Determine the largest tensile stress developed in regions AB and DC of the arm.

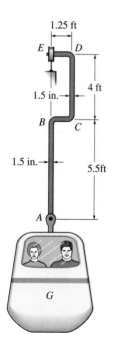

Prob. 14–13

14–14.   A bar having a square cross section of 30 mm by 30 mm is 2 m long and is held upward. If it has a mass of 5 kg/m, determine the largest angle θ at which it can be supported before it is subjected to a tensile stress along its axis near the grip.

14–15.   Solve Prob. 14–14 if the bar has a circular cross section of 30-mm diameter.

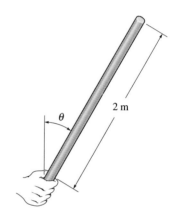

Probs. 14–14/14–15

*14–16.   The vertical force **P** acts on the bottom of the plate having a negligible weight. Determine the shortest distance d to the edge of the plate at which it can be applied so that it produces no compressive stresses on the plate at section a–a. The plate has a thickness of 10 mm and **P** acts along the center line of this thickness.

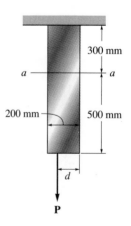

Prob. 14–16

14–17. The masonry pier is subjected to the 800-kN load. Determine the equation of the line $y = f(x)$ along which the load can be placed without causing a tensile stress in the pier. Neglect the weight of the pier.

14–18. The masonry pier is subjected to the 800-kN load. If $x = 0.25$ m and $y = 0.5$ m, determine the normal stress at each corner A, B, C, D (not shown) and plot the stress distribution over the cross section. Neglect the weight of the pier.

*14–20. The bar has a diameter of 40 mm. If it is subjected to a force of 800 N as shown, determine the stress components that act at points A and B and show the results on volume elements located at these points.

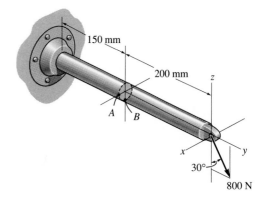

Prob. 14–20

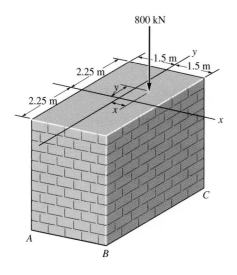

Probs. 14–17/14–18

14–19. The block is subjected to the three axial loads shown. Determine the normal stress developed at points A and B. Neglect the weight of the block.

14–21. The cylindrical post is being pulled from the ground using a sling of negligible thickness. If the rope is subjected to a vertical force of 500 N, determine the stress at points A and B. Show the results on a volume element located at each of these points.

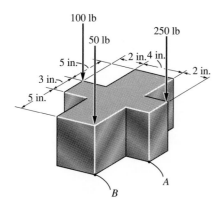

Prob. 14–19

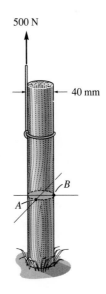

Prob. 14–21

14–22. The beam supports the loading shown. Determine the stresses at points E and F at section a–a and represent the results on a volume element located at each of these points.

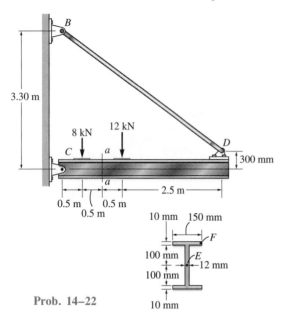

Prob. 14–22

14–23. The cable drum is locked in the position shown while supporting a horizontal cable force of 4 kN. If the drum and wound cable have a weight of 2 kN and center of gravity at G, determine the stress components in the supporting beam at point A.

*14–24. The cable drum is locked in the position shown while supporting a horizontal cable force of 4 kN. If the drum and wound cable have a weight of 2 kN and center of gravity at G, determine the stress components in the supporting beam at point B.

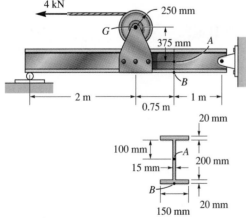

Probs. 14–23/14–24

14–25. The pliers are made from two steel parts pinned together at A. If a smooth bolt is held in the jaws and a gripping force of 10 lb is applied at the handles, determine the stress components developed in the pliers at points B and C. Here the cross section is rectangular, having the dimensions shown in the figure.

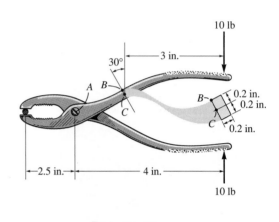

Prob. 14–25

14–26. The pin support is made from a steel rod and has a diameter of 20 mm. Determine the stress components at points A and B and represent the results on a volume element located at each of these points.

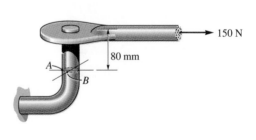

Prob. 14–26

REVIEW PROBLEMS

14–27. The uniform sign has a weight of 1500 lb and is supported by the pipe *AB,* which has an inner radius of 2.75 in. and an outer radius of 3.00 in. If the face of the sign is subjected to a uniform wind pressure of $p = 150$ lb/ft², determine the stress components at points *C* and *D*. Show the results on a volume element located at each of these points. Neglect the thickness of the sign, and assume that it is supported along the outside edge of the pipe.

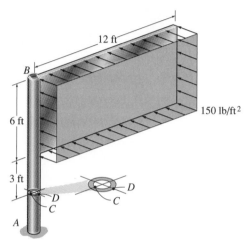

Prob. 14–27

*14–28. The column has a circular cross section of radius *c*. Determine the maximum radius *e* at which the load can be applied so that no part of the column experiences a tensile stress. Neglect the weight of the column.

Prob. 14–28

14–29. The handle of the press is subjected to a force of 20 lb. Due to internal gearing, this causes the block to be subjected to a compressive force of 80 lb. Determine the normal stress acting in the frame at points *A* and *B* caused by these loadings. *Hint:* The curved-beam formula should be used to compute the bending stress.

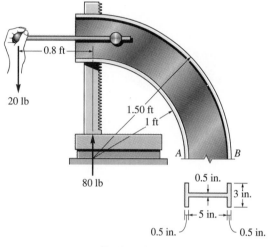

Prob. 14–29

14–30. The wide-flange beam is subjected to the loading shown. Determine the stress components at points *A* and *B* and show the results on a volume element located at each of these points. Use the shear formula to compute the shear stress.

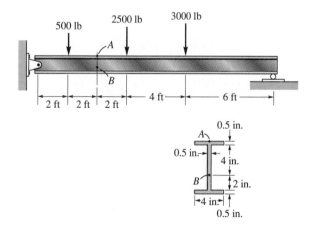

Prob. 14–30

14–31.   The beveled gear is subjected to the loads shown. Determine the stress components acting on the shaft at point A, and show the results on a volume element located at this point. The shaft has a diameter of 1 in. and is fixed to the wall at C.

*14–32.   The beveled gear is subjected to the loads shown. Determine the stress components acting on the shaft at point B, and show the results on a volume element located at this point. The shaft has a diameter of 1 in. and is fixed to the wall at C.

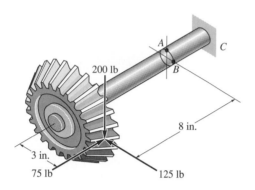

Probs. 14–31/14–32

14–33.   The frame supports a centrally applied distributed load of 1.8 kip/ft. Determine the stress components developed at points A and B on member CD and indicate the results on a volume element located at each of these points. The pins at C and D are at the same location as the neutral axis for the cross section.

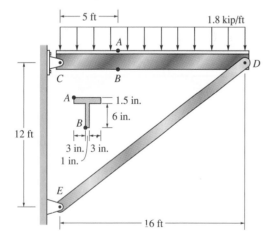

Prob. 14–33

14–34.   The C-frame is used in a riveting machine. If the force at the ram on the dolly D is 8 kN, sketch the stress distribution acting over the section a–a.

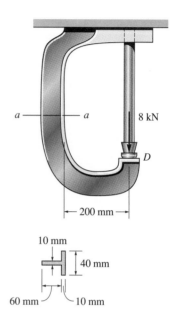

Prob. 14–34

14–35.   Plot the distribution of stress acting over the cross section a–a of the offset link.

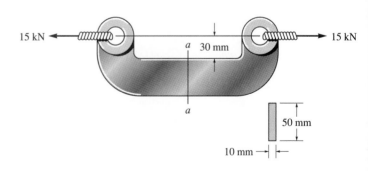

Prob. 14–35

15 Stress and Strain Transformation

In this chapter we will show how to transform the stress and strain components that are associated with a particular coordinate system, into components associated with another coordinate system. Once the transformation equations are established, we will then be able to obtain the maximum normal and shear stress and strain components at a point and find the orientation of an element on which they act. At the end of the chapter we will discuss various ways of measuring strain and develop some important material-property relationships, including a generalized form of Hooke's law.

15.1 Plane-Stress Transformation

As discussed in Sec. 8.2, in the general case, the state of stress at a point is characterized by *six* independent normal and shear stress components, which act on the faces of an element of material located at the point, Fig. 15–1*a*. This state of stress, however, is not often encountered in engineering practice. Instead, engineers frequently make approximations or simplifications in order that the stress produced in a structural member or mechanical element can be analyzed in a single plane. When this is the case, the material is said to be subjected to *plane stress,* Fig. 15–1*b*. For example, if there is no load on the surface of a body, then the normal and shear stress components will be zero on the face of an element that lies on the surface. Consequently, the corresponding stress components on the opposite face will also be zero, and so the material at the point will be subjected to plane stress.

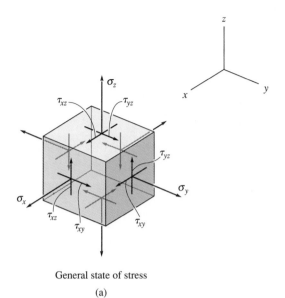

General state of stress

(a)

Fig. 15–1

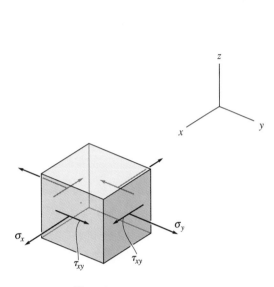

Plane stress

(b)

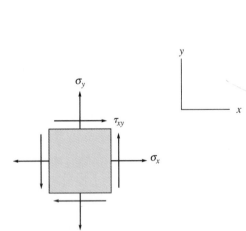

Plane stress
(two dimensional view)

(c)

The general state of *plane stress* at a point is therefore represented by a combination of two normal-stress components, σ_x, σ_y, and one shear-stress component, τ_{xy}, which act on four faces of the element as shown in Fig. 15–1*b*. For convenience, in this text we will view this state of stress in the *x–y* plane, Fig. 15–1*c*. It should be realized, however, that if one can determine these three stress components at a point, for an element oriented in the *x* and *y* directions, Fig. 15–2*a*, then the three stress components representing the *same state of stress* at the point on an element oriented in the *x'* and *y'* directions will be *different,* Fig. 15–2*b*. In other words, the state of plane stress at the point is uniquely represented by three components acting on an element that has a *specific orientation* at the point.

In this section, by using numerical examples, we will show how to *transform* the stress components from one orientation of an element to an element having a different orientation. That is, if the state of stress is defined by the components σ_x, σ_y, τ_{xy}, oriented along the *x*, *y* axes, Fig. 15–2*a*, we will show how to obtain the components $\sigma_{x'}$, $\sigma_{y'}$, $\tau_{x'y'}$, oriented along the *x'*, *y'* axes, Fig. 15–2*b*, so that they represent the *same* state of stress at the point. This is like knowing two force components, say, $\mathbf{F}_x$ and $\mathbf{F}_y$, directed along the *x*, *y* axes, that produce a resultant force $\mathbf{F}_R$, and then trying to find the force components $\mathbf{F}_{x'}$ and $\mathbf{F}_{y'}$, directed along the *x'*, *y'* axes, so they produce the *same* resultant. The transformation of stress components, however, is more difficult than that of force components, since for stress, the transformation must account for the magnitude and direction of each stress component and

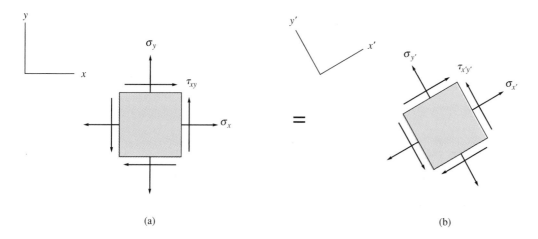

Fig. 15–2

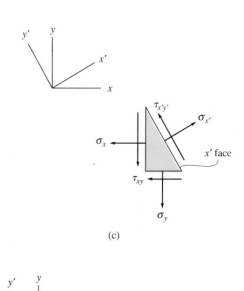

(c)

the orientation of the area upon which each component acts, whereas for force, the transformation must account only for the force component's magnitude and direction.

Recall from Sec. 8.2 that each face of an element is specified by a coordinate axis that is directed *perpendicular* to the face. For example, the $+x$ face in Fig. 15–2a is the vertical right-hand face, the $-y$ face is the bottom horizontal face, and so on. If we wish to determine the normal and shear stress components acting on the $+x'$ face of the element in Fig. 15–2b, then we must section the x, y element as shown in Fig. 15–2c. The three *known* x, y components of stress, σ_x, σ_y, τ_{xy}, acting on this segment can be related to the two unknown x', y' components, $\sigma_{x'}$, $\tau_{x'y'}$, using the equations of equilibrium. This is done by first *converting* all the stress components to forces by multiplying the stresses by their associated areas. The forces (not the stresses) are then shown on a free-body diagram of the segment and the two unknown x' and y' force components, acting on the sectioned plane ($+x'$ face), are determined from the equations of force equilibrium. Once obtained, these forces are then converted to the required normal and shear stress components, $\sigma_{x'}$ and $\tau_{x'y'}$, by dividing the forces by the sectioned area.

If $\sigma_{y'}$, acting on the $+y'$ face of the element in Fig. 15–2b, is to be determined, then it is necessary to consider a segment of the x, y element as shown in Fig. 15–2d and follow the same procedure just described. Here, however, the shear stress $\tau_{x'y'}$ does not have to be determined if it was previously calculated. Recall that shear stress is complementary, and so equilibrium of the x', y' element requires $\tau_{x'y'}$ to have the same magnitude on each of the four faces of the element, Fig. 15–2b.

The following examples illustrate this procedure numerically.

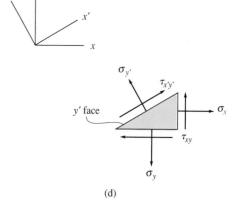

(d)

Example 15–1

An axial force of 600 N acts on the steel bar shown in Fig. 15–3*a*. Determine the stress components acting on a plane defined by section *a–a*.

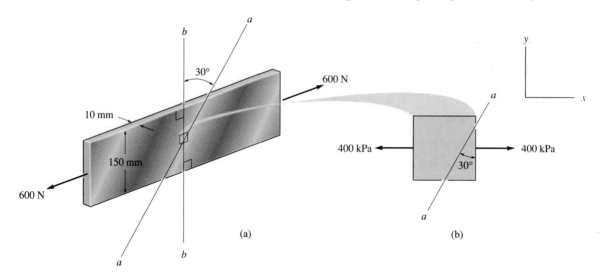

(a)

(b)

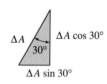

(c)

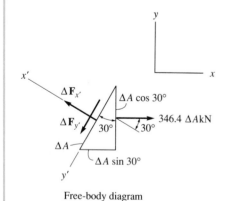

Free-body diagram

(d)

Fig. 15–3

SOLUTION I

Since the force is applied through the centroid of the cross section, the bar is subjected only to a normal stress along section *b–b*. This stress is

$$\sigma_x = \frac{P}{A} = \frac{600 \text{ N}}{(0.150 \text{ m})(0.01 \text{ m})} = 400 \text{ kPa}$$

The result is shown acting on an element of material in Fig. 15–3*b*.

Since the stress components along *a–a* are to be determined, the element will now be sectioned as shown in Fig. 15–3*b*. If we assume that the inclined face of the segment has an area ΔA, then, as shown in Fig. 15–3*c*, the horizontal and vertical faces have areas of $\Delta A \sin 30°$ and $\Delta A \cos 30°$. Using these areas, the *free-body diagram* of the segment is shown in Fig. 15–3*d*. Note that the force on the $+x$ face is

$$\Delta F_x = 400 \text{ kPa } (\Delta A \cos 30°) = 346.4 \ \Delta A \text{ kN}$$

We can obtain a direct solution for the two unknown forces $\Delta F_{x'}$ and $\Delta F_{y'}$ by applying the equations of equilibrium along the x' and y' axes.*

$$\begin{aligned}
+\nwarrow \ \Sigma F_{x'} = 0; \quad & \Delta F_{x'} - (346.4 \ \Delta A) \cos 30° = 0 \quad & \Delta F_{x'} = 300 \ \Delta A \\
+\swarrow \ \Sigma F_{y'} = 0; \quad & \Delta F_{y'} - (346.4 \ \Delta A) \sin 30° = 0 \quad & \Delta F_{y'} = 173 \ \Delta A
\end{aligned}$$

*If the equations of equilibrium are applied in the x and y directions, they will have to be solved *simultaneously* for $\Delta F_{x'}$ and $\Delta F_{y'}$.

The normal stress on section a–a is therefore

$$\sigma_{x'} = \frac{\Delta F_{x'}}{\Delta A} = \frac{300 \, \Delta A}{\Delta A} = 300 \text{ kPa} \qquad \textit{Ans.}$$

And the shear stress on the section is

$$\tau_{x'y'} = \frac{\Delta F_{y'}}{\Delta A} = \frac{173 \, \Delta A}{\Delta A} = 173 \text{ kPa} \qquad \textit{Ans.}$$

The results are independent of the area ΔA and are shown distributed uniformly over the entire section a–a in Fig. 15–3e.

SOLUTION II

A free-body diagram of the right side of the bar, sectioned along a–a, is shown in plane view, Fig. 15–3f. Applying the equations of force equilibrium in the x' and y' directions, we have

$$\begin{array}{lll} +\nwarrow \, \Sigma F_{x'} = 0; & F_{x'} - 600 \cos 30° = 0 & F_{x'} = 519.6 \text{ N} \\ +\swarrow \, \Sigma F_{y'} = 0; & F_{y'} - 600 \sin 30° = 0 & F_{y'} = 300 \text{ N} \end{array}$$

The area of the inclined section a–a is

$$A' = \left(\frac{0.150 \text{ m}}{\cos 30°}\right)(0.01 \text{ m}) = 1.732(10^{-3}) \text{ m}^2$$

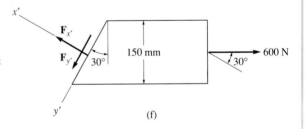

(e)

(f)

The average normal and shear stresses on the inclined plane are thus

$$\sigma_{x'} = \frac{F_{x'}}{A'} = \frac{519.6 \text{ N}}{1.732(10^{-3}) \text{ m}^2} = 300 \text{ kPa} \qquad \textit{Ans.}$$

$$\tau_{x'y'} = \frac{F_{y'}}{A'} = \frac{300 \text{ N}}{1.732(10^{-3}) \text{ m}^2} = 173 \text{ kPa} \qquad \textit{Ans.}$$

Note that this method of analysis can be used here since the formulas $\sigma = P/A$ and $\tau = V/A$ can be applied in this manner. Stresses generated by beam shear, bending moment, and torsional moment must be determined only on the *cross-sectional area* or on sections that are perpendicular to the member's axis, and as a result, the first method of analysis must be used for these cases.

Example 15–2

The state of plane stress at a point on the surface of the airplane fuselage is represented on the element oriented as shown in Fig. 15–4a. Represent the state of stress at the point on another element that is oriented 30° clockwise from the position shown.

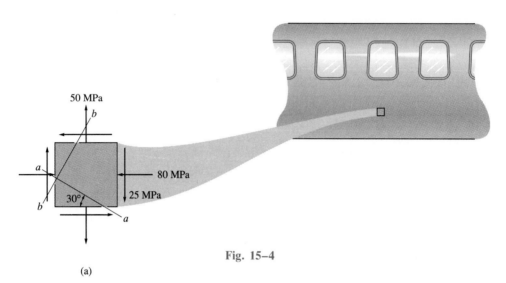

Fig. 15–4

(a)

SOLUTION

The element is sectioned by the line a–a in Fig. 15–4a, the bottom segment is removed, and assuming the sectioned (inclined) plane has an area ΔA, the horizontal and vertical planes have the areas shown in Fig. 15–4b. The free-body diagram of the segment is shown in Fig. 15–4c. Applying the equations of force equilibrium in the x' and y' directions to avoid a simultaneous solution for the two unknowns $\Delta F_{x'}$ and $\Delta F_{y'}$, we have

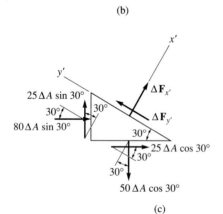

(b)

(c)

$$+\nearrow\ \Sigma F_{x'} = 0;\quad \Delta F_{x'} - (50\ \Delta A\ \cos 30°)\cos 30°$$
$$+(25\ \Delta A\ \cos 30°)\sin 30° + (80\ \Delta A\ \sin 30°)\sin 30°$$
$$+(25\ \Delta A\ \sin 30°)\cos 30° = 0$$
$$\Delta F_{x'} = -4.15\ \Delta A$$

$$+\searrow\ \Sigma F_{y'} = 0;\quad -\Delta F_{y'} + (50\ \Delta A\ \cos 30°)\sin 30°$$
$$+(25\ \Delta A\ \cos 30°)\cos 30° + (80\ \Delta A\ \sin 30°)\cos 30°$$
$$-(25\ \Delta A\ \sin 30°)\sin 30° = 0$$
$$\Delta F_{y'} = 68.8\ \Delta A$$

Since $\Delta F_{x'}$ is negative, $\Delta F_{x'}$ acts in the opposite direction of that shown in Fig. 15–4c.

The normal and shear stress components acting on the inclined face along section *a–a* are therefore

$$\sigma_{x'} = \frac{\Delta F_{x'}}{\Delta A} = \frac{4.15 \, \Delta A}{\Delta A} = 4.15 \text{ MPa} \qquad Ans.$$

$$\tau_{x'y'} = \frac{\Delta F_{y'}}{\Delta A} = \frac{68.8 \, \Delta A}{\Delta A} = 68.8 \text{ MPa} \qquad Ans.$$

These results are shown on the *top* of the element in Fig. 15–4d, since this surface is the one considered in Fig. 15–4c.

We must now repeat the procedure to obtain the stress on the *perpendicular* plane *b–b*. Sectioning the element in Fig. 15–4a in the direction of *b–b* results in a segment having sides with areas shown in Fig. 15–4e. The associated free-body diagram is shown in Fig. 15–4f. Thus,

$$+\searrow \ \Sigma F_{x'} = 0; \quad \Delta F_{x'} - (25 \, \Delta A \cos 30°) \sin 30°$$
$$+ (80 \, \Delta A \cos 30°) \cos 30° - (25 \, \Delta A \sin 30°) \cos 30°$$
$$- (50 \, \Delta A \sin 30°) \sin 30° = 0$$
$$\Delta F_{x'} = -25.8 \, \Delta A$$

$$+\nearrow \ \Sigma F_{y'} = 0; \quad -\Delta F_{y'} + (25 \, \Delta A \cos 30°) \cos 30°$$
$$+ (80 \, \Delta A \cos 30°) \sin 30° - (25 \, \Delta A \sin 30°) \sin 30°$$
$$+ (50 \, \Delta A \sin 30°) \cos 30° = 0$$
$$\Delta F_{y'} = 68.8 \, \Delta A$$

From these results, note that $\Delta F_{x'}$ acts opposite to its direction shown in Fig. 15–4f. The stress components are therefore

$$\sigma_{x'} = \frac{\Delta F_{x'}}{\Delta A} = \frac{25.8 \, \Delta A}{\Delta A} = 25.8 \text{ MPa} \qquad Ans.$$

$$\tau_{x'y'} = \frac{\Delta F_{y'}}{\Delta A} = \frac{68.8 \, \Delta A}{\Delta A} = 68.8 \text{ MPa} \qquad Ans.$$

These stress components are shown acting on the *right side* of the element in Fig. 15–4d. Here the direction and magnitude of the shear stress, which have been calculated twice, verifies the requirement for moment equilibrium discussed in Sec. 8.2. From this analysis we may therefore conclude that the state of stress at the point can, for example, be represented by choosing an element oriented as shown in Fig. 15–4a or by choosing one oriented as shown in Fig. 15–4d.

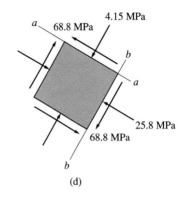

(d)

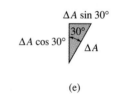

(e)

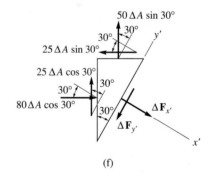

(f)

15.2 General Equations of Plane-Stress Transformation

The method of transforming the normal and shear stress components from the x, y to the x', y' coordinate axes, as discussed in the previous section, will now be developed in a general manner and expressed as a set of stress-transformation equations.

Sign Convention. Before the transformation equations can be developed, we must first establish a sign convention for the stress components. Here we will adopt the same one used in Sec. 8.2. Briefly stated, once the x, y or x', y' axes have been established, a normal or shear stress component is *positive* provided it acts in the *positive* coordinate direction on the *positive* face of the element, or it acts in the *negative* coordinate direction on the *negative* face of the element, Fig. 15–5a. For example, σ_x is positive since it acts to the right ($+x$ direction) on the right-hand vertical face (defined by the $+x$ axis), and it acts to the left ($-x$ direction) on the left-hand vertical face (defined by the $-x$ axis). The shear stress in Fig. 15–5a is shown acting in the positive direction on all four faces of the element. On the right-hand face (defined by the $+x$ axis), τ_{xy} acts upward ($+y$ direction); on the bottom face (defined by the $-y$ axis), τ_{xy} acts to the left ($-x$ direction), and so on.

All the stress components shown in Fig. 15–5a maintain equilibrium of the element as discussed in Sec. 8.2, and because of this, knowing the direction of τ_{xy} on one face of the element defines its direction on the other three faces. Hence, the above sign convention can also be remembered by simply noting that *positive normal stress* acts *outward* and *positive shear stress* acts *upward on the right-hand face* of the element.

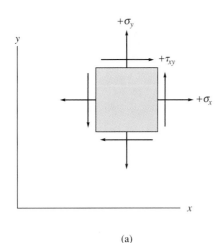

(a)

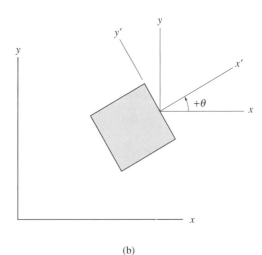

(b)

Positive Sign Convention

Fig. 15–5

Given the state of plane stress shown in Fig. 15–5a, the orientation of the inclined plane on which the normal and shear stress components are to be determined will be defined using the angle θ. To establish this angle properly, it is first necessary to establish a positive x' axis, *directed outward, perpendicular* or normal to the plane, and an associated y' axis, directed along the plane, Fig. 15–5b. Both the unprimed and primed sets of axes form right-handed coordinate systems; that is, the positive z (or z') axis is established by the right-hand rule. Curling the fingers from x (or x') toward y (or y') gives the direction for the positive z (or z') axis that points outward. The *angle θ* is measured from the positive x to the positive x' axis, such that it is *positive* provided it also follows the curl of the right-hand fingers, i.e., counterclockwise as shown in Fig. 15–5b.

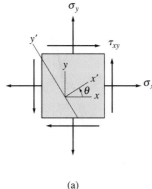

(a)

Fig. 15–6

Normal and Shear Stress Components.

Using the established sign convention, the element in Fig. 15–6a is sectioned along the inclined plane defined by $+x'$ and the segment shown in Fig. 15–6b is isolated. If the sectioned area is ΔA, then the horizontal and vertical faces of the segment have an area of $\Delta A \sin \theta$ and $\Delta A \cos \theta$, respectively.

The resulting *free-body diagram* of the segment is shown in Fig. 15–6c. Applying the equations of force equilibrium to determine the unknown normal and shear stress components $\sigma_{x'}$ and $\tau_{x'y'}$, we obtain

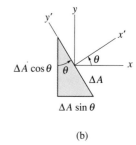

(b)

$+\nearrow \ \Sigma F_{x'} = 0;$ $\sigma_{x'} \, \Delta A - (\tau_{xy} \, \Delta A \sin \theta) \cos \theta$
$$-(\sigma_y \, \Delta A \sin \theta) \sin \theta - (\tau_{xy} \, \Delta A \cos \theta) \sin \theta$$
$$-(\sigma_x \, \Delta A \cos \theta) \cos \theta = 0$$

$$\sigma_{x'} = \sigma_x \cos^2 \theta + \sigma_y \sin^2 \theta + \tau_{xy}(2 \sin \theta \cos \theta)$$

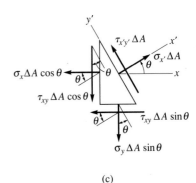

$+\nwarrow \ \Sigma F_{y'} = 0;$ $\tau_{x'y'} \, \Delta A + (\tau_{xy} \, \Delta A \sin \theta) \sin \theta$
$$-(\sigma_y \, \Delta A \sin \theta) \cos \theta - (\tau_{xy} \, \Delta A \cos \theta) \cos \theta$$
$$+(\sigma_x \, \Delta A \cos \theta) \sin \theta = 0$$

$$\tau_{x'y'} = (\sigma_y - \sigma_x) \sin \theta \cos \theta + \tau_{xy}(\cos^2 \theta - \sin^2 \theta)$$

(c)

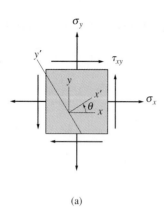

Fig. 15–6(a)

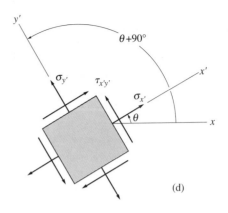

(d)   Fig. 15–6(d)

These two equations may be simplified by using the trigonometric identities $\sin 2\theta = 2\sin\theta\cos\theta$, $\sin^2\theta = (1 - \cos 2\theta)/2$, and $\cos^2\theta = (1 + \cos 2\theta)/2$, in which case,

$$\sigma_{x'} = \frac{\sigma_x + \sigma_y}{2} + \frac{\sigma_x - \sigma_y}{2}\cos 2\theta + \tau_{xy}\sin 2\theta \qquad (15\text{–}1)$$

$$\tau_{x'y'} = -\frac{\sigma_x - \sigma_y}{2}\sin 2\theta + \tau_{xy}\cos 2\theta \qquad (15\text{–}2)$$

To check these equations, note that when $\theta = 0°$, $\sigma_{x'} = \sigma_x$ and $\tau_{x'y'} = \tau_{xy}$ as required, Fig. 15–6a. Also, when $\theta = 90°$, $\sigma_{x'} = \sigma_y$ and $\tau_{x'y'} = -\tau_{xy}$. The negative sign indicates that $\tau_{x'y'}$ acts in the negative y' direction as required, Fig. 15–6a.

If the normal stress acting in the y' direction is needed, it can be obtained by simply substituting $(\theta = \theta + 90°)$ for θ into Eq. 15–1, Fig. 15–6d. The result is

$$\sigma_{y'} = \frac{\sigma_x + \sigma_y}{2} - \frac{\sigma_x - \sigma_y}{2}\cos 2\theta - \tau_{xy}\sin 2\theta \qquad (15\text{–}3)$$

If $\sigma_{y'}$ is calculated as a positive quantity, it indicates that it acts in the positive y' direction as shown in Fig. 15–6d.

PROCEDURE FOR ANALYSIS

To apply the stress transformation Eqs. 15–1 and 15–2, it is simply necessary to substitute in the known data for σ_x, σ_y, τ_{xy}, and θ in accordance with the established sign convention, Fig. 15–6a. If $\sigma_{x'}$ and $\tau_{x'y'}$ are calculated as positive quantities, then these stresses act in the positive direction of the x' and y' axes as established in Fig. 15–6c.

For convenience these equations can easily be programmed on a pocket calculator or microcomputer. The following example illustrates their numerical application.

Example 15–3

The state of plane stress at a point is represented by the element shown in Fig. 15–7a. Determine the state of stress at the point on another element oriented 30° clockwise from the position shown.

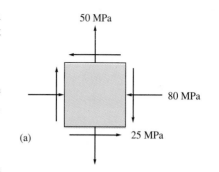

(a)

SOLUTION

This problem was solved in Example 15–2 using basic principles. Here we will apply Eqs. 15–1 and 15–2. From the established sign convention, Fig. 15–5, it is seen that

$$\sigma_x = -80 \text{ MPa} \qquad \sigma_y = 50 \text{ MPa} \qquad \tau_{xy} = -25 \text{ MPa}$$

To obtain the stress components on plane CD, Fig. 15–7b, the positive x' axis is directed outward, perpendicular to CD, and the associated y' axis is directed along CD. The angle θ measured from the x to the x' axis is $\theta = -30°$ (clockwise). Applying Eqs. 15–1 and 15–2 yields

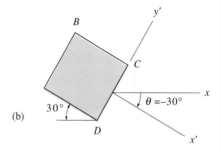

(b)

$$\sigma_{x'} = \frac{-80 + 50}{2} + \frac{-80 - 50}{2} \cos 2(-30°) + (-25) \sin 2(-30°)$$

$$= -25.8 \text{ MPa} \qquad\qquad Ans.$$

$$\tau_{x'y'} = -\frac{-80 - 50}{2} \sin 2(-30°) + (-25) \cos 2(-30°)$$

$$= -68.8 \text{ MPa} \qquad\qquad Ans.$$

The negative signs indicate that $\sigma_{x'}$ and $\tau_{x'y'}$ act in the negative x' and y' directions, respectively. The results are shown acting on the element in Fig. 15–7d.

In a similar manner, the stress components acting on face BC, Fig. 15–7b, are obtained using $\theta = 60°$, Fig. 15–7c. Applying Eqs. 15–1 and 15–2,* we get

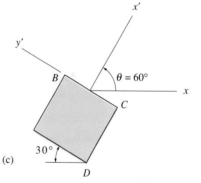

(c)

$$\sigma_{x'} = \frac{-80 + 50}{2} + \frac{-80 - 50}{2} \cos 2(60°) + (-25) \sin 2(60°)$$

$$= -4.15 \text{ MPa} \qquad\qquad Ans.$$

$$\tau_{x'y'} = -\frac{-80 - 50}{2} \sin 2(60°) + (-25) \cos 2(60°)$$

$$= 68.8 \text{ MPa} \qquad\qquad Ans.$$

Here $\tau_{x'y'}$ has been computed twice in order to provide a check. The negative sign for $\sigma_{x'}$ indicates that this stress acts in the negative x' direction shown in Fig. 15–7c. The results are shown on the element in Fig. 15–7d.

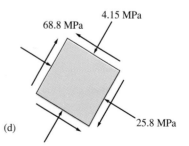

(d)

*Alternatively, we could apply Eq. 15–3 with $\theta = -30°$ rather than Eq. 15–1.

Fig. 15–7

15.3 Principal Stresses and Maximum In-Plane Shear Stress

From Eqs. 15–1 and 15–2, it can be seen that $\sigma_{x'}$ and $\tau_{x'y'}$ depend on the angle of inclination θ of the planes on which these stresses act. In engineering practice it is often important to determine the orientation of the planes that causes the normal stress to be a maximum and a minimum and the orientation of the planes that causes the shear stress to be a maximum. In this section each of these problems will be considered.

In-Plane Principal Stresses. To determine the maximum and minimum normal stress we must differentiate Eq. 15–1 with respect to θ and set the result equal to zero. Thus,

$$\frac{d\sigma_{x'}}{d\theta} = -\frac{\sigma_x - \sigma_y}{2}(2 \sin 2\theta) + 2\tau_{xy} \cos 2\theta = 0$$

Solving this equation we obtain the orientation $\theta = \theta_p$ of the planes of maximum and minimum normal stress. At $\theta = \theta_p$,

$$\tan 2\theta_p = \frac{\tau_{xy}}{(\sigma_x - \sigma_y)/2} \tag{15–4}$$

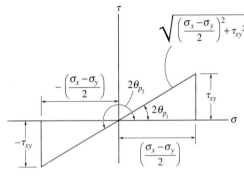

Fig. 15–8

The solution of this equation has two roots, θ_{p_1} and θ_{p_2}. Specifically, the values of $2\theta_{p_1}$ and $2\theta_{p_2}$ are 180° apart, so θ_{p_1} and θ_{p_2} are 90° apart.

The values of θ_{p_1} and θ_{p_2} must be substituted into Eq. 15–1 if we are to obtain the required normal stresses. Rather than doing this, we can instead obtain the necessary sine and cosine of $2\theta_{p_1}$ and $2\theta_{p_2}$ from the triangles shown in Fig. 15–8. The construction of these triangles is based on Eq. 15–4, assuming that τ_{xy} and $(\sigma_x - \sigma_y)$ are both positive or both negative quantities. We have,

for θ_{p_1}, $\sin 2\theta_{p_1} = \tau_{xy} \Big/ \sqrt{\left(\dfrac{\sigma_x - \sigma_y}{2}\right)^2 + \tau_{xy}{}^2}$

$\cos 2\theta_{p_1} = \left(\dfrac{\sigma_x - \sigma_y}{2}\right) \Big/ \sqrt{\left(\dfrac{\sigma_x - \sigma_y}{2}\right)^2 + \tau_{xy}{}^2}$

for θ_{p_2}, $\sin 2\theta_{p_2} = -\tau_{xy} \Big/ \sqrt{\left(\dfrac{\sigma_x - \sigma_y}{2}\right)^2 + \tau_{xy}{}^2}$

$\cos 2\theta_{p_2} = -\left(\dfrac{\sigma_x - \sigma_y}{2}\right) \Big/ \sqrt{\left(\dfrac{\sigma_x - \sigma_y}{2}\right)^2 + \tau_{xy}{}^2}$

If either of these two sets of trigonometric relations are substituted into Eq. 15–1 and simplified, we obtain

$$\sigma_{1,2} = \frac{\sigma_x + \sigma_y}{2} \pm \sqrt{\left(\frac{\sigma_x - \sigma_y}{2}\right)^2 + \tau_{xy}{}^2} \tag{15–5}$$

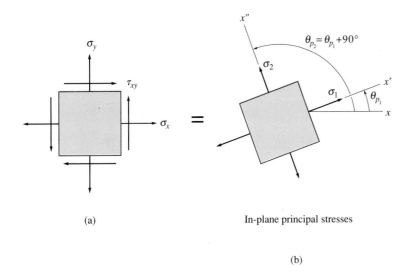

(a)

In-plane principal stresses

(b)

Fig. 15–9

Depending on the sign chosen, this result gives the maximum or minimum in-plane normal stress acting at a point, where $\sigma_1 \geq \sigma_2$. This particular set of values are called in-plane *principal stresses,* and the corresponding planes on which they act are called the *principal planes* of stress, Fig. 15–9*b*. Furthermore, if the above trigonometric relations for θ_{p_1} and θ_{p_2} are substituted into Eq. 15–2, it can be seen that $\tau_{x'y'} = 0$; that is, *no shear stress acts on the principal planes.*

To summarize, then, the state of stress at a point, if initially represented by the stress components acting on an element oriented as shown in Fig. 15–9*a*, can instead be represented by the normal stress components acting on an element oriented as shown in Fig. 15–9*b*. The faces of this element represent principal planes of stress defined by the angles θ_{p_1} and θ_{p_2}. In this position, only the maximum and minimum in-plane normal stresses (principal stresses) σ_1 and σ_2 act on the element; that is, the element in this position is *not* subjected to shear stress.

Maximum In-Plane Shear Stress. The orientation of an element that is subjected to maximum shear stress on its faces can be determined by taking the derivative of Eq. 15–2 with respect to θ and setting the result equal to zero. This gives

$$\tan 2\theta_s' = \frac{-(\sigma_x - \sigma_y)/2}{\tau_{xy}} \qquad (15\text{–}6)$$

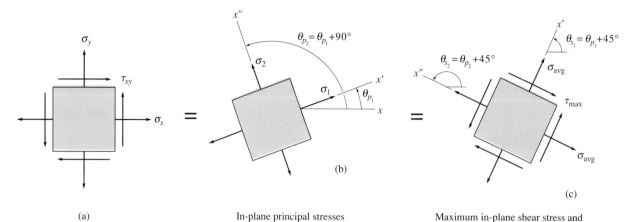

(a)

(b)

In-plane principal stresses

(c)

Maximum in-plane shear stress and associated average normal stress

Fig. 15–9

The two roots of this equation, θ_{s_1} and θ_{s_2}, can be determined from the triangles shown in Fig. 15–10. By comparison with Fig. 15–8, each root of $2\theta_s$ is 90° from $2\theta_p$. Thus, the roots θ_s and θ_p are 45° apart, and as a result the planes for maximum shear stress can be determined by orienting an element 45° from the position of an element that defines the planes of principal stress.

Using either one of the roots θ_{s_1} or θ_{s_2}, the maximum shear stress can be determined by taking the trigonometric values of $\sin 2\theta_s$ and $\cos 2\theta_s$ from Fig. 15–10 and substituting them into Eq. 15–2. The result is

$$\tau_{max} = \sqrt{\left(\frac{\sigma_x - \sigma_y}{2}\right)^2 + \tau_{xy}^2} \tag{15–7}$$

In Sec. 15.5 we will show that the absolute maximum shear stress on the element may, in some cases, lie in a plane that is perpendicular to the one considered here. For this reason, the value of τ_{max} as calculated by Eq. 15–7 will be referred to as the *maximum in-plane shear stress* because it acts on the element in the x–y plane.

Substituting the values for $\sin 2\theta_s$ and $\cos 2\theta_s$ into Eq. 15–1, we see that there is also a normal stress on the planes of maximum in-plane shear stress. We get

$$\sigma_{avg} = \frac{\sigma_x + \sigma_y}{2} \tag{15–8}$$

To summarize, the planes of maximum in-plane shear stress are oriented 45° from the planes of principal stress, Fig. 15–9c. The shear stress on these planes is defined by Eq. 15–7; furthermore, on these planes there is also an associated average normal stress σ_{avg}, Eq. 15–8.

Like the stress-transformation equations, it may be convenient to program all the above equations for use on a pocket calculator or a microcomputer. The following examples illustrate their numerical application.

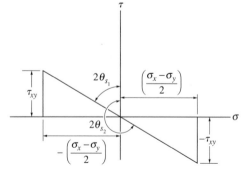

Fig. 15–10

The state of plane stress at a point on a body is shown on the element in Fig. 15–11a. Represent the stress state at the point in terms of the principal stresses.

SOLUTION

From the sign convention established in Fig. 15–5a, we have

$$\sigma_x = -20 \text{ MPa} \qquad \sigma_y = 90 \text{ MPa} \qquad \tau_{xy} = 60 \text{ MPa}$$

The orientation of the principal planes of stress is determined from Eq. 15–4.

$$\tan 2\theta_p = \frac{\tau_{xy}}{(\sigma_x - \sigma_y)/2} = \frac{60}{(-20 - 90)/2}$$

Solving, and referring to this root as θ_{p_2}, for the reason to be stated below, gives

$$2\theta_{p_2} = -47.49° \qquad \theta_{p_2} = -23.7°$$

Since the difference between $2\theta_{p_1}$ and $2\theta_{p_2}$ is 180°, we have

$$2\theta_{p_1} = 180° + 2\theta_{p_2} = 132.51° \qquad \theta_{p_1} = 66.3°$$

Recall that an angle θ is measured positive *counterclockwise* from the x axis to the outward normal (x' axis) on the face of the element, Fig. 15–11b.

The principal stresses are determined from Eq. 15–5. We have

$$\sigma_{1,2} = \frac{\sigma_x + \sigma_y}{2} \pm \sqrt{\left(\frac{\sigma_x - \sigma_y}{2}\right)^2 + \tau_{xy}^2}$$

$$= \left(\frac{-20 + 90}{2}\right) \pm \sqrt{\left(\frac{-20 - 90}{2}\right)^2 + (60)^2}$$

$$= 35.0 \pm 81.4$$

$$\sigma_1 = 116 \text{ MPa} \qquad\qquad\qquad Ans.$$

$$\sigma_2 = -46.4 \text{ MPa} \qquad\qquad\qquad Ans.$$

The principal plane on which each of these normal stresses acts can be determined by applying Eq. 15–1 with, say, $\theta = \theta_{p_2} = -23.7°$. We have

$$\sigma_{x'} = \frac{\sigma_x + \sigma_y}{2} + \frac{\sigma_x - \sigma_y}{2} \cos 2\theta + \tau_{xy} \sin 2\theta$$

$$= \left(\frac{-20 + 90}{2}\right) + \left(\frac{-20 - 90}{2}\right) \cos 2(-23.7°) + 60 \sin 2(-23.7°)$$

$$= -46.4 \text{ MPa}$$

Hence, $\sigma_2 = -46.4$ MPa acts on the plane defined by $\theta_{p_2} = -23.7°$, whereas $\sigma_1 = 116$ MPa acts on the plane defined by $\theta_{p_1} = 66.3°$. The principal stresses are shown on the element in Fig. 15–11c. As expected, no shear stress acts on this element.

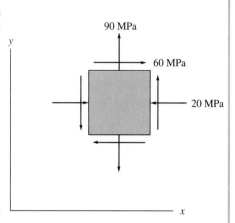

(a)

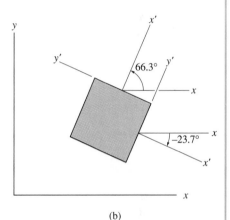

(b)

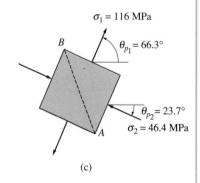

(c)

Fig. 15–11

621

Example 15–5

The state of plane stress at a point on a body is represented on the element shown in Fig. 15–12a. Represent the stress state at the point in terms of the maximum in-plane shear stress and average normal stress.

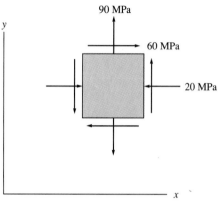

Fig. 15–12

(a)

(b)

SOLUTION

The orientation of the planes for maximum in-plane shear stress is determined from Eq. 15–6. Since $\sigma_x = -20$ MPa, $\sigma_y = 90$ MPa, and $\tau_{xy} = 60$ MPa, we have

$$\tan 2\theta_s = \frac{-[(\sigma_x - \sigma_y)/2]}{\tau_{xy}}$$

$$= \frac{-(-20 - 90)/2}{60}$$

$$2\theta_{s_2} = 42.5° \qquad \theta_{s_2} = 21.3°$$
$$2\theta_{s_1} = 180° + 2\theta_{s_2} \qquad \theta_{s_1} = 111.3°$$

These angles can also be obtained in a more direct manner if the directions of the principal planes of stress are known, Example 15–4. Realizing that the planes of maximum shear stress are 45° away from the principal planes of stress, then, from Example 15–4, $\theta_{s_1} = \theta_{p_1} + 45° = 111.3°$ and $\theta_{s_2} = \theta_{p_2} + 45° = 21.3°$, Fig. 15–12b.

The maximum shear stress on these planes is determined from Eq. 15–7; that is,

$$\tau_{max} = \sqrt{\left(\frac{\sigma_x - \sigma_y}{2}\right)^2 + \tau_{xy}^2}$$

$$= \sqrt{\left(\frac{-20 - 90}{2}\right)^2 + (60)^2} = 81.4 \text{ MPa}$$

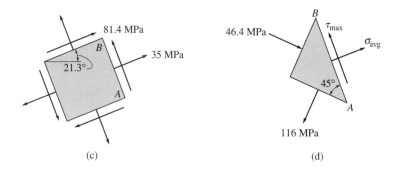

(c)   (d)

The proper direction of τ_{max} on the element can be determined in one of two ways. Applying Eq. 15–2 with, say, $\theta = \theta_{s_2} = 21.3°$, we have

$$\tau_{x'y'} = -\left(\frac{\sigma_x - \sigma_y}{2}\right) \sin 2\theta + \tau_{xy} \cos 2\theta$$

$$= -\left(\frac{-20 - 90}{2}\right) \sin 2(21.3°) + 60 \cos 2(21.3°)$$

$$= 81.4 \text{ MPa}$$

Thus, $\tau_{max} = \tau_{x'y'}$ acts in the *positive y'* direction on this face ($\theta = 21.3°$), Fig. 15–12b. The shear stresses on the other three faces are directed as shown in Fig. 15–12c.

Assuming that the principal stress and the direction of the principal planes of stress are known, it is also possible to determine the direction of τ_{max} using equilibrium principles. The element in Fig. 15–11c is sectioned along one of its diagonals, say AB (a plane of maximum shear stress), and one of the segments is then isolated, Fig. 15–12d. By observation, to preserve force equilibrium along AB, τ_{max} must create a force that acts upward on this face of the element in order to balance the downward force components produced by the principal stresses 46.4 MPa and 116 MPa.

Besides the maximum shear stress, as calculated above, the element is also subjected to an average normal stress determined from Eq. 15–8; that is,

$$\sigma_{avg} = \frac{\sigma_x + \sigma_y}{2}$$

$$= \frac{-20 + 90}{2} = 35 \text{ MPa} \qquad\qquad Ans.$$

This is a tensile stress, since it is calculated as a positive quantity. The results are shown in Fig. 15–12c.

■ Example 15–6 ■

Due to the applied loading, the element at point A on the cantilevered beam in Fig. 15–13a is subjected to the state of stress shown. Determine the principal stresses acting at point A.

SOLUTION

According to our sign convention,

$$\sigma_x = 20 \text{ ksi} \qquad \sigma_y = 0 \qquad \tau_{xy} = 3 \text{ ksi}$$

Orientation of the principal planes of stress is determined from Eq. 15–4; that is,

$$\tan 2\theta_p = \frac{\tau_{xy}}{(\sigma_x - \sigma_y)/2}$$

$$= \frac{3}{(20 - 0)/2}$$

So that

$$2\theta_{p_1} = 16.70° \qquad\qquad \theta_{p_1} = 8.35°$$
$$2\theta_{p_2} = 180° + 16.70° = 196.7° \qquad \theta_{p_2} = 98.4°$$

The principal stresses are found from Eq. 15–5,

$$\sigma_{1,2} = \left(\frac{\sigma_x + \sigma_y}{2}\right) \pm \sqrt{\left(\frac{\sigma_x - \sigma_y}{2}\right)^2 + \tau_{xy}^2}$$

$$= \left(\frac{20 + 0}{2}\right) \pm \sqrt{\left(\frac{20 - 0}{2}\right)^2 + (3)^2}$$

$$= 10 \pm 10.440$$

$$\sigma_1 = 20.4 \text{ ksi} \qquad\qquad\qquad\qquad Ans.$$
$$\sigma_2 = -0.440 \text{ ksi} \qquad\qquad\qquad\qquad Ans.$$

To determine the face over which each of these stresses acts, apply Eq. 15–1, with $\theta = \theta_{p_1} = 8.35°$. We have

$$\sigma_{x'} = \left(\frac{\sigma_x + \sigma_y}{2}\right) + \left(\frac{\sigma_x - \sigma_y}{2}\right) \cos 2\theta + \tau_{xy} \sin 2\theta$$

$$= \left(\frac{20 + 0}{2}\right) + \left(\frac{20 - 0}{2}\right) \cos 2(8.35°) + 3 \sin 2(8.35°)$$

$$= 20.4 \text{ ksi}$$

Thus, $\sigma_{x'} = \sigma_1 = 20.4$ ksi acts on the principal plane defined by $\theta_{p_1} = 8.35°$. In other words, the notation using θ_{p_1} to represent the angle 8.35° is correct. The final results are shown in Fig. 15–13b.

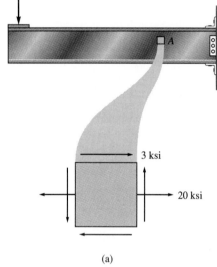

(a)

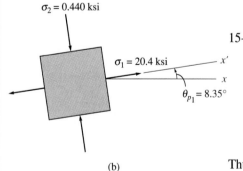

(b)

Fig. 15–13

PROBLEMS

15–1. Prove that the sum of the normal stresses $\sigma_x + \sigma_y = \sigma_{x'} + \sigma_{y'}$ is constant.

15–2. The state of stress at a point in a member is shown on the element. Determine the stress components acting on the inclined plane AB. Solve the problem using the method of equilibrium described in Sec. 15.1.

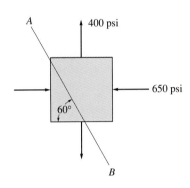

Prob. 15–2

15–3. The state of stress at a point in a member is shown on the element. Determine the stress components acting on the inclined plane AB. Solve the problem using the method of equilibrium described in Sec. 15.1.

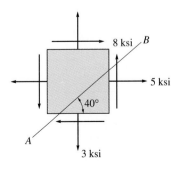

Prob. 15–3

*15–4. The state of stress at a point in a member is shown on the element. Determine the stress components acting on the inclined plane AB. Solve the problem using the method of equilibrium described in Sec. 15.1.

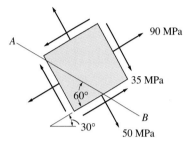

Prob. 15–4

15–5. Solve Prob. 15–2 using the stress-transformation equations developed in Sec. 15.2.

15–6. Solve Prob. 15–3 using the stress-transformation equations developed in Sec. 15.2.

15–7. Solve Prob. 15–4 using the stress-transformation equations developed in Sec. 15.2.

*15–8. Determine the equivalent state of stress on an element if the element is oriented 60° clockwise from the element shown. Use the stress-transformation equations.

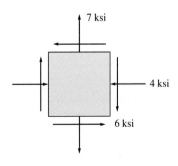

Prob. 15–8

15–9.   Determine the equivalent state of stress on an element if the element is oriented 30° clockwise from the element shown. Use the stress-transformation equations.

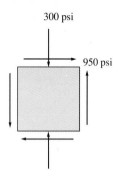

300 psi

950 psi

Prob. 15–9

15–10.   Determine the equivalent state of stress on an element if the element is oriented 50° counterclockwise from its position shown.

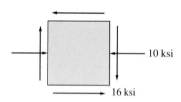

10 ksi

16 ksi

Prob. 15–10

15–11.   The wood beam is subjected to a load of 12 kN. If grains of wood in the beam at point A make an angle of 25° with the horizontal as shown, determine the normal and shear stress that act perpendicular and parallel to the grains due to the loading.

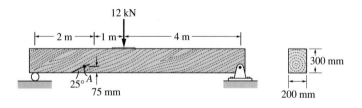

12 kN

2 m   1 m   4 m

25° A
75 mm

300 mm

200 mm

Prob. 15–11

*15–12.   The grains of wood in the board make an angle of 20° with the horizontal as shown. Determine the normal and shear stress that act perpendicular and parallel to the grains if the board is subjected to an axial load of 250 N.

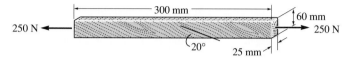

300 mm

60 mm

250 N   250 N

20°

25 mm

Prob. 15–12

15–13.   The state of stress at a point is shown on the element. Determine (a) the principal stresses and (b) the maximum in-plane shear stress and average normal stress at the point. Specify the orientation of the element in each case.

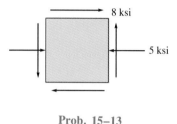

8 ksi

5 ksi

Prob. 15–13

15–14.   The state of stress at a point is shown on the element. Determine (a) the principal stresses and (b) the maximum in-plane shear stress and average normal stresses at the point. Specify the orientation of the element in each case.

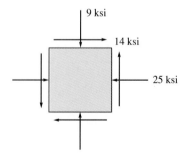

9 ksi

14 ksi

25 ksi

Prob. 15–14

15–15. The clamp bears down on the smooth surfaces at C and D by tightening the bolt. If the tensile force in the bolt is 40 kN, determine the principal stress at points A and B and show the results on elements located at each of these points. The cross-sectional area at A and B is shown in the adjacent figure.

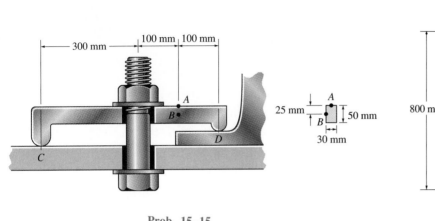

Prob. 15–15

15–17. Determine the principal stresses acting at point A of the supporting frame. Show the results on an element located at the point.

Prob. 15–17

*15–16.** The cantilevered rectangular bar is subjected to the force of 5 kip. Determine the principal stresses at points A and B.

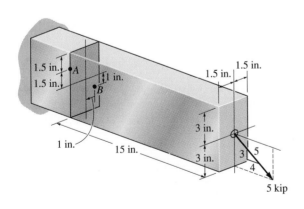

Prob. 15–16

15–18. The clamp exerts a force of 150 lb on the boards at G. Determine the axial force in each screw, AB and CD, and then compute the principal stresses at points E and F. Show the results on elements located at these points. The section through EF is rectangular and is 1 in. wide.

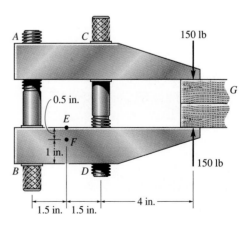

Prob. 15–18

15–19. The bell crank is pinned at A and supported by a short link BC. If it is subjected to the force of 80 N, determine the principal stresses at (a) point D and (b) point E. The crank is constructed from an aluminum plate having a thickness of 20 mm.

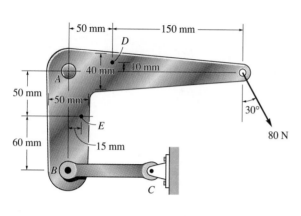

Prob. 15–19

15–21. The cantilevered beam is subjected to the load at its end. Determine the principal stresses in the beam at points A and B.

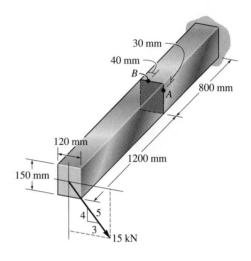

Prob. 15–21

*15–20. The box beam is subjected to the 26-kN force that is applied at the center of its width, 75 mm from each side. Determine the principal stresses at points A and B and show the results on elements located at each of these points. Use the shear formula to compute the shear stress.

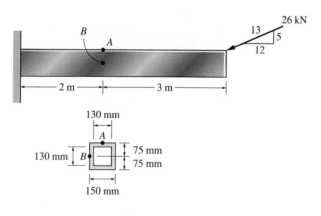

Prob. 15–20

15–22. The wide-flange beam is subjected to the 50-kN force. Determine the principal stress in the beam at point A located on the *web* at the bottom of the upper flange and at point B located on the *web* at the top of the bottom flange. Although not very accurate, use the shear formula to compute the shear stress.

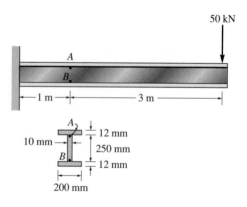

Prob. 15–22

15–23. The bolt is fixed to its support at C. If a force of 18 lb is applied to the wrench to tighten it, determine the principal stresses developed in the bolt shank at points A and B. Represent the results on an element located at each of these points. The shank has a diameter of 0.25 in.

15–26. The internal loadings at a cross section through the 6-in.-diameter drive shaft of a turbine consist of an axial force of 2500 lb, a bending moment of 800 lb · ft, and a torsional moment of 1500 lb · ft. Determine the principal stresses at point B. Also compute the maximum in-plane shear stress at this point.

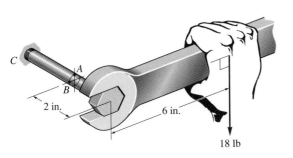

Prob. 15–23

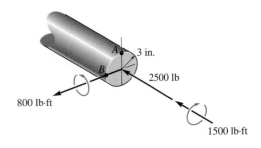

Probs. 15–25/15–26

*****15–24.** The solid shaft is subjected to a torque, bending moment, and shear force as shown. Determine the principal stresses acting at points A and B.

15–27. The box beam is subjected to the loading shown. Determine the principal stresses in the beam at points A and B.

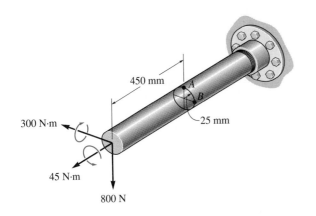

Prob. 15–24

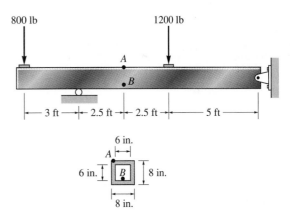

Prob. 15–27

15–25. The internal loadings at a cross section through the 6-in.-diameter drive shaft of a turbine consist of an axial force of 2500 lb, a bending moment of 800 lb · ft, and a torsional moment of 1500 lb · ft. Determine the principal stresses at point A. Also compute the maximum in-plane shear stress at this point.

*****15–28.** A bar has a circular cross section with a diameter of 1 in. It is subjected to a torque and a bending moment. At the point of maximum bending stress the principal tensile stresses are 20 ksi and −10 ksi. Determine the torque and the bending moment.

15.4 Mohr's Circle—Plane Stress

In this section we will show that the equations for plane stress transformation have a graphical solution that is often convenient to use and easy to remember. Furthermore, this approach will allow us to "visualize" how the normal and shear stress components vary as the plane on which they act is orientated in different directions.

Equations 15–1 and 15–2 can be rewritten in the form

$$\sigma_{x'} - \left(\frac{\sigma_x + \sigma_y}{2}\right) = \left(\frac{\sigma_x - \sigma_y}{2}\right) \cos 2\theta + \tau_{xy} \sin 2\theta \qquad (15\text{–}9)$$

$$\tau_{x'y'} = -\left(\frac{\sigma_x - \sigma_y}{2}\right) \sin 2\theta + \tau_{xy} \cos 2\theta \qquad (15\text{–}10)$$

The parameter θ can be *eliminated* by squaring each equation and adding the equations together. The result is

$$\left[\sigma_{x'} - \left(\frac{\sigma_x + \sigma_y}{2}\right)\right]^2 + \tau_{x'y'}^2 = \left(\frac{\sigma_x - \sigma_y}{2}\right)^2 + \tau_{xy}^2$$

For a specific problem, σ_x, σ_y, and τ_{xy} are *known constants*. Thus the above equation can be written in a more compact form as

$$(\sigma_{x'} - \sigma_{avg})^2 + \tau_{x'y'}^2 = R^2 \qquad (15\text{–}11)$$

where

$$\sigma_{avg} = \frac{\sigma_x + \sigma_y}{2}$$

$$R = \sqrt{\left(\frac{\sigma_x - \sigma_y}{2}\right)^2 + \tau_{xy}^2} \qquad (15\text{–}12)$$

If we establish coordinate axes, σ *positive to the right* and τ *positive downward*, and then plot Eq. 15–11, it will be seen that this equation repre-

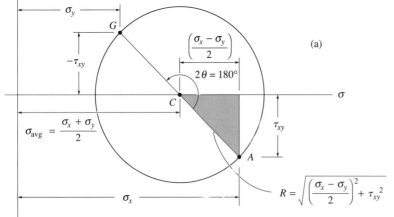

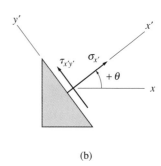

Fig. 15–14

sents the equation of a *circle* having a radius R and center on the σ axis at point $C(\sigma_{avg}, 0)$, Fig. 15–14a. This circle is called *Mohr's circle,* because it was developed by the German engineer Otto Mohr.

Stress Components on an Arbitrary Plane. Since Mohr's circle represents the stress-transformation equations graphically, it can be used to determine the normal and shear stress components $\sigma_{x'}$ and $\tau_{x'y'}$ acting on any arbitrary plane, Fig. 15–14b. For example, consider the case when $\theta = 0°$, that is, when the x' axis is coincident with the x axis, Fig. 15–14c. Here the stress-transformation Eqs. 15–1 and 15–2 give the expected values of $\sigma_{x'} = \sigma_x$ and $\tau_{x'y'} = \tau_{xy}$. This "reference" point $A(\sigma_x, \tau_{xy})$ is plotted in Fig. 15–14a. The radius of the circle is therefore CA. Its value can be calculated from the shaded triangle by using the Pythagorean theorem, which gives the expected result of Eq. 15–12. Now, when $\theta = 90°$, Fig. 15–14d, Eqs. 15–1 and 15–2 give $\sigma_{x'} = \sigma_y$ and $\tau_{x'y'} = -\tau_{xy}$. This point is plotted as point $G(\sigma_y, -\tau_{xy})$ on the circle, Fig. 15–14a. By inspection we see that the radial line CG is $2\theta = 180°$ away from the radial line CA.

As a general statement, then, the normal and shear stress components $\sigma_{x'}$, $\tau_{x'y'}$ acting on any specified plane defined by the x' axis and orientated at an angle θ from the x axis, Fig. 15–15a, can be found from the coordinates of point P on the circle, which is orientated at an angle 2θ measured *from* the radial "reference" line CA ($\theta = 0°$) *to* the radial line CP, Fig. 15–15b. It is important to realize that if the τ axis is constructed *positive downward* as shown, then the angle 2θ on the circle is measured in the *same direction* as the angle θ for the orientation of the plane.*

*If instead the τ axis is constructed *positive upwards,* then the angle 2θ on the circle would be measured in the *opposite direction* of the orientation θ of the plane.

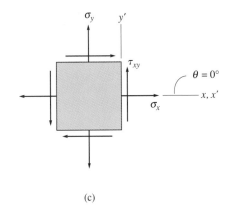

(c)

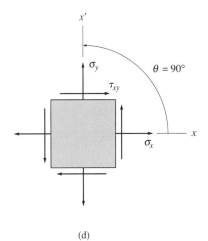

(d)

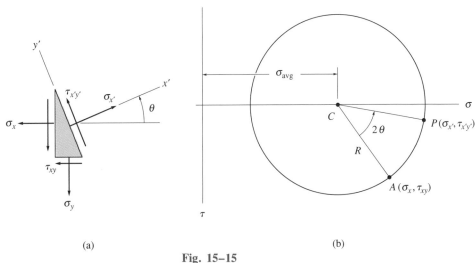

(a) (b)

Fig. 15–15

Principal Stresses. The points B and D, where the circle intersects the σ axis, Fig. 15–16a, give the two values of the principal stress since these points represent the maximum and minimum normal stresses on the element. Note that, as required, the shear stress is zero at these points. From the construction, $\sigma_1 = \sigma_{\text{avg}} + R$ and $\sigma_2 = \sigma_{\text{avg}} - R$, which is the same as applying Eq. 15–5.

The *counterclockwise* angle $2\theta_{p_1}$ is shown measured from the reference radial line CA ($\theta = 0°$) to the radial line CB. From the shaded triangle, the angle θ_{p_1} is then found from $\tan 2\theta_{p_1} = \tau_{xy} / [(\sigma_x - \sigma_y)/2]$, which is the same as Eq. 15–4. This angle, θ_{p_1}, is then measured in the same direction, i.e., *counterclockwise* from the x (reference) axis to the plane on which σ_1 acts, Fig. 15–16b. Note also on the circle that the angle between the radial lines CB and CD is 180°, and as expected the planes on which σ_1 and σ_2 act are $180°/2 = 90°$ apart.

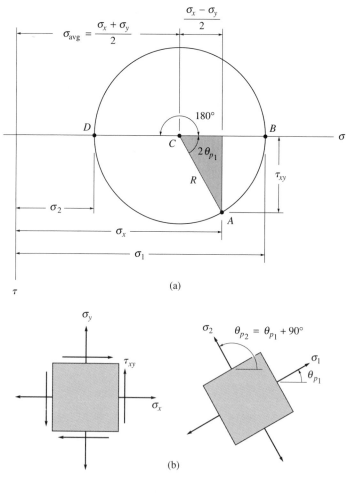

(a)

(b)

Fig. 15–16

Maximum In-Plane Shear Stress. The maximum in-plane shear stress is identified on the circle as either point E or F, Fig. 15–17a. These points have coordinates $E(\sigma_{avg}, R)$ and $F(\sigma_{avg}, -R)$, which represent the same values computed from Eqs. 15–7 and 15–8. Since CE is $90°$ from CB, Fig. 15–17a, the planes of maximum shear stress are $90°/2 = 45°$ from the principal planes of stress, as expected. Also, note that the *clockwise* angle, measured from CA to CE, is $2\theta_{s_1}$. The angle θ_{s_1} can be computed from trigonometry using the data on the circle. This angle is then measured *clockwise* from the x (reference) axis to the x' axis, which identifies the plane that has σ_{avg} and $(\tau_{x'y'})_{max}$ acting on it, Fig. 15–17b. Once these components are established, the stress components on the other three faces of the element can then be constructed as shown.

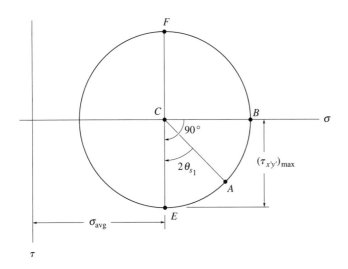

(a)

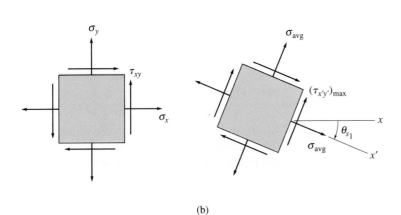

(b)

Fig. 15–17

PROCEDURE FOR ANALYSIS

Mohr's circle represents graphically the state of stress at a point. Once established, it can be used to determine the principal stresses, the maximum in-plane shear stress, or the normal and shear stress components acting on an arbitrary inclined plane. The following steps are required to draw and use the circle.

Construction of the Circle.   Establish a coordinate system such that the abscissa represents the normal stress σ, with *positive to the right*, and the ordinate represents the shear stress τ, with *positive downward*, Fig. 15–18c.

Using the positive sign convention for σ_x, σ_y, τ_{xy}, as shown in Fig. 15–18a, determine the center of the circle C, which is located on the σ axis at a distance $\sigma_{avg} = (\sigma_x + \sigma_y)/2$ from the origin, Fig. 15–18c.

Plot the reference point A having coordinates $A(\sigma_x, \tau_{xy})$. This point represents the normal and shear stress components on the element's right-hand vertical face, and since the x' axis coincides with the x axis, this represents $\theta = 0°$, Fig. 15–18a.

Connect point A with the center C of the circle and determine the hypotenuse CA by trigonometry. This distance represents the radius R of the circle, Fig. 15–18c.

Once R has been determined, sketch the circle.

By constructing the circle in this manner, the stress components $\sigma_{x'}$, $\tau_{x'y'}$ on any plane oriented at an angle θ, Fig. 15–18b, are represented by the coordinates of point P on the circle, measured 2θ from the reference line CA, Fig. 15–18c. Both angles are measured in the *same direction*.

Principal Stresses.   The principal stresses σ_1 and σ_2 are determined from the two points of intersection of the circle with the σ axis, i.e., the points B and D where $\tau = 0$, Fig. 15–16a. The principal stresses act on planes defined by the angles θ_{p_1} and θ_{p_2} which can be determined from the circle using trigonometry. They are represented by angles $2\theta_{p_1}$ and $2\theta_{p_2}$ and are measured *from* the radial reference line CA *to* lines CB and CD. Actually, only one of these angles needs to be calculated, since they are 90° apart. Remember that the direction of rotation $2\theta_p$ on the circle represents the *same* direction of rotation θ_p from the reference axis $(+x)$ to the principal plane $(+x')$. The angles $2\theta_{p_1}$ and θ_{p_1} are shown in Fig. 15–16a and 15–16b, respectively.

Maximum In-Plane Shear Stress.   The average normal stress and maximum in-plane shear stress components are determined from the circle as the coordinates of points E and F, Fig. 15–17a. In this case the angles θ_{s_1} and θ_{s_2} give the orientation of the planes that contain these components. The calculations from the circle are performed in the same manner as for finding the principal stresses and the principal planes. The results are shown in Fig. 15–17b for the element oriented on the basis of calculating both θ_{s_1} and the stress components acting on the plane associated with point E on the circle, Fig. 15–17a.

Stresses on Arbitrary Plane. The normal and shear stress components $\sigma_{x'}$ and $\tau_{x'y'}$ acting on a specified plane defined by the angle θ, Fig. 15–18b, can be obtained from the circle using trigonometry to determine the coordinates of point P, Fig. 15–18c. To locate P, the known angle θ for the plane (in this case counterclockwise) is measured on the circle in the *same direction* 2θ (counterclockwise), *from* the radial reference line CA to the radial line CP. Also, if the value of $\sigma_{y'}$ is required, it can be determined by calculating the σ coordinate of point Q in Fig. 15–18c, since Q lies 180° away from P, and thus represents a rotation of 90° of the x' axis on the element.

The following examples illustrate how to apply this procedure to problems that involve each of the three cases just discussed.

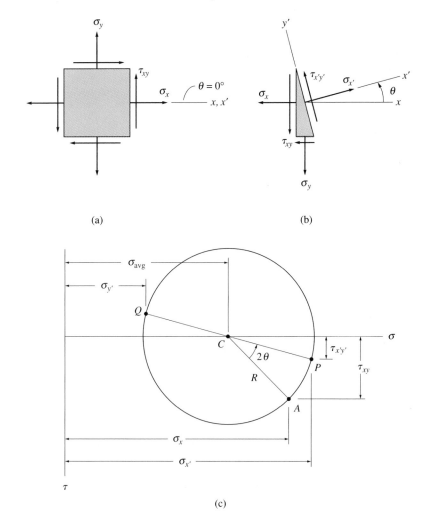

(a)

(b)

(c)

Fig. 15–18

Example 15–7

The state of plane stress at a point is shown on the element in Fig. 15–19a. Determine the principal stresses and the orientation of the element at the point. Also, determine the maximum in-plane shear stresses and the orientation of the element upon which they act.

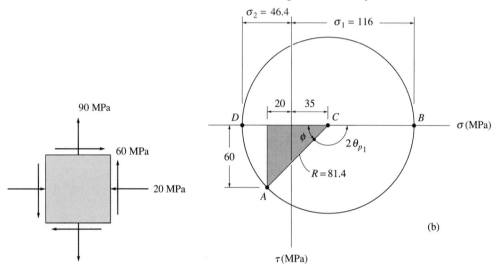

(a)

Fig. 15–19

SOLUTION

Construction of the Circle.   From the problem data,

$$\sigma_x = -20 \text{ MPa} \qquad \sigma_y = 90 \text{ MPa} \qquad \tau_{xy} = 60 \text{ MPa}$$

The σ, τ axes are established in Fig. 15–19b. Note that the positive τ axis must be directed *downward,* so that rotations of the element correspond to rotations in the same direction around the circle. The center of the circle C is located on the σ axis, at the point

$$\sigma_{\text{avg}} = \frac{-20 + 90}{2} = 35 \text{ MPa}$$

The stress components on the right-hand face of the element are the coordinates of the reference point A $(-20, 60)$, $\theta = 0°$, Fig. 15–19b. Applying the Pythagorean theorem to the shaded triangle to determine the circle's radius CA, we have

$$R = \sqrt{(60)^2 + (55)^2} = 81.4 \text{ MPa}$$

Principal Stresses.   The principal stresses are represented by points B and D in Fig. 15–19b. The σ coordinates of these points are

$$\sigma_1 = 35 + 81.4 = 116 \text{ MPa} \qquad\qquad \textit{Ans.}$$
$$\sigma_2 = 35 - 81.4 = -46.4 \text{ MPa} \qquad\qquad \textit{Ans.}$$

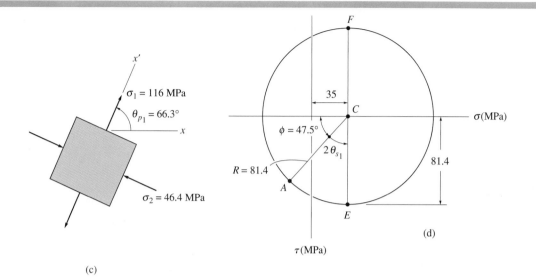

(c)

(d)

The orientation of the element is determined by computing the *counterclockwise* angle $2\theta_{p_1}$ in Fig. 15–19b, which defines the direction of σ_1 ($\sigma_1 > \sigma_2$) and its associated principal plane. Thus,

$$2\theta_{p_1} = 180° - \phi = 180° - \tan^{-1}\frac{60}{55} = 180° - 47.5° = 132.5°$$

$$\theta_{p_1} = 66.3° \hspace{3cm} \textit{Ans.}$$

The element is oriented such that the x' axis or σ_1 is directed at a *counterclockwise* angle of 66.3° with the horizontal as shown in Fig. 15–19c.

Maximum In-Plane Shear Stress.   The maximum in-plane shear stress and the average normal stress are identified by point E or F on the circle in Fig. 15–19d. The coordinates of these points are $E(35, 81.4)$ and $F(35, -81.4)$. Thus,

$$\tau_{max} = 81.4 \text{ MPa} \hspace{2cm} \textit{Ans.}$$

$$\sigma_{avg} = 35 \text{ MPa} \hspace{2cm} \textit{Ans.}$$

The *counterclockwise* angle θ_{s_1} can be computed from the circle, identified as $2\theta_{s_1}$. Using the result of $\phi = 47.5°$, computed above, we have

$$2\theta_{s_1} = 90° - 47.5° = 42.5°$$

$$\theta_{s_1} = 21.3° \hspace{3cm} \textit{Ans.}$$

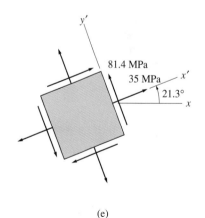

(e)

This *counterclockwise* angle defines the direction of σ_{avg} (or the x' axis), Fig. 15–19e. Since point E has *positive* coordinates, then on the associated face of the element, the average normal stress and the maximum in-plane shear stress act in the *positive* x' and y' directions as shown.

Example 15–8

Due to the applied loading, the element at point A on the solid cylinder in Fig. 15–20a is subjected to the state of stress shown. Determine the principal stresses acting at this point.

SOLUTION

Construction of the Circle. From the element's right-hand face,

$$\sigma_x = -12 \text{ ksi} \qquad \sigma_y = 0 \qquad \tau_{xy} = -6 \text{ ksi}$$

The center of the circle is at

$$\sigma_{avg} = \frac{-12 + 0}{2} = -6 \text{ ksi}$$

The initial point $A(-12, -6)$ and the center $C(-6, 0)$ are plotted in Fig. 15–20b. The circle is constructed having a radius of

$$R = \sqrt{(12 - 6)^2 + (6)^2} = 8.49 \text{ ksi}$$

Principal Stresses. The principal stresses are indicated by the coordinates of points B and D. We have, for $\sigma_1 > \sigma_2$,

$$\sigma_1 = 8.49 - 6 = 2.49 \text{ ksi} \hspace{3em} \textit{Ans.}$$
$$\sigma_2 = -6 - 8.49 = -14.5 \text{ ksi} \hspace{3em} \textit{Ans.}$$

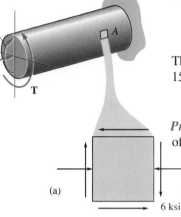

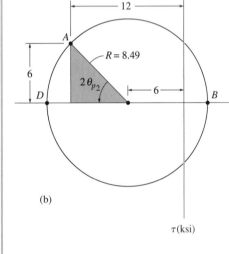

12 ksi

6 ksi

(a)

The orientation of the element is determined by computing the *counterclockwise* angle $2\theta_{p_2}$ in Fig. 15–20b, which defines the direction θ_{p_2} of σ_2 and its associated principal plane. We have

$$2\theta_{p_2} = \tan^{-1} \frac{6}{(12 - 6)} = 45.0°$$

$$\theta_{p_2} = 22.5°$$

The element is oriented such that the x' axis or σ_2 is directed *counterclockwise* 22.5° from the horizontal (x axis) as shown in Fig. 15–20c.

As an exercise, use the circle to show that the maximum in-plane shear stress is shown on the element in Fig. 15–20d.

Fig. 15–20

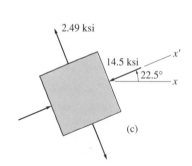

(c)

(d)

■ Example 15–9

The state of plane stress at a point is shown on the element in Fig. 15–21a. Represent this state of stress on an element oriented 30° counterclockwise from the position shown.

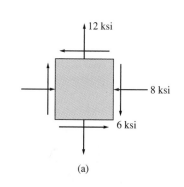

(a)

SOLUTION

Construction of the Circle. From the problem data,

$$\sigma_x = -8 \text{ ksi} \qquad \sigma_y = 12 \text{ ksi} \qquad \tau_{xy} = -6 \text{ ksi}$$

The σ and τ axes are established in Fig. 15–21b. The center of the circle C is on the σ axis at

$$\sigma_{\text{avg}} = \frac{-8 + 12}{2} = 2 \text{ ksi}$$

The initial point for $\theta = 0°$ has coordinates $A(-8, -6)$. Hence from the shaded triangle the radius CA is

$$R = \sqrt{(10)^2 + (6)^2} = 11.66$$

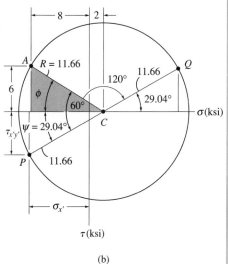

(b)

Stresses on 30° Element. Since the element is to be rotated 30° *counterclockwise*, we must construct a radial line CP, $2(30°) = 60°$ *counterclockwise*, measured from CA ($\theta = 0°$), Fig. 15–21b. The coordinates of point P ($\sigma_{x'}$, $\tau_{x'y'}$) must now be obtained. From the geometry of the circle,

$$\phi = \tan^{-1}\frac{6}{10} = 30.96° \qquad \psi = 60° - 30.96° = 29.04°$$

$$\sigma_{x'} = 2 - 11.66 \cos 29.04° = -8.20 \text{ ksi} \qquad \textit{Ans.}$$

$$\tau_{x'y'} = 11.66 \sin 29.04° = 5.66 \text{ ksi} \qquad \textit{Ans.}$$

Using established x', y' axes, these stress components act on face BD of the element shown in Fig. 15–21c.

The stress components acting on an adjacent face are represented by the coordinates of point Q on the circle. This point lies on the radial line CQ, which is 180° from CP (90° from face BD on the element, as expected). The coordinates of point Q are

$$\sigma_{x'} = 2 + 11.66 \cos 29.04° = 12.2 \text{ ksi} \qquad \textit{Ans.}$$

$$\tau_{x'y'} = -(11.66 \sin 29.04) = -5.66 \text{ ksi} \qquad \text{(check)}$$

Since CQ is 120° *clockwise* from CA, Fig. 15–21b, the above stress components act on face DE of the element in Fig. 15–21c since the x' axis for this face is oriented 120°/2 = 60° *clockwise* from the positive x axis.

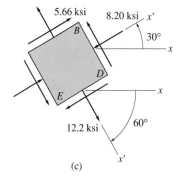

(c)

Fig. 15–21

15.5 Absolute Maximum Shear Stress

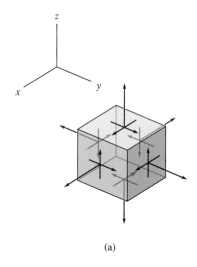

(a)

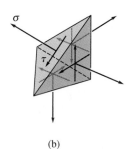

(b)

When a point in a body is subjected to a general three-dimensional state of stress, an element of material has a normal-stress and two shear-stress components acting on each of its faces, Fig. 15–22a. Like the case of plane stress, it is possible to develop stress-transformation equations that can be used to determine the normal and shear stress components σ and τ acting on *any* skewed plane of the element, Fig. 15–22b. Furthermore, at the point it is also possible to determine the unique orientation of an element having only principal stresses acting on its faces. As shown in Fig. 15–22c, these principal stresses are assumed to have magnitudes of maximum, intermediate, and minimum intensity, i.e., $\sigma_{max} \geq \sigma_{int} \geq \sigma_{min}$.

A discussion of the transformation of stress in three dimensions is beyond the scope of this text; however, it is discussed in books related to the theory of elasticity. For our purposes, we will assume that the principal orientation of the element and the principal stresses are known, Fig. 15–22c. This is a condition known as *triaxial stress*. If we view the element in two dimensions, that is, in the $y'-z'$, $x'-z'$, and $x'-y'$ planes, Fig. 15–23a, 15–23b, and 15–23c, we can then use Mohr's circle to determine the *maximum in-plane shear stress* for each case. For example, the diameter of Mohr's circle extends between the principal stresses σ_{int} and σ_{min} for the case shown in Fig. 15–23a. From this circle, Fig. 15–23d, the maximum in-plane shear stress is $(\tau_{y'z'})_{max} = (\sigma_{int} - \sigma_{min})/2$, and the associated average normal stress is $(\sigma_{int} + \sigma_{min})/2$. As shown in Fig. 15–23e, the element having these stress components on it must be oriented 45° from the position of the element shown in Fig. 15–23a. Mohr's circles for the elements in Fig. 15–23b and 15–23c have also been constructed in Fig. 15–23d. The corresponding elements having a 45° orientation and subjected to maximum in-plane shear and average normal stress components are shown in Fig. 15–23f and 15–23g, respectively.

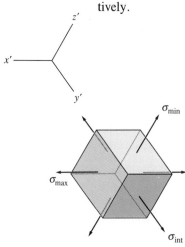

(c)

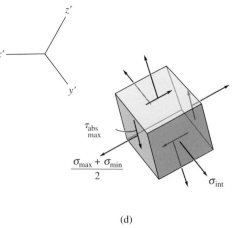

(d)

Fig. 15–22

Comparing the three circles in Fig. 15–23d, it is seen that the *absolute maximum shear stress*, $\tau_{\text{abs}}_{\text{max}}$, is defined by the circle having the largest radius, which occurs for the element shown in Fig. 15–23b. In other words, the element in Fig. 15–23f is oriented by a rotation of 45° about the y' axis from the element in Fig. 15–22c. It is shown in Fig. 15–22d. Notice that this condition can also be *determined directly* by simply choosing the maximum and minimum principal stresses from Fig. 15–22c, in which case the absolute maximum shear stress will be

$$\tau_{\text{abs}}_{\text{max}} = \frac{\sigma_{\max} - \sigma_{\min}}{2} \qquad (15\text{–}13)$$

And the associated average normal stress will be

$$\sigma_{\text{avg}} = \frac{\sigma_{\max} + \sigma_{\min}}{2} \qquad (15\text{–}14)$$

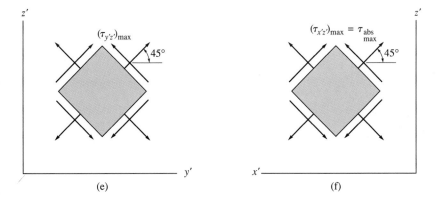

(d)

Fig. 15–23

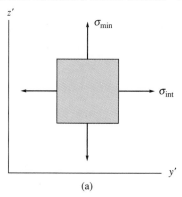

(a)

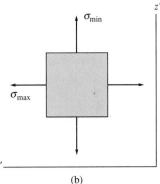

(b)

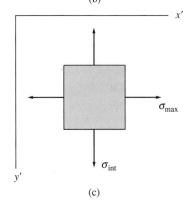

(c)

(e)   (f)   (g)

The analysis considered only the stress components acting on elements located in positions found from rotations about the x', y', or z' axis. If we had used the three-dimensional stress-transformation equations of the theory of elasticity to obtain values of the normal and shear stress components acting on any arbitrary skewed plane at the point, Fig. 15–22b, it could be shown that regardless of the orientation of the plane, specific values of the shear stress τ on the plane will *always be less* than the maximum shear stress computed from Eq. 15–13. Furthermore, the normal stress σ acting on the plane will have a value lying between the maximum and minimum principal stresses, that is, $\sigma_{max} \geq \sigma \geq \sigma_{min}$.

Plane Stress. The above results have an important implication for the case of plane stress, particularly when the in-plane principal stresses have the *same sign*, i.e., they are both tensile or both compressive. For example, consider the material to be subjected to plane stress such that the in-plane principal stresses are represented as σ_{max} and σ_{int}, in the x' and y' directions, respectively; while the out-of-plane principal stress in the z' direction is $\sigma_{min} = 0$, Fig. 15–24a. Mohr's circles that describe this state of stress for element orientations about each coordinate axis are shown in Fig. 15–24b. Here it is seen that although the maximum in-plane shear stress is $(\tau_{x'y'})_{max} = (\sigma_{max} - \sigma_{int})/2$, this value is *not* the absolute maximum shear stress to which the material is subjected. Instead, from Eq. 15–13 or Fig. 15–24b,

$$\tau_{\substack{abs \\ max}} = (\tau_{x'z'})_{max} = \frac{\sigma_{max} - 0}{2} = \frac{\sigma_{max}}{2} \qquad (15\text{–}15)$$

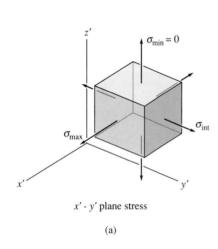

x' - y' plane stress

(a)

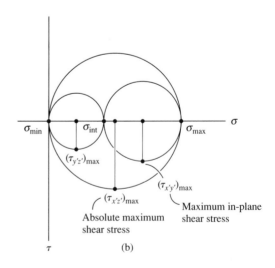

(b)

Fig. 15–24

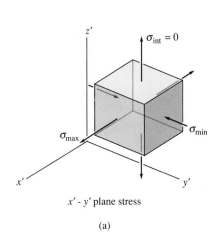

x' - y' plane stress

(a)

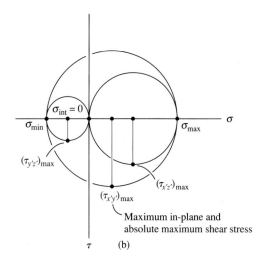

Maximum in-plane and
absolute maximum shear stress

(b)

Fig. 15–25

On the other hand, if one of the in-plane principal stresses had the *opposite sign* of that of the other, then these stresses would be represented as σ_{max} and σ_{min}, and the out-of-plane principal stress $\sigma_{int} = 0$, Fig. 15–25a. Mohr's circles that describe this state of stress for element orientations about each coordinate axis are shown in Fig. 15–25b. Clearly, in this case

$$\tau_{\substack{abs \\ max}} = (\tau_{x'y'})_{max} = \frac{\sigma_{max} - \sigma_{min}}{2} \qquad (15\text{–}16)$$

In summary, then, if the in-plane principal stresses both have the *same sign,* the *absolute maximum shear stress* will occur *out of the plane* and has a value of $\tau_{\substack{abs \\ max}} = \sigma_{max}/2$. However, if the in-plane principal stresses are of *opposite signs,* then the *absolute maximum shear stress equals the maximum in-plane shear stress;* that is, $\tau_{\substack{abs \\ max}} = (\sigma_{max} - \sigma_{min})/2$. Calculation of the absolute maximum shear stress as indicated here is important when designing members made of a ductile material, since the strength of the material depends on its ability to resist shear stress.

Example 15–10

Due to the applied loading, the element A on the wood frame in Fig. 15–26a is subjected to the state of plane stress shown. Determine the principal stresses and the absolute maximum shear stress at A.

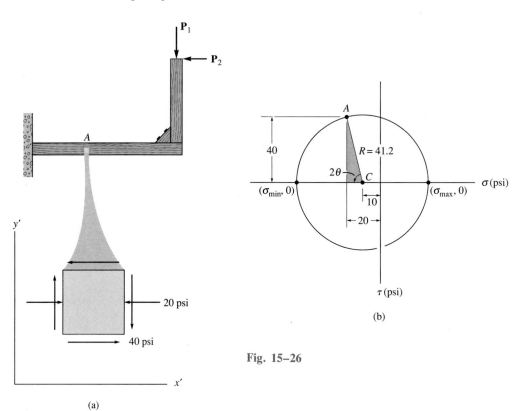

Fig. 15–26

SOLUTION

Principal Stresses. The principal planes and principal stresses can be determined from Mohr's circle. The center of the circle is on the σ axis at $\sigma_{avg} = (-20 + 0)/2 = -10$ psi. Plotting the controlling point $A(-20, -40)$, the circle can be drawn as shown in Fig. 15–26b. The radius is

$$R = \sqrt{(20 - 10)^2 + (40)^2} = 41.2 \text{ psi}$$

The principal stresses are the points where the circle intersects the σ axis; i.e.,

$$\sigma_{max} = -10 + 41.2 = 31.2 \text{ psi}$$
$$\sigma_{min} = -10 - 41.2 = -51.2 \text{ psi}$$

From the circle, the *counterclockwise* angle 2θ, measured from CA to the $-\sigma$ axis, is

$$2\theta = \tan^{-1}\left(\frac{40}{(20-10)}\right) = 76.0°$$

Thus,

$$\theta = 38.0°$$

This *counterclockwise* rotation defines the direction of the x' axis or σ_{min} and its associated principal plane, Fig. 15–26c. Since there is no principal stress on the element in the z direction, we have

$$\sigma_{max} = 31.2 \text{ psi} \qquad \sigma_{int} = 0 \qquad \sigma_{min} = -51.2 \text{ psi} \qquad Ans.$$

Notice that the notation has been selected so that $\sigma_{max} \geq \sigma_{int} \geq \sigma_{min}$.

Absolute Maximum Shear Stress. Applying Eqs. 15–13 and 15–14, we have

$$\tau_{\substack{abs \\ max}} = \frac{\sigma_{max} - \sigma_{min}}{2} = \frac{31.2 - (-51.2)}{2} = 41.2 \text{ psi} \qquad Ans.$$

$$\sigma_{avg} = \frac{\sigma_{max} + \sigma_{min}}{2} = \frac{31.2 - 51.2}{2} = -10 \text{ psi}$$

These same results can be obtained by drawing Mohr's circle for each orientation of an element about the x', y', and z' axes, Fig. 15–26d. In either case, since σ_{max} and σ_{min} are of *opposite signs,* the state of absolute maximum shear stress results from the 45° rotation of the element in Fig. 15–26c about the z' axis. In other words, the absolute maximum shear stress equals the maximum in-plane shear stress. The properly oriented element is shown in Fig. 15–26e.

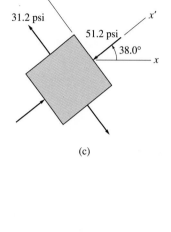

(c)

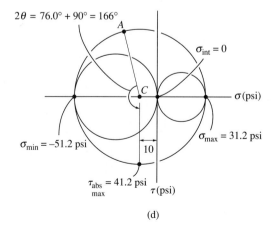

(d)

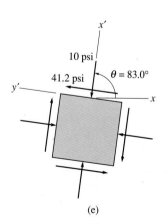

(e)

PROBLEMS

15–29.   Solve Prob. 15–2 using Mohr's circle.

15–30.   Solve Prob. 15–16 using Mohr's circle.

15–31.   Solve Prob. 15–15 using Mohr's circle.

*15–32.   Determine the equivalent state of stress if an element is oriented 60° counterclockwise from the element shown.

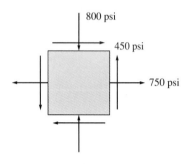

Prob. 15–32

15–33.   Determine the equivalent state of stress if an element is oriented 40° clockwise from the element shown.

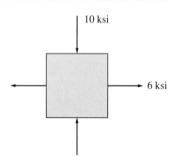

Prob. 15–33

15–34.   Determine the equivalent state of stress if an element is oriented 25° counterclockwise from the element shown.

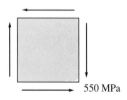

Prob. 15–34

15–35.   Determine (a) the principal stress and (b) the maximum in-plane shear stress and average normal stress. Specify the orientation of the element in each case.

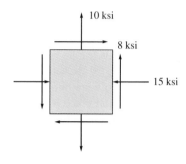

Prob. 15–35

*15–36.   Determine (a) the principal stress and (b) the maximum in-plane shear stress and average normal stress. Specify the orientation of the element in each case.

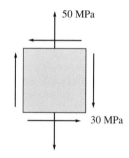

Prob. 15–36

15–37.   Determine (a) the principal stress and (b) the maximum in-plane shear stress and average normal stress. Specify the orientation of the element in each case.

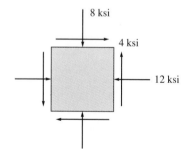

Prob. 15–37

15–38. A point on a thin plate is subjected to two successive states of stress as shown. Determine the resulting state of stress with reference to an element oriented as shown on the right.

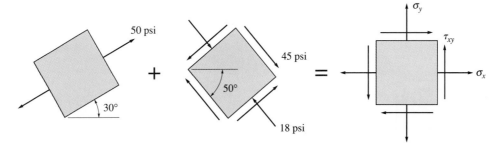

Prob. 15–38

15–39. Mohr's circle for the state of stress in Fig. 15–15a is shown in Fig. 15–15b. Show that finding the coordinates of point P ($\sigma_{x'}$, $\tau_{x'y'}$) on the circle gives the same value as the stress-transformation Eqs. 15–1 and 15–2.

*****15–40.** The post has a square cross-sectional area. If it is fixed-supported at its base and a horizontal force is applied at its end as shown, determine (a) the maximum in-plane shear stress developed at A and (b) the principal stresses at A.

15–41. The ladder is supported on the rough surface at A and by a smooth wall at B. If a man weighing 150 lb stands upright at C, determine the principal stress in one of the legs at point D. Each leg is made from a 1-in.-thick board having a rectangular cross section. Assume that the total weight of the man is exerted vertically on the rung at C and is shared equally by each of the ladder's two legs. Neglect the weight of the ladder and the forces developed by the man's arms.

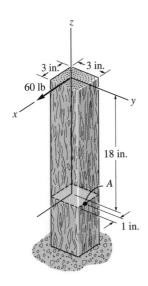

Prob. 15–40

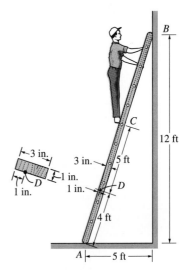

Prob. 15–41

15–42.   Determine (*a*) the principal stress and (*b*) the maximum in-plane shear stress and average normal stress. Specify the orientation of the element in each case.

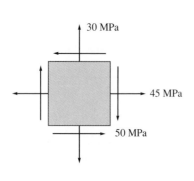

Prob. 15–42

*15–44.   The stress at a point is shown on the element. Determine the principal stresses and the absolute maximum shear stress.

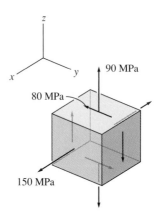

Prob. 15–44

15–43.   The crane is used to support the 350-lb load. Determine the principal stresses acting in the boom at points *A* and *B*. The cross section is rectangular and has a width of 6 in. and a thickness of 3 in.

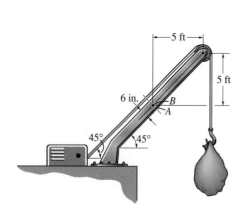

Prob. 15–43

15–45.   The frame is subjected to a horizontal force and couple moment at its end. Determine the principal stresses and the absolute maximum shear stress at point *A*. The cross-sectional area at this point is shown.

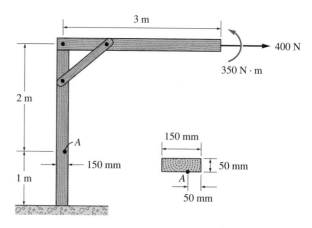

Prob. 15–45

15–46.   The stress at a point is shown on the element. Determine the principal stresses and the absolute maximum shear stress.

15–47.   The state of stress at a point is shown on the element. Determine the principal stresses and the absolute maximum shear stress.

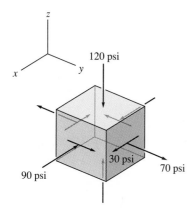

Prob. 15–46

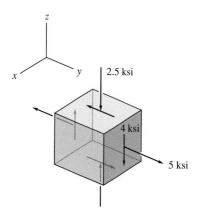

Prob. 15–47

15.6 Plane Strain

As outlined in Sec. 8.8, the general state of strain at a point in a body is represented by a combination of three components of normal strain, ϵ_x, ϵ_y, ϵ_z, and three components of shear strain, γ_{xy}, γ_{xz}, γ_{yz}. These six components tend to deform each face of an element of the material, and like stress, the normal and shear strain *components* at the point will vary according to the orientation of the element. The *strain* components at a point are often determined by using strain gauges, which measure these components in *specified directions*. For both analysis and design, however, engineers must sometimes transform this data in order to obtain the strain components in other directions.

To understand how this is done, we will first confine our attention to a study of *plane strain*. Specifically, we will not consider the effects of the components ϵ_z, γ_{xz}, and γ_{yz}. In general, then, a plane-strained element is subjected to two components of normal strain, ϵ_x, ϵ_y, and one component of shear strain, γ_{xy}. The deformations of an element caused by each of these strains are shown graphically in Fig. 15–27. Note that the normal strains are produced by *changes in length* of the element in the x and y directions, and the

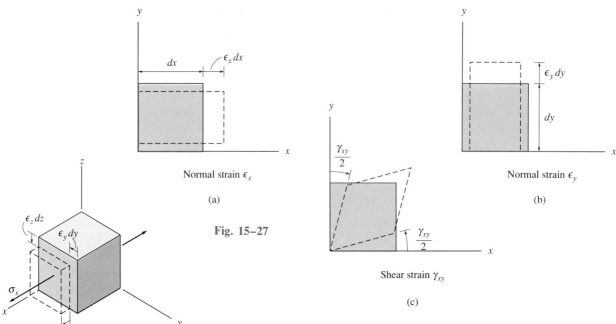

Normal strain ϵ_x

(a)

Fig. 15–27

Normal strain ϵ_y

(b)

Shear strain γ_{xy}

(c)

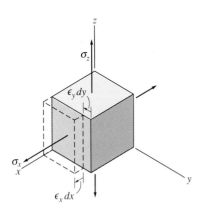

Plane stress, σ_x, does not cause plane
strain in the x–y plane since $\epsilon_z \neq 0$

(a)

Plane strain, ϵ_x, ϵ_y, does not cause plane
stress in the x–y plane since $\sigma_z \neq 0$

(b)

Fig. 15–28

shear strain is produced by the *relative rotation* of two adjacent sides of the element.

Although plane strain and plane stress both have three components lying in the same plane, realize that plane stress *does not* necessarily cause plane strain or vice versa. The reason for this has to do with the Poisson effect discussed in Sec. 9.6. For example, if the element in Fig. 15–28a is subjected to a uniaxial stress σ_x, not only is a normal strain ϵ_x produced, but there are *also* two associated normal strains, $\epsilon_y = -\nu\epsilon_x$ and $\epsilon_z = -\nu\epsilon_x$. This is obviously *not* a case of plane strain in the x–y plane, since $\epsilon_z \neq 0$. On the other hand, a uniaxial strain of ϵ_x requires application of σ_x, Fig. 15–28b. However, this will also produce contractions of the element in the y and z directions due to the Poisson effect. In order to *require* $\epsilon_z = 0$, and thereby produce plane strain in the x–y plane, it is therefore *necessary* that a normal stress of σ_z *also* be applied to the element, Fig. 15–28b. As a result, plane stress does not exist in the x–y plane. In general, then, unless $\nu = 0$, the Poisson effect will *prevent* the simultaneous occurrence of plane strain and plane stress. It should also be pointed out that since shear stress and shear strain are *not* affected by Poisson's ratio, a condition of $\tau_{xz} = \tau_{yz} = 0$ requires $\gamma_{xz} = \gamma_{yz} = 0$.

As stated above, the measurement of strain at a point is generally determined using strain gauges. These measurements are usually made on the load-free surface of the body, and as a result, only the strain within the surface is determined. Realize, however, that such strains are a consequence of *plane stress*. Although this is the case, the analysis of plane strain, which is to be discussed in this chapter, can also be used to analyze the in-plane strains caused by plane stress.

15.7 General Equations of Plane-Strain Transformation

It is important in plane-strain analysis to establish transformation equations that can be used to determine the x', y' components of normal and shear strain at a point, provided the x, y components of strain are known. Essentially this problem is one of geometry and requires relating the deformations and rotations of differential line segments, which represent the sides of differential elements that are parallel to each set of axes.

Sign Convention. Before the strain-transformation equations can be developed, we must first establish a sign convention for the strains. This convention is the same as that established in Sec. 8.8 and will be restated here for the condition of plane strain. With reference to the differential element shown in Fig. 15–29a, *normal strains* ϵ_x and ϵ_y are *positive* if they cause *elongation* along the x and y axes, respectively, and the *shear strain* γ_{xy} is *positive* if the interior angle *AOB becomes smaller* than 90°. This sign convention also follows the corresponding one used for plane stress, Fig. 15–5a, that is, positive σ_x, σ_y, τ_{xy} will cause the element to *deform* in the positive ϵ_x, ϵ_y, γ_{xy} directions, respectively.

The problem here will be to determine at a point the normal and shear strains $\epsilon_{x'}$, $\epsilon_{y'}$, $\gamma_{x'y'}$, measured relative to the x', y' axes, if we know ϵ_x, ϵ_y, γ_{xy}, measured relative to the x, y axes. If the angle between the x and x' axes is θ, then, like the case of plane stress, θ will be *positive* provided it follows the curl of the right-hand fingers, i.e., counterclockwise, as shown in Fig. 15–29b.

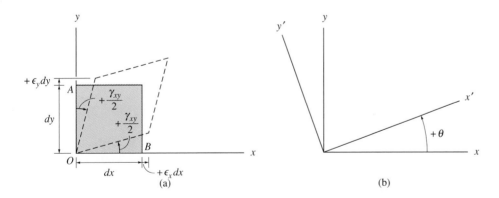

(a)   (b)

Positive sign convention

Fig. 15–29

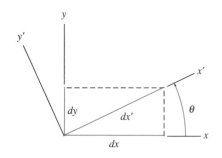

Before deformation

(a)

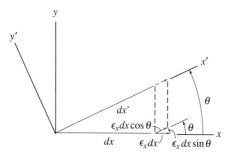

Normal strain ϵ_x

(b)

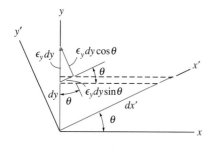

Normal strain ϵ_y

(c)

Fig. 15–30(a–c)

Normal and Shear Strains. In order to develop the strain-transformation equation for determining $\epsilon_{x'}$, we must determine the elongation of a line segment dx' that lies along the x' axis and is subjected to strain components ϵ_x, ϵ_y, γ_{xy}. As shown in Fig. 15–30a, the components of the line dx' along the x and y axes are

$$dx = dx' \cos \theta$$
$$dy = dx' \sin \theta \tag{15–17}$$

When the positive normal strain ϵ_x occurs, Fig. 15–30b, the line dx is elongated $\epsilon_x\, dx$, which causes line dx' to elongate $\epsilon_x\, dx \cos \theta$. Likewise, when ϵ_y occurs, Fig. 15–30c, line dy elongates $\epsilon_y\, dy$, which causes line dx' to elongate $\epsilon_y\, dy \sin \theta$. Lastly, assuming that dx remains fixed in position, the shear strain γ_{xy}, which is the change in angle between dx and dy, causes the top of line dy to be displaced $\gamma_{xy}\, dy$ to the right, as shown in Fig. 15–30d. This causes dx' to elongate $\gamma_{xy}\, dy \cos \theta$. If all three of these elongations are added together, the resultant elongation of dx' is then

$$\delta x' = \epsilon_x\, dx \cos \theta + \epsilon_y\, dy \sin \theta + \gamma_{xy}\, dy \cos \theta$$

From Eq. 8–11, the normal strain along the line dx' is $\epsilon_{x'} = \delta x'/dx'$. Using Eqs. 15–17, we therefore have

$$\epsilon_{x'} = \epsilon_x \cos^2 \theta + \epsilon_y \sin^2 \theta + \gamma_{xy} \sin \theta \cos \theta \tag{15–18}$$

The strain-transformation equation for determining $\gamma_{x'y'}$ can be developed by considering the amount of rotation each of the line segments dx' and dy' undergo when subjected to the strain components ϵ_x, ϵ_y, γ_{xy}. First we will consider the rotation of dx', which is defined by the counterclockwise angle α shown in Fig. 15–30e. It can be determined from the displacement $\delta y'$ using $\alpha = \delta y'/dx'$. To obtain $\delta y'$, consider the following three displacement components acting in the y' direction: one from ϵ_x, giving $-\epsilon_x\, dx \sin \theta$, Fig. 15–30b; another from ϵ_y, giving $\epsilon_y\, dy \cos \theta$, Fig. 15–30c; and the last from γ_{xy}, giving $-\gamma_{xy}\, dy \sin \theta$, Fig. 15–30d. Thus, $\delta y'$, as caused by all three strain components, is

$$\delta y' = -\epsilon_x\, dx \sin \theta + \epsilon_y\, dy \cos \theta - \gamma_{xy}\, dy \sin \theta$$

Using Eq. 15–17, with $\alpha = \delta y'/dx'$, we have

$$\alpha = (-\epsilon_x + \epsilon_y) \sin \theta \cos \theta - \gamma_{xy} \sin^2 \theta \tag{15–19}$$

As shown in Fig. 15–30e, the line dy' rotates by an amount β. We can determine this angle by a similar analysis, or by simply substituting $\theta + 90°$ for θ into Eq. 15–19. Using the identities $\sin (\theta + 90°) = \cos \theta$, $\cos(\theta + 90°) = -\sin \theta$, we have

$$\beta = (-\epsilon_x + \epsilon_y) \sin (\theta + 90°) \cos (\theta + 90°) - \gamma_{xy} \sin^2 (\theta + 90°)$$
$$= -(-\epsilon_x + \epsilon_y) \cos \theta \sin \theta - \gamma_{xy} \cos^2 \theta$$

Since α and β represent the rotation of the sides dx' and dy' of a differential element whose sides were originally oriented along the x' and y' axes, and

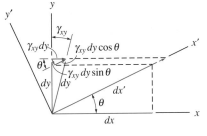

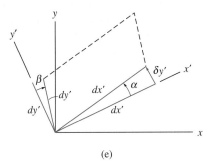

(e)

Shear strain γ_{xy}

(d)

Fig. 15–30(d, e)

β is in the opposite direction to α, Fig. 15–30e, the element is then subjected to a shear strain of

$$\gamma_{x'y'} = \alpha - \beta = -2(\epsilon_x - \epsilon_y) \sin\theta \cos\theta + \gamma_{xy}(\cos^2\theta - \sin^2\theta) \quad (15\text{–}20)$$

Using the trigonometric identities $\sin 2\theta = 2\sin\theta\cos\theta$, $\cos 2\theta = (1 + \cos^2\theta)/2$, and $\sin^2\theta + \cos^2\theta = 1$, we can rewrite Eqs. 15–18 and 15–20 in the final form

$$\epsilon_{x'} = \frac{\epsilon_x + \epsilon_y}{2} + \frac{\epsilon_x - \epsilon_y}{2}\cos 2\theta + \frac{\gamma_{xy}}{2}\sin 2\theta \quad (15\text{–}21)$$

$$\frac{\gamma_{x'y'}}{2} = -\frac{\epsilon_x - \epsilon_y}{2}\sin 2\theta + \frac{\gamma_{xy}}{2}\cos 2\theta \quad (15\text{–}22)$$

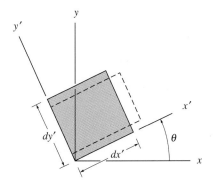

Positive normal strain, $\epsilon_{x'}$

(a)

These strain-transformation equations give the normal strain $\epsilon_{x'}$ in the x' direction and the shear strain $\gamma_{x'y'}$ of an element oriented at an angle θ, as shown in Fig. 15–31. According to the established sign convention, if $\epsilon_{x'}$ is *positive*, the element *elongates* in the positive x' direction, Fig. 15–31a, and if $\gamma_{x'y'}$ is positive, the element deforms as shown in Fig. 15–31b. Note again, as stated earlier, that these deformations occur due to positive normal stress $\sigma_{x'}$ and positive shear stress $\tau_{x'y'}$ acting on the element.

If the normal strain in the y' direction is required, it can be obtained from Eq. 15–21 by simply substituting $(\theta + 90°)$ for θ. The result is

$$\epsilon_{y'} = \frac{\epsilon_x + \epsilon_y}{2} - \frac{\epsilon_x - \epsilon_y}{2}\cos 2\theta - \frac{\gamma_{xy}}{2}\sin 2\theta \quad (15\text{–}23)$$

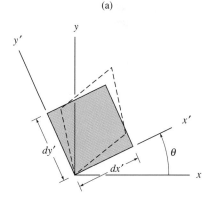

Positive shear strain, $\gamma_{x'y'}$

(b)

The similarity between the above three equations and those for plane-stress transformation, Eqs. 15–1, 15–2, and 15–3, should be noted. By comparison, σ_x, σ_y, $\sigma_{x'}$, $\sigma_{y'}$ correspond to ϵ_x, ϵ_y, $\epsilon_{x'}$, $\epsilon_{y'}$; and τ_{xy}, $\tau_{x'y'}$ correspond to $\gamma_{xy}/2$, $\gamma_{x'y'}/2$.

Fig. 15–31

Principal Strains. Like stress, the orientation of an element at a point can be determined such that the element's deformation is represented by normal strains, with *no* shear strain. When this occurs the normal strains are referred to as *principal strains,* and if the material is isotropic, the axes along which these strains occur coincide with the axes that define the planes of principal stress.

From Eqs. 15–4 and 15–5, and the correspondence between stress and strain mentioned above, the direction of the axes and the two values of the principal strains ϵ_1 and ϵ_2 are determined from

$$\tan 2\theta_p = \frac{\gamma_{xy}}{\epsilon_x - \epsilon_y} \tag{15–24}$$

and

$$\epsilon_{1,2} = \frac{\epsilon_x + \epsilon_y}{2} \pm \sqrt{\left(\frac{\epsilon_x - \epsilon_y}{2}\right)^2 + \left(\frac{\gamma_{xy}}{2}\right)^2} \tag{15–25}$$

Maximum In-Plane Shear Strain. Carrying the stress–strain analogy further, the axes along which maximum in-plane shear strain occurs are 45° away from those that define the principal strains. Using Eq. 15–6, they can be found from

$$\tan 2\theta_s = -\frac{(\epsilon_x - \epsilon_y)}{\gamma_{xy}} \tag{15–26}$$

By analogy to Eqs. 15–7 and 15–8, the maximum in-plane shear strain and associated average normal strain are determined from the following equations:

$$\frac{(\gamma_{x'y'})_{max}}{2} = \sqrt{\left(\frac{\epsilon_x - \epsilon_y}{2}\right)^2 + \left(\frac{\gamma_{xy}}{2}\right)^2} \tag{15–27}$$

and

$$\epsilon_{avg} = \frac{\epsilon_x + \epsilon_y}{2} \tag{15–28}$$

Application of the above equations is illustrated numerically in the following examples.

■ Example 15–11

A differential element of material at a point is subjected to a state of plane strain $\epsilon_x = 500(10^{-6})$, $\epsilon_y = -300(10^{-6})$, $\gamma_{xy} = 200(10^{-6})$, which tends to distort the element as shown in Fig. 15–32a. Determine the equivalent strains acting on an element oriented at the point, *clockwise* 30° from the original position.

SOLUTION

The strain-transformation Eqs. 15–21 and 15–22 will be used to solve the problem. Since θ is *positive counterclockwise*, then for this problem $\theta = -30°$. Thus,

$$\epsilon_{x'} = \frac{\epsilon_x + \epsilon_y}{2} + \frac{\epsilon_x - \epsilon_y}{2} \cos 2\theta + \frac{\gamma_{xy}}{2} \sin 2\theta$$

$$= \left[\frac{500 + (-300)}{2}\right](10^{-6}) + \left[\frac{500 - (-300)}{2}\right](10^{-6}) \cos(2(-30))$$

$$+ \frac{200(10^{-6})}{2} \sin(2(-30))$$

$$\epsilon_{x'} = 213(10^{-6}) \qquad\qquad\qquad Ans.$$

$$\frac{\gamma_{x'y'}}{2} = -\frac{\epsilon_x - \epsilon_y}{2} \sin 2\theta + \frac{\gamma_{xy}}{2} \cos 2\theta$$

$$= -\left[\frac{500 - (-300)}{2}\right](10^{-6}) \sin(2(-30)) + \frac{200(10^{-6})}{2} \cos(2(-30))$$

$$\gamma_{x'y'} = 793(10^{-6}) \qquad\qquad\qquad Ans.$$

We can obtain $\epsilon_{y'}$ using Eq. 15–21 with $\theta = 60°$ ($\theta = -30° + 90°$), Fig. 15–32b. We have with $\epsilon_{y'}$ replacing $\epsilon_{x'}$,

$$\epsilon_{y'} = \frac{\epsilon_x + \epsilon_y}{2} + \frac{\epsilon_x - \epsilon_y}{2} \cos 2\theta + \frac{\gamma_{xy}}{2} \sin 2\theta$$

$$= \left[\frac{500 + (-300)}{2}\right](10^{-6}) + \left[\frac{500 - (-300)}{2}\right](10^{-6}) \cos(120°)$$

$$+ \frac{200(10^{-6})}{2} \sin(120°)$$

$$\epsilon_{y'} = -13.4(10^{-6}) \qquad\qquad\qquad Ans.$$

These results tend to distort the element as shown in Fig. 15–32c.

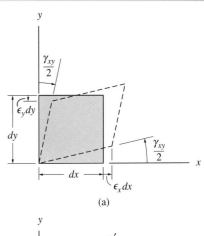

(a)

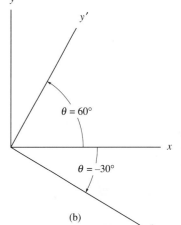

(b)

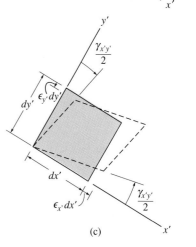

(c)

Fig. 15–32

Example 15–12

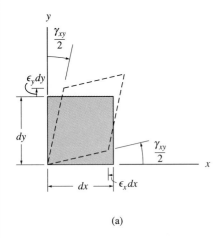

(a)

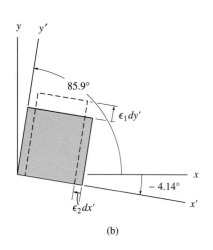

(b)

Fig. 15–33

A differential element of material at a point is subjected to a state of plane strain defined by $\epsilon_x = -350(10^{-6})$, $\epsilon_y = 200(10^{-6})$, $\gamma_{xy} = 80(10^{-6})$, which tends to distort the element as shown in Fig. 15–33a. Determine at the point the orientation of an element and the associated principal strains, and the orientation of an element and the maximum in-plane shear strains.

SOLUTION

Principal Strains. The orientation of the principal planes of strain is determined from Eq. 15–24. We have

$$\tan 2\theta_p = \frac{\gamma_{xy}}{\epsilon_x - \epsilon_y}$$

$$= \frac{80(10^{-6})}{(-350 - 200)(10^{-6})}$$

Thus, $2\theta_p = -8.28°$ and $-8.28° + 180° = 172°$, so that

$$\theta_p = -4.14° \text{ and } 85.9° \qquad Ans.$$

Each of these angles is measured *positive counterclockwise,* from the x axis to the outward normals on each face of the element, Fig. 15–33b.

The principal strains are determined from Eq. 15–25. We have

$$\epsilon_{1,2} = \frac{\epsilon_x + \epsilon_y}{2} \pm \sqrt{\left(\frac{\epsilon_x - \epsilon_y}{2}\right)^2 + \left(\frac{\gamma_{xy}}{2}\right)^2}$$

$$= \frac{(-350 + 200)(10^{-6})}{2} \pm \left[\sqrt{\left(\frac{-350 - 200}{2}\right)^2 + \left(\frac{80}{2}\right)^2}\right](10^{-6})$$

$$= -75.0(10^{-6}) \pm 277.9(10^{-6})$$

$$\epsilon_1 = 203(10^{-6}) \qquad \epsilon_2 = -353(10^{-6}) \qquad Ans.$$

We can determine which of these two strains deforms the element in the x' direction by applying Eq. 15–21 with $\theta = -4.14°$. Thus,

$$\epsilon_{x'} = \frac{\epsilon_x + \epsilon_y}{2} + \frac{\epsilon_x - \epsilon_y}{2} \cos 2\theta + \frac{\gamma_{xy}}{2} \sin 2\theta$$

$$= \left(\frac{-350 + 200}{2}\right)(10^{-6}) + \left(\frac{-350 - 200}{2}\right)(10^{-6}) \cos 2(-4.14°)$$

$$+ \frac{80(10^{-6})}{2} \sin 2(-4.14°)$$

$$\epsilon_{x'} = -353(10^{-6})$$

Hence $\epsilon_{x'} = \epsilon_2$. When subjected to the principal strains, the element is distorted as shown in Fig. 15–33b.

Maximum In-Plane Shear Strain. The element for maximum in-plane shear strain is oriented 45° from the one shown in Fig. 15–33b. In other

656

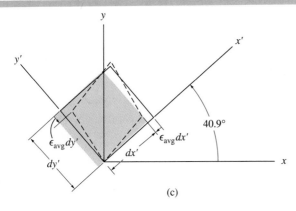

(c)

words, measured from the x axis, $\theta_s = \theta_p + 45° = -4.14° + 45° = 40.9°$, Fig. 15–33c. These same results can also be obtained from Eq. 15–26.

The maximum in-plane shear strains are determined from Eq. 15–27; that is,

$$\frac{(\gamma_{x'y'})_{max}}{2} = \sqrt{\left(\frac{\epsilon_x - \epsilon_y}{2}\right)^2 + \left(\frac{\gamma_{xy}}{2}\right)^2}$$

$$= \left[\sqrt{\left(\frac{-350 - 200}{2}\right)^2 + \left(\frac{80}{2}\right)^2}\right](10^{-6})$$

$$(\gamma_{x'y'})_{max} = 556(10^{-6}) \qquad\qquad Ans.$$

Also, there are associated average normal strains imposed on the element that are determined from Eq. 15–28:

$$\epsilon_{avg} = \frac{\epsilon_x + \epsilon_y}{2} = \frac{-350 + 200}{2}(10^{-6})$$

$$\epsilon_{avg} = -75(10^{-6})$$

The proper sign of $(\gamma_{x'y'})_{max}$ can be obtained by applying Eq. 15–22 with $\theta_s = 40.9°$. We have

$$\frac{\gamma_{x'y'}}{2} = -\frac{\epsilon_x - \epsilon_y}{2}\sin 2\theta + \frac{\gamma_{xy}}{2}\cos 2\theta$$

$$= -\left(\frac{-350 - 200}{2}\right)(10^{-6})\sin 2(40.9°) + \frac{80(10^{-6})}{2}\cos 2(40.9°)$$

$$\gamma_{x'y'} = 556(10^{-6})$$

Thus, $(\gamma_{x'y'})_{max}$ tends to distort the element in such a way so that the right angle between dx' and dy' is decreased (positive sign convention). Note that this is in agreement with the distortion caused by positive shear stress acting on the x' face of the element and corresponding shear stresses acting on the element's adjacent faces. The distorted element is shown in Fig. 15–33c.

15.8 Mohr's Circle—Plane Strain

Since the equations of plane-strain transformation are mathematically similar to the equations of plane-stress transformation, we can also solve problems involving the transformation of strain using Mohr's circle. This approach has the advantage of making it possible to see graphically how the normal and shear strain components at a point vary from one orientation of the element to the next.

Like the case for stress, the parameter θ in Eqs. 15–21 and 15–22 can be eliminated and the result rewritten in the form

$$(\epsilon_x - \epsilon_{avg})^2 + \left(\frac{\gamma_{xy}}{2}\right)^2 = R^2 \qquad (15\text{–}29)$$

where

$$\epsilon_{avg} = \frac{\epsilon_x + \epsilon_y}{2}$$

$$R = \sqrt{\left(\frac{\epsilon_x - \epsilon_y}{2}\right)^2 + \left(\frac{\gamma_{xy}}{2}\right)^2}$$

Equation 15–29 represents the equation of Mohr's circle for strain. It has a center on the ϵ axis at point $C(\epsilon_{avg}, 0)$ and a radius R.

PROCEDURE FOR ANALYSIS

The procedure for drawing Mohr's circle for strain follows the same one established for stress. The following steps are required to draw and use the circle.

Construction of the Circle. Establish a coordinate system such that the abscissa represents the normal strain ϵ, with *positive to the right,* and the ordinate represents *half* the value of the shear strain, $\gamma/2$, with *positive downward,* Fig. 15–34.

Using the positive sign convention for ϵ_x, ϵ_y, γ_{xy}, as shown in Fig. 15–34, determine the center of the circle C, which is located on the ϵ axis at a distance $\epsilon_{avg} = (\epsilon_x + \epsilon_y)/2$ from the origin, Fig. 15–34.

Plot the reference point A having coordinates $A(\epsilon_x, \gamma_{xy}/2)$. This point represents the case for which the x' axis coincides with the x axis. Hence $\theta = 0°$, Fig. 15–34.

Connect point A with the center C of the circle and determine the hypotenuse CA of the shaded triangle by trigonometry. This distance represents the radius R of the circle, and CA is referred to as the radial reference line, Fig. 15–34.

Once R has been determined, sketch the circle.

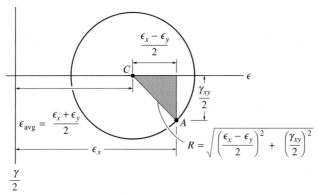

Fig. 15–34

Principal Strains. The principal strains ϵ_1 and ϵ_2 are determined from the points of intersection of the circle with the ϵ axis, i.e., where $\gamma/2 = 0$, Fig. 15–35a. These strains act in two directions, defined by the angles θ_{p_1} and θ_{p_2}, which can be determined from the circle using trigonometry. They are represented on the circle by the angles $2\theta_{p_1}$ and $2\theta_{p_2}$ and measured *from* the radial reference line *CA to* lines *CB* and *CD*, Fig. 15–35a. Actually only one of these angles needs to be calculated, since θ_{p_1} and θ_{p_2} are 90° apart. Remember that a *rotation* of 2θ on the *circle* represents a *rotation* of θ in the *same direction,* from the reference axis x to the x' axis.* The angle θ_{p_1} is shown in Fig. 15–35b. Notice that since ϵ_1 and ϵ_2 are indicated as being positive in Fig. 15–35a, the element in Fig. 15–35b will elongate in the x' and y' directions as shown by the dashed outline. This deformation is the result of positive principal stresses σ_1 and σ_2 acting on the element.

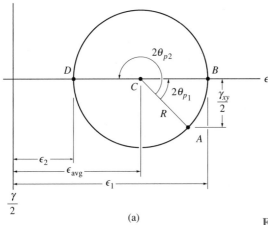

(a)

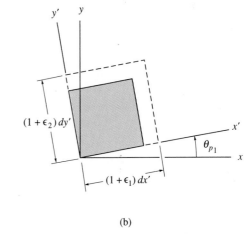

(b)

Fig. 15–35

*If instead the $\gamma/2$ axis were constructed *positive upward,* then the angle 2θ on the circle would be measured in the *opposite direction* as the orientation θ of the plane.

Maximum In-Plane Shear Strain. The average normal strain and maximum in-plane shear strain components are determined from the circle as the coordinates of points E and F, Fig. 15–36a. In this case the angles θ_{s_1} and θ_{s_2} define the two directions for orienting the planes that experience the strain components. The calculations from the circle are performed in the same manner as for finding the principal strains. The results are shown in Fig. 15–36b for the element oriented on the basis of calculating θ_{s_1} and the strain components associated with point E on the circle. Here $(\gamma_{x'y'})_{max}$ and ϵ_{avg} are positive quantities, and so the element deforms as shown by the dashed outline.

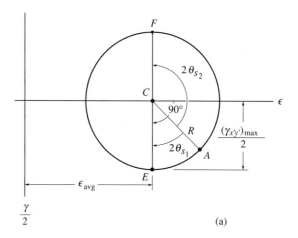

(a)

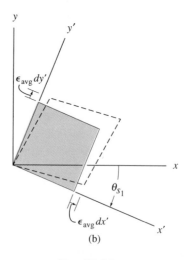

(b)

Fig. 15–36

Strains on Arbitrary Plane.   The normal and shear strain components $\epsilon_{x'}$ and $\gamma_{x'y'}$ for a specified plane oriented at an angle θ, Fig. 15–37a, can be obtained from the circle using trigonometry to determine the coordinates of point P, Fig. 15–37b. To locate P, the known angle θ for the x' axis (in this case counterclockwise) is measured in the *same direction* as 2θ (counterclockwise) on the circle. The measurement of 2θ is *from* the radial reference line CA to the radial line CP. Also, if the value of $\epsilon_{y'}$ is required, it can be determined by calculating the ϵ coordinate of point Q in Fig. 15–37b, since CQ lies 180° away from CP and thus represents a rotation of 90° of the x' axis.

   The following examples illustrate numerically how to apply this procedure to problems that involve each of the three cases just discussed.

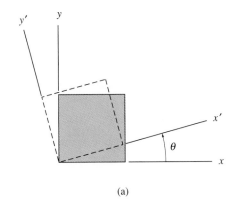

(a)

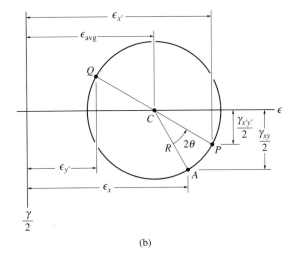

(b)

Fig. 15–37

Example 15–13

The state of plane strain at a point is represented by the components $\epsilon_x = 250(10^{-6})$, $\epsilon_y = -150(10^{-6})$, and $\gamma_{xy} = 120(10^{-6})$. Determine the orientation of an element and the associated principal strains, and the orientation of an element and the associated maximum in-plane shear strains.

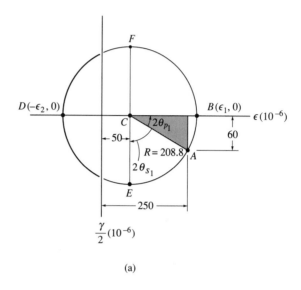

(a)

Fig. 15–38

SOLUTION

Construction of the Circle. The ϵ and $\gamma/2$ axes are established in Fig. 15–38a. Remember that the *positive* $\gamma/2$ axis must be directed *downward* so that *counterclockwise* rotations of the element correspond to *counterclockwise* rotation around the circle, and vice versa. The center of the circle C is located on the ϵ axis at

$$\epsilon_{avg} = \frac{250 + (-150)}{2}\ (10^{-6}) = 50(10^{-6})$$

Since $\gamma_{xy}/2 = 60(10^{-6})$, the reference point A ($\theta = 0°$) has coordinates $A(250(10^{-6}), 60(10^{-6}))$. From the shaded triangle in Fig. 15–38a, the radius of the circle is CA; that is,

$$R = \sqrt{(250 - 50)^2 + (60)^2} = 208.8$$

Principal Strains. The ϵ coordinates of points B and D represent the principal strains. They are

$$\epsilon_1 = (50 + 208.8)(10^{-6}) = 259(10^{-6}) \qquad \textit{Ans.}$$

$$\epsilon_2 = (50 - 208.8)(10^{-6}) = -159(10^{-6}) \qquad \textit{Ans.}$$

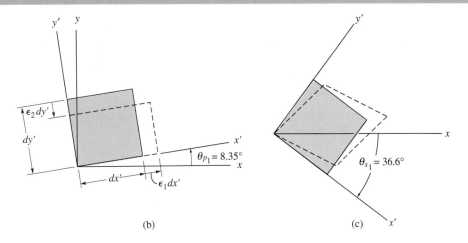

(b)

(c)

The direction of the positive principal strain ϵ_1 is defined by the *counterclockwise* angle $2\theta_{p_1}$, measured from the radial reference line *CA* to the line *CB*. We have

$$\tan 2\theta_{p_1} = \frac{60}{(250 - 50)}$$

$$\theta_{p_1} = 8.35° \qquad\qquad Ans.$$

Hence the side dx' of the element is oriented *counterclockwise* 8.35° as shown in Fig. 15–38b. This also defines the direction of ϵ_1. The deformation of the element is also shown in the figure.

Maximum In-Plane Shear Strain.   The maximum in-plane shear strain is represented by points *E* and *F* on the circle, Fig. 15–38a. From the coordinates of point *E*,

$$\frac{(\gamma_{x'y'})_{\max}}{2} = 208.8(10^{-6})$$

$$(\gamma_{x'y'})_{\max} = 418(10^{-6}) \qquad\qquad Ans.$$

$$\epsilon_{\text{avg}} = 50(10^{-6})$$

To orient the element, we can determine the clockwise angle $2\theta_{s_1}$ from the circle.

$$2\theta_{s_1} = 90° - 2(8.35°)$$

$$\theta_{s_1} = 36.6° \qquad\qquad Ans.$$

This angle is shown in Fig. 15–38c. Since the shear strain defined from point *E* on the circle has a positive value and the average normal strain is also positive, corresponding positive shear and positive average normal stress deform the element into the dashed shape shown in Fig. 15–38c.

■ Example 15–14 ■

The state of plane strain at a point is represented on an element having components $\epsilon_x = -300(10^{-6})$, $\epsilon_y = -100(10^{-6})$, and $\gamma_{xy} = 100(10^{-6})$. Determine the state of strain on an element oriented 20° clockwise from this reported position.

SOLUTION

Construction of the Circle. The ϵ and $\gamma/2$ axes are established in Fig. 15–39a. The center of the circle is on the ϵ axis at

$$\epsilon_{avg} = \left(\frac{-300 - 100}{2}\right)(10^{-6}) = -200(10^{-6})$$

The reference point A has coordinates $A(-300(10^{-6}), 50(10^{-6}))$. The radius CA determined from the shaded triangle is therefore

$$R = \left[\sqrt{(300 - 200)^2 + (50)^2}\right](10^{-6}) = 111.8(10^{-6})$$

Strains on Inclined Element. Since the element is to be oriented 20° *clockwise*, we must establish a radial line CP, $2(20°) = 40°$ *clockwise*, measured from CA ($\theta = 0°$), Fig. 15–39a. The coordinates of point P ($\epsilon_{x'}$, $-\gamma_{x'y'}/2$) are obtained from the geometry of the circle. Note that

$$\theta = \tan^{-1}\left(\frac{50}{(300 - 200)}\right) = 26.57°, \qquad \phi = 40° - 26.57° = 13.43°$$

Thus,

$$\epsilon_{x'} = -(200 + 111.8 \cos 13.43°)(10^{-6})$$
$$= -309(10^{-6}) \qquad\qquad Ans.$$

$$\frac{\gamma_{x'y'}}{2} = -(111.8 \sin 13.43°)(10^{-6})$$

$$\gamma_{x'y'} = -51.9(10^{-6}) \qquad\qquad Ans.$$

The normal strain $\epsilon_{y'}$ can be determined from the ϵ coordinate of point Q on the circle, Fig. 15–39a. Why?

$$\epsilon_{y'} = -(200 - 111.8 \cos 13.43°)(10^{-6}) = -91.3(10^{-6}) \quad Ans.$$

As a result of these strains, the element deforms relative to x', y' axes as shown in Fig. 15–39b.

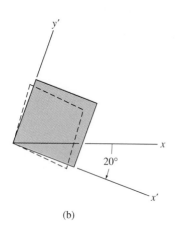

(a)

(b)

Fig. 15–39

PROBLEMS

*15–48.   Prove that the sum of the normal strains in perpendicular directions is constant.

15–49.   The state of strain at the point on the bracket has components $\epsilon_x = -130(10^{-6})$, $\epsilon_y = 280(10^{-6})$, $\gamma_{xy} = 75(10^{-6})$. Use the strain-transformation equations to determine (a) the in-plane principal strains and (b) the maximum in-plane shear strain and average normal strain. In each case specify the orientation of the element and show how the strains deform the element within the x–y plane.

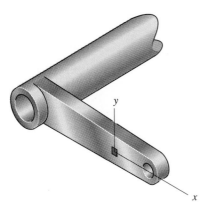

Prob. 15–49

15–50.   The state of strain at the point on the bracket has components $\epsilon_x = 350(10^{-6})$, $\epsilon_y = -860(10^{-6})$, $\gamma_{xy} = 250(10^{-6})$. Use the strain-transformation equations to determine the equivalent in-plane strains on an element oriented at an angle of $\theta = 45°$ clockwise from the original position. Sketch the deformed element within the x–y plane due to these strains.

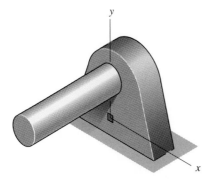

Prob. 15–50

15–51.   The state of strain at the point on the arm has components of $\epsilon_x = 250(10^{-6})$, $\epsilon_y = -450(10^{-6})$, $\gamma_{xy} = -825(10^{-6})$. Use the strain-transformation equations to determine (a) the in-plane principal strains and (b) the maximum in-plane shear strain and average normal strain. In each case specify the orientation of the element and show how the strains deform the element within the x–y plane.

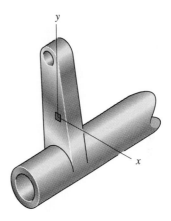

Prob. 15–51

*15–52.   The state of strain at the point on the gear tooth has components $\epsilon_x = 850(10^{-6})$, $\epsilon_y = 480(10^{-6})$, $\gamma_{xy} = 650(10^{-6})$. Use the strain-transformation equations to determine (a) the in-plane principal strains and (b) the maximum in-plane shear strain and average normal strain. In each case specify the orientation of the element and show how the strains deform the element within the x–y plane.

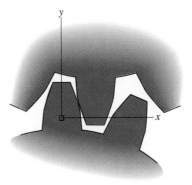

Prob. 15–52

15–53. The state of strain at the point on the bracket has components $\epsilon_x = 400(10^{-6})$, $\epsilon_y = -250(10^{-6})$, $\gamma_{xy} = 310(10^{-6})$. Use the strain-transformation equations to determine the equivalent in-plane strains on an element oriented at an angle of $\theta = 30°$ clockwise from the original position. Sketch the deformed element due to these strains within the x–y plane.

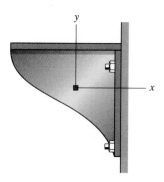

Prob. 15–53

15–54. The state of strain at the point on the bracket has components $\epsilon_x = -200(10^{-6})$, $\epsilon_y = -650(10^{-6})$, $\gamma_{xy} = -175(10^{-6})$. Use the strain-transformation equations to determine the equivalent in-plane strains on an element oriented at an angle of $\theta = 20°$ counterclockwise from the original position. Sketch the deformed element due to these strains within the x–y plane.

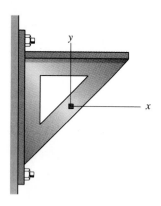

Prob. 15–54

15–55. Solve Prob. 15–49 using Mohr's circle.

*15–56. Solve Prob. 15–52 using Mohr's circle.

15–57. Solve Prob. 15–50 using Mohr's circle.

15–58. Solve Prob. 15–51 using Mohr's circle.

15–59. Solve Prob. 15–53 using Mohr's circle.

*15–60. Solve Prob. 15–54 using Mohr's circle.

15.9 Strain Rosettes

It was mentioned in Sec. 9.1 that the normal strain in a tension-test specimen can be measured using an *electrical-resistance strain gauge,* which consists of a wire grid or piece of metal foil bonded to the specimen. However, for a general loading on a body, the *normal strains* at a point on its free surface are often determined using a cluster of three electrical-resistance strain gauges, arranged in a specified pattern. This pattern is referred to as a *strain rosette,* and once the readings on the three gauges are made, the data can then be used to specify the state of strain at the point. It should be noted, however, that these strains are measured *only* in the plane of the gauges, and since the body is stress-free in a direction perpendicular to the gauges, the gauges may be subjected to *plane stress* but *not* plane strain. In this regard, the normal line to the free surface is a principal axis of strain, and so the principal normal strain along this axis is *not* measured by the strain rosette. What is important here is that the out-of-plane displacement caused by this principal strain will *not* affect the in-plane measurements of the gauges.

In the general case, the axes of the three gauges are arranged at the angles θ_a, θ_b, θ_c as shown in Fig. 15–40a. If the readings ϵ_a, ϵ_b, ϵ_c are taken, we can determine the strain components ϵ_x, ϵ_y, γ_{xy} at the point by applying the strain-transformation equation, Eq. 15–18, for each gauge. We have,

$$\epsilon_a = \epsilon_x \cos^2 \theta_a + \epsilon_y \sin^2 \theta_a + \gamma_{xy} \sin \theta_a \cos \theta_a$$
$$\epsilon_b = \epsilon_x \cos^2 \theta_b + \epsilon_y \sin^2 \theta_b + \gamma_{xy} \sin \theta_b \cos \theta_b \qquad (15\text{–}30)$$
$$\epsilon_c = \epsilon_x \cos^2 \theta_c + \epsilon_y \sin^2 \theta_c + \gamma_{xy} \sin \theta_c \cos \theta_c$$

The values of ϵ_x, ϵ_y, γ_{xy} are determined by solving these three equations simultaneously.

Strain rosettes are often arranged in 45° or 60° patterns. In the case of the 45° or "rectangular" strain rosette shown in Fig. 15–40b, $\theta_a = 0°$, $\theta_b = 45°$, $\theta_c = 90°$, so that Eq. 15–30 gives

$$\epsilon_x = \epsilon_a$$
$$\epsilon_y = \epsilon_c \qquad (15\text{–}31)$$
$$\gamma_{xy} = 2\epsilon_b - (\epsilon_a + \epsilon_c)$$

And for the 60° strain rosette in Fig. 15–40c, $\theta_a = 0°$, $\theta_b = 60°$, $\theta_c = 120°$. Here Eq. 15–30 gives

$$\epsilon_x = \epsilon_a$$
$$\epsilon_y = \frac{1}{3}\left(2\epsilon_b + 2\epsilon_c - \epsilon_a\right) \qquad (15\text{–}32)$$
$$\gamma_{xy} = \frac{2}{\sqrt{3}}\left(\epsilon_b - \epsilon_c\right)$$

Once ϵ_x, ϵ_y, γ_{xy} are determined, the transformation equations of Sec. 15.6 or Mohr's circle can then be used to determine the principal in-plane strains and the maximum in-plane shear strain at the point.

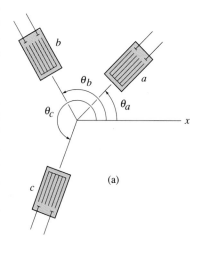

(a)

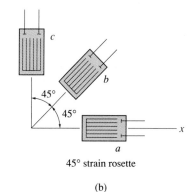

45° strain rosette

(b)

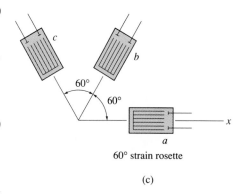

60° strain rosette

(c)

Fig. 15–40

Example 15–15

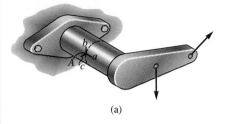

(a)

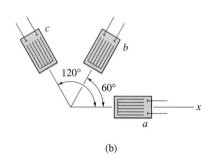

(b)

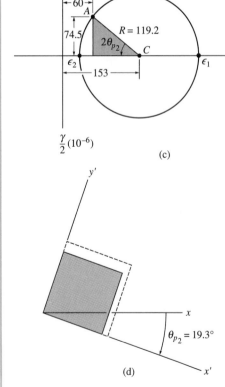

(c)

(d)

The state of strain at point A on the bracket in Fig. 15–41a is measured using the strain rosette shown in Fig. 15–41b. Due to the loadings, the readings from the gauges give $\epsilon_a = 60(10^{-6})$, $\epsilon_b = 135(10^{-6})$, and $\epsilon_c = 264(10^{-6})$. Determine the in-plane principal strains at the point and the directions in which they act.

SOLUTION

We will use Eq. 15–30 for the solution. Establishing an x axis as shown in Fig. 15–41b and measuring the angles from the $+x$ axis to the center-lines of each gauge, we have $\theta_a = 0°$, $\theta_b = 60°$, and $\theta_c = 120°$. Substituting these results, along with the problem data, into Eq. 15–30 gives

$$60(10^{-6}) = \epsilon_x \cos^2 0° + \epsilon_y \sin^2 0° + \gamma_{xy} \sin 0° \cos 0°$$

$$= \epsilon_x \qquad (1)$$

$$135(10^{-6}) = \epsilon_x \cos^2 60° + \epsilon_y \sin^2 60° + \gamma_{xy} \sin 60° \cos 60°$$

$$= 0.25\epsilon_x + 0.75\epsilon_y + 0.433\gamma_{xy} \qquad (2)$$

$$264(10^{-6}) = \epsilon_x \cos^2 120° + \epsilon_y \sin^2 120° + \gamma_{xy} \sin 120° \cos 120°$$

$$= 0.25\epsilon_x + 0.75\epsilon_y - 0.433\gamma_{xy} \qquad (3)$$

Using Eq. 1 and solving Eqs. 2 and 3 simultaneously, we get

$$\epsilon_x = 60(10^{-6}) \qquad \epsilon_y = 246(10^{-6}) \qquad \gamma_{xy} = -149(10^{-6})$$

These same results can also be obtained in a more direct manner from Eq. 15–32.

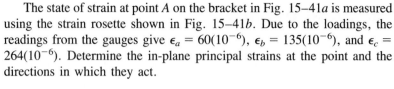

The in-plane principal strains can be determined using Mohr's circle. The reference point on the circle is at A $(60(10^{-6}), -74.5(10^{-6}))$ and the center of the circle, C, is on the ϵ axis at $\epsilon_{avg} = 153(10^{-6})$, Fig. 15–41c. From the shaded triangle, the radius is

$$R = \sqrt{(153 - 60)^2 + (74.5)^2} = 119.2$$

The in-plane principal strains are thus

$$\epsilon_1 = 153(10^{-6}) + 119.2(10^{-6}) = 272(10^{-6}) \qquad Ans.$$

$$\epsilon_2 = 153(10^{-6}) - 119.2(10^{-6}) = 33.8(10^{-6}) \qquad Ans.$$

$$2\theta_{p_2} = \tan^{-1}\frac{74.5}{(153 - 60)} = 38.7°$$

$$\theta_{p_2} = 19.3° \qquad Ans.$$

The deformed element is shown in the dashed position in Fig. 15–41d.

Fig. 15–41

PROBLEMS

15–61. The 45° strain rosette is mounted on a machine element. The following readings are obtained for each gauge: $\epsilon_a = 650(10^{-6})$, $\epsilon_b = -300(10^{-6})$, $\epsilon_c = 480(10^{-6})$. Determine (a) the in-plane principal strains and (b) the maximum in-plane shear strain and associated average normal strain. In each case show the deformed element due to these strains.

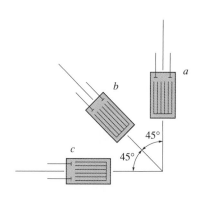

Prob. 15–61

15–63. The 60° strain rosette is mounted on the surface of an aluminum plate. The following readings are obtained for each gauge: $\epsilon_a = 950(10^{-6})$, $\epsilon_b = 380(10^{-6})$, $\epsilon_c = 220(10^{-6})$. Determine the in-plane principal strains and their orientation developed at a point on the plate.

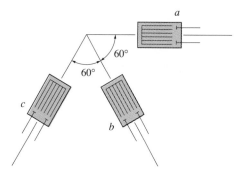

Prob. 15–63

15–62. The 60° strain rosette is mounted on a beam. The following readings are obtained for each gauge: $\epsilon_a = 250(10^{-6})$, $\epsilon_b = -400(10^{-6})$, $\epsilon_c = 280(10^{-6})$. Determine (a) the in-plane principal strains and their orientation, and (b) the maximum in-plane shear strain and average normal strain. In each case show the deformed element due to these strains.

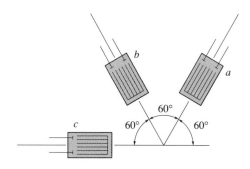

Prob. 15–62

*15–64. The 45° strain rosette is mounted on a steel shaft. The following readings are obtained for each gauge: $\epsilon_a = 800(10^{-6})$, $\epsilon_b = 520(10^{-6})$, $\epsilon_c = -450(10^{-6})$. Determine the in-plane principal strains and their orientation developed at a point on the shaft.

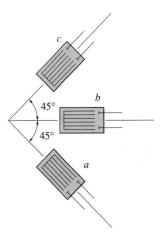

Prob. 15–64

15.10 Material-Property Relationships

Now that the general principles of multiaxial stress and strain have been presented, we will use these principles to develop some important relationships involving the material's properties. To do so we will assume that the material is homogeneous and isotropic and behaves in a linear-elastic manner.

Generalized Hooke's Law. If the material at a point is subjected to a state of triaxial stress, σ_x, σ_y, σ_z, Fig. 15–42a, associated normal strains ϵ_x, ϵ_y, ϵ_z are developed in the material. The stresses can be related to the strains by using the principle of superposition, Poisson's ratio, $\epsilon_{lat} = -\nu\epsilon_{long}$, and Hooke's law, as it applies in the uniaxial direction, $\epsilon = \sigma/E$. To show how this is done we will first consider the normal strain of the element in the x direction, caused by separate application of each normal stress. When σ_x is applied, Fig. 15–42b, the element elongates in the x direction and the strain ϵ'_x in this direction is

$$\epsilon'_x = \frac{\sigma_x}{E}$$

Application of σ_y causes the element to contract with a strain ϵ''_x in the x direction, Fig. 15–42c. Here

$$\epsilon''_x = -\nu\frac{\sigma_y}{E}$$

Likewise, application of σ_z, Fig. 15–42d, causes a contraction in the x direction such that

$$\epsilon'''_x = -\nu\frac{\sigma_z}{E}$$

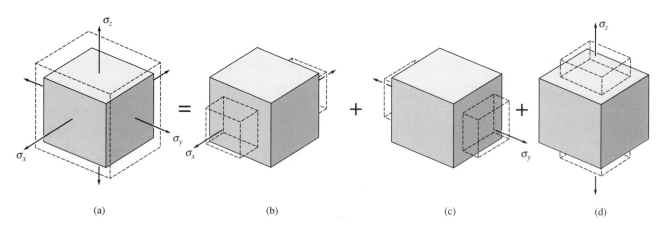

(a)   (b)   (c)   (d)

Fig. 15–42

When these three normal strains are superimposed, the normal strain ϵ_x is determined for the state of stress in Fig. 15–42a. Similar equations can be developed for the normal strains in the y and z directions. The final results can be written as

$$\epsilon_x = \frac{1}{E}\left[\sigma_x - \nu(\sigma_y + \sigma_z)\right]$$

$$\epsilon_y = \frac{1}{E}\left[\sigma_y - \nu(\sigma_x + \sigma_z)\right] \qquad (15\text{–}33)$$

$$\epsilon_z = \frac{1}{E}\left[\sigma_z - \nu(\sigma_x + \sigma_y)\right]$$

These three equations express Hooke's law in general form for a triaxial state of stress. As noted in the derivation, they are valid only if the principle of superposition applies, which requires a *linear-elastic* response of the material and application of strains that do not severely alter the shape of the material—i.e., small deformations are required. When applying these equations, note that tensile stresses are considered positive quantities, and compressive stresses are negative. If a resulting normal strain is *positive,* it indicates that the material *elongates,* whereas a *negative* normal strain indicates the material *contracts.*

Since the material is isotropic, the element in Fig. 15–42a will *remain rectangular* when subjected to the normal stresses; i.e., *no shear strains* will be produced in the material. If we now apply a shear stress τ_{xy} to the element, Fig. 15–43a, experimental observations indicate that the material will deform *only* due to a shear strain γ_{xy}; that is, τ_{xy} will not cause other strains in the material. Likewise, τ_{yz} and τ_{zx} will only cause shear strains γ_{yz} and γ_{zx}, respectively. Hooke's law for shear stress and shear strain can therefore be written as

$$\gamma_{xy} = \frac{1}{G}\tau_{xy} \qquad \gamma_{yz} = \frac{1}{G}\tau_{yz} \qquad \gamma_{zx} = \frac{1}{G}\tau_{zx} \qquad (15\text{–}34)$$

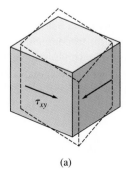

(a)

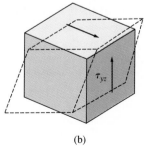

(b)

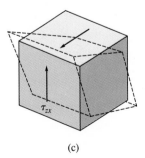

(c)

Fig. 15–43

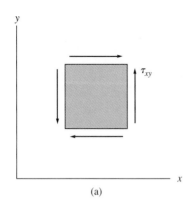

(a)

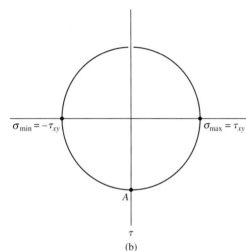

(b)

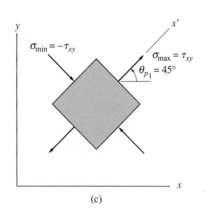

(c)

Fig. 15–44

Relationship Involving E, ν, and G.

In Sec. 3.7 we stated that the modulus of elasticity E is related to the shear modulus G by Eq. 9–11, namely,

$$G = \frac{E}{2(1 + \nu)} \qquad (15\text{–}35)$$

One way to derive this relationship is to consider an element of the material to be subjected to pure shear ($\sigma_x = \sigma_y = \sigma_z = 0$), Fig. 15–44a. In accordance with the discussion of Sec. 15.4, Mohr's circle for the element is shown in Fig. 15–44b. The center of the circle is at $\sigma_{avg} = (\sigma_x + \sigma_y)/2 = 0$, and the reference point A has coordinates $A(0, \tau_{xy})$. Hence, the radius of this circle is $R = \tau_{xy}$, and the principal stresses are therefore $\sigma_{max} = \tau_{xy}$ and $\sigma_{min} = -\tau_{xy}$. Since $2\theta_{p_1} = 90°$ counterclockwise, Fig. 15–44b, the element must be oriented $\theta_{p_1} = 45°$ counterclockwise from the x axis in order to define the direction of the plane on which σ_{max} acts, Fig. 15–44c. If the three principal stresses $\sigma_{max} = \tau_{xy}$, $\sigma_{int} = 0$, and $\sigma_{min} = -\tau_{xy}$ are substituted into the first of Eq. 15–33, the principal strain ϵ_{max} can be related to the shear stress τ_{xy}. The result is

$$\epsilon_{max} = \frac{\tau_{xy}}{E}(1 + \nu) \qquad (15\text{–}36)$$

This strain, which deforms the element along the x' axis, can also be related to the shear strain γ_{xy} using the strain transformation equations or Mohr's circle of strain. To do this, first note that from the first two equations of Hooke's law, Eq. 15–33, since $\sigma_x = \sigma_y = \sigma_z = 0$, then $\epsilon_x = \epsilon_y = 0$, Fig. 15–44a. Substituting these results into Eq. 15–25, we get

$$\epsilon_1 = \epsilon_{max} = \frac{\gamma_{xy}}{2}$$

By Hooke's law, $\gamma_{xy} = \tau_{xy}/G$, so that $\epsilon_{max} = \tau_{xy}/2G$. Substituting this result into Eq. 15–36 and rearranging terms gives the final result, namely, Eq. 15–35.

Dilatation and Bulk Modulus.

When an elastic material is subjected to normal stress, its volume will change. In order to compute this change, consider the volume element shown in Fig. 15–45, which is subjected to the principal stresses σ_x, σ_y, σ_z. The sides of the element are originally dx, dy, dz, Fig. 15–45a; however, after application of the stress they become $(1 + \epsilon_x)\,dx$, $(1 + \epsilon_y)\,dy$, $(1 + \epsilon_z)\,dz$, respectively, Fig. 15–45b. The change in volume of the element is therefore

$$\delta V = (1 + \epsilon_x)(1 + \epsilon_y)(1 + \epsilon_z)\,dx\,dy\,dz - dx\,dy\,dz$$

Neglecting the products of the strains since the strains are very small as discussed in Sec. 8.8, we have

$$\delta V = (\epsilon_x + \epsilon_y + \epsilon_z) \, dx \, dy \, dz$$

The change in volume per unit volume is called the "volumetric strain" or the *dilatation e*. It can be written as

$$e = \frac{\delta V}{V} = \epsilon_x + \epsilon_y + \epsilon_z \qquad (15\text{–}37)$$

By comparison, the shear strains will *not* change the volume of the element, rather they will only change its rectangular shape.

If we use the generalized Hooke's law, as defined by Eq. 15–33, we can write the dilatation in terms of the applied stress. After substitution and simplification, we have

$$e = \frac{1 - 2\nu}{E}(\sigma_x + \sigma_y + \sigma_z) \qquad (15\text{–}38)$$

When a volume element of material is subjected to the uniform pressure p of a liquid, the pressure on the body is the same in all directions and is always normal to any surface on which it acts. Shear stresses are *not present*, since the shear resistance of a liquid is zero. This state of "hydrostatic" loading requires the normal stresses to be equal in any and all directions, and therefore an element of the body is subjected to principal stresses $\sigma_x = \sigma_y = \sigma_z = -p$, Fig. 15–46. Substituting into Eq. 15–38 and rearranging terms yields

$$\frac{p}{e} = \frac{E}{3(1 - 2\nu)} \qquad (15\text{–}39)$$

The term on the right consists *only* of the material's properties E and ν. It is equal to the ratio of the uniform normal stress p to the dilatation or "volumetric strain." Since this ratio is *similar* to the ratio of linear-elastic stress to strain, which defines E, i.e., $\sigma/\epsilon = E$, the terms on the right are called the *volume modulus of elasticity* or the *bulk modulus*. It has the same units as stress and will be symbolized by the letter k; that is,

$$k = \frac{E}{3(1 - 2\nu)} \qquad (15\text{–}40)$$

Note that for most metals $\nu \approx \frac{1}{3}$ so $k \approx E$. Also, a *rigid material* requires k to be infinite, since $\delta V = 0$. Hence, from Eq. 15–40, the theoretical *maximum* value for Poisson's ratio is $\nu = 0.5$. It may also be added that during yielding, no actual volume change of the material is observed, and so $\nu = 0.5$ is used when plastic yielding occurs.

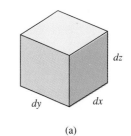

(a)

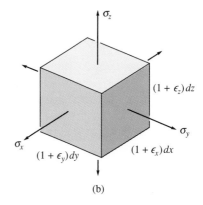

(b)

Fig. 15–45

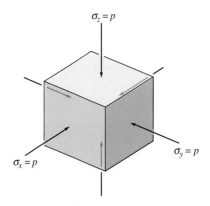

Hydrostatic stress

Fig. 15–46

Example 15–16

The bracket in Example 15–15, Fig. 15–47a, is made of steel for which $E_{st} = 200$ GPa, $\nu_{st} = 0.3$. Determine the principal stresses at point A.

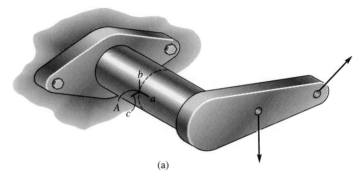

(a)

Fig. 15–47

SOLUTION I

From Example 15–15 the principal strains have been determined as

$$\epsilon_1 = 272(10^{-6})$$
$$\epsilon_2 = 33.8(10^{-6})$$

Since point A is on the *surface* of the bracket for which there is no loading, the stress on the surface is zero, and so point A is subjected to plane stress. Applying Hooke's law with $\sigma_z = 0$, we have

$$\epsilon_1 = \frac{\sigma_1}{E} - \frac{\nu}{E}\sigma_2; \qquad 272(10^{-6}) = \frac{\sigma_1}{200(10^9)} - \frac{0.3}{200(10^9)}\sigma_2$$
$$54.4(10^6) = \sigma_1 - 0.3\sigma_2 \qquad (1)$$

$$\epsilon_2 = \frac{\sigma_2}{E} - \frac{\nu}{E}\sigma_1; \qquad 33.8(10^{-6}) = \frac{\sigma_2}{200(10^9)} - \frac{0.3}{200(10^9)}\sigma_1$$
$$6.76(10^6) = \sigma_2 - 0.3\sigma_1 \qquad (2)$$

Solving Eqs. 1 and 2 simultaneously yields

$$\sigma_1 = 62.0 \text{ MPa} \qquad\qquad\qquad Ans.$$
$$\sigma_2 = 25.4 \text{ MPa} \qquad\qquad\qquad Ans.$$

SOLUTION II

It is also possible to solve the problem using the given state of strain,

$$\epsilon_x = 60(10^{-6}) \qquad \epsilon_y = 246(10^{-6}) \qquad \gamma_{xy} = -149(10^{-6})$$

as specified in Example 15–15. Applying Hooke's law in the $x - y$ plane, we have

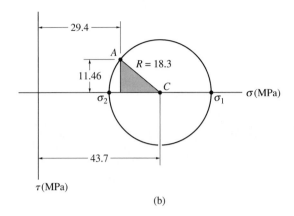

(b)

$$\epsilon_x = \frac{\sigma_x}{E} - \frac{\nu}{E}\sigma_y; \qquad 60(10^{-6}) = \frac{\sigma_x}{200(10^9)} - \frac{0.3\sigma_y}{200(10^9)}$$

$$\epsilon_y = \frac{\sigma_y}{E} - \frac{\nu}{E}\sigma_x; \qquad 246(10^{-6}) = \frac{\sigma_y}{200(10^9)} - \frac{0.3\sigma_x}{200(10^9)}$$

$$\sigma_x = 29.4 \text{ MPa} \qquad \sigma_y = 58.0 \text{ MPa}$$

The shear stress is computed using Hooke's law for shear. First, however, we must calculate G.

$$G = \frac{E}{2(1 + \nu)} = \frac{200 \text{ GPa}}{2(1 + 0.3)} = 76.9 \text{ GPa}$$

Thus,

$$\tau_{xy} = G\gamma_{xy}; \quad \tau_{xy} = 76.9(10^9)[-149(10^{-6})] = -11.46 \text{ MPa}$$

The Mohr's circle for this state of plane stress has a reference point $A(29.4 \text{ MPa}, -11.46 \text{ MPa})$ and center at $\sigma_{avg} = 43.7$ MPa, Fig. 15–47b. The radius is determined from the shaded triangle.

$$R = \sqrt{(43.7 - 29.4)^2 + (11.46)^2} = 18.3 \text{ MPa}$$

Therefore,

$$\sigma_1 = 43.7 + 18.3 = 62.0 \text{ MPa} \qquad \qquad Ans.$$

$$\sigma_2 = 43.7 - 18.3 = 25.4 \text{ MPa} \qquad \qquad Ans.$$

Note that each of these solutions is valid provided the material is both linear-elastic and isotropic, since then the principal planes of stress and strain coincide.

Example 15–17

The copper bar in Fig. 15–48 is subjected to a uniform loading along its edges as shown. If it has a length $a = 300$ mm, width $b = 50$ mm, and thickness $t = 20$ mm before the load is applied, determine its new length, width, and thickness after application of the load. Take $E_{cu} = 120$ GPa, $\nu_{cu} = 0.34$.

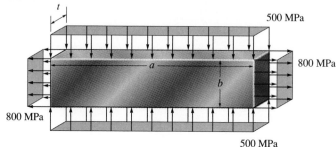

Fig. 15–48

SOLUTION

By inspection, the bar is subjected to a state of plane stress. From the loading we have

$$\sigma_x = 800 \text{ MPa} \qquad \sigma_y = -500 \text{ MPa} \qquad \tau_{xy} = 0 \qquad \sigma_z = 0$$

The associated normal strains are determined from the generalized Hooke's law, Eq. 15–33; that is,

$$\epsilon_x = \frac{\sigma_x}{E} - \frac{\nu}{E}(\sigma_y + \sigma_z)$$

$$= \frac{800 \text{ MPa}}{120(10^3) \text{ MPa}} - \frac{0.34}{120(10^3) \text{ MPa}}(-500 \text{ MPa}) = 0.00808$$

$$\epsilon_y = \frac{\sigma_y}{E} - \frac{\nu}{E}(\sigma_x + \sigma_z)$$

$$= \frac{-500 \text{ MPa}}{120(10^3) \text{ MPa}} - \frac{0.34}{120(10^3) \text{ MPa}}(800 \text{ MPa} + 0) = -0.00643$$

$$\epsilon_z = \frac{\sigma_z}{E} - \frac{\nu}{E}(\sigma_x + \sigma_y)$$

$$= 0 - \frac{0.34}{120(10^3) \text{ MPa}}(800 \text{ MPa} - 500 \text{ MPa}) = -0.000850$$

The new bar length, width, and thickness are therefore

$$a' = 300 \text{ mm} + 0.00808(300 \text{ mm}) = 302.4 \text{ mm} \qquad \textit{Ans.}$$

$$b' = 50 \text{ mm} - 0.00643(50 \text{ mm}) = 49.68 \text{ mm} \qquad \textit{Ans.}$$

$$t' = 20 \text{ mm} - 0.000850(20 \text{ mm}) = 19.98 \text{ mm} \qquad \textit{Ans.}$$

Example 15-18

If the rectangular rubber block shown in Fig. 15–49 is subjected to a uniform pressure of $p = 20$ psi, determine the dilatation and the change in length of each side. Take $E_r = 600$ psi, $\nu_r = 0.45$.

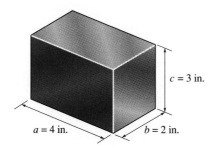

$c = 3$ in.

$a = 4$ in. $b = 2$ in.

Fig. 15–49

SOLUTION

Dilatation. The dilatation can be determined using Eq. 15–38 with $\sigma_x = \sigma_y = \sigma_z = -20$ psi. We have

$$e_r = \frac{1 - 2\nu}{E}(\sigma_x + \sigma_y + \sigma_z)$$

$$= \frac{1 - 2(0.45)}{600 \text{ psi}}[3(-20 \text{ psi})]$$

$$= -0.01 \text{ in}^3/\text{in}^3 \qquad\qquad Ans.$$

Change in Length. The normal strain on each side can be determined from Hooke's law, Eq. 15–33; that is,

$$\epsilon = \frac{1}{E}[\sigma - \nu(\sigma + \sigma)]$$

$$= \frac{1}{600 \text{ psi}}[-20 \text{ psi} - (0.45)(-20 \text{ psi} - 20 \text{ psi})] = -0.00333 \text{ in./in.}$$

Thus the change in length of each side is

$$\Delta a = -0.00333(4 \text{ in.}) = -0.0133 \text{ in.} \qquad Ans.$$

$$\Delta b = -0.00333(2 \text{ in.}) = -0.00667 \text{ in.} \qquad Ans.$$

$$\Delta c = -0.00333(3 \text{ in.}) = -0.0100 \text{ in.} \qquad Ans.$$

The negative signs indicate that each dimension is decreased.

PROBLEMS

15–65. Use Hooke's law, Eq. 15–33, to develop the strain-transformation equations, Eqs. 15–21 and 15–22, from the stress-transformation equations, Eqs. 15–1 and 15–2.

15–66. Determine the bulk modulus for each of the following materials: (a) rubber, $E_r = 0.4$ ksi, $\nu_r = 0.48$, and (b) glass, $E_g = 8(10^3)$ ksi, $\nu_g = 0.24$.

15–67. A uniform edge load of $w_1 = 60$ N/m and $w_2 = 45$ N/m is applied to the rubber membrane. If it is originally square and has dimensions of $a = 100$ mm and $b = 100$ mm and a thickness of $t = 2$ mm, determine its new dimensions a', b', and t' after the load is applied. $E_r = 4$ MPa, $\nu_r = 0.42$.

*15–68. A uniform edge load of $w_1 = 60$ N/m and $w_2 = 0$ is applied to the rubber membrane. If it is originally square and has dimensions of $a = 100$ mm and $b = 100$ mm and a thickness of $t = 2$ mm, determine its new dimensions a', b', and t' after the load is applied. $E_r = 4$ MPa, $\nu_r = 0.42$.

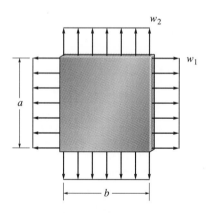

Probs. 15–67/15–68

15–69. The aluminum bar is subjected to an axial force of 15 kip. If it has the original dimensions shown, determine the *change* in the angle θ after the load is applied. $E_{al} = 10(10^3)$ ksi, $\nu_{al} = 0.33$.

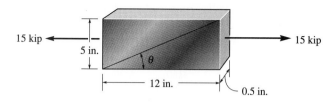

Prob. 15–69

15–70. Determine the principal strains that occur at a point on a steel member where the principal stresses are $\sigma_{max} = 22$ ksi, $\sigma_{int} = 0$, $\sigma_{min} = -14$ ksi. $E_{st} = 29(10^3)$ ksi, $\nu_{st} = 0.3$.

15–71. A bar of copper alloy is loaded in a tension machine and it is determined that $\epsilon_x = 940(10^{-6})$ and $\sigma_x = 14$ ksi, $\sigma_y = 0$, $\sigma_z = 0$. Determine the modulus of elasticity, E_{cu}, and the dilatation, e_{cu}, of the copper. $\nu_{cu} = 0.35$.

*15–72. The principal plane stresses and associated strains in a plane at a point are $\sigma_1 = 36$ ksi, $\sigma_2 = 16$ ksi, $\epsilon_1 = 1.02(10^{-3})$, $\epsilon_2 = 0.180(10^{-3})$. Determine the modulus of elasticity and Poisson's ratio.

15–73. From experiment, the principal strains in a plane at a point on a steel shell are $\epsilon_1 = 350(10^{-6})$ and $\epsilon_2 = -250(10^{-6})$. If $E_{st} = 200$ GPa and $\nu_{st} = 0.28$, determine the principal plane stresses in this plane.

15–74. The principal stresses at a point are shown in the figure. If the material is nylon for which $E_n = 2.5$ GPa and $\nu_n = 0.4$, determine the principal strains.

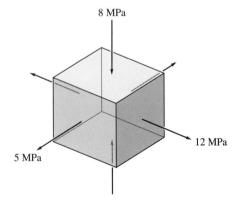

Prob. 15–74

15–75.   For the case of plane stress, show that Hooke's law can be written as

$$\sigma_x = \frac{E}{(1 - \nu^2)}(\epsilon_x + \nu\epsilon_y), \qquad \sigma_y = \frac{E}{(1 - \nu^2)}(\epsilon_y + \nu\epsilon_x)$$

*15–76.   The principal strains in a plane, measured experimentally at a point on the aluminum fuselage of a jet aircraft, are $\epsilon_1 = 780(10^{-6})$ and $\epsilon_2 = 400(10^{-6})$. Determine the associated principal stresses at the point in the same plane. $E_{al} = 10(10^3)$ ksi, $\nu_{al} = 0.33$. *Hint:* See Prob. 15–75.

15–77.   A thin-walled spherical pressure vessel has an inner radius r, thickness t, and is subjected to an internal pressure p. If the material constants are E and ν, determine the strain in the circumferential direction in terms of the stated parameters.

15–78.   The spherical pressure vessel has an inner diameter of 2 m and a thickness of 10 mm. A strain gauge having a length of 20 mm is attached to the vessel and it is observed to increase in length by 0.012 mm as the pressure in the vessel is increased. Determine the change in pressure causing this deformation and also compute the maximum in-plane shear stress and the absolute maximum shear stress. The material is steel, for which $E_{st} = 200$ GPa and $\nu_{st} = 0.3$.

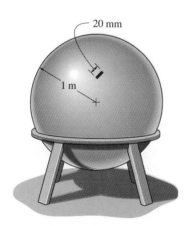

20 mm

1 m

Prob. 15–78

15–79.   A strain gauge, placed on the outer surface and at an angle of 45° to the axis of the copper pipe, gives a reading at point A of $\epsilon_a = 250(10^{-6})$. Determine the force P applied to the wrench. The pipe has an outer diameter of 1 in. and an inner diameter of 0.6 in. $E_{cu} = 18(10^3)$ ksi, $\nu_{cu} = 0.33$.

*15–80.   A strain gauge, placed on the outer surface and at an angle of 45° to the axis of the copper pipe, gives a reading at point A of $\epsilon_a = 250(10^{-6})$. Determine the principal strains in the pipe at point A. The pipe has an outer diameter of 1 in. and an inner diameter of 0.6 in. $E_{cu} = 18(10^3)$ ksi, $\nu_{cu} = 0.33$.

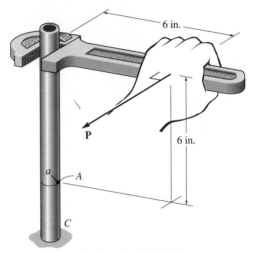

6 in.

P

6 in.

a

A

C

Probs. 15–79/15–80

15–81.   The cross section of the rectangular beam is subjected to the bending moment **M**. Determine an expression for the increase in length of lines AB and CD. The material has a modulus of elasticity E and Poisson's ratio is ν.

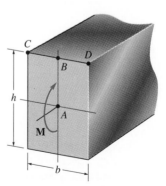

C

B

D

h

A

M

b

Prob. 15–81

15–82. The aluminum beam has the rectangular cross section shown. If it is subjected to a bending moment of $M = 60$ kip · in., determine the increase in the 2-in. dimension at the top of the beam and the decrease in this dimension at the bottom. $E_{al} = 10(10^3)$ ksi, $\nu_{al} = 0.3$.

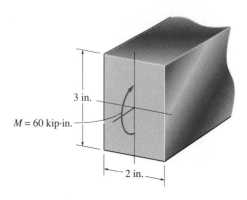

3 in.

$M = 60$ kip·in.

2 in.

Prob. 15–82

15–83. The thin-walled cylindrical pressure vessel of inner radius r and thickness t is subjected to an internal pressure p. Determine the maximum x,y in-plane shear strain at point A, where the principal stresses are σ_{max} and σ_{int}, and $\sigma_{min} = 0$. Also compute the absolute maximum shear strain at A. The material properties are E and ν.

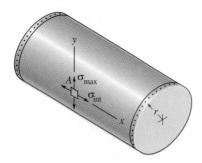

Prob. 15–83

*15–84. Air is pumped into the steel thin-walled pressure vessel at C. If the ends of the vessel are closed using two pistons connected by a rod AB, determine the increase in the diameter of the pressure vessel when the internal gauge pressure is 5 MPa. Also, what is the tensile stress in rod AB if it has a diameter of 100 mm? The inner radius of the vessel is 400 mm, and its thickness is 10 mm. $E_{st} = 200$ GPa and $\nu_{st} = 0.3$.

15–85. Determine the increase in the diameter of the pressure vessel in Prob. 15–84 if the pistons are replaced by walls connected to the vessel.

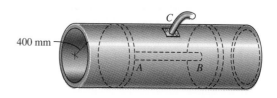

400 mm

Probs. 15–84/15–85

15–86. The strain gauge is placed on the surface of a thin-walled steel boiler as shown. If it is 0.5 in. long, determine the pressure in the boiler when the gauge elongates $0.2(10^{-3})$ in. The boiler has a thickness of 0.5 in. and inner diameter of 60 in. Also, determine the maximum x, y in-plane shear strain in the material. $E_{st} = 29(10^3)$ ksi, $\nu_{st} = 0.3$.

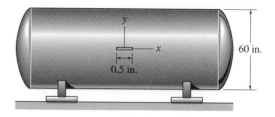

60 in.

0.5 in.

Prob. 15–86

15–87. The steel shaft has a radius of 15 mm. Determine the torque T in the shaft if the two strain gauges, attached to the surface of the shaft, report strains of $\epsilon_{x'} = -80(10^{-6})$ and $\epsilon_{y'} = 80(10^{-6})$. Also, compute the strains acting in the x and y directions. $E_{st} = 200$ GPa, $\nu_{st} = 0.3$.

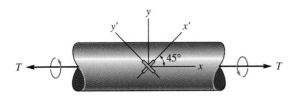

Prob. 15–87

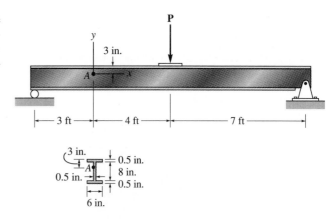

Prob. 15–89

15–90. A thin-walled cylindrical pressure vessel has an inner radius r, thickness t, and length L. If it is subjected to an internal pressure p, show that the increase in its inner radius is $dr = r\epsilon_1 = pr^2(1 - \frac{1}{2}\nu)/Et$ and the increase in its length is $\Delta L = pLr(\frac{1}{2} - \nu)/Et$. Using these results, show that the change in internal volume becomes $dV = \pi r^2(1 + \epsilon_1)^2(1 + \epsilon_2)L - \pi r^2 L$. Since ϵ_1 and ϵ_2 are small quantities, show further that the change in volume per unit volume, called *volumetric strain*, can be written as $dV/V = Pr(2.5 - 2\nu)/Et$.

15–91. The cylindrical pressure vessel is fabricated using hemispherical end caps in order to reduce the bending stress that would occur if flat ends were used. The bending stresses at the seam where the caps are attached can be eliminated by proper choice of the thickness t_h and t_c of the caps and cylinder, respectively. This requires the radial expansion to be the same for both the hemispheres and cylinder. Show that this ratio is $t_c/t_h = (2 - \nu)/(1 - \nu)$. Assume that the vessel is made of the same material and both the cylinder and hemispheres have the same inner radius. If the cylinder is to have a thickness of 0.5 in., what is the required thickness of the hemispheres? Take $\nu = 0.3$.

***15–88.** Two strain gauges are attached to the 6-in.-diameter steel shaft. The gauges are placed at 45° with the axis of the shaft. If the readings are $\epsilon_{x'} = 320(10^{-6})$ and $\epsilon_{y'} = -320(10^{-6})$, determine the torque T developed in the shaft. $E_{st} = 29(10^3)$ ksi, $\nu_{st} = 0.3$.

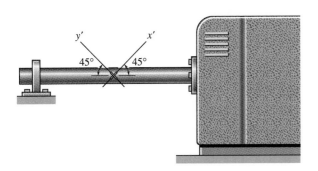

Prob. 15–88

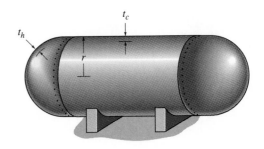

Prob. 15–91

15–89. The strain in the x direction at point A on the steel beam is measured and found to be $\epsilon_x = -100(10^{-6})$. Determine the applied load P. What is the shear strain γ_{xy} at point A? $E_{st} = 29(10^3)$ ksi, $\nu_{st} = 0.3$.

REVIEW PROBLEMS

*15–92.   A rod has a circular cross section with a diameter of 2 in. It is subjected to a torque of 12 kip · in. and a bending moment **M**. The greater principal stress at the point of maximum flexural stress is 15 ksi. Determine the magnitude of the bending moment.

Prob. 15–92

15–93.   A steel pipe has an inner diameter of 2.75 in. and an outer diameter of 3 in. If it is fixed at C and subjected to the horizontal force acting on the handle of the pipe wrench at its end, determine the principal stresses in the pipe at point A.

Prob. 15–93

15–94.   The state of stress at a point on the upper surface of the wing is shown on the element. Determine (a) the principal stresses and (b) the maximum in-plane shear stress and average normal stress at the point. Specify the orientation of the element in each case.

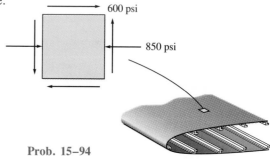

Prob. 15–94

15–95.   The 60° strain rosette is mounted on the surface of a dome. The following readings are obtained for each gauge: $\epsilon_a = -780(10^{-6})$, $\epsilon_b = 400(10^{-6})$, $\epsilon_c = 500(10^{-6})$. Determine (a) the principal strains and (b) the maximum in-plane shear strain and associated average strain. In each case, specify the orientation of the element and show how the strain deforms the element.

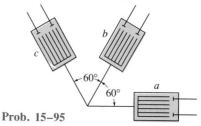

Prob. 15–95

*15–96.   The state of stress at a point is shown on the element. Determine (a) the principal stresses and (b) the maximum in-plane shear stress and average normal stress on the element. Specify the orientation of the element in each case. Use the stress transformation equations.

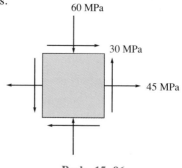

Prob. 15–96

15–97. The drill pipe has an outer diameter of 3 in. and a wall thickness of 0.25 in. If it is subjected to a torque and axial load as shown, determine (a) the principal stresses and (b) the maximum in-plane shear stress at a point on its surface.

1500 lb

800 lb·ft

Prob. 15–97

15–98. The beam is subjected to the two forces shown. Determine the principal stresses at point A.

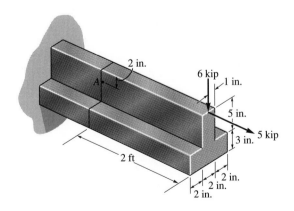

2 in.

6 kip

A

1 in.

5 in.

3 in.   5 kip

2 ft

2 in.
2 in.
2 in.

Probs. 15–98

15–99. Determine the equivalent state of stress if the element is oriented 30° clockwise from its position shown. Use Mohr's circle.

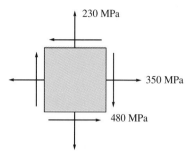

230 MPa

350 MPa

480 MPa

Prob. 15–99

*15–100. The state of stress at a point in shown on the element. Determine (a) the principal stress and (b) the maximum in-plane shear stress and the average normal stress at the point. Specify the orientation of the element in each case.

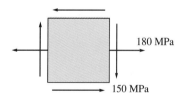

180 MPa

150 MPa

Prob. 15–100

15–101. The square steel plate has a thickness of 0.5 in. and is subjected to the edge loading shown. Determine the principal stresses developed in the steel.

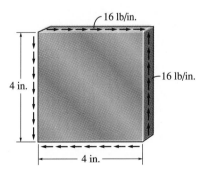

16 lb/in.

4 in.

16 lb/in.

4 in.

Prob. 15–101

16 Design of Beams

A beam is an important structural element in engineering, and in the first part of this chapter we will discuss how to design a beam so that it is able to resist bending and shear loads. Often limits must also be placed on the amount of deflection a beam may undergo when it is subjected to a load, and we will discuss various methods for determining the deflection and slope at specific points on beams. The analytical methods include the integration method, the use of discontinuity functions, and the method of superposition. Also, a semi-graphical technique, called the moment-area method, will be presented. At the end of the chapter, we will use the superposition method to solve for the support reactions on a beam that is statically indeterminate.

16.1 Basis for Beam Design

As stated in Sec. 8.2, beams are structural members designed to support loadings applied perpendicular to their longitudinal axes. Certainly beams may be considered among the most important of all structural elements. Examples include members used to support the floor of a building, the deck of a bridge, or the wing of an aircraft. Also, the axle of an automobile, the boom of a crane, even many bones of the body, act as beams.

Because of the applied loadings, beams develop an internal shear force and bending moment that, in general, vary from point to point along the axis of the beam. Some beams may also be subjected to an internal axial force; however, the effects of this force are often neglected in design, since the axial stress is generally much smaller than the stresses developed by shear and bending. A beam that is chosen to resist both shear and bending stresses is said to be designed on the *basis of strength*. In the first part of this chapter we will show how engineers establish design criteria based on the use of the shear and flexure formulas developed in Chapters 12 and 13. Application of these formulas, however, is limited to beams made of a homogeneous material that has linear-elastic behavior. Also, the cross-sectional area must have an axis of symmetry in the plane of the loading. The design approach to be discussed is in a way approximate, as opposed to a more exact theoretical method; however, it does provide an adequate means of obtaining both a safe and economical design.

Although beams are designed mainly for strength, they must also be braced properly along their sides so that they do not buckle or suddenly become unstable. Furthermore, in some cases beams must be designed to resist a limited amount of *deflection,* as when they support ceilings made of brittle materials such as plaster. Methods for computing beam deflections will be discussed in the last part of this chapter.

16.2 Stress Variations Throughout a Prismatic Beam

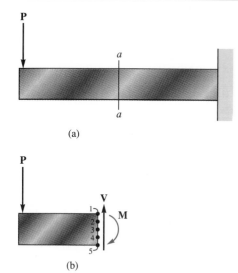

(a)

(b)

Fig. 16–1(a–b)

In Sec. 16.1 we stated that the stress analysis of a beam requires application of the shear and flexure formulas, since beams resist both internal shear and moment loadings. Specific application of these formulas has been treated in Chapters 12 and 13. Here we will discuss the general results obtained when these equations are applied to various points in a cantilevered beam that has a rectangular cross section and supports a load **P** at its end, Fig. 16–1a.

In general, at an arbitrary section *a–a* along the beam's axis, Fig. 16–1b, the internal shear V and moment M are developed by a *parabolic* shear-stress distribution, Fig. 16–1c, and a *linear* normal-stress distribution, Fig. 16–1d. If these results are applied to specific elements located at points 1 through 5 along the section, Fig. 16–1b, the stresses acting on these elements will be as shown in Fig. 16–1e. In particular, elements 1 and 5 are subjected only to the maximum normal stress, whereas element 3, which is on the neutral axis, is subjected only to the maximum shear stress. The intermediate elements 2 and 4 resist *both* normal and shear stress.

In each case the state of stress can be transformed into *principal stresses,* using either the stress-transformation equations or Mohr's circle. The results are shown in Fig. 16–1f. Notice that each successive element, 1 through 5, undergoes a slight counterclockwise orientation. Specifically, relative to element 1, considered to be at the 0° position, element 3 is oriented at 45° and element 5 is oriented at 90°. Also, the *maximum tensile stress* acting on the vertical faces of element 1 becomes smaller on the corresponding faces of each of the successive elements, until it is zero on the horizontal faces of element 5. In a similar manner, the *maximum compressive stress* on the vertical faces of element 5 reduces to zero on the horizontal faces of element 1.

If this analysis is extended to many vertical sections along the beam other than *a–a*, a profile of the results can be represented by curves called *stress trajectories*. Each of these curves indicates the *direction* of a principal stress having a constant magnitude. Some of them are shown for the cantilevered beam in Fig. 16–2. Here the solid lines represent the direction of the tensile principal stresses and the dashed lines represent the direction of the compressive principal stresses. As expected, the lines intersect the neutral axis at 45° angles, and the solid and dashed lines always intersect at 90°. Why? Knowing the direction of these lines can help engineers decide where to reinforce a beam so that it does not crack or become unstable.

Localized Stresses.

The above stress analysis neglects the effects caused by external distributed loadings and concentrated forces applied to the beam. As shown in Fig. 16–3, these loadings will create additional stresses in the beam directly under the load. Notably, a compressive stress σ_y will be developed, in addition to the bending stress σ_x and shear stress τ_{xy} discussed previously. Using advanced methods of analysis, as treated in the theory of elasticity, it can be shown, however, that the stress σ_y diminishes rapidly throughout the beam's depth, and for *most* beam span-to-depth ratios used in engineering practice, the maximum value of σ_y generally represents only a small percentage of the bending stress σ_x, that is, $\sigma_x >> \sigma_y$. Furthermore, the direct application of concentrated loads is generally avoided in beam design. Instead, *bearing pads* or plates are used to spread these loads more evenly along the surface of the beam.

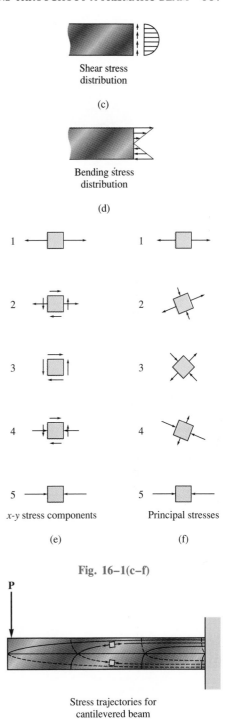

Shear stress distribution

(c)

Bending stress distribution

(d)

x-y stress components

(e)

Principal stresses

(f)

Fig. 16–1(c–f)

Stress trajectories for cantilevered beam

Fig. 16–2

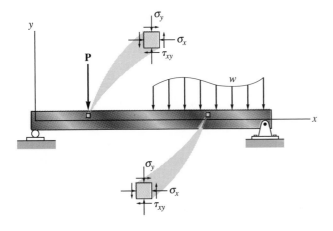

Fig. 16–3

The wide-flange beam shown in Fig. 16–4a is subjected to the distributed loading of $w = 120$ kN/m. Determine the principal stresses in the beam at points 1 through 5. Neglect the fillets and stress concentrations at points 2 and 4, which lie at the top and bottom of the web at the web–flange junction.

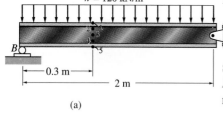

$w = 120$ kN/m

B

├─0.3 m─┤

├──────── 2 m ────────┤

(a)

SOLUTION

Equilibrium Equation. The support reaction on the beam at B is determined, and equilibrium of the sectioned beam shown in Fig. 16–4b yields

$$V = 84 \text{ kN} \qquad M = 30.6 \text{ kN} \cdot \text{m}$$

Section Properties. The beam's moment of inertia about the neutral axis is

$$I = \left[\frac{1}{12}(0.01)(0.2)^3\right]$$
$$+ 2\left[\frac{1}{12}(0.175)(0.015)^3 + (0.175)(0.015)(0.1075)^2\right]$$
$$= 67.4(10^{-6}) \text{ m}^4$$

Stresses. At points 1 and 5,

$$\sigma_1 = \sigma_5 = \frac{Mc}{I} = \frac{30.6(10^3) \text{ N} \cdot \text{m} \, (0.115 \text{ m})}{67.4(10^{-6}) \text{ m}^4} = 52.2 \text{ MPa} \qquad \textit{Ans.}$$

$$\tau_1 = \tau_5 = 0 \qquad \textit{Ans.}$$

At points 2 and 4,

$$\sigma_2 = \sigma_4 = \frac{My}{I} = \frac{30.6(10^3) \text{ kN} \cdot \text{m} \, (0.100 \text{ m})}{67.4(10^{-6}) \text{ m}^4} = 45.4 \text{ MPa} \qquad \textit{Ans.}$$

$$\tau_2 = \tau_4 = \frac{VQ}{It} = \frac{84(10^3) \text{ N} \, [(0.1075 \text{ m})(0.175 \text{ m})(0.015 \text{ m})]}{67.4(10^{-6}) \text{ m}^4(0.010 \text{ m})}$$
$$= 35.2 \text{ MPa} \qquad \textit{Ans.}$$

At point 3,

$$\sigma_3 = 0 \qquad \textit{Ans.}$$

$$\tau_3 = \frac{VQ}{It}$$
$$= \frac{84(10^3) \text{ N} \, [(0.1075 \text{ m})(0.175 \text{ m})(0.015 \text{ m}) + (0.050 \text{ m})(0.100 \text{ m})(0.010 \text{ m})]}{67.4(10^{-6}) \text{ m}^4(0.010 \text{ m})}$$
$$= 41.4 \text{ MPa} \qquad \textit{Ans.}$$

These results are shown in Fig. 16–4c.

15 mm

200 mm N ─┼─ A
              10 mm

15 mm 175 mm

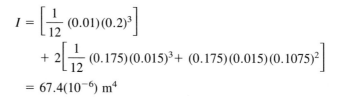

36 kN
0.15 m

$M = 30.6$ kN·m

─0.3 m─ $V = 84$ kN

120 kN

(b)

Fig. 16–4

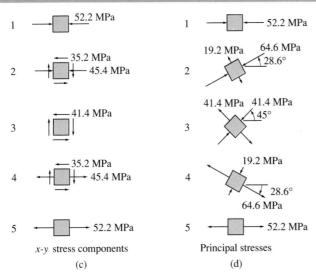

1   52.2 MPa   1   52.2 MPa

19.2 MPa   64.6 MPa

2   35.2 MPa   2   28.6°
     45.4 MPa

41.4 MPa   41.4 MPa

3   41.4 MPa   3   45°

4   35.2 MPa   4   19.2 MPa
     45.4 MPa   28.6°
                              64.6 MPa

5   52.2 MPa   5   52.2 MPa

x-y stress components   Principal stresses

(c)   (d)

Using Mohr's circle, the principal stresses at 2 (and 4) can be determined. As shown in Fig. 16–4e, for element 2 the center of the circle is at $(-45.4 + 0)/2 = -22.7$, and the radius is calculated to be $R = 41.9$. Thus $\sigma_1 = (41.9 - 22.7) = 19.2$ MPa, $\sigma_2 = -(22.7 + 41.9) = -64.6$ MPa. The element is rotated counterclockwise $2\theta_{p_2} = 57.2°$, so that $\theta_{p_2} = 28.6°$. In a similar manner, the principal stresses for element 3, $\sigma_1 = 41.4$ MPa and $\sigma_2 = -41.4$ MPa, are calculated from Mohr's circle, Fig. 16–4f. Here $2\theta_{p_2} = 90°$, so that $\theta_{p_2} = 45°$ counterclockwise.

The results for each element are shown in Fig. 16–4d. By comparison, it can be seen that the elements at points 2 and 4 located at the web-flange junction are subjected to the largest principal stress (64.6 MPa). This occurs because at the section we have considered, the shear stress (35.2 MPa) is *significant* compared with the normal stress (45.4 MPa), Fig. 16–4c. In practice, however, beams are relatively long and are designed to resist the maximum internal moment. At this "critical" section, where the internal moment is a maximum, the shear stress at the web–flange junction *will often be rather small* compared with the bending stress.* Consequently, the maximum principal stress at these points will *not* be greater than the maximum bending stress, which occurs at the beam's outer fibers. For this reason, design codes generally require the beam's maximum bending stress to be compared with the allowable bending stress as stated in the codes. In other words, the allowable bending stress is thought to have enough of a factor of safety to compensate for cases when the maximum principal stress at points between the beam's web and flange exceeds the maximum bending stress at points on the beam's top or bottom surface.

*For the beam in Fig. 16–4a, the shear and moment diagrams reveal that maximum bending occurs at the *center* of the beam, and the internal shear force (and shear stress) is actually zero at this point. See Fig. 7–12.

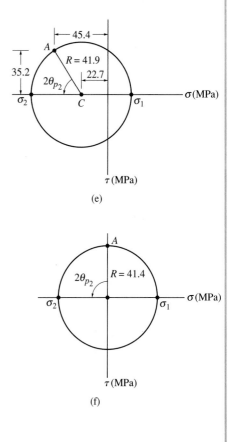

16.3 Prismatic Beam Design

The design of a prismatic beam requires selecting its cross-sectional area to have a practical size and shape so that the material is not overstressed or the beam does not deflect excessively or buckle when subjected to a given loading. Here we will discuss the design of a beam based only on its *strength*. This requires that the actual bending and shear stresses in the beam do not exceed allowable bending and shear stresses for the material as defined by structural or mechanical codes.

If the suspended span of the beam is relatively long, so that the internal moments become large, the engineer will first consider a design based upon bending. This requires a determination of the beam's *section modulus*, which is the ratio of the unknowns I and c, that is, $S = I/c$. Using the flexure formula, $\sigma = Mc/I$, we have

$$S_{\text{req'd}} = \frac{M}{\sigma_{\text{allow}}} \qquad\qquad (16\text{–}1)$$

Here M is determined from the beam's moment diagram, and the allowable bending stress, σ_{allow}, is specified in a design code. In most cases the beam's unknown weight will be small and can be neglected in comparison with the loads the beam must carry. However, if the additional moment caused by the weight is to be included in the design, a selection for S is made so that it slightly *exceeds* $S_{\text{req'd}}$.

Once $S_{\text{req'd}}$ is known, if the beam has a simple cross-sectional shape, such as a square, a circle, or a rectangle of known width-to-height proportions, its *dimensions* can be determined directly from $S_{\text{req'd}}$, since by definition $S_{\text{req'd}} = I/c$. However, if the cross section is made from several elements, such as a wide-flange section, then an infinite number of web and flange dimensions can be determined that satisfy the value of $S_{\text{req'd}}$. In practice, however, engineers choose a particular beam meeting the requirement that $S > S_{\text{req'd}}$ from a handbook that lists the standard shapes available from manufacturers. Often several beams that have the same section modulus can be selected from these tables. If deflections are not restricted, usually the beam having the smallest cross-sectional area is chosen, since it is made of less material and is therefore both lighter and more economical than the others.

The above discussion assumes that the material's allowable bending stress is the *same* for both tension and compression. If this is the case, then a beam having a cross section that is *symmetric* with respect to the neutral axis should be chosen. However, if the allowable tensile and compressive bending stresses are *not* the same, then the choice of an unsymmetric cross section may be more efficient. Under these circumstances the beam must be designed to resist *both* the largest positive and the largest negative moment in the span.

Once the beam has been selected, the shear formula $\tau_{\text{allow}} \geq VQ/It$ can then be used to check that the allowable shear stress is not exceeded. Often this requirement will not present a problem. However, if the beam is "short" and supports large concentrated loads, the shear-stress limitation may dictate the size of the beam. This limitation is particularly important in the design of wood beams, because wood tends to split along its grain due to shear (see Fig. 13–6).

Fabricated Beams. Since beams are often made of steel and wood, we will now discuss some of the tabulated properties of beams made of these materials.

Steel Sections. Most manufactured steel beams are produced by rolling a hot ingot of steel until the desired shape is formed. These so-called *rolled shapes* have properties that are tabulated in the American Institute of Steel Construction (AISC) manual. A representative listing taken from this manual is given in Appendix *D*. As noted in this appendix, the wide-flange shapes are designated by their depth and weight per unit length; for example, $W18 \times 46$ indicates a wide-flange cross section (W) having a depth of 18 in. and a weight of 46 lb/ft, Fig. 16–5. For any given section, the weight per length, dimensions, cross-sectional area, moment of inertia, and section modulus are reported. Also included is the radius of gyration r, which is a geometric property related to the section's buckling strength. This will be discussed in Chapter 17.

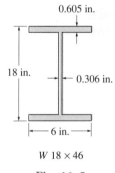

$W\ 18 \times 46$

Fig. 16–5

Wood Sections. Most beams made of wood have rectangular cross sections because such beams are easy to manufacture and handle. Manuals, such as that of the American Forest Products Association, list the dimensions of lumber often used in the design of wood beams. Often both the nominal and net dimensions are reported. Lumber is identified by its *nominal* dimensions, such as 2×4 (2 in. by 4 in.); however, its actual or "dressed" dimensions are smaller, being 1.5 in. by 3.5 in. The reduction in the dimensions occurs due to the requirement of obtaining smooth surfaces from lumber that is rough-sawn. Obviously, the *actual dimensions* must be used whenever stress calculations are performed on wood beams.

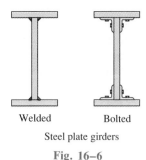

Welded   Bolted

Steel plate girders

Fig. 16–6

Wood box–beam

(a)

Glulam beam

(b)

Fig. 16–7

Built-up Sections.   A *built-up section* is constructed from two or more parts joined together to form a single unit. As indicated by Eq. 16–1, the capacity of the beam to resist a moment will vary directly with its section modulus S, and since $S = I/c$, then S is *increased* if I is *increased*. In order to increase I, *most of the material* should be placed as far *away* from the neutral axis as practical. This, of course, is what makes a deep wide-flange beam so efficient in resisting a moment. For very large loads, however, an available rolled-steel section may not have a section modulus great enough to support a given moment. Rather than using several available beams to support the load, engineers will usually ''build up'' a beam made from plates and angles. A deep I-shaped section having this form is called a *plate girder*. For example, the steel plate girder in Fig. 16–6 has two flange plates that are either welded or, using angles, bolted to the web plate.

Wood beams are also ''built up,'' usually in the form of a box-beam section, Fig. 16–7a. They may be made having plywood webs and larger boards for the flanges. For very large spans, *glulam beams* are used. These members are made from several boards glue-laminated together to form a single unit, Fig. 16–7b.

Just as in the case of rolled sections or beams made from a single piece, the design of built-up sections requires that the bending and shear stresses be checked. In addition the shear stress in the fasteners, such as weld, glue, nails, etc., must be checked to be certain the beam acts as a single unit. The principles for doing this were outlined in Sec. 13.4.

PROCEDURE FOR ANALYSIS

Based on the previous discussion, the following procedure provides a rational method for the design of a beam on the basis of strength.

Shear and Moment Diagrams.   Determine the maximum shear and moment in the beam. Often this is done by constructing the beam's shear and moment diagrams. For built-up beams these diagrams are also useful for identifying *regions* where the shear and moment are excessively large and may require additional structural reinforcement or fasteners to hold the beam together.

Bending Stress. If the beam is relatively long, it is designed on the basis of determining its section modulus using the flexure formula, $S_{req'd} = M_{max}/\sigma_{allow}$, where M_{max} is the maximum moment in the beam and σ_{allow} is the allowable bending stress for the material. Once $S_{req'd}$ is determined, the cross-sectional dimensions for simple shapes can then be computed, since $S_{req'd} = I/c$. If rolled-steel sections are to be used, several possible values of S may be selected from the tables in Appendix D. Of these, choose the one having the smallest cross-sectional area, since this beam has the least weight and is therefore the most economical. Make sure that the selected section modulus, S, is *slightly greater* than $S_{req'd}$, so that the additional moment created by the beam's weight is considered.

Shear Stress. Using the shear formula, check to see that the allowable shear stress is not exceeded; that is, use $\tau_{allow} \geqq V_{max} Q/It$. Here V_{max} is the maximum internal shear as determined from the shear diagram. In particular, if the beam has a solid *rectangular* cross section, this equation becomes $\tau_{allow} \geqq 1.5(V_{max}/A)$, Eq. 13–5, and if the cross section is a *wide flange,* it is generally appropriate to assume that the shear stress is *constant* over the cross-sectional area of the beam's web so that $\tau_{allow} \geqq V_{max}/A_{web}$, where A_{web} is determined from the product of the beam's depth and the web's thickness. See Sec. 13.4.

When a beam that has been designed on the basis of bending stress fails to meet the criterion for shear, it must be redesigned to resist the shear stress. Normally beams that are short and carry large loads, especially those made of wood, are first designed to resist shear and then later checked against the allowable-bending-stress requirements.

Adequacy of Fasteners. Built-up beams made from several elements must be fastened together properly so the beam acts as a single unit when it is loaded. In this regard, the adequacy of a fastener depends on the shear stress it can resist. Specifically, the required spacing of nails or bolts of a particular size is determined from the allowable shear flow, $q_{allow} = VQ/I$, calculated at points on the cross section where the fasteners are located. See Sec. 13.4.

The following examples illustrate application of these principles. Although this procedure for analysis is generally followed by most codes in structural and mechanical design, it should be mentioned that often further analysis must be performed to be certain that a selected beam is adequate. A complete analysis would *also* include checks to see that the beam is properly braced to prevent unstable sidesway, that built-up elements, such as the web and flanges, are not too thin so that they may buckle, that the beam does not severely deflect under loading, and that the localized stresses at points of stress concentrations and concentrated loads are reduced.

Example 16–2

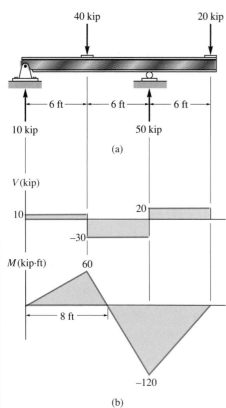

40 kip   20 kip

6 ft — 6 ft — 6 ft

10 kip   50 kip

(a)

V(kip)

10

20

−30

x(ft)

M(kip·ft)   60

8 ft

x(ft)

−120

(b)

Fig. 16–8

A beam is to be made of steel that has an allowable bending stress of $\sigma_{\text{allow}} = 24$ ksi and an allowable shear stress of $\tau_{\text{allow}} = 14.5$ ksi. Select an appropriate W shape that will carry the loading shown in Fig. 16–8a.

SOLUTION

Shear and Moment Diagrams. The support reactions have been calculated in Fig. 16–8a, and the shear and moment diagrams are shown in Fig. 16–8b. From these diagrams, $V_{\text{max}} = 30$ kip and $M_{\text{max}} = 120$ kip · ft.

Bending Stress. The required section modulus for the beam is determined from the flexure formula,

$$S_{\text{req'd}} = \frac{M_{\text{max}}}{\sigma_{\text{allow}}} = \frac{120 \text{ kip} \cdot \text{ft}(12 \text{ in./ft})}{24 \text{ kip/in}^2} = 60 \text{ in}^3$$

Using the table in Appendix D, the following beams are adequate:

W18 × 40	$S = 68.4 \text{ in}^3$
W16 × 45	$S = 72.7 \text{ in}^3$
W14 × 43	$S = 62.7 \text{ in}^3$
W12 × 50	$S = 64.7 \text{ in}^3$
W10 × 54	$S = 60.0 \text{ in}^3$
W8 × 67	$S = 60.4 \text{ in}^3$

The beam having the least weight per foot is chosen, i.e.,

$$W18 \times 40$$

The *actual* maximum moment M_{max}, which includes the weight of the beam, can be computed and the adequacy of the selected beam can be checked. In comparison with the applied loads, however, the beam's weight, $(0.040 \text{ kip/ft})(18 \text{ ft}) = 0.720$ kip, will only *slightly increase* $S_{\text{req'd}}$. In spite of this,

$$S_{\text{req'd}} = 60 \text{ in}^3 < 68.4 \text{ in}^3 \qquad \text{OK}$$

Shear Stress. Since the beam is a *wide-flange section*, the *average shear stress* within the web will be considered. Here the web is assumed to extend from the very top to the very bottom of the beam. From Appendix D, for a W18 × 40, $d = 17.90$ in., $t_w = 0.315$ in. Thus,

$$\tau_{\text{avg}} = \frac{V_{\text{max}}}{A_w} = \frac{30 \text{ kip}}{(17.90 \text{ in.})(0.315 \text{ in.})} = 5.32 \text{ ksi} < 14.5 \text{ ksi} \qquad \text{OK}$$

Use a W18 × 40. *Ans.*

Example 16–3

The laminated wooden beam shown in Fig. 16–9a supports a uniform distributed loading of 12 kN/m. If the beam is to have a height-to-width ratio of 1.5, determine its smallest width. The allowable bending stress is $\sigma_{allow} = 9$ MPa and the allowable shear stress is $\tau_{allow} = 0.6$ MPa. Neglect the weight of the beam.

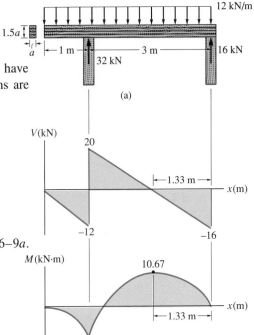

(a)

SOLUTION

Shear and Moment Diagrams.   The support reactions at A and B have been calculated in Fig. 16–9a, and the shear and moment diagrams are shown in Fig. 16–9b. Here $V_{max} = 20$ kN, $M_{max} = 10.7$ kN · m.

Bending Stress.   Applying the flexure formula yields

$$S_{req'd} = \frac{M_{max}}{\sigma_{allow}} = \frac{10.67 \text{ kN} \cdot \text{m}}{9(10^3) \text{ kN/m}^2} = 0.00119 \text{ m}^3$$

Assuming that the width is a, then the height is $h = 1.5a$, Fig. 16–9a. Thus,

$$S_{req'd} = \frac{I}{c} = \frac{\frac{1}{12}(a)(1.5a)^3}{(0.75a)} = 0.00119 \text{ m}^3$$

$$a^3 = 0.003160 \text{ m}^3$$

$$a = 0.147 \text{ m}$$

(b)

Fig. 16–9

Shear Stress.   Applying the shear formula for rectangular sections, we have

$$\tau_{max} = \frac{3}{2} \frac{V_{max}}{A} = \frac{3}{2} \frac{20 \text{ kN}}{(0.147 \text{ m})(1.5)(0.147 \text{ m})} = 0.926 \text{ MPa} > 0.6 \text{ MPa}$$

Since the shear criterion fails, the beam must be redesigned on the basis of shear.

$$\tau_{allow} = \frac{3}{2} \frac{V_{max}}{A}$$

$$600 \text{ kN/m}^2 = \frac{3}{2} \frac{20 \text{ kN}}{(a)(1.5a)}$$

$$a = 0.183 \text{ m} = 183 \text{ mm} \qquad \qquad \textit{Ans.}$$

This larger section will also adequately resist the normal stress.

Example 16–4

The wooden T-beam shown in Fig. 16–10a is made from two 200 mm × 30 mm boards. If the allowable bending stress is $\sigma_{allow} = 12$ MPa and the allowable shear stress is $\tau_{allow} = 0.8$ MPa, determine if the beam can safely support the loading shown. Also specify the maximum spacing of nails needed to hold the two boards together if each nail can safely resist 1.50 kN in shear.

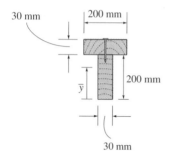

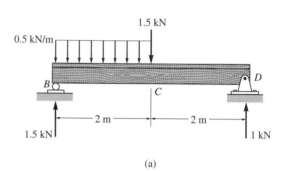

(a)

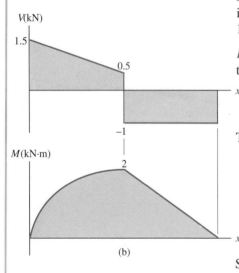

Fig. 16–10

SOLUTION

Shear and Moment Diagrams. The reactions on the beam are shown in Fig. 16–10a, and the shear and moment diagrams are drawn in Fig. 16–10b. Here $V_{max} = 1.5$ kN, $M_{max} = 2$ kN · m.

Bending Stress. The neutral axis (centroid) will be located from the bottom of the beam. Working in units of meters, we have

$$\bar{y} = \frac{\Sigma \bar{y}A}{\Sigma A} = \frac{(0.1)(0.03)(0.2) + 0.215(0.03)(0.2)}{0.03(0.2) + 0.03(0.2)} = 0.1575 \text{ m}$$

Thus

$$I = \left[\frac{1}{12}(0.03)(0.2)^3 + (0.03)(0.2)(0.1575 - 0.1)^2 \right]$$
$$+ \left[\frac{1}{12}(0.2)(0.03)^3 + (0.03)(0.2)(0.215 - 0.1575)^2 \right]$$
$$= 60.125(10^{-6}) \text{ m}^4$$

Since $c = 0.1575$ m (not 0.230 m − 0.1575 m = 0.0725 m), we require

$$\sigma_{allow} \geq \frac{M_{max}c}{I}$$

$$12(10^3) \text{ kPa} \geq \frac{2 \text{ kN} \cdot \text{m}(0.1575 \text{ m})}{60.125(10^{-6}) \text{ m}^4} = 5.24(10^3) \text{ kPa} \qquad \text{OK}$$

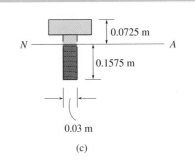

(c)

Shear Stress.   Maximum shear stress in the beam depends upon the magnitude of Q and t. It occurs at the neutral axis, since Q is a maximum there and the neutral axis is in the web, where the thickness $t = 0.03$ m is smallest for the cross section. For simplicity we will use the rectangular area below the neutral axis to calculate Q, rather than a two-part composite area above this axis, Fig. 16–10c. We have

$$Q = \bar{y}'A' = \left(\frac{0.1575}{2}\right)[(0.1575)(0.03)] = 0.372(10^{-3})\ \text{m}^3$$

So that

$$\tau_{allow} \geq \frac{V_{max}Q}{It}$$

$$800\ \text{kPa} \geq \frac{1.5\ \text{kN}[0.372(10^{-3})]\ \text{m}^3}{60.125(10^{-6})\ \text{m}^4(0.03\ \text{m})} = 309\ \text{kPa} \qquad \text{OK}$$

Nail Spacing.   From the shear diagram it is seen that the shear varies over the entire span. Since the nail spacing depends on the magnitude of shear in the beam, for simplicity (and to be conservative), we will design the spacing on the basis of $V = 1.5$ kN for region BC and $V = 1$ kN for region CD. Since the nails join the flange to the web, Fig. 16–10d, we have

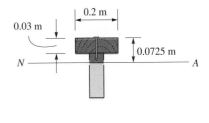

(d)

$$Q = \bar{y}'A' = (0.0725 - 0.015)[(0.2)(0.03)] = 0.345(10^{-3})\ \text{m}^3$$

The shear flow for each region is therefore

$$q_{BC} = \frac{V_{BC}Q}{I} = \frac{1.5\ \text{kN}[0.345(10^{-3})]\ \text{m}^3}{60.125(10^{-6})\ \text{m}^4} = 8.61\ \text{kN/m}$$

$$q_{CD} = \frac{V_{CD}Q}{I} = \frac{1\ \text{kN}[0.345(10^{-3})]\ \text{m}^3}{60.125(10^{-6})\ \text{m}^4} = 5.74\ \text{kN/m}$$

One nail can resist 1.50 kN in shear, so the spacing becomes

$$s_{BC} = \frac{1.50\ \text{kN}}{8.61\ \text{kN/m}} = 0.174\ \text{m}$$

$$s_{CD} = \frac{1.50\ \text{kN}}{5.74\ \text{kN/m}} = 0.261\ \text{m}$$

For ease of measuring, use

$$s_{BC} = 150\ \text{mm} \qquad\qquad \textit{Ans.}$$

$$s_{CD} = 250\ \text{mm} \qquad\qquad \textit{Ans.}$$

PROBLEMS

Neglect the weight of the beam in the following problems.

16–1. The simply-supported beam is made of timber that has an allowable bending stress of $\sigma_{allow} = 6.5$ MPa and an allowable shear stress of $\tau_{allow} = 500$ kPa. Determine its dimensions if it is to be rectangular and have a height-to-width ratio of 1.25.

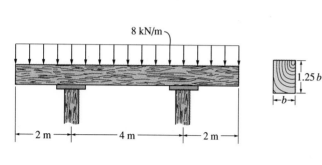

Prob. 16–1

16–2. The beam is made of cypress having an allowable bending stress of $\sigma_{allow} = 850$ psi and an allowable shear stress of $\tau_{allow} = 80$ psi. Determine the width b of the beam if the height $h = 1.5b$.

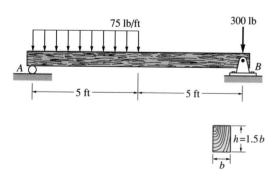

Prob. 16–2

16–3. The beam is made of Douglas fir having an allowable bending stress of $\sigma_{allow} = 1.1$ ksi and an allowable shear stress of $\tau_{allow} = 0.70$ ksi. Determine the width b of the beam if the height $h = 2b$.

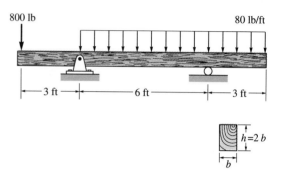

Prob. 16–3

*16–4. The joists of a floor in a warehouse are to be selected using square timber beams made of oak. If each beam is to be designed to carry 90 lb/ft over a simply-supported span of 25 ft, determine the dimension a of its square cross section. The allowable bending stress is $\sigma_{allow} = 4.5$ ksi and the allowable shear stress is $\tau_{allow} = 125$ psi.

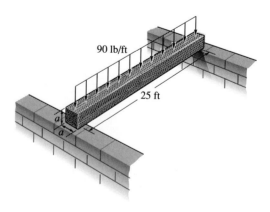

Prob. 16–4

16–5.   The simply-supported beam is made of timber that has an allowable bending stress of $\sigma_{allow} = 960$ psi and an allowable shear stress of $\tau_{allow} = 75$ psi. Determine its dimensions if it is to be rectangular and have a height-to-width ratio of 1.25.

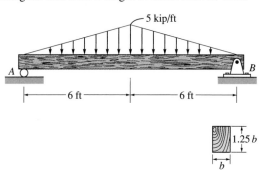

Prob. 16–5

16–6.   The timber beam has a rectangular cross section. If the width of the beam is 6 in., determine its height h so that it simultaneously reaches its allowable bending stress of $\sigma_{allow} = 1.50$ ksi and an allowable shear stress of $\tau_{allow} = 50$ psi. Also, what is the maximum load P that the beam can then support?

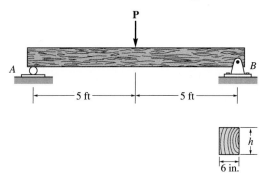

Prob. 16–6

16–7.   Select the lightest-weight steel wide-flange beam from Appendix D that will safely support the loading shown. The allowable bending stress is $\sigma_{allow} = 24$ ksi and the allowable shear stress is $\tau_{allow} = 14$ ksi.

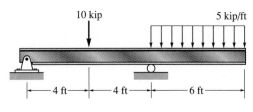

Prob. 16–7

*16–8.   Select the lightest-weight steel wide-flange beam from Appendix D that will safely support the loading shown. The allowable bending stress is $\sigma_{allow} = 24$ ksi and the allowable shear stress is $\tau_{allow} = 14$ ksi.

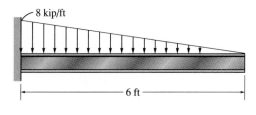

Prob. 16–8

16–9.   Select the lightest-weight steel wide-flange beam from Appendix D that will safely support the loading shown. The allowable bending stress is $\sigma_{allow} = 24$ ksi and the allowable shear stress is $\tau_{allow} = 14$ ksi.

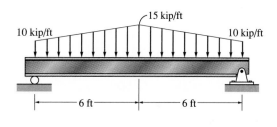

Prob. 16–9

16–10.   Select the lightest-weight steel wide-flange beam from Appendix D that will safely support the loading shown. The allowable bending stress is $\sigma_{allow} = 22$ ksi and the allowable shear stress is $\tau_{allow} = 12$ ksi.

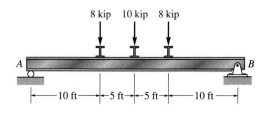

Prob. 16–10

16–11. The simply-supported beam is composed of two $W12 \times 22$ sections built up as shown. Determine the maximum uniform loading w the beam will support if the allowable bending stress is $\sigma_{allow} = 22$ ksi and the allowable shear stress is $\tau_{allow} = 14$ ksi.

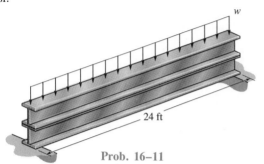

24 ft

Prob. 16–11

*16–12. The beam is to be used to support the machine, which exerts the forces of 6 kip and 8 kip as shown. If the maximum bending stress is not to exceed $\sigma_{allow} = 22$ ksi, determine the required width b of the flanges.

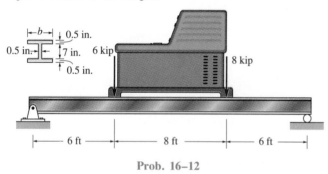

Prob. 16–12

16–13. The steel cantilevered beam is made from two 150-mm by 15-mm plates welded together as shown. Determine the maximum loads P that can be safely supported on the beam if the allowable bending stress is $\sigma_{allow} = 170$ MPa and the allowable shear stress is $\tau_{allow} = 95$ MPa.

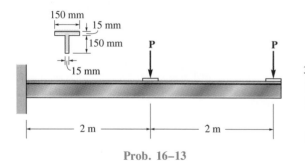

Prob. 16–13

16–14. The steel beam has an allowable bending stress of $\sigma_{allow} = 140$ MPa and an allowable shear stress of $\tau_{allow} = 90$ MPa. Determine the maximum load that it can safely support.

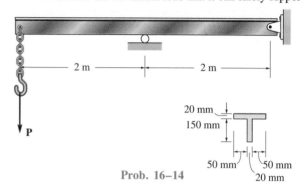

Prob. 16–14

16–15. The simply-supported joist is used in the construction of a floor for a building. In order to keep the floor low with respect to the sill beams C and D, the ends of the joists are notched as shown. If the allowable shear stress for the wood is $\tau_{allow} = 350$ psi and the allowable bending stress is $\sigma_{allow} = 1500$ psi, determine the height h that will cause the beam to reach both allowable stresses at the same time. Also, what load P causes this to happen? Neglect the stress concentration at the notch.

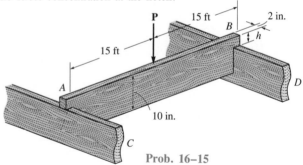

Prob. 16–15

*16–16. The wooden box beam has an allowable bending stress of $\sigma_{allow} = 10$ MPa and an allowable shear stress of $\tau_{allow} = 775$ kPa. Determine the maximum intensity w of the distributed loading that it can safely support. Also, determine the maximum safe nail spacing for each third of the length of the beam. Each nail can resist a shear force of 200 N.

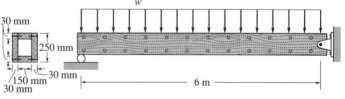

Prob. 16–16

16–17.   The joist AB used in housing construction is to be made from 8-in. by 1.5-in. Southern-pine boards. If the design loading on each board is placed as shown, determine the largest room width L that the boards can span. The allowable bending stress for the wood is σ_{allow} = 2 ksi and the allowable shear stress is τ_{allow} = 180 psi. Assume that the beam is simply-supported from the walls at A and B.

16–18.   The overhang beam is constructed using two 2-in. by 4-in. pieces of wood braced as shown. If the allowable bending stress is σ_{allow} = 600 psi, determine the largest load P that can be applied. Also, determine the associated maximum spacing of nails, s, along the beam section AC if each nail can resist a shear force of 800 lb. Assume that the beam is pin-connected at A, B, and D. Neglect the axial force developed in the beam along DA.

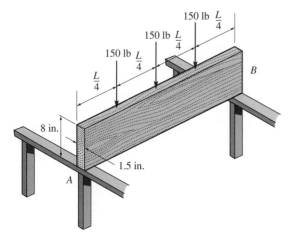

Prob. 16–17

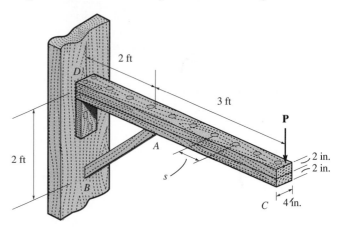

Prob. 16–18

16.4 The Elastic Curve

Having considered the method for designing a beam on the basis of strength, we will now turn our attention to explaining various methods used to determine the slope and deflection at various points along the axis of the beam. Before doing this, it is often helpful to sketch the deflected shape of the beam when it is loaded, in order to "visualize" any computed results and thereby partially check these results. The deflection diagram of the longitudinal axis that passes through the centroid of each cross-sectional area of the beam is called the *elastic curve*. For most beams the elastic curve can be sketched without much difficulty. When doing so, however, it is necessary to know the type of restrictions to slope or displacement that often occur at a support. In general, supports that resist a *force*, such as a pin, restrict *displacement*, and those that resist a *moment*, such as a fixed wall, restrict *rotation* or *slope*. With this in mind, two typical examples of the elastic curves for loaded beams (or shafts), sketched to a greatly exaggerated scale, are shown in Fig. 16–11.

If the elastic curve for a beam seems difficult to establish, it is suggested that the moment diagram for the member be drawn first. Using the beam sign

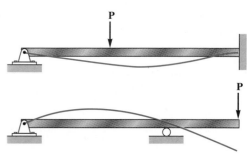

Fig. 16–11

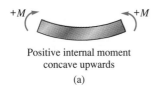

Positive internal moment
concave upwards

(a)

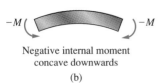

Negative internal moment
concave downwards

(b)

Fig. 16–12

convention established in Chapter 13, a positive internal moment tends to bend the beam concave upward, Fig. 16–12a. Likewise, a negative moment tends to bend the beam concave downward, Fig. 16–12b. Therefore, if the moment diagram is *known,* it will be easy to construct the elastic curve. For example, consider the beam in Fig. 16-13a with its associated moment diagram shown in Fig. 16–13b. Due to the roller and pin supports, the displacement at B and D must be zero. Within the region of negative moment, AC, Fig. 16–13b, the elastic curve must be concave downward, and within the region of positive moment, CD, the elastic curve must be concave upward. Hence, there must be an *inflection point* at point C, where the curve changes from concave up to concave down, since this is a point of zero moment. Using these facts, the beam's elastic curve is sketched to a greatly exaggerated scale in Fig. 16–13c. It should also be noted that the displacements Δ_A and Δ_E are especially critical. At point E the *slope* of the elastic curve is *zero,* and there the beam's *deflection* may be a *maximum.* Whether Δ_E is actually greater than Δ_A depends on the relative magnitudes of $\mathbf{P}_1$ and $\mathbf{P}_2$ and the location of the roller at B.

Following these sample principles, note how the elastic curve is Fig. 16–14 was constructed. Here the beam is cantilevered from a fixed support at A and therefore the elastic curve must have both zero displacement and zero slope at this point. Also, the largest displacement will occur either at D, where the slope is zero, or at C.

Moment–Curvature Relationship. We will now develop an important relation between the internal moment in the beam and the radius of curvature ρ (rho) of the elastic curve at a point. The resulting equation will be used throughout the chapter as a basis for establishing each of the methods for finding the slope and deflection of the elastic curve for a beam.

The following analysis, here and in the next section, will require the use of three coordinates. As shown in Fig. 16–15a, the x axis extends positive to

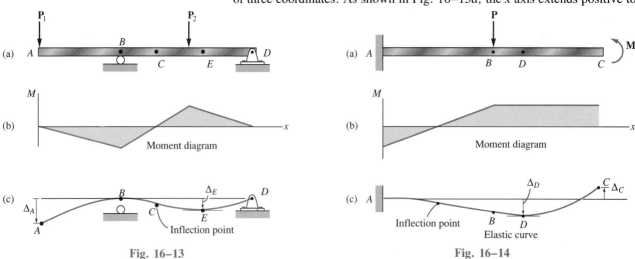

Fig. 16–13

Fig. 16–14

the right, along the initially straight longitudinal axis of the beam. It is used to locate the differential element, having an undeformed width dx. The v axis extends *positive upward* from the x axis. It measures the *displacement* of the centroid on the cross-sectional area of the element. With these two coordinates, we will later define the equation of the elastic curve, v, as a function of x. Lastly, a "localized" y coordinate is used to specify the position of a fiber of the beam element. It is measured *positive upward* from the neutral axis, as shown in Fig. 16–15b. Recall that this same sign convention for x and y was used in the derivation of the flexure formula.

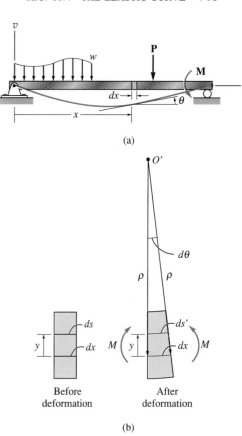

(a)

(b)

Fig. 16–15

In order to derive the relationship between the internal moment and ρ, we will limit the analysis to the most common case of an initially straight beam that is elastically deformed by loads applied perpendicular to the beam's x axis and lying in the x–v plane of symmetry for the beam's cross-sectional area. Due to the loading, the deformation of the beam is caused by both the internal shear force and bending moment. If the beam has a length that is much greater than its depth, the greatest deformation will be caused by bending, and therefore we will direct our attention to its effects.

In Sec. 12.1 we described what happens when the internal moment M deforms the element of the beam in Fig. 16–15b. Recall that the cross sections of the element remain plane, such that the angle between them becomes $d\theta$. The arc that represents a portion of the elastic curve intersects the neutral axis for each cross section, and since it does *not* stretch, it maintains a length dx. The *radius of curvature* for this arc is defined as the distance ρ, which is measured from the *center of curvature O'* to dx. Any arc on the element other than dx is subjected to a normal strain. For example, it was shown in Sec. 12.1 that the strain in arc ds, located at a position y from the neutral axis, can be determined from Eq. 12–1, namely,

$$\frac{1}{\rho} = -\frac{\epsilon}{y} \qquad (16\text{–}2)$$

If the material is homogeneous and behaves in a linear-elastic manner, then $\epsilon = \sigma/E$. Also, since the flexure formula applies, $\sigma = -My/I$. Combining these equations and substituting into Eq. 16–2, we have

$$\boxed{\frac{1}{\rho} = \frac{M}{EI}} \qquad (16\text{–}3)$$

where

ρ = the radius of curvature at a specific point on the elastic curve ($1/\rho$ is referred to as the *curvature*)
M = the internal moment in the beam at the point where ρ is to be determined
E = the material's modulus of elasticity
I = the beam's moment of inertia computed about the neutral axis

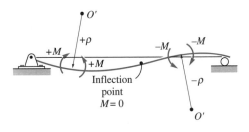

Fig. 16–16

The product EI in this equation is referred to as the *flexural rigidity,* and it is always a positive quantity. The sign for ρ therefore depends on the direction of the moment. As shown in Fig. 16–16, when M is *positive,* ρ extends *above* the beam, i.e., in the positive v direction; when M is *negative,* ρ extends *below* the beam, or in the negative v direction.

Using the flexure formula, $\sigma = -My/I$, we can also express the curvature in terms of the stress in the beam, namely,

$$\frac{1}{\rho} = -\frac{\sigma}{Ey} \tag{16–4}$$

Both Eqs. 16–3 and 16–4 are valid for either small or large radii of curvature. However, the value of ρ is almost always calculated as a *very large quantity.* For example, consider a steel beam made from a $W14 \times 53$ (Appendix D), where $E_{st} = 29(10^3)$ ksi and $\sigma_Y = 36$ ksi. When the material at the outer fibers, $y = \pm 7$ in., is about to *yield,* then, from Eq. 16–3, $\rho = \pm 5639$ in. Values of ρ calculated at other points along the beam's elastic curve may be even *larger,* since σ cannot exceed σ_Y at the outer fibers.

16.5 Slope and Displacement by Integration

The elastic curve for a beam can be expressed mathematically as $v = f(x)$. To obtain this equation, we must first represent the curvature $(1/\rho)$ in terms of v and x. In most calculus books it is shown that this relationship is

$$\frac{1}{\rho} = \frac{d^2v/dx^2}{[1 + (dv/dx)^2]^{3/2}}$$

Substituting into Eq. 16–3, we get

$$\frac{d^2v/dx^2}{[1 + (dv/dx)^2]^{3/2}} = \frac{M}{EI} \tag{16–5}$$

This equation represents a nonlinear second-order differential equation. Its solution, which is called the *elastica,* gives the exact shape of the elastic curve, assuming, of course, that beam deflections occur only due to bending. Through the use of higher mathematics, elastica solutions have been obtained only for simple cases of beam geometry and loading.

In order to facilitate the solution of a greater number of deflection problems, Eq. 16–5 can be modified. Most engineering design codes specify *limitations* on deflections for tolerance or esthetic purposes, and as a result the elastic deflections for the majority of beams and shafts form a shallow curve. Consequently, the slope of the elastic curve which is determined from dv/dx will be *very small,* and its square will be negligible compared with unity.* Therefore the curvature, as defined above, can be approximated by $1/\rho = d^2v/dx^2$. Using this simplification, Eq. 16–5 can now be written as

$$\frac{d^2v}{dx^2} = \frac{M}{EI} \tag{16–6}$$

*See Example 12–1.

It is also possible to write this equation in two alternative forms. If we differentiate each side with respect to x and substitute $V = dM/dx$ (Eq. 7–2), we get

$$\frac{d}{dx}\left(EI\,\frac{d^2v}{dx^2}\right) = V(x) \qquad (16\text{–}7)$$

Differentiating again, using $-w = dV/dx$ (Eq. 7–1), yields

$$\frac{d^2}{dx^2}\left(EI\,\frac{d^2v}{dx^2}\right) = -w(x) \qquad (16\text{–}8)$$

For most problems the flexural rigidity will be constant along the length of the beam. Assuming this to be the case, the above results may be reordered into the following set of equations:

$$EI\,\frac{d^4v}{dx^4} = -w(x) \qquad (16\text{–}9)$$

$$EI\,\frac{d^3v}{dx^3} = V(x) \qquad (16\text{–}10)$$

$$EI\,\frac{d^2v}{dx^2} = M(x) \qquad (16\text{–}11)$$

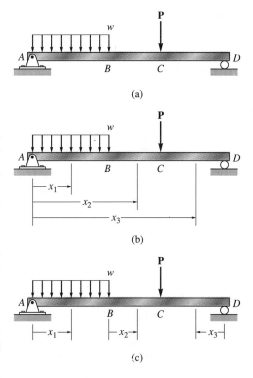

(a)

(b)

(c)

Fig. 16–17

Solution of any of these equations requires successive integrations to obtain the deflection v of the elastic curve. For each integration it is necessary to introduce a "constant of integration" and then solve for all the constants to obtain a unique solution for a particular problem. For example, if the distributed load is expressed as a function of x and Eq. 16–9 is used, then four constants of integration must be evaluated; however, if the internal moment M is determined and Eq. 16–11 is used, only two constants of integration must be found. The choice of which equation to start with depends on the problem. Generally, however, it is easier to determine the internal moment M as a function of x, integrate twice, and evaluate only two integration constants.

Recall from Sec. 13.2 that if the loading on a beam is discontinuous, that is, consists of a series of several distributed and concentrated loads, then several functions must be written for the internal moment, each valid within the region between the discontinuities. Also, for convenience in writing each moment expression, *the origin for each x coordinate can be selected arbitrarily.* For example, consider the beam shown in Fig. 16–17a. The internal moment in regions *AB, BC,* and *CD* can be written in terms of the x_1, x_2, and x_3 coordinates selected, as shown in either Fig. 16–17b or 16–17c, or in fact in any manner that will yield $M = f(x)$ in as simple a form as possible. Once these functions are integrated through the use of Eq. 16–11 and the constants of integration determined, the functions will give the slope and deflection (elastic curve) for each region of the beam for which they are valid.

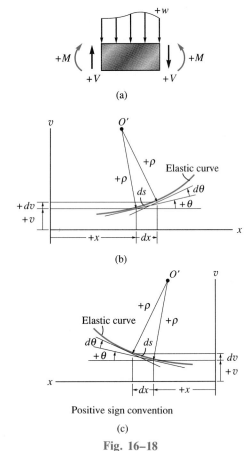

Fig. 16–18

Sign Convention and Coordinates.

When applying Eqs. 16-9 through 16-11, it is important to use the proper signs for M, V, or w as established by the sign convention that was used in the derivation of these equations. For review, these terms are shown in their *positive directions* in Fig. 16–18a. Furthermore, recall that *positive deflection, v*, is *upward*, and as a result, the *positive slope angle θ* will be measured *counterclockwise* from the x axis when x is *positive to the right*. The reason for this is shown in Fig. 16–18b. Here positive increases dx and dv in x and v create an increased θ that is counterclockwise. On the other hand, if *positive x* is directed to the *left*, then θ will be *positive clockwise*, Fig. 16–18c.

It should be pointed out that by assuming dv/dx to be very small, the original horizontal length of the beam's axis and the arc of its elastic curve will be about the same. In other words, ds in Fig. 16–18b and 16–18c is approximately equal to dx, since $ds = \sqrt{(dx)^2 + (dv)^2} = \sqrt{1 + (dv/dx)^2}\ dx \approx dx$. As a result, points on the elastic curve will only be *displaced vertically*, and not horizontally. Also, since the *slope angle θ* will be *very small*, its value in radians can be determined *directly* from $\theta \approx \tan\theta = dv/dx$.

Boundary and Continuity Conditions.

The constants of integration are determined by evaluating the functions for shear, moment, slope, or displacement at a particular point on the beam where the value of the function is known. These values are called *boundary conditions*. Several possible boundary conditions that are often used to solve beam deflection problems are listed in Table 16–1. For example, if the beam is supported by a roller or pin (1, 2, 3, 4), then it is required that the displacement be *zero* at these points. Furthermore, if these supports are located at the *ends of the beam* (1, 2), the internal moment in the beam must also be zero. At the fixed support (5) the slope and displacement are both zero, whereas the free-ended beam (6) has both zero moment and zero shear. Lastly, if two segments of a beam are connected by an "internal" pin or hinge (7), the moment must be zero at this connection.

If a single x coordinate cannot be used to express the equation for the beam's slope or the elastic curve, then *continuity conditions* must be used to evaluate some of the integration constants. For example, consider the beam in Fig. 16–19a. Here the x coordinates are both chosen with origins at A. Each is valid only within the regions $0 \le x_1 \le a$ and $a \le x_2 \le (a + b)$. Once the functions for the slope and deflection are obtained, they must give the *same values* for the slope and deflection at point B so the elastic curve is physically *continuous*. Expressed mathematically, this requires that $\theta_1(a) = \theta_2(a)$ and $v_1(a) = v_2(a)$. These equations can then be used to evaluate two constants of integration. The elastic curve can also be expressed in terms of the coordinates $0 \le x_1 \le a$ and $0 \le x_2 \le b$, shown in Fig. 16–19b. Here continuity of slope and deflection at B requires $\theta_1(a) = -\theta_2(b)$ and $v_1(a) = v_2(b)$. In this particular case, a *negative* sign is necessary to match the slopes at B since x_1 extends positive to the right, whereas x_2 extends positive to the left. Consequently, θ_1 is positive counterclockwise, and θ_2 is positive clockwise. See Fig. 16–18b and 16–18c.

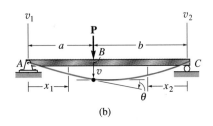

Fig. 16–19

PROCEDURE FOR ANALYSIS

The following procedure provides a method for determining the slope and deflection of a beam using the method of integration. It should be realized that this method is suitable only for *elastic deflections* such that the beam's slope is very small. Furthermore, the method considers *only deflections due to bending*. Additional deflection due to shear generally represents only a few percent of the bending deflection, and so it is usually neglected in engineering practice.

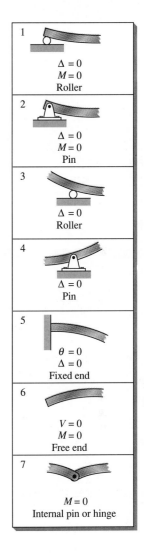

Table 16–1

Elastic Curve. Draw an exaggerated view of the beam's elastic curve. Recall that points of zero slope and zero displacement occur at a fixed support, and zero displacement occurs at all pin and roller supports.

Establish the x and v coordinate axes. The x axis must be parallel to the undeflected beam and can have an origin at any point along the beam, with a positive direction either to the right or to the left. If several discontinuous loads are present, establish x coordinates that are valid for each region of the beam between the discontinuities. Choose these coordinates so that they will simplify subsequent algebraic work. In all cases, the associated positive v axis should be directed upward.

Load or Moment Function. For each region in which there is an x coordinate, express the loading w or the internal moment M as a function of x. In particular, always assume that M acts in the *positive direction* when applying the equation of moment equilibrium to determine $M = f(x)$.

Slope and Elastic Curve. Provided EI is constant, apply either the load equation $EI\,d^4v/dx^4 = -w(x)$, which requires four integrations to get $v = v(x)$, or the moment equation $EI\,d^2v/dx^2 = M(x)$, which requires only two integrations. For each integration it is important to include a constant of integration. The constants are evaluated using the boundary conditions for the supports (Table 16-1) and the continuity conditions that apply to slope and displacement at points where two functions meet. Once the constants are evaluated and substituted back into the slope and deflection equations, the slope and displacement at *specific points* on the elastic curve can then be determined. The numerical values obtained can be checked graphically by comparing them with the sketch of the elastic curve. Realize that *positive* values for *slope* are *counterclockwise* if the x axis extends *positive* to the *right*, and *clockwise* if the x axis extends *positive* to the *left*. In either of these cases, *positive displacement* is *upward*, Fig. 16–18.

The following examples illustrate application of this procedure.

Example 16–5

The cantilevered beam shown in Fig. 16–20a is subjected to a vertical load **P** at its end. Determine the equation of the elastic curve. *EI* is constant.

SOLUTION I

Elastic Curve.   The load tends to deflect the beam as shown in Fig. 16–20a. By inspection, the internal moment can be represented throughout the beam using a single x coordinate.

Moment Function.   From the free-body diagram, with **M** acting in the *positive direction,* Fig. 16–20b, we have

$$M = -Px$$

Slope and Elastic Curve.   Applying Eq. 16–11 and integrating twice, yields

$$EI\frac{d^2v}{dx^2} = -Px \tag{1}$$

$$EI\frac{dv}{dx} = -\frac{Px^2}{2} + C_1 \tag{2}$$

$$EI\,v = -\frac{Px^3}{6} + C_1x + C_2 \tag{3}$$

Using the boundary conditions $dv/dx = 0$ at $x = L$ and $v = 0$ at $x = L$, Eqs. 2 and 3 become

$$0 = -\frac{PL^2}{2} + C_1$$

$$0 = -\frac{PL^3}{6} + C_1L + C_2$$

Thus, $C_1 = PL^2/2$ and $C_2 = -PL^3/3$. Substituting these results into Eqs. 2 and 3 with $\theta = dv/dx$, we get

$$\theta = \frac{P}{2EI}(L^2 - x^2)$$

$$v = \frac{P}{6EI}(-x^3 + 3L^2x - 2L^3) \qquad\qquad Ans.$$

Maximum slope and displacement occur at $A(x = 0)$, for which

$$\theta_A = \frac{PL^2}{2EI} \tag{4}$$

$$v_A = -\frac{PL^3}{3EI} \tag{5}$$

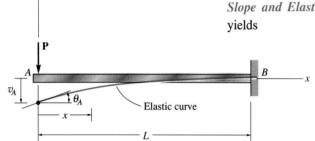

v

P

A

v_A

θ_A

Elastic curve

x

B

x

L

(a)

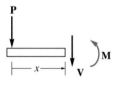

P

x

V

M

(b)

Fig. 16–20

The *positive* result for θ_A indicates *counterclockwise* rotation and the *negative* result for v_A indicates that v_A is *downward*. This agrees with the results sketched in Fig. 16–20a.

In order to obtain some idea as to the actual *magnitude* of the slope and displacement at the end *A*, consider the beam in Fig. 16–20a to have a length of 15 ft, support a load of $P = 6$ kip, and be made of steel having $E_{st} = 29(10^3)$ ksi. Using the methods of Sec. 16–3, if this beam was designed without a factor of safety by assuming the allowable normal stress is equal to the yield stress $\sigma_{\text{allow}} = 36$ ksi, then a $W16 \times 26$ would be found to be adequate ($I = 204$ in^4). From Eqs. 4 and 5 we get

$$\theta_A = \frac{6 \text{ kip}(15 \text{ ft})^2(16 \text{ in./ft})^2}{2[29(10^3) \text{ kip/in}^2](204 \text{ in}^4)} = 0.0164 \text{ rad}$$

$$v_A = -\frac{6 \text{ kip}(15 \text{ ft})^3(16 \text{ in./ft})^3}{3[29(10^3) \text{ kip/in}^2](204 \text{ in}^4)} = -1.97 \text{ in.}$$

Since $\theta_A^2 = 0.000269 \text{ rad}^2 \ll 1$, this justifies the use of Eq. 16–5, rather than applying the more exact Eq. 16–4, for computing the deflection of beams. Also, since this numerical application is for a *cantilevered beam*, we have obtained *larger values* for θ and v than would have been obtained if the beam was supported using pins, rollers, or other fixed supports.

SOLUTION II

This problem can also be solved using Eq. 16–9, $EI \, d^4v/dx^4 = -w(x)$. Here $w(x) = 0$ for $0 \le x \le L$, Fig. 16–20a, so that upon integrating once we get the form of Eq. 16–10, i.e.,

$$EI \frac{d^4v}{dx^4} = 0$$

$$EI \frac{d^3v}{dx^3} = C_1' = V$$

The shear constant C_1' can be evaluated at $x = 0$, since $V_A = -P$ (negative according to the beam sign convention, Fig. 16–18a). Thus, $C_1' = -P$. Integrating again yields the form of Eq. 16–11, i.e.,

$$EI \frac{d^3v}{dx^3} = -P$$

$$EI \frac{d^2v}{dx^2} = -Px + C_2' = M$$

Here $M = 0$ at $x = 0$, so $C_2' = 0$, and as a result one obtains Eq. 1 and the solution proceeds as before.

Example 16–6

The simply-supported beam shown in Fig. 16–21a supports the triangular distributed loading. Determine its maximum deflection. EI is constant.

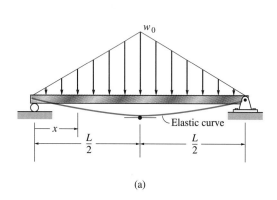

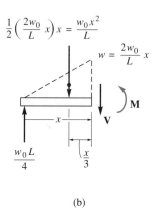

(a)

(b)

Fig. 16–21

SOLUTION I

Elastic Curve. Due to symmetry, only one x coordinate is needed for the solution. The beam deflects as shown in Fig. 16–21a. Notice that maximum deflection occurs at the center since the slope is zero at this point.

Moment Function. The distributed load acts downward, and therefore it is positive according to our sign convention. A free-body diagram of the segment on the left is shown in Fig. 16–21b. The equation for the distributed loading is

$$w = \frac{2w_0}{L}x$$

Hence,

$$\zeta + \Sigma M_{NA} = 0; \qquad M + \frac{w_0 x^2}{L}\left(\frac{x}{3}\right) - \frac{w_0 L}{4}(x) = 0$$

$$M = \frac{-w_0 x^3}{3L} + \frac{w_0 L}{4}x$$

Slope and Elastic Curve. Using Eq. 16–11 and integrating twice, we have

$$EI\frac{d^2v}{dx^2} = M = -\frac{w_0}{3L}x^3 + \frac{w_0L}{4}x \tag{1}$$

$$EI\frac{dv}{dx} = -\frac{w_0}{12L}x^4 + \frac{w_0L}{8}x^2 + C_1$$

$$EIv = -\frac{w_0}{60L}x^5 + \frac{w_0L}{24}x^3 + C_1x + C_2$$

The constants of integration are obtained by applying the boundary condition $v = 0$ at $x = 0$ and the symmetry condition that $dv/dx = 0$ at $x = L/2$. This leads to

$$C_1 = -\frac{5w_0L^3}{192} \qquad C_2 = 0$$

Hence for $0 \le x \le L/2$,

$$EI\frac{dv}{dx} = -\frac{w_0}{12L}x^4 + \frac{w_0L}{8}x^2 - \frac{5w_0L^-}{192}$$

$$EIv = -\frac{w_0}{60L}x^5 + \frac{w_0L}{24}x^3 - \frac{5w_0L^3}{192}x$$

Computing the maximum deflection at $x = L/2$, we have

$$v_{max} = -\frac{w_0L^4}{120EI} \qquad\qquad \textit{Ans.}$$

SOLUTION II
Starting with the distributed loading, Eq. 1, and applying Eq. 16–9, we have

$$EI\frac{d^4v}{dx^4} = -\frac{2w_0}{L}x$$

$$EI\frac{d^3v}{dx^3} = V = -\frac{w_0}{L}x^2 + C_1'$$

Since $V = +w_0L/4$ at $x = 0$, then $C_1' = w_0L/4$. Integrating again yields

$$EI\frac{d^3v}{dx^3} = V = -\frac{w_0}{L}x^2 + \frac{w_0L}{4}$$

$$EI\frac{d^2v}{dx^2} = M = -\frac{w_0}{3L}x^3 + \frac{w_0L}{4}x + C_2'$$

Here $M = 0$ at $x = 0$, so $C_2' = 0$. This yields Eq. 1. The solution now proceeds as before.

Example 16–7

The simply-supported beam shown in Fig. 16–22a is subjected to the concentrated force **P.** Determine the maximum deflection of the beam. *EI* is constant.

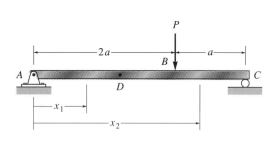

(a)

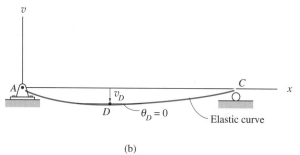

(b)

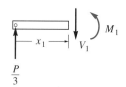

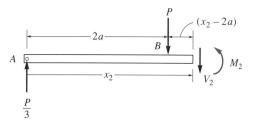

(c)

Fig. 16–22

SOLUTION

Elastic Curve. The beam deflects as shown in Fig. 16–22b. Two coordinates must be used, since the moment becomes discontinuous at *P*. Here we will take x_1 and x_2 having the *same origin* at *A*, so that $0 \leq x_1 < 2a$ and $2a < x_2 \leq 3a$.

Moment Function. From the free-body diagrams shown in Fig. 16–22c,

$$M_1 = \frac{P}{3}x_1$$

$$M_2 = \frac{P}{3}x_2 - P(x_2 - 2a) = \frac{2P}{3}(3a - x_2)$$

Slope and Elastic Curve. Applying Eq. 16–11 for M_1 and integrating twice yields

$$EI\frac{d^2v_1}{dx_1^2} = \frac{P}{3}x_1$$

$$EI\frac{dv_1}{dx_1} = \frac{P}{6}x_1^2 + C_1 \tag{1}$$

$$EIv_1 = \frac{P}{18}x_1^3 + C_1x_1 + C_2 \tag{2}$$

Likewise for M_2,

$$EI\frac{d^2v_2}{dx_2^2} = \frac{2P}{3}(3a - x_2)$$

$$EI\frac{dv_2}{dx_2} = \frac{2P}{3}\left(3ax_2 - \frac{x_2^2}{2}\right) + C_3 \tag{3}$$

$$EIv_2 = \frac{2P}{3}\left(\frac{3}{2}ax_2^2 - \frac{x_2^3}{6}\right) + C_3x_2 + C_4 \tag{4}$$

The four constants are evaluated using *two* boundary conditions, namely, $x_1 = 0$, $v_1 = 0$ (Eq. 2) and $x_2 = 3a$, $v_2 = 0$ (Eq. 4). Also, *two* continuity conditions must be applied at B, that is, $dv_1/dx_1 = dv_2/dx_2$ at $x_1 = x_2 = 2a$ (Eqs. 1 and 3) and $v_1 = v_2$ at $x_1 = x_2 = 2a$ (Eqs. 2 and 4). Substitution as specified results in the following four equations:

$v_1 = 0$ at $x_1 = 0$; $0 = 0 + 0 + C_2$

$v_2 = 0$ at $x_2 = 3a$; $0 = \dfrac{2P}{3}\left(\dfrac{3}{2}a(3a)^2 - \dfrac{(3a)^3}{6}\right) + C_3(3a) + C_4$

$\dfrac{dv_1(2a)}{dx} = \dfrac{dv_2(2a)}{dx}$; $\dfrac{P}{6}(2a)^2 + C_1 = \dfrac{2P}{3}\left(3a(2a) - \dfrac{(3a)^2}{2}\right) + C_3$

$v_1(2a) = v_2(2a)$; $\dfrac{P}{18}(2a)^3 + C_1(2a) + C_2 = \dfrac{2P}{3}\left(\dfrac{3}{2}a(2a)^2 - \dfrac{(2a)^3}{6}\right) + C_3(2a) + C_4$

Solving these equations we get

$$C_1 = -\frac{4}{9}Pa^2 \qquad C_2 = 0$$

$$C_3 = -\frac{22}{9}Pa^2 \qquad C_4 = \frac{4}{3}Pa^3$$

Thus Eqs. 1–4 become

$$\frac{dv_1}{dx_1} = \frac{P}{6EI}x_1{}^2 - \frac{4}{9}\frac{Pa^2}{EI} \tag{5}$$

$$v_1 = \frac{P}{18EI}x_1{}^3 - \frac{4}{9}\frac{Pa^2}{EI}x_1 \tag{6}$$

$$\frac{dv_2}{dx_2} = \frac{2Pa}{EI}x_2 - \frac{P}{3EI}x_2{}^2 - \frac{22}{9}\frac{Pa^2}{EI} \tag{7}$$

$$v_2 = \frac{Pa}{EI}x_2{}^2 - \frac{P}{9EI}x_2{}^3 - \frac{22}{9}\frac{Pa^2}{EI}x_2 + \frac{4}{3}\frac{Pa^3}{EI} \tag{8}$$

By inspection of the elastic curve, Fig. 16–22b, the maximum deflection occurs at D, somewhere within region AB. Here the slope must be zero. From Eq. 5,

$$\frac{1}{6}x_1{}^2 - \frac{4}{9}a^2 = 0$$

$$x_1 = 1.633a$$

Substituting into Eq. 6,

$$v_{\text{max}} = -0.484\frac{Pa^3}{EI} \qquad\qquad Ans.$$

The negative sign indicates that the deflection is downward.

Example 16–8

The beam in Fig. 16–23a is subjected to a load **P** at its end. Determine the displacement at *C*. *EI* is constant.

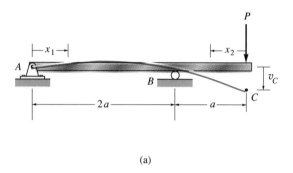

(a)

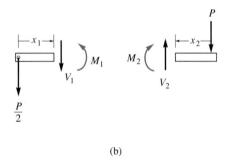

(b)

Fig. 16–23

SOLUTION

Elastic Curve.   The beam deflects into the shape shown in Fig. 16–23a. Due to the loading, two *x* coordinates may be considered. x_2 is directed to the left from *C*, since the internal moment is easy to formulate.

Moment Functions.   Using the free-body diagrams shown in Fig. 16–23b, we have

$$M_1 = -\frac{P}{2}x_1 \qquad M_2 = -Px_2$$

Slope and Elastic Curve.   Applying Eq. 16–11,

for x_1,

$$EI \frac{d^2 v_1}{dx_1^2} = -\frac{P}{2} x_1$$

$$EI \frac{dv_1}{dx_1} = -\frac{P}{4} x_1^2 + C_1 \tag{1}$$

$$EIv_1 = -\frac{P}{12} x_1^3 + C_1 x + C_2 \tag{2}$$

For x_2,

$$EI \frac{d^2 v_2}{dx_2^2} = -P x_2$$

$$EI \frac{dv_2}{dx_2} = -\frac{P}{2} x_2^2 + C_3 \tag{3}$$

$$EIv_2 = -\frac{P}{6} x_2^3 + C_3 x_2 + C_4 \tag{4}$$

The *four* constants of integration are determined using *three* boundary conditions, namely, $v_1 = 0$ at $x_1 = 0$, $v_1 = 0$ at $x_1 = 2a$, and $v_2 = 0$ at $x_2 = a$ and *one* continuity equation. Here the continuity of slope at the roller requires $dv_1/dx_1 = -dv_2/dx_2$ at $x_1 = 2a$ and $x_2 = a$. Why is there a negative sign in this equation? Note that continuity of displacement at B has been indirectly considered in the boundary conditions, since $v_1 = v_2 = 0$ at $x_1 = 2a$ and $x_2 = a$. Applying these four conditions yields

$v_1 = 0$ at $x_1 = 0$; $0 = 0 + 0 + C_2$

$v_1 = 0$ at $x_1 = 2a$; $0 = -\dfrac{P}{12}(2a)^3 + C_1(2a) + C_2$

$v_2 = 0$ at $x_2 = a$; $0 = -\dfrac{P}{6} a^3 + C_3 a + C_4$

$\dfrac{dv_1(2a)}{dx_1} = -\dfrac{dv_2(a)}{dx_2}$; $-\dfrac{P}{4}(2a)^2 + C_1 = -\left(-\dfrac{P}{2}(a)^2 + C_3 \right)$

Solving, we obtain

$$C_1 = \frac{Pa^2}{3} \qquad C_2 = 0 \qquad C_3 = \frac{7}{6} Pa^2 \qquad C_4 = -Pa^3$$

Substituting C_3 and C_4 into Eq. 4 gives

$$v_2 = -\frac{P}{6EI} x_2^3 + \frac{7Pa^2}{6EI} x_2 - \frac{Pa^3}{EI}$$

The displacement at C is determined by setting $x_2 = 0$. We get

$$v_C = -\frac{Pa^3}{EI} \qquad\qquad\qquad \textit{Ans.}$$

PROBLEMS

16–19.   A steel strap having a thickness of 0.125 in. and a width of 2 in. is bent into a circular arc of radius $\rho = 600$ in. Determine the maximum bending stress in the strap. $E_{st} = 29(10^3)$ ksi.

*16–20.   The steel blade of the band saw wraps around the pulley having a radius of 12 in. Determine the maximum normal stress in the blade. The blade is made of steel having a width of 0.75 in. and a thickness of 0.0625 in. $E_{st} = 29(10^3)$ ksi.

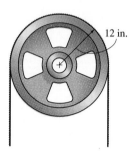

Prob. 16–20

16–21.   Determine the equation of the elastic curve for the beam using the x coordinate that is valid for $0 < x < L/2$. Specify the slope at A and the maximum deflection. EI is constant.

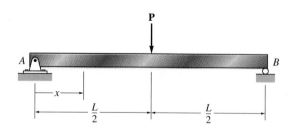

Prob. 16–21

16–22.   The shaft is supported at A by a journal bearing that exerts only vertical reactions on the shaft, and at B by a thrust bearing that exerts horizontal and vertical reactions on the shaft. Draw the bending-moment diagram for the shaft and then, from this diagram, sketch the deflection or elastic curve for the shaft's centerline. Determine the equations of the elastic curve using the coordinates x_1 and x_2.

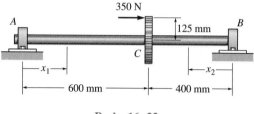

Prob. 16–22

16–23.   The steel beam has a depth of 10 in. and is subjected to a constant moment $\mathbf{M}_0$, which causes the stress at the outer fibers to become $\sigma_Y = 36$ ksi. Determine the radius of curvature of the beam and the deflection at its end B. $E_{st} = 29(10^3)$ ksi. *Comment:* Notice how *small* the maximum elastic deflection actually is.

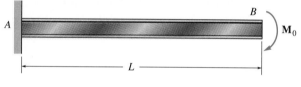

Prob. 16–23

*16–24.   Determine the equations of the elastic curve for the beam using the x_1 and x_2 coordinates. Specify the slope at A and the maximum deflection. EI is constant.

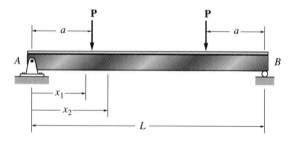

Prob. 16–24

16–25. The shaft is supported at A by a journal bearing that exerts only vertical reactions on the shaft, and at B by a thrust bearing that exerts horizontal and vertical reactions on the shaft. Draw the bending-moment diagram for the shaft and then, from this diagram, sketch the deflection or elastic curve for the shaft's centerline. Determine the equations of the elastic curve using the coordinates x_1 and x_2.

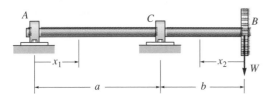

Prob. 16–25

16–26. The shaft is supported at A by a journal bearing that exerts only vertical reactions on the shaft, and at B by a thrust bearing that exerts both horizontal and vertical reactions on the shaft. Draw the bending-moment diagram for the shaft and then, from this diagram, sketch the deflection or elastic curve for the shaft's centerline. Determine the equations of the elastic curve using the coordinates x_1 and x_2. EI is constant.

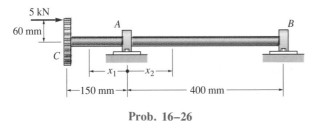

Prob. 16–26

16–27. The shaft is supported by a journal bearing at A, which exerts only vertical reactions on the shaft, and by a thrust bearing at B, which exerts both horizontal and vertical reactions on the shaft. Draw the bending-moment diagram for the shaft and then, from this diagram, sketch the deflection or elastic curve for the shaft's centerline. Determine the equations of the elastic curve using the coordinates x_1 and x_2. EI is constant.

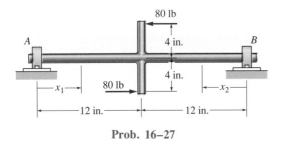

Prob. 16–27

16–28. Determine the elastic curve for the cantilevered beam, which is subjected to the couple moment $\mathbf{M}_0$. Also compute the maximum slope and maximum deflection of the beam. EI is constant.

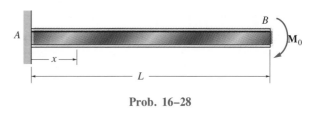

Prob. 16–28

16–29. Determine the elastic curve for the simply-supported beam, which is subjected to a couple moment $\mathbf{M}_0$. Also, compute the slope at each end A and B and the maximum deflection of the beam. EI is constant.

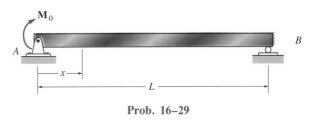

Prob. 16–29

16–30. Determine the elastic curve for the simply-supported beam, which is subjected to the couple moments $\mathbf{M}_0$. Also, compute the maximum slope and the maximum deflection of the beam. EI is constant.

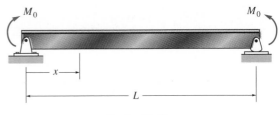

Prob. 16–30

16–31.   Determine the equations of the elastic curve for the beam using the x_1 and x_2 coordinates. Specify the slope at A and the deflection at C. EI is constant.

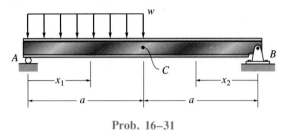

Prob. 16–31

*16–32.   Determine the equations of the elastic curve for the beam using the x_1 and x_2 coordinates. Specify the slope at A and the maximum deflection. EI is constant.

16–33.   Determine the maximum deflection between the supports A and B. EI is constant.

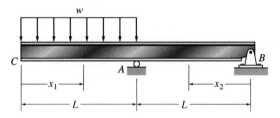

Probs. 16–32/16–33

16–34.   Determine the equations of the elastic curve using the coordinate x and specify the deflection and slope at point C. EI is constant.

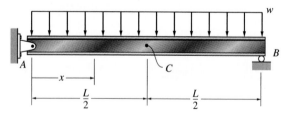

Prob. 16–34

16–35.   The floor beam of the airplane is subjected to the loading shown. Assuming that the fuselage exerts only vertical reactions on the ends of the beam, determine the maximum deflection of the beam. EI is constant.

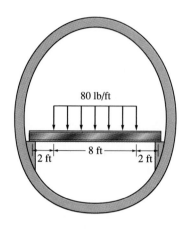

Prob. 16–35

*16–36.   Determine the equations of the elastic curve for the beam using the x coordinate. Specify the slope and deflection at A. EI is constant.

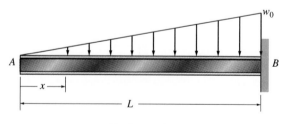

Prob. 16–36

16–37. Wooden posts used for a retaining wall have a diameter of 3 in. If the soil pressure along a post varies uniformly from zero at the top A to a maximum of 300 lb/ft at the bottom B, determine the slope and deflection at the top of the post. $E_w = 1.6(10^3)$ ksi.

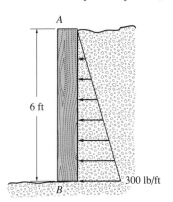

Prob. 16–37

16–38. At what distance a should the bearing supports at A and B be placed so that the deflection at the center of the shaft is equal to the deflection at its ends? EI is constant.

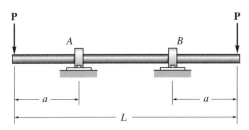

Prob. 16–38

16–39. The two wooden meter sticks are separated at their centers by a smooth rigid cylinder having a diameter of 50 mm. Determine the force F that must be applied at each end in order to just make their ends touch. Each stick has a width of 20 mm and a thickness of 5 mm. $E_w = 11$ GPa.

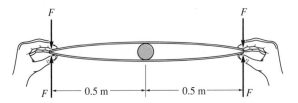

Prob. 16–39

*16–40. The fence board weaves between the three smooth fixed posts, each of which has a diameter of 3 in. If the posts remain along the same line, determine the maximum bending stress in the board. The board has a width of 6 in. and a thickness of 0.5 in. $E_w = 1.60(10^3)$ ksi. Assume the deflection of each end of the board relative to its center is 3 in.

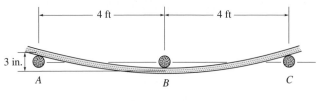

Prob. 16–40

16–41. The tapered beam has a rectangular cross section. Determine the deflection of its end in terms of the load P, length L, modulus of elasticity E, and the moment of inertia I_0 of its end A.

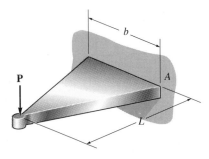

Prob. 16–41

16–42. The tapered beam has a rectangular cross section. Determine the deflection of its center in terms of the load P, length L, modulus of elasticity E, and the moment of inertia I_c of its center.

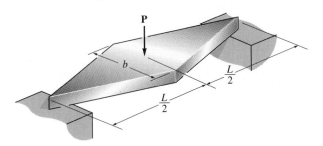

Prob. 16–42

*16.6 Discontinuity Functions

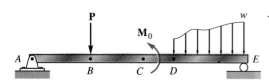

Fig. 16–24

The method of integration, to find the equation of the elastic curve for a beam or shaft, is convenient if the internal moment can be expressed as a *continuous function* throughout the beam's *entire length*. If several loadings act on the beam, however, the method becomes more tedious to apply, because separate moment functions must be written for each region between these loadings. Furthermore, integration of these moment functions requires the evaluation of integration constants using boundary conditions and/or continuity conditions. For example, the beam shown in Fig. 16–24 requires *four* moment functions to be written. They describe the moment in regions *AB*, *BC*, *CD*, and *DE*. Applying the moment-curvature relationship, $EI\,d^2v/dx^2 = M$, and integrating each moment equation twice, requires the evaluation of eight constants of integration. These are evaluated using *two* boundary conditions that require zero displacement at points *A* and *E*, and *six* continuity conditions of both slope and displacement at points *B*, *C*, and *D*.

In this section we will discuss a method for finding the equation of the elastic curve for a multiply-loaded beam using a *single expression* to define the beam's internal moment. When this expression is substituted into the moment-curvature relationship and integrated twice, the *two* constants of integration will be determined *only* from the *boundary conditions*. Since the continuity equations are not involved, the analysis will be greatly simplified.

In order to express the internal moment in the beam as a single function, we will use a special mathematical operator known as a *discontinuity function*, which for purposes of beam deflection may be written as

$$\langle x - a\rangle^n = \begin{cases} 0 & \text{for } x < a \\ (x - a)^n & \text{for } x \geq a \end{cases} \qquad (16\text{–}12)$$
$$n \geq 0$$

Here x represents the coordinate position of a point along the beam and a is the location on the beam where a "discontinuity" occurs, such as the point of application of an external force or couple moment or the start of a distributed loading. Note that the discontinuity function $\langle x - a\rangle^n$ is written with angle brackets to distinguish it from the ordinary function $(x - a)^n$, written with parentheses. As stated in Eq. 16–12, only when $x \geq a$ is $\langle x - a\rangle^n = (x - a)^n$, otherwise it is zero. In particular, integration of the discontinuity function follows the same rules as for ordinary functions, i.e.,

$$\int \langle x - a\rangle^n \, dx = \frac{\langle x - a\rangle^{n+1}}{n + 1} + C \qquad (16\text{–}13)$$

Discontinuity functions can be developed for various types of loadings applied to a beam. We will now consider four that are often encountered in engineering practice.

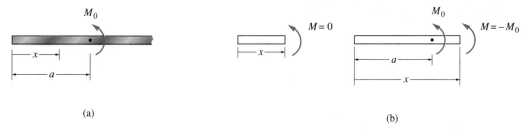

(a)

(b)

Fig. 16–25

Couple Moment M_0.   When an *external* couple moment $\mathbf{M_0}$ is applied to a beam at $x = a$, Fig. 16–25a, the *internal moment* M in the beam can be represented by the discontinuity function

$$M = -M_0\langle x - a\rangle^0 \qquad\qquad (16\text{--}14)$$

This equation states that when $x < a$, $M = 0$, and when $x \ge a$, $M = -M_0(x - a)^0 = -M_0$. The validity of these values can be checked by using the method of sections, shown in Fig. 16–25b. Note that application of the equation indicates that a *counterclockwise couple moment* $\mathbf{M_0}$ yields a *negative* expression for M, whereas M would be *positive* if $\mathbf{M_0}$ is *clockwise*.

Concentrated Force P.   If a concentrated force $\mathbf{P}$ is applied to the beam at $x = a$, Fig. 16–26a, then the discontinuity function for the internal moment in the beam is

$$M = -P\langle x - a\rangle \qquad\qquad (16\text{--}15)$$

Here when $\mathbf{P}$ acts *downward*, it yields a *negative* expression for M, whereas if $\mathbf{P}$ is upward, then M is *positive*. Equation 16–15 states that when $x < a$, $M = 0$, and when $x \ge a$, $M = -P(x - a)$. Indeed this is the case, as shown by the method of sections, Fig. 16–26b.

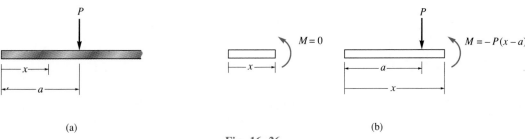

(a)

(b)

Fig. 16–26

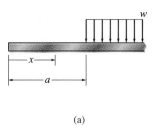

(a)

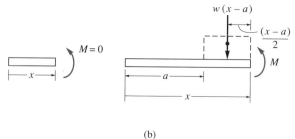

(b)

Fig. 16–27

Uniform Distributed Loading, w. If a uniform distributed loading acts on the beam, starting at $x = a$ and extending all the way to the right end of the beam, Fig. 16–27a, then the discontinuity function that describes the internal moment is

$$M = -\frac{1}{2}w\langle x - a \rangle^2 \tag{16–16}$$

This equation states that when $x < a$, $M = 0$, and when $x \geq a$, $M = -\frac{1}{2}w(x - a)^2$. These values may be checked using the method of sections as shown in Fig. 16–27b. Here w acting *downward* on the beam yields a negative expression for M, whereas if w acts *upward*, then M is *positive*.

Linear Distributed Loading. If the load on the beam is triangular and has a slope m, Fig. 16–28a, then the internal moment in the beam is defined by the discontinuity function

$$M = -\frac{1}{6}m\langle x - a \rangle^3 \tag{16–17}$$

Here the values $M = 0$ for $x < a$ and $M = -\frac{1}{6}m(x - a)^3$ for $x \geq a$ can be verified by the method of sections, Fig. 16–28b. As above, when the *distributed loading* acts *downward* on the beam it yields a *negative* expression for M, whereas if w acts upward, then M is positive.

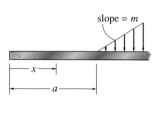

(a)

(b)

Fig. 16–28

This analysis can be extended to distributed loadings that have other forms. Also, it is possible to use superposition with the uniform and triangular loadings to create the discontinuity function for a trapezoidal loading.

Application of Eqs. 16–14 through 16–17 provides a rather direct means for writing the internal moment in a beam as a function of x. When doing so, close attention must be paid to the signs of the *external loadings*. As stated above, *concentrated forces* and *distributed loads* are *positive downward,* and *couple moments* are *positive counterclockwise*. If this sign convention is followed, then the *internal moment* **M** is in accordance with the beam sign convention established in Sec. 16.5; i.e., positive moment bends the beam concave upward.

As an example of how to apply the above functions, consider the beam loaded as shown in Fig. 16–29a. Here the reactive force $\mathbf{R}_1$ created by the pin, Fig. 16–29b, is negative since it acts upward, and $\mathbf{M}_0$ is negative since it acts clockwise. Applying Eqs. 16–14 through 16–16, the internal moment at *any point* x in the beam, Fig. 16–29a, is therefore

$$M = R_1\langle x - 0 \rangle - P\langle x - a \rangle + M_0\langle x - b \rangle^0 - \frac{1}{2}w\langle x - c \rangle^2 \qquad (16\text{–}18)$$

The validity of this expression may be checked by using the method of sections, say, within the region $b < x < c$, Fig. 16–19b. Moment equilibrium requires that

$$M = R_1 x - P(x - a) + M_0$$

This result agrees with that of Eq. 16–18, since by Eq. 16–12 only the last term in Eq. 16–18 is zero when $x < c$.

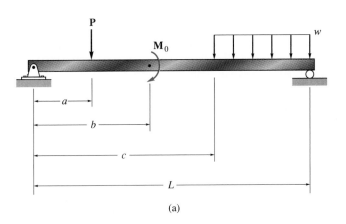

(a)

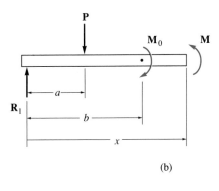

(b)

Fig. 16–29

PROCEDURE FOR ANALYSIS

The following procedure provides a method for using discontinuity functions to determine a beam's elastic curve. This method is particularly advantageous for solving problems involving beams or shafts subjected to *several loadings,* since the two constants of integration can be evaluated by using *only* the boundary conditions.

Elastic Curve.　Sketch the beam's elastic curve and identify the boundary conditions at the supports. Recall that zero displacement occurs at all pin and roller supports, and zero slope and zero displacement occur at fixed supports. Establish the x axis so that it extends to the right and has its origin at the beam's left end.

Moment Function.　Calculate the support reactions and then use the discontinuity functions to express the internal moment M as a function of x. When doing so, make sure to follow the sign convention for each loading as it applies for Eqs. 16–14 through 16–17. Also, note that the distributed loadings must extend all the way to the beam's right end for Eqs. 16–16 and 16–17 to be valid. If this does not occur, use the method of superposition, which is illustrated in Example 16–9.

Slope and Elastic Curve.　Substitute M into the moment-curvature relation $EI\,d^2v/dx^2 = M$, and integrate twice to obtain the equations for the beam's slope and deflection. Evaluate the two constants of integration using the boundary conditions and substitute the constants into the slope and deflection equations to obtain the final results. When these functions are evaluated at any point on the beam, a *positive slope* is *counterclockwise,* and a *positive displacement* is *upward.*

The following examples illustrate application of this procedure.

Example 16–9

Determine the equation of the elastic curve for the cantilevered beam shown in Fig. 16–30a. EI is constant.

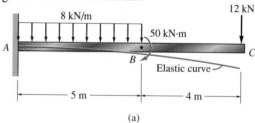

Fig. 16–30

(a)

SOLUTION

Elastic Curve.　The loads cause the beam to deflect as shown in Fig. 16–30a. The boundary conditions require zero slope and displacement at A.

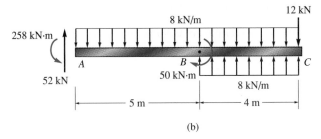

(b)

Moment Function.   The support reactions at A have been calculated by statics and are shown on the free-body diagram in Fig. 16–30b. Since the distributed loading in Fig. 16–30a does not extend to C as required, we can use the superposition of loadings shown in Fig. 16–30b to represent the same effect. By our sign convention, the 50-kN · m couple moment, the 52-kN force at A, and the portion of distributed loading from B to C on the bottom of the beam are all negative. With reference to Fig. 16–30b, applying Eqs. 16–14 through 16–16, the beam's internal moment is therefore

$$M = -258\langle x - 0\rangle^0 - (-52)\langle x - 0\rangle - \frac{1}{2}(8)\langle x - 0\rangle^2$$

$$-\frac{1}{2}(-8)\langle x - 5\rangle^2 - (-50)\langle x - 5\rangle^0$$

$$= -258 + 52x - 4x^2 + 4\langle x - 5\rangle^2 + 50\langle x - 5\rangle^0$$

The moment of the 12-kN load at C is *not included* here, since x cannot be greater than 9 m.

Slope and Elastic Curve.   Applying Eq. 16–11 and integrating twice, using Eq. 16–13, we have

$$EI \frac{d^2v}{dx^2} = -258 + 52x - 4x^2 + 4\langle x - 5\rangle^2 + 50\langle x-5\rangle^0$$

$$EI \frac{dv}{dx} = -258x + 26x^2 - \frac{4}{3}x^3 + \frac{4}{3}\langle x - 5\rangle^3 + 50\langle x - 5\rangle + C_1$$

$$EIv = -129x^2 + \frac{26}{3}x^3 - \frac{1}{3}x^4 + \frac{1}{3}\langle x - 5\rangle^4 + 25\langle x - 5\rangle^2 + C_1x + C_2$$

Since $dv/dx = 0$ at $x = 0$, $C_1 = 0$; and $v = 0$ at $x = 0$, so $C_2 = 0$. Thus,

$$v = \frac{1}{EI}\left(-129x^2 + \frac{26}{3}x^3 - \frac{1}{3}x^4 + \frac{1}{3}\langle x - 5\rangle^4 + 25\langle x - 5\rangle^2\right) \; Ans.$$

Note: If we *did not* use discontinuity functions, two x coordinates would be needed for regions AB and BC, and the problem would require evaluating two more constants of integration involving the continuity of slope and displacement at point B.

■ **Example 16–10** ▬▬▬▬▬

Determine the maximum deflection of the beam shown in Fig. 16–31a. *EI* is constant.

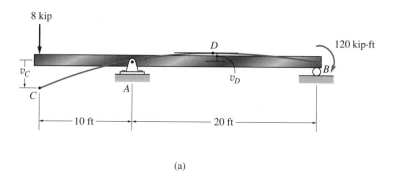

(a)

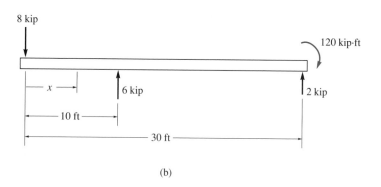

(b)

Fig. 16–31

SOLUTION

Elastic Curve. The beam deflects as shown in Fig. 16–31a. The boundary conditions require zero displacement at *A* and *B*.

Moment Function. The reactions have been calculated and are shown on the free-body diagram in Fig. 16–31b. Using Eqs. 16–14 and 16–15, we have

$$M = -8\langle x - 0 \rangle - (-6)\langle x - 10 \rangle$$
$$= -8x + 6\langle x - 10 \rangle$$

The couple moment and force at *B* are not included here, since they are located at the right end of the beam, and *x* cannot be greater than 30 ft.

Slope and Elastic Curve. Integrating Eq. 16–11 twice yields

$$EI\,\frac{d^2v}{dx^2} = -8x + 6\langle x - 10\rangle$$

$$EI\,\frac{dv}{dx} = -4x^2 + 3\langle x - 10\rangle^2 + C_1$$

$$EIv = -\frac{4}{3}x^3 + \langle x - 10\rangle^3 + C_1x + C_2 \qquad (1)$$

From Eq. 1, the boundary condition $v = 0$ at $x = 10$ ft and $v = 0$ at $x = 30$ ft gives

$$0 = -1333 + (10 - 10)^3 + C_1(10) + C_2$$
$$0 = -36,000 + (30 - 10)^3 + C_1(30) + C_2$$

Solving these equations simultaneously for C_1 and C_2, we get $C_1 = 1333$ and $C_2 = -12,000$. Thus,

$$EI\,\frac{dv}{dx} = -4x^2 + 3\langle x - 10\rangle^2 + 1333 \qquad (2)$$

$$EIv = -\frac{4}{3}x^3 + \langle x - 10\rangle^3 + 1333x - 12,000 \qquad (3)$$

From Fig. 16–31a, maximum displacement may occur at either C or D, where the slope $dv/dx = 0$. To obtain the displacement of C, set $x = 0$ in Eq. 3. We get

$$v_C = -\frac{12,000\ \text{kip} \cdot \text{ft}^3}{EI}$$

The *negative* sign indicates that the displacement is *downward* as shown in Fig. 16–31a. To locate point D, use Eq. 2 with $x > 10$ and $dv/dx = 0$. This gives

$$0 = -4x_D^2 + 3(x_D - 10)^2 + 1333$$
$$x_D^2 + 60x_D - 1633 = 0$$

Solving for the positive root,

$$x_D = 20.3\ \text{ft}$$

Hence, from Eq. 3,

$$EIv_D = -\frac{4}{3}(20.3)^3 + (20.3 - 10)^3 + 1333(20.3) - 12,000$$

$$v_D = \frac{29,000\ \text{kip} \cdot \text{ft}^3}{EI} \qquad\qquad\qquad \textit{Ans.}$$

Comparing this value with v_C, we see that $v_{\max} = v_D$.

PROBLEMS

16–43.   The beam is subjected to the load shown. Using discontinuity functions, determine the equation of the elastic curve. *EI* is constant.

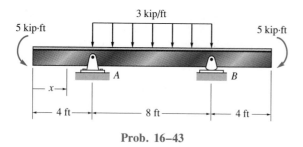

Prob. 16–43

*16–44.   The beam is subjected to the load shown. Using discontinuity functions, determine the equations of the slope and elastic curve. *EI* is constant.

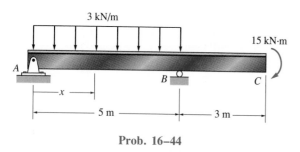

Prob. 16–44

16–45.   The beam is subjected to the load shown. Using discontinuity functions, determine the equation of the elastic curve. *EI* is constant.

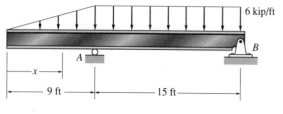

Prob. 16–45

16–46.   The beam is subjected to the load shown. Using discontinuity functions, determine the equation of the elastic curve. *EI* is constant.

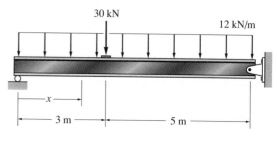

Prob. 16–46

16–47.   The wooden beam is subjected to the load shown. Using discontinuity functions, determine the equation of the elastic curve. If $E_w = 12$ GPa, determine the deflection and the slope at end *B*.

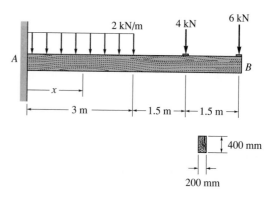

Prob. 16–47

*16–48.   The shaft supports the two pulley loads shown. Using discontinuity functions, determine the equation of the elastic curve. The bearings at *A* and *B* exert only vertical reactions on the shaft. *EI* is constant.

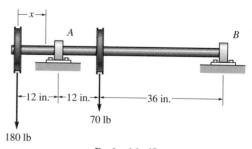

Prob. 16–48

16–49. The beam is subjected to the load shown. Using discontinuity functions, determine the equation of the elastic curve. EI is constant.

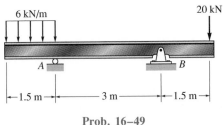

Prob. 16–49

16–50. The shaft supports the two pulley loads shown. Using discontinuity functions, determine the equation of the elastic curve. The bearings at A and B exert only vertical reactions on the shaft. EI is constant.

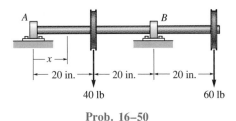

Prob. 16–50

16–51. The beam is subjected to the load shown. Using discontinuity functions, determine the equation of the elastic curve. EI is constant.

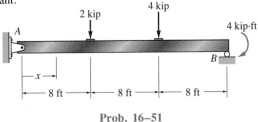

Prob. 16–51

*16–52. The wooden beam is subjected to the load shown. Using discontinuity functions, determine the equation of the elastic curve. Specify the deflection at the end C. $E_w = 1.6(10^3)$ ksi.

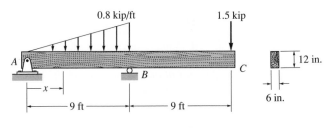

Prob. 16–52

16–53. Determine the slope at B and the deflection at C for the $W10 \times 45$. Solve the problem using discontinuity functions. $E_{st} = 29(10^3)$ ksi.

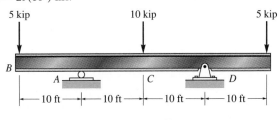

Prob. 16–53

16–54. Using discontinuity functions, determine the deflection at each of the pulleys C, D, and E. The shaft is made of steel and has a diameter of 30 mm. The bearings at A and B exert only vertical reactions on the shaft. $E_{st} = 200$ GPa.

16–55. Using discontinuity functions, determine the slope of the shaft at the bearings at A and B. The shaft is made of steel and has a diameter of 30 mm. The bearings at A and B exert only vertical reactions on the shaft. $E_{st} = 200$ GPa.

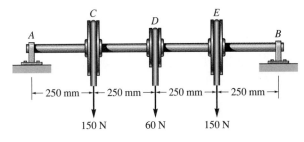

Probs. 16–54/16–55

*16–56. The shaft is made of steel and has a diameter of 15 mm. Using discontinuity functions, determine its maximum deflection. The bearings at A and B exert only vertical reactions on the shaft. $E_{st} = 200$ GPa.

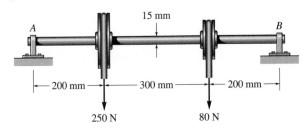

Prob. 16–56

*16.7 Slope and Displacement by the Moment-Area Method

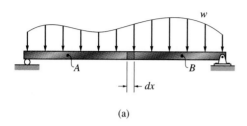

(a)

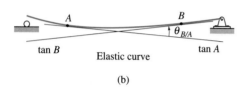

tan B   Elastic curve   tan A

(b)

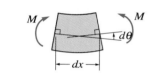

(c)

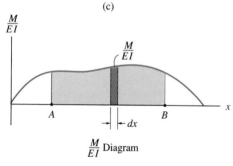

$\frac{M}{EI}$ Diagram

(d)

Fig. 16–32

The moment-area method provides a semigraphical technique for finding the slope and displacement at *specific points* on the elastic curve of a beam or shaft. Application of the method requires computing areas associated with the beam's moment diagram; so if this diagram consists of simple shapes, the method is very convenient to use. Normally this is the case when the beam is loaded with concentrated forces or couple moments or regions of the beam have different moments of inertia.

To develop the moment-area method we will make the same assumptions as we used for the method of integration: The beam is initially straight, it is elastically deformed by the loads, such that the slope and deflection of the elastic curve are very small, and the deformations are caused by bending. Before we can apply the moment-area method we must first develop two theorems. The first theorem provides a means of obtaining the slope of the elastic curve, and the second theorem provides the means of obtaining the displacement of a point.

Theorem 1. To develop the first moment-area theorem, consider the simply-supported beam shown in Fig. 16–32a with its associated elastic curve, Fig. 16–32b. A differential segment dx of the beam is isolated in Fig. 16–32c. It is seen that the beam's internal moment M deforms the element such that the *tangents* to the elastic curve at each side of the element intersect at an angle $d\theta$. This angle can be determined from Eq. 16–11, written as

$$EI \frac{d^2v}{dx^2} = EI \frac{d}{dx}\left(\frac{dv}{dx}\right) = M$$

Since the *slope* is *small*, $\theta = dv/dx$, and therefore

$$d\theta = \frac{M}{EI} dx \qquad (16-19)$$

If the moment diagram for the beam is constructed and divided by both the beam's moment of inertia I and modulus of elasticity E, Fig. 16–32d, Eq. 16–19 indicates that $d\theta$ is equal to the *area* under the "M/EI diagram" for the beam segment dx. Integrating from a selected point A on the elastic curve to another point B, Fig. 16–32b, we have

$$\theta_{B/A} = \int_A^B \frac{M}{EI} dx \qquad (16-20)$$

This equation forms the basis for the first moment-area theorem.

Theorem 1: *The angle between the tangents at any two points on the elastic curve equals the area under the M/EI diagram between these two points.*

The notation $\theta_{B/A}$ is referred to as the angle of the tangent at B measured *with respect to* the tangent at A. From the proof it should be evident that this angle is measured *counterclockwise*, from tangent A to tangent B, if the area under the M/EI diagram is *positive*, Fig. 16–32d. Conversely, if the area is *negative*, or lies below the x axis, the angle $\theta_{B/A}$ is measured clockwise from tangent A to tangent B. Furthermore, from the dimensions of Eq. 16–20, $\theta_{B/A}$ will be *measured* in radians.

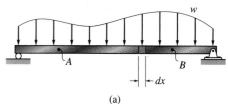

(a)

Theorem 2. The second moment-area theorem is based on the relative deviation of tangents to the elastic curve. Shown in Fig. 16–33b is a greatly exaggerated view of the vertical deviation dt of the tangents on each side of the differential element dx. This deviation is caused by the curvature of the element and has been measured along a vertical line passing through point A located on the elastic curve. Since the slope of the elastic curve and its deflection are assumed to be very small, it is satisfactory to approximate the length of each tangent line by x and the arc ds' by dt. Using the circular-arc formula $s = \theta r$, where r is the length x and s is dt, we can write $dt = x\, d\theta$. Substituting Eq. 16–19 into this equation and integrating from A to B, the vertical deviation of the tangent at A *with respect to* the tangent at B can then be determined; i.e.,

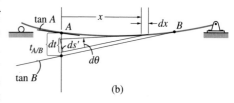

(b)

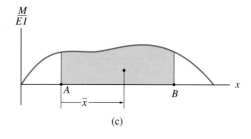

(c)

$$t_{A/B} = \int_A^B x\, \frac{M}{EI}\, dx \qquad (16\text{–}21)$$

Since the centroid of an area is found from $\bar{x} \int dA = \int x\, dA$, and $\int (M/EI)dx$ represents the area under the M/EI diagram, we can also write

$$t_{A/B} = \bar{x} \int_A^B \frac{M}{EI}\, dx \qquad (16\text{–}22)$$

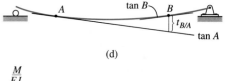

(d)

Here $\bar{x}$ is the distance from the vertical axis through A to the *centroid* of the area under the M/EI diagram between A and B, Fig. 16–22c.

The second moment-area theorem can now be stated as follows in reference to Fig. 16–33b.

Theorem 2: *The vertical deviation of the tangent at a point (A) on the elastic curve with respect to the tangent extended from another point (B) equals the moment of the area under the M/EI diagram between the two points (A and B). This moment is computed about a vertical axis passing through the point (A) where the vertical deviation ($t_{A/B}$) is to be determined.*

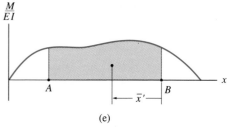

(e)

Fig. 16–33

The distance $t_{A/B}$ used in the theorem can also be interpreted as the vertical displacement from the point located on the extended tangent drawn from B to the point A on the elastic curve. Note that $t_{A/B}$ is *not* equal to $t_{B/A}$, which is shown in Fig. 16–33d. Specifically, the moment of the area under the M/EI diagram between A and B is computed about a vertical axis through point A to determine $t_{A/B}$, Fig. 16–33c, and it is computed about a vertical axis through point B to determine $t_{B/A}$, Fig. 16–33e.

If the moment of a *positive M/EI* area from *A* to *B* is computed for $t_{A/B}$, as in Fig. 16–33c, it indicates that point *A* is *above* the tangent extended from point *B*, Fig. 16–33b. Similarly, *negative M/EI* areas indicate that point *A* is *below* the tangent extended from point *B*.

PROCEDURE FOR ANALYSIS

The following procedure provides a method that may be used to determine the slope and displacement at a specific point on a beam using the two moment-area theorems.

M/EI Diagram. Determine the support reactions and draw the beam's *M/EI* diagram. If the beam is loaded with concentrated forces, the *M/EI* diagram will consist of a series of straight line segments, and the areas and their moments required for the moment-area theorems will be relatively easy to compute. If the loading consists of a series of distributed loads, the *M/EI* diagram will consist of parabolic or perhaps higher-order curves, and it is suggested that the table in Appendix *B* be used to locate the area and centroid under each curve.

Elastic Curve. Draw an exaggerated view of the beam's elastic curve. Recall that points of zero slope and zero displacement always occur at a fixed support, and zero displacement occurs at all pin and roller supports. If it becomes difficult to draw the general shape of the elastic curve, use the moment (or *M/EI*) diagram. Realize that when the beam is subjected to a *positive moment*, the beam bends *concave up*, whereas *negative moment* bends the beam *concave down*, Fig. 16–12. Furthermore, an inflection point or change in curvature occurs where the moment in the beam (or *M/EI*) is zero.

The unknown displacement and slope to be determined should be indicated on the curve. Since the moment-area theorems apply *only between two tangents,* attention should be given as to which tangents should be constructed on the curve so that the angles or deviations between them will lead to the solution of the problem. In this regard, *the tangents at the points of unknown slope and displacement and at the supports should be considered,* since the beam usually has zero displacement and/or zero slope at the supports.

Moment-Area Theorems. Apply Theorem 1 to determine the *angle* between any two tangents on the elastic curve and Theorem 2 to determine the *tangential deviation*. The algebraic sign of the answer can be checked from the angle or deviation indicated on the elastic curve. Specifically, a *positive* $\theta_{B/A}$ represents a *counterclockwise* rotation of the tangent at *B* with respect to the tangent at *A*, and a *positive* $t_{A/B}$ indicates that point *A* on the elastic curve lies *above* the extended tangent from point *B*.

The following examples illustrate application of this procedure.

Example 16–11

Determine the slope of the beam shown in Fig. 16–34*a* at points *B* and *C*. *EI* is constant.

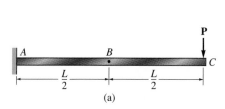

(a)

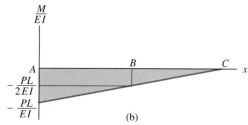

(b)

Fig. 16–34

SOLUTION

M/EI Diagram. This diagram is shown in Fig. 16–34*b*.

Elastic Curve. The force **P** causes the beam to deflect as shown in Fig. 16–34*c*. (The elastic curve is concave downward, since *M/EI* is negative.) The tangents at *B* and *C* are indicated since we are required to find θ_B and θ_C. Also, the tangent at the support (*A*) is shown. This tangent has a *known* zero slope. By the construction, the angle between tan *A* and tan *B*, that is, $\theta_{B/A}$, is equivalent to θ_B, or

$$\theta_B = \theta_{B/A}$$

Also

$$\theta_C = \theta_{C/A}$$

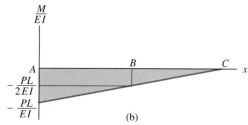

(c)

Moment-Area Theorem. Applying Theorem 1, $\theta_{B/A}$ is equal to the area under the *M/EI* diagram between points *A* and *B*; that is,

$$\theta_B = \theta_{B/A} = \left(-\frac{PL}{2EI}\right)\left(\frac{L}{2}\right) + \frac{1}{2}\left(-\frac{PL}{2EI}\right)\left(\frac{L}{2}\right)$$

$$= -\frac{3PL^2}{8EI} \qquad \qquad Ans.$$

The *negative sign* indicates that the angle measured from the tangent at *A* to the tangent at *B* is *clockwise*. This checks since the beam slopes downward at *B*.

In a similar manner, the area under the *M/EI* diagram between points *A* and *C* equals $\theta_{C/A}$. We have

$$\theta_C = \theta_{C/A} = \frac{1}{2}\left(-\frac{PL}{EI}\right)L$$

$$= -\frac{PL^2}{2EI} \qquad \qquad Ans.$$

Example 16–12

Determine the displacement of points B and C of the beam shown in Fig. 16–35a. EI is constant.

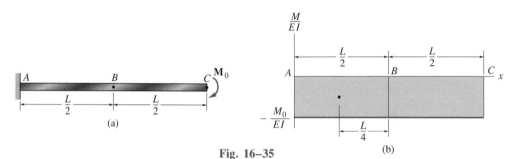

(a)

(b)

Fig. 16–35

SOLUTION

M/EI Diagram. See Fig. 16–35b.

Elastic Curve. The couple moment at C causes the beam to deflect as shown in Fig. 16–35c. The tangents at B and C are indicated since we are required to find Δ_B and Δ_C. Also, the tangent at the support (A) is shown since it is horizontal. The required displacements can now be related directly to the deviations between the tangents at B and A and C and A. Specifically, Δ_B is equal to the deviation of tan A from tan B; that is,

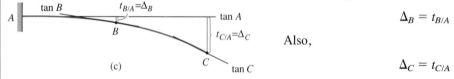

(c)

$$\Delta_B = t_{B/A}$$

Also,

$$\Delta_C = t_{C/A}$$

Moment-Area Theorem. Applying Theorem 2, $t_{B/A}$ is equal to the moment of the shaded area under the M/EI diagram between A and B computed about a vertical axis passing through point B (the point on the elastic curve), since this is the point where the tangential deviation is to be determined. Hence, from Fig. 16–35b,

$$\Delta_B = t_{B/A} = \left(\frac{L}{4}\right)\left[\left(-\frac{M_0}{EI}\right)\left(\frac{L}{2}\right)\right] = -\frac{M_0 L^2}{8EI} \qquad Ans.$$

Likewise, for $t_{C/A}$ we must compute the moment of the area under the *entire* M/EI diagram from A to C about a vertical axis passing through point C (the point on the elastic curve). We have

$$\Delta_C = t_{C/A} = \left(\frac{L}{2}\right)\left[\left(-\frac{M_0}{EI}\right)(L)\right] = -\frac{M_0 L^2}{2EI} \qquad Ans.$$

Since both answers are *negative*, they indicate that points B and C lie *below* the tangent at A. This checks with Fig. 16–35c.

Example 16–13

Determine the slope at point C of the beam in Fig. 16–36a. EI is constant.

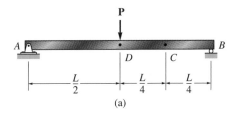

(a)

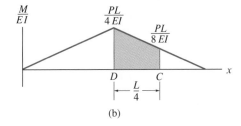

(b)

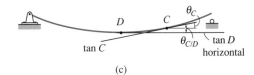

(c)

Fig. 16–36

SOLUTION

M/EI Diagram.   See Fig. 16–36b.

Elastic Curve.   Since the loading is applied symmetrically to the beam, the elastic curve is symmetric and the tangent at D is horizontal as shown in Fig. 16–36c. Tangents are drawn at C, since we must find the slope θ_C, and at D. By the construction, the angle $\theta_{C/D}$, between tan D and tan C, is equal to θ_C; that is,

$$\theta_C = \theta_{C/D}$$

Moment-Area Theorem.   Using Theorem 1, $\theta_{C/D}$ is equal to the shaded area under the M/EI diagram between points D and C. We have

$$\theta_C = \theta_{C/D} = \left(\frac{PL}{8EI}\right)\left(\frac{L}{4}\right) + \frac{1}{2}\left(\frac{PL}{4EI} - \frac{PL}{8EI}\right)\left(\frac{L}{4}\right) = \frac{3PL^2}{64EI} \qquad \textit{Ans.}$$

What does the positive result indicate?

◼ Example 16–14

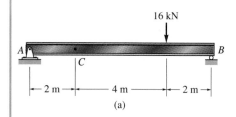

16 kN

A · C · B

├─ 2 m ─┼─── 4 m ───┼─ 2 m ─┤

(a)

$\dfrac{M}{EI}$

$\dfrac{24}{EI}$

$\dfrac{8}{EI}$

A · C · B · x

├─ 2 m ─┼─── 4 m ───┼─ 2 m ─┤

(b)

Determine the slope at point C for the steel beam in Fig. 16–37a. Take $E_{st} = 200$ GPa, $I = 17(10^6)$ mm^4.

SOLUTION

M/EI Diagram.   See Fig. 16–37b.

Elastic Curve.   The elastic curve is shown in Fig. 16–37c. The tangent at C is shown since we are required to find θ_C. To do this, tangents at the *supports*, A and B, are also constructed as shown. Angle $\theta_{C/A}$ is the angle between the tangents at A and C. The slope at A, θ_A, in Fig. 16–37c can be found using $|\theta_A| = |t_{B/A}|/L_{AB}$. This equation is valid since $t_{B/A}$ is actually very small, so that θ_A in radians can be approximated by the length of a circular arc defined by a radius of $L_{AB} = 8$ m and a sweep of θ_A. (Recall that $s = \theta r$.) From the geometry of Fig. 16–37c, we have

$$|\theta_C| = |\theta_A| - |\theta_{C/A}| = \left|\frac{t_{B/A}}{8}\right| - |\theta_{C/A}| \tag{1}$$

Moment-Area Theorems.   Using Theorem 1, $\theta_{C/A}$ is equivalent to the area under the M/EI diagram between points A and C; that is,

$$\theta_{C/A} = \frac{1}{2}(2)\left(\frac{8}{EI}\right) = \frac{8}{EI}$$

Applying Theorem 2, $t_{B/A}$ is equivalent to the moment of the area under the M/EI diagram between B and A about a vertical axis passing through point B (the point on the elastic curve), since this is the point where the tangential deviation is to be determined. We have

$$t_{B/A} = \left(2 + \frac{1}{3}(6)\right)\left[\frac{1}{2}(6)\left(\frac{24}{EI}\right)\right] + \left(\frac{2}{3}(2)\right)\left[\frac{1}{2}(2)\left(\frac{24}{EI}\right)\right]$$

$$= \frac{320}{EI}$$

Substituting these results into Eq. 1, we get

$$\theta_C = \frac{320}{8EI} - \frac{8}{EI} = \frac{32}{EI} \downarrow$$

We have computed this result in units of kN and m, so converting EI into these units, we have

$$\theta_C = \frac{32 \text{ kN} \cdot \text{m}^2}{200(10^6) \text{ kN/m}^2 \ 17(10^{-6}) \text{ m}^4} = 0.00941 \text{ rad} \downarrow \qquad Ans.$$

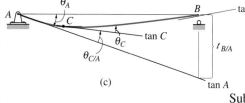

θ_A

A · C · tan C · B · tan B

θ_C

$\theta_{C/A}$

$t_{B/A}$

tan A

(c)

Fig. 16–37

Example 16–15

Determine the displacement at C for the beam shown in Fig. 16–38a. EI is constant.

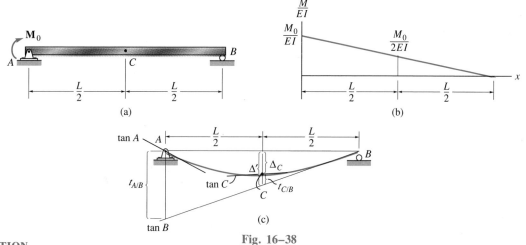

Fig. 16–38

SOLUTION

M/EI Diagram.   See Fig. 16–38b.

Elastic Curve.   The tangent at C is drawn on the elastic curve since we are required to find Δ_C, Fig. 16–38c. Note that C is *not* the location of the maximum deflection of the beam, because the loading and hence the elastic curve are *not symmetric*. Also indicated in Fig. 16–38c are the tangents at the supports A and B. It is seen that $\Delta_C = \Delta' - t_{C/B}$. If $t_{A/B}$ is determined, then Δ' can be found from proportional triangles, that is, $\Delta'/(L/2) = t_{A/B}/L$ or $\Delta' = t_{A/B}/2$. Hence,

$$\Delta_C = \frac{t_{A/B}}{2} - t_{C/B} \qquad (1)$$

Moment-Area Theorem.   Applying Theorem 2 to determine $t_{A/B}$ and $t_{C/B}$, we have

$$t_{A/B} = \left(\frac{1}{3}(L)\right)\left[\frac{1}{2}(L)\left(\frac{M_0}{EI}\right)\right] = \frac{M_0 L^2}{6EI}$$

$$t_{C/B} = \left(\frac{1}{3}\left(\frac{L}{2}\right)\right)\left[\frac{1}{2}\left(\frac{L}{2}\right)\left(\frac{M_0}{2EI}\right)\right] = \frac{M_0 L^2}{48EI}$$

Substituting these results into Eq. 1 gives

$$\Delta_C = \frac{1}{2}\left(\frac{M_0 L^2}{6EI}\right) - \left(\frac{M_0 L^2}{48EI}\right)$$

$$= \frac{M_0 L^2}{16EI} \downarrow \qquad\qquad\qquad Ans.$$

▪ Example 16–16

Determine the displacement at point C for the steel overhanging beam shown in Fig. 16–39a. Take $E_{st} = 39(10^3)$ ksi, $I = 125$ in⁴.

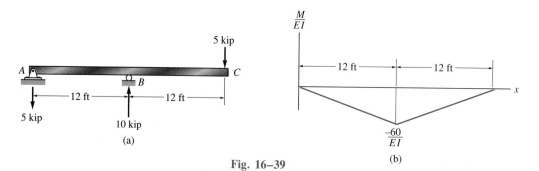

Fig. 16–39

SOLUTION

M/EI Diagram.   See Fig. 16–39b.

Elastic Curve.   The loading causes the beam to deflect as shown in Fig. 16–39c. We are required to find Δ_C. By constructing tangents at C and at the supports A and B, it is seen that $\Delta_C = |t_{C/A}| - \Delta'$. However, Δ' can be related to $t_{B/A}$ by proportional triangles; that is, $\Delta'/24 = |t_{B/A}|/12$ or $\Delta' = 2|t_{B/A}|$. Hence

$$\Delta_C = |t_{C/A}| - 2|t_{B/A}| \qquad (1)$$

Moment-Area Theorem.   Applying Theorem 2 to determine $t_{C/A}$ and $t_{B/A}$, we have

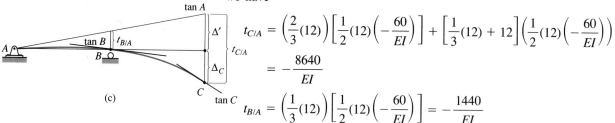

$$t_{C/A} = \left(\frac{2}{3}(12)\right)\left[\frac{1}{2}(12)\left(-\frac{60}{EI}\right)\right] + \left[\frac{1}{3}(12) + 12\right]\left(\frac{1}{2}(12)\left(-\frac{60}{EI}\right)\right)$$

$$= -\frac{8640}{EI}$$

$$t_{B/A} = \left(\frac{1}{3}(12)\right)\left[\frac{1}{2}(12)\left(-\frac{60}{EI}\right)\right] = -\frac{1440}{EI}$$

Why are these terms negative? Substituting the results into Eq. 1 yields

$$\Delta_C = \frac{8640}{EI} - 2\left(\frac{1440}{EI}\right) = \frac{5760}{EI} \ \downarrow$$

Realizing that the computations were made in units of kip and ft, we have

$$\Delta_C = \frac{5760 \text{ kip} \cdot \text{ft}^3 \ (1728 \text{ in}^3/\text{ft}^3)}{[29(10^3) \text{ kip/in}^2](125 \text{ in}^4)} = 2.75 \text{ in. } \downarrow \qquad \textit{Ans.}$$

PROBLEMS

16–57. Determine the slope and deflection at *B*. *EI* is constant.

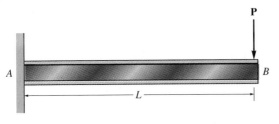

Prob. 16–57

16–58. Determine the slope and deflection at *C*. *EI* is constant.

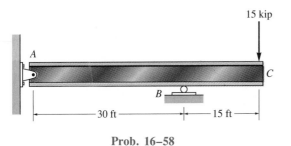

Prob. 16–58

16–59. If the bearings at *A* and *B* exert only vertical reactions on the shaft, determine the slope at *B* and the deflection at *C*. *EI* is constant.

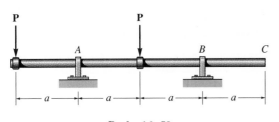

Prob. 16–59

*16–60. If the bearings at *A* and *B* exert only vertical reactions on the shaft, determine the slope at *B* and the deflection at *C*. *EI* is constant.

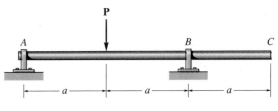

Prob. 16–60

16–61. The composite simply-supported steel shaft is subjected to a force of 10 kN at its center. Determine its maximum deflection. $E_{st} = 200$ GPa.

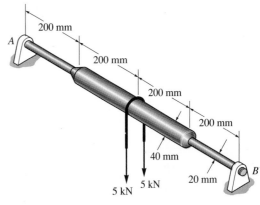

Prob. 16–61

16–62. Determine the value of *a* so that the slope at *A* is equal to zero. *EI* is constant.

16–63. Determine the value of *a* so that the deflection at *C* is equal to zero. *EI* is constant.

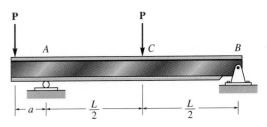

Probs. 16–62/16–63

*16–64. Determine the deflection and slope at C. EI is constant.

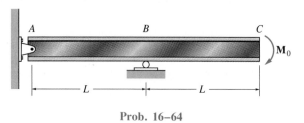

Prob. 16–64

16–65. Determine the maximum deflection of the beam and the slope at A. EI is constant.

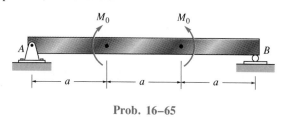

Prob. 16–65

16–66. Determine the slope at B and the deflection at C. EI is constant.

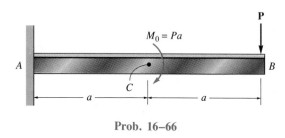

Prob. 16–66

16–67. The beam is subjected to the loading shown. Determine the slope at C and deflection at B. EI is constant.

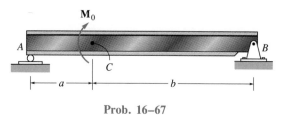

Prob. 16–67

*16–68. If the bearings at A and B exert only vertical reactions on the shaft, determine the slope at A and the maximum deflection.

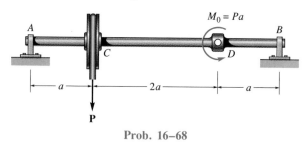

Prob. 16–68

16–69. Determine the slope at B and deflection at C. EI is constant.

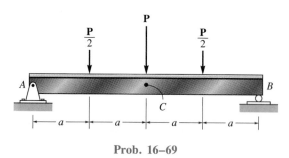

Prob. 16–69

16–70. The two steel bars have a thickness of 1 in. and a width of 4 in. They are designed to act as a spring for the machine, which exerts a force of 4 kip on them at A and B. If the supports exert only vertical forces on the bars, determine the maximum deflection of the bottom bar. $E_{st} = 29(10^3)$ ksi.

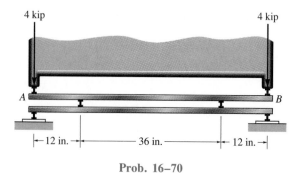

Prob. 16–70

16–71. The beam is subjected to the loading shown. Determine the slope at B and deflection at C. EI is constant.

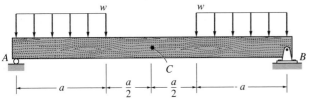

Prob. 16–71

*16–72. Determine the slope at A and the deflection at C. EI is constant.

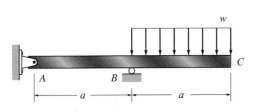

Prob. 16–72

16–73. Determine the maximum deflection of the beam. EI is constant.

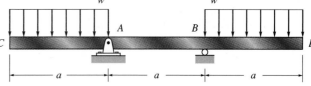

Prob. 16–73

16–74. The two bars are pin-connected at D. Determine the slope at A and the deflection at D. EI is constant.

Prob. 16–74

16.8 Method of Superposition

The differential equation $EId^4v/dx^4 = -w(x)$ satisfies the two necessary requirements for applying the principle of superposition as stated in Sec. 10.3. To be specific, the load $w(x)$ is linearly related to the deflection $v(x)$, and the load is assumed not to change significantly the original geometry of the beam or shaft, since the slope is considered very small for this equation to apply. As a result, the deflections for a series of separate loadings acting on a beam may be superimposed. For example, if v_1 is the deflection for one load and v_2 is the deflection for another load, the total deflection for both loads acting together is the algebraic sum $v_1 + v_2$. Using tabulated results for various beam loadings, such as the ones listed on the inside back cover of this book, or those found in various engineering handbooks, it is therefore possible to find the slope and displacement at a point on a beam once the actual loading has been represented by its various component parts.

The following examples illustrate how to use the method of superposition to solve deflection problems, where the deflection is caused not only by beam deformations, but also by rigid-body displacements, which can occur when the beam is supported by springs or portions of a segmented beam are supported by hinges.

Example 16–17

Determine the displacement at point C and the slope at the support A of the beam shown in Fig. 16–40a. EI is constant.

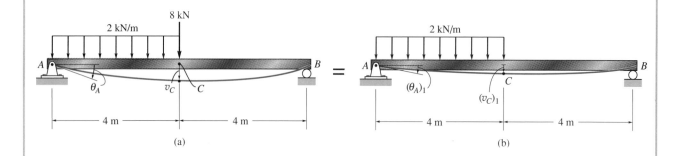

(a)   (b)

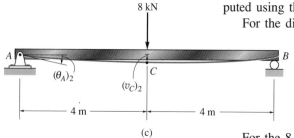

(c)

Fig. 16–40

SOLUTION

The loading can be separated into two component parts as shown in Fig. 16–40b and 16–40c. The displacement at C and slope at A are computed using the table on the inside back cover for each part.

For the distributed loading, Fig. 16–40b,

$$(\theta_A)_1 = \frac{3wL^3}{128EI} = \frac{3(2 \text{ kN/m})(8 \text{ m})^3}{128EI} = \frac{24}{EI} \downarrow$$

$$(v_C)_1 = \frac{5wL^4}{768EI} = \frac{5(2 \text{ kN/m})(8 \text{ m})^4}{768EI} = \frac{53.33}{EI} \downarrow$$

For the 8-kN concentrated force, Fig. 16–40c,

$$(\theta_A)_2 = \frac{PL^2}{16EI} = \frac{8 \text{ kN}(8 \text{ m})^2}{16EI} = \frac{32}{EI} \downarrow$$

$$(v_C)_2 = \frac{PL^3}{48EI} = \frac{8 \text{ kN}(8 \text{ m})^3}{48EI} = \frac{85.33}{EI} \downarrow$$

The total displacement at C and the slope at A, Fig. 16–40a, are the algebraic sums of these components. Hence

$$(^+\downarrow) \qquad \qquad \theta_A = (\theta_A)_1 + (\theta_A)_2 = \frac{56}{EI} \downarrow \qquad \qquad Ans.$$

$$(+\downarrow) \qquad \qquad v_C = (v_C)_1 + (v_C)_2 = \frac{139}{EI} \downarrow \qquad \qquad Ans.$$

Example 16–18

Determine the displacement at the end C of the overhanging beam shown in Fig. 16–41a. EI is constant.

SOLUTION

Since the table on the back cover *does not* include beams with overhangs, the beam will be separated into a simply-supported and a cantilevered portion. First we will calculate the slope at B, as caused by the distributed load acting on the simply-supported span, Fig. 16–41b.

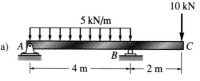

$$(\theta_B)_1 = \frac{wL^3}{24EI} = \frac{5 \text{ kN/m}(4 \text{ m})^3}{24EI} = \frac{13.33}{EI} \text{↰}$$

Since this angle is *small*, $\theta_B \approx \tan \theta_B$, and the vertical displacement at point C is

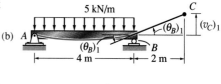

$$(v_C)_1 = (2 \text{ m})\left(\frac{13.33}{EI}\right) = \frac{26.67}{EI} \uparrow$$

Next, the 10-kN load on the overhang causes a statically equivalent force of 10 kN and couple moment of 20 kN · m at the support B of the simply-supported span, Fig. 16–41c. The 10-kN force does not cause a displacement or slope at B; however, the 20-kN · m couple moment does. The slope at B due to this moment is

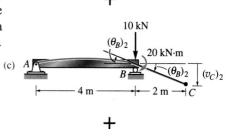

$$(\theta_B)_2 = \frac{M_0 L}{3EI} = \frac{20 \text{ kN · m}(4 \text{ m})}{3EI} = \frac{26.67}{EI} \text{↲}$$

So that the extended point C is displaced

$$(v_C)_2 = (2 \text{ m})\left(\frac{26.7}{EI}\right) = \frac{53.33}{EI} \downarrow$$

Finally, the cantilevered portion BC is displaced by the 10-kN force, Fig. 16–41d. We have

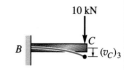

Fig. 16–41

$$(v_C)_3 = \frac{PL^3}{3EI} = \frac{10 \text{ kN}(2 \text{ m})^3}{3EI} = \frac{26.67}{EI} \downarrow$$

Summing these results algebraically, we obtain the final displacement of point C.

$$(+\downarrow) \qquad v_C = -\frac{26.7}{EI} + \frac{53.3}{EI} + \frac{26.7}{EI} = \frac{53.3}{EI} \downarrow \qquad \qquad Ans.$$

Example 16–19

The steel bar shown in Fig. 16–42a is supported by two springs at its ends A and B. Each spring has a stiffness of $k = 15$ kip/ft and is originally unstretched. If the bar is loaded with a force of 3 kip at point C, determine the vertical displacement of the force. Neglect the weight of the bar and take $E_{st} = 29(10^3)$ ksi, $I = 12$ in^4.

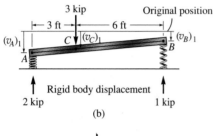

(a)

$\parallel$

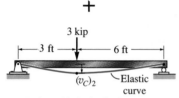

(b)

$+$

(c)

Fig. 16–42

SOLUTION

The end reactions at A and B are computed and shown in Fig. 16–42b. Each spring deflects by an amount

$$(v_A)_1 = \frac{2 \text{ kip}}{15 \text{ kip/ft}} = 0.1333 \text{ ft}$$

$$(v_B)_1 = \frac{1 \text{ kip}}{15 \text{ kip/ft}} = 0.0667 \text{ ft}$$

If the bar is considered to be rigid, these displacements cause it to move into the position shown in Fig. 16–42b. For this case, the vertical displacement at C is

$$(v_C)_1 = (v_B)_1 + \frac{6 \text{ ft}}{9 \text{ ft}} [(v_A)_1 - (v_B)_1]$$

$$= 0.0667 \text{ ft} + \frac{2}{3} [0.1333 \text{ ft} - 0.0667 \text{ ft}] = 0.1111 \text{ ft} \qquad \downarrow$$

We can compute the displacement at C caused by the *deformation* of the bar, Fig. 16–42c, by using the table on the inside back cover. We have

$$(v_C)_2 = \frac{Pab}{6EIL}(L^2 - b^2 - a^2)$$

$$= \frac{3 \text{ kip}(6 \text{ ft})(3 \text{ ft})[(9 \text{ ft})^2 - (6 \text{ ft})^2 - (3 \text{ ft})^2]}{6(9 \text{ ft})[29(10^3)] \text{ kip/in}^2(144 \text{ in}^2/1 \text{ ft}^2) 12 \text{ in}^4(1 \text{ ft}^4/20{,}736 \text{ in}^4)}$$

$$= 0.0149 \text{ ft} \qquad \downarrow$$

Adding the two displacement components, we get

$$(+\downarrow) \quad v_C = 0.1111 \text{ ft} + 0.0149 \text{ ft} = 0.126 \text{ ft} = 1.51 \text{ in.} \qquad \downarrow \qquad \textit{Ans.}$$

PROBLEMS

16–75. The $W8 \times 48$ steel cantilevered beam is subjected to the loading shown. Using the method of superposition, determine the deflection at its end A. $E_{st} = 29(10^3)$ ksi.

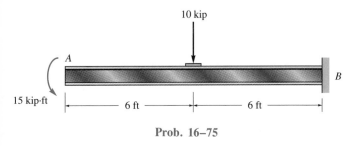

Prob. 16–75

16–77. The wide-flange beam acts as a cantilever. Due to an error it is installed at an angle θ with the vertical. Determine the ratio of its deflection in the x direction to its deflection in the y direction at A when a load $\mathbf{P}$ is applied at this point. The moments of inertia are I_x and I_y. For the solution, resolve $\mathbf{P}$ into components and use the method of superposition. *Note:* The result indicates that large lateral deflections (x direction) can occur in narrow beams, $I_y \ll I_x$, when they are improperly installed in this manner. To show this numerically, compute the deflections in the x and y directions for a $W10 \times 15$, with $P = 1.5$ kip, $\theta = 10°$, and $L = 12$ ft. $E_{st} = 29(10^3)$ ksi.

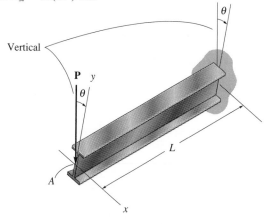

Prob. 16–77

16–78. The $W10 \times 30$ steel cantilevered beam is subjected to unsymmetrical bending caused by the applied moment. Determine the deflection of the centroid at its end A due to the loading. $E_{st} = 29(10^3)$ ksi. *Hint:* Resolve the moment into components and use superposition.

*16–76. The $W12 \times 45$ simply-supported beam is made of steel and is subjected to the loading shown. Using the method of superposition, determine the deflection at its center C. $E_{st} = 29(10^3)$ ksi.

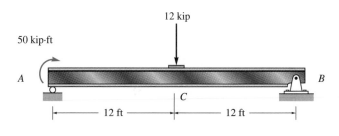

Prob. 16–76

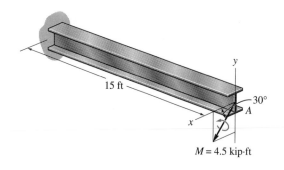

Prob. 16–78

16–79.   The $W8 \times 24$ simply-supported beam is subjected to the loading shown. Using the method of superposition, determine the deflection at its center C. The beam is made of steel for which $E_{st} = 29(10^3)$ ksi.

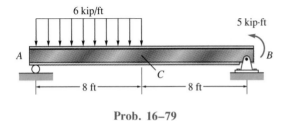

Prob. 16–79

*16–80.   Using the method of superposition, determine the magnitude of M_0 in terms of the distributed load w and dimension a so that the deflection at the center of the beam is zero. EI is constant.

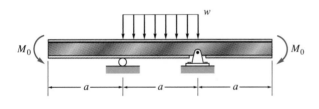

Prob. 16–80

16–81.   The shaft for an electric motor and generator supports the weights of the rotor, the armature, and the commutator. Using the method of superposition, determine the deflection at C of the shaft due to these loadings. The bearings exert vertical forces on the shaft at A and B. EI is constant.

16–82.   The shaft for an electric motor and generator supports the weights of the rotor, the armature, and the commutator. Using the method of superposition, determine the slope of the shaft at the bearings at A and B. The bearings exert vertical forces on the shaft at A and B. EI is constant.

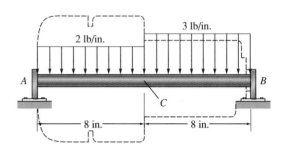

Probs. 16–81/16–82

16.9 Statically Indeterminate Beams

In this section we will illustrate a general method for determining the reactions on statically indeterminate beams. Specifically, a member of any type is classified as *statically indeterminate* if the number of unknown reactions *exceeds* the available number of equilibrium equations.

The additional support reactions on the beam or shaft that are *not needed* to keep it in stable equilibrium are called *redundants*. The number of these redundants is referred to as the *degree of indeterminacy*. For example, consider the beam shown in Fig. 16–43a. If the free-body diagram is drawn, Fig. 16–43b, there will be four unknown support reactions, and since three equilibrium equations are available for solution, the beam is classified as being indeterminate to the first degree. Either A_y, B_y, or M_A can be classified as the redundant, for if any one of these reactions is removed, the beam remains stable and in equilibrium. (A_x cannot be classified as the redundant, for if it were removed, $\Sigma F_x = 0$ would not be satisfied.) In a similar manner, the *continuous beam* in Fig. 16–44a is indeterminate to the second degree, since there are five unknown reactions and only three available equilibrium equations, Fig. 16–44b. Here the two redundant support reactions can be chosen among A_y, B_y, C_y, and D_y.

To determine the reactions on a beam that is statically indeterminate, it is first necessary to specify the redundant reactions. We can determine these redundants from conditions of geometry known as *compatibility conditions*. Once determined, the redundants are then applied to the beam, and the remaining reactions are determined from the equations of equilibrium.

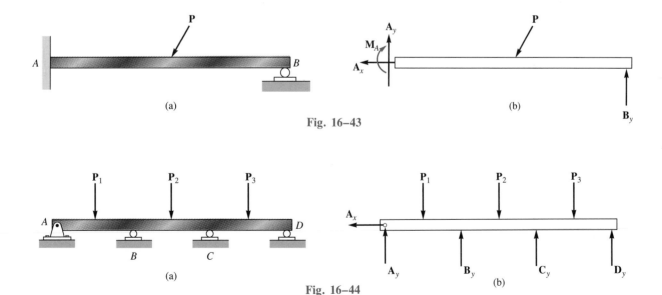

(a)

(b)

Fig. 16–43

(a)

(b)

Fig. 16–44

16.10 Statically Indeterminate Beams— Method of Superposition

In order to apply the method of superposition to the solution of statically indeterminate beams (or shafts), it is first necessary to identify the redundant support reactions. By *removing* them from the beam we obtain the so-called *primary beam,* which is statically determinate and stable, and is subjected *only* to the external load. If we add to this beam a succession of similarly supported beams, each loaded with a *separate* redundant, then by the principle of superposition, we obtain the actual loaded beam. Finally, in order to solve for the redundants, we must write the *conditions of compatibility* that exist at the supports where each of the redundants act. Since the redundant forces are determined directly in this manner, this method of analysis is sometimes called the *force method.* Once the redundants are obtained, the other reactions on the beam are then determined from the three equations of equilibrium.

To clarify these concepts, consider the beam shown in Fig. 16–45a. If we choose the reaction $\mathbf{B}_y$ at the roller as the redundant, then the primary beam is shown in Fig. 16–45b, and the beam with the redundant $\mathbf{B}_y$ acting on it is shown in Fig. 16–45c. The displacement at the roller is to be zero, and since the displacement of point B on the primary beam is v_B, and $\mathbf{B}_y$ causes point B to be displaced upward v_B', we can write the compatibility equation at B as

$$(+\uparrow) \qquad\qquad 0 = -v_B + v_B'$$

The displacements v_B, Fig. 16–45b, and v_B', Fig. 16–45c, can be obtained using any one of the methods discussed in Sec. 16.5 through 16.7. Here we will obtain them directly from the table on the inside back cover. We have

$$v_B = \frac{5PL^3}{48EI} \qquad \text{and} \qquad v_B' = \frac{B_y L^2}{3EI}$$

Substituting into the compatibility equation, we get

$$0 = -\frac{5PL^3}{48EI} + \frac{B_y L^3}{3EI}$$

$$B_y = \frac{5}{16}P$$

Now that B_y is known, the reactions at the wall are determined from the three equations of equilibrium applied to the entire beam, Fig. 16–45d. The results are

$$A_x = 0$$

$$A_y = \frac{11}{16}P$$

$$M_A = \frac{3}{16}PL$$

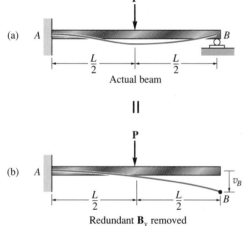

(a)   A ____ B

Actual beam

‖

(b)   A ____ B

Redundant $\mathbf{B}_y$ removed

+

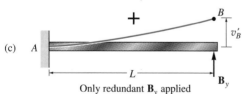

(c)   A ____ B

Only redundant $\mathbf{B}_y$ applied

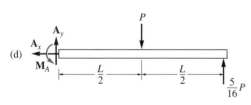

(d)

Fig. 16–45

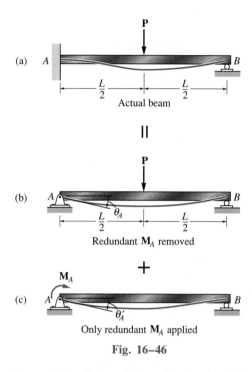

(a)

Actual beam

||

(b)

Redundant M_A removed

+

(c)

Only redundant M_A applied

Fig. 16–46

As stated in Sec. 16.9, choice of the redundant is *arbitrary,* provided the primary beam remains stable. For example, the moment at A for the beam in Fig. 16–46a can also be chosen as the redundant. In this case the capacity of the beam to resist M_A is removed, and so the primary beam is then pin-supported at A, Fig. 16–46b. Also, the redundant at A acts alone on this beam, Fig. 16–46c. Referring to the slope at A caused by the load P as θ_A, Fig. 16–46b, and the slope at A caused by the redundant M_A as θ'_A, Fig. 16–46c, the compatibility equation for slope at A requires

($\curvearrowleft$+) $$0 = \theta_A + \theta'_A$$

Again using the table on the inside back cover, we have

$$\theta_A = \frac{PL^2}{16EI} \quad \text{and} \quad \theta'_A = \frac{M_A L}{3EI}$$

Thus

$$0 = \frac{PL^2}{16EI} + \frac{M_A L}{3EI}$$

$$M_A = -\frac{3}{16}PL$$

This is the same result computed previously. Here the negative sign for M_A simply means that M_A acts in the opposite sense of direction of that shown in Fig. 16–46c.

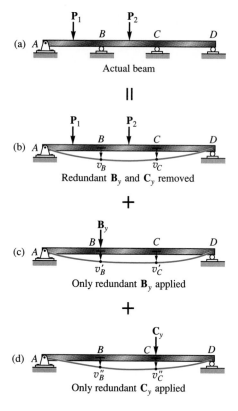

Fig. 16–47

Another example that illustrates this method is given in Fig. 16–47a. In this case the beam is indeterminate to the second degree and therefore *two* compatibility equations will be necessary for the solution. We will choose the forces at the roller supports B and C as redundants. The primary (statically determinate) beam deflects as shown in Fig. 16–47b when the redundants are removed. Each redundant force deflects this beam as shown in Fig. 16–47c and 16–47d, respectively. By superposition, the compatibility equations for the displacements at B and C are

$$(+\downarrow) \qquad 0 = v_B + v_B' + v_B''$$
$$(+\downarrow) \qquad 0 = v_C + v_C' + v_C'' \qquad (16\text{–}23)$$

Here the displacement components v_B' and v_C' will be expressed in terms of the unknown B_y, and the components v_B'' and v_C'' will be expressed in terms of the unknown C_y. When these displacements have been computed and substituted into Eq. 16–23, these equations may then be solved simultaneously for the two unknowns B_y and C_y.

PROCEDURE FOR ANALYSIS

The following procedure provides a means for applying the method of superposition (or the force method) to determine the reactions on statically indeterminate beams.

Principle of Superposition.   Specify the unknown redundant forces or moments that must be removed from the beam in order to make it statically determinate and stable. Then, using the principle of superposition, draw the statically indeterminate beam and show it equal to a sequence of corresponding *statically determinate beams*. The first of these beams, the primary beam, supports the same external loads as the statically indeterminate beam, and each of the other beams "added" to the primary beam shows the beam loaded with a separate redundant force or moment. Also, sketch the deflection curve on each beam and indicate symbolically the displacement or slope at the point of each redundant force or moment. (See Figs. 16–45 through 16–47.)

Compatibility Equations.   Write a compatibility equation for the displacement or slope at each point where there is a redundant force or moment. Determine all the displacements or slopes using an appropriate method as explained in Sec. 16.2 through 16.5. Substitute the results into the compatibility equations and solve for the unknown redundants. If a numerical value for a redundant is *positive,* it has the *same sense of direction* as originally assumed. Similarly, a *negative* numerical value indicates the redundant acts *opposite* to its assumed *sense of direction*.

Equilibrium Equations.   Once the redundant forces and/or moments have been determined, the remaining unknown reactions can be found from the equations of equilibrium applied to the loadings shown on the beam's free-body diagram.

The following examples illustrate application of this procedure. For brevity, all displacements and slopes have been computed using the table on the inside back cover.

Example 16–20

Determine the reactions at the roller support B of the beam shown in Fig. 16–48a, then draw the shear and moment diagrams. EI is constant.

SOLUTION

Principle of Superposition. By inspection, the beam is statically indeterminate to the first degree. The roller support at B will be chosen as the redundant so that $\mathbf{B}_y$ will be determined *directly*. Figures 16–48b and 16–48c show application of the principle of superposition. Here we have assumed that $\mathbf{B}_y$ acts upward on the beam.

Compatibility Equation. Taking positive displacement as downward, the compatibility equation at B is

$$(+\downarrow) \qquad\qquad 0 = v_B - v_B' \qquad\qquad (1)$$

These displacements can be obtained directly from the table on the inside back cover.

$$v_B = \frac{wL^4}{8EI} + \frac{5PL^3}{48EI} = \frac{2(10)^4}{8EI} + \frac{5(8)(10)^3}{48EI} = \frac{3{,}333}{EI} \qquad \downarrow$$

$$v_B' = \frac{PL^3}{3EI} = \frac{B_y(10^3)}{3EI} = \frac{333.3B_y}{EI} \qquad \uparrow$$

Substituting into Eq. 1 and solving yields

$$0 = \frac{3{,}333}{EI} + \frac{333.3B_y}{EI}$$

$$B_y = 10 \text{ kip} \qquad\qquad \textit{Ans.}$$

Equilibrium Equations. Using this result and applying the three equations of equilibrium, we obtain the results shown on the beam's free-body diagram in Fig. 16–48d. The shear and moment diagrams are shown in Fig. 16–48e.

(a)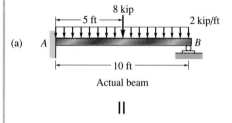

Actual beam

||

(b)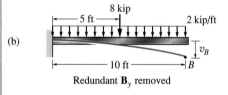

Redundant $\mathbf{B}_y$ removed

+

(c)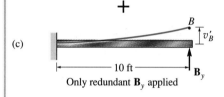

Only redundant $\mathbf{B}_y$ applied

(d), (e)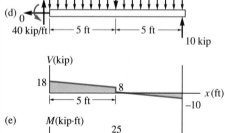

Fig. 16–48

Example 16–21

Determine the reactions on the beam shown in Fig. 16–49a. Due to the loading and poor construction, the roller support at B settles 12 mm. Take $E = 200$ GPa and $I = 80 \, (10^6)$ mm^4.

SOLUTION

Principle of Superposition. By inspection, the beam is indeterminate to the first degree. The roller support at B will be chosen as the redundant. The principle of superposition is shown in Fig. 16–49b and 16–49c. Here $\mathbf{B}_y$ is assumed to act upward on the beam.

Compatibility Equation. With reference to point B, using units of meters, we require

$$(+\downarrow) \qquad\qquad 0.012 = v_B - v_B' \qquad\qquad (1)$$

Using the table on the inside back cover, the displacements are

$$v_B = \frac{5wL^4}{768EI} = \frac{5(24)(8)^4}{768EI} = \frac{640}{EI} \qquad \downarrow$$

$$v_B' = \frac{PL^3}{48EI} = \frac{B_y(8)^3}{48EI} = \frac{10.67B_y}{EI} \qquad \uparrow$$

Thus Eq. 1 becomes

$$0.012EI = 640 - 10.67B_y$$

Expressing E and I in units of kN/m^2 and m^4, respectively, we have

$$0.012(200)(10^6)[80(10^{-6})] = 640 - 10.67B_y$$
$$B_y = 42.0 \text{ kN} \qquad \uparrow \qquad\qquad Ans.$$

Equilibrium Equations. Applying this result to the beam, Fig. 16–49c, we can compute the reactions at A and C using the equations of equilibrium. We obtain

$$\zeta^+ \ \Sigma M_A = 0; \qquad -96(2) + 42.0(4) + C_y(8) = 0$$
$$C_y = -3.00 \text{ kN} = 3.00 \text{ kN} \qquad \downarrow \qquad Ans.$$
$$+\uparrow \ \Sigma F_y = 0; \qquad A_y - 96 + 42.0 - 3.00 = 0$$
$$A_y = 57.0 \text{ kN} \qquad \uparrow \qquad Ans.$$

(a)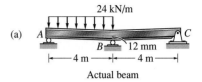

Actual beam

$\|$

(b)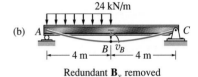

Redundant $\mathbf{B}_y$ removed

$+$

(c)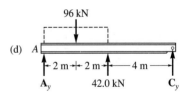

Only redundant $\mathbf{B}_y$ applied

(d)

Fig. 16–49

Example 16–22

The beam in Fig. 16–50a is fixed-supported to the wall at A and pin-connected to a $\frac{1}{2}$-in.-diameter rod BC. If $E = 29(10^3)$ ksi for both members, determine the force developed in the rod due to the loading. The moment of inertia of the beam about its neutral axis is $I = 475$ in^4.

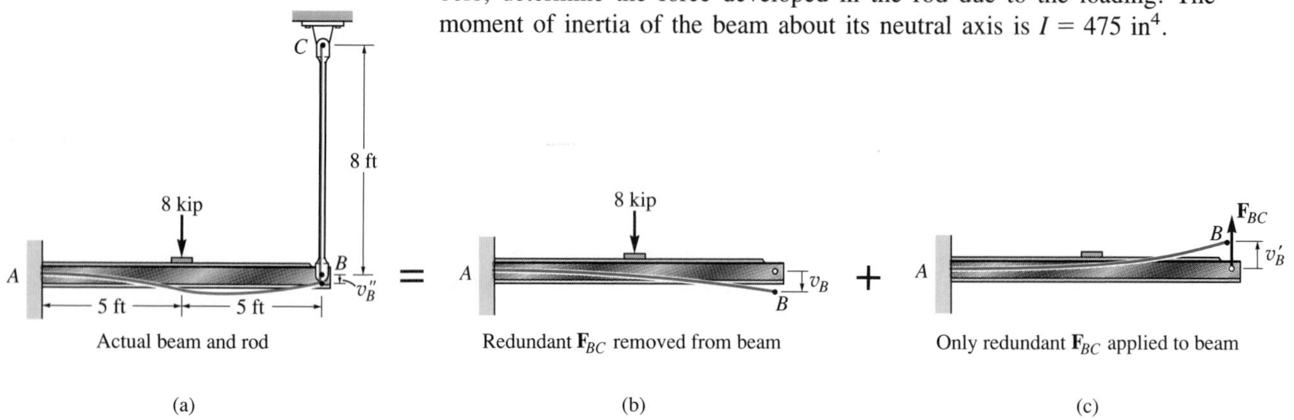

Actual beam and rod

(a)

Redundant $\mathbf{F}_{BC}$ removed from beam

(b)

Only redundant $\mathbf{F}_{BC}$ applied to beam

(c)

Fig. 16–50

SOLUTION I

Principle of Superposition.   By inspection, this problem is indeterminate to the first degree. Here B will undergo an unknown displacement v_B'', since the rod will stretch. The rod will be treated as the redundant and hence the force of the rod is removed from the beam at B, Fig. 16–50b, and then reapplied, Fig. 16–50c.

Compatibility Equation.   At point B we require

$$(+\downarrow) \qquad\qquad v_B'' = v_B - v_B' \qquad\qquad (1)$$

The displacements v_B and v_B' are determined from the table on the back cover. v_B'' is computed from Eq. 10-2. Working in kilopounds and inches, we have

$$v_B'' = \frac{PL}{AE} = \frac{F_{BC}(8\text{ ft})(12\text{ in./ft})}{(\pi/4)(\frac{1}{2}\text{ in.})^2[29(10^3)\text{ kip/in}^2]} = 0.01686F_{BC} \;\downarrow$$

$$v_B = \frac{5PL^3}{48EI} = \frac{5(8\text{ kip})(10\text{ ft})^3(12\text{ in./ft})^3}{48[29(10^3)\text{kip/in}^2](475\text{ in}^4)} = 0.1045\text{ in.} \;\downarrow$$

$$v_B' = \frac{PL^3}{3EI} = \frac{F_{BC}(10\text{ ft})^3(12\text{ in./ft})^3}{3[29(10^3)\text{ kip/in}^2](475\text{ in}^4)} = 0.04181F_{BC} \;\uparrow$$

Thus, Eq. 1 becomes

$$(+\downarrow) \qquad\qquad 0.01686F_{BC} = 0.1045 - 0.04181F_{BC}$$

$$F_{BC} = 1.78\text{ kip} \qquad\qquad\qquad Ans.$$

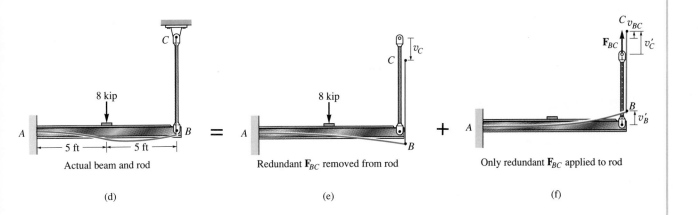

Actual beam and rod   Redundant $\mathbf{F}_{BC}$ removed from rod   Only redundant $\mathbf{F}_{BC}$ applied to rod

(d)   (e)   (f)

SOLUTION II

Principle of Superposition. We can also solve this problem by removing the pin support at C and keeping the rod attached to the beam. In this case the 8-kip load will cause points B and C to be displaced downward the *same amount* v_C, Fig. 16–50e, since no force exists in rod BC. When the redundant force $\mathbf{F}_{BC}$ is applied at point C, it causes the end C of the rod to be displaced upward v'_C and the end B of the beam to be displaced upward v'_B, Fig. 16–50f. The difference in these two displacements, v_{BC}, represents the stretch of the rod due to $\mathbf{F}_{BC}$, so that $v'_C = v_{BC} + v'_B$. Hence, from Fig. 16–50d, 16–50e, and 16–50f, the compatibility of displacement at point C is

$(+\downarrow)$ $\qquad\qquad\qquad 0 = v_C - (v_{BC} + v'_B)$ $\qquad\qquad$ (2)

From Solution I, we have

$$v_C = v_B = 0.1045 \text{ in.} \quad \downarrow$$
$$v_{BC} = v''_B = 0.01686 F_{BC} \quad \uparrow$$
$$v'_B = 0.04181 F_{BC} \quad \uparrow$$

Therefore Eq. 2 becomes

$(+\downarrow)$ $\qquad 0 = 0.1045 - (0.01686 F_{BC} + 0.04181 F_{BC})$

$$F_{BC} = 1.78 \text{ kip} \qquad\qquad\qquad Ans.$$

Example 16–23

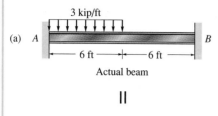

(a) A   B

3 kip/ft

6 ft ——┼—— 6 ft

Actual beam

||

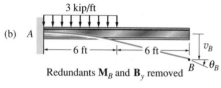

(b) A

3 kip/ft

6 ft ——┼—— 6 ft

v_B

θ_B

Redundants M_B and B_y removed

+

(c) A

$\mathbf{B}_y$

v_B'

θ_B'

——— 12 ft ———

Only redundant B_y applied

+

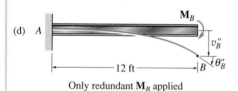

(d) A

$\mathbf{M}_B$

v_B''

θ_B''

——— 12 ft ———

Only redundant M_B applied

Fig. 16–51

Determine the moment at B for the beam shown in Fig. 16–51a. EI is constant. Neglect the effects of axial load.

SOLUTION

Principle of Superposition.   If the axial load on the beam is neglected, there will be a vertical force and moment at A and B. Since there are only two available equations of equilibrium ($\Sigma M = 0$, $\Sigma F_y = 0$), the problem is indeterminate to the second degree. We will assume that $\mathbf{B}_y$ and $\mathbf{M}_B$ are redundant, so that by the principle of superposition, the beam is represented as a cantilever, loaded *separately* by the distributed load and reactions $\mathbf{B}_y$ and $\mathbf{M}_B$, Fig. 16–51b, 16–51c, and 16–51d.

Compatibility Equations.   Referring to the displacement and slope at B in Fig. 16–51b, 16–51c, and 16–51d, we require

$$(\curvearrowleft +) \qquad 0 = \theta_B + \theta_B' + \theta_B'' \qquad (1)$$

$$(+\downarrow) \qquad 0 = v_B + v_B' + v_B'' \qquad (2)$$

Using the table on the inside back cover to compute the slopes and displacements, we have

$$\theta_B = \frac{wL^3}{48EI} = \frac{3(12)^3}{48EI} = \frac{108}{EI} \;\; \downarrow$$

$$v_B = \frac{7wL^4}{384EI} = \frac{7(3)(12)^4}{384EI} = \frac{1134}{EI} \;\; \downarrow$$

$$\theta_B' = \frac{PL^2}{2EI} = \frac{B_y(12)^2}{2EI} = \frac{72B_y}{EI} \;\; \downarrow$$

$$v_B' = \frac{PL^3}{3EI} = \frac{B_y(12)^3}{3EI} = \frac{576B_y}{EI} \;\; \downarrow$$

$$\theta_B'' = \frac{ML}{EI} = \frac{M_B(12)}{EI} = \frac{12M_B}{EI} \;\; \downarrow$$

$$v_B'' = \frac{ML^2}{2EI} = \frac{M_B(12)^2}{2EI} = \frac{72M_B}{EI} \;\; \downarrow$$

Substituting these values into Eqs. 1 and 2 and canceling out the common factor EI, we get

$$(\curvearrowleft +) \qquad 0 = 108 + 72B_y + 12M_B$$

$$(+\downarrow) \qquad 0 = 1134 + 576B_y + 72M_B$$

Solving these equations simultaneously gives

$$B_y = -3.375 \text{ kip}$$

$$M_B = 11.25 \text{ kip} \cdot \text{ft}$$

Ans.

PROBLEMS

16–83. The beam has a moment of inertia I and is supported at its ends by a pin and roller and at its center by a rod having a cross-sectional area A and length L. Determine the tension in the rod when a uniform load w is placed on the beam. The modulus of elasticity is E.

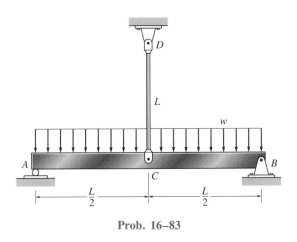

Prob. 16–83

*16–84. The beam AB has a moment of inertia $I = 475$ in^4 and rests on the smooth supports at its ends. A 0.75-in.-diameter rod CD is welded to the center of the beam and to the fixed support at D. If the temperature of the rod is decreased by 150°F, determine the force developed in the rod. The beam and rod are both made of steel, for which $E_{st} = 29(10^3)$ ksi and $\alpha_{st} = 6.5(10^{-6})$/°F.

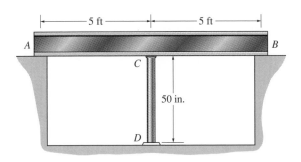

Prob. 16–84

16–85. Determine the reactions at the supports A and B. EI is constant.

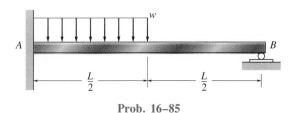

Prob. 16–85

16–86. Determine the deflection at the end B of the clamped steel strip. The spring has a stiffness of $k = 2$ N/mm. The strip is 5 mm wide and 10 mm thick. Also, draw the shear and moment diagrams for the strip. $E_{st} = 200$ GPa.

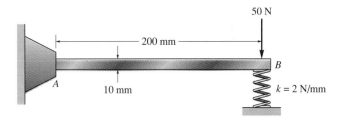

Prob. 16–86

16–87. The assembly consists of a steel and an aluminum bar, each of which is 1 in. thick, fixed at its ends A and B, and pin-connected to the *rigid* short link CD. If a horizontal force of 80 lb is applied to the link as shown, determine the moments created at A and B. $E_{st} = 29(10^3)$ ksi, $E_{al} = 10(10^3)$ ksi.

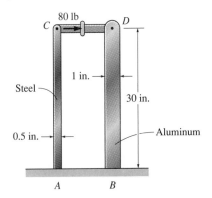

Prob. 16–87

*16–88.   Determine the reaction at support C. EI is constant for both beams.

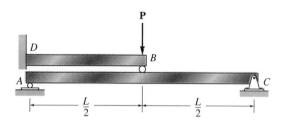

Prob. 16–88

16–89.   Determine the reactions at A and B. The support at A can exert only a moment on the beam. EI is constant.

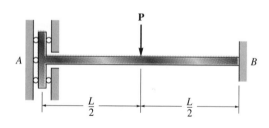

Prob. 16–89

16–90.   The 1-in.-diameter steel shaft is supported by unyielding bearings at A and C. The bearing at B rests on a simply-supported steel wide-flange beam having a moment of inertia of $I = 500$ in⁴. If the belt loads on the pulley are 400 lb each, determine the vertical reactions at A, B, and C. $E_{st} = 29(10^3)$ ksi.

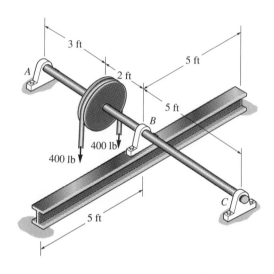

Prob. 16–90

16–91.   Determine the reactions on the beam. EI is constant.

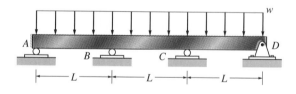

Prob. 16–91

REVIEW PROBLEMS

*16–92. Determine the equations of the elastic curve for the beam using the method of integration and the x_1 and x_2 coordinates. Specify the slope at A and the maximum deflection. EI is constant.

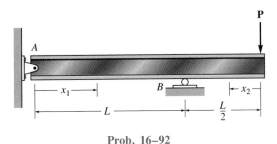

Prob. 16–92

16–94. The beam is subjected to the load shown. Using discontinuity functions, determine the equation of the elastic curve. EI is constant.

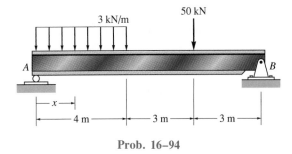

Prob. 16–94

16–95. Determine the equation of the elastic curve for the beam using the x coordinate. Specify the slope at A and the maximum deflection. EI is constant.

Prob. 16–95

16–93. The compound beam is made from two sections, which are pinned together at B. Use Appendix D and select the lightest wide-flange beam that would be safe for each section if the allowable bending stress is $\sigma_{allow} = 24$ ksi and the allowable shear stress is $\tau_{allow} = 14$ ksi. The beam supports a pipe loading of 4 kip and 6 kip as shown.

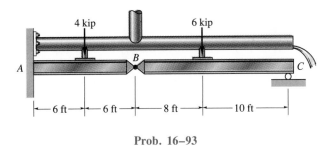

Prob. 16–93

*16–96. The beam is subjected to the load **P** as shown. Determine the magnitude of force **F** that must be applied at the end of the overhang C so that the deflection at C is zero. Use the moment-area method. EI is constant.

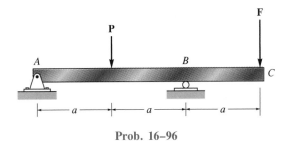

Prob. 16–96

16–97. The shaft supports the two pulley loads shown. Using discontinuity functions, determine the equation of the elastic curve. The bearings at A and B exert only vertical reactions on the shaft. What is the maximum deflection? EI is constant.

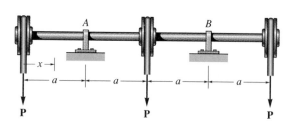

Prob. 16–97

16–98. The steel shaft is subjected to the loadings developed in the belts passing over the two pulleys. If the bearings at A and B exert vertical reactions on the shaft, determine the slope at A. The shaft has a diameter of 0.75 in. Use the moment-area method. $E_{st} = 29(10^3)$ ksi.

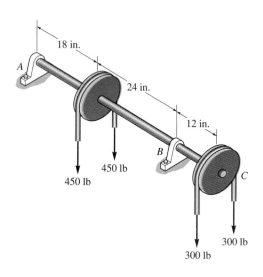

Prob. 16–98

16–99. Determine the deflection at C and the slope of the beam at A, B, and C. Use the moment-area method. EI is constant.

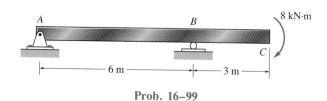

Prob. 16–99

*16–100. The beam is subjected to the linearly varying distributed load. Determine the maximum deflection of the beam. Use the method of integration. EI is constant.

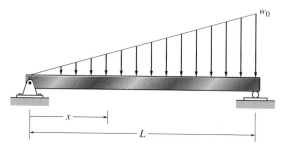

Prob. 16–100

16–101. The 25-mm-diameter steel shaft is supported at A and B by bearings. If the tension in the belt on the pulley at C is 0.75 kN, determine the largest belt tension T on the pulley at D so that the slope of the shaft at A or B does not exceed 0.02 rad. The bearings exert only vertical reactions on the shaft. Use the moment-area method. $E_{st} = 200$ GPa.

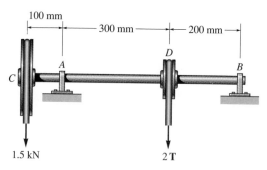

Prob. 16–101

17 Buckling of Columns

In this chapter we will discuss the behavior of columns and indicate some of the methods used for their design. The chapter begins with a general discussion of buckling, followed by a determination of the axial load needed to buckle a so-called ideal column. Inelastic buckling of a column is presented as a special topic.

17.1 Critical Load

Whenever a member is designed, it is necessary that it satisfy specific strength, deflection, and stability requirements. In the preceding chapters we have discussed some of the methods used to determine a member's strength and deflection, while assuming that the member was always in stable equilibrium. Some members, however, may be subjected to compressive loadings, and if these members are long and slender the loading may be large enough to cause the member to deflect laterally or sideway. To be specific, long slender members subjected to an axial compressive force are called *columns,* and the lateral deflection that occurs is called *buckling.* Quite often the buckling of a column can lead to a sudden and dramatic failure of a structure or mechanism, and as a result, special attention must be given to the design of columns so that they can safely support their intended loadings without buckling.

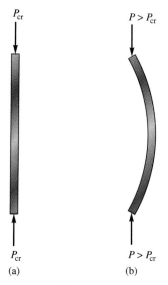

P_{cr}

P_{cr}

(a)

$P > P_{cr}$

$P > P_{cr}$

(b)

Fig. 17–1

The maximum axial load that a column can support when it is on the *verge* of buckling is called the *critical load, P_{cr}* Fig. 17–1a. Any additional loading will cause the column to buckle and therefore deflect laterally as shown in Fig. 17–1b. In order to better understand the nature of this instability, consider a two-bar mechanism consisting of weightless bars that are rigid and pin-connected at their ends, Fig. 17–2a. When the bars are in the vertical position, the spring, having a stiffness k, is unstretched, and a *small* vertical force **P** is applied at the top of one of the bars. We can upset this equilibrium position by displacing the pin at A by a small amount Δ, Fig. 17–2b. As shown on the free-body diagram of the pin, Fig. 17–2c, the spring will produce a restoring force $F = k\Delta$, while the applied load **P** develops two horizontal components, $P_x = P \tan \theta$, which tend to push the pin (and the bars) further out of equilibrium. Since θ is small, $\Delta = \theta(L/2)$ and $\tan \theta \approx \theta$. Thus the restoring spring force becomes $F = k\theta L/2$, and the disturbing force is $2P_x = 2P\theta$.

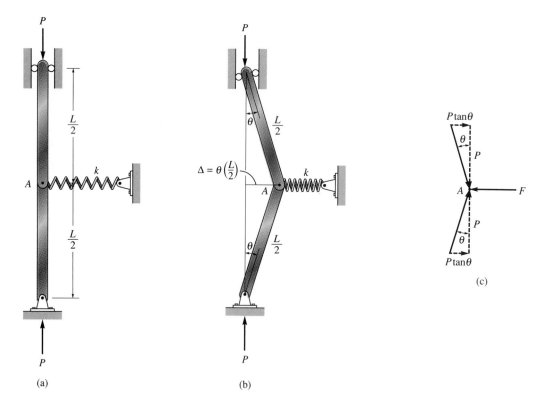

(a)

(b)

(c)

Fig. 17–2

If the restoring force is greater than the disturbing force, that is, $k\theta L/2 > 2P\theta$, then, noticing that θ cancels out, we can solve for P, which gives

$$P < \frac{kL}{4} \qquad \text{stable equilibrium}$$

This is a condition for *stable equilibrium,* since the force developed by the spring would be adequate in restoring the bars back to their vertical position. On the other hand, if $kL\theta/2 < 2P\theta$, or

$$P > \frac{kL}{4} \qquad \text{unstable equilibrium}$$

then the mechanism would be in *unstable equilibrium.* In other words, if this load P is applied, and a slight displacement occurs at A, the mechanism will tend to move out of equilibrium and not be restored to its original position.

The intermediate value of P, defined by requiring $kL\theta/2 = 2P\theta$, is the *critical load.* Here

$$P_{cr} = \frac{kL}{4} \qquad \text{neutral equilibrium}$$

This loading represents a case of the mechanism being in *neutral equilibrium.* Since P_{cr} is *independent* of the (small) displacement θ of the bars, any slight disturbance given to the mechanism will not cause it to move further out of equilibrium, nor will it be restored to its original position. Instead, the bars will *remain* in the deflected position.

These three different states of equilibrium are represented graphically in Fig. 17–3. The transition point where the load is equal to the critical value $P = P_{cr}$ is called the *bifurcation point.* At this point the mechanism will be in equilibrium for any *small value* of θ, measured either to the right or to the left of the vertical. Physically, P_{cr} represents the load for which the mechanism is on the verge of buckling. It is quite valid to determine this value by assuming *small displacements* as done here; however, it should be understood that P_{cr} may *not* be the largest value of P that the mechanism can support. Indeed, if a larger load is placed on the bars, then the mechanism may have to undergo a further deflection before the spring is compressed or elongated enough to hold the mechanism in equilibrium.

Like the two-bar mechanism just discussed, the critical buckling loads on columns supported in various ways can be obtained, and the method used to do this will be explained in the next section. Although in engineering design the critical load may be considered to be the largest load the column can support, realize that, like the two-bar mechanism in the deflected or buckled position, a column may actually support an even greater load than P_{cr}. Unfortunately, however, this loading may require the column to undergo a *large* deflection, which is generally not tolerated in engineering structures or machines. For example, it may take only a few newtons of force to buckle a meterstick, but the additional load it may support can be applied only after the stick undergoes a relatively large lateral deflection.

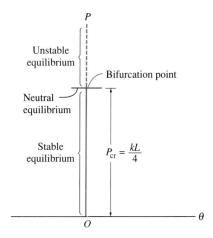

Fig. 17–3

17.2 Ideal Column with Pin Supports

In this section we will determine the critical buckling load for a column that is pin-supported as shown in Fig. 17–4a. The column to be considered is an *ideal column,* meaning one that is perfectly straight before loading, is made of homogeneous material, and upon which the load is applied through the centroid of the column's cross section. It is further assumed that the material behaves in a linear-elastic manner and that the column buckles or bends in a single plane. In reality, the conditions of column straightness and load application are never accomplished; however, the analysis to be performed on an "ideal column" is similar to that used to analyze initially crooked columns or those having an eccentric load application. These more realistic cases will be discussed later in this chapter.

Since an ideal column is straight, theoretically the axial load P could be increased until failure occurs by either fracture or yielding of the material. However, when the critical load P_{cr} is reached, the column is on the verge of becoming unstable, so that a small lateral force F, Fig. 17–4b, will cause the column to remain in the deflected position when F is removed, Fig. 17–4c. Any slight reduction in the axial load P from P_{cr} will allow the column to straighten out, and any slight increase in P, beyond P_{cr}, will cause further increases in lateral deflection.

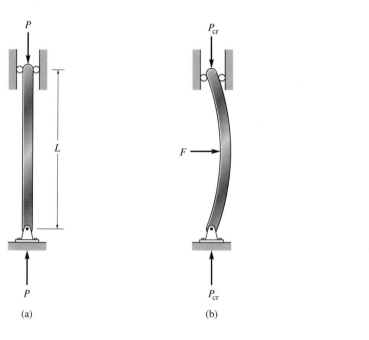

Fig. 17–4

Whether or not a column will remain stable or become unstable when subjected to an axial load will depend on its ability to restore itself, which is based on its resistance to bending. Hence, in order to determine the critical load and the buckled shape of the column, we will apply Eq. 16–11, which relates the internal moment in the column to its deflected shape, i.e.,

$$EI \frac{d^2v}{dx^2} = M \qquad (17\text{–}1)$$

Recall that this equation assumes that the slope of the elastic curve is small* and that deflections occur only by bending. When the column is in its deflected position, Fig. 17–5a, the internal bending moment can be determined by using the method of sections. The free-body diagram shown in Fig. 17–5b will be considered. Here both the deflection v and the internal moment M are shown in the *positive direction* according to the sign convention used to establish Eq. 17–1. Summing moments, the internal moment is $M = -Pv$. Thus Eq. 17–1 becomes

$$EI \frac{d^2v}{dx^2} = -Pv$$

$$\frac{d^2v}{dx^2} + \left(\frac{P}{EI}\right)v = 0 \qquad (17\text{–}2)$$

This is a homogeneous, second-order, linear differential equation with constant coefficients. It can be shown by using the methods of differential equations, or by direct substitution into Eq. 17–2, that the general solution is

$$v = C_1 \sin\left(\sqrt{\frac{P}{EI}}\,x\right) + C_2 \cos\left(\sqrt{\frac{P}{EI}}\,x\right) \qquad (17\text{–}3)$$

The two constants of integration are determined from the boundary conditions at the ends of the column. Since $v = 0$ at $x = 0$, then $C_2 = 0$. And since $v = 0$ at $x = L$, then

$$C_1 \sin\left(\sqrt{\frac{P}{EI}}\,L\right) = 0$$

This equation is satisfied if $C_1 = 0$; however, then $v = 0$, which is a *trivial solution* that requires the column to always remain straight, even though the load causes the column to become unstable. The other possibility is for

$$\sin\left(\sqrt{\frac{P}{EI}}\,L\right) = 0$$

which is satisfied if

$$\sqrt{\frac{P}{EI}}\,L = n\pi$$

*If large deflections are to be considered, the more accurate differential equation, Eq. 16–5, $EI(d^2v/dx^2)/[1 + (dv/dx)^2]^{3/2} = M$ must be used.

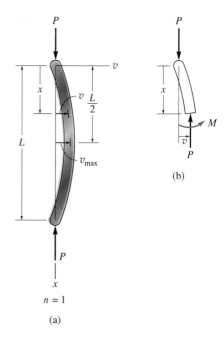

Fig. 17–5

or

$$P = \frac{n^2 \pi^2 EI}{L^2} \qquad n = 1, 2, 3, \ldots \qquad (17-4)$$

The *smallest value* of P is obtained when $n = 1$, so the *critical load* for the column is therefore

$$P_{cr} = \frac{\pi^2 EI}{L^2}$$

This load is sometimes referred to as the *Euler load,* after the Swiss mathematician Leonhard Euler, who originally solved this problem in 1757. The corresponding buckled shape is defined by the equation

$$v = C_1 \sin \frac{\pi x}{L}$$

Here the constant C_1 represents the maximum deflection, v_{max}, which occurs at the midpoint of the column, Fig. 17–5a. Specific values for C_1 cannot be obtained, since the exact deflected form for the column is unknown once it has buckled. It has been assumed, however, that this deflection is small.

Realize that n in Eq. 17–4 represents the number of waves in the deflected shape of the column. For example, if $n = 2$, then from Eqs. 17–3 and 17–4, *two waves* will appear in the buckled shape, Fig. 17–5c, and the column will support a critical load that is $4P_{cr}$ just prior to buckling. Since this value is

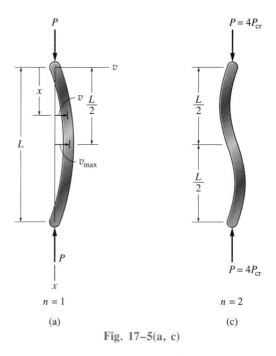

Fig. 17–5(a, c)

four times the critical load, and the deflected shape is unstable, this form of buckling, practically speaking, will not exist, unless the column has a lateral brace at its center.

Like the two-bar mechanism discussed in Sec. 17–1, we can represent the load-deflection characteristics of the ideal column by the graph shown in Fig. 17–6. The bifurcation point represents the state of *neutral equilibrium,* at which point the *critical load* acts on the column. Here the column is on the verge of impending buckling.

It should be noted that the critical load is independent of the strength of the material; rather it depends only on the column's dimensions (*I* and *L*) and the material's stiffness or modulus of elasticity *E*. For this reason, as far as elastic buckling is concerned, columns made, for example, of high-strength steel offer no advantage over those made of lower-strength steel, since the modulus of elasticity for both is approximately the same. Also note that the load-carrying capacity of a column will increase as the moment of inertia of the cross section increases. Thus, efficient columns are designed so that most of the column's cross-sectional area is located as far away as possible from the principal centroidal axes for the section. This is why hollow sections such as tubes are more economical than solid sections. Furthermore, wide-flange sections, and columns that are "built up" from channels, angles, plates, etc., are better than sections that are solid and rectangular.

It is also important to realize that a column will buckle about the principal axis of the cross section having the *least moment of inertia* (the weakest axis). For example, a column having a rectangular cross section, like a meter stick, as shown in Fig. 17–7, will buckle about the *a–a* axis, not the *b–b* axis. As a result, engineers usually try to achieve a balance, keeping the moments of inertia the same in all directions. Geometrically, then, circular tubes would make excellent columns. Also, square tubes or those shapes having $I_x \approx I_y$ are often selected for columns.

Summarizing the above discussion, the buckling equation for a pin-supported column can be rewritten and the terms defined as follows:

$$P_{cr} = \frac{\pi^2 EI}{L^2} \tag{17–5}$$

where

P_{cr} = critical or maximum axial load on the column just before it begins to buckle. This load must *not* cause the stress in the column to exceed the proportional limit

E = modulus of elasticity for the material

I = *least* moment of inertia for the column's cross-sectional area

L = unsupported length of the column, whose ends are pinned

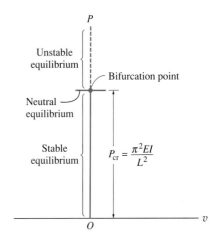

Fig. 17–6

Fig. 17–7

For purposes of design, Eq. 17–5 can also be written in a more useful form by expressing $I = Ar^2$, where A is the cross-sectional area and r is the *radius of gyration* of the cross-sectional area. Thus,

$$P_{cr} = \frac{\pi^2 E(Ar^2)}{L^2}$$

$$\left(\frac{P}{A}\right)_{cr} = \frac{\pi^2 E}{(L/r)^2}$$

or

$$\boxed{\sigma_{cr} = \frac{\pi^2 E}{(L/r)^2}} \qquad (17\text{–}6)$$

Here

$\sigma_{cr} =$ critical stress, which is an *average stress* in the column just before the column buckles. This stress is an *elastic stress* and therefore $\sigma_{cr} \leq \sigma_Y$

$E =$ modulus of elasticity for the material

$L =$ unsupported length of the column, whose ends are pinned

$r =$ *smallest* radius of gyration of the column, determined from $r = \sqrt{I/A}$, where I is the *least* moment of inertia of the column's cross-sectional area A

The geometric ratio L/r in Eq. 17–6 is known as the *slenderness ratio*. It is a measure of the column's flexibility, and as will be discussed later, it serves as a means of classifying columns as long, intermediate, or short.

It is possible to graph Eq. 17–6 using axes that represent the critical stress versus the slenderness ratio. Examples of this graph for columns made of a typical structural steel and aluminum alloy are shown in Fig. 17–8. Note that the curves are hyperbolic and are valid only for critical stresses below the material's yield point (proportional limit), since the material must behave elastically. For the steel the yield stress is $(\sigma_Y)_{st} = 36$ ksi $[E_{st} = 29(10^3)$ ksi$]$ and for the aluminum it is $(\sigma_Y)_{al} = 27$ ksi $[E_{al} = 10(10^3)$ ksi$]$. Substituting $\sigma_{cr} = \sigma_Y$ into Eq. 17–6, the *smallest* acceptable slenderness ratios for the steel and aluminum columns are therefore $(L/r)_{st} = 89$ and $(L/r)_{al} = 60.5$. Thus, for a steel column, if $(L/r)_{st} \geq 89$, Euler's formula can be used to determine the buckling load since the stress in the column remains elastic. On the other hand, if $(L/r)_{st} < 89$, the column's stress will exceed the yield point before buckling can occur, and therefore the Euler formula is not valid in this case.

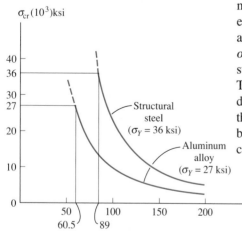

Fig. 17–8

Example 17-1

A 24-ft-long steel tube having the cross section shown in Fig. 17–9 is to be used as a pin-ended column. If $E_{st} = 29(10^3)$ ksi, determine the maximum allowable load the column can support so that it does not buckle. Take $\sigma_Y = 36$ ksi.

SOLUTION
Using Eq. 17–5 to obtain the critical load,

$$P_{cr} = \frac{\pi^2 EI}{L^2}$$

$$= \frac{\pi^2 [29(10^3) \text{ kip/in}^2](\frac{1}{4}\pi(3)^4 - \frac{1}{4}\pi(2.75)^4) \text{ in}^4}{[24 \text{ ft}(12 \text{ in./ft})]^2}$$

$$= 64.5 \text{ kip} \hspace{3cm} \textit{Ans.}$$

This force creates an average compressive stress in the column of

$$\sigma_{cr} = \frac{P_{cr}}{A} = \frac{64.5 \text{ kip}}{[\pi(3)^2 - \pi(2.75)^2] \text{ in}^2} = 14.3 \text{ ksi}$$

Since $\sigma_{cr} < \sigma_Y = 36$ ksi, application of Euler's equation is appropriate.

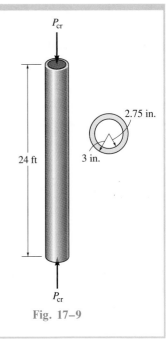

Fig. 17–9

Example 17-2

The steel $W \, 8 \times 31$ shown in Fig. 17–10 is to be used as a pin-connected column. Determine the largest load it can support before it either begins to buckle or the steel yields. $E_{st} = 29(10^3)$ ksi, and $\sigma_Y = 36$ ksi.

SOLUTION
From the table in Appendix D, the column's cross-sectional area and moments of inertia are $A = 9.13 \text{ in}^2$, $I_x = 110 \text{ in}^4$, and $I_y = 37.1 \text{ in}^4$. By inspection, buckling will occur about the y–y axis. Why? Applying Eq. 17–5, we have

$$P_{cr} = \frac{\pi^2 EI}{L^2} = \frac{\pi^2 [29(10^3) \text{ kip/in}^2](37.1 \text{ in}^4)}{[12 \text{ ft}(12 \text{ in./ft})]^2} = 512 \text{ kip}$$

When fully loaded, the average compressive stress in the column is

$$\sigma_{cr} = \frac{P_{cr}}{A} = \frac{512 \text{ kip}}{9.13 \text{ in}^2} = 56.1 \text{ ksi}$$

Since this stress exceeds the yield stress (36 ksi), the load P is determined from simple compression:

$$36 \text{ ksi} = \frac{P}{9.13 \text{ in}^2}; \hspace{2cm} P = 329 \text{ kip} \hspace{2cm} \textit{Ans.}$$

In actual practice, a factor of safety would be placed on this loading.

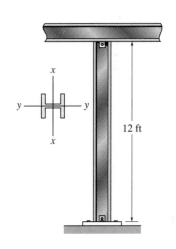

Fig. 17–10

17.3 Columns Having Various Types of Supports

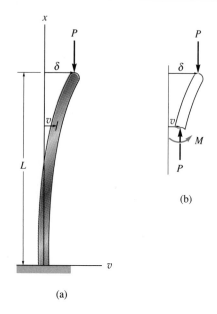

x

P

δ

v

L

P

δ

v

M

P

(b)

(a)

Fig. 17–11

In Sec. 17.2 we derived the Euler load for a column that is pin-connected or free to rotate at its ends. Oftentimes, however, columns may be supported in some other way. For example, consider the case of a column fixed at its base and free at the top, Fig. 17–11a. Determination of the buckling load on this column follows the same procedure as that used for the pinned column. From the free-body diagram in Fig. 17–11b, the internal moment at the arbitrary section is $M = P(\delta - v)$. Consequently, the differential equation for the deflection curve is

$$EI \frac{d^2v}{dx^2} = P(\delta - v)$$

$$\frac{d^2v}{dx^2} + \frac{P}{EI}v = \frac{P}{EI}\delta \qquad (17\text{–}7)$$

Unlike Eq. 17–2, this equation is nonhomogeneous because of the nonzero term on the right side. The solution consists of both a complementary and particular solution, namely,

$$v = C_1 \sin\left(\sqrt{\frac{P}{EI}}\, x\right) + C_2 \cos\left(\sqrt{\frac{P}{EI}}\, x\right) + \delta$$

The constants are determined from the boundary conditions. At $x = 0$, $v = 0$, so that $C_2 = -\delta$. Also,

$$\frac{dv}{dx} = C_1 \sqrt{\frac{P}{EI}} \cos\left(\sqrt{\frac{P}{EI}}\, x\right) - C_2 \sqrt{\frac{P}{EI}} \sin\left(\sqrt{\frac{P}{EI}}\, x\right)$$

At $x = 0$, $dv/dx = 0$, so that $C_1 = 0$. The deflection curve is therefore

$$v = \delta\left[1 - \cos\left(\sqrt{\frac{P}{EI}}\, x\right)\right] \qquad (17\text{–}8)$$

Since the deflection at the top of the column is δ, that is, at $x = L$, $v = \delta$, we require

$$\delta \cos\left(\sqrt{\frac{P}{EI}}\, L\right) = 0$$

The trivial solution $\delta = 0$ indicates that no buckling occurs, regardless of the load P. Instead,

$$\cos\left(\sqrt{\frac{P}{EI}}\, L\right) = 0 \qquad \text{or} \qquad \sqrt{\frac{P}{EI}}\, L = \frac{n\pi}{2}$$

The smallest critical load occurs when $n = 1$, so that

$$P_{cr} = \frac{\pi^2 EI}{4L^2} \qquad (17\text{–}9)$$

By comparison with Eq. 17–5 it is seen that a column fixed-supported at its base will carry only one-fourth the critical load that can be applied to a pin-supported column.

Other types of supported columns are analyzed in much the same way and will not be covered in detail here. Instead, we will tabulate the results for the most common types of column support and show how to apply these results by writing Euler's formula in a general form.

Effective Length. As stated previously, the Euler formula, Eq. 17–5, was developed for the case of a column having ends that are pinned or free to rotate. In other words, L in the equation represents the unsupported distance between the points of zero moment. If the column is supported in other ways, then Euler's formula can be used to determine the critical load provided "L" represents the distance between the zero-moment points. This distance is called the column's *effective length*, L_e. Obviously, for a pin-ended column $L_e = L$, Fig. 17–12a. For the fixed and free-ended column analyzed above, the deflection curve was found to be one-half that of a column that is pin-connected and has a length of $2L$, Fig. 17–12b. Thus the effective length between the points of zero moment is $L_e = 2L$. Examples for two other columns with different end supports are also shown in Fig. 17–12. The column fixed at its ends, Fig. 17–12c, has inflection points or points of zero moment $L/4$ from each support. The effective length is therefore represented by the middle half of its length, that is, $L_e = 0.5L$. Lastly, the pin- and fixed-ended column, Fig. 17–12d, has an inflection point at approximately $0.7L$ from its pinned end, so that $L_e = 0.7L$.

Rather than specifying the column's effective length, many design codes provide column formulas that employ a dimensionless coefficient K called the *effective-length factor*. K is defined from

$$L_e = KL \qquad (17\text{–}10)$$

Specific values of K are also given in Fig. 17–12. Based on this generality, we can therefore write Euler's formula as

$$\boxed{P_{cr} = \frac{\pi^2 EI}{(KL)^2}} \qquad (17\text{–}11)$$

or

$$\boxed{\sigma_{cr} = \frac{\pi^2 E}{(KL/r)^2}} \qquad (17\text{–}12)$$

Here (KL/r) is the column's *effective-slenderness ratio*. For example, note that for the column fixed at its base and free at its end, $K = 2$, and therefore Eq. 17–11 gives the same result as Eq. 17–9.

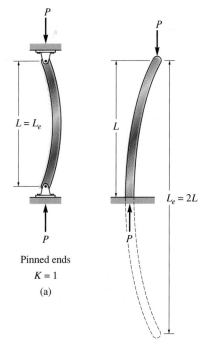

Pinned ends
$K = 1$
(a)

Fixed and free ends
$K = 2$
(b)

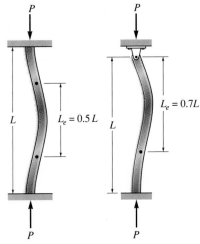

Fixed ends
$K = 0.5$
(c)

Pinned and fixed ends
$K = 0.7$
(d)

Fig. 17–12

Example 17–3

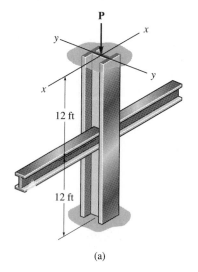

P

12 ft

12 ft

(a)

A *W* 6 × 15 steel column is 24 ft long and is fixed at its ends as shown in Fig. 17–13*a*. Its load-carrying capacity is increased by bracing it about the *y–y* (weak) axis using struts that are assumed to be pin-connected to its midheight. Determine the load it can support so that the column does not buckle nor the material exceed the yield stress. Take $E_{st} = 29(10^3)$ ksi and $\sigma_Y = 60$ ksi.

SOLUTION

The buckling behavior of the column will be *different* about the *x* and *y* axes due to the *y–y* axis bracing. The buckled shape for each of these cases is shown in Fig. 17–13*b* and 17–13*c*. From Fig. 17–13*b*, the effective length for the *x–x* axis is $(KL)_x = 0.5(24 \text{ ft}) = 12 \text{ ft} = 144$ in., and from Fig. 17–13*c*, for the *y–y* axis, $(KL)_y = 0.7(24 \text{ ft}/2) = 8.40 \text{ ft} = 100.8$ in. The moments of inertia for a *W* 6 × 15 are determined from the table in Appendix D. We have $I_x = 29.1 \text{ in}^4$, $I_y = 9.32 \text{ in}^4$.

Applying Eq. 17–12, we have

$$(P_{cr})_x = \frac{\pi^2 E I_x}{(KL)_x^2} = \frac{\pi^2 [29(10^3) \text{ ksi}]29.1 \text{ in}^4}{(144 \text{ in.})^2} = 401.7 \text{ kip} \qquad (1)$$

$$(P_{cr})_y = \frac{\pi^2 E I_y}{(KL)_y^2} = \frac{\pi^2 [29(10^3) \text{ ksi}]9.32 \text{ in}^4}{(100.8 \text{ in.})^2} = 262.5 \text{ kip} \qquad (2)$$

By comparison, buckling will occur about the *y–y* axis, Fig. 17–13*c*.

The area of the cross section is 4.43 in², so the average compressive stress in the column will be

$$\sigma_{cr} = \frac{P_{cr}}{A} = \frac{262.5 \text{ kip}}{4.43 \text{ in}^2} = 59.3 \text{ ksi}$$

Since this stress is less than the yield stress, buckling will occur before the material yields. Thus,

$$P_{cr} = 263 \text{ kip} \qquad \qquad \textit{Ans.}$$

Note: From Eq. 17–11 it can be seen that buckling will always occur about the column axis having the *largest* slenderness ratio, since a large slenderness ratio will give a small critical load. Thus, using the data for the radius of gyration from the table in Appendix *D*, we have

$$\left(\frac{KL}{r}\right)_x = \frac{144}{2.56} = 56.2$$

$$\left(\frac{KL}{r}\right)_y = \frac{100.8}{1.46} = 69.0$$

Hence, *y–y* axis buckling will occur, the same conclusion as reached by comparing Eqs. 1 and 2.

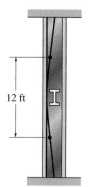

12 ft

x–x axis buckling

(b)

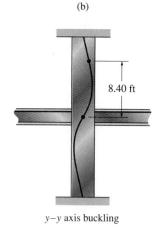

8.40 ft

y–y axis buckling

(c)

Fig. 17–13

The aluminum column is fixed at its bottom and is braced at its top by cables so as to prevent movement at the top in the x–z plane, Fig. 17–14a. If it is assumed to be fixed at its base, determine the largest allowable load P that can be applied. Use a factor of safety for buckling of F.S. = 3.0. Take E_{al} = 70 GPa, σ_Y = 215 MPa, A = 7.5(10^{-3}) m^2, I_x = 61.3(10^{-6}) m^4, I_y = 23.2(10^{-6}) m^4.

SOLUTION
Buckling about the x and y axes is shown in Fig. 17–14b and 17–14c, respectively. Using Fig. 17–12, for x–x axis buckling, K = 2, so $(KL)_x$ = 2(5 m) = 10 m. Also, for y–y axis buckling, K = 0.7, so $(KL)_y$ = 0.7(5 m) = 3.5 m.

Applying Eq. 17–11, the critical loads for each case are

$$(P_{cr})_x = \frac{\pi^2 E I_x}{(KL)_x^2} = \frac{\pi^2 [70(10^9)\text{ N/m}^2](61.3(10^{-6})\text{ m}^4)}{(10\text{ m})^2}$$

$$= 424\text{ kN}$$

$$(P_{cr})_y = \frac{\pi^2 E I_y}{(KL)_y^2} = \frac{\pi^2 [70(10^9)\text{ N/m}^2](23.2(10^{-6})\text{ m}^4)}{(3.5\text{ m})^2}$$

$$= 1.31\text{ MN}$$

By comparison, as P is increased the column will buckle about the x–x axis. The allowable load is therefore

$$P_{allow} = \frac{P_{cr}}{\text{F.S.}} = \frac{424\text{ kN}}{3} = 141\text{ kN} \qquad \textit{Ans.}$$

Since

$$\sigma_{cr} = \frac{P_{cr}}{A} = \frac{424\text{ kN}}{7.5(10^{-3})\text{ m}^2} = 56.5\text{ MPa} < 215\text{ MPa}$$

Euler's equation can be applied.

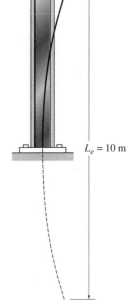

(a)

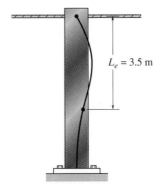

L_e = 3.5 m

y–y axis buckling

(c)

Fig. 17–14

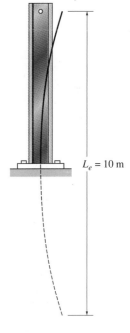

L_e = 10 m

x–x axis buckling

(b)

PROBLEMS

17–1.   The aircraft link is made from a steel rod. Determine the smallest diameter of the rod, to the nearest $\frac{1}{16}$ in., that will support the load of 4 kip without buckling. The ends are pin-connected. $E_{st} = 29(10^3)$ ksi, $\sigma_Y = 36$ ksi.

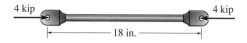

Prob. 17–1

17–2.   A steel column has a length of 5 m and is fixed at both ends. If the cross-sectional area has the dimensions shown, determine the critical load. $E_{st} = 200$ GPa, $\sigma_Y = 360$ MPa.

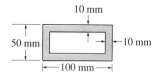

Prob. 17–2

17–3.   A steel column has a length of 15 ft and is pinned at both ends. If the cross-sectional area has the dimensions shown, determine the critical load. $E_{st} = 29(10^3)$ ksi, $\sigma_Y = 36$ ksi.

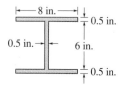

Prob. 17–3

*17–4.   The steel pipe is fixed-supported at its ends. If it is 4 m long and has an outer diameter of 50 mm and a thickness of 10 mm, determine the maximum axial load P that it can carry without buckling. Use a factor of safety with respect to buckling of F.S. = 1.5. $E_{st} = 200$ GPa, $\sigma_Y = 360$ MPa.

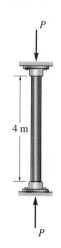

Prob. 17–4

17–5.   The handle is used to operate a machine. If the machine is jammed, however, determine the maximum force P that can be applied to the handle so that the steel control rod AB does not buckle. The rod is made of steel and has a diameter of 1.25 in. It is pin-connected at its ends. $E_{st} = 29(10^3)$ ksi, $\sigma_Y = 36$ ksi.

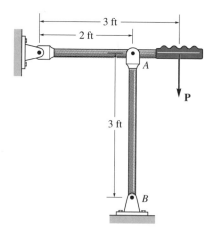

Prob. 17–5

17–6.   The $W\ 12 \times 87$ column has a length of 12 ft. If its bottom end is fixed-supported while its top is free, and it is subjected to an axial load of 380 kip, determine the factor of safety with respect to buckling. $E_{st} = 29(10^3)$ ksi, $\sigma_Y = 36$ ksi.

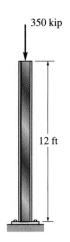

350 kip

12 ft

Prob. 17–6

*17–8.   An acrobat weighing 150 lb sits in the chair at the top of the 18-ft-long wooden pole, which is assumed fixed at the ground. If his center of gravity is positioned directly over the pole's longitudinal axis, determine the smallest diameter of the pole so that it will not buckle. $E_w = 1.6(10^3)$ ksi, $\sigma_Y = 5$ ksi.

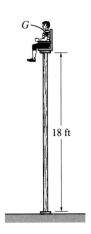

G

18 ft

Prob. 17–8

17–7.   The 12-ft pipe column has an outer diameter of 3 in. and a thickness of 0.25 in. Determine the critical load if (*a*) the ends are assumed to be pin-connected, (*b*) the bottom is fixed and the top is pinned. $E_{st} = 29(10^3)$ ksi, $\sigma_Y = 36$ ksi.

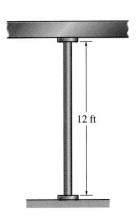

12 ft

Prob. 17–7

17–9.   The steel pipe has an outer diameter of 2 in. and a thickness of 0.5 in. If it is held in place by a guywire, determine the largest horizontal force P that can be applied without causing the pipe to buckle. Assume that the ends of the pipe are pin-connected. $E_{st} = 29(10^3)$ ksi, $\sigma_Y = 36$ ksi.

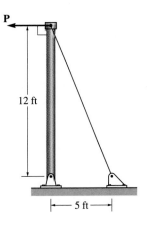

P

12 ft

5 ft

Prob. 17–9

17–10. The members of the truss are assumed to be pin-connected. If member BD is a steel rod of radius $r = 2$ in., determine the maximum load P that can be supported by the truss without causing the member to buckle. $E_{st} = 29(10^3)$ ksi, $\sigma_Y = 36$ ksi.

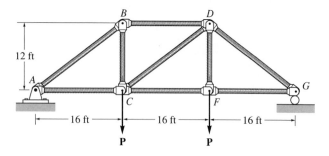

Prob. 17–10

17–11. The truss is made from steel bars, each of which has a circular cross section with a diameter of 1.5 in. Determine the maximum force P that can be applied without causing any of the members to buckle. The members are pin-supported at their ends. $E_{st} = 29(10^3)$ ksi, $\sigma_Y = 36$ ksi.

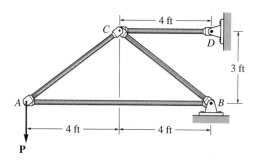

Prob. 17–11

*17–12. The column is supported at B by a support that does not permit rotation but allows vertical deflection. Determine the critical load P_{cr}. Assume elastic buckling.

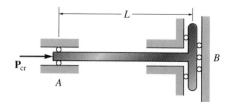

Prob. 17–12

17–13. The linkage is made using two steel rods, each having a circular cross section. Determine the diameter of each rod to the nearest $\frac{1}{8}$ in. that will support the 6-kip load. Assume that the rods are pin-connected at their ends. Use a factor of safety with respect to buckling of F.S. $= 1.8$. $E_{st} = 29(10^3)$, $\sigma_Y = 36$ ksi.

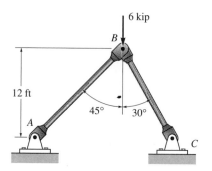

Prob. 17–13

17–14. The steel bar AB of the frame is pin-connected at its ends. Determine the factor of safety with respect to buckling about the y–y axis due to the applied loading. $E_{st} = 200$ GPa, $\sigma_Y = 360$ MPa.

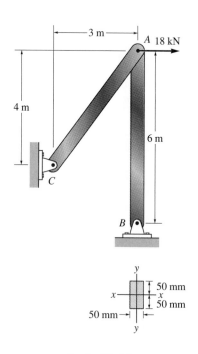

Prob. 17–14

17–15. The steel bar *AB* has a square cross section. If it is pin-connected at its ends, determine the maximum allowable load *P* that can be applied to the frame. Use a factor of safety with respect to buckling of F.S. = 2. $E_{st} = 29(10^3)$ ksi, $\sigma_Y = 36$ ksi.

17–17. The steel bar *AB* has a rectangular cross section. If it is pin-connected at its ends, determine the maximum allowable intensity *w* of the distributed load that can be applied to *BC* without causing bar *AB* to buckle. Use a factor of safety with respect to buckling of F.S. = 1.5. $E_{st} = 200$ GPa, $\sigma_Y = 360$ MPa.

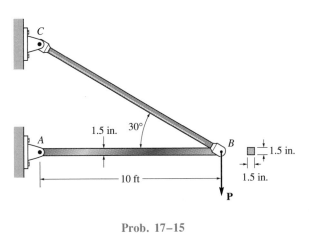

Prob. 17–15

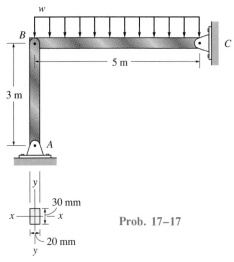

Prob. 17–17

17–18. Determine the maximum allowable intensity *w* of the distributed load that can be applied to member *BC* without causing member *AB* to buckle. Assume that *AB* is made of steel and is pinned at its ends for *x–x* axis buckling and fixed at its ends for *y–y* axis buckling. Use a factor of safety with respect to buckling of F.S. = 3. $E_{st} = 200$ GPa, $\sigma_Y = 360$ MPa.

*17–16. The steel bar *AB* of the frame is pin-connected at its ends. Determine the factor of safety with respect to buckling about the *y–y* axis due to the applied loading. $E_{st} = 29(10^3)$ ksi, $\sigma_Y = 36$ ksi.

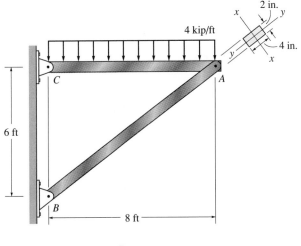

Prob. 17–16

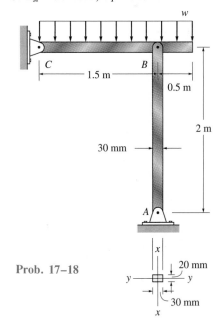

Prob. 17–18

*17.4 Inelastic Buckling

In engineering practice, columns are generally classified according to the type of stresses developed within the column at the time of failure. *Long slender columns* will become unstable when the compressive stress remains elastic. The failure that occurs is referred to as *elastic instability*. *Intermediate columns* fail due to *inelastic instability,* meaning that the compressive stress at failure is greater than the material's yield stress. And *short columns,* sometimes called *posts,* do not become unstable; rather the material simply yields or fractures.

Application of the Euler equation requires that the stress in the column remain *below* the material's yield point (actually the proportional limit) when the column buckles, and so this equation applies only to long columns. In practice, however, most columns are selected to have intermediate lengths. The behavior of these columns can be studied by modifying the Euler equation so that it applies for inelastic buckling. To show how this can be done, consider the material to have a stress–strain diagram as shown in Fig. 17–15a. Here the yield stress is σ_Y, and the modulus of elasticity, or slope of the line AB, is E. A plot of Euler's hyperbola, Fig. 17–8, is shown in Fig. 17–15b. This equation is valid for a column having a slenderness ratio as small as $(KL/r)_Y$, since at this point the axial stress in the column becomes $\sigma_{cr} = \sigma_Y$. If the column has a slenderness ratio that is *less* than $(KL/r)_Y$, then the critical stress in the column must be greater than σ_Y. For example, suppose a column has a slenderness ratio of $(KL/r)_1 < (KL/r)_Y$, with corresponding critical stress $\sigma_D > \sigma_Y$ needed to cause instability. When the column is *about to buckle,* the change in strain that occurs in the column is within a *small range* $\Delta\epsilon$, so that the modulus of elasticity or stiffness for the material can be taken as the *tangent modulus* E_t, defined as the slope of the σ–ϵ diagram at point D, Fig. 17–15a. In other words, at the time of failure, the column

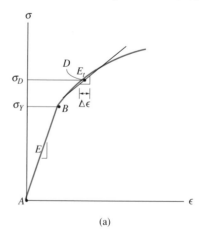

(a)

Fig. 17–15(a)

behaves as if it was made from a material that has a *lower stiffness* than when it behaves elastically, $E_t < E$. In general, therefore, as the slenderness ratio decreases, the *critical stress* for a column continues to rise; and from the σ–ϵ diagram, the *tangent modulus* for the material *decreases*. Using this idea, we can modify Euler's equation to include these cases of inelastic buckling by substituting the material's tangent modulus E_t for E, so that

$$\sigma_{cr} = \frac{\pi^2 E_t}{(KL/r)^2} \qquad (17\text{–}13)$$

This is the so-called *tangent modulus* or *Engesser equation,* proposed by F. Engesser in 1889. A plot of this equation for intermediate and short-length columns of a material defined by the σ–ϵ diagram in Fig. 17–15*a* is shown in Fig. 17–15*b*.

No *actual column* can be considered to be either perfectly straight or loaded along its centroidal axis, as assumed here, and therefore it is indeed very difficult to develop an expression that will provide a full analysis of this phenomenon. It should also be pointed out that other methods of describing the inelastic buckling of columns have been considered. One of these methods was developed by the aeronautical engineer F. R. Shanley and is called the *Shanley theory* of inelastic buckling. Although it provides a better description of the phenomenon than the tangent modulus theory, as explained here, experimental testing of a large number of columns, each of which approximates the ideal column, has shown that Eq. 17–13 is *reasonably accurate* in predicting the column's critical stress. Furthermore, the tangent modulus approach to modeling inelastic column behavior is relatively easy to apply.

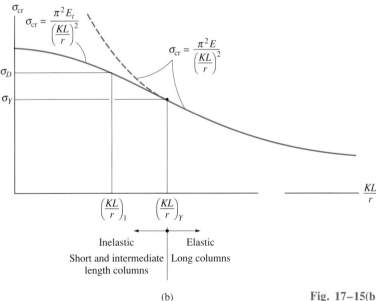

(b)

Fig. 17–15(b)

■ Example 17–5

A solid rod has a diameter of 30 mm and is 600 mm long. It is made of a material that can be modeled by the stress–strain diagram shown in Fig. 17–16. If it is used as a pin-supported column, determine the critical load.

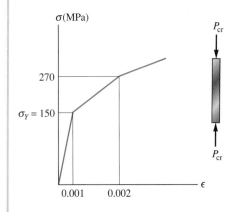

Fig. 17–16

SOLUTION

The radius of gyration is

$$r = \sqrt{\frac{I}{A}} = \sqrt{\frac{(\pi/4)(15)^4}{\pi(15)^2}} = 7.5 \text{ mm}$$

and therefore the slenderness ratio is

$$\frac{KL}{r} = \frac{1(600)}{7.5} = 80$$

Applying Eq. 17–13 yields

$$\sigma_{cr} = \frac{\pi^2 E_t}{(KL/r)^2} = \frac{\pi^2 E_t}{(80)^2} = 1.542(10^{-3})E_t \qquad (1)$$

First we will assume that the critical stress is elastic. From Fig. 17–16,

$$E = \frac{150 \text{ MPa}}{0.001} = 150 \text{ GPa}$$

Thus, Eq. 1 becomes

$$\sigma_{cr} = 1.542(10^{-3})[150(10^3)] \text{ MPa} = 231.3 \text{ MPa}$$

Since $\sigma_{cr} > \sigma_Y = 150$, inelastic buckling occurs.

From the second line segment of the σ–ϵ diagram, Fig. 17–16, we have

$$E_t = \frac{\Delta\sigma}{\Delta\epsilon} = \frac{270 - 150}{0.002 - 0.001} = 120 \text{ GPa}$$

Applying Eq. 1 yields

$$\sigma_{cr} = 1.542(10^{-3})[120(10^3)] \text{ MPa} = 185 \text{ MPa}$$

Since this value falls within the limits of 150 MPa and 270 MPa, it is indeed the critical stress.

The critical load on the rod is therefore

$$P_{cr} = \sigma_{cr}A = 185 \text{ MPa}[\pi(0.015 \text{ m})^2] = 131 \text{ kN} \qquad Ans.$$

PROBLEMS

17–19. The stress–strain diagram for a material can be approximated by the two line segments shown. If a bar having a diameter of 80 mm and a length of 1.5 m is made of this material, determine the critical load provided the ends are pinned. Assume that the load acts through the axis of the bar. Use Engesser's equation.

*17–20. The stress–strain diagram for a material can be approximated by the two line segments shown. If a bar having a diameter of 80 mm and a length of 1.5 m is made of this material, determine the critical load provided the ends are fixed. Assume that the load acts through the axis of the bar. Use Engesser's equation.

17–21. The stress–strain diagram for a material can be approximated by the two line segments shown. If a bar having a diameter of 80 mm and length of 1.5 m is made of this material, determine the critical load provided one end is pinned and the other is fixed. Assume that the load acts through the axis of the bar. Use Engesser's equation.

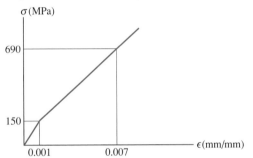

Probs. 17–19/17–20/17–21

17–22. A column of intermediate length buckles when the compressive stress is 40 ksi. If the slenderness ratio is 60, determine the tangent modulus. The material's stress–strain curve is shown in the figure.

17–23. Construct the buckling curve, P/A versus L/r, for a column that has a bilinear stress–strain curve in compression as shown.

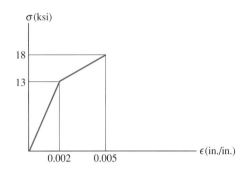

Probs. 17–22/17–23

REVIEW PROBLEMS

*17–24. The steel pipe is constrained between the walls at A and B. If there is a gap of 3 mm at B when $T_1 = 4°C$, determine the temperature required to cause the pipe to become unstable and begin to buckle. The pipe has an outer diameter of 40 mm and a wall thickness of 10 mm. Assume that the collars at A and B provide fixed connections for the pipe. Neglect their size. $\alpha_{st} = 12(10^{-6})/°C$, $E_{st} = 200$ GPa, $\sigma_Y = 250$ MPa.

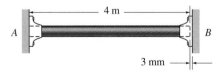

Prob. 17–24

17–25. Construct the buckling curve, P/A versus L/r, for a column that has a bilinear stress–strain curve in compression as shown.

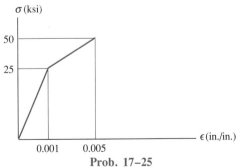

Prob. 17–25

A Mathematical Expressions

Quadratic Formula

If $ax^2 + bx + c = 0$, then $x = \dfrac{-b \pm \sqrt{b^2 - 4ac}}{2a}$

Hyperbolic Functions

$\sinh x = \dfrac{e^x - e^{-x}}{2}$, $\cosh x = \dfrac{e^x + e^{-x}}{2}$, $\tanh x = \dfrac{\sinh x}{\cosh x}$

Trigonometric Identities

$\sin^2 \theta + \cos^2 \theta = 1$

$\sin (\theta \pm \phi) = \sin \theta \cos \phi \pm \cos \theta \sin \phi$

$\sin 2\theta = 2 \sin \theta \cos \theta$

$\cos (\theta \pm \phi) = \cos \theta \cos \phi \mp \sin \theta \sin \phi$

$\cos 2\theta = \cos^2 \theta - \sin^2 \theta$

$\cos \theta = \pm \sqrt{\dfrac{1 + \cos 2\theta}{2}}$, $\sin \theta = \pm \sqrt{\dfrac{1 - \cos 2\theta}{2}}$

$\tan \theta = \dfrac{\sin \theta}{\cos \theta}$

$1 + \tan^2 \theta = \sec^2 \theta \qquad 1 + \cot^2 \theta = \csc^2 \theta$

Power-Series Expansions

$\sin x = x - \dfrac{x^3}{3!} + \dfrac{x^5}{5!} - \dfrac{x^7}{7!} + \cdots$

$\cos x = 1 - \dfrac{x^2}{2!} + \dfrac{x^4}{4!} - \dfrac{x^6}{6!} + \cdots$

$\sinh x = x + \dfrac{x^3}{3!} + \dfrac{x^5}{5!} + \cdots$

$\cosh x = 1 + \dfrac{x^2}{2!} + \dfrac{x^4}{4!} + \cdots$

Derivatives

$\dfrac{d}{dx}(u^n) = nu^{n-1} \dfrac{du}{dx}$

$\dfrac{d}{dx}(uv) = u\dfrac{dv}{dx} + v\dfrac{du}{dx}$

$\dfrac{d}{dx}\left(\dfrac{u}{v}\right) = \dfrac{v\dfrac{du}{dx} - u\dfrac{dv}{dx}}{v^2}$

$\dfrac{d}{dx}(\sin u) = \cos u \dfrac{du}{dx}$

$\dfrac{d}{dx}(\cos u) = -\sin u \dfrac{du}{dx}$

$\dfrac{d}{dx}(\tan u) = \sec^2 u \dfrac{du}{dx}$

$\dfrac{d}{dx}(\cot u) = -\csc^2 u \dfrac{du}{dx}$

$\dfrac{d}{dx}(\sec u) = \tan u \sec u \dfrac{du}{dx}$

$\dfrac{d}{dx}(\csc u) = -\csc u \cot u \dfrac{du}{dx}$

$\dfrac{d}{dx}(\sinh u) = \cosh u \dfrac{du}{dx}$

$\dfrac{d}{dx}(\cosh u) = \sinh u \dfrac{du}{dx}$

Integrals

$$\int x^n \, dx = \frac{x^{n+1}}{n+1} + C, \; n \neq -1$$

$$\int \frac{dx}{a+bx} = \frac{1}{b} \ln (a+bx) + C$$

$$\int \frac{dx}{a+bx^2} = \frac{1}{2\sqrt{-ba}} \ln \left[\frac{\sqrt{a} + 2\sqrt{-b}}{\sqrt{a} - x\sqrt{-b}} \right] + C,$$
$$a > 0, \; b < 0$$

$$\int \frac{x \, dx}{a+bx^2} = \frac{1}{2b} \ln (bx^2 + a) + C$$

$$\int \frac{x^2 \, dx}{a+bx^2} = \frac{x}{b} - \frac{a}{b\sqrt{ab}} \tan^{-1} \frac{x\sqrt{ab}}{a} + C$$

$$\int \frac{dx}{a^2 - x^2} = \frac{1}{2a} \ln \left[\frac{a+x}{a-x} \right] + C, \; a^2 > x^2$$

$$\int \sqrt{a+bx} \, dx = \frac{2}{3b} \sqrt{(a+bx)^3} + C$$

$$\int x\sqrt{a+bx} \, dx = \frac{-2(2a-3bx)\sqrt{(a+bx)^3}}{15b^2} + C$$

$$\int x^2\sqrt{a+bx} \, dx =$$
$$\frac{2(8a^2 - 12abx + 15b^2x^2)\sqrt{(a+bx)^3}}{105b^3} + C$$

$$\int \sqrt{a^2 - x^2} \, dx = \frac{1}{2} \left[x\sqrt{a^2 - x^2} + a^2 \sin^{-1} \frac{x}{a} \right] + C,$$
$$a > 0$$

$$\int x\sqrt{a^2 - x^2} \, dx = -\frac{1}{3} \sqrt{(a^2 - x^2)^3} + C$$

$$\int x^2\sqrt{a^2 - x^2} \, dx = -\frac{x}{4} \sqrt{(a^2 - x^2)^3}$$
$$+ \frac{a^2}{8} \left(x\sqrt{a^2 - x^2} + a^2 \sin^{-1} \frac{x}{a} \right) + C, \; a > 0$$

$$\int \sqrt{x^2 \pm a^2} \, dx =$$
$$\frac{1}{2} \left[x\sqrt{x^2 \pm a^2} \pm a^2 \ln (x + \sqrt{x^2 \pm a^2}) \right] + C$$

$$\int x\sqrt{x^2 \pm a^2} \, dx = \frac{1}{3} \sqrt{(x^2 \pm a^2)^3} + C$$

$$\int x^2\sqrt{x^2 \pm a^2} \, dx = \frac{x}{4} \sqrt{(x^2 \pm a^2)^3}$$
$$\pm \frac{a^2}{8} x\sqrt{x^2 \pm a^2} - \frac{a^4}{8} \ln (x + \sqrt{x^2 \pm a^2}) + C$$

$$\int \frac{dx}{\sqrt{a+bx}} = \frac{2\sqrt{a+bx}}{b} + C$$

$$\int \frac{x \, dx}{\sqrt{x^2 \pm a^2}} = \sqrt{x^2 \pm a^2} + C$$

$$\int \frac{dx}{\sqrt{a+bx+cx^2}} = \frac{1}{\sqrt{c}} \ln \left[\sqrt{a+bx+cx^2} + \right.$$
$$\left. x\sqrt{c} + \frac{b}{2\sqrt{c}} \right] + C, \; c > 0$$
$$= \frac{1}{\sqrt{-c}} \sin^{-1} \left(\frac{-2cx - b}{\sqrt{b^2 - 4ac}} \right) + C, \; c < 0$$

$$\int \sin x \, dx = -\cos x + C$$

$$\int \cos x \, dx = \sin x + C$$

$$\int x \cos (ax) \, dx = \frac{1}{a^2} \cos (ax) + \frac{x}{a} \sin (ax) + C$$

$$\int x^2 \cos (ax) \, dx = \frac{2x}{a^2} \cos (ax) + \frac{a^2x^2 - 2}{a^3} \sin (ax) + C$$

$$\int e^{ax} \, dx = \frac{1}{a} e^{ax} + C$$

$$\int x e^{ax} \, dx = \frac{e^{ax}}{a^2} (ax - 1) + C$$

$$\int \sinh x \, dx = \cosh x + C$$

$$\int \cosh x \, dx = \sinh x + C$$

B Average Mechanical Properties of Typical Engineering Materials[a]

(SI Units)

Materials	Density (Mg/m³)	Modulus of Elasticity (GPa)	Modulus of Rigidity (GPa)	Yield Strength (MPa)[b] Tens.	Yield Strength (MPa)[b] Comp.	Yield Strength (MPa)[b] Shear	Ultimate Strength (MPa)[b] Tens.	Ultimate Strength (MPa)[b] Comp.	Ultimate Strength (MPa)[b] Shear	% Elongation in 50 mm specimen	Poisson's Ratio	Coef. of Therm. Expansion (10⁻⁶)/°C
Metallic												
Aluminum Wrought Alloys ⌈ 2014-T6	2.79	73.1	27	414	414	172	469	469	290	10	0.35	23
⌊ 6061-T6	2.71	68.9	26	255	255	131	290	290	186	12	0.35	24
Cast Iron ⌈ Gray ASTM 20	7.19	68.9	27	—	—	—	179	669	—	0.6	0.28	12
Alloys ⌊ Malleable ASTM A-197	7.28	172	68	—	—	—	276	572	—	5	0.28	12
Copper ⌈ Red Brass C83400	8.74	101	37	68.9	68.9	—	241	241	—	35	0.35	18
Alloys ⌊ Bronze C86100	8.83	103	38	345	345	—	655	655	—	20	0.34	17
Magnesium Alloy [Am 1004-T61]	1.83	44.7	18	152	152	—	276	276	152	1	0.30	26
Steel ⌈ Structural A36	7.85	200	75	250	250	—	400	400	—	30	0.32	12
Alloys ⌊ Stainless 304	7.86	193	75	207	207	—	517	517	—	40	0.27	17
Tool L2	8.16	200	78	703	703	—	800	800	—	22	0.32	12
Titanium Alloy [Ti-6Al-4V]	4.43	120	44	924	924	—	1,000	1,000	—	16	0.36	9.4
Nonmetallic												
Concrete ⌈ Low Strength	2.38	22.1	—	—	—	12	—	—	—	—	0.15	11
⌊ High Strength	2.38	29.0	—	—	—	38	—	—	—	—	0.15	11
Plastic ⌈ Kevlar 49	1.45	131	—	—	—	—	717	483	20.3	2.8	0.34	—
Reinforced ⌊ 30% Glass	1.45	72.4	—	—	—	—	90	131	—	—	0.34	—
Wood Select Structural ⌈ Douglas Fir	0.47	13.1	—	—	—	—	2.1[c]	26[d]	6.2[d]	—	0.29[e]	—
Grade ⌊ White Spruce	3.60	9.65	—	—	—	—	2.5[c]	36[d]	6.7	—	0.31[e]	—

[a] Specific values may vary for a particular material due to alloy or mineral composition, mechanical working of the specimen, or heat treatment. For a more exact value reference books for the material should be consulted.

[b] The yield and ultimate strengths for ductile materials can be assumed equal for both tension and compression.

[c] Measured perpendicular to the grain.

[d] Measured parallel to the grain.

[e] Deformation measured perpendicular to the grain when the load is applied along the grain.

(U.S. Customary Units)

Materials	Specific Weight (lb/in³)	Modulus of Elasticity (10³) ksi	Modulus of Rigidity (10³) ksi	Yield Strength (ksi) Tens.	Comp.[b]	Shear	Ultimate Strength (ksi) Tens.	Comp.[b]	Shear	% Elongation in 2 in. specimen	Poisson's Ratio	Coef. of Therm. Expansion (10⁻⁶)/°F
Metallic												
Aluminum Wrought Alloys — 2014-T6	0.101	10.6	3.9	60	60	25	68	68	42	10	0.35	12.8
Aluminum Wrought Alloys — 6061-T6	0.098	10.0	3.7	37	37	19	42	42	27	12	0.35	13.1
Cast Iron Alloys — Gray ASTM 20	0.260	10.0	3.9	—	—	—	26	97	—	0.6	0.28	6.70
Cast Iron Alloys — Malleable ASTM A-197	0.263	25.0	9.8	—	—	—	40	83	—	5	0.28	6.60
Copper Alloys — Red Brass C83400	0.316	14.6	5.4	10	10	—	35	35	—	35	0.35	9.80
Copper Alloys — Bronze C86100	0.319	15.0	5.6	50	50	—	95	95	—	20	0.34	9.60
Magnesium Alloy [Am 1004-T61]	0.066	6.48	2.5	22	22	—	40	40	22	1	0.30	14.3
Steel Alloys — Structural A36	0.284	29.0	11.0	36	36	—	58	58	—	30	0.32	6.60
Steel Alloys — Stainless 304	0.284	28.0	11.0	30	30	—	75	75	—	40	0.27	9.60
Steel Alloys — Tool L2	0.295	29.0	11.0	102	102	—	116	116	—	22	0.32	6.50
Titanium Alloy [Ti-6A1-4V]	0.160	17.4	6.4	134	134	—	145	145	—	16	0.36	5.20
Nonmetallic												
Concrete — Low Strength	0.086	3.20	—	—	—	1.8	—	—	—	—	0.15	6.0
Concrete — High Strength	0.086	4.20	—	—	—	5.5	—	—	—	—	0.15	6.0
Plastic Reinforced — Kevlar 49	0.0524	19.0	—	—	—	—	104	70	10.2	2.8	0.34	—
Plastic Reinforced — 30% Glass	0.0524	10.5	—	—	—	—	13	19	—	—	0.34	—
Wood Select Structural Grade — Douglas Fir	0.017	1.90	—	—	—	—	0.30[c]	3.78[d]	0.90[d]	—	0.29[e]	—
Wood Select Structural Grade — White Spruce	0.130	1.40	—	—	—	—	0.36[c]	5.18[d]	0.97[d]	—	0.31[e]	—

[a] Specific values may vary for a particular material due to alloy or mineral composition, mechanical working of the specimen, or heat treatment. For a more exact value reference books for the material should be consulted.

[b] The yield and ultimate strengths for ductile materials can be assumed equal for both tension and compression.

[c] Measured perpendicular to the grain.

[d] Measured parallel to the grain.

[e] Deformation measured perpendicular to the grain when the load is applied along the grain.

Geometric Properties of An Area and Volume

Centroid Location	Centroid Location	Area Moment of Inertia

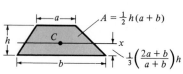

$A = \frac{1}{2}h(a+b)$

$\frac{1}{3}\left(\frac{2a+b}{a+b}\right)h$

Trapezoidal area

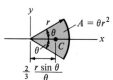

$A = \theta r^2$

$\frac{2}{3}\frac{r \sin \theta}{\theta}$

Circular sector area

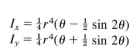

$I_x = \frac{1}{4}r^4(\theta - \frac{1}{2}\sin 2\theta)$
$I_y = \frac{1}{4}r^4(\theta + \frac{1}{2}\sin 2\theta)$

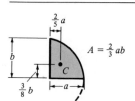

$A = \frac{2}{3}ab$

Semiparabolic area

$A = \frac{\pi r^2}{4}$

$\frac{4r}{3\pi}$

$\frac{4r}{3\pi}$

Quarter circular area

$I_x = \frac{1}{16}\pi r^4$
$I_y = \frac{1}{16}\pi r^4$

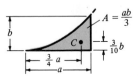

$A = \frac{ab}{3}$

$\frac{3}{10}b$

Exparabolic area

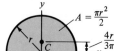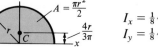

$A = \frac{\pi r^2}{2}$

$\frac{4r}{3\pi}$

Semicircular area

$I_x = \frac{1}{8}\pi r^4$
$I_y = \frac{1}{8}\pi r^4$

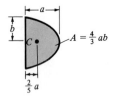

$A = \frac{4}{3}ab$

Parabolic area

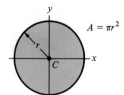

$A = \pi r^2$

Circular area

$I_x = \frac{1}{4}\pi r^4$
$I_y = \frac{1}{4}\pi r^4$

(continued)

Centroid Location	**Centroid Location**	**Area Moment of Inertia**

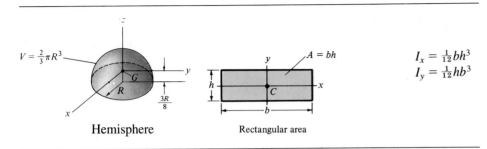

$V = \frac{2}{3}\pi R^3$

Hemisphere

$A = bh$

Rectangular area

$I_x = \frac{1}{12}bh^3$
$I_y = \frac{1}{12}hb^3$

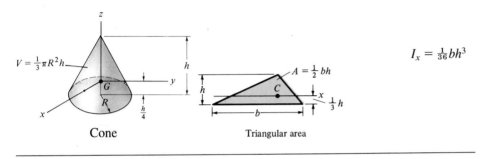

$V = \frac{1}{3}\pi R^2 h$

Cone

$A = \frac{1}{2}bh$

Triangular area

$I_x = \frac{1}{36}bh^3$

D Properties of Wide-Flange Sections

Designation	Area A in²	Depth d in.	Web thickness t_w in.	Flange width b in.	Flange thickness t_f in.	x-x axis I in⁴	x-x axis S in³	x-x axis r in.	y-y axis I in⁴	y-y axis S in³	y-y axis r in.
W24 × 104	30.6	24.06	0.500	12.750	0.750	3100	258	10.1	259	40.7	2.91
W24 × 94	27.7	24.31	0.515	9.065	0.875	2700	222	9.87	109	24.0	1.98
W24 × 84	24.7	24.10	0.470	9.020	0.770	2370	196	9.79	94.4	20.9	1.95
W24 × 76	22.4	23.92	0.440	8.990	0.680	2100	176	9.69	82.5	18.4	1.92
W24 × 68	20.1	23.73	0.415	8.965	0.585	1830	154	9.55	70.4	15.7	1.87
W24 × 62	18.2	23.74	0.430	7.040	0.590	1550	131	9.23	34.5	9.80	1.38
W24 × 55	16.2	23.57	0.395	7.005	0.505	1350	114	9.11	29.1	8.30	1.34
W18 × 65	19.1	18.35	0.450	7.590	0.750	1070	117	7.49	54.8	14.4	1.69
W18 × 60	17.6	18.24	0.415	7.555	0.695	984	108	7.47	50.1	13.3	1.69
W18 × 65	16.2	18.11	0.390	7.530	0.630	890	98.3	7.41	44.9	11.9	1.67
W18 × 50	14.7	17.99	0.355	7.495	0.570	800	88.9	7.38	40.1	10.7	1.65
W18 × 46	13.5	18.06	0.360	6.060	0.605	712	78.8	7.25	22.5	7.43	1.29
W18 × 40	11.8	17.90	0.315	6.015	0.525	612	68.4	7.21	19.1	6.35	1.27
W18 × 35	10.3	17.70	0.300	6.000	0.425	510	57.6	7.04	15.3	5.12	1.22
W16 × 57	16.8	16.43	0.430	7.120	0.715	758	92.2	6.72	43.1	12.1	1.60
W16 × 50	14.7	16.26	0.380	7.070	0.630	659	81.0	6.68	37.2	10.5	1.59
W16 × 45	13.3	16.13	0.345	7.035	0.565	586	72.7	6.65	32.8	9.34	1.57
W16 × 36	10.6	15.86	0.295	6.985	0.430	448	56.5	6.51	24.5	7.00	1.52
W16 × 31	9.12	15.88	0.275	5.525	0.440	375	47.2	6.41	12.4	4.49	1.17
W16 × 26	7.68	15.69	0.250	5.500	0.345	301	38.4	6.26	9.59	3.49	1.12
W14 × 53	15.6	13.92	0.370	8.060	0.660	541	77.8	5.89	57.7	14.3	1.92
W14 × 43	12.6	13.66	0.305	7.995	0.530	428	62.7	5.82	45.2	11.3	1.89
W14 × 38	11.2	14.10	0.310	6.770	0.515	385	54.6	5.87	26.7	7.88	1.55
W14 × 34	10.0	13.98	0.285	6.745	0.455	340	48.6	5.83	23.3	6.91	1.53
W14 × 30	8.85	13.84	0.270	6.730	0.385	291	42.0	5.73	19.6	5.82	1.49
W14 × 26	7.69	13.91	0.255	5.025	0.420	245	35.3	5.65	8.91	3.54	1.08
W14 × 22	6.49	13.74	0.230	5.000	0.335	199	29.0	5.54	7.00	2.80	1.04

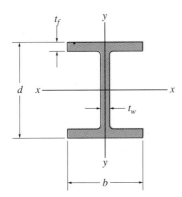

Designation	Area A in^2	Depth d in.	Web thickness t_w in.	Flange width b in.	Flange thickness t_f in.	x-x axis I in^4	x-x axis S in^3	x-x axis r in.	y-y axis I in^4	y-y axis S in^3	y-y axis r in.
W12 × 87	25.6	12.53	0.515	12.125	0.810	740	118	5.38	241	39.7	3.07
W12 × 50	14.7	12.19	0.370	8.080	0.640	394	64.7	5.18	56.3	13.9	1.96
W12 × 45	13.2	12.06	0.335	8.045	0.575	350	58.1	5.15	50.0	12.4	1.94
W12 × 26	7.65	12.22	0.230	6.490	0.380	204	33.4	5.17	17.3	5.34	1.51
W12 × 22	6.48	12.31	0.260	4.030	0.425	156	25.4	4.91	4.66	2.31	0.847
W12 × 16	4.71	11.99	0.220	3.990	0.265	103	17.1	4.67	2.82	1.41	0.773
W12 × 14	4.16	11.91	0.200	3.970	0.225	88.6	14.9	4.62	2.36	1.19	0.753
W10 × 100	29.4	11.10	0.680	10.340	1.120	623	112	4.60	207	40.0	2.65
W10 × 54	15.8	10.09	0.370	10.030	0.615	303	60.0	4.37	103	20.6	2.56
W10 × 45	13.3	10.10	0.350	8.020	0.620	248	49.1	4.32	53.4	13.3	2.01
W10 × 30	8.84	10.47	0.300	5.810	0.510	170	32.4	4.38	16.7	5.75	1.37
W10 × 39	11.5	9.92	0.315	7.985	0.530	209	42.1	4.27	45.0	11.3	1.98
W10 × 19	5.62	10.24	0.250	4.020	0.395	96.3	18.8	4.14	4.29	2.14	0.874
W10 × 15	4.41	9.99	0.230	4.000	0.270	68.9	13.8	3.95	2.89	1.45	0.810
W10 × 12	3.54	9.87	0.190	3.960	0.210	53.8	10.9	3.90	2.18	1.10	0.785
W8 × 67	19.7	9.00	0.570	8.280	0.935	272	60.4	3.72	88.6	21.4	2.12
W8 × 58	17.1	8.75	0.510	8.220	0.810	228	52.0	3.65	75.1	18.3	2.10
W8 × 48	14.1	8.50	0.400	8.110	0.685	184	43.3	3.61	60.9	15.0	2.08
W8 × 40	11.7	8.25	0.360	8.070	0.560	146	35.5	3.53	49.1	12.2	2.04
W8 × 31	9.13	8.00	0.285	7.995	0.435	110	27.5	3.47	37.1	9.27	2.02
W8 × 24	7.08	7.93	0.245	6.495	0.400	82.8	20.9	3.42	18.3	5.63	1.61
W8 × 15	4.44	8.11	0.245	4.015	0.315	48.0	11.8	3.29	3.41	1.70	0.876
W6 × 25	7.34	6.38	0.320	6.080	0.455	53.4	16.7	2.70	17.1	5.61	1.52
W6 × 20	5.87	6.20	0.260	6.020	0.365	41.4	13.4	2.66	13.3	4.41	1.50
W6 × 15	4.43	5.99	0.230	5.990	0.260	29.1	9.72	2.56	9.32	3.11	1.46
W6 × 16	4.74	6.28	0.260	4.030	0.405	32.1	10.2	2.60	4.43	2.20	0.966
W6 × 12	3.55	6.03	0.230	4.000	0.280	22.1	7.31	2.49	2.99	1.50	0.918
W6 × 9	2.68	5.90	0.170	3.940	0.215	1.64	5.56	2.47	2.19	1.11	0.905

Answers*

*Note: Answers to every fourth problem are omitted.

2–35. $\mathbf{F} = \{289\mathbf{i} + 225\mathbf{j} - 261\mathbf{k}\}$ lb

2–37. $\alpha = 52.2°$, $\beta = 52.2°$, $\gamma = 120°$

2–38. 166 N, $\alpha = 97.5°$, $\beta = 63.7°$, $\gamma = 27.5°$

2–39. $\alpha_1 = 36.9°$, $\beta_1 = 90°$, $\gamma_1 = 53.1°$, $\alpha_R = 69.3°$, $\beta_R = 52.2°$, $\gamma_R = 45°$

2–41. 180 N, $\alpha_2 = 147°$, $\beta_2 = 119°$, $\gamma_2 = 75.0°$

2–42. $F_x = 6$ kN, $F_y = 6$ kN, $F_z = 8.49$ kN

2–43. $F = 2.02$ kN, $F_y = 0.523$ kN

2–45. $\mathbf{r}_{AB} = \{2\mathbf{i} - 7\mathbf{j} - 5\mathbf{k}\}$ m, $r_{AB} = 8.83$ m, $\alpha = 76.9°$, $\beta = 142°$, $\gamma = 124°$

2–46. $\alpha = 69.6°$, $\beta = 116°$, $\gamma = 34.4°$

2–47. 7.55 m, $\alpha = 105°$, $\beta = 22.0°$, $\gamma = 74.6°$

2–49. $\mathbf{r} = \{9.48\mathbf{i} - 1.24\mathbf{j} + 2.30\mathbf{k}\}$ m, $r = 9.83$ m, $\alpha = 15.4°$, $\beta = 97.2°$, $\gamma = 76.5°$

2–50. $\mathbf{r} = \{5\mathbf{i} - 4\mathbf{j} - 5\mathbf{k}\}$ m, 8.12 m, $\alpha = 52.0°$, $\beta = 119$, $\gamma = 128°$

2–51. 19.0 in.

2–53. 98.0 ft, 114°

2–54. $r_{AD} = 1.5$ m, $r_{BD} = 1.5$ m, $r_{CD} = 1.73$ m

2–55. $\mathbf{F} = \{-2\mathbf{i} + 1\mathbf{j} + 2\mathbf{k}\}$ kN, $\alpha = 132°$, $\beta = 70.5°$, $\gamma = 48.2°$

2–57. $\mathbf{F} = \{-5.54\mathbf{i} + 22.2\mathbf{j} - 7.38\mathbf{k}\}$ kN, $\alpha = 103°$, $\beta = 22.6°$, $\gamma = 108°$

2–58. $\mathbf{F}_1 = \{38.0\mathbf{i} + 104\mathbf{j} - 101\mathbf{k}\}$ N, $\mathbf{F}_2 = \{119\mathbf{i} - 19.9\mathbf{j} - 159\mathbf{k}\}$ N

2–59. $\mathbf{F}_1 = \{-3.79\mathbf{i} + 11.4\mathbf{k}\}$ lb, $\mathbf{F}_2 = \{-6.65\mathbf{i} - 11.8\mathbf{j} + 11.8\mathbf{k}\}$ lb, $F_R = 28.1$ lb, $\alpha = 112°$, $\beta = 115°$, $\gamma = 34.2°$

2–61. $\mathbf{F}_1 = \{75.5\mathbf{i} - 43.6\mathbf{j} - 122\mathbf{k}\}$ lb, $\mathbf{F}_2 = \{26.8\mathbf{i} + 33.5\mathbf{j} - 90.4\mathbf{k}\}$ lb, $F_R = 236$ lb, $\alpha = 64.3°$, $\beta = 92.5°$, $\gamma = 154°$

2–62. 10.0 ft, $\mathbf{F} = \{-19.1\mathbf{i} - 14.9\mathbf{j} + 43.7\mathbf{k}\}$ lb, $\alpha = 112°$, $\beta = 107°$, $\gamma = 29.0°$

2–63. $(-1.79$ ft, 1.34 ft, 4.47 ft$)$

2–65. $\mathbf{F}_A = \{-1.46\mathbf{i} + 5.82\mathbf{k}\}$ kN, $\mathbf{F}_C = \{0.857\mathbf{i} + 0.857\mathbf{j} + 4.85\mathbf{k}\}$ kN, $\mathbf{F}_B = \{0.970\mathbf{i} - 1.68\mathbf{j} + 7.76\mathbf{k}\}$ kN, $F_R = 18.5$ kN, $\alpha = 88.8°$, $\beta = 92.6°$, $\gamma = 2.81°$

2–66. 672 N, $\alpha = 82.9°$, $\beta = 93.4°$, $\gamma = 172°$

2–67. $x = 1.96$ m, $y = 2.34$ m, 704 N

2–69. 85.9°

2–70. $(\mathbf{r}_1)_2 = \{0.143\mathbf{i} + 0.214\mathbf{j} - 0.0714\mathbf{k}\}$ m

2–71. 78.1°

2–73. 2.67 m

2–74. 3.87 m

2–75. 128°

2–77. $F_1 = 18.3$ lb, $F_{per} = 35.6$ lb

2–78. $F_{AC} = 26.5$ lb, $F_{BA} = 66.4$ lb

2–79. 85.5 lb

2–81. 34.2°

2–82. $F_x = 47.8$ lb, $F_{AC} = 45.5$ lb

2–83. $\theta = 52.4°$, $\phi = 68.2°$

2–85. 5.43 lb

2–86. 100°

2–87. $\mathbf{F}_1 = 19.4$ N, $\mathbf{F}_2 = 53.4$ N

2–89. $\mathbf{F}_1 = \{20\mathbf{i} + 34.6\mathbf{j}\}$ lb, $\mathbf{F}_2 = \{-42.4\mathbf{i} + 42.4\mathbf{j}\}$ lb, $F_R = 80.3$ lb, $\theta = 106°$

2–90. 80.3 lb, 106°

2–91. $F_v = 80.4$ lb, $F_u = 75.0$ lb

2–93. $x = 8.67$ ft, $y = 1.89$ ft

2–94. 0.667 kN

2–95. $\mathbf{F}_1 = \{70\mathbf{i} - 121\mathbf{j}]$ lb, $\mathbf{F}_2 = \{-180\mathbf{j}\}$ lb, $\mathbf{F}_3 = \{-88.4\mathbf{i} - 88.4\mathbf{j}\}$ lb, $F_R = 390$ lb

2–97. 90°

Chapter 3

3–1. 3.15 kN · m ↙

3–2. 858 lb · in. ↙

3–3. 1.17 kip · in. ↙

3–5. 4.47 kN · m ↙

3–6. 14.5 lb · ft ↙

3–7. 39.2°

3–9. 11.2 kip · ft ↙

3–10. 6.16 kip · ft ↙

3–11. 66.1 N

3–13. $M_A = 80 \cos\theta + 320 \sin\theta$

3–14. $M_P = \{538 \cos\theta + 75 \sin\theta\}$ lb · ft

3–15. 239 lb

3–17. $\mathbf{M}_P = \{153\mathbf{i} - 29.7\mathbf{j} + 174\mathbf{k}\}$ lb · ft

3–18. $(\mathbf{M}_R)_O = \{-160\mathbf{i} + 200\mathbf{j} + 510\mathbf{k}\}$ N · m

3–19. $(\mathbf{M}_R)_p = \{80\mathbf{i} + 660\mathbf{j} + 530\mathbf{k}\}$ N · m

3–21. $\mathbf{M}_1 = \{-52\mathbf{i} + 24\mathbf{j} - 36\mathbf{k}\}$ N · m, $\mathbf{M}_2 = \{-34\mathbf{i} + 34\mathbf{j} - 68\mathbf{k}\}$ N · m, 147 N · m, $\alpha = 126°$, $\beta = 66.7°$, $\gamma = 135°$

3–22. $\mathbf{M}_1 = \{-36\mathbf{i} - 36\mathbf{k}\}$ N · m, $\mathbf{M}_2 = \{-42\mathbf{i} - 14\mathbf{j} - 68\mathbf{k}\}$ N · m, 131 N · m, $\alpha = 127°$, $\beta = 96.1°$, $\gamma = 143°$

3–23. $\mathbf{M}_c = \{343\mathbf{i} + 263\mathbf{j} + 198\mathbf{k}\}$ lb · ft

3–25. $(\mathbf{M}_R)_A = \{36\mathbf{i} - 78\mathbf{j}\}$ lb · in.,
$(\mathbf{M}_R)_B = \{36\mathbf{i} + 192\mathbf{j} + 90\mathbf{k}\}$lb · in.,
$(\mathbf{M}_R)_C = \{-396\mathbf{i} + 192\mathbf{j} - 222\mathbf{k}\}$ lb · in.

3–26. 165 lb · ft, $\alpha = 91.3°$, $\beta = 62.5°$, $\gamma = 152°$

3–27. $\alpha = 77.8°$, $\beta = 29.6°$, $\gamma = 63.4°$

3–29. $(\mathbf{M}_R)_{aa} = \{-41.3\mathbf{i} - 44.7\mathbf{j} + 13.8\mathbf{k}\}$ kN · m

3–30. $\mathbf{M}_{aa} = \{0.869\mathbf{i} - 2.17\mathbf{j} + 1.74\mathbf{k}\}$kN · m

3–31. $M_x = -72$ N · m, $M_y = 12$ N · m, $M_z = -52$ N · m

3–33. 942 lb · in.

3–34. 62 lb · in.

3–35. 75 lb · in.

3–37. 3.75 N · m

3–38. $\mathbf{M}_y = \{0.828\mathbf{j}\}$ N · m

3–39. $\mathbf{M}_y = \{87.4\mathbf{j}\}$ N · m

3–41. 19.8 lb

3–42. 14.8 N · m

3–43. 20.2 N

3–45. 21.9 kN · m

3–46. 3.12 kip · ft

3–47. 133 N, 267 N

3–49. 28.9 lb · ft

3–50. 45°, 8.99 lb

3–51. 139 lb

3–53. 134 lb · ft

3–54. $M_R = 39.6$ lb · ft ↖

3–55. 9.69 kN · m ↓

3–57. 78.1 N · m, $\alpha = 48.3°$, $\beta = 50.2°$, $\gamma = 67.4°$

3–58. $\mathbf{M}_C = \{-50\mathbf{i} + 60\mathbf{j}\}$ lb · ft, $M_C = 78.1$ lb · ft

3–59. $\mathbf{M}_R = \{-12.1\mathbf{i} - 10\mathbf{j} - 17.3\mathbf{k}\}$ N · m

3–61. $F = 3$ lb, $P = 4$ lb

3–62. 22.6° ∠θ, 204 lb · ft ↓

3–63. 30° ↘θ, 2.46 kN · m ↓

3–65. 1.19 kN, 85.9° θ↗, 21.6 kN · m ↓

3–66. 375 lb, 15.5 ft

3–67. 52.1 lb, 20.9° ↘θ, 60.9 ft

3–69. 1.08 kN, 68.2° θ↗, 0.901 kN · m ↖

3–70. 60 N

3–71. 29.9 lb, 78.4° ∠θ, 214 lb · in. ↓

3–73. 342 N, 43.0° ↗, 150 N · m ↓

3–74. 342 N, 43.0° θ↗, 99.5 N · m ↓

3–75. 464 lb, 45.1° ↘θ, 756 lb · ft ↓

3–77. 110 lb, 67.5° θ↗, 5.95 ft

3–78. 110 lb, 67.5° θ↗, 705 lb · ft ↖

3–79. 26.6°, 8.94 kN, 4.5 m

3–81. 219 lb, 79.5° θ↗, 1.19 kip · ft ↓

3–82. 219 lb, 79.5° θ↗, 1.39 kip · ft ↓

3–83. 98.5 lb, 24.0° ∠θ, 390 lb · ft ↓

3–85. 98.5 lb, 24.0° ∠θ, 6.0 ft

3–86. 4.5 kN ↓, 2.22 m

3–87. $\mathbf{F} = \{50\mathbf{i} - 30\mathbf{j} + 80\mathbf{k}\}$ N,
$\mathbf{M}_{RO} = \{410\mathbf{i} + 230\mathbf{j} - 170\mathbf{k}\}$ N · m

3–89. $\mathbf{F} = \{8\mathbf{i} + 6\mathbf{j} + 8\mathbf{k}\}$ kN,
$\mathbf{M}_{RO} = \{-10\mathbf{i} + 18\mathbf{j} - 56\mathbf{k}\}$ kN · m

3–90. $\mathbf{F} = \{8\mathbf{i} + 6\mathbf{j} + 8\mathbf{k}\}$ kN,
$\mathbf{M}_{RP} = \{-46\mathbf{i} + 66\mathbf{j} - 56\mathbf{k}\}$ kN · m

3–91. $\mathbf{F}_R = \{20\mathbf{i} + 30\mathbf{j} - 60\mathbf{k}\}$ lb,
$\mathbf{M}_{R_o} = \{-2700\mathbf{i} + 600\mathbf{j} - 600\mathbf{k}\}$ lb · ft

3–93. 140 kN ↓, $y = 7.14$ m, $x = 5.71$ m

3–94. 40 lb ↑, 0.5 ft

3–95. $\mathbf{F}_R = \{506\mathbf{i} - 23.2\mathbf{j} + 1236\mathbf{k}\}$ N,
$\mathbf{M}_{R_o} = \{-17.7\mathbf{i} - 600\mathbf{j} + 26.5\mathbf{k}\}$ N · m

3–97. $(M_1)_B = 3.96$ kip · ft ↓, $(M_2)_B = 4.80$ kip · ft ↓,
$(M_3)_B = 1.30$ kip · ft ↓, $M_{RB} = 10.1$ kip · ft ↓

3–98. $\mathbf{M}_{BA} = \{-3.39\mathbf{i} + 2.54\mathbf{j} - 2.54\mathbf{k}\}$ N · m

3–99. $\mathbf{M}_{OC} = \{-4.61\mathbf{i} - 3.46\mathbf{j}\}$N · m

3–101. $\mathbf{M}_O = \{-128\mathbf{i} + 128\mathbf{j} - 257\mathbf{k}\}$ N · m

3–102. $\mathbf{M}_B = \{-37.6\mathbf{i} + 90.7\mathbf{j} - 155\mathbf{k}\}$ N · m

3–103. 10.75 kip ↓, 99.5 kip · ft ↓

3–105. $\mathbf{F}_R = \{-40\mathbf{j} - 40\mathbf{k}\}$ N, $(\mathbf{M}_R)_A = \{-12\mathbf{j} + 12\mathbf{k}\}$ N · m

3–106. $\mathbf{F}_R = \{-28.3\mathbf{j} - 68.3\mathbf{k}\}$N,
$(\mathbf{M}_R)_A = \{-20.5\mathbf{j} + 8.49\mathbf{k}\}$ N · m

Chapter 4

(no answers for 4–1—4–10)

4–11. $F_1 = 259$ N, $F_2 = 366$ N

4–13. $F_1 = 3.59$ kN, $F_2 = 0.536$ kN

4–14. $F_{AC} = 267$ N, $F_{AB} = 98.6$ N

4–15. $\tan \theta = \tan 62.99° = \dfrac{1.5 + d}{2}$, $d = 2.42$ m

4–17. $x_{AB} = 0.467$ m, $x_{AC} = 0.793$ m, $x_{AD} = 0.490$ m

4–18. $A_x = 0$, $A_y = 3.83$ kN, $B_y = 11.2$ kN

4–19. $A_x = 0$, $A_y = 200$ N, $B_y = 400$ N

4–21. $A_x = 0$, $A_y = 6.75$ kN, $M_A = 46.25$ kN · m

4–22. $F_A = 105$ lb, $B_x = 97.4$ lb, $B_y = 269$ lb

4–23. 9.08 lb

4–25. $T_{BC} = 16.4$ kN, $A_x = 13.1$ kN, $A_y = 14.6$ kN

4–26. $F_{BD} = 41.6$ kN, $A_x = 41.6$ kN, $A_y = 8$ kN

4–27. $N_A = 425$ N, $N_B = 245$ N

4–29. $A_x = 4.00$ kN, $A_y = 8.71$ kN, $B_y = 78.2$ kN

4–30. $F_{BC} = 574$ lb, $A_x = 1.08$ kip, $A_y = 637$ lb

4–31. (a) $N_E = 1.16$ kip, $N_D = 2.19$ kip, (b) 4.74 kip

4–33. $T_{BD} = 76.8$ kN, $A_x = 62.9$ kN, $A_y = 65.6$ kN

4–34. $N_B = -4.38$ lb

4–35. $A_x = 33.3$ lb, $B_x = 33.3$ lb, $B_y = 100$ lb

4–37. (a) 1.64 kN, (b) 213 N

4–38. 53.1°, 196 N

4–39. $F_{BC} = 3.76$ kip, $A_x = 3.26$ kip, $A_y = 3.12$ kip

4–41. $F_{BC} = 410$ lb, $A_x = 71.2$ lb, $A_y = 396$ lb

4–42. $A_x = 4.61$ lb, $A_y = B_y = 43.3$ lb

4–43. $F_A = 432$ lb, $F_B = 0$, $F_C = 432$ lb

4–45. $w_1 = 3.22$ kN/m, $w_2 = 1.63$ kN/m

4–46. $P = F_A = 0.01$ N

4–47. 250 N/m

4–49. 12.8°

4–50. $F_A = (W/2)\sqrt{(4l^2 - 3d^2)/(l^2 - d^2)}$

4–51. $F_C = \dfrac{2}{5}wL$, $F_B = \dfrac{1}{5}wL$, $\Delta_C = \dfrac{2wL}{5K}$

4–53. $A_x = -100$ N, $A_y = -20$ N, $A_z = 45$ N,
$M_{A_x} = 35$ N $\cdot$ m, $M_{A_y} = -15$ N $\cdot$ m, $M_{A_z} = 240$ N $\cdot$ m

4–54. $F_A = A_{z'} = 125$ lb, $M_A = 206$ lb $\cdot$ ft

4–55. $A_x = -8$ lb, $A_y = 4$ lb, $A_z = 0$, $(M_A)_y = 44$ lb $\cdot$ ft,
$(M_A)_z = -32$ lb $\cdot$ ft

4–57. $C_z = 600$ N, $B_z = 500$ N, $A_y = 700$ N, $B_x = 0$, $B_y = 0$

4–58. $F_{DE} = 21.3$ kip, $F_{BC} = F_{AC} = 8.89$ kip

4–59. $F_C = 147$ N, $F_B = 81.8$ N, $F_A = 164$ N

4–61. $A_y = 0$, $F = 34.6$ N, $B_z = 129$ N, $A_z = -66.0$ N,
$B_x = 95.0$ N, $A_x = 75.0$ N

4–62. $P = 75$ lb, $B_z = 75$ lb, $B_x = 112$ lb, $A_y = 0$,
$A_x = 37.5$ lb, $A_z = 75$ lb

4–63. $B_z = 312$ N, $M_{B_y} = 104$ N $\cdot$ m, $B_x = 180$ N,
$A_x = 480$ N, $B_y = 0$, $A_z = 831$ N

4–65. $A_x = -56.6$ lb, $A_z = 28.3$ lb, $B_y = -28.3$ lb,
$B_z = 72.4$ lb, $C_y = 28.3$ lb, $C_z = 15.9$ lb

4–66. $A_x = 0$, $T_{CD} = 88.0$ N, $B_z = 141$ N, $B_y = -11.4$ N,
$A_y = 34.1$ N, $A_z = -29.3$ N

4–67. $F_{BC} = 175$ lb, $A_x = 130$ lb, $A_y = -10$ lb,
$M_{Ax} = -300$ lb $\cdot$ ft, $M_{Ay} = 0$, $M_{Az} = -720$ lb $\cdot$ ft

4–69. $T_{DE} = 721$ lb, $T_{BC} = 2.16$ kip, $A_x = -309$ lb,
$A_y = 1.55$ kip, $A_z = -1.21$ kip

4–70. $W = 275$ lb, $A_z = -392$ lb, $A_x = -100$ lb, $A_y = 500$ lb

4–71. 344 lb, $\mu_s \geq 0.70$

4–73. 76.4 lb $\leq P \leq$ 144 lb

4–74. $x = 1.6$ ft > 1.5 ft, 75 lb

4–75. the spool will not remain in equilibrium

4–77. 6.67 in.

4–78. 21.8°

4–79. 45°, 1

4–81. 72.0 lb

4–82. 21.7 lb

4–83. 20 lb

4–85. 447 lb

4–86. 4.05 kip

4–87. 106 lb

4–89. wedges are self-locking

4–90. 6.97 kN, 15.3 kN

4–91. 1.25 kN, 6.89 kN

4–93. 2.50 kip $\cdot$ ft

4–94. 2.21 kip $\cdot$ ft

4–95. he can push cabinet to the left

4–97. 0.268

4–98. 377 lb

4–99. 33.8°

4–101. 1.17 kN

4–102. he cannot move the crate

4–103. 26.6°

4–105. 326 N

4–106. 40.2 N

4–107. 46.4°

4–109. $P = \dfrac{W \sin (\alpha + \theta)}{\cos (\phi - \theta)}$ (Q.E.D.)

4–110. $P = W \sin(\alpha + \theta)$, $\phi = \theta$

4–111. 1.02°

4–113. $P = 100$ lb, $B_z = 40$ lb, $B_x = -35.7$ lb, $A_x = 136$ lb,
$B_y = 0$, $A_z = 40$ lb

4–114. $F_{BD} = 171$ N, $F_{BC} = 145$ N

4–115. $T = 1.01$ kN, $F_D = 982$ N

4–117. $A_y = 7.36$ kip, $B_x = 0.5$ kip, $B_y = 16.6$ kip

4–118. $T_{BA} = 715$ N, $T_{BC} = 1.04$ kN, $D_x = 490$ N,
$D_y = 654$ N, $D_z = 2.29$ kN

Chapter 5

5–1. $F_{BC} = 374$ lb (C), $F_{BA} = 143$ lb (T), $F_{CA} = 264$ lb (T)

5–2. $F_{BA} = 1.18$ kip (C), $F_{AC} = 833$ lb (T), $F_{BC} = 373$(C)

5–3. $F_{DE} = 0$, $F_{DC} = 16.5$ kip (C), $F_{AB} = 4.21$ kip (C),
$F_{AE} = 4.33$ kip (T), $F_{CE} = 7.81$ kip (T),
$F_{CB} = 2.33$ kip (C), $F_{BE} = 3.50$ kip (T)

5–5. $F_{EC} = 9.01$ kN (C), $F_{ED} = 7.50$ kN (T),
$F_{DC} = 5$ kN (C), $F_{DA} = 7.50$ kN (T), $F_{CA} = 4.51$ kN (T),
$F_{CB} = 13.5$ kN (C)

5–6. $F_{GC} = F_{FE} = 0$, $F_{CD} = F_{ED} = 0$,
$F_{GB} = F_{FA} = 4.24$ kN (C),
$F_{AB} = 3.0$ kN (T), $F_{GD} = F_{FD} = 4.24$ kN (C),
$F_{CB} = F_{EA} = 4$ kN (C)

5–7. $F_{CD} = 400$ lb (T), $F_{CB} = 346$ lb (C), $F_{BD} = 0$,
$F_{BA} = 346$ lb (C), $F_{DA} = 400$ lb (C), $F_{DE} = 800$ lb (T),
$F_{AE} = 200$ lb (T)

5–9. $F_{EC} = 1.20$ P (T), $F_{ED} = 0$,
$F_{AB} = F_{AD} = F = 0.373$ P (C), $F_{DC} = 0.373$ P (C),
$F_{DB} = 0.333$ P (T), $F_{BC} = 0.373$ P (C)

5–10. $F_{FC} = 1.39$ kip (T), $F_{GF} = 19.0$ kip (C),
$F_{CD} = 16.7$ kip (T)

5–11. $F_{HG} = 15$ kN (C), $F_{HC} = 7.07$ kN (T), $F_{GC} = 0$

5–13. $F_{BE} = 21.2$ kN(T), $F_{EF} = 25$ kN(C)

5–14. $F_{DF} = 2.26$ kN (T), $F_{CE} = 4.53$ kN (C),
$F_{ED} = 400$ N (C)

5–15. $F_{FE} = 4.50$ kip (C), $F_{CD} = 1.41$ kip (C),
$F_{CE} = 2.0$ kip (T)

5–17. $F_{FE} = 1.0$ kip (T), $F_{BC} = 2.24$ kip (C), $F_{FB} = 0$

5–18. $F_{CD} = 671$ N (C), $F_{KJ} = 800$ N (T), $F_{CJ} = 283$ N (C)

5–19. $F_{BC} = F_{AB} = F_{CD} = F_{DE} = 0$, $F_{IC} = 6.88$ kN (C),
$F_{CG} = 11.0$ kN (T)

5–21. $F_{BC} = 10.4$ kN (C), $F_{HC} = 2.24$ kN (T),
$F_{HG} = 9.15$ kN (T)

5–22. $F_{CD} = 11.2$ kN (C), $F_{CF} = 3.21$ kN (T),
$F_{CG} = 6.80$ kN (C)

5–23. $F_{CD} = 2.62$ kip (T), $F_{KJ} = 3.02$ kip (C), $F_{JN} = 0$

5–25. $F_{CD} = 3.69$ kip (C), $F_{KJ} = 3.69$ kip (T)

5–26. $F_C = 220$ lb, $F_A = 434$ lb, $\theta = 64.0° \angle \theta$

5–27. $A_x = 300$ N, $A_y = 300$ N, $C_x = 300$ N, $C_y = 300$ N

5–29. $F_{AB} = 1.13$ kN, $F_{AC} = 511$ N

5–30. $F_{BC} = 576$ N, $F_{AB} = 360$ N, $D_x = 360$ N,
$D_y = 450$ N, $M_D = 450$ N · m

5–31. $F_{AB} = 8.60$ MN, $F_C = 9.78$ MN

5–33. $B_x = 1600$ lb, $B_y = 800$ lb, $D_x = 800$ lb, $D_y = 800$ lb

5–34. 667 N

5–35. 111 N

5–37. $F_{BC} = 297$ lb, $D_y = 210$ lb, $D_x = 60$ lb

5–38. $A_y = 2$ kN, $B_x = 0$, $B_y = 12$ kN, $M_B = 32$ kN · m

5–39. $B_y = 30$ kip, $D_x = 0$, $D_y = 30$ kip, $C_y = 135$ kip,
$A_x = 0$, $A_y = 75$ kip, $F_y = 135$ kip, $E_x = 0$, $E_y = 75$ kip

5–41. $T_{AB} = 357$ lb, $N_D = 70$ lb, $C_x = C_y = 0$

5–42. 36.0 lb

5–43. 120 lb

5–45. $C_y = 285$ lb, $C_x = 326$ lb, $A_x = 176$ lb, $A_y = 135$ lb

5–46. $D_x = 50$ lb, $D_y = 50$ lb, $B_x = 112$ lb, $A_x = 62.5$ lb,
$B_y = 117$ lb, $A_y = 167$ lb

5–47. $F_{EF} = 3.53$ kN, $A_x = 1.96$ kN, $A_y = 571$ N

5–49. $D_x = 300$ lb, $D_y = 300$ lb, $E_y = 600$ lb, $B_y = 300$ lb,
$E_x = 225$ lb, $B_x = 75$ lb

5–50. $F_{AB} = 1.71$ kip, $F_C = 2.28$ kip

5–51. $B_x = 2.06$ kN, $A_y = 0$, $A_x = 2.06$ kN, $F_C = 2.43$ kN

5–53. $C_x = 19.2$ lb, $C_y = 25.6$ lb, $B_x = 305$ lb, $B_y = 0$,
$A_x = 206$ lb, $A_y = 25.6$ lb

5–54. $C_y = 25.5$ lb, $A_y = 144$ lb, $B_x = 12.8$ lb, $B_y = 25.5$ lb

5–55. $T = 147$ N, $\theta = 14.6°$

5–57. 170 lb · in.

5–58. $F_{DE} = W \cot \theta$

5–59. 850 N

5–61. $G_x = 333$ lb, $G_y = 250$ lb, $F_x = G_x = 333$ lb,
$F_y = G_y = 250$ lb

5–62. 71.7 lb

5–63. $D_y = 2.5$ kip, $B_y = 12$ kip, $A_y = 1.5$ kip, $A_x = 0$

5–65. $F_{AD} = 990$ lb (C), $F_{AB} = 700$ lb (T), $F_{DB} = 495$ lb (C),
$F_{DC} = 1.48$ kip (C), $F_{CB} = 1.05$ kip (T)

5–66. $F_{HD} = 7.07$ kN (C), $F_{CD} = 50$ kN (T), $F_{GD} = 5$ kN (T)

5–67. $F_{HB} = 21.2$ kN (C), $F_{BC} = 50$ kN (T), $F_{HI} = 35$ kN (C)

5–69. $B_y = 449$ lb, $A_x = 92.3$ lb, $A_y = 186$ lb,
$M_A = 359$ lb · ft

5–70. $B_x = 84.9$ lb, $B_y = 84.9$ lb, $A_x = 84.9$ lb, $A_y = 265$ lb,
$M_A = 953$ lb · ft

5–71. $P = 25$ lb, $P = 33.3$ lb, $P = 11.1$ lb

Chapter 6

6–1. $\bar{x} = (2/3)a$, $\bar{y} = (m/3)a$

6–2. $\bar{x} = (2/7)a$, $\bar{y} = (2/5)a$

6–3. $\bar{x} = [(1 + n)/2(2 + n)]a$, $\bar{y} = [(1 + n)/(1 + 2n)]h$

6–5. $\bar{x} = 1.08$ in., $\bar{y} = 0.541$ in.

6–6. $\bar{x} = (5/8)a$, $\bar{y} = (2k/5)a$

6–7. 1.70 ft, $T_B = 26.7$ lb, $T_A = 36.2$ lb

6–9. $\bar{x} = (\pi/2)a$, $\bar{y} = (\pi/8)a$

6–10. $\bar{x} = 0.45$ m, $\bar{y} = 0.45$ m

6–11. $\bar{x} = 0.4$ ft, $\bar{y} = 1.0$ ft

6–13. $\bar{x} = 1.26$ m, $\bar{y} = 0.143$ m

6–14. $\bar{y} = \frac{3}{4} h$

6–15. $(4/5)h$

6–17. 1.33 ft

6–18. 2.67 m

6–19. 84.7 mm

6–21. $\bar{y} = \frac{3}{8} a$

6–22. 95 mm

6–23. 87.5 mm

6–25. $\bar{x} = 2.5$ in., $\bar{y} = 4$ in.

6–26. $\bar{y} = (b^2c + 3abc + 3a^2c)/3(a + b)(2a + b)$, $\bar{x} = b(b + 3a)/3(2a + b)$

6–27. $\bar{x} = 432$ mm, $\bar{y} = -339$ mm

6–29. $\bar{x} = 1.82$ m, $\bar{y} = 1.28$ m

6–30. 142 mm

6–31. $\bar{x} = 2.73$ in., $\bar{y} = 1.42$ in.

6–33. 11.5 ft

6–34. 7.18 ft, $B_y = 2633$ lb, $A_y = 667$ lb

6–35. 4.28 ft, $B_y = 620$ lb, $A_y = 830$ lb

6–37. 5.32 in.

6–38. $\bar{z} = 52.5$ mm, $\bar{x} = \bar{y} = 0$

6–39. 2.95 in.

6–41. $\bar{z} = \frac{3}{8} h$

6–42. $\bar{z} = \frac{1}{2}a$

6–43. 323 mm

6–45. 1.95 kip · ft $\downarrow$

6–46. 27 kN, 2.11 m

6–47. 18.0 kip $\downarrow$, 11.7 ft

6–49. 3.25 kip, 67.2° $\theta\nearrow$, 3.86 ft

6–50. $b = 9.0$ ft, $a = 7.50$ ft

6–51. 107 kN $\leftarrow$, 1.60 m

6–53. $F = 24$ kN, $\bar{x} = 2$ m, $\bar{y} = 1.33$ m

6–54. 133 lb, $\bar{y} = 7.50$ ft, $\bar{x} = 0$

6–55. 42.7 kN, $\bar{y} = 2.40$ m, $\bar{x} = 0$, $C_z = B_z = 12.8$ kN, $A_x = 0$, $A_y = 0$

6–57. 1.41 MN, 4 m

6–58. 157 kN, 235 kN, 4.22 m

6–59. 3.65 m

6–61. $F_h = 488$ kip, $F_v = 260$ kip

6–62. 4.92 kip, 24.0° $\searrow\theta$

6–63. 6.26 kip

6–65. 13.8 kip, 61.4 kip

6–66. 356 in^4

6–67. $I_x = \frac{1}{12}bh^3$

6–69. $I_x = \dfrac{nbh^3}{1 + 3n}$, $I_y = \dfrac{nhb^3}{3(n + 3)}$

6–70. 0.165 in^4

6–71. $I_x = 2.13$ in^4, $I_y = 4.57$ in^4

6–73. $114(10)^6$ mm^4

6–74. 0.167 in^4

6–75. 0.3 in^4

6–77. 2.13 ft^4

6–78. 0.610 ft^4

6–79. 0.305 m^4

6–81. $I_x = 141$ in^4, $I_y = 21.3$ in^4

6–82. $147(10)^6$ mm^4

6–83. $\bar{y} = 2$ in., $I_{\bar{x}\bar{x}} = 128$ in^4

6–85. $\bar{y} = 1.29$ in., $I_{\bar{x}\bar{x}} = 6.74$ in^4

6–86. $\bar{y} = 15.7$ in., $I_{x'x'} = 2.16(10)^3$ in^4

6–87. 86 in^4

6–89. $\bar{y} = 170$ mm, $I_{x'} = 722(10)^6$ mm^4

6–90. $91.7(10)^6$ mm^4

6–91. $\bar{y} = 2.20$ in., $I_{y'} = 57.9$ in^4

6–93. $\bar{x} = 35$ mm, $\bar{y} = 85$ mm, $I_{x'} = 45.1(10)^6$ mm^4, $I_{y'} = 7.60(10)^6$ mm^4

6–94. 90.5 mm

6–95. 0.4 m

6–97. $\bar{z} = 0.422$ m, $\bar{x} = \bar{y} = 0$

6–98. $I_x = 246$ in^4, $I_y = 61.5$ in^4

6–99. 10.8 kN, 2.26 m

6–101. $\bar{x} = 3.30$ in., $\bar{y} = 3.30$ in.

6–102. $\bar{x} = -0.262$ in., $\bar{y} = 0.262$ in.

6–103. $0.0954\ d^4$

Chapter 7

7–1. (a) $F_A = 13.8$ kip,   (b) $F_A = 34.9$ kN

7–2. (a) $T_C = 250$ N · m, $T_D = 0$,   (b) $T_D = 500$ lb · ft, $T_C = 150$ lb · ft

7–3. (a) $N_a = 500$ lb, $V_a = 0$,   (b) $V_b = 250$ lb, $N_b = 433$ lb

7–5. $N_D = 131$ N, $V_D = 175$ N, $M_D = 8.75$ N · m

7–6. $N_D = 1.20$ kip, $V_D = 0.625$ kip, $M_D = 0.769$ kip · ft, $N_E = 2.00$ kip, $V_E = 0$, $M_E = 0$

7–7. $N_D = 1.44$ kN, $V_D = 0$, $M_D = 3.75$ kN · m, $N_E = 2.89$ kN, $V_E = 0$, $M_E = 0$

7–9. (a) $N_C = 3.60$ kip, $V_C = 1.20$ kip, $M_C = 9.60$ kip · ft, (b) $N_C = 1.70$ kip, $V_C = 3.39$ kip, $M_C = 9.60$ kip · ft

7–10. $N_C = 18.2$ N, $V_C = 10.5$ N, $M_C = 9.46$ N · m

7–11. $V_C = 2.94$ kN, $N_C = 2.94$ kN, $M_C = 1.47$ kN · m

7–13. $(N_B)_x = 0$, $(V_B)_y = 0$, $(V_B)_z = 70.6$ N, $(T_B)_x = 9.42$ N · m, $(M_B)_y = 6.23$ N · m, $(M_B)_z = 0$

7–14. $(V_B)_x = 105$ lb, $(V_B)_y = (N_B)_z = (M_B)_x = 0$, $(M_B)_y = 788$ lb · ft, $(T_B)_z = 52.5$ lb · ft

7–15. $(V_C)_x = 250$ N, $(N_C)_y = 0$, $(V_C)_z = 240$ N, $(M_C)_x = 108$ N · m, $(T_C)_y = 0$, $(M_C)_z = 138$ N · m

7–17. $V_B = 0.785wr$, $N_B = 0$, $T_B = 0.0783wr^2$, $M_B = 0.293wr^2$

7–18. $V = 20$ lb, $M = \{20x\}$ lb · in., $V = -15$ lb, $M = \{525 - 15x\}$ lb · in.

7–19. -300 lb, $\{-300x\}$ lb · ft, 300 lb, $M = \{300x - 900\}$ lb · ft

7–21. $V = -2M_O/L$, $M = -(2M_O/L)x$, $V = -2M_O/L$, $M = M_O - (2M_O/L)x$, $V = -2M_O/L$, $M = 2M_O - (2M_O/L)x$

7–22. $V = \{2250 - 300x\}$ lb, $M = \{2250x - 150x^2\}$ lb · ft

7–23. 0.75 kN, $\{0.75x\}$ kN · m, $V = \{3.75 - 1.5x\}$ kN, $M = \{3.75x - 0.75x^2 - 3\}$ kN · m

7–25. $V = \{250 - 300x\}$ N, $M = \{250x - 150x^2\}$ N · m, $V = \{550 - 600x\}$ N, $M = \{-300x^2 + 550x - 150\}$ N · m

7–26. $V = \{1350 - 60x\}$ lb, $M = \{1350x - 30x^2 - 8250\}$ lb · ft, $M = \{600x - 30x^2 - 3000\}$ lb · ft

7–27. $V = (w/2)(3L - 2x)$, $M = (w/2)(3Lx - x^2 - 2L^2)$

7–29. $V = \{-50x^2 - 650\}$ N, $M = \{-16.7x^3 - 650x\}$ N · m, $V = \{-300x + 2100\}$ N, $M = \{-150x^2 + 2100x - 7350\}$ N · m

7–30. $V = \{-66.7x^2\}$ lb, $M = \{-22.2x^3\}$ lb · ft

7–31. $V = \{3000 - 41.7x^2\}$ lb, $M = \{-13.9x^3 + 3000x - 18000\}$ lb · ft, $V = \{41.7x^2 - 1000x + 6000\}$ lb, $M = \{-13.9(12 - x)^3\}$ lb · ft

7–33. See Prob. 7–37

7–34. See Prob. 7–38

7–35. $V = \{-66.7x^2\}$ lb, $M = \{-22.2x^3\}$ lb · ft

7–37. $x = 0$, $V = 358$, $M = -1458$; $x = 25$, $V = -41.7$, $M = -500$

7–38. $x^- = 4$, $V = -400$, $M = -1600$; $x^+ = 16$, $V = -950$, $M = 3800$

7–39. $x = 0$, $V = 0$, $M = 0$; $x = 12$, $V = 0$, $M = 2700$

7–41. $x = 0$, $V = 1.52$, $M = 0$; $x^+ = 800$, $V = -1.98$, $M = 0.395$

7–42. $x = 0$, $V = 0$, $M = 0$; $x = 6^+$, $V = 800$, $M = -1200$

7–43. $x = 20$, $V = 12$, $M = 0.08$; $x = 50$, $V = 0$, $M = 0.26$

7–45. $x = 0$, $V = 1.10$, $M = -10.4$; $x = 12$, $V = 0.4$, $M = 0$

7–46. $x = 0$, $V = wL/3$, $M = 0$; $x = L^+$, $V = wL/2$, $M = -wL^2/6$

7–47. $x = 0$, $V = 0$, $M = 24$; $x = 3$, $V = -12$, $M = 12$

7–49. $x = 0$, $V = 52.8$, $M = 0$; $x = 4.36$, $V = 0$, $M = 153$

7–50. $x = 0$, $V = 0$, $M = 0$; $x = 18$, $V = 0$, $M = 0$

7–51. $N_D = 6.08$ kN, $V_D = 2.60$ kN, $M_D = 13.0$ kN · m

7–53. $V_D = M_D = 0$, $N_D = F_{CD} = 86.6$ lb, $N_E = 0$, $V_E = 28.9$ lb, $M_E = 86.6$ lb · ft

7–54. $x = 2$, $V = 7.5$, $M = 15$; $x = 2.75$, $V = 0$, $M = 17.8$

7–55. $x = 0$, $V = 2.5$, $M = 0$; $x = 1.25$, $V = 0$, $M = 1.56$

Chapter 8

8–1. $\sigma = 1.82$ MPa

8–2. $\sigma_{40} = 3.98$ MPa, $\sigma_{30} = 7.07$ MPa, $\tau_{avg} = 5.09$ MPa

8–3. $\tau_B = \tau_C = 324$ MPa, $\tau_A = 324$ MPa

8–5. $\bar{x} = 4$ in., $\bar{y} = 4$ in., $\sigma = 9.26$ psi

8–6. $\sigma = 8$ MPa, $\tau_{avg} = 4.62$ MPa

8–7. $\sigma = 1.90$ psi, $\tau_{avg} = 7.08$ psi

8–9. $\sigma_B = 151$ kPa, $\sigma_C = 32.5$ kPa, $\sigma_D = 25.5$ kPa

8–10. $\sigma = (238 - 22.6z)$ kPa

8–11. 3.39 psi

8–13. $\sigma_s = 208$ MPa, $(\tau_{avg})_a = 4.72$ MPa, $(\tau_{avg})_b = 45.5$ MPa

8–14. $d = 1.20$ m

8–15. $\tau_{avg} = 61.3$ MPa

8–17. 21.8 kN/m

8–18. $w = w_1 e^{(w_1^2\gamma/2P)z}$

8–19. $d = 1$ in.

8–21. $d_{AB} = 19.9$ mm, $d_{AC} = 19.1$ mm

8–22. $h = 1.5$ in.

8–23. 3 in. $\times$ 3 in., $4\frac{1}{2}$ in. $\times$ $4\frac{1}{2}$ in.

8–25. $d = 144$ mm

8–26. $A_{BC} = 0.577$ in^2, $d_A = 0.743$ in., $d_B = 0.525$ in.

8–27. $P = 5.11$ kN

8–29. $P = 3.26$ kip

8–30. $d_b = 6.11$ mm, $d_w = 15.4$ mm

8–31. F.S. = 2.71, F.S. = 1.53

8–33. $\tau = 23.9$ ksi, F.S. = 1.05, $\sigma = 30.6$ ksi, F.S. = 1.24

8–34. 0.452 kip/ft

8–35. $\epsilon = 0.0472$ in./in.

8–37. $\epsilon_{CE} = 0.00250$ mm/mm, $\epsilon_{BD} = 0.00107$ mm/mm

8–38. $\Delta L = \dfrac{kL^2}{2}$

8–39. $\epsilon_{AB} = 0.0343$

8–41. 0.00578 mm/mm

8–42. $\epsilon = 2kx$

8–43. $\gamma_{xy} = 0.00880$ rad

8–45. 0.00884 mm/mm

8–46. $(\epsilon_x)_{avg} = \dfrac{kL^2}{3}$

8–47. $\gamma_{xy} = 0.0502$ rad, $\epsilon_x = -0.03$ in./in., $\epsilon_y = 0.02$ in./in.

8–49. $\gamma_{xy} = 0.02$ rad

8–50. $\sigma_{AB} = 417$ psi (C), $\sigma_{BC} = 469$ psi (T), $\sigma_{AC} = 833$ psi (T)

8–51. 267 kPa

8–53. $N_D = 17.6$ kN, $V_D = 2.61$ kN, $M_D = 5.22$ kN $\cdot$ m, $N_E = 10.1$ kN, $V_E = 4.89$ kN, $M_E = 15.3$ kN $\cdot$ m

8–54. $N_C = 0$, $V_C = 3.50$ kip, $M_C = 47.5$ kip $\cdot$ ft, $N_D = 0$, $V_D = 0.240$ kip, $M_D = 0.360$ kip $\cdot$ ft

8–55. $N_B = 0$, $V_{BR} = 22.5$ kN, $M_B = 102$ kN $\cdot$ m, $V_{BL} = 62.1$ kN

8–57. $\tau = 7.50$ MPa

8–58. $\epsilon_{BE} = 3.82(10^{-3})$ in./in., $\epsilon_{CF} = 7.31(10^{-3})$ in./in., $\epsilon_{AD} = 0.333(10^{-3})$ in./in.

8–59. $\gamma_{xy} = 0.00125$ rad

Chapter 9

9–1. $E = 286$ GPa, $U_r = 91.6$ kJ/m^3

9–2. $E = 20.0(10^3)$ ksi, $U_r = 25.6$ in. $\cdot$ lb/in^3

9–3. $E_{approx} = 3.13(10^3)$ ksi

9–5. $U_T = 16.3$ in. $\cdot$ kip/in^3

9–6. $(\sigma_{ult})_{approx} = 110$ ksi, $\sigma_R = 93.1$ ksi, $(\sigma_Y)_{approx} = 55$ ksi, $E_{approx} = 32.0(10^3)$ ksi

9–7. $E = 28.7(10^3)$ ksi, $\sigma_{pl} = 43$ ksi, $\sigma_{ult} = 85$ ksi, $U_{r(approx)} = 32.2$ in. $\cdot$ lb/in^3, elastic recovery = 0.00251 in./in., permanent set = 0.0475 in./in.

9–9. $\Delta P = 50.1$ kip

9–10. $\Delta L = 0.304$ in.

9–11. $\epsilon_b = 0.00227$ mm/mm, $\epsilon_s = 0.000884$ mm/mm

9–13. $\epsilon_{DE} = 0.00116$ in./in., $W = 112$ lb, $\epsilon_{BC} = 0.00193$ in./in.

9–14. $A = 0.209$ in^2, $P = 1.62$ kip

9–15. $P = 2.46$ kN

9–17. (a) $\Delta L = -0.611(10^{-3})$ in., (b) $D' = 0.500067$ in.

9–18. $\nu = 0.300$

9–19. $\epsilon_x = 0.0075$ in./in., $\epsilon_y = -0.00375$ in./in., $\gamma_{xy} = 0.0122$ rad

9–21. 8.33 mm

9–22. $\Delta_h = 3.02$ mm

9–23. $\delta = \dfrac{Pa}{2bhG}$

9–25. $\nu = 0.33$, $h' = 2.000176$ in.

9–26. $L = 2.792$ in.

9–27. $d_{AB} = 3.54$ mm, $d_{AC} = 3.23$ mm, $l_{AB} = 750.488$ mm

9–29. $x = 1.53$ m, $d'_A = 30.00782$ mm

9–30. 2.45 kip

9–31. $E = 28.6(10^3)$ ksi

Chapter 10

10–1. $\sigma_{AB} = 22.2$ ksi (T), $\sigma_{BC} = 41.7$ ksi (C), $\sigma_{CD} = 25.0$ ksi (C), $\Delta_{A/D} = -0.00157$ in.

10–2. $\Delta_{B/C} = -0.0278$ in.

10–3. $\Delta_B = 1.59$ mm, $\Delta_A = 6.14$ mm

10–5. $\Delta_{A/D} = 0.111$ in.

10–6. $\Delta_B = 2.31$ mm, $\Delta_A = 2.64$ mm

10–7. $\Delta_D = 1.17$ mm

10–9. $\Delta_{C_h} = 0.0975$ mm

10–10. $x = 10$ in.

10–11. 0.281 in.

10–13. $\Delta = \dfrac{PL}{AE} + \dfrac{\gamma L^2}{2E}$

10–14. $\Delta = \dfrac{Ph}{tE(d_2 - d_1)} \ln \dfrac{d_2}{d_1}$

10–15. $\Delta = \dfrac{PL}{\pi E r_1 r_2} + \dfrac{\gamma L^2 (r_2 + r_1)}{6E(r_2 - r_1)} - \dfrac{\gamma L^2 r_1^2}{3E r_2 (r_2 - r_1)}$

10–17. $\sigma_{st} = 48.8$ MPa, $\sigma_{con} = 5.85$ MPa

10–18. $\sigma_{st} = 65.9$ MPa, $\sigma_{con} = 8.24$ MPa

10–19. $\sigma_{st} = 4.05$ ksi, $\sigma_{con} = 0.488$ ksi

10–21. $\sigma_{BC} = 14.1$ MPa (T), $\sigma_{AB} = 88.1$ MPa (T), $\sigma_{AD} = 113$ MPa (C)

10–22. $\sigma_{br} = 0.340$ ksi, $\sigma_{st} = 0.657$ ksi

10–23. $T_{AB} = 1.12$ kip, $T_{AC} = 1.68$ kip

10–25. $P_t = P_b = 116$ kN

10–26. 145 ksi

10–27. $T_{AB} = T_{CD} = 16.7$ kN, $T_{EF} = 33.3$ kN

10–29. $F_{AB} = F_{AD} = 465$ N, $F_{AC} = 727$ N

10–30. $\sigma_{AB} = \dfrac{7P}{12A}$, $\sigma_{CD} = \dfrac{P}{3A}$, $\sigma_{EF} = \dfrac{P}{12A}$

10–31. $x = \dfrac{E_B h}{E_B + E_A}$

10–33. $\Delta_B = 0.0489$ in.

10–34. $F_A = 5.79$ kN, $F_B = 9.64$ kN, $F_C = 11.6$ kN

10–35. $T_B = 86.6$ lb, $T_C = 195$ lb

10–37. 0.312 in., 24.0 k

10–38. $F = 67.2$ kip

10–39. $F = 65.7$ kip

10–41. 11.6 ksi, 116°F

10–42. $F = 16.5$ kip

10–43. 4.23 kN

10–45. $F_B = 3.30$ kip, $F_A = 3.32$ kip

10–46. $\sigma_b = 29.5$ MPa, $\sigma_s = 40.1$ MPa

10–47. $F = \dfrac{\alpha AE(T_B - T_A)}{2}$

10–49. $\sigma_{\max} = 190$ MPa

10–50. $P = 5.05$ kN

10–51. 1.21 kip

10–53. $w = 0.836$ in.

10–54. $P = 19$ kN, $K = 1.26$

10–55. $L = 463.41$ ft

10–57. $\sigma_{BC} = 9.55$ MPa, $\sigma_D = 13.4$ MPa

10–58. $x = 2.12$ ft, $P = 4.33$ kip

10–59. $F = 2.39$ kip

10–61. $\Delta = 0.129$ mm, $h' = 49.9988$ mm, $w' = 59.9986$ mm

10–62. 0.130 kip/ft, 0.0596 in.

10–63. $F = 11.8$ kN

Chapter 11

11–1. (a) $T = 7.95$ kip · in., (b) $T = 6.38$ kip · in.

11–2. $\tau_{\max} = 26.7$ MPa

11–3. $\tau_A = 14.1$ MPa, $\tau_B = 8.06$ MPa

11–5. $\tau_C = 3.91$ ksi, $\tau_D = 1.56$ ksi

11–6. $\tau_{AB} = 7.82$ ksi, $\tau_{BC} = 2.36$ ksi

11–7. $(\tau_{BC})_{\max} = 5.79$ ksi, $(\tau_{DE})_{\max} = 8.26$ ksi

11–9. $\tau_A = 6.88$ MPa, $\tau_B = 10.3$ MPa

11–10. $F = 154$ N

11–11. $T' = 125$ N · m, $(\tau_{CD})_{\max} = 14.8$ MPa, $(\tau_{AB})_{\max} = 9.43$ MPa

11–13. $T = 71.5$ N · m, $(\tau_s)_{\max} = 23.3$ MPa

11–14. $t_0 = 133$ N · m/m, $\tau_A = 255$ kPa, $\tau_B = 141$ kPa

11–15. $T_B = T_A + \frac{1}{2}t_A L$, $(\tau_s)_{\max} = \dfrac{r_o(2T_A + t_A L)}{\pi(r_o{}^4 - r_i{}^4)}$

11–17. $d = 1.25$ in.

11–18. 216 psi

11–19. $T = 1.51$ lb · ft, $\tau_{\max} = 219$ psi

11–21. $t = 0.174$ in.

11–22. 0.104 in.

11–23. $\omega = 296$ rad/s

11–25. 1.72°

11–26. $\phi_{A/C} = 0.362°$

11–27. $\gamma_{C/2} = \dfrac{Tc}{2JG}$

11–29. 0.243°

11–30. $\phi_{A/D} = 0.879°$

11–31. $d = 2.75$ in.

11–33. 60.3 N, 7.20 mm

11–34. $\tau_{\max} = 64.0$ MPa

11–35. $t = 7.53$ mm

11–37. $1\frac{1}{4}$ in.

11–38. $\phi_A = 1.92°$

11–39. $\phi_B = 3.06°$

11–41. $\phi_B = \dfrac{t_0 L^2}{\pi c^4 G}$

11–42. $\phi = \dfrac{2L(t_0 L + 3T_A)}{3\pi(r_o{}^4 - r_i{}^4)G}$, $T_B = \dfrac{1}{2}t_0 L + T_A$

11–43. $\phi_A = \dfrac{7LT}{12\pi r^4 G}$

11–45. 583 psi

11–46. $d = 42.7$ mm

11–47. $x = \dfrac{17}{32} L$

11–49. $\phi_C = 0.116°$, $(\tau_{st})_{\max BC} = 395$ psi, $(\gamma_{st})_{\max} = 0.0343(10^{-3})$ rad, $(\tau_{br})_{\max} = 96.1$ psi, $(\gamma_{br})_{\max} = 0.0172(10^{-3})$ rad

11–50. $d = 403$ mm

11–51. $T_A = 127$ N · m, $T_B = 424$ N · m

11–53. $(\tau_{AC})_{\max} = 2.17$ ksi, $(\tau_{BD})_{\max} = 4.35$ ksi

11–54. $T_A = 371$ lb $\cdot$ ft, $T_B = 429$ lb $\cdot$ ft

11–55. $T_B = \dfrac{7t_0 L}{12}$, $T_A = \dfrac{3t_0 L}{4}$

11–57. 90.2 kW

11–58. $T = 8.16$ N $\cdot$ m

11–59. $T = 193$ lb $\cdot$ ft, $\phi = 17.2°$

11–61. $T = 19.2$ kN $\cdot$ m, $\phi = 24.9°$, $\phi_r = 6.72°$

11–63. $T_Y = 1.26$ kN $\cdot$ m, $\phi_Y = 3.58°$, $\phi = 4.86°$

11–65. $T_p = 13.6$ kN $\cdot$ m, $\phi_r = 12.3°$

11–66. 71.8 kN $\cdot$ m, 7.47°

11–67. 110 lb $\cdot$ ft.

11–69. $\tau_{max} = \dfrac{19T}{12\pi r^3}$

11–70. $\tau_{max} = 3.18$ MPa, $\tau_{min} = 1.59$ MPa

11–71. $T' = 510$ N $\cdot$ m

11–73. (a) 6.67%   (b) 6.67%

11–74. $r' = 0.841r$

11–75. (a) $0.707r$,   (b) $0.707r$

11–77. $\phi_{A/B} = \dfrac{T}{2a\pi G} (1 - e^{-4aL})$

11–78. $\tau_{max} = \dfrac{T}{2\pi r_i^2 h}$

11–79. $T = 331$ lb $\cdot$ ft

Chapter 12

12–1. (a) $(\sigma_z)_{max} = 11.3$ psi,   (b) $(\sigma_y)_{max} = 22.7$ psi

12–2. $\sigma_{max} = 6.44$ ksi

12–3. $\sigma_A = 115$ psi (C), $\sigma_B = 68.9$ psi (T)

12–5. $\sigma_{max} = 1.35$ ksi

12–6. $F_R = 5.88$ kN

12–7. $\sigma_B = 3.61$ MPa, $\sigma_C = 1.55$ MPa

12–9. $F_{R_A} = 0$, $F_{R_B} = 1.50$ kN

12–10. $\%\left(\dfrac{M'}{M}\right) = 84.6\%$

12–11. $M = 36.5$ kN $\cdot$ m, $\sigma_{max} = 40$ MPa

12–13. $\sigma_{max} = 635$ kPa

12–14. $\sigma_A = 2.05$ ksi, $\sigma_B = 1.63$ ksi

12–15. $F_R = 11.8$ kip

12–17. $P = 7.29$ kip

12–18. $P = 5.97$ kip

12–19. 10.4 kN

12–21. $w = 1.65$ kip/ft

12–22. $\sigma_{max} = 66.8$ ksi

12–23. $\sigma_A = 0$, $\sigma_B = 462$ kPa, $\sigma_D = -462$ kPa, $\sigma_E = 0$

12–25. $\sigma_{max} = 2.90$ MPa, $\alpha = -66.6°$

12–26. 3.33 ksi, $-63.1°$

12–27. $\bar{y} = 4.83$ in., $\sigma_A = -2.35$ ksi, $\sigma_B = -2.62$ ksi

12–29. $\sigma_A = 164$ psi, $\alpha = -62.1°$, $\sigma_A = 125$ psi

12–30. 8.10 kN

12–31. $M = 24.0$ N $\cdot$ m

12–33. $\sigma_{max} = 768$ psi

12–34. $570(10^{-6})$ m^3, 1.16

12–35. $M_Y = 193$ kip $\cdot$ ft, $M_P = 342$ kip $\cdot$ ft

12–37. $M_Y = 9.44$ kN $\cdot$ m, $M_P = 17.1$ kN $\cdot$ m

12–38. 2.0, $667(10^{-6})$ m^3

12–39. $K = 1.70$, $Z = 36.0$ in^3

12–41. (a) $P = 66.7$ kN,   (b) $P = 100$ kN

12–42. (a) 18.0 kip/ft,   (b) 22.8 kip/ft

12–43. $\sigma_{max} = 5.0$ ksi, $F_A = 17.7$ kip, $F_B = 13.7$ kip

12–45. $\sigma_{max} = 2.06$ MPa

12–46. $M = 1.25$ kip $\cdot$ ft

12–47. $(\sigma_{max})_t = 18.4$ ksi, $(\sigma_{max})_c = 19.2$ ksi

12–49. $M_P = 172$ kip $\cdot$ ft

12–50. $\sigma_{max} = 12.2$ ksi

12–51. $w_2 = 800$ lb/in., $w_1 = 533$ lb/in., $\sigma_{max} = 45.1$ ksi

Chapter 13

13–1. $\tau_A = 1.99$ MPa, $\tau_B = 1.65$ MPa

13–2. $\tau_A = \tau_B = 108$ psi

13–3. 928 psi

13–5. $\tau_{max} = 276$ psi, shear-stress jump = 156 psi

13–6. 1.36 ksi

13–7. $\tau_{max} = 2.44$ MPa

13–9. 1.96 MPa

13–10. $\tau_B = 795$ psi, $\tau_C = 596$ psi

13–11. $\tau_{max} = 928$ psi

13–13. 4.85 MPa

13–14. $V = 9.96$ kip

13–15. $V = 131$ kN

13–17. $V = 50.3$ kN

13–18. $\tau = \dfrac{2V}{bh^3} (h^2 + 2yh - 8y^2)$

13–19. $\tau_{max} = 281$ psi

13–21. $\tau_{avg} = 97.2$ MPa

13–22. $s_t = 1.42$ in., $s_b = 1.69$ in.

13–23. $V = 4.97$ kip, $s_t = 1.14$ in., $s_b = 1.36$ in.

13–25. $F_C = 197$ lb, $F_D = 1.38$ kip, $\tau_{max} = 495$ psi

13–26. $s' = 1.49$ in., $s = 9.88$ in.

13–27. $s = 5.53$ in.

13–29. (a) $V = 4.10$ kip,   (b) $V = 749$ lb

13–30. $\dfrac{4V}{3A}$ Q.E.D.

13–31. $\tau = \dfrac{4V\,(c^2 - y^2)}{3\pi c^4}$

13–33. $\tau_{max} = 4.48$ ksi

13–34. $V = 317$ lb

13–35. 1.47 kip

13–37. 7.38 ksi

Chapter 14

14–1. $t = 18.8$ mm

14–2. $\sigma_1 = 209$ psi, $\sigma_2 = 0$

14–3. $p = 3.60$ MPa, $n = 113$

14–5. (a) $\sigma_1 = 127$ MPa,   (b) $\sigma_1 = 63.3$ MPa, $\tau_{avg} = 322$ MPa

14–6. $\sigma_h = 432$ psi, $\sigma_b = 8.80$ ksi

14–7. $s = 32.7$ in.

14–9. $\sigma_{max} = 1.07$ MPa

14–10. $(\sigma_{max})_c = 1.12$ ksi, $(\sigma_{max})_t = 1.28$ ksi

14–11. $w = 79.7$ mm

14–13. $(\sigma_{AB})_{max} = 667$ psi, $(\sigma_{CD})_{max} = 40.7$ ksi

14–14. $\theta = 0.286°$

14–15. $\theta = 0.215°$

14–17. $y = 0.750 - 1.50x$

14–18. $\sigma_A = 9.88$ kPa, $\sigma_B = -49.4$ kPa, $\sigma_C = -128$ kPa, $\sigma_D = -69.1$ kPa

14–19. $\sigma_A = -21.3$ psi, $\sigma_B = -12.2$ psi

14–21. $\sigma_A = 0.398$ MPa, $\sigma_B = -1.19$ MPa

14–22. $\sigma_F = -30.7$ MPa, $\tau_F = 0$, $\sigma_E = -1.23$ MPa, $\tau_E = 3.09$ MPa

14–23. $\sigma_A = 444$ kPa, $\tau_A = 130$ kPa

14–25. $\sigma_C = -62.5$ psi, $\tau_C = 162$ psi, $\sigma_B = 5.56$ ksi, $\tau_B = 0$

14–26. $\sigma_A = 15.3$ MPa, $\tau_A = 0$, $\sigma_B = 0$, $\tau_B = 0.637$ MPa

14–27. $(\sigma_D)_z = 124$ ksi, $(\tau_D)_{zx} = 62.4$ ksi, $(\sigma_C)_z = 15.6$ ksi, $(\tau_C)_{zy} = 52.4$ ksi

14–29. $\sigma_A = 94.4$ psi, $\sigma_B = -59.0$ psi

14–30. $\sigma_A = -9.41$ ksi, $\tau_A = 0$, $\sigma_B = 2.69$ ksi, $\tau_B = 0.869$ ksi

14–31. $(\sigma_A)_y = 16.2$ ksi, $(\tau_A)_{yx} = 2.84$ ksi, $(\tau_A)_{yz} = 0$

14–33. $\tau_A = 0$, $\sigma_A = -15.9$ ksi, $\tau_B = 0$, $\sigma_B = 44.6$ ksi

14–34. $(\sigma_{max})_t = 106$ MPa, $(\sigma_{max})_c = 159$ MPa

14–35. $(\sigma_{max})_t = 228$ MPa, $(\sigma_{max})_c = 168$ MPa

Chapter 15

15–1. $\sigma_{x'} + \sigma_{y'} = \sigma_x + \sigma_y$ Q.E.D.

15–2. $\sigma_{x'} = -388$ psi, $\tau_{x'y'} = 455$ psi

15–3. $\sigma_{x'} = -4.05$ ksi, $\tau_{x'y'} = -0.404$ ksi

15–5. $\sigma_{x'} = -388$ psi, $\tau_{x'y'} = 455$ psi

15–6. $\sigma_{x'} = -4.05$ ksi, $\tau_{x'y'} = -0.404$ ksi

15–7. $\sigma_{x'} = 49.7$ MPa, $\tau_{x'y'} = -34.8$ MPa

15–9. $\sigma_{x'} = -898$ psi, $\tau_{x'y'} = 605$ psi, $\sigma_{y'} = 598$ psi

15–10. $\sigma_{x'} = -19.9$ ksi, $\tau_{x'y'} = 7.70$ ksi, $\sigma_{y'} = 9.89$ ksi

15–11. $\sigma_{x'} = 0.507$ MPa, $\tau_{x'y'} = 0.958$ MPa

15–13. (a) $\sigma_1 = 5.88$ ksi, $\sigma_2 = -10.9$ ksi,
(b) $\tau_{\substack{max \\ in\text{-}plane}} = 8.38$ ksi, $\sigma_{avg} = -2.5$ ksi, $\theta_{p_1} = 53.7°$, $\theta_{p_2} = -36.3°$, $\theta_s = 8.68°$ or $98.7°$

15–14. (a) $\sigma_1 = -0.875$ ksi, $\sigma_2 = -33.1$ ksi,
(b) $\tau_{\substack{max \\ in\text{-}plane}} = 16.1$ ksi, $\sigma_{avg} = -17.0$ ksi, $\theta_{p_1} = 59.9°$, $\theta_{p_2} = -30.1°$, $\theta_s = 14.9°$ or $-75.1°$

15–15. point A: $\sigma_1 = 0$, $\sigma_2 = -192$ MPa, point B: $\sigma_1 = 24.0$ MPa, $\sigma_2 = -24.0$ MPa

15–17. $\sigma_1 = 0$, $\sigma_2 = -85.7$ MPa

15–18. $\sigma_1 = 800$ psi, $\sigma_2 = 0$, $\sigma_1 = 356$ psi, $\sigma_2 = -88.9$ psi

15–19. point D: $\sigma_1 = 1.03$ MPa, $\sigma_2 = -0.00919$ MPa, point E: $\sigma_1 = 0.0257$ MPa, $\sigma_2 = -0.570$ MPa

15–21. point A: $\sigma_1 = 0.0378$ MPa, $\sigma_2 = -10.8$ MPa, point B: $\sigma_1 = 42.0$ MPa, $\sigma_2 = -0.0106$ MPa

15–22. $\sigma_1 = 198$ MPa, $\sigma_2 = -1.37$ MPa, $\sigma_1 = 1.37$ MPa, $\sigma_2 = -198$ MPa

15–23. point A: $\sigma_1 = 48.8$ ksi, $\sigma_2 = -25.4$ ksi, point B: $\sigma_1 = 34.7$ ksi, $\sigma_2 = -34.7$ ksi

15–25. $\sigma_1 = 233$ psi, $\sigma_2 = -774$ psi, $\tau_{\substack{max \\ in\text{-}plane}} = 503$ psi

15–26. $\sigma_1 = 383$ psi, $\sigma_2 = -471$ psi, $\gamma_{\substack{max \\ in\text{-}plane}} = 427$ psi

15–27. point A: $\sigma_1 = 61.7$ psi, $\sigma_2 = 0$, point B: $\sigma_1 = 0$, $\sigma_2 = -46.3$ psi

15–29. $\sigma_{x'} = -388$ psi, $\tau_{x'y'} = 455$ psi

15–30. $\theta_p = 27.3°$ ↲, $\theta_s = 17.7°$ ↰, $\sigma_1 = 310$ psi, $\sigma_2 = -1160$ psi, $\gamma_{\substack{max \\ in\text{-}plane}} = 735$ psi, $\sigma_{avg} = -425$ psi

15–31. $\sigma_1 = 265$ MPa, $\sigma_2 = -84.9$ MPa, $\tau_{\max_{\text{in-plane}}} =$ 175 MPa, $\sigma_{\text{avg}} = 90$ MPa, $\theta_p = 29.5°$ ↓, $\theta_s = 15.5°$ ↑

15–33. $\sigma_{x'} = -0.611$ ksi, $\sigma_{y'} = -3.39$ ksi, $\tau_{x'y'} = 7.88$ ksi

15–34. $\sigma_{x'} = -421$ MPa, $\tau_{x'y'} = -354$ MPa, $\sigma_{y'} = 421$ MPa

15–35. $\sigma_1 = 12.3$ ksi, $\sigma_2 = -17.3$ ksi, $\tau_{\max_{\text{in-plane}}} = 14.8$ ksi, $\sigma_{\text{avg}} = -2.50$ ksi

15–37. $\sigma_1 = -5.53$ ksi, $\sigma_2 = -14.5$ ksi, $\tau_{\max_{\text{in-plane}}} = 4.47$ ksi, $\sigma_{\text{avg}} = -10$ ksi

15–38. $\sigma_x = 74.4$ psi, $\sigma_y = -42.4$ psi, $\tau_{xy} = 22.7$ psi

15–39. $\sigma_{x'} = \dfrac{\sigma_x + \sigma_y}{2} + \dfrac{\sigma_x - \sigma_y}{2} \cos 2\theta + \tau_{xy} \sin 2\theta$ Q.E.D.

$\tau_{x'y'} = \tau_{xy} \cos 2\theta - \dfrac{\sigma_x - \sigma_y}{2} \sin 2\theta$ Q.E.D.

15–41. $\sigma_1 = 0.129$ psi, $\sigma_2 = -121$ psi

15–42. (a) $\sigma_1 = 88.1$ MPa, $\sigma_2 = -13.1$ MPa,
(b) $\gamma_{\max_{\text{in-plane}}} = 50.6$ MPa, $\sigma_{\text{avg}} = 37.5$ MPa

15–43. point A: $\sigma_2 = -1.20$ ksi, $\sigma_1 = 0$,
point B: $\sigma_1 = 9.88$ psi, $\sigma_2 = -43.1$ psi

15–45. $\sigma_1 = 6.27$ kPa, $\sigma_2 = 0$, $\sigma_3 = -806$ kPa,
$\gamma_{\max_{\text{abs}}} = 406$ kPa

15–46. $\sigma_1 = 75.4$ psi, $\sigma_2 = -95.4$ psi, $\sigma_3 = -120$ psi,
$\tau_{\max_{\text{abs}}} = 97.7$ psi

15–47. $\sigma_1 = 6.73$ ksi, $\sigma_2 = 0$, $\sigma_3 = -4.23$ ksi, $\tau_{\max_{\text{abs}}} = 5.48$ ksi

15–49. $\epsilon_1 = 283(10^{-6})$, $\epsilon_2 = -133(10^{-6})$, $\gamma_{\max_{\text{in-plane}}} = 417(10^{-6})$, $\epsilon_{\text{avg}} = 75.0(10^{-6})$, $\theta_{p_1} = 84.8°$, $\theta_{p_2} = -5.18°$, $\theta_s = 39.8°$ or $130°$

15–50. $\epsilon_{x'} = -380(10^{-6})$, $\epsilon_{y'} = -130(10^{-6})$, $\gamma_{x'y'} = 1.21(10^{-3})$

15–51. (a) $\epsilon_1 = 441(10^{-6})$, $\epsilon_2 = -641(10^{-6})$,
(b) $\gamma_{\max_{\text{in-plane}}} = 1.08(10^{-3})$, $\epsilon_{\text{avg}} = -100(10^{-6})$,
$\theta_{p_1} = -24.8°$, $\theta_{p_2} = 65.2°$, $\theta_s = 20.2°$, $110°$

15–53. $\epsilon_{x'} = 103(10^{-6})$, $\epsilon_{y'} = 46.7(10^{-6})$, $\gamma_{x'y'} = 718(10^{-16})$

15–54. $\epsilon_{x'} = -309(10^{-6})$, $\epsilon_{y'} = -541(10^{-6})$, $\gamma_{x'y'} = -423(10^{-6})$

15–61. (a) $\epsilon_1 = 1434(10^{-6})$, $\epsilon_2 = -304(10^{-6})$,
(b) $\gamma_{\max_{\text{in-plane}}} = 1738(10^{-6})$ $\epsilon_{\text{avg}} = 565(10^{-6})$

15–62. (a) $\epsilon_1 = 487(10^{-6})$, $\epsilon_2 = -400(10^{-6})$, (b) $\epsilon_{\text{avg}} = 43.3(10^{-6})$, $\gamma_{\max_{\text{in-plane}}} = 887(10^{-6})$

15–63. $\epsilon_1 = 1.05(10^{-3})$, $\epsilon_2 = -306(10^{-6})$, $\theta_{p_1} = 15.4°$ ↓

15–65. $\epsilon_{x^1} = \dfrac{\epsilon_x + \epsilon_y}{2} + \dfrac{\epsilon_x - \epsilon_y}{2} \cos 2\theta + \dfrac{\gamma_{xy}}{2} \sin 2\theta$,

$\dfrac{\gamma_{x'y'}}{2} = -\dfrac{\epsilon_x - \epsilon_y}{2} \sin 2\theta + \dfrac{\gamma_{xy}}{2} \cos 2\theta$

15–66. (a) $K = 3.33$ ksi,   (b) $K = 5.13$ ksi

15–67. $a' = 100.248$ mm, $b' = 100.514$ mm, $t' = 1.989$ mm

15–69. $-0.0162°$

15–70. $\epsilon_{\max} = 0.903(10^{-3})$, $\epsilon_{\text{int}} = -0.0828(10^{-3})$, $\epsilon_{\min} = -0.710(10^{-3})$

15–71. $E = 14.9(10^3)$ ksi, $e = 0.282(10^{-3})$

15–73. $\sigma_3 = 0$, $\sigma_1 = 60.8$ MPa, $\sigma_2 = -33.0$ MPa

15–74. $\epsilon_{\max} = 5.28(10^{-3})$, $\epsilon_{\text{int}} = 1.36(10^{-3})$, $\epsilon_{\min} = -5.92(10^{-3})$

15–75. $\sigma_x = \dfrac{E(\epsilon_x + \nu\epsilon_y)}{1 - \upsilon^2}$ Q.E.D.,

$\sigma_y = \dfrac{E(\upsilon\epsilon_x + \epsilon_y)}{1 - \nu^2}$ Q.E.D.

15–77. $\epsilon = \dfrac{Pr}{2tE}(1 - \nu)$

15–78. $p = 3.43$ MPa, $\tau_{\text{abs}_{\max}} = 85.7$ MPa

15–79. $P = 86.9$ lb

15–81. $\Delta L_{AB} = \dfrac{3\nu M}{2Ebh}$, $\Delta L_{CD} = \dfrac{6\nu M}{Eh^2}$

15–82. $\Delta L_{\text{top}} = 1.2(10^{-3})$ in., $\Delta L_{\text{bottom}} = -1.2(10^{-3})$ in.

15–83. $\gamma_{\max_{\text{in-plane}}} = \dfrac{Pr}{2tE}(1 + \nu)$, $\gamma_{\text{abs}_{\max}} = \dfrac{Pr}{tE}(1 + \nu)$

15–85. 0.680 mm

15–86. $p = 0.967$ ksi, $\gamma_{\max_{\text{in-plane}}} = 1.30(10^{-3})$

15–87. $\gamma_{xy} = -160(10^{-6})$, $\epsilon_x = \epsilon_y = 0$, $T = 65.2$ N · m

15–89. 13.9 kip, $0.156(10^{-3})$ rad

15–90. $\dfrac{dV}{V} = \dfrac{Pr}{Et}(2.5 - 2\nu)$

15–91. $t_h = 0.206$ in.

15–93. $\sigma^1 = 236$ psi, $\sigma_2 = -236$ psi

15–94. (a) $\sigma_1 = 310$ psi, $\sigma_2 = -1160$ psi,   (b) $\gamma_{\max_{\text{in-plane}}} = 735$ psi, $\sigma_{\text{avg}} = -425$ psi, $\theta_{p_1} = 62.7°$, $\theta_{p_2} = -27.3°$, $\theta_s = 17.7°$ and $-72.3°$

15–95. (a) $\epsilon_1 = 862(10^{-6})$, $\epsilon_2 = -782(10^{-6})$, $\theta_{p_2} = 2.01°$ ↓, (b) $\epsilon_{\text{avg}} = 40(10^{-6})$, $\gamma_{\max_{\text{in-plane}}} = 1.64(10^{-3})$, $\theta_s = 43.0°$ ↑

15–97. $\sigma_1 = 3.17$ ksi, $\sigma_2 = -3.86$ ksi, $\tau_{\max_{\text{in-plane}}} = 3.51$ ksi

15–98. $\sigma_1 = 4.00$ ksi, $\sigma_2 = -0.0317$ ksi

15–99. $\sigma_{x'} = 736$ MPa, $\sigma_{y'} = -156$ MPa, $\tau_{x'y'} = 188$ MPa

15–101. $\sigma_1 = 32$ psi, $\sigma_2 = -32$ psi

Chapter 16

16–1. $b = 211$ mm

16–2. $b = 2.71$ in.

16–3. $b = 3.40$ in.

16–5. $b = 15.5$ in.

16–6. $P = 3.20$ kip, $h = 8.0$ in.

16–7. $W\ 16 \times 31$

16–9. $W\ 24 \times 62$

16–10. $W\ 18 \times 50$

16–11. $w = 1.66$ kip/ft

16–13. $P = 2.90$ kN

16–14. $P = 9.52$ kN

16–15. $P = 6.67$ kip, $h = 7.14$ in.

16–17. 35.6 ft

16–18. $P = 178$ lb, $s = 12.0$ in.

16–19. $\sigma = 3.02$ ksi

16–21. $\theta_A = -\dfrac{PL^2}{16EI}$, $v_{max} = -\dfrac{PL^3}{48EI}$,

$$v = \frac{Px}{48EI}(4x^2 - 3L^2)$$

16–22. $v_1 = \dfrac{1}{EI}(-7.29x_1^3 + 3.79x_1)$ N $\cdot$ m^3

$$v_2 = \frac{1}{EI}(7.29x_2^3 + 0.583x_2)\ \text{N} \cdot \text{m}^3$$

16–23. $\rho = 336$ ft, $v_B = -\dfrac{M_0 L^2}{2EI}$

16–25. $v_1 = \dfrac{wb}{6aEI}[x_1^3 - a^2 x_1]$,

$$v_2 = \frac{w}{6EI}[-x_2^3 + (2ab + 3b^2)x_2 - 2(b^3 + ab^2)]$$

16–26. $v_1 = \dfrac{1}{EI}(150x_1^2 + 40x_1)$ N $\cdot$ m^3

$$v_2 = \frac{1}{EI}(-125x_2^3 + 150x_2^2 - 40x_2)\ \text{N} \cdot \text{m}^3$$

16–27. $v_1 = \dfrac{1}{EI}(4.44x_1^3 - 640x_1)$ lb $\cdot$ in^3,

$$v_2 = \frac{1}{EI}(-4.44x_2^3 + 640x_2)\ \text{lb} \cdot \text{in}^3$$

16–29. $\theta_A = -\dfrac{M_0 L}{3EI}$, $\theta_B = \dfrac{M_0 L}{6EI}$,

$$v = \frac{M_0}{6EIL}(3Lx^2 - x^3 - 2L^2 x),$$

$$v_{max} = \frac{-0.0642\ M_0 L^2}{EI}$$

16–30. $|\theta_{max}| = \left| \dfrac{M_0 L}{2EI} \right|$, $v = \dfrac{M_0 x}{2EI}(x - L)$,

$$v_{max} = -\frac{M_0 L^2}{8EI}$$

16–31. $\theta_A = \dfrac{3wa^3}{16EI}$, $v_1 = \dfrac{wx_1}{48EI}[6ax_1^2 - 2x_1^3 - 9a^3]$,

$$v_2 = \frac{wax_2}{48EI}[2x_2^2 - 7a^2], \quad v_C = -\frac{5wa^4}{48EI}$$

16–33. $v_{max} = \dfrac{wL^4}{18\sqrt{3}EI}$

16–34. $\theta_c = 0$, $v = \dfrac{wx}{24EI}(2\,Lx^2 - x^3 - L^3)$,

$$v_c = -\frac{5wL^4}{384EI}$$

16–35. $v_{max} = -\dfrac{18.8\ \text{kip} \cdot \text{ft}^3}{EI}$

16–37. $\theta_A = 0.0611$ rad, $v_A = 3.52$ in.

16–38. $0.152L$

16–39. $F = 1.375$ N

16–41. $v_B = -\dfrac{PL^3}{2EI_0}$

16–42. $v_c = -\dfrac{PL^3}{32EI_c}$

16–43. $v = \dfrac{1}{EI}[-2.5x^2 + 2 < x - 4 >^3 - \frac{1}{8} < x - 4 >^4 +$
$2 < x - 12 >^3 + \frac{1}{8} < x - 12 >^4 - 24x + 136]$
kip $\cdot$ ft^3

16–45. $v = \dfrac{1}{EI}[-0.00556x^5 + 12.9 < x - 9 >^3 +$
$0.00556 < x - 9 >^5 - 256x + 2637]$ kip $\cdot$ ft^3

16–46. $v = \dfrac{1}{EI}[11.1x^3 - 0.5x^4 - 5 < x - 3 >^3 - 378x]$
kN $\cdot$ m^3

16–47. $v = \dfrac{1}{EI}\left[-31.5x^2 + \dfrac{8}{3}x^3 - \dfrac{x^4}{12} + \dfrac{1}{12} < x - 3 >^4 \right.$
$\left. - \dfrac{2}{3} < x - 4.5 >^3 \right]$ kN $\cdot$ m^3, $\theta_B = -0.705°$,
$v_B = -51.7$ mm

16–49. $v = \dfrac{1}{EI} [-0.25x^4 + 0.208 <x - 1.5>^3 +$
$0.25 <x - 1.5>^4 + 4.625 <x - 4.5>^3 +$
$25.1x - 36.4]$ kN $\cdot$ m^3

16–50. $v = \dfrac{1}{EI} [-1.67x^3 - 6.67 <x - 20>^3 + 18.3 <$
$x - 40>^3 + 4000x]$ lb $\cdot$ in^3

16–51. $v = \dfrac{1}{EI} [0.417x^3 - 0.333 <x - 8>^3$
$-0.667 <x - 16>^3 - 169x]$ kip $\cdot$ ft^3

16–53. $\theta_B = 0.574°$, $v_C = 0.200$ in.

16–54. $v_c = -0.501$ mm, $v_D = -0.698$ mm,
$v_E = -0.501$ mm

16–55. $\theta_A = 0.128°$ ↓, $\theta_B = 0.128°$ ↑

16–57. $\theta_B = \dfrac{PL^2}{2EI}$ ↓, $\Delta_B = \dfrac{PL^3}{3EI}$ ↓

16–58. $\theta_c = \dfrac{3937.5}{EI}$ ↘, $\Delta_c = \dfrac{50625}{EI}$ ↓

16–59. $\theta_B = \dfrac{Pa^2}{12EI}$ ↑, $\Delta_C = \dfrac{Pa^3}{12EI}$ ↓

16–61. $\Delta_{max} = 12.2$ mm ↓

16–62. $a = \dfrac{3L}{16}$

16–63. $a = \dfrac{L}{3}$

16–65. $\theta_A = \dfrac{M_0 a}{2EI}$ ↓, $\Delta_{max} = \dfrac{5M_0 a^2}{8EI}$ ↓

16–66. $\theta_B = \dfrac{M_0(b^3 + 3ab^2 - 2a^3)}{6EI(a + b)^2}$
$\Delta_c = \dfrac{M_0 ab(b - a)}{3EI(a + b)}$

16–67. $\theta_C = \dfrac{5Pa^2}{2EI}$ ↓, $\Delta_B = \dfrac{25Pa^3}{6EI}$ ↓

16–69. $\theta_B = \dfrac{7Pa^2}{4EI}$ ↓, $\Delta_C = \dfrac{9Pa^3}{4EI}$ ↓

16–70. 2.12 in ↓

16–71. $\theta_B = \dfrac{7wa^3}{12EI}$ ↓, $\Delta_C = \dfrac{25wa^4}{48EI}$ ↓

16–73. $\Delta_{max} = \dfrac{3wa^4}{8EI}$ ↓

16–74. $\theta_A = \dfrac{Pa^2}{4EI}$ ↓, $\Delta_D = \dfrac{Pa^3}{4EI}$ ↑

16–75. $\Delta_A = 0.933$ in. ↓

16–77. $\dfrac{x_{max}}{y_{max}} = \dfrac{I_x}{I_y} \tan \theta$, $y_{max} = 0.736$ in., $x_{max} = 3.09$ in.

16–78. $\Delta_A = 0.916$ in.

16–79. $\Delta_C = 1.90$ in. ↓

16–81. $\Delta_C = \dfrac{2133}{EI}$ ↓

16–82. $\Delta_G = 5.82$ in. ↓

16–83. $T_{CD} = \dfrac{5AwL^3}{8(48I + L^2A)}$

16–85. $B_y = \dfrac{7wL}{128}$, $A_y = \dfrac{57wL}{128}$, $M_A = \dfrac{9wL^2}{128}$, $A_x = 0$

16–86. $\Delta_B = 1.50$ mm ↓

16–87. $M_A = 0.639$ kip $\cdot$ in., $M_B = 1.76$ kip $\cdot$ in.

16–89. $M_A = \dfrac{PL}{8}$, $M_B = \dfrac{3PL}{8}$, $B_y = P$

16–90. $B_y = 634$ lb, $A_y = 243$ lb, $C_y = 76.8$ lb

16–91. $B_y = C_y = \dfrac{11wL}{10}$, $A_y = D_y = \dfrac{2wL}{5}$

16–93. Segment AB: W 16 × 26, Segment BC: W 12 × 14

16–94. $v = \dfrac{1}{EI} [4.10x^3 - 0.125x^4 + 0.125 <x - 4>^4$
$-8.33 <x - 7>^3 - 279x]$ kN $\cdot$ m^3

16–95. $\theta_A = \dfrac{M_0 L}{3EI}$ ↓, $v = \dfrac{M_0}{6EIL} (3Lx^2 - x^3 - 2L^2x)$,
$v_{max} = \dfrac{0.0642M_0 L^2}{EI}$ ↓

16–97. $v = \dfrac{P}{12EI} [-2x^3 + 3 <x - a>^3 - 2 <x - 2a>^3$
$+3 <x - 3a>^3 + 15a^2x - 13a^3]$, $v_{max} = \dfrac{Pa^3}{3EI}$ ↑

16–98. $\theta_A = 6.54°$ ↓

16–99. $\theta_A = \dfrac{8}{EI}$ ↑, $\theta_B = \dfrac{16}{EI}$ ↓, $\theta_C = \dfrac{40}{EI}$ ↓, $\Delta_C = \dfrac{84}{EI}$

16–101. $s = 26.8$ mm, $s' = 10.7$ mm

Chapter 17

17–1. $d = \frac{9}{16}$ in.

17–2. $P_{cr} = 272$ kN

17–3. $P_{cr} = 377$ kip

17–5. $P = 17.6$ kip

17–6. $F.S. = 2.19$

17–7. (a) $P_{cr} = 28.4$ kip, (b) $P_{cr} = 58.0$ kip

17–9. $P = 4.23$ kip

17–10. $P = 73.2$ kip

17–11. $P = 5.79$ kip

17–13. $d_{AB} = 2\frac{1}{8}$ in., $d_{BC} = 2$ in.

17–14. $F.S. = 2.38$

17–15. $P = 2.42$ kip

17–17. $w = 1.17$ kN/m

17–18. $w = 5.55$ kN/m

17–19. $P_{cr} = 794$ kN

17–21. 1.62 MN

17–22. $E_t = 14.6(10^3)$ ksi

17–23. $0 < L/r < 35.6$, $P/A = 16\ 449/(L/r)^2$;
$35.6 < L/r < 70.2$, $P/A = 13$; $70.2 < L/r$,
$P/A = 64\ 152/(L/r)^2$

17–25. $0 < L/r < 49.7$, $P/A = 61.7(10^3)/(L/r)^2$;
$49.7 < L/r < 99.3$, $P/A = 25$ ksi, $99.3 < L/r$,
$P/A = 247(10^3)/(L/r)^2$

Index

Simply Supported Beam Slopes and Deflections

Beam	Slope	Deflection	Elastic Curve
	$\theta_{max} = \dfrac{PL^2}{16EI}$	$v_{max} = \dfrac{-PL^3}{48EI}$	$v = \dfrac{-Px}{48EI}(3L^2 - 4x^2)$ $0 \le x \le L/2$
	$\theta_1 = \dfrac{Pab(L+b)}{6EIL}$ $\theta_2 = \dfrac{Pab(L+a)}{6EIL}$	$v\Big|_{x=a} = \dfrac{-Pba}{6EIL}(L^2 - b^2 - a^2)$	$v = \dfrac{-Pbx}{6EIL}(-x^2 - b^2 + L^2)$ $0 \le x \le a$
	$\theta_1 = \dfrac{ML}{3EI}$ $\theta_2 = \dfrac{-ML}{6EI}$	$v_{max} = \dfrac{-ML^2}{\sqrt{243}\ EI}$	$v = \dfrac{-Mx}{6LEI}(x^2 - 3Lx + 2L^2)$ $0 \le x \le L$
	$\theta_{max} = \dfrac{wL^3}{24EI}$	$v_{max} = \dfrac{-5wL^4}{384EI}$	$v = \dfrac{-wx}{24EI}(x^3 - 2Lx^2 + L^3)$ $0 \le x \le L$
	$\theta_1 = \dfrac{3wL^3}{128EI}$ $\theta_2 = \dfrac{-7wL^3}{384EI}$	$v\Big|_{x=L/2} = \dfrac{-5wL^4}{768EI}$	$v = \dfrac{-wx}{384EI}(16x^3 - 24Lx^2 + 9L^3)$ $0 \le x \le L/2$ $v = \dfrac{-wL}{384EI}(8x^3 - 24Lx^2$ $+ 17L^2x - L^3)$ $L/2 \le x \le L$
	$\theta_1 = \dfrac{7w_0L^3}{360EI}$ $\theta_2 = \dfrac{-w_0L^3}{45EI}$		$v = \dfrac{-w_0x}{360LEI}(3x^4 - 10L^2x^2 + 7L^4)$ $0 \le x \le L$

SP04